Reproductive Endocrinology

Editor
Pedro J. Chedrese
Department of Biology
College of Arts and Science
University of Saskatchewan
Saskatoon, Saskatchewan
Canada
pjc108@mail.usask.ca

ISBN 978-0-387-88185-0 e-ISBN 978-0-387-88186-7
DOI 10.1007/978-0-387-88186-7

Library of Congress Control Number: 2008935613

Printed on acid-free paper

springer.com

Pedro J. Chedrese
Editor

Reproductive Endocrinology

A Molecular Approach

Almost all aspects of life are organized at the molecular level, and if we do not understand molecules, our comprehension of life will be very incomplete

Francis Crick

Preface

Molecular biology is a relatively new science that emerged from the advances in biochemistry of the 1940s and 1950s, when the structure of the nucleic acids and proteins were elucidated. With progress in the 1970s in nucleic acid enzymology and the discovery of the restriction enzymes, the tools of molecular biology became widely available and were applied to cell biology to study how genes are regulated. This new knowledge impacted endocrinology and reproductive biology, since it was largely known that the secretion of the internal glands affected phenotypes, and therefore expression of genes. In 1982 the first edition of *"Molecular Cloning: A Laboratory Manual,"* (Maniatis, Fritch and Sambrook) was published, offering researchers the first organized collection of protocols for molecular techniques. It was the application of these techniques by endocrinologists that gave birth to molecular endocrinology. From then on, the use of these modern tools not only shed light on fundamental questions in endocrinology and reproductive biology, but also has become central to scientific studies in all living matter, impacting disciplines such as human and veterinary medicine and related health sciences, including animal, agriculture and environmental sciences. Thus, modern reproductive biology encompasses every level of biological study from genomics to ecology, taking in a great deal of cell biology, biochemistry, endocrinology and general physiology. All of these disciplines require a basic knowledge, both as a tool and as essential aid to fundamental understanding the principles of life in health and disease.

Graduate students enter into reproductive biology programs with very varied backgrounds and in many cases their knowledge requires updates to allow them to appreciate the full potential of the field. This is particularly true for the students with professional degrees because the need to concentrate on information that often leads to the conclusion that molecular biology is all too esoteric. Hence the purpose of this book is to serve the needs of students and professionals to encourage them to integrate the extensive knowledge that they already have with a field that they often find off-putting because of its unfamiliar concepts and terminology. In this book, the basic biochemistry of nucleic acids and proteins are reviewed. Methodologies used to study signaling and gene regulation in the endocrine/reproductive system are also discussed. Topics include the basis and tools of molecular biology regulation of reproductive hormones, organs and tissues, and several endocrine disorders affecting the reproductive system. We believe that graduate students and professionals in the medical, veterinary and animal sciences fields will find this book an exciting and stimulating material that will enhance the breadth and quality of research.

Saskatoon, SK

Pedro J. Chedrese

Acknowledgments

My warm thanks to Christine Meaden and Victoria Fachal for their friendship, constant support, help and editing skills. Special acknowledgement to Dr. Mike Henson, Dr. Loro Kujjo, Dr. Holly LaVoie, Dr. Tracy Marchant, Dr. Gloria Perez and Dr. Martina Piasek who collaborated with great enthusiasm. Thanks for your friendship and for sharing your scholarship and advice. As always when last minute glitches come up there are special people in the wings who are willing to lend a hand. I am grateful to Dr. Charles Blake for his help and Dr. Abraham Kierszenbaum who graciously in very short time contributed a key chapter to this book. Gratitude is expressed to Dr. Bruce D. Murphy for his support and mentorship.

Contents

Part III Molecular Regulation of Reproductive Hormones

Part IV Molecular Regulation of the Reproductive Organs and Tissues

Contributors

Dalhia Abramovich Instituto de Biología y Medicina Experimental-CONICET, Buenos Aires, Argentina abramovich@dna.uba.ar

Ayman Al-Hendy Department of Obstetrics and Gynecology, Meharry Medical College Center for Women Health Research, Nashville, TN, USA, ahendy@MMC.edu

Luz Andreone Centro de Investigaciones Endocrinológicas, Hospital de Niños Ricardo Gutiérrez, Buenos Aires, Argentina landreone@cedie.org.ar

Christy Barlund Department of Large Animal Clinical Sciences, University of Saskatchewan Western College of Veterinary Medicine, Saskatoon, SK, Canada, csb272@mail.usask.ca

Alejandro M. Bertorello Department of Medicine Membrane Signaling Networks, Karolinska Institutet, Stockholm, Sweden alejandro.bertorello@ki.se

Gheorghe T. Braileanu Department of Pediatrics, University of Maryland at Baltimore School of Medicine, Baltimore, MD, USA, gbrai001@umaryland.edu

Stella Campo Centro de Investigaciones Endocrinológicas, Hospital de Niños Ricardo Gutiérrez, Buenos Aires, Argentina scampo@cedie.org.ar

V. Daniel Castracane Department of Obstetrics & Gynecology Health Science Center, Texas Tech University School of Medicine, Lubbox, TX, USA daniel.castracane@ttuhsc.edu

Javier S. Castresana Brain Tumor Biology Unit, University of Navarra, Pamplona, Spain, jscastresana@unav.es

Stella M. Celuch Instituto de Investigaciones Farmacológicas, CONICET, Buenos Aires, Argentina, sceluch@ffyb.uba.ar

Pedro J. Chedrese Department of Biology, University of Saskatchewan College of Arts and Science, Saskatoon, SK, Canada, jorge.chedrese@usask.ca

Indrajit Chowdhury Department of Physiology, Morehouse School of Medicine, Atlanta, GA, USA indrajitchowdhury@yahoo.co.uk

M. Victoria Fachal Department of Biology, University of Saskatchewan College of Art and Sciences, Saskatoon, SK, Canada, mvf259@mail.usask.ca

Jorge A. Flores Department of Biology, West Virginia University, Morgantown, WV, USA, jflores@wvu.edu

Lisa C. Freeman Department of Anatomy and Physiology, Kansas State University College of Veterinary Medicine, Manhattan, KS, USA, freeman@vet.k-state.edu

Michael Furlan Novozymes Biologicals, Saskatoon, SK, Canada, maf131@mail.usask.ca

Michael C. Henson Department of Biological Sciences, Purdue University Calumet, Hammond, IN, USA, henson@calumet.purdue.edu

Griselda Irusta Instituto de Biologìa y Medicina Experimental-CONICET, Buenos Aires, Argentina girusta@dna.uba.ar

Abraham L. Kierszenbaum Department of Cell Biology and Anatomy, The Sophie Davis School of Biomedical Education/The City University of New York Medical School, New York NY, USA, kier@med.cuny.edu

Steven R. King Scott Department of Urology Baylor College of Medicine Houston, TX, USA, srking100@yahoo.com

Loro L. Kujjo Department of Physiology, Biomedical and Physical Sciences, Michigan State University, East Lansing, MI, USA kujjo@msu.edu

Paula Lázcoz Department of Health Sciences, University of Navarra, Pamplona, Spain paula.lazcoz@unavarra.es

Holly A. LaVoie Deptartment of Cell Biology and Anatomy, University of South Carolina School of Medicine, Columbia, SC, USA, HLAVOIE@gw.med.sc.edu

Yonghai Li Department of Anatomy & Physiology, Kansas State University College of Veterinary Medicine, Manhattan, KS, USA yhli@vet.k-state.edu

Nazareth Loreti Centro de Investigaciones Endocrinológicas, Hospital de Niños Ricardo Gutiérrez, Buenos Aires, Argentina lnazareth@cedie.org.ar

Christine Meaden Saskatchewan Health, Saskatoon, SK, Canada, cmeaden@shaw.ca

Rena J. Okrainetz Cytogenetics Laboratory, Royal University Hospital, Saskatoon, SK, Canada rjc401@mail.usask.ca

Fernanda Parborell Instituto de Biología y Medicina Experimental-CONICET, Buenos Aires, Argentina fparbo@dna.uba.ar

Gloria I. Perez Department of Physiology, Biomedical and Physical Sciences, Michigan State University, East Lansing, MI, USA, perezg@msu.edu

Tim G. Rozell Department of Animal Sciences & Industry, Kansas State University, Manhattan, KS, USA, trozell@ksu.edu

Salama A. Salama Department of Obstetrics & Gynecology, University of Texas Medical Branch, Galveston, TX, USA, sasalama@utmb.edu

Rajagopala Sridaran Department of Physiology, Morehouse School of Medicine, Atlanta, GA, USA, rsridaran@msm.edu

Carlos Stocco Department of Obstetrics Gynecology and Reproductive Sciences, Yale University School of Medicine, New Haven, CT, USA, carlos.stocco@yale.edu

Marta Tesone Institute of Experimental Biology and Medicine-CONICET, Buenos Aires, Argentina mtesone@dna.uba.ar

Laura L. Tres Department of Cell Biology and Anatomy, The Sophie Davis School of Biomedical Education, The City University of New York Medical School, New York NY, USA, tres@med.cuny.edu

Part I
The Basis for Molecular Reproductive Biology

Chapter 1

Introduction to the Molecular Organization of the Endocrine/Reproductive System

Pedro J. Chedrese

1.1 Introduction

The endocrine system developed during evolution from single cells to multi-cellular organisms as a means of communication through chemical signals that coordinates multiple organic functions. The investigation of these chemical signals in the early years of the twentieth century gave origin to the science of *endocrinology*, which refers to the study of a group of specialized secretory organs called *endocrine glands* that deliver their products, called *hormones* or *first messengers*, directly into the interstitial space and enter the circulatory system. Endocrine glands are called *internal glands* to differentiate from *exocrine glands*, which deliver their products through ducts into the gastrointestinal tract or outside the body.

In the early 1930 s, anatomical and functional studies demonstrated that *hormones affect the nervous system, and that the endocrine system* is both directly and indirectly controlled by the nervous system. The nervous system is organized as a cellular network, *with* axons directing information via chemical mediators called *neurotransmitters* that are secreted into the closed space of the synapse and act on targets located on the efferent side of the cleft. Neurotransmitters can also diffuse into the extracellular space, enter the circulatory system, and act outside of the synaptic cleft as classical hormones on distant target organs. The neurotransmitters that act outside of the synaptic cleft are called *neurohormones*. It is now known that neurohormones secreted by the hypothalamus, which is anatomically related to the pituitary gland, are the signals that integrate the endocrine system into one functional unit controlled by the central nervous system (CNS). Hence, the term *neuroendocrinology* was incorporated into endocrinology and is defined as the study of the relationship between the nervous system and the endocrine system.

From the evolutionary perspective, the endocrine system provided an internal stable environment to the primitive organism. Thus, the development of the endocrine system preceded the development of the CNS that requires conditions of internal constancy for normal functioning of its highly specialized cells. The CNS further evolved developing a sensorial system of visual, olfactory, auditory, and tactile perceptions, which are linked to the endocrine system. Accordingly, the endocrine glands respond to the nervous system by secreting hormones and other extracellular signaling molecules that interact with chemically defined cellular structures called *receptors* located within target organs. Recognition of hormones by specific receptors triggers a cascade of reactions that involve the synthesis and/or mobilization of a second group of molecules called *intracellular second messengers,* which ultimately affect expression of genes and elicit biological responses. From studies on the molecular mechanisms by which endocrine signaling molecules affect expression of genes emerged the science of *molecular endocrinology*, which is defined as the study of hormone action at the cellular and molecular levels.

In this introductory chapter the classic and current concepts of endocrinology will be reviewed, focusing on the molecules that regulate reproduction and how they are organized within the context of the endocrine system.

1.2 The Endocrine System: Classical and Current Concepts

The classically described endocrine glands include the hypothalamus, pituitary, pineal, thyroid, parathyroid, adrenals, pancreatic Islets of Langerhans, and gonads. Although some of these glands would seem unrelated to the reproductive system, most of them, directly or indirectly, affect reproduction. Important concepts to consider are that some cells and non-glandular tissues can produce, convert, or secrete hormones or

P.J. Chedrese (✉)
Department of Biology, University of Saskatchewan College of Arts and Science, Saskatoon, SK, Canada
e-mail: jorge.chedrese@usask.ca

P.J. Chedrese (ed.), *Reproductive Endocrinology*,
DOI 10.1007/978-0-387-88186-7_1, © Springer Science+Business Media, LLC 2009

hormone-like substances. In addition, virtually every organ in the body have endocrine capabilities related to their primary functions, including the brain, placenta, thymus, kidney, heart, blood vessels, skin and adipose tissue.

1.2.1 Hormones and Hormone-Targets

The term *hormone* derives from the Greek verb *hormaein,* which means "to excite or to set in motion." The classical endocrine definition is that hormones are *extracellular signaling molecules synthesized and secreted by specialized cells that are released into the circulatory system to exert specific biochemical actions on target cells located at distant sites* (Fig. 1.1A). The main functional features that characterize hormones are that they

- do not initiate new functions, but regulate already established cellular processes;
- have relatively brief effects and fluctuate in concentration, responding to immediate physiological requirements. Albeit, in some cases hormones may cause permanent changes, such as sex differentiation, that persist even when the hormone is no longer required;
- have specific and sensitive effects that influence unique cellular responses at very low concentrations;
- trigger a cascade of amplifying mechanisms, meaning that a small amount of hormone induces the release of a larger amount of hormone and/or intracellular metabolites in the target cells until physiological demands are met.

Once hormones reach target tissues, they interact with receptor molecules that are hormone specific. Receptors have high affinity for the hormone and the capacity to discriminate between hundreds of other regulatory molecules that

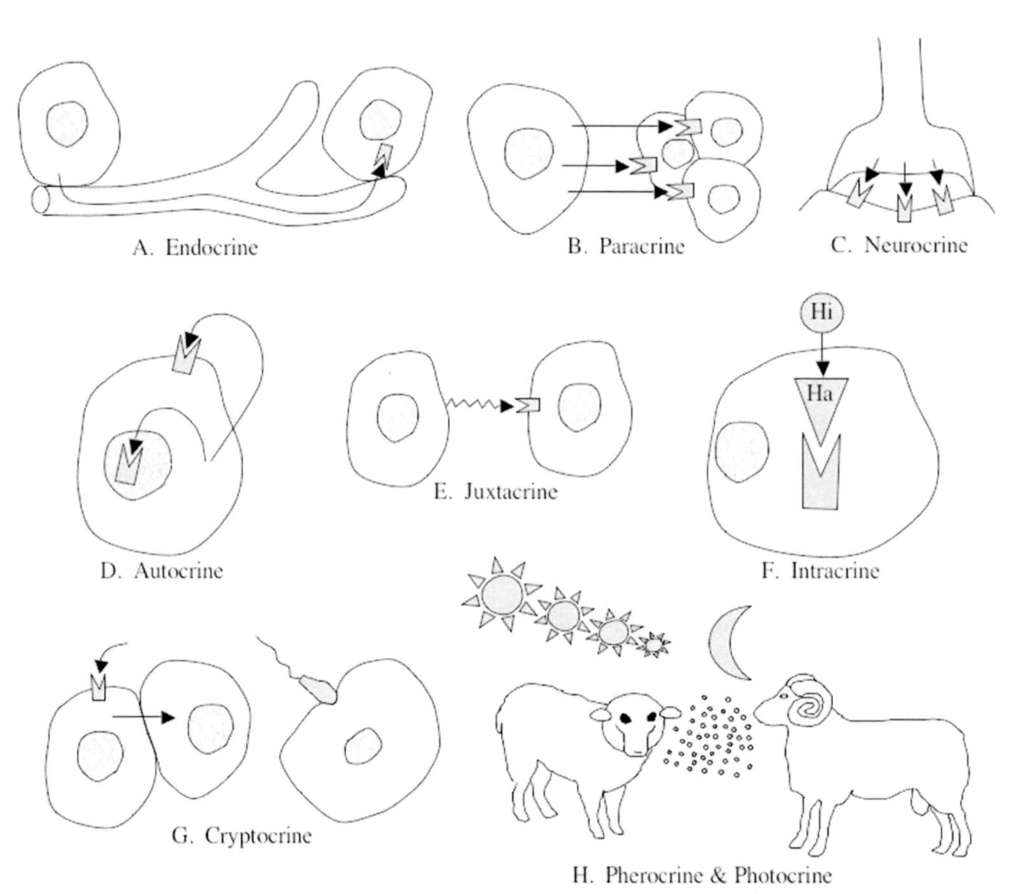

Fig. 1.1 Signaling in the endocrine/reproductive system. (**A**) Endocrine signaling: signals are released into the blood and exert actions at distant sites. (**B**) Paracrine signaling: signals diffuse from one cell type to a target cell located in proximity within the same organ. (**C**) Neurocrine (or synaptic) signaling: signals can be secreted either by neurons, as a neurotransmitter, or by an endocrine gland as a classic hormone. (**D**) Autocrine signaling: signals are produced by a cell population and regulate the same population of cells. (**E**) Juxtacrine signaling: the signal is a plasma membrane-bound peptide that binds to a receptor in a cell of close proximity. (**F**) Intracrine signaling: an endocrine inactive precursor (Hi) generated in one organ reach a target organ where it is transformed into an endocrine active compound (Ha). (**G**) Cryptocrine signaling: the signaling molecule is synthesized and exerts its action into a closed cellular environment without extracellular release. (**H**) Pherocrine and photocrine signaling: in the pherocrine system, pheromones are secreted into the environment targeting individuals of the same species. In the photocrine system, daylight affects structures within the eye and signals are transmitted to the brain influencing the reproductive cycles

circulate in the blood stream and are diffused throughout the body. The receptor affinity for the hormone must be high enough in relation to hormone concentration in the blood; thus, the amount of hormone bound to its receptor will change in response to changes in the level of hormone in circulation.

The endocrine system can further ensure specificity by directly secreting hormones into confined areas, such as the interstitial spaces or synaptic clefts. In this way, bloodstream dilution is avoided and effective hormonal concentrations are established at the receptor level. Therefore, hormones with affinity for widely distributed receptors are able to elicit specific biological responses within limited spaces without affecting other tissues. Accordingly, the classical definition of hormone has been extended to include other *extracellular signaling molecules,* such as regulatory peptides, growth factors, and neurotransmitters that are locally synthesized and do not act by the classical endocrine signaling mode. In general, *while all hormones are considered extracellular signaling molecules, not all extracellular signaling molecules are considered hormones*, although the terms are commonly used interchangeably in the literature. Therefore, in addition to the classical *endocrine* signaling mode, the interaction between extracellular signaling molecules and their targets include other signaling modes termed *paracrine, neurocrine, autocrine, juxtacrine, intracrine,* and *cryptocrine*.

Paracrine signaling (Fig. 1.1B): the cells that produce the extracellular signaling molecules are in close proximity to the target cells. Examples of paracrine signaling include: the androgenic steroid hormone testosterone that is synthesized by the Leydig cells of the testis and stimulates spermatogenesis in the adjacent seminiferous tubules; and the polypeptide hormone insulin that is synthesized in the β-cells of the pancreatic Islets of Langerhans and stimulates secretion of glucagon produced by the adjacent α-cells.

Neurocrine signaling (Fig. 1.1C): the extracellular signaling molecules are peptides or amines that are secreted by the neurons as either a neurotransmitter or a neurohormone (Table 1.1). A classic example of neurosecretion is the hypothalamic-releasing hormone, gonadotropin-releasing hormone (GnRH) that reaches the pituitary gland through the porta-hypophyseal circulation and stimulates gonadotrope cells to synthesize and release the follicle-stimulating hormone (FSH) and the luteinizing hormone (LH). Other examples of neurosecretory regulation include the neuropeptides secreted by the posterior lobe of the pituitary gland. They include vasopressin, also called antidiuretic hormone (ADH), and oxytocin, which are both synthesized by hypothalamic neurons and transported to the pituitary neural lobe by axoplasmic flow. They are then released as hormones into the general circulation to regulate functions in remote organs. ADH increases water reabsorption by the kidneys and oxytocin regulates contraction of the uterus during parturition and milk ejection in the mammary gland during breastfeeding.

Autocrine signaling (Fig. 1.1D): the extracellular signaling molecules produced by a cell population also regulate the same cell population. Within the reproductive system, 17β-estradiol, synthesized by the granulosa cells of the ovarian follicle, regulates progesterone synthesis and cell growth in the same population of granulosa cells. Tumor cells acquire autocrine growth control by producing their own hormonal growth factors, making the cells independent of the physiological regulators of proliferation.

Juxtacrine signaling (Fig. 1.1E): the extracellular signaling molecules are membrane-bound peptides that bind to cell receptors located in close proximity. Intercellular communication is transmitted via membrane components and signaling may affect either the emitting cell or transmembrane-linked adjacent cells. Unlike other modes of cell signaling, juxtacrine signaling requires

Table 1.1 Endocrine signaling molecules of the CNS

Signaling molecule	Structure	Action
Thyrotropin-releasing hormone (TRH)	Tripeptide	↑TSH, PRL
Gonadotropin-releasing hormone (GnRH or LHRH)	Decapeptide	↑FSH and LH
Growth hormone inhibiting hormone (GHIH) or somatostatin	14-amino acid peptide	↓GH
Corticotropin-releasing factor (CRF)	41-amino acid peptide	↑ACTH
GH releasing factor (GHRF)	44-amino acid peptide	↑GH
Pituitary adenylate cyclase-activating peptide (PACAP)	27- or 38-amino acid peptide	↑LH, PRL, GH, ACTH
Dopamine	Amine	↑GH; ↓PRL
Epinephrine	Amine	↑GnRH
Norepinephrine	Amine	↑GnRH
γ-aminobutyric acid (GABA)	Amine	↓PRL
Opioids	Peptides	↓GnRH
Vasointestinal peptide (VIP)	Peptide	↑ PRL

ACTH: adrenocorticotropic hormone; FSH: follicle-stimulating hormone; GH: growth hormone; LH: luteinizing hormone; PRL: prolactin; TSH: thyroid-stimulating hormone.

physical contact between the cells involved. The peptide must remain attached to the membrane; otherwise if it is cleaved to yield a free molecule, the signal activity is lost. Juxtacrine extracellular signaling molecules include many growth factors, such as epidermal growth factors (EGF), transforming growth factor-α (TGF-α), tumor necrosis factor-α (TNF-α), and colony stimulating factor-1 (CSF-1).

Intracrine signaling (Fig. 1.1F): endocrine-active enzymes in target cells transform the extracellular signaling molecules. Endocrine-active target cells not only express enzymes that synthesize active hormones, but also express enzymes that inactivate them. Therefore, intracrine signaling modulates hormone concentrations according to the needs of the tissues. The overall contribution of intracrine signaling to the regulation of steroid synthesis is significant, since 40% of all androgens in males and 75–100% of estrogens in postmenopausal females originate from adrenal steroids that are transformed in target tissues. Intracrine signaling occurs in two ways. First, the extracellular signaling molecule is enzymatically converted into a different hormone. For example, the androgen androstenedione is converted into the estrogen 17β-estradiol in granulosa cells by cytochrome P450 aromatase (P450arom). Second, the extracellular signaling molecule is enzymatically transformed into a version that has increased or decreased biological activity. Intracrine signaling occurs during fetal development when testosterone is transformed by 5α-reducase type 2 into a much more potent androgen, 5α-dihydrotestosterone (5α-DHT), which is responsible for the normal masculinization of the external genitalia. Another example of increased biological activity is the hormone thyroxine (T_4), which is secreted by the thyroid gland and transformed by iodothyronine deiodinases type 1 and type 2 into its more active form, triiodothyronine (T_3). Endocrine-active peripheral tissues also express enzymes that inactivate extracellular signaling biological activity, such as the sulfotransferases. An example of inactivation is the transformation of T_3 and T_4 into inactive metabolites by iodothyronine deiodinase type 3, which is found mainly in fetal tissue and placenta.

Cryptocrine signaling (Fig. 1.1G): the extracellular signaling molecules are synthesized into a closed cellular environment without extracellular release. This system requires an intimate association between the cell that produces the signal and the target cell. Thus, extracellular signaling molecules can influence the function of cells lacking specific receptors. Examples include the relationship between Sertoli cells and spermatids and the transfer of second messengers, such as cyclic nucleotides or inositol triphosphate, through gap junctions between adjacent cells.

Pherocrine and Photocrine signaling (Fig. 1.1H): the endocrine system also responds to external environmental endocrine signals that include air-borne hormones or hormone-like molecules present in the environment and light signals, termed pherocrine and photocrine, respectively.

Pherocrine signaling refers to the effect of pheromones secreted into the environment, which are extracellular signaling molecules associated with sex pairing. Pherocrine signals have been described in mice, rats, pigs, sheep, and monkeys and is suggested to also exist in humans. Pheromones are sensed by olfaction by opposite members of the species. Examples of male pherocrine signaling include the androgenic metabolite 3α-androstenol, produced in the submaxillary salivary glands, and 5α-androstenone that is present in fat and other tissues of the domesticated boar pig. In many species the bacterial flora of the vagina produces female pheromones in the form of short-chain fatty acids. Activity of the vaginal flora changes along with changes in the circulating levels of estrogens and progesterone, and it is believed that these fatty acids are the olfactory stimuli emitted during estrus, providing information to the male about the physiological status of the female.

Photocrine signaling refers to the effect of the length of daylight, or photoperiod, on structures within the eye. Daylight signals are then transmitted to the brain and ultimately influence the reproductive cycles in a number of species. The CNS pathway involved in the translation of light includes the retina, the suprachiasmatic nucleus, the superior cervical ganglion, and the pineal gland that produces the hormone melatonin in response to darkness. Although melatonin has been described as an *anti*-gonadal hormone, its function in photoperiodicity is controversial, since both short and long phases of darkness elicit different effects on reproductive cycles of many species, including horse, sheep, goat, and cat. These species have annual periods of ovarian activity that are interrupted by periods of inactivity, termed anoestrus. Although reproductive cycles are dependent on the length of exposure to daylight, the response to photoperiods differs among species. Thus, increasing day length induces ovulation in cats and horses, while decreasing day length induces ovulation in goats and sheep.

While pherocrine and photocrine signaling are important components of neuroendocrine communication between the external and internal environments, the definition of these signals as hormonal remains controversial and some authors think that they should not be considered as such.

It can be concluded that the endocrine signaling system provides a means of communication between

- specific cell populations;
- neighboring cell populations;
- distant organs via the circulatory system; and
- the external and internal environments.

Overall, the endocrine system coordinates fundamental aspects of animal life such as maintenance of the internal environment, metabolism, reproduction, development, and growth.

1.2.2 Organization of the Extracellular Signaling Molecules

Extracellular signaling molecules can be divided into six main groups:

1) Peptides, proteins, and glycoproteins, which are classified according to their size into the following:
   - small peptides include three amino acid molecules, such as the thyrotropin releasing hormone (TRH), through to ten amino acid molecules, such as GnRH;
   - large peptides and proteins include insulin, growth factors, adrenocorticotropic hormone (ACTH), growth hormone (GH), and prolactin (PRL); and
   - glycoproteins such as thyroid stimulating hormone (TSH); the gonadotropins, FSH and LH; the placental gonadotropins, human chorionic gonadotropin (hCG), and equine chorionic gonadotropin (eCG).
2) Amino acid derivatives, such as the neurotransmitters: epinephrine, serotonin and dopamine; and iodinated derivatives of the amino acid tyrosine, the thyroid hormones, T_4 and T_3.
3) Steroid hormones, which are derivatives of cholesterol that can be divided into two groups:
   - steroids with an intact sterol group, such as the gonadal androgens, estrogens, and progestins, and adrenal steroids, including glucocorticoids and mineralocorticoids; and
   - steroids with a broken sterol nucleus, such as the vitamin-D_2 derivatives cholecalciferol and ergocalciferol.
4) Eicosanoids, which are derivatives of long-chain polyunsaturated fatty acids, such as prostaglandins, thromboxanes, and leukotrienes.
5) Retinoids, such as the vitamin-A derivative, retinoic acid.
6) Dissolved gases, such as nitric oxide (NO) and carbon monoxide (CO).

1.2.3 Sex-Specific Hormone Actions

Typically, sex hormones are classified into two groups: female hormones that include estrogens and progestins and male hormones, androgens. However, it must be emphasized that *the sex hormones that have been characterized to date are not exclusive to either gender*. All sex hormones are present in both males and females, and both genders have receptors that bind and respond to all of the sex hormones. *Gender is characterized by the amount of individual sex hormones and their genetically programmed patterns of secretion.* Enzymes appropriately expressed by the gonads (the ovaries or the testis), at critical stages of embryo development, are what define patterns of sex hormone secretion in each gender.

1.2.4 Growth Factors and Cytokines

Growth factors are hormone-like peptides that stimulate cell division by inducing differentiation and cell growth or inhibit cell division by inducing cellular hypertrophy. Growth factors do not originate in the classically defined endocrine glands but in many different tissues of the body and exert their actions in an autocrine and/or paracrine manner. Table 1.2 [1] is a summary of the main growth factors expressed in the gonads. The main growth factors are

- the family of somatomedins, including insulin-like growth factor-I (IGF-I or somatomedin-C), insulin-like growth factor-II (IGF-II), and proinsulin;
- the family of epidermal growth factors, including EGF, also known as urogastrone, keratinocyte autocrine factor (KAF), or amphiregulin, and heparin-binding-EGF (HB-EGF);
- the family of transforming growth factor-β (TGF-β), including TGF-β1, TGF-β2, TGF-β3, activins (activin A, activin B, and activin AB), inhibin, Müllerian-inhibiting substance (MIS), also known as antimüllerian hormone (AMH);
- the family of platelet-derived growth factors (PDGF), including PDFG-AA, PDFG-BB, and PDGF-AB;
- the family of fibroblast growth factors (FGF) or heparin-binding growth factors, including acidic FGF (aFGF) and basic FGF (bFGF); and
- the family of nerve growth factors, including nerve growth factor (NGF), brain-derived neurotrophic factor (BDNF), neurotropin-3 (NT-3), and ciliary neurotrophic factor (CNTF).

The nervous system and the endocrine system are also integrated with the immune system through a communication network that recognizes internal (i.e., tumor) or external (i.e., bacterial, viral, and fungal) antigens. In response to these antigens, the immune system secretes powerful signaling peptides called *cytokines*, which are growth factors produced by inflammatory cells. Mononuclear phagocytes, lymphocytes, epithelial cells, fibroblasts, endothelial cells, and chondrocytes produce cytokines. Cytokines play key roles in a

Table 1.2 Cell-to-cell communication growth factors in the testis and ovary

Growth factor	Cell origin	Cell target	Effects
Ovary			
BMP	Oocyte theca	Granulosa theca	Regulates cellular differentiation
CSF	Theca	Granulosa theca	Regulates cellular growth
FGFs	Granulosa theca oocyte	Granulosa theca	Regulates cellular growth and differentiation
GDF-9	Oocyte	Granulosa theca	Regulates steroideogenesis and cellular differentiation
HGF	Theca	Granulosa	Stimulates cellular proliferation and differentiation
Fas ligand	Granulosa theca oocyte	Granulosa theca Oocyte	Stimulates apoptosis
IGF-1	Granulosa theca	Granulosa theca oocyte	Stimulates cellular growth and differentiation
Inhibin	Granulosa	Oocyte theca granulosa	Regulates cellular differentiation
Interleukins	Granulosa theca	Granulosa theca	Regulates cellular differentiation
KGF	Theca	Granulosa	Regulates cellular growth
SCF/KL	Granulosa	Oocyte theca	Recruitment of primordial follicles and regulation of theca cells growth
LIF	Granulosa	Oocyte theca	Stimulates cellular growth and differentiation
NGF	Theca	Granulosa theca	Ovulation
TNF	Granulosa theca oocyte	Oocyte granulosa Theca	Induces apoptosis and regulates cell growth
TGFα	Theca	Granulosa theca	Stimulates cellular growth
TGFβ	Theca granulosa	Theca granulosa	Inhibits cellular growth and regulates cellular differentiation
VEGF	Theca granulosa	Endothelium granulosa	Promote angiogenesis
Testis			
FGFs	Sertoli germ Leydig	Germ peritubular sertoli Leydig	Regulates cellular growth and differentiation
HGF	Peritubular	Leydig peritubular sertoli	Stimulates cellular growth and seminiferous tubule formation
IGF-1	Sertoli peritubular Leydig	Sertoli peritubular Leydig germ	Homeostasis and DNA synthesis
Inhibin	Sertoli	Germ Leydig	Regulates cellular differentiation
Interleukins	Sertoli Leydig	Sertoli Leydig germ	Regulates cellular growth and differentiation
SCF/KL	Sertoli	Germ	Regulates spermatogonial cells proliferation
LIF	Peritubular sertoli Leydig	Germ	Stimulates cellular growth and survival
Neurotropins	Germ sertoli	Sertoli peritubular	Stimulates cellular growth, migration and differentiation
PDGFs	Sertoli	Peritubular Leydig	Stimulates cellular growth and regulates differentiation
TNF	Germ Leydig	Sertoli germ	Regulates apoptosis of germ cells and cellular differentiation of sertoli cells
TGFα	Sertoli peritubular Leydig	Sertoli peritubular Leydig germ	Stimulates cellular growth
TGFβ	Sertoli peritubular Leydig	Sertoli peritubular Leydig germ	Inhibits cellular growth and stimulates cellular differentiation

Modified from Skinner [1] BMP: bone morphogenic protein; CSF: colony stimulating factor; FGFs: fibroblast growth factors; GDF-9: growth differentiation factor-9; HGF: hepatocyte growth factor; IGF-I: insulin growth factor-I; KGF: keratocyte growth factor; LIF: leukemia inhibitor factor; NGF: nerve growth factor; PDGFs: platelet-derived growth factors; SCF/KL: stem cell factor/Kit ligand; TNF: tumor necrosis factor; TGFα: transforming growth factor-α; TGFβ: transforming growth factor-β; VEGF: vascular endothelial growth factor

variety of biological processes, including cell growth and activation, inflammation, immunity, hematopoiesis, tumorigenesis, tissue repair, fibrosis, and morphogenesis. As growth factors, cytokines operate mainly in autocrine/paracrine signaling, but can also exert classical endocrine signaling affecting distant cell targets. The main cytokines include

- interferons;
- interleukins;
- tumor necrosis factors; and
- colony stimulating factors.

In addition, a variety of other growth factors and hormones of the endocrine/reproductive system have cytokine activity, including

- PRL, GH, and leptin;
- erythropoietin and thrombopoietin;
- the neurotrophins: BDNF, NGF, NT-3, NT-6, and the glial-derived neurotrophic factor;
- the neuropoietic factors, CNTF, oncostatin-M, and leukemia inhibitory factor (LIF); and
- the TGF-β family, which includes the TGF-β, bone morphogenic proteins (BMP), and activins.

Thus, the immune system, long considered to function autonomously, is now recognized as a regulated system that is subject to a reciprocal relationship with the neuroendocrine system.

1.2.5 Dissolved Gases as Signaling Molecules

Many types of cells use dissolved gases, such as nitric oxide (NO) and carbon monoxide (CO), as extracellular signaling molecules. NO is produced by deamination of the amino acid *arginine*, catalyzed by the enzyme NO-synthase. NO dissolves rapidly and diffuses out of the cell and enters into neighboring cells. The effects of NO are local and very short. It has a half-life of about 5–10 s, and then is converted into nitrates and nitrites in the extracellular space.

Nitric oxide plays a crucial regulatory function as an extracellular signaling molecule in the reproductive system. In the penis, NO is released by autonomic nerves causing dilatation of the local blood vessels responsible for erection. NO binds to the active site of the enzyme guanylate cyclase, stimulating it to produce the second messenger, cyclic guanosine monophosphate (cGMP). This effect lasts only seconds because cGMP has a high turnover rate due to its rapid degradation by cGMP-phosphodiesterase. The drug sildenafil (Viagra®) is a cGMP-phosphodiesterase inhibitor that reduces the cGMP turnover rate, keeping the penile blood vessels dilated and therefore, the penis erect.

1.3 Regulation of Synthesis and Actions of the Signaling Molecules: The Concept of Feedback

Most endocrine cells are both producers and targets of extracellular signaling molecules and act in concert to generate an integrated response to the changing external and internal environments. The neuroendocrine network possesses the ability to perceive changes and then selectively generates signals that control and coordinate hormone synthesis. The simplest mechanisms of hormone regulation are *negative feedback* and *positive feedback.* In negative feedback, a hormone secreted by the target gland signals the producer gland to decrease its activity. In positive feedback, the hormone produced by the target gland signals the producer gland to increase secretion.

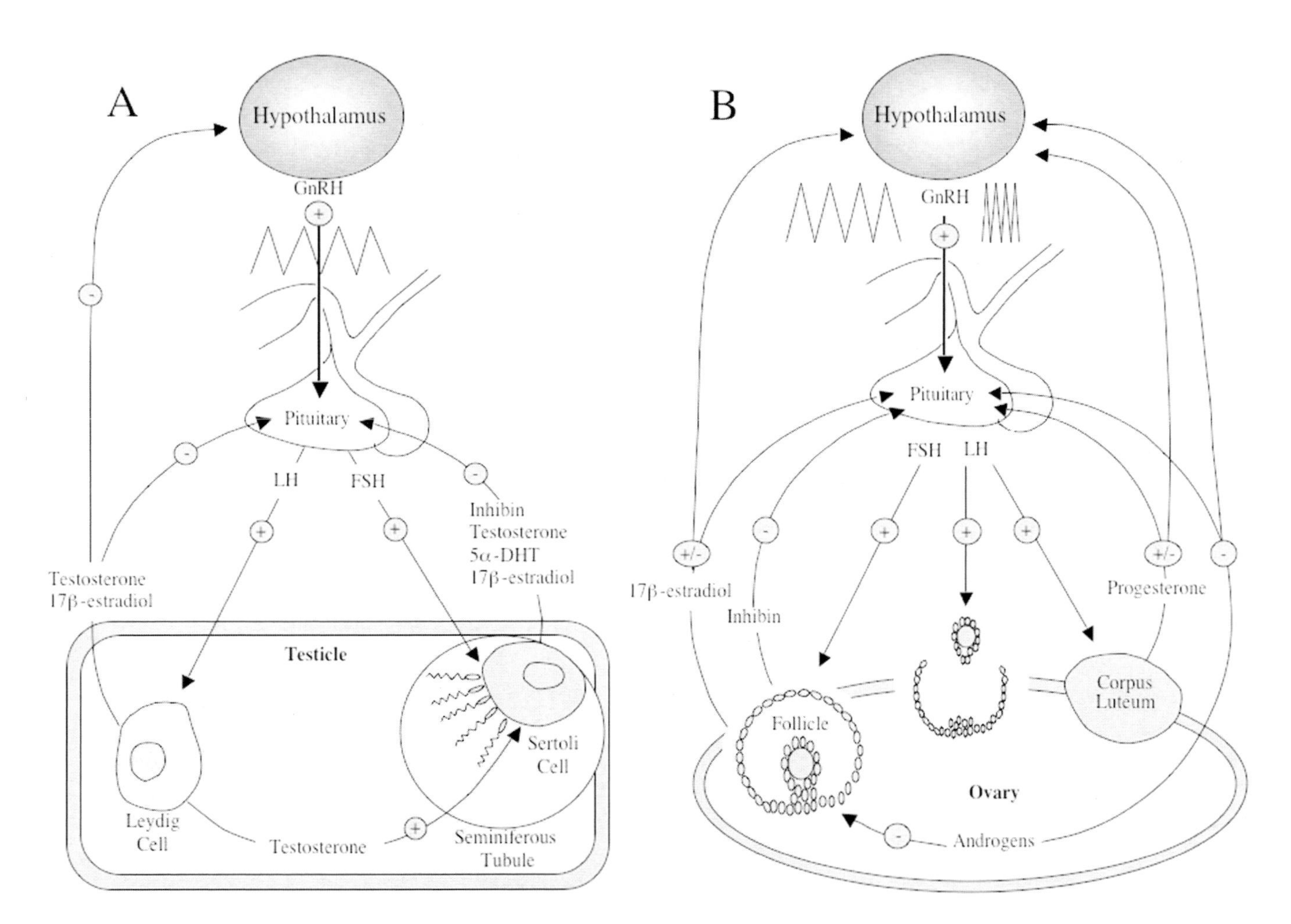

Fig. 1.2 Feedback regulation of testicular (**A**) and ovarian (**B**) functions

Regulation of gonadal activity by the hypothalamic pituitary axis is an example of both negative and positive feedback. Hypothalamic neurons secrete GnRH in pulses with a frequency of approximately one pulse every 90–120 minutes. Pulses of GnRH are followed by pulses of gonadotropins, mainly LH and to a lesser extent FSH. Both gonadotropins bind to specific receptors in the gonads regulating synthesis of steroid hormones.

In the male, LH stimulates synthesis of androgens, which bind to androgen receptors in both the hypothalamus and pituitary glands and exert negative feedback by reducing the frequency of LH pulses. Estradiol, which is produced locally in the hypothalamus by aromatization of androgens, binds to estrogen receptors in the hypothalamus and pituitary glands and reduces the amplitude of the LH pulses (Fig. 1.2A).

In the female, gonadotropins stimulate estrogen synthesis and induce ovulation. At the beginning of the follicular phase, estrogens produced by ovarian granulosa cells exert negative feedback on the hypothalamic pituitary axis. However, at the end of the follicular phase, elevated levels of estrogens exert positive feedback on the hypothalamic pituitary axis, which responds with an increase in the frequency of GnRH pulses followed by the release of an ovulatory surge of gonadotropins (Fig. 1.2B).

1.4 Summary

The *endocrine system* coordinates fundamental organic physiological maintenance of the body's internal environment, metabolism, reproduction, development, and growth. *Molecular endocrinology*, defined as the study of hormone action at the cellular and molecular levels, emerged from studies on the molecular mechanisms by which hormones and other extracellular signaling molecules affect expression of genes. *Hormones* are synthesized by specialized cells of the endocrine organs/glands and are secreted into the blood stream to exert their specific biochemical effects on target cells at distant sites, which is known as *endocrine signaling*. The classical definition of hormone was extended to include other extracellular signaling molecules (i.e., growth factors, neurotransmitters) that do not originate from the endocrine organs/glands and act by other signaling modes including paracrine, neurocrine, autocrine, juxtacrine, intracrine, cryptocrine, pherocrine, and photocrine. Thus, the endocrine and extracellular signaling systems provide a means of communication between distant organs via the circulatory system, specific cell populations, neighboring cell populations, and the external and internal environments.

Glossary of Terms and Acronyms

5α-DHT: 5α-dihydrotestosterone

ACTH: adrenocorticotropic hormone

ADH: antidiuretic hormone

aFGF: acidic fibroblast growth factor

AMH: antimüllerian hormone or Müllerian-inhibiting substance (MIS)

AR: androgen receptors

BDNF: brain-derived neurotrophic factor

bFGF: basic fibroblast growth factor

BMP: bone morphogenic proteins

cGMP: cyclic guanosine monophosphate

CNS: central nervous system

CNTF: ciliary neurotrophic factor

CO: carbon monoxide

CRF: corticotropin-releasing factor

CSF: colony-stimulating factor

eCG: equine chorionic gonadotropin

EGF: epidermal growth factor also known as urogastrone

ER: estrogen receptors

FGF: fibroblast growth factor or heparin-binding growth factors

FSH: follicle-stimulating hormone

GABA: γ-aminobutyric acid

GDF-9: growth differentiation factor-9

GH: growth hormone

GHIH: GH inhibitory hormone

GHRF: growth hormone releasing factor

GMP: guanosine monophosphate

GnRH: gonadotropin releasing hormone, also called LH releasing hormone, LHRH HB-

HB-EGF: heparin-binding EGF

hCG: human chorionic gonadotropin

HGF: hepatocyte growth factor

IGF-I: insulin-like growth factor-I or somatomedin-C

IGF-II: insulin-like growth factor-II

KAF: keratinocyte autocrine factor or amphiregulin

KGF: keratocyte growth factor

LH: luteinizing hormone

LIF: leukemia inhibitory factor

MIS: Müllerian-inhibiting substance or antimüllerian hormone, AMH

NGFs: the family of nerve growth factors, including nerve growth factor

NO: nitric oxide

NT-3: neurotropin-3

NT-6: neurotropin-6

P450arom: cytochrome P450 aromatase

PACAP: pituitary adenylate cyclase-activating polypeptide

PDGFs: platelet-derived growth factors

PRL: prolactin

SCF/KL: stem cell factor/Kit ligand

SRIF: somatotropin release-inhibiting factor or somatostatins also called GHIH

T_3: triiodothyronine

T_4: thyroxine

TGF: transforming growth factor

TNF: tumor necrosis factor

TRH: thyrotropin releasing hormone

TSH: thyroid stimulating hormone

VEGF: vascular endothelial growth factor

VIP: vasointestinal peptide

Bibilography

1. Baulieu E-E, Kelly PA. Hormones: From Molecules to Disease. Paris, New York: Hermann, Chapman and Hall, 1990.
2. Braunstein GD. Testes. In: Greenspan FS, Gardner DG, editors. Basic & Clinical Endocrinology, sixth edition. New York: McGraw-Hill, 2001:422–52.
3. Bolander F. Molecular Endocrinology, third edition. San Diego: Academic Press, 2004.
4. Diamond N. Perspectives in Reproduction and Sexual Behavior. Bloomington: Indiana University Press, 1968.
5. Grumbach MM, Conte FA. Disorders of sex differentiation. In: Wilson JD, Foster DW, Kronenberg HM, Larsen PR, editors. W. B Williams Textbook of Endocrinology, ninth edition. Philadelphia: W.B Saunders Company, 1998:1303–425.
6. Wilson JD, Foster DW, Kronenberg H, et al. Principles of endocrinology. In: Wilson JD, Foster DW, Kronenberg HM, Larsen PR., editors. Williams Textbook of Endocrinology, ninth edition. Philadelphia: W. B. Saunders Company, 1998:1–10.

References

1. Skinner MK. Cell–cell signaling in the testis and ovary. In: Bradshaw RA, Dennis EA, editors. Handbook of Cell Signaling, third volume. San Diego: Academic Press, 2003:531–43.

Chapter 2

Extracellular Signaling Receptors

Pedro J. Chedrese and Stella M. Celuch

2.1 Introduction

Classically, *receptors* were defined as cellular structures that recognize and bind hormones. In present days this definition is expanded to include receptors for a variety of other extracellular regulatory signaling molecules, such as growth factors and neurotransmitters. Accordingly, the term *ligand* was introduced to encompass all the extracellular signaling molecules that bind to receptors. The main attribute of receptors is their ability to specifically recognize a ligand among the many different molecules present in the environment surrounding the cells. Ligand-specific binding results in activation of intracellular signaling pathways, which amplify the signal and affect gene expression. Thus, receptors not only receive extracellular signals but also increase and transmit those signals to the genome and ultimately elicit a biological response.

2.2 Cell Surface and Nuclear Receptors

Extracellular signaling receptors are classified into two main groups according to their location in the cell: *cell surface receptors,* located in the plasma membrane, and *nuclear transcription factor receptors* located inside the cell. A common feature of the system is that ligands that evolved from identical or similar precursors are grouped into families that interact with families of related receptors (Table 2.1).

2.2.1 Cell Surface Receptors

Cell surface receptors are usually glycoproteins with three defined structural and functional domains:

- the *extracellular domain,* which is hydrophilic and contains the ligand-binding site;
- the *transmembrane domain* that anchors the receptor into the plasma membrane by one or more stretches of approximately 23 hydrophobic amino acids. This is the number of residues required to form an α-helix across the cell membrane; and
- the *intracellular domain,* which contains amino acids that are targets for phosphorylating enzymes that regulate receptor activity.

Cell surface receptors activate regulatory molecules called *effectors* starting a cascade of reactions that transform extracellular signals into intracellular signals. Effectors are either part of the membrane receptor itself or a complex group of molecules linked to the receptor (Fig 2.1). Cell surface receptors can be classified into three major groups.

Seven transmembrane domain receptors—include receptors for hormonal peptides, glycoproteins, neurotransmitters, and eicosanoids. They are anchored into the plasma membrane by seven hydrophobic stretches of amino acids, with their amino terminal facing the extracellular space and the carboxy terminal in the cytoplasm (Fig. 2.1A). A distinct characteristic of these receptors is that they are associated with a group of regulatory proteins called G-proteins that function as effectors by binding guanine nucleotides. Because these receptors span the cell membrane seven times and are associated with G-proteins, they are also called seven transmembrane segment (7-TMS) receptors, serpentine receptors, or G-protein-coupled receptors (GPCR).

Enzyme-linked receptors—include receptors for growth factors and cytokines. They possess intrinsic enzymatic activity, indicating that the receptor and the effector reside within the same molecule (Fig. 2.1B). These receptors have a single transmembrane domain that anchors the receptor into the membrane and an intracellular domain that contain the catalytic subunit of the enzyme. They are classified into four main groups:

P.J. Chedrese (✉)
Department of Biology, University of Saskatchewan College of Arts and Science, Saskatoon, SK, Canada
e-mail: jorge.chedrese@usask.ca

P.J. Chedrese (ed.), *Reproductive Endocrinology*,
DOI 10.1007/978-0-387-88186-7_2,

Table 2.1 Hormones and receptors grouped by families

Hormones	Type of receptor
PRL, GH, and placental lactogen	Tyrosine kinase associated
Insulin, IGF-I, and IGF-II	Tyrosine kinase
Secretin, glucagon, and VIP	Seven transmembrane
Gastrin and cholecystokinin	Seven transmembrane
LH, FSH, hCG, and eCG	Seven transmembrane
ACTH, MSH, and derivatives of the pro-opiomelanocortin	Seven transmembrane
Androgens, estrogens, progestins, glucocorticoids, mineralocorticoids, thyroid hormones, vitamin D, and retinoic acid	Zinc finger nuclear receptors

ACTH: adrenocorticotropic hormone; eCG: equine chorionic gonadotropic hormone; FSH: follicle-stimulating hormone; GH: growth hormone; hCG: human chorionic gonadotropic hormone; IGF-I and IGF-II: insulin-like growth factor-I and -II; MSH: melanocyte-stimulating hormone; PRL: prolactin; LH: luteinizing hormone; VIP: vasointestinal peptide.

i) *tyrosine kinase (TK) receptors* that transmit signals by autophosphorylation of their intracellular *tyrosine* residues;
ii) *tyrosine-kinase associated receptors* that are devoid of intrinsic enzymatic activity, but are associated with cytoplasmic tyrosine kinases;
iii) *serine-threonine kinase (STK) receptors* that transmit signals by autophosphorylation of their intracellular *serine* and *threonine* residues;
iv) *guanylate cyclase associated receptors* that mediate their effect by activating the enzyme guanylate cyclase.

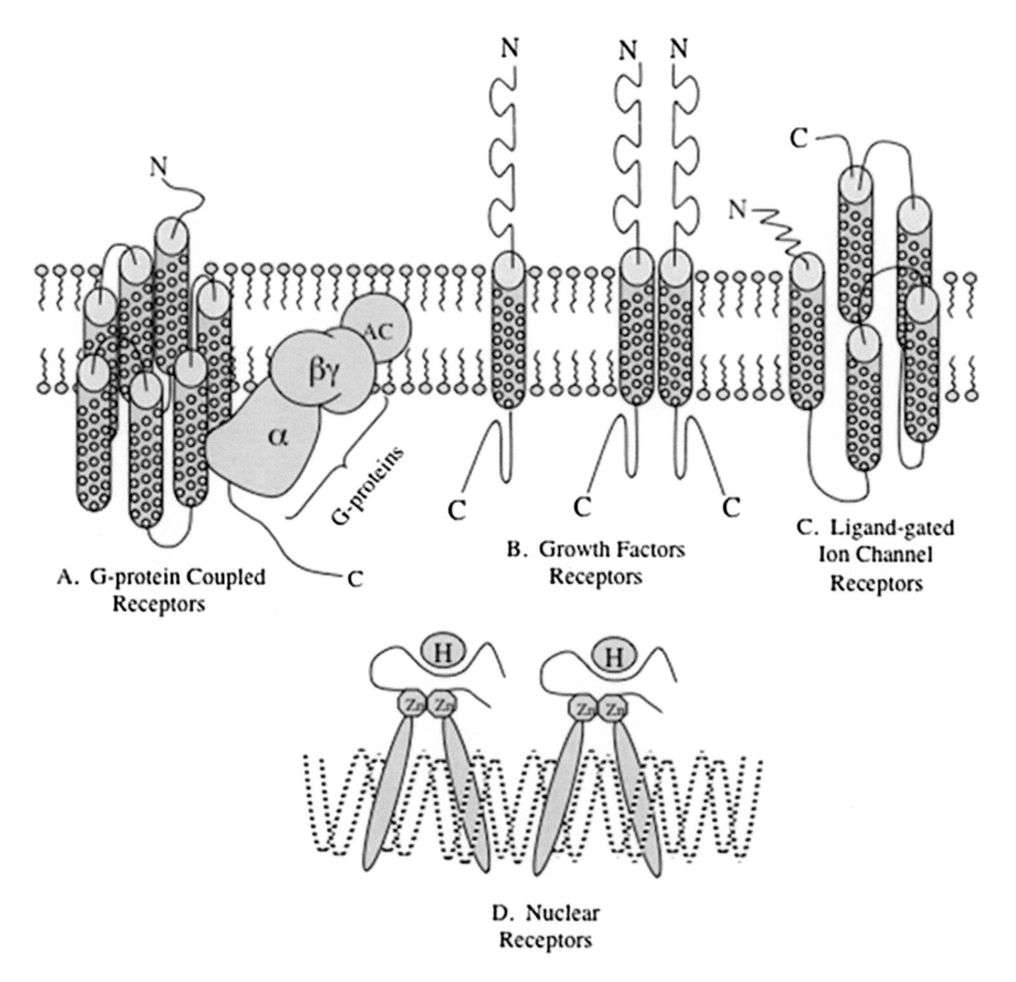

Fig. 2.1 Structural organization of the membrane and nuclear receptors. AC: adenylate cyclase; Greek symbols stand for the G-proteins associated to the seven transmembrane receptors; H: hormone; N and C stand for the amino and carboxy end of the molecules, respectively

Ligand-gated ion channel (LGIC) receptors or ionotropic receptors are ion-selective transmembrane channels formed by multi-subunit proteins (Fig. 2.1C). LGIC are activated and open upon binding of specific ligands, such as the neurotransmitters acetylcholine (nicotinic cholinergic receptors), glutamate (ionotropic glutamate receptors), and serotonin (5-HT_3 receptors).

2.2.2 Nuclear Transcription Factor Receptors

Nuclear transcription factor receptors compile a large family of intracellular single subunit phosphoproteins that interact with membrane-soluble signaling molecules and directly affect gene transcription (Fig. 2.1D). They are also called *ligand-activated zinc (Zn)-finger nuclear receptor transcription factors* because they contain one atom of Zn that interacts via coordination bonds with four cysteine residues and recognize specific DNA consensus sequences in the regulatory regions of their target genes. Members of this family include the receptors for

- progesterone (PR)
- estrogens (ER);
- glucocorticoids (GR);
- mineralocorticoids (MR);
- vitamin-D (VDR);
- thyroid hormone (TR);
- retinoic acid and 9-*cis*-retinoic acid (RAR); and
- orphan receptors, which are a group of receptors with unknown ligands.

2.3 Functional Properties of Receptors

Ligands bind to receptors by reversible *non-covalent* interactions, implying that the ligand can dissociate from the receptor. Overall, receptors are characterized by functional properties, including limited binding capacity, ligand specificity, cellular specificity, high affinity, and correlation with biological responses.

Limited binding capacity—cells have a limited number of receptors. This property can be demonstrated experimentally by incubating receptor samples with a labeled ligand (*H) in the presence of increasing concentrations of the unlabeled ligand (H). Since H and *H compete for the same receptor sites, H will displace *H until all receptor sites are occupied by the unlabeled ligand (Fig. 2.2).

Ligand specificity—under adequate conditions of salt concentration, temperature, and pH, receptors have the capacity to recognize specific ligands among hundreds

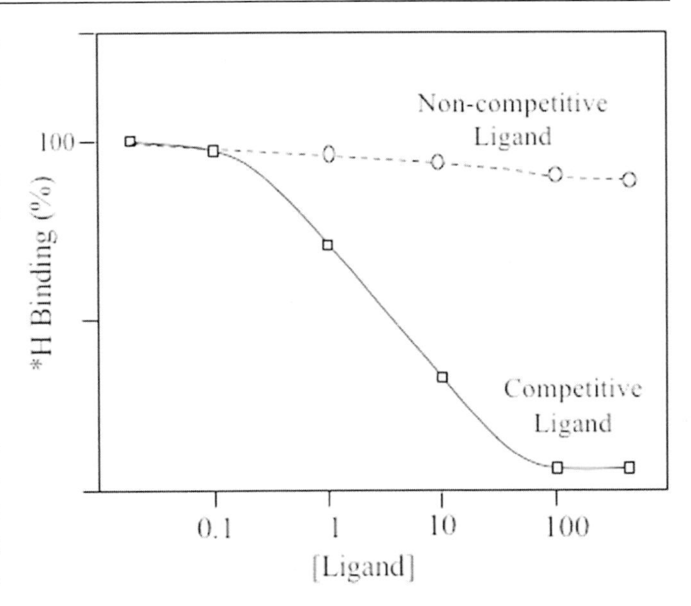

Fig. 2.2 Competition binding analysis. Receptors are incubated with a labeled ligand (*H) in the presence of increasing concentrations of either the unlabeled ligand (H), or other chemically related competitive ligand, or a non-competitive ligand. Competitive ligands including H displace *H from receptors while non-competitive ligand are unable to displace *H

present in the body's circulation. Approximately 500 different receptor ligands are expressed in high mammals, which circulate in the range of 10^{-11} to 10^{-9} M. In addition, many related molecules—such as steroids, amino acids, peptides, and proteins—circulate in the 10^{-5} to 10^{-3} M range. Therefore, target cells must bind ligands present at small concentrations while at the same time distinguish them from similar molecules circulating at higher concentrations. Overall, ligand specificity is a crucial property that accounts for the efficiency of the endocrine system. Ligand specificity is illustrated in Fig. 2.2, which shows that the competitive unlabeled ligand H displaces *H from its receptor, while a non-competitive ligand is unable to displace *H.

Cellular specificity—many different receptors are present in a wide variety of tissues. However, to be functional, receptors must be expressed in sufficient amounts within the target tissue in order to elicit a biological response.

High affinity—affinity refers to the intermolecular force that binds the ligand to its receptor. To compete with a large number of related molecules that may be present in the cellular environment, ligands must bind to the receptor with high affinity. Receptor affinity is represented by the dissociation constant at equilibrium (K_d) for the reaction:

$$\text{Ligand} + \text{Receptor} \rightleftarrows \text{Ligand-receptor Complex}$$

The K_d is unique for each ligand-receptor complex. K_d is expressed in molar (M) concentration units and defined as the concentration of ligand occupying 50% of the receptor

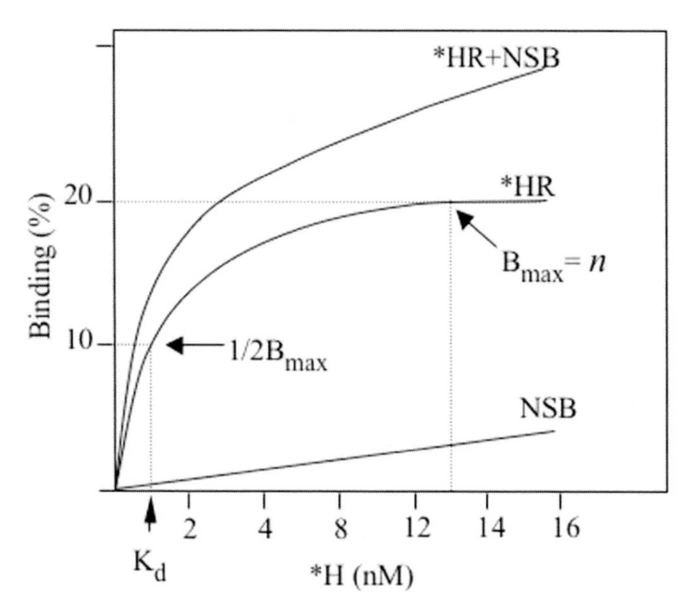

Fig. 2.3 Graphic representation of the equilibrium saturation-binding assay data. The *top* asymptotic hyperbola represents the total binding (*H+NSB). The *middle* asymptotic hyperbola represents the specific binding (SB), which is the relationship between increased concentration of *H and receptor occupancy, minus the value of NSB. B_{max}: maximum binding, which is a measure of *n*; *H: labeled ligand; K_d: dissociation constant, which is equivalent to the molar concentration of *H occupying 50% of the R sites; NSB: non-specific binding; R: receptor

binding sites (Fig. 2.3). The lower the numerical value of K_d, the higher the ligand affinity for its receptor. K_d values for hormones fall into the range of their plasma concentrations, usually 10^{-11} to 10^{-9} M. Thus, minor variations in hormone concentration will significantly change the number of occupied receptors.

Affinity distinguishes receptors from plasma transport proteins that bind and carry hormones (Table 2.2). Plasma transport proteins are much more abundant than receptors and bind steroid and thyroid hormones with lower affinity and specificity through weak hydrophobic and electrostatic interactions that are easily reversible. Thus, transport plasma proteins provide a readily available circulating reservoir of hormones. In addition, binding to plasma transport proteins protects steroid and thyroid hormones from metabolic degradation. Therefore, plasma half-life of steroid and thyroid hormones is much longer than the plasma half-life of protein hormones (hours versus seconds and minutes), which do not bind to transport proteins and circulate as free ligands.

Correlation with biological responses—ligand-receptor binding must occur within a ligand concentration range to cause a biological response. Ligand-receptor binding activates intracellular signaling pathways and gene expression, which leads to changes in protein synthesis, hormone secretion, and cell growth. Examples of correlation with biological responses are described below.

Table 2.2 Comparison between receptors and plasma transport proteins

	Receptors	Transport proteins
Concentration	Thousands/cell	Billions/ml
Binding affinity	10^{-11} to 10^{-9} M	10^{-7} to 10^{-5} M
Binding specificity	High	Low
Saturability	Yes	No
Reversibility	Yes	Yes
Signal transduction	Yes	No

2.4 Receptor Assays

Four main techniques are used to study receptors: quantitative analysis of receptor mRNA, immunoassays, radioligand binding, and functional assays.

2.4.1 Quantitative Analysis of Receptor mRNA

Receptor mRNA levels can be measured by nucleic acid hybridization, either in situ or in nucleic acid extracts, and by quantitative reverse transcription polymerase chain reaction, termed quantitative RT-PCR, real-time RT-PCR, or kinetic RT-PCR. These techniques involve the reverse transcription of a defined portion of mRNA followed by cycles of automated amplification and quantification of the resulting cDNA. Therefore, it is necessary to know the nucleotide sequence of the receptor mRNA in order to design specific oligonucleotides to be used as probes and primers.

Interpretation of the information emerging from quantitative RT-PCR is based on the assumption that mRNA is rapidly degraded and that the number of mRNA transcripts measured at a particular time is directly proportional to the number of receptor protein molecules. However, these results must be interpreted with caution, since the half-life of transcripts can vary. Moreover, since not all transcripts are necessarily translated into a functional protein, the amount of mRNA may not correspond with the number of functional receptors.

2.4.2 Immunoassays

Receptors can be studied by immunoassay techniques based on the same principles used to measure other proteins. A specific antibody against the receptor is required for efficient identification and quantification of the receptor-antibody complex. Types of immunoassays are as follows:

- Enzyme-linked immunoabsorbent assays (ELISAs) measure receptors in solubilized samples by using specific antibodies complexed on a solid matrix. These techniques employ either antigens (receptor proteins) or antibodies labeled with enzymes (i.e., horseradish peroxidase), which generate colored products that are quantified by spectrophotometry.
- Radioimmunoassay (RIA) is based on the competition between receptors labeled with ^{125}I and unlabeled receptors for binding sites at specific antibodies. Receptor number is determined by measuring the radioactivity bound to the receptor-antibody complex.
- Western Blot analysis involves the separation of proteins by gel electrophoresis followed by transfer to a membrane matrix and incubation with antibodies labeled with enzymes such as horseradish peroxidase. The use of chemiluminescent substrates allows identification and quantification of specific bands on a photographic film by densitometry.
- Immunohistochemistry and immunocytochemistry involve incubation of either tissue sections (immunohistochemistry) or cultured cells (immunocytochemistry) in the presence of specific antibodies. Antigen–antibody complexes labeled with color or fluorescent tags are localized by microscopy.

ELISA and RIA are sensitive and accurate quantitative methods; whereas Western Blot, immunohistochemistry, and immunocytochemistry are semi-quantitative methods. Overall, the information obtained is limited because they only provide estimation of receptor number, which may not correspond to function. In addition, these techniques do not provide information on either receptor affinity or hormone specificity.

2.4.3 Radioligand Binding Assays

This methodology allows accurate measurements of specificity, affinity, and number of receptors [1]. Radioligand binding assays are performed on either purified membranes or solubilized receptor preparations. Radioligand binding assays require

- ligand (H) and the same, or functionally similar, ligand (*H) labeled with a radioactive isotope (^{125}I for protein ligands, ^{3}H or ^{14}C for steroid ligands);
- samples containing the receptors (R), either particulated cell membranes or cytosolic fraction or nuclear preparations;
- a method to separate the ligand bound to receptors (B) from the free ligand (F) in solution, such as centrifugation or membrane filtration; and
- a detection system for γ- (^{125}I) or β- (^{3}H or ^{14}C) radiation.

The basic procedure consists of adding increasing concentrations of labeled ligand (*H) to a fixed amount of receptor sample and determining the specific binding (SB) of *H to R. Labeled ligands can bind to many components of the reaction, including the test tube, which are non-specific binding (NSB) sites. Therefore, NSB is determined by conducting the same reaction in the presence of an excess of H, which will displace *H from the receptors, but not from the NSB sites.

As shown in Fig. 2.3, NSB is linear with respect to ligand concentration because it is unsaturable and related to low affinity/high capacity binding sites. SB is calculated by subtracting NSB from total binding. Percentage of SB is the ratio of SB to total labeled hormone added.

$$\%SB = (^{*}HR - NSB)100/^{*}H$$

Binding of a ligand to a receptor can be analyzed as predicted by the *law of mass action:*

$$k_a H + R \rightleftarrows HR k_d \quad (2.1)$$

where the *rate of association* between H and R is equal to

$$k_a[H][R] \quad (2.2)$$

and the *rate of dissociation* of HR is equal to

$$k_d[HR] \quad (2.3)$$

B: Bound ligand;
B$_{max}$: maximun binding
F: Free ligand (H or *H);
[H]: Molar concentration (M) of free ligand;
***H:** Labeled ligand;
[HR]: Concentration of hormone-receptor complex;
k_a: *Association Rate Constant* (M^{-1} sec^{-1});
K$_a$: *Association Constant at Equilibrium* (M^{-1});
k_d: *Dissociation Rate Constant* (M);
K$_d$: *Dissociation Constant at Equilibrium* (M);
***n*:** Number of receptors;
NSB: Non-specific binding
[R]: Concentration of free receptors;
R$_{max}$: maximal response
SB: Specific binding.

When the reaction reaches the equilibrium, the *rate of association* is equal to the *rate of dissociation*.

Therefore,

$$k_a[\mathrm{H}][\mathrm{R}] = \mathrm{k_d}[\mathrm{HR}], \quad (2.4)$$

which can be rearranged as

$$[\mathrm{H}][\mathrm{R}] = [\mathrm{HR}]k_d/k_a \quad (2.5)$$

or

$$k_d/k_a = [\mathrm{H}][\mathrm{R}]/[\mathrm{HR}] \quad (2.6)$$

The ratio of the two rate constants is the *dissociation constant at equilibrium*, K_d, equal to

$$[\mathrm{H}][\mathrm{R}]/[\mathrm{HR}] = k_d/k_a \quad (2.7)$$

Or its reciprocal, the *association constant at equilibrium*, K_a, that measure ligand affinity for the receptor and is equal to

$$[\mathrm{HR}]/[\mathrm{H}][\mathrm{R}] = \mathrm{k_a/k_d} \quad (2.8)$$

2.4.4 Equilibrium Saturation-Binding Assay

A commonly used approach for studying receptors is the equilibrium saturation-binding assay (Fig. 2.3), which provides information on ligand affinity, receptor number, and specificity of ligand-receptor interaction.

Equilibrium saturation binding assay requires the incubation of samples containing equal amount of protein in the presence of increasing concentrations of *H. The incubation time must be long enough to allow for equilibrium, which means that the rate of formation and dissociation of the ligand-receptor complex are constant. Moreover, *H must reach concentrations high enough to occupy all receptors in the sample. The graphic representation of the equilibrium saturation-binding assay data is an asymptotic hyperbola, in which the maximum binding (B_{max}) is a measure of *n*, and the dissociation constant at equilibrium (K_d) is the molar concentration of *H occupying 50% of the total receptor population (Fig. 2.3).

2.5 Interpretation of the Equilibrium Saturation Binding Assay Data

A variety of derivations of the *law of mass action* have been used to linearize and simplify the interpretation of the saturation-binding assay data. A traditional method of linearization is the *Scatchard* plot [2] (Fig. 2.4A), by which *n* and *affinity* can be estimated on the basis of the saturation-binding assay. In the *Scatchard plot*, [*H] plus [H] are considered the free ligand (F), while [HR] is considered the bound ligand (B). Therefore, when the reaction reaches the equilibrium, [*n*–B] will be equal to [R].

$$n = [\mathrm{R}] + [\mathrm{HR}], \text{ thus } [\mathrm{R}] = \mathrm{n} - [\mathrm{HR}] \quad (2.9)$$

Substituting these terms in equation (2.8):

$$\mathrm{K_a} = [\mathrm{HR}]/[\mathrm{H}](\mathrm{n} - [\mathrm{HR}]) \quad (2.10)$$

Since [H] is equal to F and [HR] equal to B,

$$\mathrm{K_a} = \mathrm{B/F(n - B)} \quad (2.11)$$

To solve this equation, both sides have to be multiplied by (*n*–B):

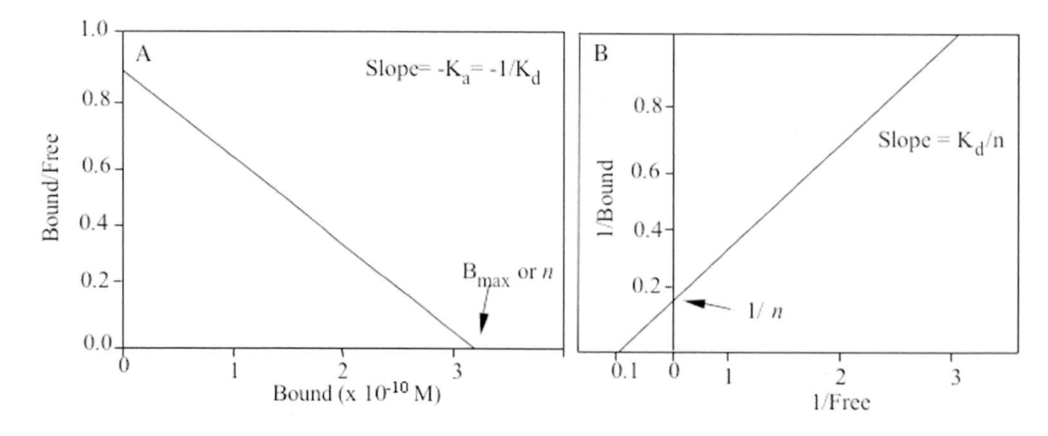

Fig. 2.4 Graphic linear representation of equilibrium saturation-binding assay data by the Scatchard plot (**A**) or the double reciprocal plot of Lineweaver-Burk (**B**). B_{max}: maximum binding at equilibrium, which is a measure of the total number of receptors (n); Bound = [HR]; Free + [*H]+[H]; K_a: association constant; K_d: dissociation constant

$$(n - \mathrm{B})K_a = (n - \mathrm{B})\mathrm{B}/\mathrm{F}(n - \mathrm{B}) \tag{2.12}$$

Which is equal to

$$(n - \mathrm{B})\mathrm{K_a} = \mathrm{B}/\mathrm{F} \tag{2.13}$$

The data from the equilibrium saturation-binding assay can be plotted using equation (2.13), which is the equation for a straight line (Fig. 2.4A):

- B/F is the ordinate;
- B is the abscissa;
- The x intercept at B/F = 0 is the B_{max} or total number of receptors (n); and
- The slope is the negative value of the association constant ($-K_a$), also defined in terms of the dissociation constant K_d (slope = $-1/K_d$).

In addition to the Scatchard plot, equilibrium saturation binding assay data from equation (2.11) can rearranged to obtain other linear plots, such as the Lineweaver-Burk depicted in Fig. 2.4B, in which $1/B = K_d\, n\, F + 1/n$.

Scatchard analysis for a one-site binding assay yields a straight line. However, there are cases in which the Scatchard analysis yields curvilinear plots (Fig. 2.5) [3], suggesting that there are at least two components, one of high affinity/low capacity and the other with low affinity/high capacity. Curvilinear plots are observed when ligands bind to multiple forms of the same receptor, such as the β_1- and β_2-adrenoceptors. Curvilinear plots are also observed when the receptor exhibits negative cooperativity, such as the case of the insulin receptor, in which affinity decreases when the hormone is added at high concentrations. Actually, linear plots are very useful for easy visualization of changes in B_{max}, K_d, and K_a, but they fail to satisfy rigorous statistical criteria. Computer analysis is now available and linearization of the data may no longer be needed.

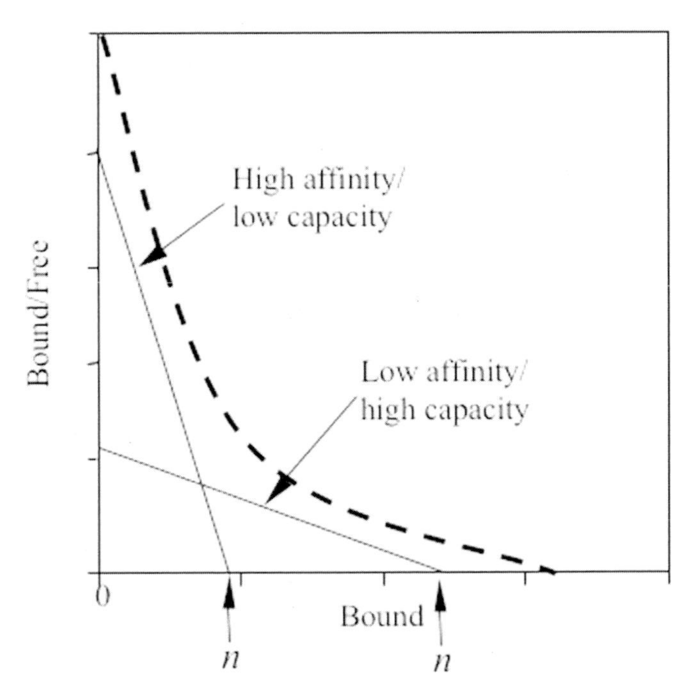

Fig. 2.5 Scatchard plot for a receptor sample with two independent binding sites. The *curvilinear dashed line* is interpreted as the sum of two binding sites. *Straight-lines* represent a single population of high affinity receptors with low number of binding sites, or a single population of low affinity receptors with high number of binding sites. Modified from De Meyts [3]

2.6 Functional Receptor Assays

The previously described methods are useful to analyze receptor expression and ligand affinity. However, functional assays that analyze biological responses elicited through the activation of the receptor, either in vivo or in vitro, are also required to completely understand how receptors operate. Functional assays are usually interpreted on the grounds of *Receptor Occupancy Theory* that describes the relationship between the degree of receptor occupation (binding) and the biological response [4,5]. The results of functional assays are plotted as log of the ligand (or agonist) dose versus biological response, which yield a sigmoid curve (Figs. 2.6 and 2.7). The effective dose/concentration of agonist that produces 50% of the maximal response (R_{max}) is termed effective dose/concentration 50 (ED_{50} / EC_{50}; Fig. 2.6A).

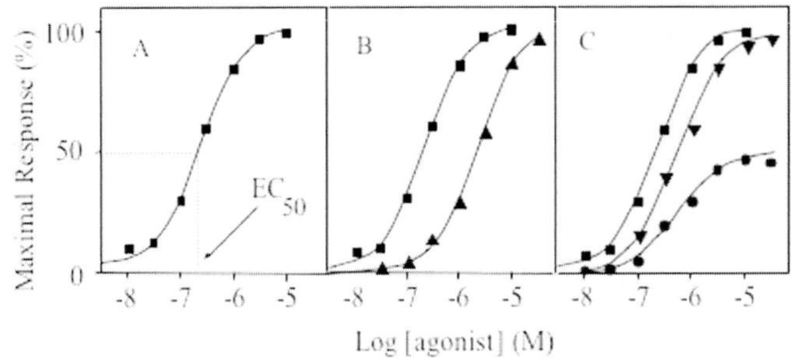

Fig. 2.6 Functional studies: relationship between ligand-receptor binding and biological response. Data is expressed as percentages of the maximal response (R_{max}). Molar concentrations of the agonist are represented as log values. (**A**): Agonist EC_{50} (). (**B**): A competitive antagonist shifts the curve to the right (). (**C**): Low concentration of non-competitive antagonist does not alter the R_{max} but change the EC_{50} by shifting the curve to the right (▼). In the same graph, high concentration of non-competitive antagonist decreases R_{max} (). ED_{50}/EC_{50}: Effective dose/concentration of agonist that produces 50% of the maximal response; R_{max}: maximal response

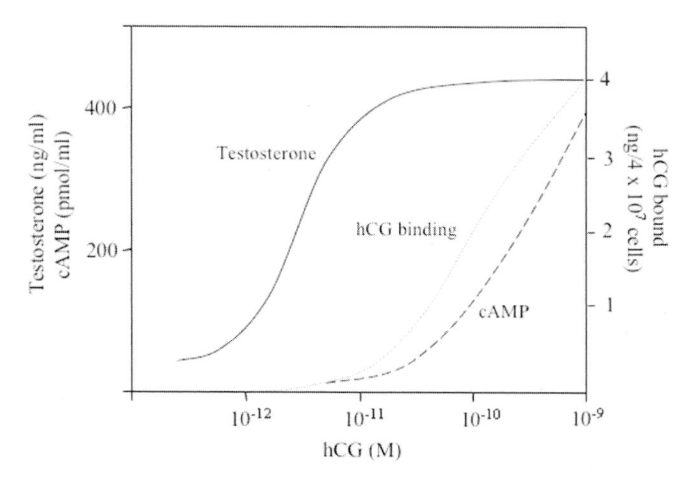

Fig. 2.7 Relationship between receptor occupancy, stimulation of adenylate cyclase and synthesis of testosterone in rat Leydig cell. Modified from Catt and Dufau [6]

2.7 Regulation of Receptors

Receptors, like other proteins, are continuously synthesized and recycled or degraded. Since these processes occur at different rates, the number of available receptors could change over time. In addition, the intrinsic activity of the receptors may change as well. These changes modify the sensitivity to ligands and hence affect the biological responses.

Down-regulation or desensitization is the decrease in the sensitivity to a hormone due to a decrease in the number of functional receptors. Down-regulation may occur in response to high plasma concentrations of a hormone. An example of this is the decrease of β_2-adrenoceptor population in limphocytes of patients with pheochromocytoma, a catecholamine-secreting tumor [7]. Administration of pharmacological doses of hormone can also cause down-regulation. Forexample, gonadotropin-releasing hormone (GnRH), which is normally secreted in pulses by the hypothalamus, stimulates gonadotropins synthesis and release, but when administered in large doses, it causes down-regulation of the GnRH receptor and hence inhibits the synthesis and release of gonadotropins [8,9].

Up-regulation is positive regulation of the number of receptors. An example of up-regulation is the observation that high plasma concentrations of estrogens and FSH increase the number of LH receptors in the ovarian granulosa cells.

Receptor cross-talk refers to the interaction between intracellular signaling pathways linked to independent receptors. In this way, the activation of a receptor could regulate or even trigger signaling processes linked to other receptors expressed in the same cell. An example of cross-talk is the activation of estrogen-, progesterone-, and androgen-receptor signaling by growth factors [10].

2.8 Receptor Terminology

- *Endogenous agonists* are ligands synthesized within the organism, such as hormones, neurotransmitters, and growth factors, which bind receptors and elicit biological responses.
- *Exogenous agonists* are ligands generated outside the organism that can occupy the same receptors as the endogenous agonists and produce the same biological responses.
- *Full agonists* are ligands that induce maximal biological responses. In some cases this can be achieved without occupation of all the available receptors. The fact that only a fraction of the total receptor population could be required to elicit a full biological response supports the concept of amplification of the ligand-receptor induced signal.
- *Spare receptors* or reserve receptors are the receptors that remain unoccupied when the maximal biological response to a full agonist has been reached. For example, there is a close relationship between LH binding to LH receptors and synthesis of the second messenger, cyclic AMP, in rat Leydig cells. However, cAMP-dependent maximal stimulation of steroidogenesis is achieved when less than 1% of the LH receptors are occupied (Fig. 2.7) [6].
- *Partial agonist* is a ligand with lower intrinsic activity than a full agonist, thereby inducing a lower maximal response.
- *Competitive antagonist* is a ligand with no intrinsic activity that competes with the agonist to occupy the same receptor site, but does not activate the receptor. Since both ligands, the agonist and the competitive antagonist, compete for the same binding site, there is a parallel shift to the right in the concentration/dose-response curve for the agonist, without changes in the maximal effect (Fig. 2.6B).
- *Non-competitive antagonist* is an antagonist that occupies the receptor at a site other than the agonist-binding site. Therefore, the non-competitive antagonist does not impair the binding of the agonist but blocks the receptor activation. A low concentration of a non-competitive antagonist may produce a shift to the right in the concentration/dose-response curve without changing the maximal response, due to the existence of reserve receptors. However, as reserve receptors become occupied by higher concentrations of antagonist, the maximal response to the agonist will decrease (Fig. 2.6C).
- *Receptor constitutive activity* means that the receptor elicits a biological response in the absence of agonist. Constitutive activity was first described in in vitro overexpressed-receptor systems, but there is now evidence of constitutive signaling in native G-protein-coupled receptors in vivo. An example of this condition is the melanocortin receptor MC1 and its endogenous inverse agonist agouti [11].

- *Inverse agonists* are agonists with negative intrinsic activity. Inverse agonists occupy the same receptor-sites as the agonists, but inhibit receptor constitutive activity causing an opposite effect. In fact, many ligands that have been classified as pure antagonists in the past are now considered as inverse agonists for some tissues [12,13].
- *Cryptic receptors* are inactive forms of receptors. It is believed that cryptic receptors are a ready reservoir of receptors available for rapid cellular response. Examples of cryptic receptors are the receptors for GnRH, neuropeptide-Y, insulin, FSH, and LH, which are present in the plasma membrane and can be rapidly induced into their active forms by phosphorylation, sulfhydrylation, cleavage, or deglycosylation.
- *Isoreceptors* are receptors with different structures and functions that are recognized by the same endogenous ligands. Isoreceptors may have different tissue distribution and can be linked to independent intracellular signaling pathways, thus eliciting different, even opposite, effects. Isoreceptors may account for the diverse effects of the same ligand in a tissue. For instance, β_1-, β_2-, and β_3-adrenoceptors mediate lipolysis with different sensitivities, allowing response to a wide range of concentrations of catecholamines.

2.9 Summary

Extracellular signal-induced effects are mediated by activation of either cell surface receptors, such as G-protein coupled and enzyme-linked receptors, or the intracellular receptors, such as the nuclear transcription factor receptors. These binding sites are characterized by limited population size, ligand and cellular specificity, high affinity, and interaction with intracellular signaling pathways and gene expression regulatory mechanisms. Extracellular signaling receptors can be analyzed by quantification of specific mRNA tissue levels as well as by immunoassays of receptor protein. Other methods to study receptors are radioligand-binding and functional assays that allow getting information about receptor kinetics and biological responses. Extracellular signaling receptor-mediated biological responses are affected by regulatory mechanisms such as down- and up-regulation and receptor cross-talk.

Glossary of Terms and Acronyms

5-HT$_3$: 5-hydroxytriptamine

7-TMS: seven transmembrane segment

ACTH: adrenocorticotropic hormone

cDNA: complementary DNA

eCG: equine chorionic gonadotropin

ELISA: enzyme-linked immunoabsorbent assays

ER: estrogen

FSH: follicle stimulating hormone

GH: growth hormone

GnRH: gonadotropin-releasing hormone

GPCR: G-protein-coupled receptor

GR: glucocorticoids

hCG: human chorionic gonadotropin

IGF: insuline-like growth factor

LGIC: ligand-gated ion channel

LH: luteinizing hormone

MC1: melanocortin receptor

MR: mineralocorticoids

mRNA: messenger RNA

MSH: melanocyte-stimulating hormone

PR: progesterone

PRL: prolactin

RAR: retinoic acid and 9-*cis*-retinoic acid

RIA: radioimmunoassay

RT-PCR: reverse transcriptase polymerase chain reaction

STK: serine-threonine kinase

TK: tyrosine kinase

TR: thyroid hormone

VDR: vitamin-D

VIP: vasointestinal peptide

Bibilography

1. Baulieu EE, Kelly PA. Hormones: From Molecules to Disease. Paris, New York: Harmann, Chapman & Hall, 1990.
2. Bolander F. Molecular Endocrinology, third edition. San Diego: Academic Press, 2004.
3. Gospodarowicz D. Properties of the luteinizing hormone receptor of isolated bovine corpus luteum plasma membranes. J Biol Chem 1973; 248:5042–9.
4. Ketelslegers JM, Knott GD, Catt KJ. Kinetics of gonadotropin binding by receptors of the rat testis. Analysis by a non-linear curve-fitting method. Biochemistry 1975; 14:3075–83.

5. Lehman FPA. Stereoselective molecular recognition in biology. In: Cuatrecasas P, Greaves MF, editors. Receptors and Recognition. London: Chapman & Hall, 1978:18–46.
6. Schrader WT, O'Malley BW. Laboratory Methods Manual for Hormone Action and Molecular Biology, eleventh edition. Houston: Department of Cell Biology Baylor College of Medicine, 1987.
7. Wilson JD, Foster DW, Kronenberg H, et al. Principles of endocrinology. In: Wilson DW, Foster HM, Kronenberg H, editors. Williams Textbook of Endocrinology, ninth edition. Philadelphia: W. B. Saunders Company, 1998:1–10.

References

1. Winzor DJ, Sawyer WH. Quantitative Characterization of Ligand Binding. New York: Wiley-Liss, 1995.
2. Scatchard G. An attraction of proteins for small molecules and ions. Ann NY Acad Sci 1949; 51:660–72.
3. De Meyts P, Roth J, Neville DM Jr, et al. Insulin interactions with the receptors: experimental evidence for negative cooperativity. Biochem Biophys Res Comm 1973; 55:154–61.
4. Ariens EJ, Van Rossum JM, Simonis AM. Affinity, intrinsic activity and drug interactions. Pharmacol Rev 1957; 9:218–36.
5. Furchgott RF. Receptor mechanisms. Ann Rev Pharmacol 1964; 4:21–50.
6. Catt KJ, Dufau ML. Interactions of LH and hCG with testicular gonadotropin receptors. In: O'Malley BW, Means AR, editors. Receptors for Reproductive Hormones. Advances in Experimental Medicine and Biology. New York: Academic Press, 1979: 379–418.
7. Cases A, Bono M, Gaya J, et al. Reversible decrease of surface beta 2-adrenoceptor number and response in lymphocytes of patients with pheochromocytoma. Clin Exp Hypertens 1995; 17: 537–49.
8. Chedrese PJ, Kay TWH, Jameson JL. GnRH stimulates of glycoprotein hormone α-subunit mRNA levels by increasing transcription and mRNA stability. Endocrinology 1994; 134: 2475–81.
9. Kay TWH, Chedrese PJ, Jameson JL. GnRH causes transcriptional activation followed by desensitization in transfected α-T3 gonadotrope cells. Endocrinology 1994; 134:568–73.
10. Picard D. Molecular mechanisms of cross-talk between growth factors and nuclear receptor signaling. Pure Appl Chem 2003; 75:1743–56.
11. Adan RAH. Constitutive receptor activity series. Endogenous inverse agonists and constitutive receptor activity in the melanocortin system. Trends Pharmacol Sci 2006; 27:183–6.
12. Strange PG. Mechanisms of inverse agonism at G-protein-coupled-receptors. Trends Pharmacol Sci 2002; 23:89–95.
13. Bond RA, IJzerman AP. Recent developments in constitutive receptor activity and inverse agonism, and their potential for GPCR drug discovery. Trends Pharmacol Sci 2006; 27:92–6.

Chapter 3

The Molecules That Transmit Information into the Cell: The Intracellular Signaling Pathways

Pedro J. Chedrese and Alejandro M. Bertorello

3.1 Introduction

Extracellular regulatory molecules convey information into the cell through a fast low energy complex of signals known as *intracellular signaling pathways*. The function of the pathways is to organize and amplify the signals in a way that a small number of ligands bound to receptors affect the activity of a large number of intracellular molecules. Signaling molecules can be divided into two main groups: the *intracellular messengers* and *homology domain* proteins.

The notion of intracellular messengers derives from the concept of ligands as extracellular messengers; ligands were termed *first messengers*, therefore, the signaling molecules produced in response to ligand-activated receptors were termed *second messengers*. They include 3′,5′-monophosphate monophosphate (cyclic AMP or cAMP), cyclic guanosine monophosphate (cGMP), inositol-1,4,5-triphosphate (IP_3), 1,2-diacylglycerol (DAG), phosphatidylinositols (PI), arachidonic acid, calcium ions (Ca^{2+}), nitric oxide (NO), and carbon monoxide (CO). These messengers initiate cascade of intracellular reactions leading to activation of the *transcription factors* that regulate gene expression. Thus, some authors consider transcription factors to be the *third messengers* of the signaling pathways.

A hierarchical integration of such signals provides the cell with appropriate compartmentalization of information in a defined time and space avoiding deleterious cross talk between pathways. This is largely achieved by the capacity of signaling molecules to recognize each other through specific sites (i.e., protein modules, domains), forming an intracellular net of communication that connects the receptor with the intracellular targets. This is a higher level of organization in which scaffolding proteins, also termed adaptors that contain non-catalytic regions, termed *homology domains*, provide recognition and binding properties to the signaling molecules. They include, among others, the *Ras Homology (RH) domain*, a region originally described in proteins of the rat sarcoma oncogene (*ras*) present in the G-proteins that recruit other proteins of the same group and operate as molecular switches, the *Src Homology (SH) regions, SH2 and SH3*, originally described in proteins of the Rous sarcoma virus (*src*) oncogene family of *tyrosine* kinases, the *Pleckstrin Homology (PH) domain*, a region of approximately 120 amino acids that recognize and bind membrane proteins and phospholipids, and polyproline regions that specifically recognize SH3 domains. Scaffolding proteins connect receptors with effectors and provide secure levels of integration within the pathway, and can also interconnect different pathways.

3.2 Signaling Through Seven Transmembrane Receptors

The seven transmembrane receptors are the largest family of cell membrane-bound receptors. Their molecular structure, depicted in Fig. 3.1, is characterized by the following:

- extracellular glycosylated amino-terminal regions of varying lengths that form the ligand binding sites;
- three extracellular and three intracellular hydrophilic loops that connect seven transmembrane domains, which are composed predominantly of hydrophobic α-helices of 21–23 amino acids. The amino acids of the transmembrane domains interact with the membrane lipids, while the intracellular domains carry amino acid recognition sequences for the G-proteins; and
- cytoplasmic C-terminal domains that contain phosphorylation sites on specific amino acid residues, which regulate receptor activity.

Signaling through seven transmembrane receptors involves the participation of different molecular entities,

P.J. Chedrese (✉)
Department of Biology, University of Saskatchewan College of Arts and Science, Saskatoon, SK, Canada
e-mail: jorge.chedrese@usask.ca

P.J. Chedrese (ed.), *Reproductive Endocrinology*,
DOI 10.1007/978-0-387-88186-7_3, © Springer Science+Business Media, LLC 2009

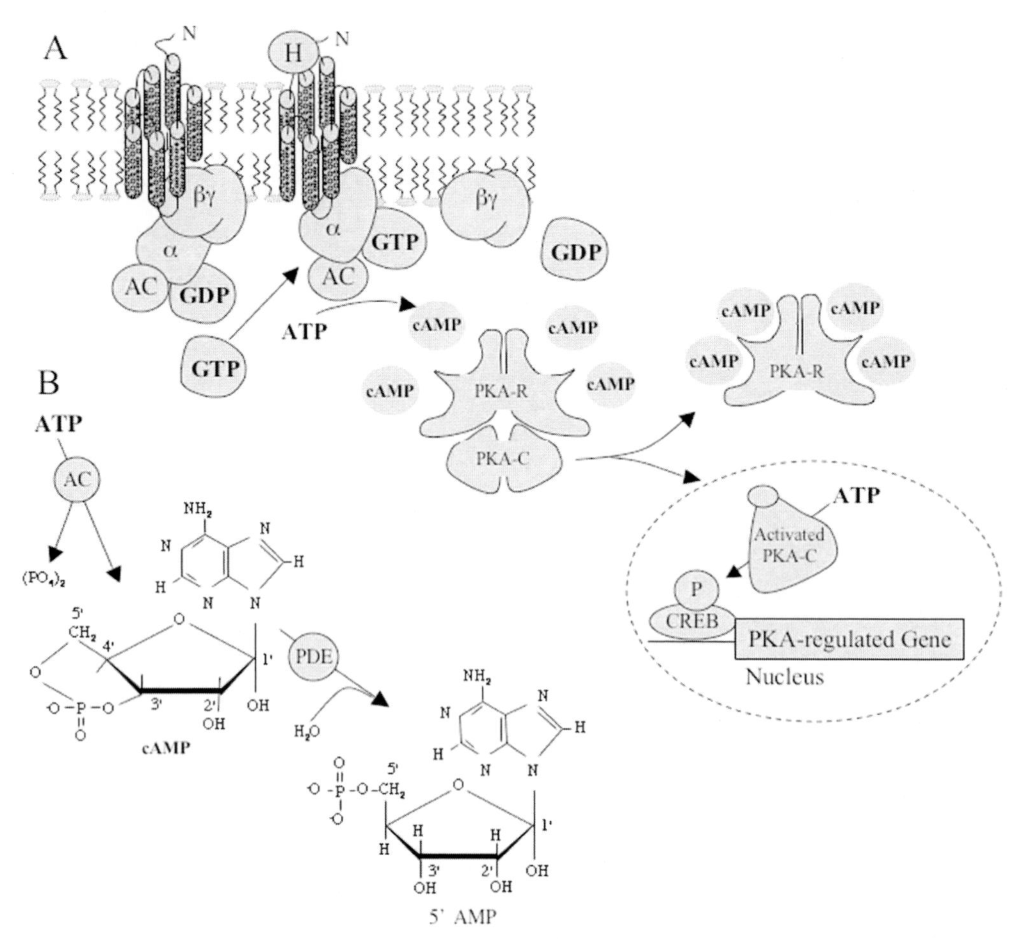

Fig. 3.1 (**A**) Organization of the seven transmembrane receptor and hormonal activation of the PKA pathway. The structure on the *left* represents an inactive receptor bound to the G-proteins. The structure on the *right* represents the reactions upon ligand binding to the receptor, including dissociation of the β/γ subunits from the G-protein complex, synthesis of cAMP, and activation of PKA. Activated PKA phosphorylates a variety of intracellular proteins including the transcription factor CREB. (**B**) Synthesis and degradation of cAMP. AC converts ATP into cAMP, which is immediately converted into 5′AMP by PDE. AC: Adenylate cyclase; 5′AMP: 5′ adenosine monophosphate; ATP: adenosine triphosphate; cAMP: cyclic 3′, 5′-AMP; CREB: cAMP regulatory element binding protein; GDP: guanosine diphosphate; GTP: guanosine triphosphate; H: hormone; P: indicates phosphorylation; PDE: phosphodiesterase; PKA: protein kinase-A; PKA-C: catalytic subunit of PKA; PKA-R: regulatory subunit of PKA; Greek symbols α, β, γ represent G-protein subunits

including G-protein molecular switches, signal effectors and second messengers, and second messenger's metabolizing enzymes.

3.2.1 G-Protein Molecular Switches

G-protein molecular switches belong to the extensive family of G-proteins that are characterized by the ability of promoting hydrolysis of guanine ribonucleotides. They couple receptors with "amplifier" enzymes, termed *signal effectors*, located on the internal side of the cell membrane that serve as allosteric regulators of receptor activity cycling between active and inactive states. Thus, they function as molecular switches and are divided into two groups, *large G-proteins complexes* and *small G-proteins.*

Large G-protein complexes are heterodimers formed by three different subunits, α– (39–46 kDa), β– (37 kDa), and γ– (8 kDa), anchored to the hydrophobic amino acids of the plasma membrane by lipid prenylation and myristolation. Small G-proteins are monomeric and range between 20 and 25 kDa. They are also called *ras*-like G-proteins for their similarities with the oncogene *ras* product. Small G-proteins are involved in the tyrosine kinase receptor signaling pathways and a variety of other cellular processes, including vesicular transport, membrane ruffling, stress fiber formation, and focal adhesion.

In the basal state, large G-proteins remain bound to guanosine diphosphate (GDP) in a trimeric complex, αβγ (Fig. 3.1A). Upon ligand-receptor binding, the G-protein α-subunit develops affinity for guanosine triphosphate (GTP), thus displacing GDP and dissociates from the complex to become activated α-GTP. Both, α-GTP and the dissociated βγ complex can stimulate a variety of effectors, including enzymes and ion channels, generating second messengers, which either stimulate or inhibit cellular responses. The α-subunit possesses intrinsic enzymatic activity (GTPase) making α-GTP unstable because hydrolyzes GTP into GDP and inorganic phosphate (P_i), causing the proteins to rapidly re-associate into the inactive basal state form, αβγ. In summary, ligand binding to seven transmembrane receptors involves reduction in receptor affinity for the ligand and dissociation of the α-subunit from the αβγ complex.

The α-subunit either activates or inactivates effectors depending on GTPase activity, which switches the reaction off and deactivates the effector. There are more than 20 known G-protein α-subunits involved in coupling seven transmembrane receptors with effectors, some of which are summarized in Table 3.1. They can be divided into four major groups according to amino acid sequence similarities: (1) *G-stimulatory α-subunit ($G_{\alpha s}$)* activated by the gonadotropins FSH and LH, ACTH, and hormones and neurotransmitters that increase cAMP levels by stimulating adenylate cyclase; (2) *G-inhibitory α-subunit ($G_{\alpha i}$)*, activated by opiates, somatostatins, angiotensin-II, β_2-adrenergic agonists, the light-activated rhodopsin, and signaling molecules that reduce cAMP by inhibiting adenylate cyclase; (3) *$G_q\alpha$-subunit*, comprised of six members associated to the muscarinic acetylcholine receptor and the α_1-adrenergic receptor stimulate phospholipase-Cβ; and (4) *G_{12}-α-subunit,* comprised of two members that stimulate phospholipase-Cβ.

The β- and γ-subunits of the G-proteins are also a large family of proteins. There are at least four β-subunits and six γ-subunits. The γ-subunit is prenylated and its main function is to anchor the G-protein complex to the plasma membrane. The βγ dimmer plays a regulatory role by inactivating $G_{\alpha s}$ and activating phospholipase-$C_{\beta 2}$, and also participates in agonist-induced receptor phosphorylation and desensitization.

3.2.2 Signal Effectors and Second Messengers

Three major intracellular pathways convey the G-protein activated signals: *the cAMP dependent protein kinase-A (PKA) pathway, the phosphatidyl inositol/protein kinase-C (PI/PKC) pathway,* and *the phospholipase-A2 (PLA_2) pathway.*

The PKA pathway—the effector of the PKA pathway is the enzyme *adenylate cyclase*, also termed *adenylyl* or *adenyl cyclase,* which is a membrane protein of approximately 1,100 amino acids, organized in two clusters of six transmembrane domains separated by two similar catalytic domains (Fig. 3.1A). There are at least nine isoforms of adenylyl cyclase regulated by the membrane's G-proteins with $G_{\alpha s}$ as the main stimulator and $G_{\alpha i}$ as the main inhibitor, and by a complex of Ca^{2+} bound to calmodulin (Ca^{2+}-CaM). Adenylate cyclase converts adenosine triphosphate (ATP) into the second-messenger cAMP that activates the tetrameric PKA expressed in all eukaryotic cells (Fig. 3.1B). PKA is composed of two regulatory (R) and two catalytic (C) subunits bound to each other with high affinity. The amino terminal domain of the R-subunits is responsible for the protein–protein interactions involved in dimerization between both subunits and binding to the C-subunit, while the carboxy terminal domain that consists of two tandems of repeated sequences form two binding sites for cAMP termed A-cAMP and B-cAMP sites. When cAMP binds to the R-subunits it releases the C-subunits, which transfer the terminal phosphate group from ATP to specific *serine* or *threonine* residues

Table 3.1 General features of selected G-proteins, their effectors and intracellular messengers

Receptor	G-protein	Effectors	Signal
FSH, LH, hCG, TSH, GHRH, PTH, PACAP, β-adrenergic, glucagon, calcitonin, vasopressin V_2, VIP, CRF, dopamine D_1, serotonin 5-HT_3	G_α-stimulatory ($G_{\alpha s}$)	Adenylate cyclase(+)L-type channels	↑cAMP↑Ca^{2+} influx
Odorants	$G_{\alpha olf}$	Adenylate cyclase(+)	↑cAMP
α_2-adrenergic, somatostatin, opioids, muscarinic-M_2 and M_4, adenosine A_1, angiotensin-II, serotonin 5-HT_1	G_α-inhibitory ($G_{\alpha i}$)	Adenylate cyclase(-)	↓cAMP
Rhodopsin (Rods)	$G_{\alpha t1}$	Guanylate cyclase & phosphodiesterase	↓cGMP (light/vision)
Rhodopsin (Cones)	$G_{\alpha t2}$	Guanylate cyclase & phosphodiesterase	↓cGMP (light/color)
Muscarinic-M_2 and M_4	$G_{\alpha 0}$	Phospholipase-CL-type channels	↑IP_3, DAG, Ca^{2+},↓Ca^{2+} influx, ↓cAMP
A1-adrenergic Angiotensin AT1	$G_{\alpha q}$	Phospholipase-$C_{\beta 1}$	↑IP_3, DAG, Ca^{2+} ↑Ca^{2+} influx

5-HT_3: 5-hydroxytryptamine; Ca^{2+}: calcium ions; cAMP: cyclic adenosine monophosphate; cGMP: cyclic guanosine monophosphate; CRF: corticotropin releasing factor; DAG: 1-2 diacylglycerol; FSH: follicle stimulating hormone; GHRH: growth hormone releasing hormone; hCG: human chorionic gonadotropin; IP_3: inositol triphosphate; LH: luteinizing hormone; PACAP: pituitary adenylate cyclase-activating polypeptide; PTH: parathyroid hormone; cyclase-activating polypeptide; TSH: thyroid stimulating hormone; VIP: vasoactive intestinal peptide.

of selected intracellular proteins of the PKA pathway. Kinetic analysis of the dissociation reaction between the R- and the C-subunits supports the concept that cAMP operates as an allosteric regulator of PKA. cAMP binds first to the B-cAMP site on the R-subunit, which leads to conformational changes. As a result, affinity of the R-subunit decreases and dissociates from the C-subunit liberating a second A-cAMP site that binds to another molecule of cAMP. Once both sites are occupied by cAMP the reaction reaches maximum velocity. The free R-subunits are recycled and remain as dimers for re-association with free C-subunits.

Two different classes of PKA R-subunits have been described, RI and RII, each of which have two isotypes—RIα and RIβ and RIIα and RIIβ. The C-subunit is expressed in three isotypes: C_α, C_β and C_γ. The C_α isotypes may have a common pattern of expression, since the RIα, RIIα, and C_α are widely distributed in a variety of tissues, whereas the expression of the β isotypes, RIβ, RIIβ, and C_β, are predominant in the brain, neuroendocrine, and endocrine tissues. Expression of C_β is mainly limited to the nervous system. The RIα and RIβ subunits are mainly found in the cytoplasm, while RIIα and RIIβ are in the cell membrane. RIIβ is also localized into the nucleus and perinuclear region, suggesting that this subunit may be associated with a C-subunit that targets nuclear phosphorylation substrates. Overall, the specific functions of the different PKA subunits are not completely understood. However, it is believed that responses to the cAMP pathway result in a differential subcellular distribution of the enzyme depending on the relative expression of the different isoforms and the type of tissue involved. The PKA C-subunit phosphorylates a variety of cytoplasmic proteins initiating a cascade of kinase reactions that amplify the original extracellular signal's magnitude by several orders. In addition, the PKC C-subunit phosphorylates nuclear proteins, including the transcription factor CREB involved in transcriptional regulation of several genes (Chapter 5).

The cAMP metabolizing enzymes, termed nucleotide phosphodiesterases (PDEs), hydrolyze cAMP into the inactive metabolite 5′-adenosine monophosphate (5′AMP) and regulate the overall activity of the PKA pathway (Fig. 3.1B). The PDEs belong to a large family of mammalian proteins, constituted of at least 20 different members that contain a central catalytic domain and an N-terminal regulatory domain that binds Ca^{2+}-CaM and the second messenger cGMP. They are classified according to their physical and functional properties into five main types: Type-I, stimulated by Ca^{2+}-CaM and by Ca^{2+}-mobilizing agonists is inhibited by methylxanthines; Type-II, stimulated by cAMP and the atrial natriuretic peptides (ANP); Type-III, inhibited by cGMP in competition with cAMP, is regulated by insulin, glucagon, and dexamethasone; Type-IV is activated specifically by cAMP-stimulating agonists; and Type-V that includes different isoforms expressed in rods and cones that are specifically activated by cGMP and *transducin* during visual transduction.

Positive and negative regulation of the PKA pathway through the PDEs has been reported in several hormone-receptor systems. Positive regulation by PDEs was observed in systems in which ligand binding-activated adenylate cyclase, followed by cAMP-induced phosphorylation of Type-III PDE, decreases the activity of Ca^{2+}-CaM PDEs. Negative regulation through the PDEs was reported in the testicular Sertoli cells, in which FSH activation of the PKA pathway also increases the expression of the high-affinity cAMP-specific PDE. It is believed that this effect may account for the progressive loss of responsiveness to FSH observed during sexual maturation.

The PI/PKC pathway—the PI/PKC pathway is activated by the phosphatidyl inositol-specific phospholipase-C (PLC), also termed phosphoinositide phospholipase-C, which is a cell membrane bound enzyme associated with G-proteins and seven transmembrane receptors (Table 3.2). PLC belongs to a family of hormone-dependent enzymes that specifically hydrolyzes membrane phospholipids containing the sugar *myo*-inositol as a polar group, such as phosphatidylinositol 4,5-biphosphate or $PI(4,5)P_2$ (Fig. 3.2). This family consists of 13 isozymes divided into six subfamilies, PLC-δ, -β, -γ, -ε, -ζ, and -η, which require Ca^{2+} for catalytic activity, except the -δ and -ζ isoforms that are inactive at basal intracellular Ca^{2+} concentrations. All members of the family contain two catalytic domains, X and Y, and mediate extracellular signaling effects but with different mechanisms of action:

- PLC-β contains the PH domain and a long C-terminal extension, which is required for activation by the Gαq subunit by a mechanism analogous to receptor activation of adenyl cyclase;
- PLC-γ contains PH, SH2, and SH3 domains and is activated by *tyrosine* kinase receptors, including the receptors for epidermal growth factor (EGF), platelet derived growth factor (PDGF), and fibroblast growth factor (FGF) that do not couple to G-proteins;
- PLC-δ is activated by binding to PIP_2 through its RH domain and by high intracellular levels of Ca^{2+};

Table 3.2 Phosphatidylinositol/protein kinase-C (PI/PKC) pathway-activated receptors and their effectors

Receptor	Effectors
TRH, GnRH, Angiotensin-II Muscarinic-M_2	Activates PLCβ
Neuropeptide YY_1, TSH	Activates PI-PLCβ and inhibits AC
PTH, Calcitonin, LH, EGF	Activates PI-PLCβ and AC
PDGF	Activates PI-PLCγ

AC: adenylate cyclase; EGF: epidermal growth factor; GnRH: gonadotropin-releasing hormone; LH: luteinizing hormone; PDGF: platelet-derived growth factor; PI-PLCβ and PI-PLCγ: phosphatidylinositol-specific phospholipase-C-β and -γ, respectively; PLC: phospholipase-C-β; PTH: parathyroid hormone; TSH: thyroid-stimulating hormone; TRH: thyroid-stimulating-releasing hormone.

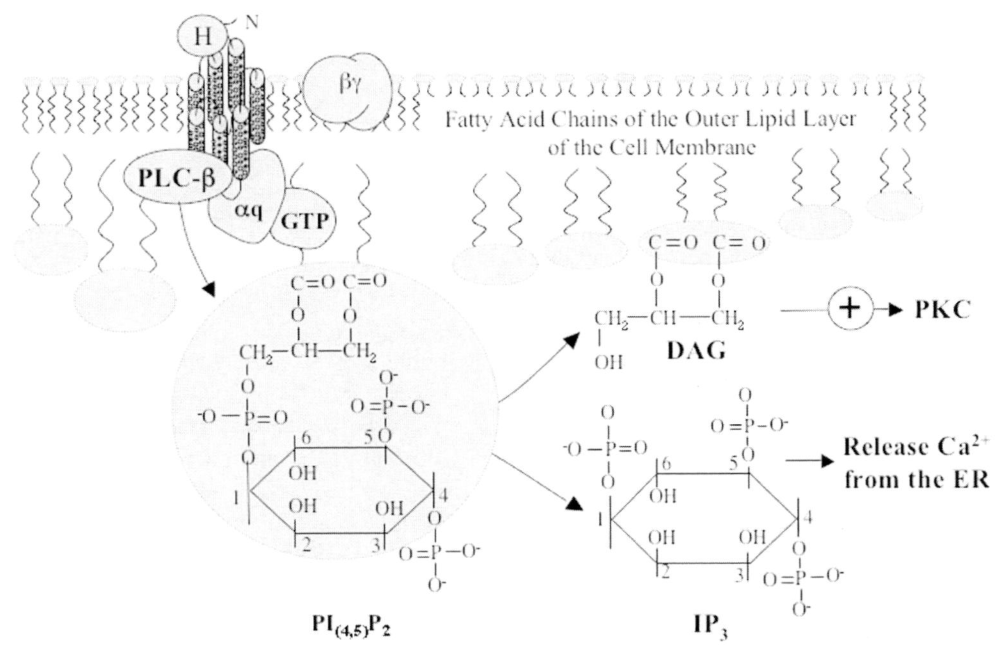

Fig. 3.2 PI cleavage and activation of the PKC pathway. Hormone binding to a PLC-β-associated receptor induces cleavage of the cell membrane phospholipid, PI, which is converted into the second messengers IP_3 and DAG. IP_3 binds to IP_3-R in the endoplasmic reticulum causing liberation of Ca^{2+}, while DAG stimulates PKC. Ca^{2+}: calcium ions; DAG: 1,2-diacylglycerol; ER: endoplasmic reticulum; H: hormone; IP_3: Inositol 1-4-5-triphosphate; $PI_{(4,5)}P_2$: PI 4-5-biphosphate; PI: phosphatidylinositol; PKC: protein kinase C; PLC-β: Phospholipase C-β; Greek symbols α, β, γ represent G-protein subunits

- PLC-ε contains the RH domain and is activated by Ras;
- PLC-ζ is involved in the fertilization process, although its mode of action is currently unknown; and
- PLC-η is involved in neuronal functioning.

The PI/PKC pathway is activated mainly by ligand binding to receptors associated to G-proteins that activate PLCβ leading to the conversion of $PI(4,5)P_2$ into the second messengers IP_3 and DAG.

IP_3 is a small, water-soluble molecule that moves away of membranes and rapidly diffuses into the cytoplasm (Fig. 3.2). IP_3 binds to endoplasmic reticulum receptors and open Ca^{2+}-channels, inducing a rapid release of the stored Ca^{2+} into the cytoplasm, which is followed by activation of cell membrane mechanisms that rapidly pump the ion out of the cell. Additionally, IP_3 can be further phosphorylated to IP_4, which functions as another intracellular signaling molecule, or inactivated by dephosphorylation by specific phosphatases to form IP_2.

DAG, on the contrary, remains attached to the membrane by its non-polar chains during signaling where it stimulates various types of PKC (Fig. 3.2). DAG can be further hydrolyzed by a DAG-lipase to the second messenger arachidonic acid. PKC was originally termed calcium-activated phospholipid-dependent protein kinase for its dependency of Ca^{2+} and phospholipids. Now, the term PKC refers to a large family of proteins of ~13 isozymes, some of which have relative dependence on Ca^{2+} or phospholipids. They are divided into three subfamilies based on their requirements for activation: (1) conventional, or classical, the isoforms α, βI, βII, and γ, that require Ca^{2+}, DAG, and a phospholipid, such as phosphatidylcholine; (2) novel, the isoforms, δ, ε, η, and θ, that require DAG, but not Ca^{2+}; and (3) atypical, the isoforms Mζ and ι/λ, which requires neither Ca^{2+} nor diacylglycerol.

Thus, conventional and novel PKCs are activated through the same signal transduction pathway as phospholipase-C. Upon activation, PKCs are translocated to the plasma membrane by a membrane-bound receptor for activated PKC (RACK: *r*eceptor for *a*ctivated *C*-*k*inase) and remain activated after the original activation signal, or the Ca^{2+}-wave, disappears. The main functions of the PKCs are to phosphorylate intracellular signaling proteins, including *m*yristolated *a*lanine-*r*ich C-*k*inase *s*ubstrate (MARKS), a protein that participates in cell shape motility, secretion, transmembrane transport, and cell cycle regulation, *m*itogen *a*ctivated *p*rotein *k*inases (MAPK), Raf, epidermal growth factor receptor (EGF-R), and vitamin-D receptor (VD-R). It is believed that alternative activation of the different isoenzymes, mediated by the receptors of different extracellular signals, can lead to distinct cellular responses.

Calcium signaling—it has been known for some time that Ca^{2+} participate in muscle contraction and neurotransmitter secretion. Now it is recognized that Ca^{2+} fit into the criteria to be defined as a second messenger, participating as an intracellular mediator in hormone secretion, gene transcription, and apoptosis. Extracellular signals regulate cytoplasmic

Ca^{2+} concentrations by controlling the extracellular influx through the plasma membrane, or the release from the intracellular stores. Plasma membrane mechanisms of Ca^{2+} influx include voltage-sensitive Ca^2-channels (VSCCs) and receptor-operated Ca^2-channels (ROC).

The VSCCs are the best characterized, of which three types have been described: long lasting (L-type); transient (T-type), found in neuroendocrine and anterior pituitary gland cells; and neither long-lasting nor transient (N-type), primarily found in neurons. In addition, Ca^{2+}influx is regulated by the Ca^{2+} pump, Ca^{2+}-Mg^{2+} ATPase, and the Na^+-Ca^{2+} exchange mechanism dependent on the Na^+ gradient established by the Na^+-K^+ ATPase. The intracellular stores of Ca^2 include the endoplasmic reticulum (ER) and the mitochondria; both sequester free Ca^{2+} from the cytoplasm. The membranes of the ER have a high-affinity/low-capacity Ca^{2+}-Mg^{2+}-ATPase that pumps Ca^{2+} into the vesicular lumen. Thus, the ER maintains low Ca^{2+} cytoplasmic concentrations within the nM range, while the mitochondria have low-affinity/high-capacity Ca^{2+}-Mg^{2+} ATPase pumps that regulate Ca^{2+} when the intracellular concentrations reach the micromolar range.

Upon ligand binding to receptors, cytoplasmic Ca^{2+} can increase from ~100 nM up to ~1,500 nM in two phases. The first is a short elevation associated to membrane mobilization of PI and synthesis of DAG and IP_3. IP_3 binds to IP_3-receptors in the ER, which is a Ca^{2+}channel that releases the ion into the cytoplasm. The released Ca^{2+} binds with high affinity to CaM, which is a ubiquitous member of a family of small acidic proteins with multiple copies of a helix-loop-helix motif that serves as an intracellular Ca^{2+} receptor. The Ca^{2+}-activated CaM amplifies and propagates extracellular regulatory signals by phosphorylation of a broad array of proteins activating numerous enzymatic systems, including adenylate cyclase, guanylate cyclase, cyclic nucleotide phosphodiesterases, PKC isoenzymes α, βI, βII, and γ, Ca^{2+}- and Mg^{2+}-ATPase, and calcineurin, also called protein phosphatase IIB (described below), which is involved in negative feedback regulation by dephosphorylation.

Thus, by influencing activities of these enzymes, the Ca^{2+}-activated CaM links the intracellular messenger systems and regulates others enzymes involved in signaling. The second Ca^{2+} elevation is caused by influx of extracellular Ca^{2+} through VSCC, which appears to be needed to replenish intracellular stores following prolonged hormonal stimulation. The entire system is regulated by negative feedback mechanisms that extrude Ca^{2+} from cells and activate sequestration into the ER.

The phospholipase-A2 pathway—in addition to PLC, other phospholipases are activated in response to extracellular signaling, such as PLA_2. PLA_2 belongs to a group of several unrelated enzymes, including cytosolic forms that are involved in intracellular signaling, and the secreted forms, such as pancreatic PLA_2 that participate in phospholipid digestion in the intestine. Two groups of PLA_2 cytosolic forms have been described, the Ca^{2+}-dependent and Ca^{2+}-independent. The cytosolic Ca^{2+}-dependent PLA_2 is activated by phosphorylation at *serine*-505 and influx of Ca^{2+} through a mechanism regulated by PKC, *tyrosine* kinase receptors, and MAPK. Stimulated Ca^{2+}-dependent PLA_2 translocates to the membrane where it specifically hydrolyzes the acyl bond at the second position (sn-2 acyl) of phospholipids releasing arachidonic acid and lysophospholipids. The Ca^{2+}-independent PLA_2 is stimulated when Ca^{2+} levels fall and dissociate from CaM, and by a group of *cystein/aspartate*-specific proteases, termed *caspases* that are involved in regulation of apoptosis.

The secreted PLA_2 (sPLA2) is located extracellularly; however, it can be internalized and transported to the nucleus to generate arachidonic acid, which is both a second messenger and a precursor of the eicosanoids, such as the prostaglandins and leukotrienes. Prostaglandins and leukotrines are synthesized in most tissues and exert a wide variety of biological activities, including participation in pain, inflammatory responses, ovulation, and regression of the corpus luteum. Eicosanoids are also synthesized from DAG, released from the cell membranes lipids by PLC as described above.

3.3 Signaling Through Ion Channel Receptors

In addition to Ca^2-channels, there are cell membrane receptor channels that control passage of the cations, K^+, Na^+, and Mg^{2+}, and the anions Cl^- and HCO_3^-, into the cell (Fig. 3.3). They are termed ligand-gated ion channels (LGIC) and include the receptors for several neurotransmitters, among others for acetylcholine (ACh), γ-aminobutiric acid (GABA), glycine, glutamate, serotonin, and ATP. ACh has two types of receptors: the muscarinic ACh receptor (mACh-R), which is a seven transmembrane G-protein associated receptor, and the nicotinic receptor (nACh-R), also called nicotinic cholinergic, which is expressed as a ligand-gated-Na^+ channel in the muscle and a Ca^{2+}-channel in the nervous system. 5-hydroxytryptamine (5-HT_3) and GABA mediate their effects through seven transmembrane G-protein associated receptors and the LGIC. However, 5-HT_3 also binds to a divalent cation channel receptor, while GABA binds to the $GABA_A$ and $GABA_C$ receptors that are Cl^- channels. The most thoroughly studied LGIC receptor is the nACh-R, which was first characterized in the *Torpedo californica*, an electric ray native to the eastern Pacific Ocean. The nACh-R is composed of 12 different glycosylated subunits, α2 through 10 and β2 through 4. These subunits combine to form pentamers

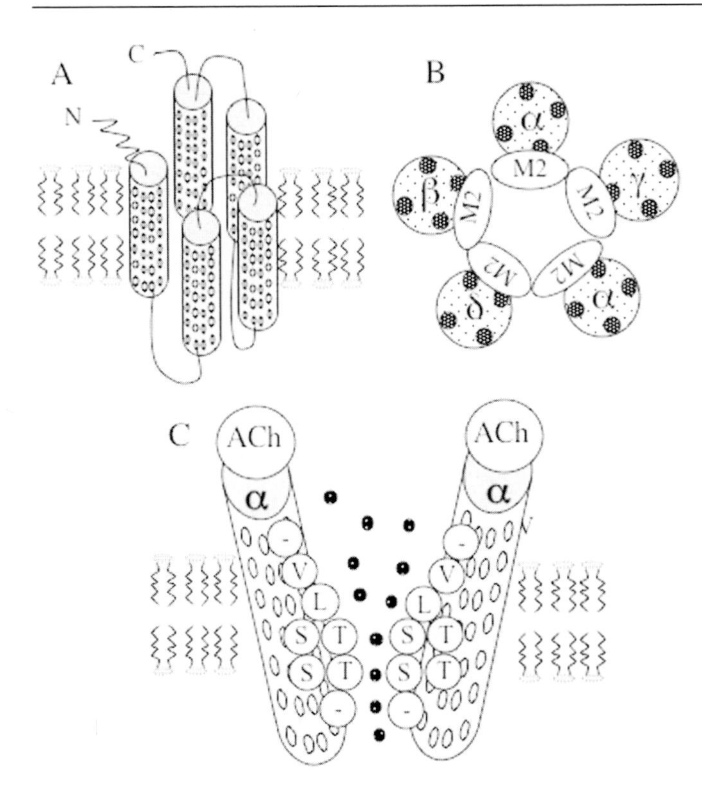

Fig. 3.3 Schematic representation of the nicotinic acetylcholine receptor (nACh-R). (**A**) Structural organization of the nACh-R across the plasma membrane. (**B**) Transverse cross section view from above at the level of the plasma membrane showing the second helix (M2) and the four transmembrane α-helices from each subunit. (**C**) Longitudinal section of two α-subunits bound to acetylcholine (ACh) showing passage of ions through the channel. V: *valine*, L: *leucine*, S: *serine*, T: *threonine*

that are similar to one another, mainly in the hydrophobic domains. The muscle nACh-R is composed of two α-subunits, one β-subunit, one δ-subunit, and either one γ-subunit or one ε-subunit. The neuronal forms of the nACh-R are more heterogeneous with a larger number of subunit combinations. The subunits are organized into a pentagonal array forming a pore that crosses the membrane (Fig. 3.3). The structure of the ligand-gated ion channel receptors can be described as follows:

- each subunit is composed of four hydrophobic α-helical domains and each domain spans the plasma membrane four times;
- the α-subunits contain the ligand-binding site, thus, activation of the receptor requires binding of two molecules of the ligand;
- a large extracellular amino-terminal domain with multiple sites for glycosylation, together with one of the α-transmembrane domains in each subunit (M2), lines the pore for the passage of ions;
- a short extracellular carboxy-terminal domain; and
- a large intracytoplasmic loop, located between transmembrane regions 3 and 4, contains multiple potential phosphorylation target sites, suggesting that protein kinases may regulate this receptor activity.

The channel part of the receptor is constituted by a structure of several rings of specific amino acids (Fig. 3.3C). An outer hydrophilic ring is followed by two hydrophobic rings, one of *valine* and the second of *leucine*, and by two *serine-threonine* rings. The *serine-threonine* rings are the narrowest part of the channel and function as a gate. Upon ACh binding, the M2 rotates in a way that leucine side chains move to the side, enlarging the pore size like a diaphragm. The structures of the other known ligand-gated ion channel are similar to that of the nACh-R, suggesting that they may have derived from a common ancestral gene.

The main functional feature of this group of receptors is that they do not depend on intracellular mediators; therefore they can convey a signal rapidly into the cell. However, the LGIC are also regulated by the G-proteins. The mACh-R-regulated K^+ channel and the β-adrenergic-gated L-type Ca^{2+}-channel found in the heart are regulated by the βγ and α_s G-proteins, respectively. The activity of LGIC is also regulated by PKA and PKC, which can phosphorylate the large intracellular loops of their G-proteins γ- and δ-subunits. PKA- and PKC-mediated phosphorylation is associated with receptor desensitization and prevention of ligand binding ion channel subunit assembly.

3.4 Signaling Through Enzyme-Linked Receptors

Enzyme-linked receptors are single-span transmembrane proteins with two functions carried by the same molecule: it binds the ligand and serves as the cell membrane signal effector. The effector is a *kinase* that transfers phosphate groups to specific amino acid residues of the receptor. Enzyme-linked receptors mediate the effects of some extracellular messengers that fit into the definition of classical hormones, however, most of their ligands are local regulators that act in a paracrine, autocrine, or juxtacrine mode of signaling (Chapter 1). Although enzyme-linked receptors were originally classified as typically slow response mediators of extracellular signals, it is now known that they can also mediate direct and rapid changes on gene expression that affects the cytoskeleton, and regulate cell growth, shape, and movements. Disorders of cell proliferation, differentiation, survival, and migration can give rise to tumors, for which it is believed that abnormalities in enzyme-linked receptors signaling pathways may be involved in the etiology of cancer. Enzyme-linked receptors can be classified into four classes:

tyrosine kinase, tyrosine kinase-associated, serine/threonine kinase, and guanylate cyclase-associated.

3.4.1 Tyrosine Kinase Receptors

Tyrosine kinase receptors are single transmembrane ligand binding proteins that are able to phosphorylate their own tyrosine residues (*autophosphorylation*) located in the cytosolic region of the molecule. Tyrosine kinase receptors are activated by many of the hormones and growth factors involved in cell growth, proliferation, differentiation, and survival (Table 3.3). Receptor activation involves interaction of two adjacent receptor molecules, forming a *dimmer* that cross-links in the cell membrane. Dimerization brings the kinase domains of both molecules into close proximity, allowing autophosphorylation (Fig. 3.4). The mechanism of dimerization varies depending on the ligand-receptor system and can be in one of the following ways.

- Monomeric ligands, such as EGF, bind to two receptors simultaneously and cross-link them directly.
- FGF form *multimers* by binding to accessory molecules, such as heparin sulfate proteoglycans, either on the target cell surface or in the extracellular matrix. These accessory molecules serve to cross-link adjacent receptors.
- PDFG, which is a dimer itself, upon binding cross-links two receptors together.
- Ephrins (Eph) and Eph receptors are a particular kind of juxtacrine mode of interaction. They act on a contact-dependent manner in the way that they serve simultaneously as both ligand and receptor. Upon binding Eph not only activate the Eph receptor, but it also it becomes self-activated creating a reciprocal bidirectional signaling that changes the behavior of both cells. This system is functional during development of the central nervous system where cells of the developing brain are maintained by mixing with neighboring cells. In addition, soluble Eph can also mediate signaling but are active only when they form clusters in the plasma membrane of the target cell. Thus, Eph activate Eph-receptors only when membrane-bound, while soluble Eph are active when it aggregates into clusters.
- Insulin and IGF-I bind to receptors that are already organized as tetramers inducing rearrangement of the transmembrane chains by which both kinase domains come together.

Tyrosine kinase receptors transduce their signals through multiple routes to different intracellular target domains. Autophosphorylation of tyrosine residues within the kinase domain induces two main effects in their signaling pathway. The receptor is an allosteric enzyme, therefore the first effect is the activation of its own kinase. Thus, ligand binding is followed by a rapid increase in the number of phosphorylated residues in its own cytoplasmatic tail that serves as recognition sites to other enzymes, including phospholipase-C-γ (PLC-γ), which function like the phospholipase-C-β activating the IP signaling pathway increasing IP_3 and releasing Ca^{2+} from the ER. The second effect is through the phosphorylation of *tyrosine* residues within proteins that binds SH2 & SH3 domains. These SH2 & SH3 domains are high-affinity docking sites for proteins that do not have their own Src homology domains, serving as *adaptors* between the receptor and other intracellular signaling pathway's proteins.

The tyrosine kinase signaling interacts with Ras, a small G-protein product of the c-*ras* proto-oncogen that belongs to a large family of monomeric GTPases. Ras, which is covalently attached by prenylation to lipids of the cytoplasmic face of the plasma membrane, couples tyrosine

Table 3.3 Tyrosine kinase receptors ligands and functions

Ligand	Typical responses
Brain-derived neurotropic factor (BDNF)	Stimulates survival, growth, and differentiation of neurons
Colony stimulating factor (CSF)	Stimulates monocytes/macrophage proliferation and differentiation;
Ephrin-A and -B	Regulate cell adhesion and repulsion, stimulate angiogenesis, and guide cells and axon migration
Epidermal growth factor (EGF)	Stimulates cell proliferation
Fibroblast growth factor-1 to -24 (FGF-1 to FGF-24)	Stimulate cell proliferation, inhibit differentiation of some precursor cells, and serve as inductive signal during development
Hepatocyte growth factor (HGF)	Induces cell dissociation and motility, influences epithelial morphology, and is involved in tissue regeneration
Insulin	Stimulates carbohydrate utilization and protein synthesis
Insulin-like growth factor-I and -II (IGF-I & IGF-II)	Stimulate cell growth and survival
Macrophage-colony-stimulating factor (M-CSF)	Stimulates monocyte/macrophage proliferation and differentiation
Nerve growth factor (NGF)	Stimulates survival growth and differentiation of neurons
Platelet-derived growth factors (PDGF AA, BB, and AB)	Stimulates cell survival, growth, and proliferation
Vascular endothelial growth factor (VEGF)	Stimulates angiogenesis

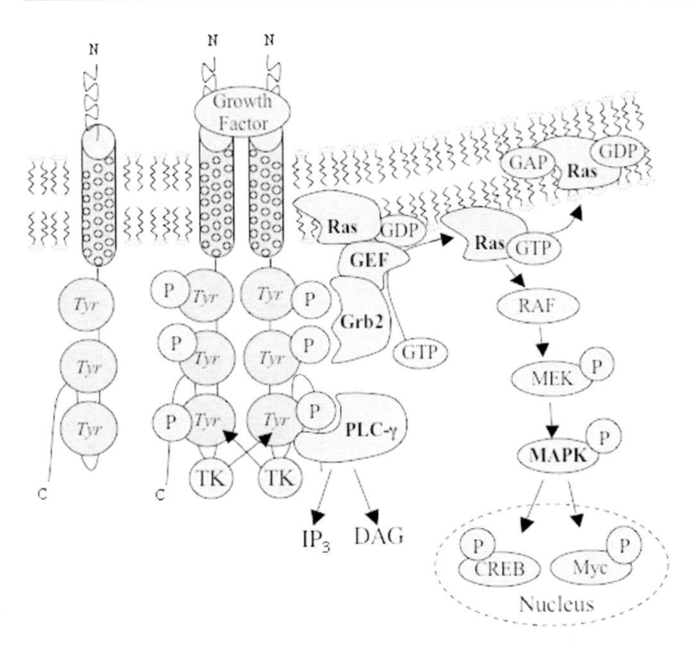

Fig. 3.4 Tyrosine kinase-stimulated receptor and activation of the MAP kinase (MAPK) pathway. The structure on the *left* represents an inactive growth factor receptor. The structure on the *right* represents the reactions upon ligand binding including receptor dimerization and cross-phosphorylation of the tyrosine (*Tyr*) residues, which create binding sites for the scaffolding protein Grb2 and for PLC-γ. These initial reactions are followed by activation of Ras and the MAPK pathway. CREB: cAMP regulatory element binding protein; DAG: 1,2-diacylglycerol; GAP: GTPase-activating protein; GDP: guanosine diphosphate; GEF: Guanine nucleotide exchange factor; Grb2: growth factor receptor-bound protein-2; GTP: guanosine triphosphate; IP_3: inositol triphosphate; MAPK: mitogen activated kinase; MEK: MAP-kinase-kinase; Myc: transcription factor encoded by the c-*myc* gene; P: indicates phosphorylation; PLC-γ: phospholipase-Cγ; RAF: MAP-kinase-kinase-kinase; Ras: small G-protein; TK: tyrosine kinase

kinase receptors with multiple downstream mitogenic effectors resembling the GTP-binding proteins that switch the seven transmembrane receptor signals. Hyperactive mutant forms of Ras resistant to GTPases have been found in approximately 30% of human tumors, which remain continuously activated in the GTP-bound state, suggesting that mutations in the Ras proto-gene may promote the development of cancer. Tyrosine kinase receptors interact with Ras through an adaptor protein termed growth factor receptor-bound protein-2 (Grb2) that carries the SH2 and SH3 domains. Grb2 binds to the proline-rich domain a guanine nucleotide exchange factor (GEF) encoded by the *son of sevenless* (*SOS*) gene. Ras also interact with GAP, a GTPase-activating protein that increases hydrolysis of bound GTP and inactivate Ras. Thus, GEF and GAP serve as signal switches modulating the GTPase activity of Ras (Fig. 3.4).

Interaction of tyrosine kinase receptors with the MAP-kinase pathway—ligand binding to tyrosine kinase receptors, autophosphorylation, and Ras activation are transient reactions. The tyrosine-specific phosphatases rapidly remove the phosphate groups from the receptor and GAP promotes GTP-Ras hydrolysis, switching off the reactions. Therefore, an alternative mechanism is activated that relays the tyrosine kinase pathway message downstream along several other pathways, prolonging and amplifying the signal. This mechanism is mediated by Ras that initiates a cascade of downstream phosphorylation on *serine* and *threonine* residues of a group of proteins termed *m*itogen-*a*ctivated *p*rotein kinase (MAP-kinase; originally termed "extracellular signal-regulated kinase," ERK, or "microtubule-associated protein kinase," MAPK), which are much more stable than tyrosine phosphorylations.

Activation of the MAP-kinase pathway involves the interaction of Ras with Raf, a *serine/threonine*-specific kinase encoded by the c-*raf* proto-oncogene, also termed MAP-kinase-kinase-kinase. This reaction initiates a cascade of serine/threonine phosphorylations that last longer than the previously described tyrosine phosphorylation (Fig. 3.4). Raf phosphorylates a second enzyme in the cascade, termed MAP-kinase-kinase (MAPKK) or MEK, which phosphorylates MAPK. Phosphorylated MAPK transmits the signal downstream by phosphorylating various other kinases and gene regulatory proteins, which modulate transcription of genes and translation of several mRNAs. The role of the Ras-MAP-kinase signaling pathway is to convey mitogenic signals from the cell membrane to the nucleus, activating several transcription factors involved in cell proliferation, including c-*myc*, c-*fos*, and CREB and regulate expression of G1-cyclins.

The MAP-kinase pathways not only regulate mitogenesis but also participate within mechanisms that control cell survival and cell growth through a signaling cascade of protein phosphorylation reactions that involve Ras-dependent phosphorylation of phosphatidylinositol-3-kinase (PI3-kinase), which is a protein that carries the SH2 domain recognized by Ras. The Ras-activated PI3-kinase then phosphorylates phosphatidylinositol at the third position of the inositol ring to generate phosphatidylinositol 3,4-biphosphate, or $PI(3,4)P_2$ and phosphatidylinositol 3,4,5-triphosphate, or $PI(3,4,5)P_3$. Both, $PI(3,4)P_2$ and $PI(3,4,5)P_3$ remain attached to the cell membrane, serving as dock sites for proteins that link the signal from the membrane to the cytoplasm. $PI(3,4,5)P_3$ docks two enzymes that carry the PH domain the protein kinase B (PKB, also termed Akt) and phosphatidylinositol-dependent protein kinase (PDKI). PDKI phosphorylates PKB, which moves to the cytoplasm and phosphorylates a variety of signaling proteins such as the S6 subunit of the ribosomes, which enhances translation efficiency. In addition, PKB inactivates by phosphorylation a protein termed Bcl-2-associated death promoter (BAD), a pro-apoptotic member of the Bcl-2 family of proteins involved in programmed cell death.

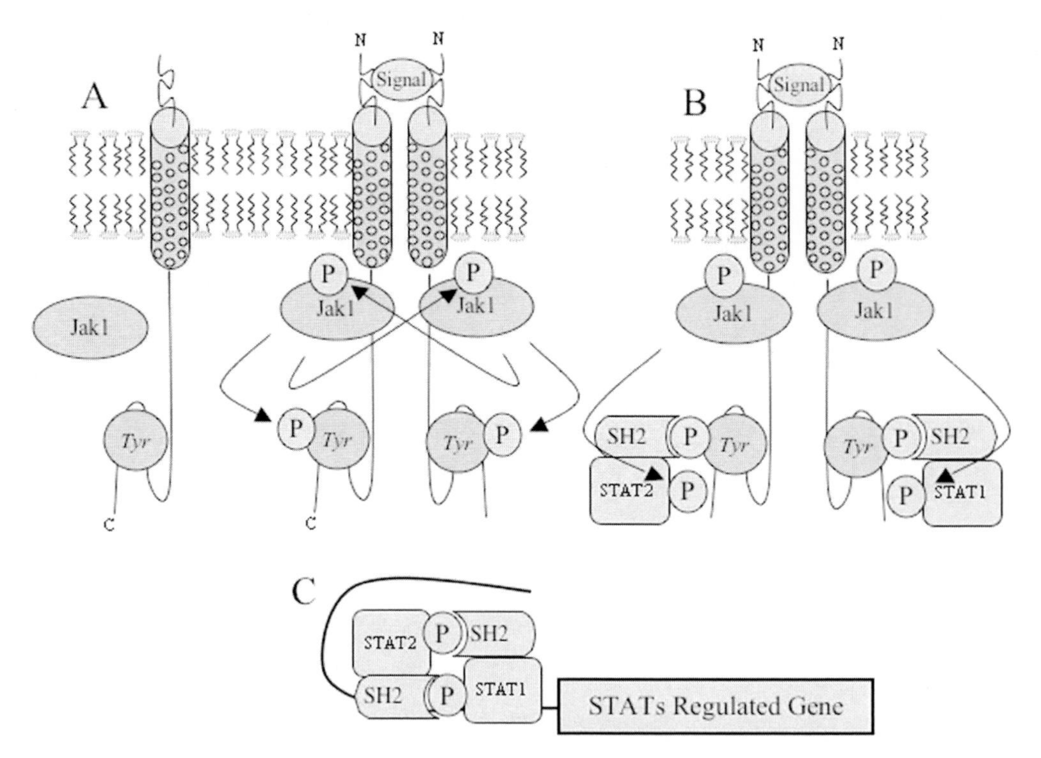

Fig. 3.5 Tyrosine kinase associated receptor. (**A**) Upon ligand binding receptors dimerize, Jaks cross phosphorylate each other and then phosphorylate the receptor. (**B**) STAT dock on the phosphorylated receptor and is phosphorylated by Jak. (**C**) STATs dissociate from the receptor, dimerize through the SH2 domains, and migrate to the nucleus where they bind to the regulatory region of activated genes. Jak: janus kinase; P: indicates phosphorylation; SH2: *src* homology region 2; STAT: signal transducers and activators of transcription; Tyr: tyrosine

3.4.2 Tyrosine Kinase Associated Receptors

Tyrosine kinase associated receptors are transmembrane proteins devoid of intrinsic enzymatic activity, but associated to cytoplasmic tyrosine kinases by non-covalent bonds (Fig. 3.5). Some hormones, such as growth hormone (GH), prolactin (PRL), and erythropoietin, and the cytokines including granulocyte-macrophage stimulating factor, granulocyte colony-stimulating factor, interferon-α, -β, and -γ, and interleukins (IL), IL-2, IL-3, IL-4, IL-5, IL-6, and IL-7 mediate their effects through tyrosine-kinase-associated receptors (Table 3.4).

Tyrosine-kinase associated receptors are composed of two or more ligand specific polypeptide chains that form dimmers. Upon ligand binding the dimmers phosphorylate their respective *tyrosine* residues, as well as the residues of other signaling proteins that activate a group of cytoplasmic enzymes termed Janus kinases (named Jaks after the two-faced Roman god). Four different Jaks have been described: Jakl, Jak2, Jak3, and Tyk2 (Table 3.4). The activated receptor induces association between two Jaks, which activate their tyrosine kinase domains and cross-phosphorylate. Phosphorylated Jaks serve as docking sites for other signaling proteins, including a group of gene regulatory proteins carrying SH2 domain, termed signal transducers and activators of transcription (STAT), which move to the nucleus and regulate transcription of selected genes. An interesting feature of STAT is that the SH2 domains enable docking with phosphorylated residues of the

Table 3.4 Extracellular ligands that activates cytokine receptors and the Jak-STAT signaling pathways

Ligand	Jaks	STATS	Typical responses
γ-interferon	Jakl and Jak2	STAT1	Activates macrophages; increases MHC protein expression
α-interferon	Tyk2 and Jak2	STAT1 and STAT2	Increases cell resistance to viral infection
Erythropoietin	Jak2	STAT5	Stimulates production of erythrocytes
PRL	Jakl and Jak2	STAT5	Stimulates milk production
GH	Jak2	STAT1 and STAT5	Stimulates growth by inducing IGF- I production
GM-CSF	Jak2	STAT5	Stimulates production of granulocytes and macrophages
IL-3	Jak2	STAT5	Stimulates early blood cell production

GH: growth hormone; GM-CSF: granulocyte-macrophage colony-stimulating factor; IL: interleukins; Jak: Janus kinase; MHC: major histocompatibility complex; PRL: prolactin; STAT: signal transducers and activators of transcription; Tyk2: tyrosine kinase 2.

activated *tyrosine* kinase receptor. STAT is phosphorylated by Jak causing dissociation from the receptor. The SH2 domains on the released STATs mediate binding to phosphorylated *tyrosine* residues with other STAT molecules, forming either STAT homodimers or heterodimers, which move into the cell nucleus where in association with other proteins regulate transcription.

3.4.3 Serine/Threonine Kinase Receptors

Serine/threonine kinases receptors are single-chain peptides with a large extracellular domain containing a cysteine-rich region, a transmembrane domain, and a cytoplasmic domain (Fig. 3.6). The cytoplasmic domain contains two separated amino acid sequences homologous to other serine-threonine kinases, such as PKA and PKC. Several members of the serine/threonine kinase receptors have been cloned and characterized, including the receptors for transforming growth factor (TGF) β-I and -II, activin, inhibin, Müllerian-inhibiting substance (MIS), and bone morphogen (BMP). Serine-threonine receptors activated by these extracellular regulatory molecules recognize and phosphorylate a group of proteins termed SMAD (*S*mall *M*others *A*gainst *D*ecapentaplegic), of which five (SMAD1, SMAD2, SMAD3, SMAD5, and SMAD9) form complexes that act as transcription factors. The effects for TGFβ and activins are mediated by SMAD2 and SMAD3, while BMP and MIS are mediated by SMAD1, SMAD5, and SMAD9. The SMAD amino terminal (MH1) has DNA-specific binding properties and binds to promoter nucleotide specific sequences, and also interacts with other transcriptional activators including members of the AP-1 family. The complex SMAD3–SMAD4 binds to a four base pair sequence 5′-GTCT-3′, or the complement 5′-AGAC-3′, termed SMAD-binding element (SFE). The carboxy terminal (MH2) recognizes the receptor and oligomerize with other SMAD.

3.4.4 Guanylate Cyclase Associated Receptors

Guanylate cyclase associated receptors mediate effects by activating the enzyme guanylate cyclase (GC), through a

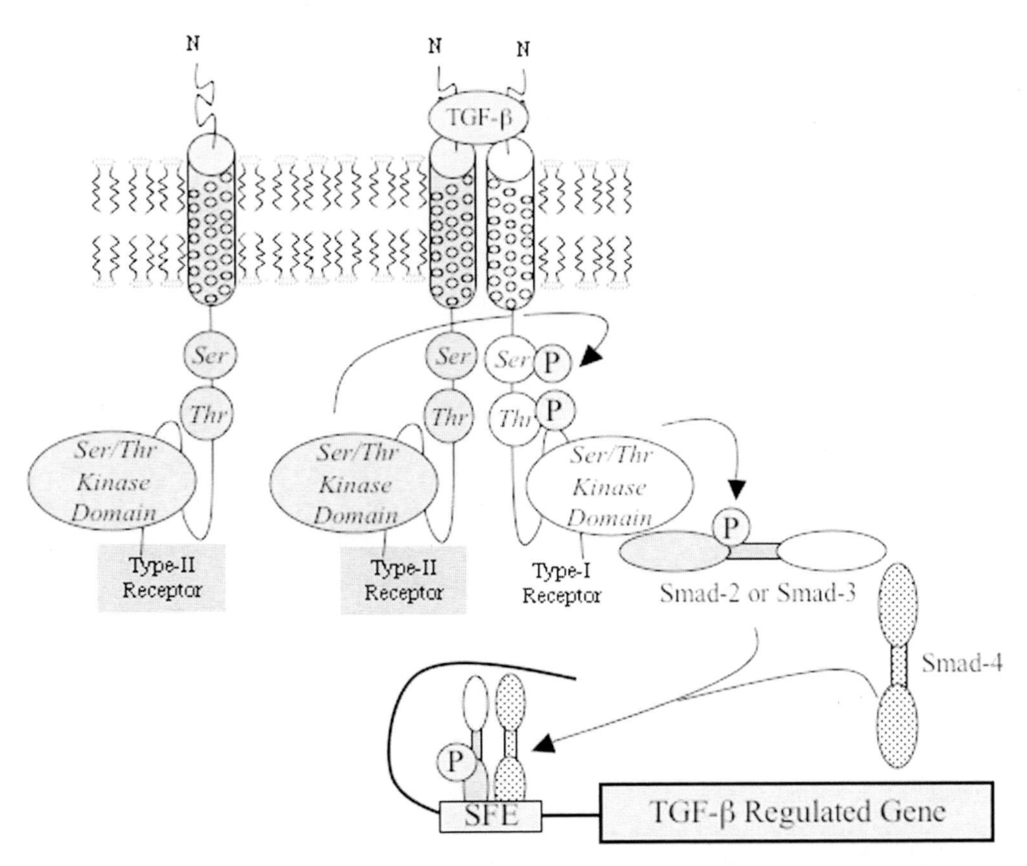

Fig. 3.6 Activation of the TGF-β *serine-threonine* receptor. TGF-β binding to type-II TGF-β receptor induces recruitment and phosphorylation of type-I TGF-β receptor. Phosphorylated type-I TGF-β receptor phosphorylates Smad2 or Smad3, which dissociates from the receptor and oligomerizes with Smad4. The oligomer migrates to the nucleus and binds the Smad binding regulatory element (SFE) in activated genes. P: indicates phosphorylation; *Ser*: serine; SMAD: group of proteins that form complexes that acts as transcription factors; TGF-β: transcription growth factor β; *Thr*: threonine

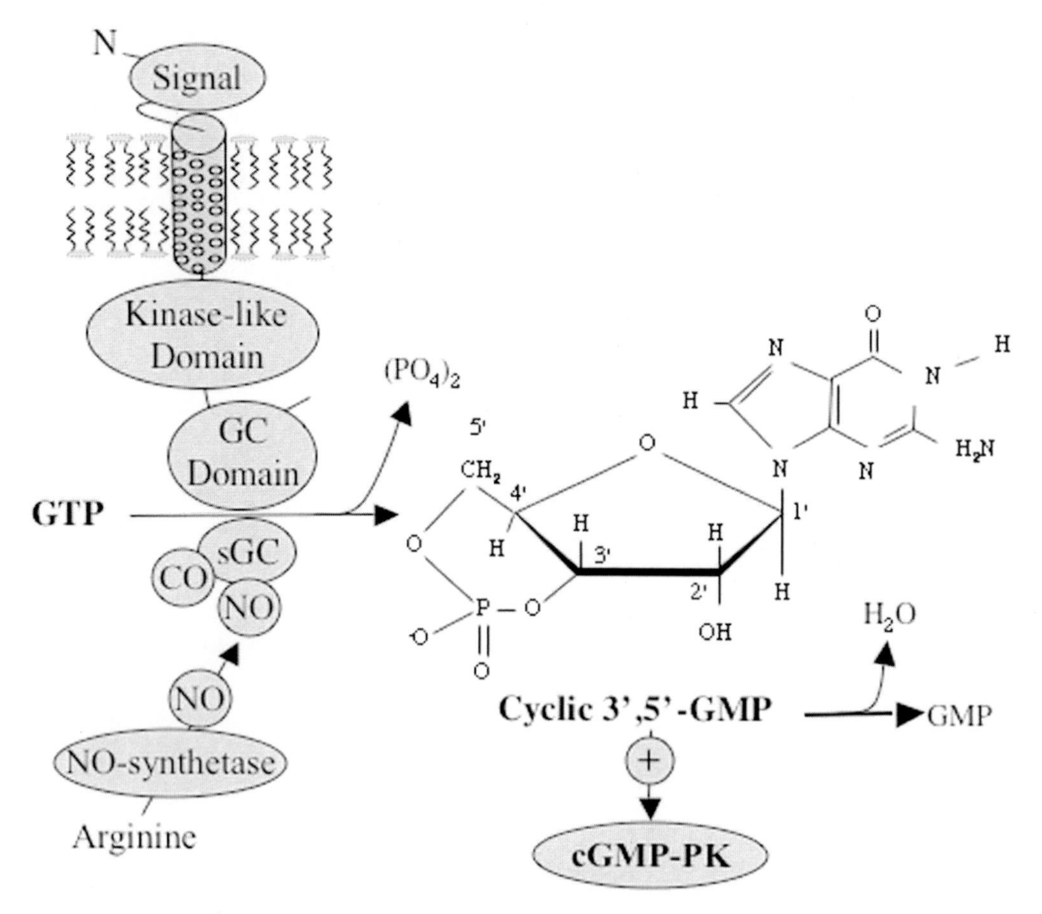

Fig. 3.7 Activation of guanylate cyclase associated receptors and synthesis of cGMP. Ligand binding to a GC-associated receptor induces conversion of GTP into cGMP. Soluble intracellular GC is activated by CO and NO. In both cases, the resulting cGMO activates cGMP-PK. cGMP: cyclic guanosine monophosphate; cGMP-PK: cGMP-dependent protein kinase; CO: carbon monoxide; GC: guanylate cyclase; GMP: guanosine monophosphate; GTP: guanosine triphosphate; NO: nitric oxide; sGC: soluble guanylate cyclase

mechanism that has some analogies with adenylate cyclase, since it catalyzes the conversion of GTP into the second messenger cyclic cGMP (Fig. 3.7). Several extracellular regulatory molecules, including ANP, also known as *atrial natriuretic factor* (ANF), the sea urchin sperm proteins *resact* and *speract* that are involved in the regulation of sperm-egg interactions, and the gases NO and CO mediate their effects by activating GC. However, like the *tyrosine* and *serine/threonine* kinases, GC can directly be both receptor and effector. Two main types of GC have been described, a membrane associated form (mGC) and a cytoplasmic soluble form (sGC) that directly catalyze the production of cGMP in the cytosol. ANP, guanylin, and the sea urchin sperm proteins interact with mGC, whereas NO and CO interact with sGC. NO is synthesized from L-arginine by the action of NO-synthetase and is induced in various tissues by cytokines and endotoxins. One of the most prominent actions of NO is vascular smooth muscle relaxation in the penis. cGMP can activate cGMP-dependent protein kinases (cGMP-PK), stimulate the actions of specific phosphodiesterases, which in turn inactivate cGMP and cAMP. The cGMP-PKs have a number of functions, including relaxation of smooth muscle, inhibition of platelet activation, reduction of endothelial permeability, and increased negative inotropic effect on the cardiac muscle.

3.5 Regulation of Receptors and Signaling Pathways

When cells are exposed to stimulus for a prolonged period of time, their response decreases through a mechanism called receptor *desensitization* or *adaptation* (Fig. 3.8). Receptor desensitization can occur in at least three ways: *inactivation* when the receptor is uncoupled from the G-proteins; *sequestration*, also termed internalization by *endocytosis*, when the receptor is temporarily moved to the interior of the cells but can be recycled into the membrane and become active; and downregulation when the internalized receptor is destroyed by the lysosomes.

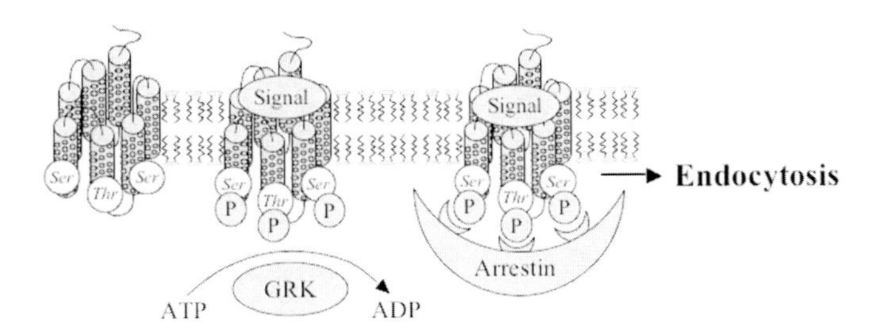

Fig. 3.8 Desensitization of G-protein linked receptors. After activation, the seven transmembrane receptor is phosphorylated by GRK at *serine* (*ser*) and *threonine* (*thr*) residues, creating docking sites for the protein arrestin that inhibits binding to the G-proteins and drives the receptor to endocytosis. ADP: adenosine diphosphate; ATP: adenosine triphosphate; GRK: G-protein-linked receptor kinase; P: indicates phosphorylation

Desensitization of the G-protein linked receptors is mediated by phosphorylation. PKA, PKC, or a kinase stimulated by another ligand-activated G-protein linked receptor (GRK) phosphorylates multiple *serine* and *threonine* residues on a targeted G protein-coupled receptor, which then binds with high affinity to a member of the *arrestin* family of proteins. Arrestins uncouple the receptor from the G-proteins and serve as adaptors to couple the receptor to clathrin-coated pits, inducing endocytosis, which is followed by intracellular sequestration and/or degradation of the ligand-receptor complex.

The second messenger pathways can also be downregulated by hormone over-exposure. Downstream of the receptor inactivation involves mainly dephosphorylation of the proteins phosphorylated by the kinases. Dephosphorylation of phorphorylated *serines* and *threonines* is catalyzed by four types of serine/threonine phosphoprotein phosphatases, which are termed protein phosphatases I, IIA, IIB or calcineurin, and IIC. Protein phosphatase-I dephosphorylate many of the PKA substrates and inactivates CREB.

In the enzyme-linked receptors the level of activation depends on the net level of phosphorylation of their own *tyrosine* residues, which is the result of two opposite reactions: phosphorylation and dephosphorylation. Phosphorylation depends on ligand binding, whereas dephosphorylation depends on the activity of phosphotyrosine phosphatases. Thus, changes in activity of the phosphotyrosine phosphatases are also determinants of enzyme-linked receptor activity. Two groups of *tyrosine* phosphatases have been described. The first are small intracellular phosphatases with a single catalytic domain. The second are the phosphotyrosine phosphatase-like receptors (PTPs), which are constituted by a large transmembrane domain with two tandems intracellular catalytic repeats that resemble membrane receptors. Although their ligands have not yet been identified, more than ten PTPs expressed within the plasma membrane have been characterized. The cellular signaling functions of the PTPs are still undefined, however, they have been described carrying SH2 domains, which suggest that they may be involved in modulating signals from cytokine and growth factor receptors. The speculative mechanisms of the SH2 domain in PTPs include the following:

- reduction of receptor activity and *tyrosine kinase* signaling by dephosphorylation;
- inactivation of other *tyrosine*-phosphorylated signaling molecules, such as phospholipase-Gγ and src *tyrosine kinase*;
- signal amplification via binding of other *tyrosine*-phosphorylated signaling proteins to one of the two SH2 domains; and
- signal transmission via interaction between another SH2 domain protein and the phosphorylated *tyrosine* in the carboxy terminus of SH2 domain PTP-1.

3.6 Interactions Between Signaling Pathways

Cells are under the influence of more than one extracellular regulatory molecule at one time, therefore, a particular signaling cascade operates under conditions where more than one signaling pathway is simultaneously activated (Fig. 3.9). In addition, the endocrine system has more than one way to achieve a given biological response, and organs may have an extra regulatory capacity in the form of overlapping controls or redundancy. Redundant systems probably developed during evolution as a fail-safe mechanism to secure vital functions at levels beyond day-to-day demand for survival. Several extracellular regulatory molecules, including hormones, neurotransmitters, and growth factors activate more than one pathway simultaneously, as summarized below:

- Adrenergic agonists bind to the α-1 receptors that activate PI hydrolysis, and to the β-receptors and α-2-receptors, that can stimulate or inhibit adenylate cyclase by coupling to different G-proteins, and regulate two different signaling pathways (Table 3.1).
- Ligand-binding to the M2-muscarinic receptor inhibits adenyl cyclase and weakly stimulates PI hydrolysis by interacting with G_i and G_q, respectively.
- Thyroid stimulating hormone (TSH) binding to the TSH receptor (TSH-R) simultaneously activate adenylate cyclase and PI hydrolysis activating both the G_s and the G_q.
- EGF and PDGF not only activate the tyrosine kinase of their respective receptors, the EGF-R and PDGF-R, but also induce PI hydrolysis by phosphorylating PI-PLC-γ.

Signaling pathways can respond to the same agonist, or to different agonists, interacting to affect similar intracellular processes. These interactions are termed *cross-talk*, in which

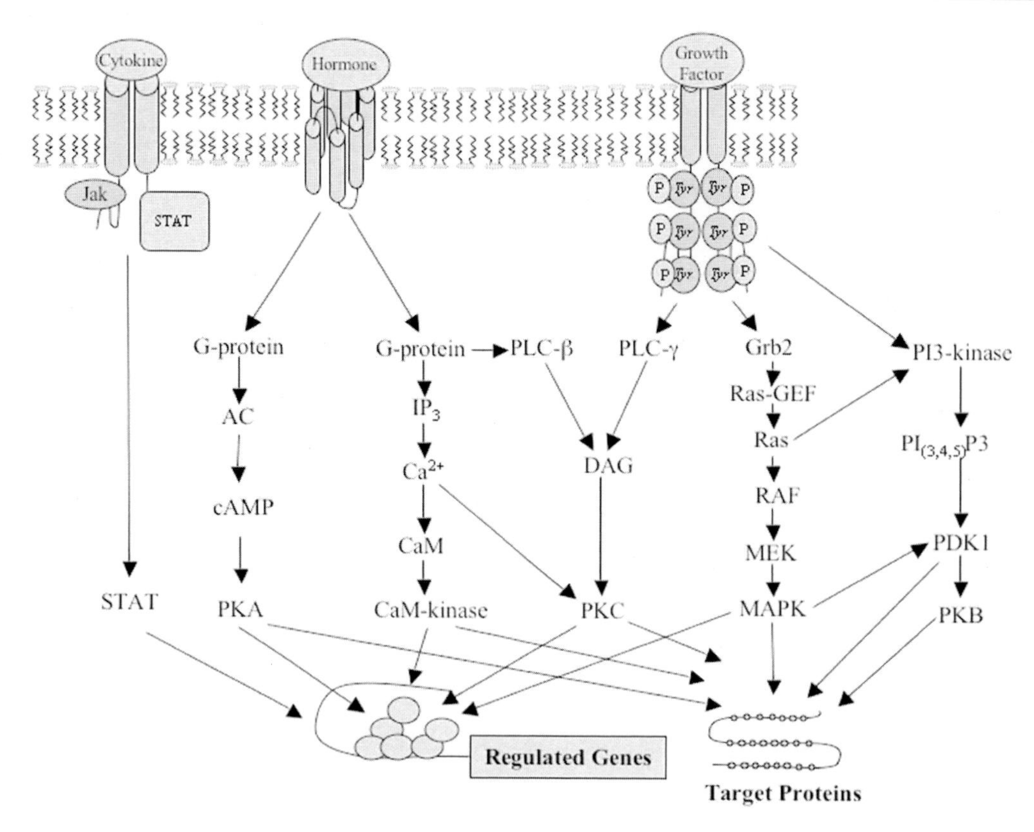

Fig. 3.9 Interactions between signaling pathways. AC: adenylate cyclase; Ca^{2+}: calcium ions; CaM: calmodulin; cAMP: cyclic AMP; DAG: 1,2-diacylglycerol; GEF: Ras guanine nucleotide exchange factor; Grb2: growth factor receptor-bound protein-2; IP_3: inositol triphosphate; Jak: janus kinase; MAPK: mitogen activated protein kinase; MEK: MAP kinase-kinase; PDK: phosphatidylinositol dependent protein kinase; $PI_{(3,4,5)}P_3$: phosphatidylinositol 3,4,5-triphosphate; PI3-Kinase: phosphatidylinositol 3-kinase; PKA: protein kinase A; PKB: protein kinase B; PKC: protein kinase C; PLC: phospholipase C; RAF: MAP-kinase-kinase-kinase; Ras: small G-protein; STAT: signal transducers and activators of transcription

two or more distinct pathways affect the same response and produce an additive or synergistic response, or that the activation of one pathway may inhibit the response stimulated by the other. Some examples of cross-talk between signaling pathways in the pituitary gland illustrate the topic:

- In the mamotroph cells of the pituitary gland, thyrotropin-releasing hormone (TRH) and the vasoactive intestinal peptide (VIP) stimulate synthesis and secretion of PRL, but through different pathways. TRH mediates its effects by elevating cAMP, whereas VIP activates the PI signaling cascade and elevations of cytoplasmic Ca^{2+}.
- In PRL-secreting cells, when TRH and VIP are added simultaneously at maximally effective concentrations, stimulation of PRL secretion is additive and the kinetics of stimulated PRL secretion is a composite of the two agonist effects. When TRH is added hours before, the amount of PRL secreted in response to VIP is increased because TRH stimulates PRL synthesis as well as secretion.
- Both angiotensin-II (AII) and TRH activates PI hydrolysis and stimulates PRL secretion in the mamotroph cells. However, if cells are pre-exposed to TRH the effect of AII on PRL is attenuated, probably by PKC-mediated phosphorylation of the AII receptor. This effect cause feedback inhibition and is believed to be a mechanism of AII receptor desensitization.

Overall, cell signaling has been classically described as the response of one pathway to one extracellular signal. However, physiology is more complex, cells are exposed to more than one external signal at a given time, and the outcome is not the consequence of unidirectional and independent signals but rather the cross-talk between multiple pathways. Thus, to understand endocrine integration at the cellular level a more holistic approach is required, in which interaction between signaling pathways must be taken into consideration to understand the mechanism behind the final biological response.

3.7 Summary

Intracellular signaling pathways organize and amplify extracellular signals in a way that a small number of ligands affect the activity of a large numberof molecules leading to the regulation of gene expression. Signaling via seven transmembrane receptors involves activation of G-protein that couple receptors with effectors. Three major intracellular pathways convey seven transmembrane activated signals PKA, PI/PKC, and PLA_2. The effector of the PKA pathway is adenylate cyclase that converts ATP into cAMP. The PI/PKC pathway is activated mainly by ligands that activates PLC leading to the conversion of PI(4,5) P_2 into IP_3 and DAG. IP_3 binds to ER receptors and open Ca^{2+}-channels, inducing a rapid release of the stored Ca^{2+} into the cytoplasm, while DAG stimulates various types of PKC. The PLA_2 is regulated by PKC, tyrosine kinase receptors, MAPK and Ca^{2+}. Stimulated PLA_2 hydrolyzes phospholipids releasing arachidonic acid and lysophospholipids. Ion channel receptors control the passage of the cations and anions into the cell. They include receptors for several neurotransmitters, among others for ACh, GABA, glycine, glutamate, serotonin, and ATP. Enzyme-linked receptors are single-span transmembrane proteins that bind the ligand, and serve as cell membrane signal effectors. Enzyme-linked receptors are classified into four classes: tyrosine kinase, tyrosine kinase-associated, serine/threonine kinases, and guanylate cyclase-associated. Most of the ligands of the enzyme-linked receptors are local regulators and growth factors. Receptor activity is regulated through mechanisms of desensitization or adaptation that involve inactivation, sequestration, and downregulation. Extracellular regulators activate more than one pathway simultaneously and respond to the same agonist, or to different agonists, interacting to affect similar intracellular processes. Interaction involves a mechanism of cross-talk, in which two or more distinct pathways affect the same response and produce an additive or synergistic response or inhibit the response stimulated by the other.

Glossary of Terms and Acronyms

5′AMP: adenosine monophosphate

AC: Adenylate cyclase

ACh: Acetylcholine

ACTH: adrenocorticotropic hormone

AII: angiotensin-II

Akt: protein kinase-B, also termed PKB

ANP: atrial natriuretic peptides

ATP: adenosine triphosphate

BAD: Bcl-2-associated death promoter

BDNF: brain-derived neurotrophic factor

BMP: bone morphogenetic protein

BNP: B-type natriuretic peptide, also known as brain natriuretic peptide or GC-B

βARK: β-adrenergic receptor kinase

Ca^{2+}: calcium ions

Ca^{2+}-CaM: Ca^{2+} bound to CaM

CaM: calmodulin

cAMP: cyclic adenosine monophosphate

c-*fos*: cellular protooncogene of the transforming gene of the FBJ and FBR osteosarcome viruses

cGMP: cyclic guanosine monophosphate

cGMP-PK: cGMP-dependent protein kinase

c-*jun*: cellular protooncogene of the transforming gene of avian sarcoma virus

CNP: C-type natriuretic peptide

CO: carbon monoxide

CREB: cAMP response element binding protein

CRF: corticotropin releasing factor

CSF: colony-stimulating factor

Cα, Cβ, Cγ: isotypes of the PKA C-subunit

EGF: epidermal growth factor

EGF-R: EGF receptor

Eph: ephrins

ER: endoplasmic reticulum

ERK: extracellular signal regulated kinase

FGF: fibroblast growth factor

Fos: transcription factor expressed by the c-*fos* gene

FSH: follicle-stimulating hormone

GABA: γ-aminobutiric acid

GAP: GTPase-activating protein

GC: guanylate cyclase

GDP: guanosine diphosphate

GEF: Ras guanine nucleotide exchange factor

GH: growth hormone

GHRH: growth hormone releasing hormone

GM-CSF: granulocyte-macrophage colony-stimulating factor

GnRH: gonadotropin releasing hormone

Grb2: growth factor receptor-bound protein-2

GRK: G-protein-linked receptor kinase

GTP: guanosine triphosphate

$G_{\alpha i}$: G-protein inhibitory α-subunit

$G_{\alpha s}$: G-protein stimulatory α-subunit

HGF: hepatocyte growth factor

HPL: human placental lactogen

IGF-I: insulin-like growth factor-I

IGF-II: insulin-like growth factor-II

IL: interleukin

IP_3: inositol triphosphate

IRS-1: insulin receptor substrate-1

Jak: Janus kinase

Jun: transcription factor expressed by the c-*jun* gene

LGIC: ligand-gated ion channels

LH: luteinizing hormone

L-type: long lasting VSCC

mAch: muscarinic acetylcholine receptor

MAPK: mitogen-activated protein kinase

MARKS: myristolated alanine-rich C-kinase substrate

M-CSF: macrophage colony-stimulating factor

MEK: MAP-kinase-kinase or MAPKK

mGC: membrane-associated guanylate cyclase

MHC: major histocompatibility complex

MIS: Müllerian-inhibiting substance

MSH: melanocyte-stimulating hormone

Myc: transcription factor encoded by the c-*myc* gene

nACh-R: nicotinic acetylcholine receptor

NGF: nerve growth factor

NO: nitric oxide

N-type: neither long-lasting nor transient VSCC

PACAP: pituitary adenylate cyclase-activating polypeptide

PDE: nucleotide phosphodiesterase

PDGF: platelet-derived growth factor

PDKI: phosphatidylinositol-dependent χ protein kinase

PH: Pleckstrin homology

PI/PKC: phosphatidylinositol/protein kinase-C

P_i: inorganic phosphate

PI: phosphatidylinositol

PI3-kinase: phosphatidylinositol-3-kinase

PI-PLC: phosphatidilinositol-specific phospholipase-C

PKA: cAMP-dependent protein kinase-A

PKB: protein kinase-B, also termed Akt

PKC: protein kinase-C

PKG: cGMP-dependent protein kinase

PLA_2: phospholipase-A2

PLC: phospholipase-C

PRL: prolactin

PTH: parathyroid-stimulating hormone

PTP: phosphotyrosine phosphatase-like receptors

RACK: receptor for activated C-kinase

RAF: MAP-kinase-kinase-kinase

***ras*:** oncogene of the Harvey (*ras*H) and Kristen (*ras*K) rat sarcoma viruses

Ras: small G-protein, first identified as product of the *ras* oncogene

RH: Ras homology domain

ROC: receptor operated Ca^{2+} channel

S6: ribosomal S6 kinase

SFE: SMAD-binding element

sGC: cytoplasmatic soluble guanylate cyclase

SH2 and SH3: src homology region-2 and -3 respectively

SMAD: Small Mothers Against Decapentaplegic

$sPLA_2$: secreted phopholipase-A_2

***src (pronounced "sark")*:** retroviral oncogene (*v-src*) from the chicken Rous sarcoma retrovirus and its precursor (*c-src*)

STAT: signal transducers and activators of transcription

TGF: transforming growth factor

TK: tyrosine kinase

TNF: tumor necrosis factor

TRH: thyrotropin-releasing hormone

Trk: tropomyosin-related kinase receptor

TSH: thyrotropin, or thyroid-stimulating hormone

TSH-R: TSH receptor

T-type: transient type VSCC

VD-R: vitamin D receptor

VEGF: vascular endothelial growth factor

VGCR: voltage-gated channel receptors

VIP: vasoactive intestinal peptide

VSCC: voltage-sensitive Ca^{2+} channel

$PI(3,4,5)P_3$: phosphatidylinositol 3,4,5-triphosphate

DAG: 1,2-diacylglycerol

$PI(3,4)P_2$: phosphatidylinositol 3,4-biphosphate

$5\text{-}HT_3$: 5-hydroxytriptamine

$PI(4,5)P_2$: phosphatidylinositol 4,5-biphosphate

Bibliography

1. Alberts B, Johnson A, Lewis J, et al. Molecular Biology of the Cell, fourth edition. New York: Garland Science, 2002.
2. Avruch J. MAP kinase pathways: The first twenty years. Biochim Biophys Acta 2007; 1773:1150–60.
3. Bolander F. Molecular Endocrinology, third edition. San Diego: Academic Press, 2004.
4. Brodde OE, Michel MC. Adrenergic and muscarinic receptors in the human heart. Pharmacol Rev 1999; 51:651–81.
5. den Hertog J, Ostman A, Böhmer FD. Protein tyrosine phosphatases: regulatory mechanisms. FEBS J 2008; 275:831–47.
6. Catt KJ. Hormone receptors. In Hillier SG, Kitchener HC, Neilson JP. Scientific Essentials of Reproductive Medicine. Philadelphia: Saunders Company, 1996:32–44.
7. Catt KJ. Intracellular signaling. In Hillier SG, Kitchener HC, Neilson JP. Scientific Essentials of Reproductive Medicine. Philadelphia: Saunders Company, 1996:45–59.
8. DeWire SM, Ahn S, Lefkowitz RJ, et al. Beta-arrestins and cell signaling. Annu Rev Physiol 2007; 69:483–510.
9. Doyle DA. Structural changes during ion channel gating. Trends Neurosci 2004; 27:298–302.
10. Itier V, Bertrand D. Neuronal nicotinic receptors: from protein structure to function. FEBS Letters 2001; 504:118–25.
11. Goldsmith EJ, Akella R, Min X, et al. Substrate and docking interactions in serine/threonine protein kinases. Chem Rev 2007; 107:5065–81.
12. Lefkowitz RJ. Seven transmembrane receptors: something old, something new. Acta Physiol (Oxf) 2007; 190:9–19.
13. Mellor H, Parker PJ. The extended protein kinase C superfamily. Biochem J 1998; 332:281–92.
14. Michel RH. Inositol derivatives: evolution and functions. Nat Rev Mol Cell Biol 2008; 9:151–61.
15. Miyazawa A, Fujiyoshi Y, Unwin N. Structure and gating mechanism of acetylcholine receptor pore. Nature 2003; 423:949–55.
16. Oldham WM, Hamm HE. How do receptors activate G proteins? Adv Protein Chem 2007; 74:67–93.
17. Oldham WM, Hamm HE. Heterotrimeric G protein activation by G-protein-coupled receptors. Nat Rev Mol Cell Biol 2008; 9: 60–71.
18. Pawson T, Nash P. Assembly of cell regulatory systems through protein interaction domains. Science 2003; 300:445–52.
19. Pawson T. Specificity in signal transduction: from phosphotyrosine-SH2 domain interactions to complex cellular systems. Cell 2004; 116:191–203.
20. Pearson G, Robinson F, Beers Gibson T, et al. Mitogen-activated protein (MAP) kinase pathways: regulation and physiological functions. Endocr Rev 2001; 22:153–83.
21. Shenolikar S. Analysis of protein phosphatases: toolbox for unraveling cell signaling networks. Methods Mol Biol 2007; 365:1–8.
22. Sutherland EW, Robinson GA. The role cyclic-3′-5′-AMP in responses to catecholamines and other hormones. Pharmacol Rev 1966; 18:145–61.
23. Willoughby D, Cooper DM. Organization and Ca2+ regulation of adenylyl cyclases in cAMP microdomains. Physiol Rev 2007; 87:965–1010.

Chapter 4

Introduction to Molecular Biology: Structure and Function of the Macromolecules of Genetic Information

Pedro J. Chedrese

4.1 Introduction

Every living organism contains two types of nucleic acids: deoxyribonucleic acid (DNA) and ribonucleic acid (RNA). DNA is the macromolecule that stores the full complement of genetic information of every individual. This information represents the chemical basis of heredity, which in the form of a code is organized into genes. RNA is the macromolecule involved in the process of conveying the genetic information from the DNA to synthesis of proteins. Understanding the basic biochemistry of nucleic acids and proteins is fundamental for complete appreciation of the mechanisms by which extracellular signals affect gene expression and ultimately the biological responses.

4.2 Structure and Functions of the Nucleic Acids

Both DNA and RNA molecules are polymers of monomeric precursors that are composed of an organic base, a sugar, and a phosphoric acid group. The organic bases found in DNA and RNA are the purines—*adenine* and *guanine*— and the pyrimidines—*cytosine, thymine,* and *uracil* (Fig. 4.1). While the sugars are deoxyribose (2-deoxy-D-ribose), found in DNA, which contains only one hydroxyl (OH) group located on the 3′ carbon, and ribose (D-ribose), found in RNA, which contains two OH groups located on the 2′ and 3′ carbons (Fig. 4.2).

Adenine, cytosine, and guanine can link to either deoxyribose or ribose, while thymine links only to deoxyribose and uracil links only to ribose. Therefore, thymine is only present as *deoxythymidine* and is exclusive to DNA and uracil is only present as *ribouridine* and is exclusive to RNA. Monomers that consist only of a sugar attached to the organic base are termed *nucleosides*. A, G, C, T, and U denote the nucleosides of adenine, guanine, cytosine, thymine, and uracil, respectively. They are termed *deoxynucleosides* when the sugar is deoxyribose and *ribonucleosides* when the sugar is ribose. Purine nucleosides end in *-osine*, while pyrimidine nucleosides end in *-idine*. Nucleosides phosphorylated on the hydroxyl groups of the sugar are termed *nucleotides*. Phosphorylated deoxyribose and ribose of the nucleosides are termed *deoxynucleotides* and *ribonucleotides*, respectively.

The position of the phosphate on the sugar molecule is indicated by a numeral followed by a prime symbol (′). The prime symbol differentiates atoms of the sugar from those of the organic bases, which are not followed by the prime symbol. The prefix "d" is used to indicate that the sugar is deoxyribose. For example, cytidine with the phosphate attached to carbon 5 of the sugar ribose is cytidine 5′-monophosphate (CMP), and dCMP when the sugar is deoxyribose. The exceptions are phosphorylated adenosine and guanosine (adenosine 5-monophosphate, AMP; guanosine 5-monophosphate, GMP), since the prime symbol in the nucleotide nomenclature is omitted when the phosphate is esterified to carbon 5 of ribose (Table 4.1; Fig. 4.2).

Additional phosphates attached to the sugar of the nucleotide form nucleotide di- and triphosphates (Fig. 4.3). The position of the phosphate group in the nucleotide is indicated by Greek symbols, α, β, and γ, with respect to being proximal, medial, or distal to the sugar molecule. Nucleotide triphosphates have high group transfer potential and participate in covalent bond formation during DNA and RNA synthesis. By forming covalent bonds, they polymerize to generate the long chains characteristic of the nucleic acids (Fig. 4.4).

- Polymerization of dTMP, dCMP, dAMP, and dGMP to form DNA;
- polymerization of UMP, CMP, AMP, and GMP form RNA.

P.J. Chedrese (✉)
Department of Biology, University of Saskatchewan College of Arts and Science, Saskatoon, SK, Canada
e-mail: jorge.chedrese@usask.ca

P.J. Chedrese (ed.), *Reproductive Endocrinology*,
DOI 10.1007/978-0-387-88186-7_4,

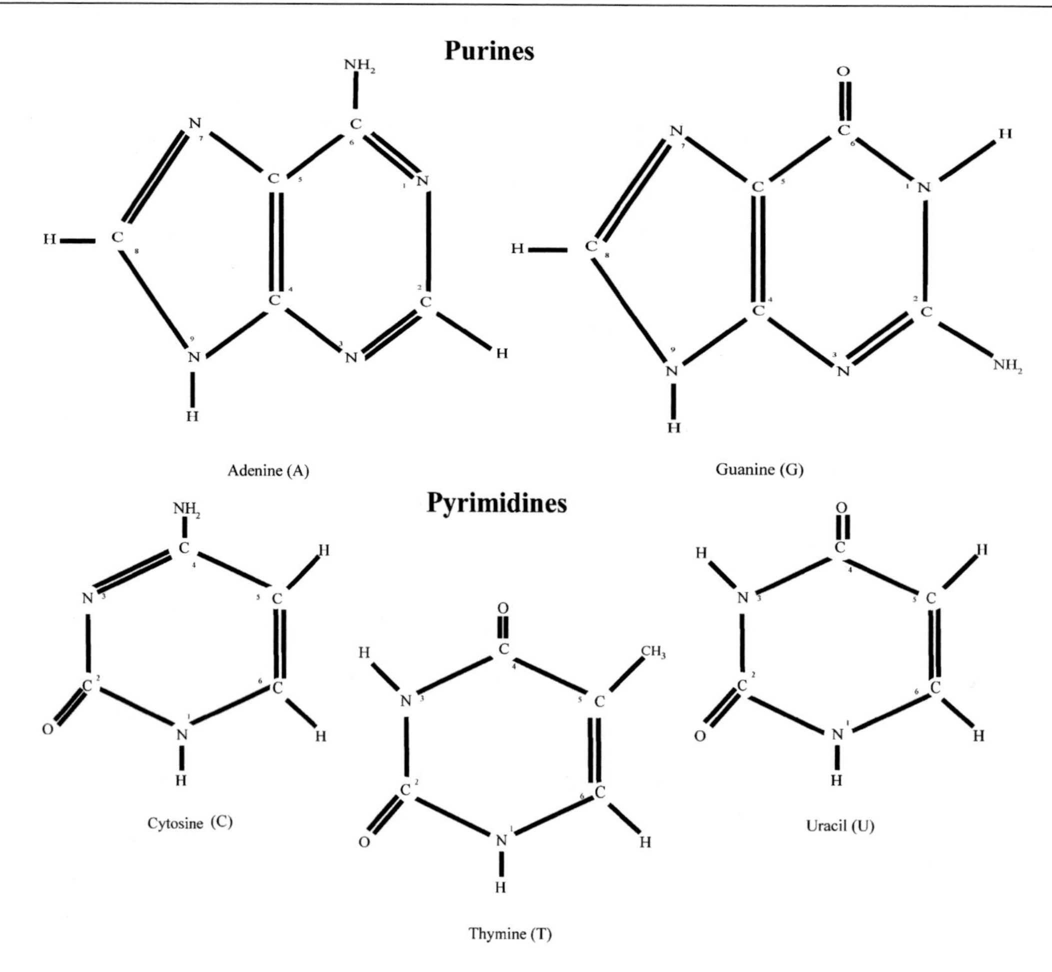

Fig. 4.1 Structure of the purines and pyrimidines

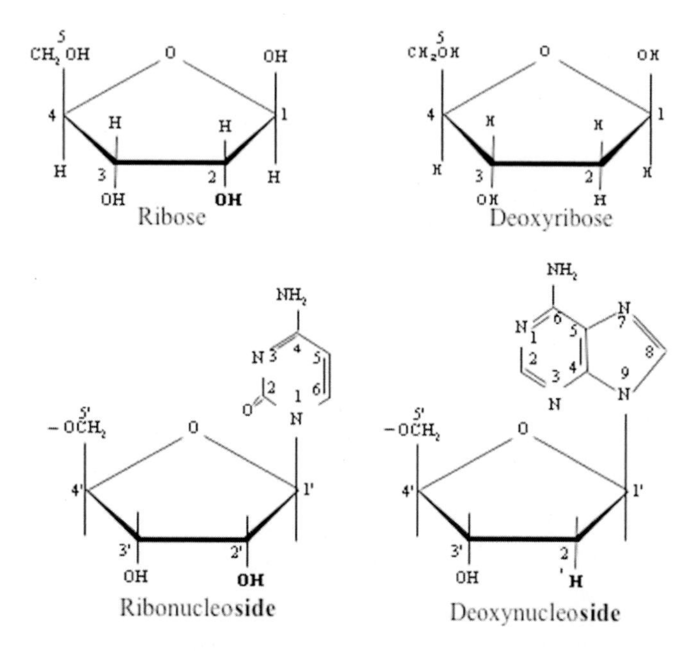

Fig. 4.2 Structure of the nucleosides of ribose and deoxyribose. The sugars are attached to the organic base by a P-N-glycosidic bond, at N^9 in the purines or N^1 in the pyrimidines

4.2.1 DNA

DNA is a very long polymer of the four deoxynucleotide A, G, C, and T that are linked together by phosphodiester bonds between the deoxyribose monomers (Fig. 4.4). Deoxynucleotides are asymmetric and each sugar phosphate chain has polarity. Thus, the phosphodiester bonds within the deoxynucleotide chains have directionality. The ends of the chains are denoted as 5′ and 3′, corresponding to the portion of the deoxynucleotide that is exposed at each end with respect to this polarity (Fig. 4.4).

The entire DNA molecule consists of two antiparallel strands that are held together by hydrogen bonds. The strands are coiled, one from 5′ to 3′ and the other from 3′ to 5′, to generate a double helix configuration. The hydrogen bonds are very selective, A links only to T by two hydrogen bonds and G links only to C by three hydrogen bonds. The specificity of the hydrogen bonds makes both strands complementary. Therefore, during replication, the deoxynucleotide sequence of one strand determines the sequence of the second strand.

Table 4.1 Organic bases, nucleosides, and nucleotides. Purine nucleosides end in *-osine*, while pyrimidine nucleosides end in *-idine*. Prefix "d" is used to indicate that the sugar is deoxyribose

Base	Ribonucleoside	Ribonucleotide
Adenine (A)	Aden*osine* (A)	Adenosine monophosphate (AMP)
Guanine (G)	Guan*osine* (G)	Guanosine monophosphate (GMP)
Cytosine (C)	Cyt*idine* (C)	Cytidine 5′-monophosphate (CMP)
Uracil (U)	Ur*idine* (U)	Uridine 5′-monophosphate (UMP)
Base	**Deoxyribonucleoside**	**Deoxyribonucleotide**
Adenine (A)	Deoxyaden*osine* (dA)	Deoxyadenosine 5′-monophosphate (dAMP)
Guanine (G)	Deoxyguan*osine* (dG)	Deoxyguanosine 5′-monophosphate (dGMP)
Cytosine (C)	Deoxycyt*idine* (dC)	Deoxycytidine 5′-monophosphate (dCMP)
Thymine (T)	Thym*idine* (dT)	Deoxythymidine 5′-monophosphate (dTMP)

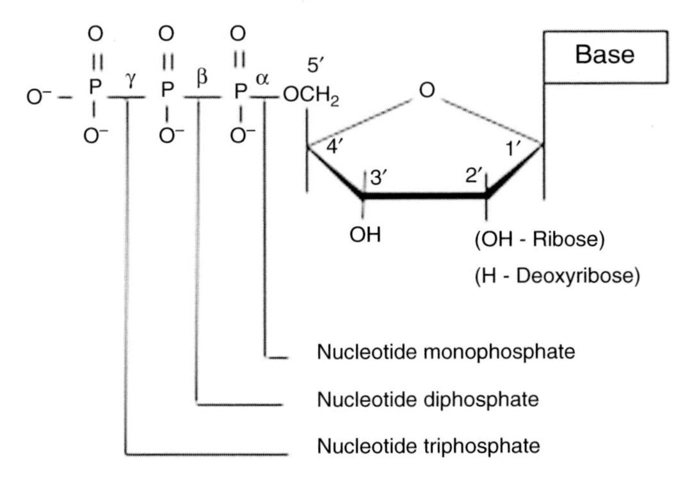

Fig. 4.3 Structure of the nucleotides

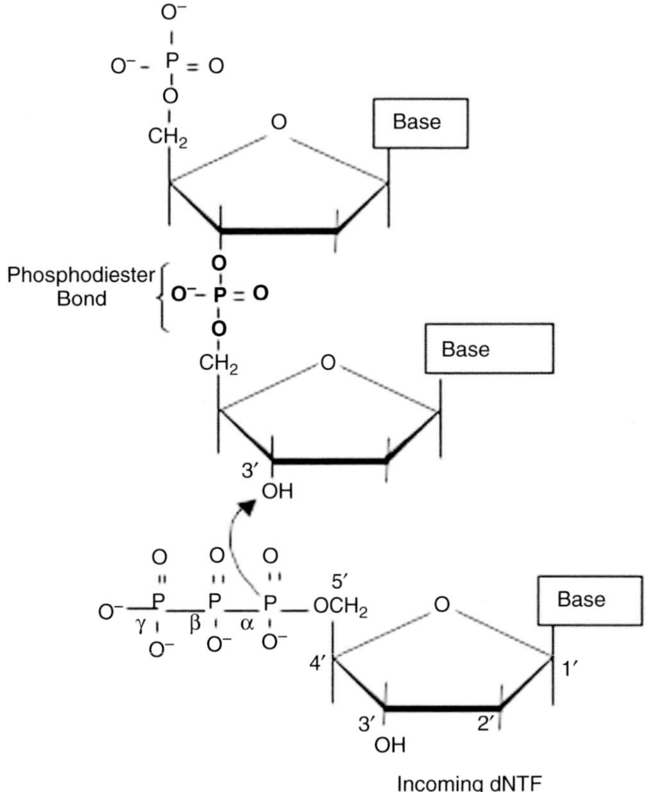

Fig. 4.4 Nucleotide triphosphate incorporation and covalent bond formation in DNA synthesis. dNTP: deoxynucleotide

The sugars of the deoxynucleotides are adjacent, while the bases are oriented toward the interior of the molecule, facing each other (Fig. 4.5). Within the double helix the hydrogen bonds form the minor grove and the adjacent deoxynucleotides form the major grove (Fig. 4.6). This structural configuration makes DNA a strong, stable molecule designed to store genetic information and transmit it from generation to generation.

DNA is organized into *genes*, which are defined as *the fundamental units of genetic information*. Genes contain the information for the synthesis of various types of RNA such as the protein coding RNA, messenger RNA (mRNA), and the non-coding RNAs, such as transfer RNA (tRNA), ribosomal RNA (rRNA), and small RNAs. The genetic information resides in the sequence of nucleotides on one strand, termed the template strand or DNA antisense strand. The opposite strand is considered the coding strand, or DNA sense strand, because it matches the RNA transcript (Fig. 4.7). In the cell, DNA is complexed with proteins called histones to form chromatin. Chromatin consists of five histones: H1, H2a, H2b, H3, and H4. Histones are basic proteins that neutralize the strong acidity of the DNA. Approximately 145 base pairs (bp) of DNA are wrapped with two molecules of histones to form the nucleosome core. Nucleosomes are compacted and condensed to form solenoids, which are further compacted to form the chromosomes (Fig. 4.8).

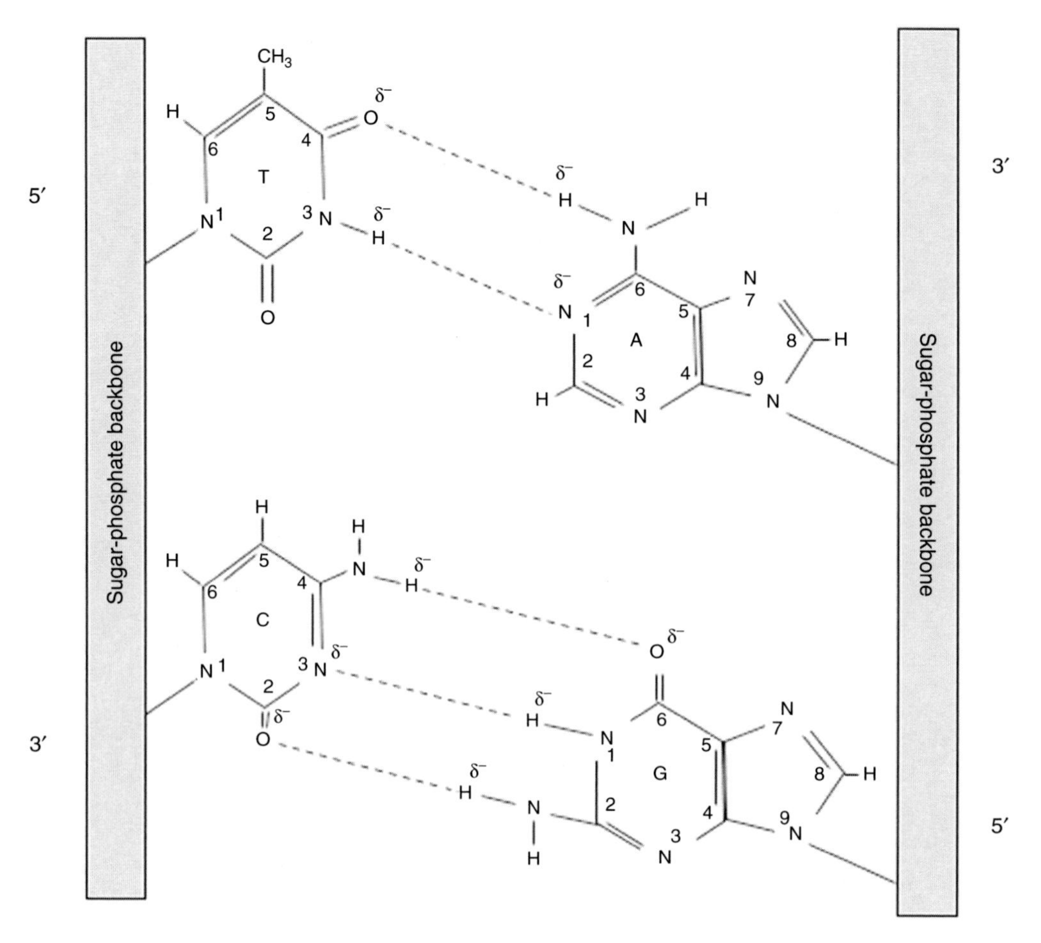

Fig. 4.5 Base-pairing between polynucleotide strands of DNA. The complementary bases are adenine (A) and thymine (T), and guanine (G) and cytosine (C). Note that between A and T there is two hydrogen bonds, and between G and C are three hydrogen bonds

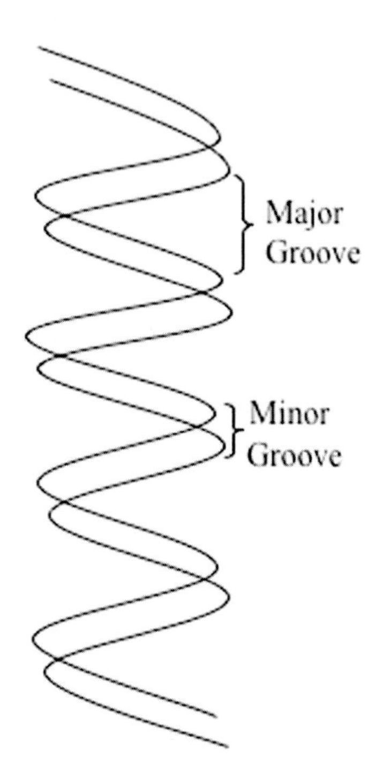

Fig. 4.6 Double helical structure of DNA

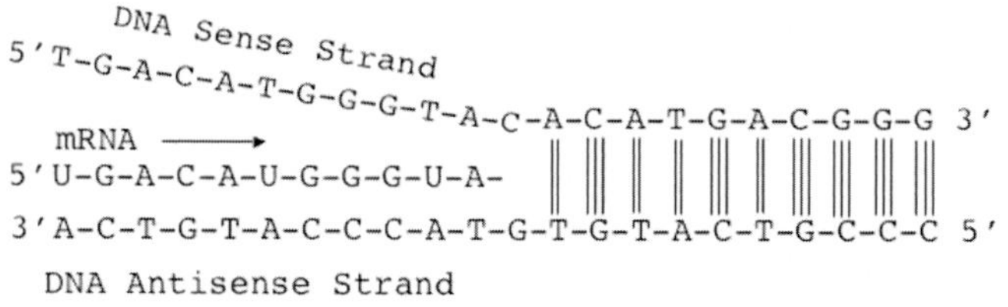

```
DNA Sense Strand
5'T-G-A-C-A-T-G-G-G-T-A-C-A-C-A-T-G-A-C-G-G-G 3'
mRNA ──→
5'U-G-A-C-A-U-G-G-G-U-A-
3'A-C-T-G-T-A-C-C-C-A-T-G-T-G-T-A-C-T-G-C-C-C 5'
DNA Antisense Strand
```

Fig. 4.7 Complementary sequence of DNA and mRNA during transcription. The *arrow* indicates the direction of transcription. mRNA is homologous to the DNA sense strand and complementary to the antisense strand

4.2.2 RNA

The structure of RNA is similar to that of DNA; except that it is a single-stranded molecule in which deoxyribose is substituted by ribose and thymine by uracil. RNA also contains minor amounts of the organic bases: pseudouridine (ψ), inosine (I), dihydrouridine (D), ribothymidine (RT), methylguanosine (MG), or methylinosine (MI). The other main difference between DNA and RNA is that RNA is a very unstable molecule that rapidly degrades. Although RNA is a

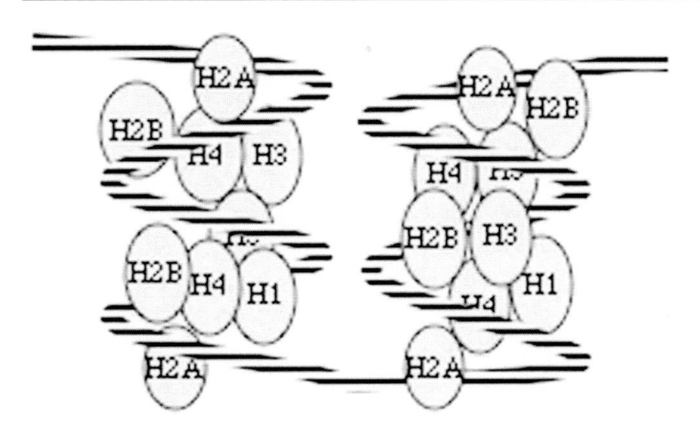

Fig. 4.8 Histone-DNA complexes

simple single-stranded molecule, it has extensive corresponding regions in which AU and GC pairs are joined by hydrogen bonds. This configuration results in molecules that fold, forming *hairpin loop* structures (Fig. 4.9).

Classically three major classes of RNA have been described: mRNA, tRNA, and rRNA. Most recently, other RNA molecules were recognized playing important roles in gene regulation, including small nuclear (snRNA), small nucleolar (snoRNA), micro RNA (microRNA), ribozymes, and small interfering RNAs (siRNAs). RNA molecules are synthesized by three different DNA-dependent RNA polymerases present in the nucleus: RNA polymerase-I, -II, and -III. In addition, other RNA polymerases have been described in the mitochondria. The three main types of RNA polymerases described in eukaryotes are as follows:

- RNA polymerase-I that synthesize the 45S pre-RNA, which matures into 28S, 18S and 5.8S rRNA and form the most abundant components of the ribosomes;
- RNA polymerase-II that synthesize the precursors of mRNAs and most snRNA and microRNAs; and
- RNA polymerase-III that synthesizes tRNAs, rRNAs 5S and other small RNAs found in the nucleus and cytosol.

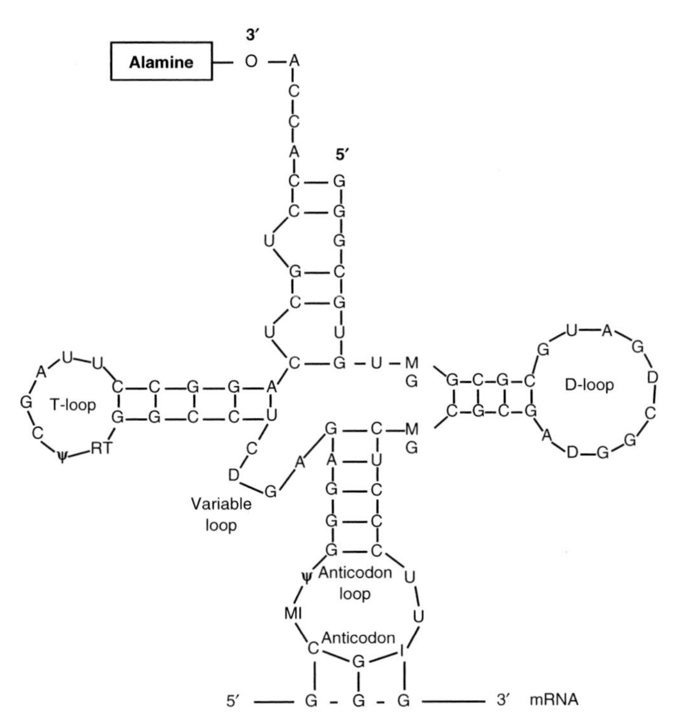

Fig. 4.9 Cloverleaf structure of the alanine-tRNA. The uncommon bases found in tRNA include: pseudouridine (ψ), inosine (I), dihydrouridine (D), ribothymidine (RT), methylguanosine (MG), and methylinosine (MI). In the anticodon of the alanine-tRNA I replaced G

The functions of the different types of RNA can be summarized as follows.

mRNA conveys the genetic information encoded in the genes and serves as a template for protein synthesis in the ribosomes. mRNA is the most heterogeneous in size and stability, and its size reflects the size of the polypeptide it encodes. Most cells produce small amounts of thousands of different mRNA molecules that can only be visualized using specific labeled nucleic acid probes. Many mRNAs are unstable molecules, being rapidly degraded by enzymes called RNAses.

tRNA identifies and transports amino acids to the ribosomes. There are 33 known different tRNAs, products of separate genes, one for each amino acid. The mature tRNA molecules are synthesized by 20 different aminoacyl-tRNA synthetases, also one for each amino acid. tRNA molecules are of uniform size, of about 73–93 ribonucleotides, and also contain inosinic acid, ribothymidylic acid, and pseudouridilic acid (ψU). Many of the bases in the chain pair with each other forming sections of double helix paired loops with the shape of a cloverleaf (Fig. 4.9). The tRNA loops include

- the acceptor loop that binds to the amino acid;
- the TψC loop that binds to the ribosomes;
- the anticodon loop that recognizes mRNA nucleotide triplets, called codons; and
- the D loop, which is the site recognized by the specific aminoacyl-tRNA synthetase.

Each tRNA carries one of the 20 amino acids at its 3′ end. Since there are more than 20 amino acids, most amino acids have more than one tRNA. In the anticodon loop the three unpaired bases pair with the complementary codon on the mRNA molecule, bringing its amino acid into the growing polypeptide chain.

rRNA is the most abundant RNA and represents the main mass of the ribosomal nucleoprotein in the cell cytoplasm. The most abundant rRNAs are the 28S and 18S that can be easily visualized on agarose gel electrophoresis stained with ethidium bromide (Fig. 4.10). One of the 18S rRNA molecules, along with 30 other proteins, forms the small subunit of the ribosomes, while one of each of the 28S, 5.8S, and

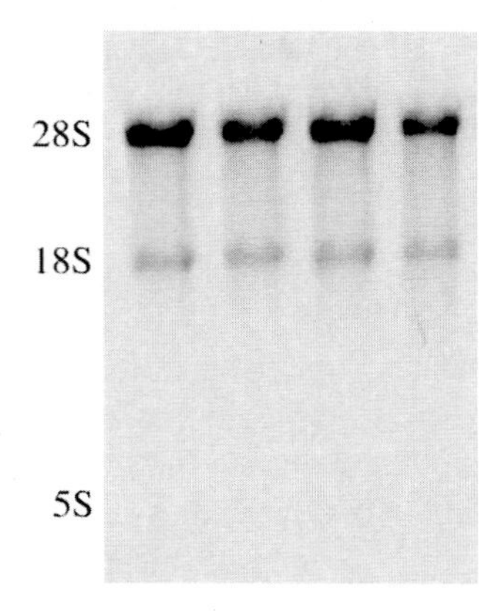

Fig. 4.10 Agarose gel electrophoresis of total RNA. Four different samples of total RNA were separated by size via formaldehyde agarose gel electrophoresis, stained with ethidium bromide, and then photographed under UV light. The above picture shows two main bands, at 28S and 18S, and indicates the position of a minor band at 5S

5S rRNA along with 45 other proteins forms the large subunit of the ribosomes.

snRNA are part of the spliceosomes that participate in mRNA processing that involve excising introns and splicing exons after transcription.

snoRNA are small RNA molecules made from the introns removed during RNA processing. snoRNA are present in the nucleolus, where they participate in a variety of functions including alternative splicing of mRNA, processing of the precursors of the 28S, 18S, and 5.8S RNA, methylation of ribose, and template for the synthesis of telomers.

microRNAs are short non-coding RNA of ~21–31 nucleotides transcribed from DNA but not translated into protein. microRNA are complementary to mRNA and can down-regulate gene expression by a mechanism known as post-transcriptional gene silencing (Chapter 5).

Ribozymes, also termed catalytic RNA, are enzymatically active RNA molecules that catalyze the hydrolysis of RNA phosphodiester bonds and also have aminotransferase activity in the ribosomes.

siRNAs are a group of RNA molecules that are not product of the RNA polymerase, but are the product of the enzyme *dicer*, which cleaves double strand RNA molecules in fragments of about 20–25 nucleotides. Like the microRNA, siRNA participate in post-transcriptional gene silencing (Chapter 5).

4.3 Flow of Genetic Information: From DNA to RNA to Proteins

The information determining a sequence of amino acids in a protein is contained in the genes, in the form of continuous sequences of nucleotides called *exons*, which are the coding regions. In eukaryotes, exons are interspersed with non-coding sections of the DNA molecule called *introns* (Fig. 4.11).

Genes are transcribed into mRNA by the enzyme RNA polymerase-II, a multisubunit enzyme complex that synthesizes a single-stranded RNA. This process starts at the first nucleotide of the first exon, called the *transcription start site* (Tx). Transcription ends when RNA polymerase reaches the *transcription stop signals, AUAAAA*. These signals indicate the post-transcriptional addition of multiple adenine ribonucleotides, called poly-adenylated (poly-A) tails, at the 3′ end of the mRNA.

The first molecule of mRNA synthesized, or first transcript, is called heteronuclear RNA (hnRNA) and is an exact copy of the DNA from the Tx to the transcription stop signals (Fig. 4.11). hnRNA is then processed within minutes after transcription into mature mRNA within the nucleus and immediately transported to the cytoplasm. The processing of hnRNA can be summarized as follows:

- The introns are spliced out and exons rejoined yielding a mature molecule that is shorter than the original transcript. Splicing of the introns and rejoining of the exons are defined by signals that include the sequences GT and AG.
- At the 5′ end of the mRNA, 7′-methylguanosine links by a unique 5′ to 5′ phosphodiester bond (Fig. 4.12) forming a structure called "CAP" (capped mRNA). CAP appears to protect mRNA from RNAses. Therefore, CAP may play an important role in mRNA stability.
- At the 3′ end, a poly-A tail attaches, which appears to play a role in ribosomal recognition of mRNA and helps regulate translation efficiency.

Not all mRNA sequences are translated into protein. The mature mRNA contains untranslated sequences at both ends of the molecule, called the 5′ and 3′ untranslated regions (5′-UT and 3′-UT, respectively). The 5′-UT extends from the CAP to the first codon AUG that encodes for *methionine*, which is the first amino acid in every protein. The 3′-UT spans from the transcription stop signal to the poly-A tail. The untranslated sequences of mRNA are regions of great variability between genes, especially the 3′-UT, which can be from a few dozen to several thousand bases in length.

4.3.1 The Genetic Code and Protein Synthesis

The genetic code is defined as *the order of the nucleotide sequences in DNA or RNA that form the basis of heredity through their role in protein synthesis.* The genetic code is organized into signals comprised of three nucleotides, or triplets, called codons, which encode for each amino

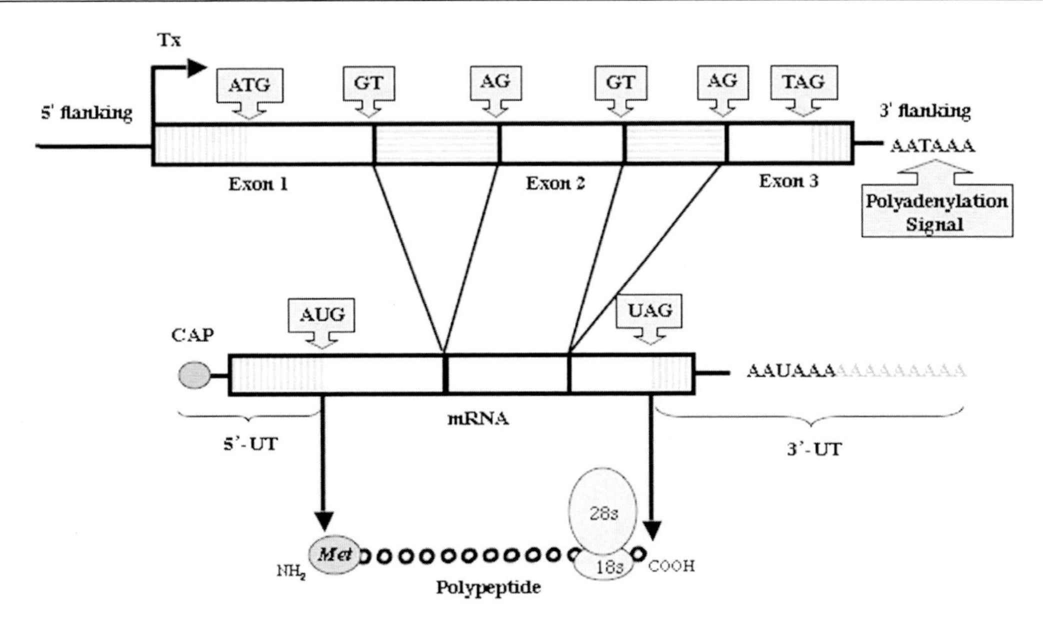

Fig. 4.11 Structure of the gene and flow of genetic information from DNA to mRNA to protein

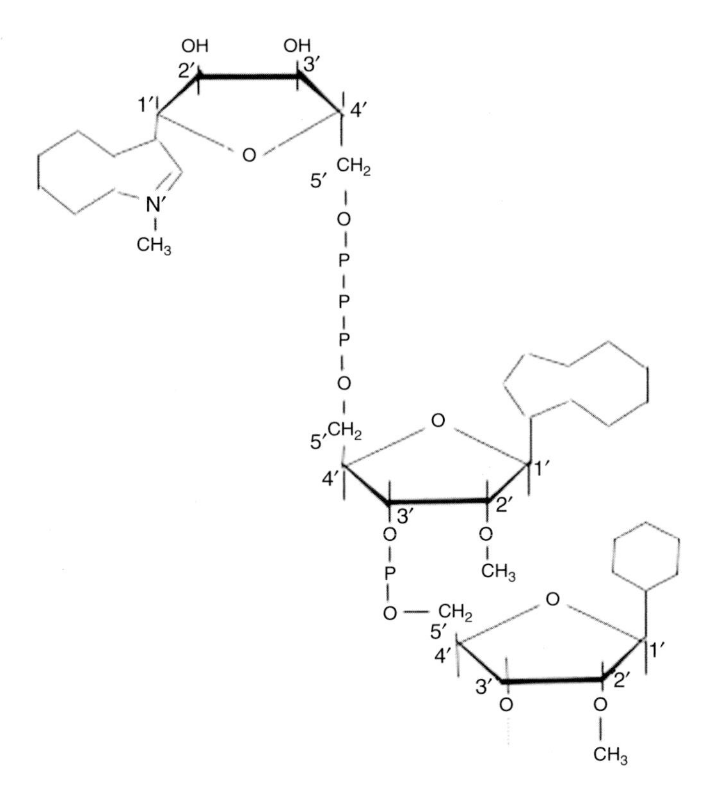

Fig. 4.12 The mRNA 5′ terminus. In most mRNA, a 7′-methyl-guanosine triphosphate, called CAP, is attached at the 5′ terminus by a unique 5′ to 5′ phosphodiester bond, which usually contains a 2′-methylpurine nucleotide

acid. Since there are only 20 amino acids and there are 64 possible combinations of the four nucleotides into triplets, amino acids are encoded by more than one codon, with the exception of *methionine* that is encoded by a unique codon (Table 4.2). The main features of the genetic code can be summarized as follows:

- It is *universal.* With a few minor exceptions, such as the mitochondria, the same genetic code is used by viruses, prokaryotes, and eukaryotes.
- It is written in *linear* form using the letters designating the ribonucleotide bases.
- *Each codon contains three nucleotides*, arranged in 16 families.
- It *does not contain comas*, meaning that when translation begins each codon is read in sequence.
- It is *non-overlapping*, meaning that any single letter is part of only one codon.
- It is *unambiguous*, meaning that each codon specifies a single amino acid.

Table 4.2 The genetic code

UUU Phe	UCU Ser	UAU Tyr	UGU Cys
UUC Phe	UCC Ser	UAC Tyr	UGC Cys
UUA Leu	UCA Ser	UAA Stop	UGA Stop
UUG Leu	UCG Ser	UAG Stop	UGG Trp
CUU Leu	CCU Pro	CAU His	CGU Arg
CUC Leu	CCC Pro	CAC His	CGC Arg
CUA Leu	CCA Pro	CAA Gln	CGA Arg
CUG Leu	CCG Pro	CAG Gln	CGG Arg
AUU Ile	ACU Thr	AAU Asn	AGU Ser
AUC Ile	ACC Thr	AAC Asn	AGC Ser
AUA Ile	ACA Thr	AAA Lys	AGA Arg
AUG Met	ACG Thr	AAG Lys	AGG Arg
GUU Val	GCU Ala	GAU Asp	GGU Gly
GUC Val	GCC Ala	GAC Asp	GGG Gly
GUA Val	GCA Ala	GAA Glu	GGA Gly
GUG Val	GCG Ala	GAG Glu	GGG Gly

- It is *degenerate*, meaning that more than one codon specifies for the same amino acid.
- It is *ordered*, meaning that degenerate codons are grouped, most often varying by only the third letter.
- It contains *start and stop* signals.

Immediately after processing, mature mRNA reaches the ribosomes, which order the interaction of all of the components of translation and protein synthesis. Translation does not start until the ribosomes recognize the *methionine* codon in the mRNA (Fig. 4.11). mRNA is translated from the 5′ to 3′ direction, and the proteins are synthesized from the N-terminus to the C-terminus. *t*RNA carries amino acids to the ribosomes. Each tRNA has an anticodon of three nucleotides that can pair with the complementary mRNA codon (Fig. 4.9). Translation ends when the ribosomes reach the first of the termination, or stop, codons—UAA, UAG, or UGA—in the mRNA molecule. No tRNA carries anticodons that recognize these sequences.

4.4 Summary

DNA and RNA are polymers of monomeric precursors composed of an organic base, a sugar, and a phosphoric acid group. The DNA molecule consists of two antiparallel strands held together by hydrogen bonds, in which the sugar of the deoxynucleotides are adjacent, while the bases face the interior of the molecule. DNA is organized into genes that are transcribed into mRNA. Transcription starts at the first nucleotide of the first exon, called the transcription start site, and ends at the transcription stop signals. The first molecule of mRNA synthesized, the hnRNA, is processed into mature mRNA, in which the introns are spliced out, capped at the 5′ end, and a poly-A tail is added at the 3′ end. Mature mRNA reaches the ribosomes, which order protein synthesis according to the genetic code, which is organized into signals comprised of three nucleotides, called codons that encode for every amino acid. Translation starts when the ribosomes recognize the methionine codon, AUG, and mRNA is read in the 5′ to 3′ direction, and proteins are synthesized sequentially from the N-terminus to the C-terminus. Translation ends when the ribosomes reach the first of the termination codons in the mRNA.

Glossary of Terms and Acronyms

2-deoxy-D-ribose: deoxyribose

3′-UT: 3′ untranslated region

5′-UT: 5′ untranslated region

A: adenine or adenosine

AMP: adenosine 5-monophosphate

bp: base pairs

C: cytosine or cytidine

CAP: capped mRNA

CMP: cytidine 5′-monophosphate

D: dihydrouridine

dAMP: deoxyadenosine monophosphate

dCMP: deoxycytidine monophosphate

dCMP: deoxyribose cytidine 5′-monophosphate

dGMP: deoxyguanosine monophosphate

DNA: deoxyribonucleic acid

D-ribose: ribose

dTMP: deoxythymidine monophosphate

G: guanine or guanosine

GMP: guanosine 5-monophosphate

hnRNA: heteronuclear RNA

I: inosine

MG: methylguanosine

MI: methylinosine

microRNA: micro RNA

mRNA: messenger RNA

OH: hydroxil group

Poly-A: poly-adenylated tail at the 3′ end of the mRNA

RNA: ribonucleic acid

RNAses: enzymes that specifically degrade RNA

rRNA: ribosomal RNA

RT: ribothymidine

siRNA: small interfering RNA

snoRNA: small nucleolar RNA

snRNA: small nuclear RNA

T: thymine or thymidine

tRNA: transfer RNA

Tx: transcription start site

U: uracil or uridine

UMP: uridine monophosphate

ψ: pseudouridine

ΨU: pseudouridilic acid

Bibilography

1. Alberts B, Johnson A, Lewis J, et al. Molecular Biology of the Cell, fourth edition. New York: Garland Science, 2002.
2. Maniatis T, Fritch EF, Sambrook J. Molecular Cloning: A Laboratory Manual. New York: Cold Spring Harbor Laboratory Press, 1982.
3. Miller WL. Basic Recombinant DNA. International Symposium on Molecular Biology for Endocrinologist. Washington: Serono Symposia USA, 1996:1–10.
4. Miller WL. Cloning Strategies. International Symposium on Molecular Biology for Endocrinologist. Washington: Serono Symposia USA, 1996:11–8.
5. Rodwell VW. Nucleotides. In: Murray RK, Granner DK, Mayes PA, et al., editors. Harper's Biochemistry, twenty-second edition. Norwalk: Appleton & Lange, 1990:333–42.
6. Rodwell VW. Metabolism of purine & pyrimidine nucleotides. In: Murray RK, Granner DK, Mayes PA, editors. Harper's Biochemistry, twenty-second edition. Norwalk: Appleton & Lange, 1990:342–56.
7. Shupnik MA. Introduction to molecular biology. In: Fauser BCJM, editor. Reproductive Medicine: Molecular, Cellular and Genetic Fundamentals. Boca Raton: Parthenon Pub Group 2003: 3–22.

Chapter 5

Regulation of Gene Expression

Pedro J. Chedrese

5.1 Introduction

All living organisms are genetically characterized by information stored in the genome as sequences of nucleotides. The genetic information required for organic development and functions is contained in the genes, which comprise only a fraction of the total sequences of nucleotides transmitted during every cell division and from generation to generation. Genes that express proteins can be divided in two main groups. The first group called *constitutive* or "housekeeping" genes encode *slow turnover* proteins required for basal functions and, therefore, are steadily expressed without regulation in most cells. The second group includes genes that encode proteins required for cell-specific biochemical activities and, therefore, are expressed only in particular cell groups. Some of these genes express *high turnover* proteins that are subject to rapid changes. Because these genes have restricted spatial patterns of expression, depending on their role, they can be highly regulated.

Genes can be regulated by extracellular signals, including hormones, growth factors, and neurotransmitters that mainly affect transcription. However, other levels of regulation are also involved, such as *epigenetic* mechanisms that include modifications in the DNA and chromatin without changes in the sequence of nucleotides. In addition, expression of genes can be regulated *post-transcriptionally* at the level of processing, transport, stability and translation of mRNA, and by RNA interference (RNAi) through small RNA molecules that target and degrade mRNA. Finally, gene expression can be regulated *post-translationally,* where modification of the synthesized protein results in changes in its biological activity.

P.J. Chedrese (✉)
Department of Biology, University of Saskatchewan College of Arts and Science, Saskatoon, SK, Canada
e-mail: jorge.chedrese@usask.ca

5.2 Transcriptional Regulation of Gene Expression: The Basic Transcription Unit

Eukaryotic genes contain the information necessary to encode proteins and the sites to initiate and terminate transcription. In addition, genes contain the information needed to regulate their own expression in the form of specific DNA nucleotide sequences called *cis*-acting regulatory elements. The *cis*-acting regulatory elements are recognized by *trans*-acting regulatory factors, such as the nuclear transcription factors that are the direct targets of the intracellular signaling pathways. Regulatory elements can be classified into five main groups: *promoters, enhancers, silencers, boundary elements, and response elements.*

5.2.1 Promoters

Promoters are classically defined as *specific regulatory regions of the gene where transcription factors bind to initiate transcription.* However, researchers generically refer to the entire 5′-flanking region of a gene as the *promoter region*. Promoters are located in the first 100–300 base pairs (bp) upstream of the transcription start site (Tx) and can be divided into two segments, the *core promoter* and the *proximal promoter.* The core promoter is involved in initiation of transcription and basal levels of gene activity (Fig. 5.1). The elements of the core promoter include the TATA box, which is an 8 bp consensus sequence rich in T and A, surrounded by a C and G rich region. The TATA box is usually found in the first 25–30 bp upstream of the transcription start site. The TATA box binds the TATA-binding protein (TBP), which is a key component of the *t*ranscription *f*actor IID (TFIID). Although some genes lack consensus TATA boxes, they have related TA-rich sequences that define the site where transcription starts. In addition, three more elements are commonly found in the core promoter: the TFII*B* *r*ecognition *e*lement (BRE) located upstream of the TATA box that binds TFIIB;

P.J. Chedrese (ed.), *Reproductive Endocrinology*,
DOI 10.1007/978-0-387-88186-7_5, © Springer Science+Business Media, LLC 2009

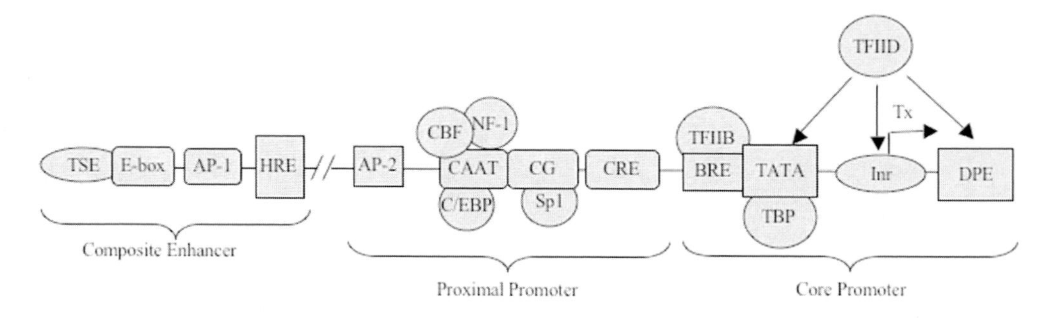

Fig. 5.1 Regulatory elements in the promoter and enhancer regions of a gene. AP-1: activating protein-1 binding site; AP-2: activating protein-2 binding site; BRE: TFIIB recognition element; CAAT: C/EBP: CAAT-enhancer binding protein; CBF: CAAT box binding factor; CRE: cAMP responsive element; CREB/AP-1: CRE binding protein/activator protein-1; DPE: downstream promoter element; E-box: binds members of the basic helix-loop-helix (bHLH) family of proteins; Inr: initiator; NF-1: nuclear factor-1; Sp1: selective promoter-1 transcription factor; TBP: TATA box-binding protein; TSE: tissue-specific element; HRE: hormone-response element; Tx: transcription start site

the initiator (Inr) sequence, located at the transcription start site that binds TFIID; and the downstream promoter element (DPE) located approximately 30 bp downstream of the transcription start site, which is recognized by TFIID.

The proximal promoter is located immediately upstream of the core promoter and may include elements also found in the core promoter (Fig. 5.1). The typical elements of the proximal promoter are the GC box and the CAAT box. The GC box contains the sequence GGGCGG found in multiple copies and can be in opposite orientation, GGCGGG. The GC box binds the selective promoter-1 (Sp1) transcription factor, and is therefore, also termed the Sp1 box. The CAAT box contains the sequence CCAAT, usually found within the first 100 bp upstream of the transcription start site. The CAAT box binds the *C*AAT box/*e*nhancer *b*inding *p*rotein (C/EBP), the *C*AAT box-*b*inding *f*actor (CBF), also termed *n*uclear *f*actor-*Y* (NF-Y), and the *C*AAT-binding *t*ranscription *f*actor (CTF), also known as *n*uclear *f*actor-1 (NF-1). The CAAT box continues to function even when located at distances further than 100 bp upstream of the Tx and when placed in opposite orientation, TAACC. In conjunction with the GC box, the CAAT box brings the RNA polymerase-II into proximity of the Tx. Overall, the CAAT box and the CG box determine efficiency of transcription. However, it must be taken into consideration that CAAT boxes and GC-boxes are associated with numerous genes, but not all genes.

5.2.2 Enhancers

Enhancers are *cis*-acting elements located in the regulatory region of the gene that improve basal transcription initiated through the core promoter (Fig. 5.1). Enhancers permit interaction of transcription factors, even when located at various distances, either upstream or downstream, of the Tx. Enhancers recognized by particular transcription factors determine tissue specific gene expression. Typical enhancers include the tissue-specific element (TSE), the E-box that binds members of the *basic helix-loop-helix* (bHLH) family of proteins, the steroidogenic factor-1 (SF-1) binding sites found in the steroidogenic enzyme gene promoters, and the pituitary-specific factor-1 (Pit-1) site found in the growth hormone gene promoter.

Both promoters and enhancers are positioned in a linear manner and although they may be located far apart from each other, DNA looping brings the enhancer-bound transcription factors close to the promoter allowing molecular interactions (Fig. 5.2). Further characterization of the enhancers, and transcription factors that they bind, reveals that there is considerable functional overlap with other regulatory elements in the promoter, such as the CAAT and GC boxes. Therefore, the difference between enhancers and promoters sometimes becomes indistinguishable.

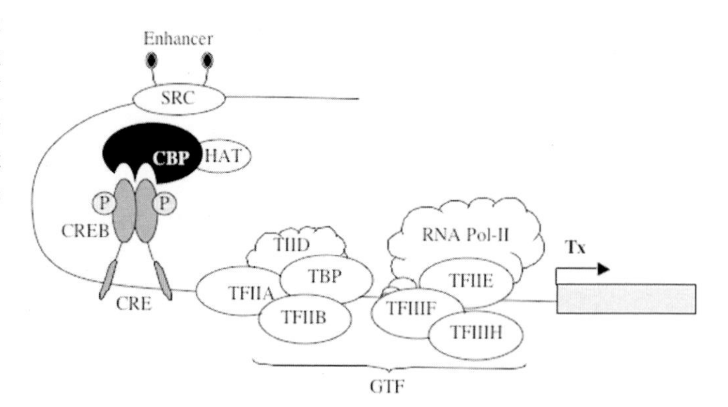

Fig. 5.2 DNA looping and interactions between enhancer bound transcription factors and components of the promoter. The diagram shows DNA looping and interaction between distant regulatory elements and the general transcription factors in the promoter. CREB: cAMP-regulatory element binding protein; CBP: CREB-binding protein; GTF: general transcription factors; HAT: histone acetyltransferase; SRC: steroid receptor co-activator; TBP: TAATA box binding protein; Tx: transcription start site

5.2.3 Silencers

Silencers, also termed repressors, are elements that reduce transcription. Silencers are located at various positions within genes, including the introns. Although they are not fully characterized, two groups have been recognized: *classical silencers* that are position-dependent and *negative regulator elements* that are position-independent.

5.2.4 Boundary Elements

Boundary elements, also termed *insulators,* are short sections of DNA that flank the regulatory elements of the gene. The function of the boundary elements is to insulate genes, from the activity of adjacent genes. For example, the nucleotide sequence CCCTC that binds an eleven zinc-finger motif protein, *CTC-binding factor* (CTCF), is found in many genes blocking the influence of other gene promoters.

5.2.5 Response Elements

Response elements are *cis*-acting DNA nucleotide sequences that modulate transcription in response to extracellular messages. They are classified into two main groups: *ligand-activated* and *second messenger-activated.*

Ligand-activated response elements mediate the effects of extracellular signaling on transcription by recognizing nuclear receptor transcription factors, such as the receptors for steroids, thyroid hormones, and retinoic acid (Table 5.1). Thus, they are termed hormone response elements (HRE) and are usually a pair of inverted repeat sequences complementary to another sequence located downstream, suggesting that the transcription factor binds as a dimer that recognizes the pair. Many different genes share the same, or very similar HREs, therefore the term "consensus sequences" has been established for nucleotide sequences that bind a common transcription factor. Conversely, consensus sequences can also be recognized by more than one transcription factor, which adds another level of complexity to gene regulation.

Second messenger-activated response elements are sequences that bind proteins activated by intracellular signaling, such as the protein kinase-A (PKA) and protein kinase-C (PKC) pathways. Second messenger-activated response element includes the sequence TGACGTCA, termed cAMP-response element (CRE), that binds the CRE-binding protein (CREB) activated by the second messenger cAMP. Some genes contain related sequences that differ from CRE by a single nucleotide deletion, TGA–GTCA, termed TPA-response element (TRE) because they respond to the PKC pathway activator phorbol ester TPA (12-*O*-tetradecanoyl-13-acetate). TRE binds the transcription factor *activating protein-1* (AP-1), a heterodimer formed by the products of the c-*jun* and c-*fos* proto-oncogene (Jun/Fos). Another common second messenger-activated response element is $GCCN_3GGC$, the site that binds the *activating protein-2* (AP-2) transcription factor (Table 5.2).

Although the signaling-activated response elements are considered enhancers, they are also often found in the core promoter of several genes. For example, CRE, which operates as a classic enhancer, is commonly found in the core promoter region of several cAMP-activated genes, such as the α-subunits of the glycoprotein hormones, FSH, LH, TSH, and hCG.

5.3 Transcription Factors

Transcription factors, also called *trans*-acting regulatory factors, are the proteins needed for initiation of transcription that are not part of the enzymatic complex *DNA-dependent RNA-polymerase-II.* Transcription factors recognize and bind

Table 5.1 Consensus sequences of ligand-activated response elements

Ligand	*trans*-acting factor	*cis*-acting element	Consensus sequence
Glucocorticoids	GR	GRE	GGTACAnnnTGTTCT
Progesterone	PR	PRE	"
Androgens	AR	ARE	"
Mineralocorticoids	MR	MRE	"
Estrogens	ER	ERE	AGGTCAnnnTGACCT
Thyroid hormones	TR	TRE	TCAGGTCA
Retinoic acid	RAR	RRE	TGACCTGA

AR: androgens receptor; ARE: androgens receptor element; ER: estrogen receptor; ERE: estrogen receptor element; GR: glucocorticoid receptor; GRE: glucocorticoid receptor element; MR: mineralocorticoid receptor; MRE: mineralocorticoid receptor element; PR: progesterone receptor; PRE: progesterone receptor element; RAR: retinoic acid receptor; RRE: retinoic acid receptor element; TR: thyroid hormone receptor; TRE: thyroid hormone receptor element.

Table 5.2 Consensus sequences of signaling-activated response elements

Signal	*Trans*-acting factor	Consensus sequences
PKA	CREB (ATF)	TGACGTCA [CRE]
PKA	Fos	TGACGTTT [CRE-like]
PKA and PKC	AP-1	TGA-GTCA [TRE]
PKA and PKC	AP-2	$GCCN_3GGC$
PKA	Jun-B	AGTGCACT
PKA	GATA	(A/T)GATA(A/G)
Serum growth factor	SRF	CC(A/T)(A/T)(A/T)(A/T)(AJT)(A/T)GG [SRE]
Heavy metals	MTF-1	TGCACAC
Interferon-γ	STAT1	TTNCNNNAAA
Heatshock	HSP70	CTNGAATNTTCTAGA

AP-1: activating protein 1; AP-2: activating protein 2; CRE: cAMP response element; CREB: CRE binding protein also called ATF or activator transcription factor; GATA: family of transcription factors; HSP70: heat shock protein-70; MTF-1: metal-responsive transcription factor-1; PKA: protein kinase-A; PKC: protein kinase-C; SRE: serum response element; SRF: serum response factor; STAT1: signal transducers and activator of transcription; TRE: TPA-response element.

cis-acting elements located in promoters through protein–DNA interaction, and can bind other transcription factor(s) through protein–protein interactions.

Initiation of transcription requires the interaction of DNA-dependent RNA-polymerase-II with the *general transcription factors* (GTF), which are the proteins necessary for basal transcriptional activity. At least seven GTFs participate in initiation of transcription, including TFIIA, TFIIB, TFIID, TFIIE, TFIIF, TFIIH, and TFIIJ that interact according to the following four steps:

1. TFIID binds to a region upstream of the TATA box, a process that requires the participation of TFIIA;
2. TFIIB allows the binding of RNA polymerase II to this complex;
3. TFIIE binds to a region downstream of RNA polymerase II; then
4. TFIIF, TFIIH and TFIIJ complete the complex and transcription begins.

Regulated gene expression requires the participation of a third group of proteins, *specific transcription factors,* which are obligated links between the extracellular signals and the gene (Table 5.1). Thus, DNA-dependent RNA-polymerase-II activity is regulated by the interaction between specific transcription factors and GTF, with the *cis*-acting elements.

5.3.1 Coregulators

Coregulators are proteins that recognize other proteins, including nuclear receptor transcription factors. Coregulators link GTF with specific transcription factors without interacting directly with DNA. *Coactivators* are coregulators that activate transcription, while *corepressors* suppress transcription.

5.4 Transcription Factors Structural Motifs

Most transcription factors have a series of common secondary structural features that facilitate interaction with DNA. They include an α-helix that fits into the DNA major groove and a free peptide strand that makes contact with the DNA minor groove. The amino-terminal region binds the bases while the carboxy-terminal region interacts with the deoxyribophosphate backbone of the DNA. According to their structural motifs, transcription factors are classified as follows: *helix-turn-helix, homeodomain proteins, high mobility group proteins, zinc-finger transcription factors, leucine zippers,* and *helix-loop-helix.*

5.4.1 Helix-Turn-Helix

Helix-Turn-Helix (HTH) motif is defined as two α-helices, each composed of 20 amino acids connected by a short amino acids chain that twist the α-helices into an "L" shape configuration (Fig. 5.3A). The carboxy terminal α-helix, referred to as the *recognition* helix, fits into the DNA major groove and interacts with specific bases. This motif is found in many DNA-binding proteins, including the catabolic activator protein (CAP), the λ-phage repressor *cro*, the *lac* repressor, and the *trp* repressor. Extensions of the HTH motif are the *winged* HTH (wHTH), which is a group of proteins characterized by a third α-helix and an adjacent β-sheet that are components of the DNA-binding motif providing additional contacts with the DNA backbone.

5.4.2 Homeodomain Proteins

In higher-order organisms the HTH motifs are more commonly found in a family of proteins termed homeodomain, which has three α-helices instead of two (Fig. 5.3B). The carboxy-terminal α-helix, which is identical to the HTH except that the length of the loop is longer, is the recognition helix and lies in the major groove of the DNA. The remaining α-helices lie on the top of the recognition helix. A variation of the homeodomain motif is found in the POU family of transcription factors, an acronym derived from *P*it-1, *O*ct, and *U*nc-86, which were the first three members

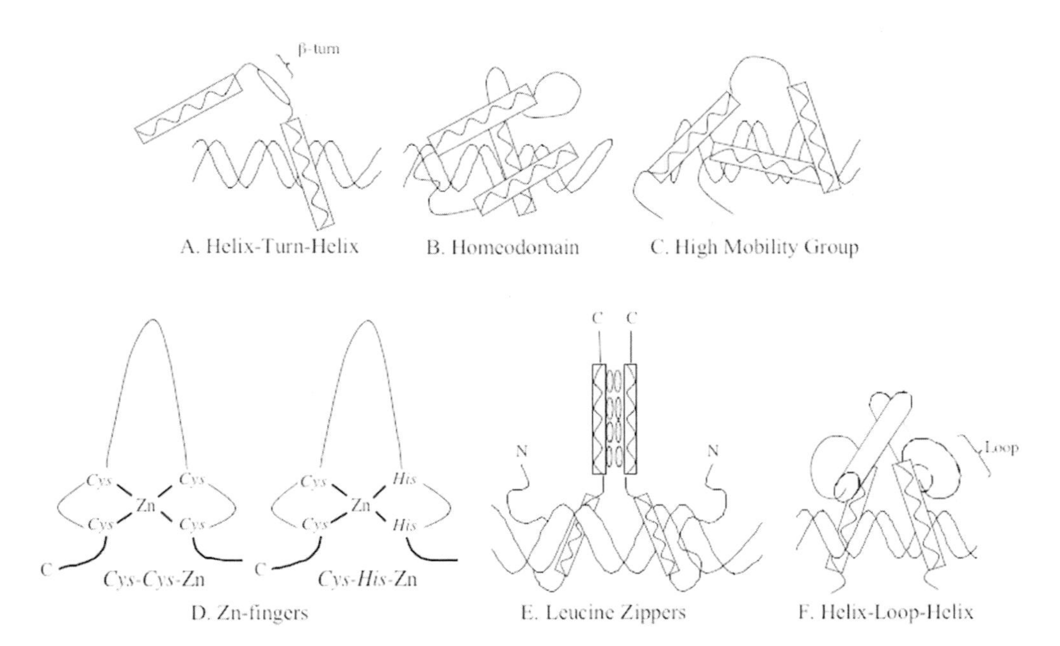

Fig. 5.3 Structural representation of the transcription factors motifs. (**A**) HTH motif with two α-helices forming an L-shape configuration. The α-helix that interacts with the DNA *cis*-acting element is termed recognition helix. (**B**) The recognition helix of the homeodomain motif fits into the DNA major groove, while the other two helices are located perpendicularly over the recognition helix. (**C**) The recognition helices in the HMG contact the minor groove of the DNA. (**D**) In the Zn-fingers the amino to carboxy ascending limb is part of a β-sheet and the descending limb forms a recognition α-helix that fits into the DNA major groove. (**E**) The *leucine* rich domain of the leucine zipper is constituted of *leucine* repeats every seventh amino acid that forms two parallel α-helices. An extended α-helix of the motif fits into the DNA major groove. (**F**) The bHLH form one helix linked by a loop to a second helix that is required to contact with the deoxyribose-phosphate backbone of the DNA. The bHLH and can form homo- and hetero-dimers through interactions between their hydrophobic residues

reported. The POU family contains a homeodomain, called POU_H, preceded by a conserved 75–82 amino acid sequence called POU_S, composed of four α-helices. Transcription factors carrying the homeodomain motif include the homeotic proteins involved in development of the fruit fly *Drosophila melanogaster*, and the thyroid transcription factor 1 (TTF1) in mammals.

5.4.3 High Mobility Group Proteins

The high mobility group (HMG) is a small group of diverse proteins, highly conserved during evolution, that bind DNA and mediate gene regulation. They are called "high mobility" because they migrate rapidly in polyacrylamide gel electrophoresis. HMG proteins have a homologous DNA-binding domain called the HMG-box and are the only group of transcription factors that bind to the DNA minor groove. The HMG proteins contain three α-helices (Fig. 5.3C). Two of the α-helices are perpendicular forming a twisted "L" over the DNA, widening the minor groove, while the third helix contacts the DNA deoxyribose-phosphate backbone. The HMG are associated with chromatin activity and two members of this group, HMG-14 and HMG-17, selectively bind to the nucleosome of active chromatin. Other members include T-cell factor/lymphoid enhancer factor (TCF/LEF), the sex-determining region-Y (SRY) and the SRY box-9 (SOX-9).

5.4.4 Zinc-Finger Transcription Factors

Zinc (Zn) finger transcription factors are a family of proteins whose amino acids are folded around Zn atoms that stabilize their three-dimensional structure (Fig. 5.3D). The first group of this family includes the transcription factors TFI-IIA, Sp1, and GATA. They are sequences of 12–14 amino acids bound to a Zn atom by a *cysteine* pair at the amino-terminus and a *histidine* pair (C_2H_2) at the carboxy-terminus. The C_2H_2 force a loop configuration of the protein, in which the amino to carboxy ascending limb is part of a β-sheet and the descending limb forms a recognition α-helix called the Zn-finger that fits into the DNA major groove. Only the tip of the Zn-finger makes contact with 3–4 bases of the DNA, therefore, many fingers are required to recognize a specific DNA sequence.

The second group of this family is the *nuclear receptor transcription factors* that bind lipophilic ligands with high affinity. This group includes the nuclear receptors for progesterone (PR), estrogens (ER), glucorticoids (GR), mineralocorticoids (MR), vitamin-D (VDR), thyroid hormone (THR), retinoic acid, 9-*cis*-retinoic acid (RAR), ecdysone, and a group of unknown ligands referred to as orphan receptors, such as the *c*hicken *o*voalbumin *u*pstream *p*romoter *t*ranscription *f*actor (COUP-TF). Nuclear receptor transcription factors bind to HRE as dimers, in which each monomer recognizes only half of the consensus sequence. Thus, the orientation and spacing between half-sites of the HRE determines receptor specificity. Three different regions can be distinguished in the nuclear receptor transcription factors (Fig. 5.4), including the following:

1. The *DNA-binding domain* is characterized by a non-conserved amino-terminal region of variable size that determines tissue specificity;
2. The *activation domain* is characterized by a conserved central DNA-binding region that contains basic amino acids and 9 conserved *cysteine* residues. Two Zn-finger substructures can be distinguished in this domain, each containing one molecule of Zn that interacts via coordination bonds with four *cysteine* residues. The two Zn-fingers are flanked by α-helices at the carboxy-terminal regions, which include a proximal box (P-box) and a distal box (D-box). Both Zn-fingers are structurally and functionally different and are encoded by two different exons. The α-helix of the amino-terminal-proximal Zn-finger interacts directly with the major groove of the DNA. Three amino acids—*glycine, serine,* and *valine*—at the amino-terminus of the α-helix, are responsible for the recognition of a DNA responsive element. The second Zn-finger appears to be involved in protein–protein interactions such as receptor dimerization and stabilization of complexes; and
3. The *hormone-binding domain*, characterized by a carboxy-terminal region, of approximately 250 amino acids that recognize and bind the ligand with high affinity. The carboxy-terminal is the second most conserved region of the nuclear receptor transcription factor family of proteins.

Steroid hormone receptors are complexed to the chaperone *heat shock protein* (HSP). Upon ligand binding, the steroid receptor undergoes phosphorylation of its *serine*

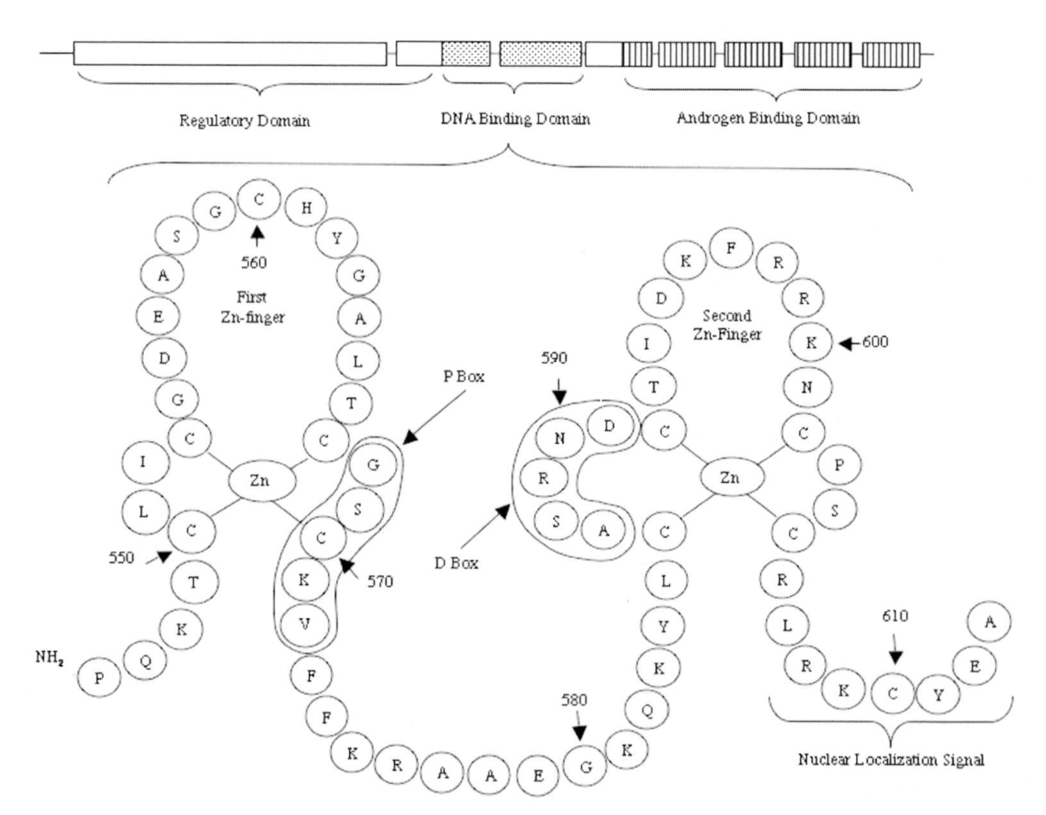

Fig. 5.4 The androgen receptor. *Top panel*: linear representation of the eight exons encoding the three main domains of the androgen receptor gene. *Bottom panel*: sequence and structure of the Zn-finger binding domain of the human AR. The amino acid sequences are represented in the single-letter code. The first Zn-finger contains the proximal box (P-box) sequences that define the specificity of receptor binding to DNA. Thus, receptors that bind to the same hexameric DNA binding site have the same or similar P-box sequences. The second Zn-finger contain the distal box (D-box) sequences, in which are located the amino acid residues involved in protein–protein interaction with a second receptor molecule, to form an active homodimer complex

residues at the amino-terminal activation domain, dissociates from the HSP and becomes activated. The activated receptors form dimers and bind to palindromic steroid-responsive *cis*-acting elements located upstream of the CAAT and TATA boxes (Table 5.1). DNA-dependent RNA-polymerase-II and other transcription factors and coactivators are then recruited to initiate transcription of the steroid response genes.

5.4.5 Leucine Zippers

Leucine zippers or bZip are a family of heterodimeric proteins with two well-defined domains, a basic DNA-binding domain rich in basic amino acids, and a *leucine* rich domain (Fig. 5.3E). The basic DNA-binding domain has an extended α-helix motif that fits into the DNA major groove. The *leucine* rich domain is constituted of *leucine* repeats every seventh amino acid that forms two parallel α-helices. This molecular configuration creates a hydrophobic interface allowing homo- and hetero-dimeric interactions with other proteins of the family. A number of transcription factors belong to this group including CREB, the CREB-binding proteins CBP and p300, the cAMP-responsive element modulator (CREM), Fos and Jun, C/EBP, and the yeast factor GCN4.

5.4.6 Helix-Loop-Helix

Helix-loop-helix (HLH), also termed *basic HLH* (bHLH), is an extension of the bZip motif. All members of this group share a common sequence of 40–50 amino acids that form one short helix and one long helix that are linked by a loop of 12–28 amino acids (Fig. 5.3F). The loop is required to flip one helix over the other, forming a parallel structure that makes contact with the deoxyribose-phosphate backbone of the DNA. Both α-helices have hydrophobic residues on one side and charged residues on the other side. Molecules with bHLH motifs can form both homo- and hetero-dimers through interactions between the hydrophobic residues on the corresponding faces of the two helices. Overall, the bHLH motif has similar characteristics to the bZIP motif; both are involved in gene transcription during development and differentiation, and bind DNA in a similar way. Transcription factors carrying the bHLH motif bind to the sequence CACGTG, termed E-box, or related consensus sequences located in the promoter region. bHLH motif transcription factors include Myc, MAD/MAX proteins, and the *u*pstream *s*timulating *f*actors USF-1 and USF-2. The transcription factor AP-2 belongs to an extension of this family, the *helix-span-helix* group, which has a longer loop.

5.5 Signaling Regulated Transcription

Every gene has a distinct temporal pattern of expression and response to specific extracellular signals. Intracellular activated signaling pathways and nuclear transcription factors trigger a cascade of reactions in the genome that regulate transcription. These reactions include changes in chromatin structure, post-translational modifications of transcription factors, and epigenetic modifications. Chromatin is the DNA complexed with histones to form the chromosomes within the nucleus of the cell. The DNA wraps around the histone to form the nucleosomes that are further compacted into chromosomes. The phosphate components of the DNA backbone are negatively charged, while the histones are rich in positively charged *lysine* residues. These opposite charges are responsible for the strong attraction and tight binding between DNA and histones that makes DNA inaccessible to most transcription factors, maintaining transcription in a repressed state. Thus, gene regulation requires controlled *chromatin decondensation,* which occurs by acetylation/deacetylation of histones.

Coactivators, which either have intrinsic histone acetyltransferase (HAT) catalytic activity or recruit HAT to the promoter region, acetylates the amino group of *lysine* residues and neutralizes the positive charges of the histones. In this way coactivators weakens binding of histones to DNA, decondense chromatin, unwinds DNA, and allow transcription factors access to the regulatory region of the gene activating transcription.

Corepressors silence transcription by reversing the reaction of histone deacetylation. They interact with DNA-bound histones and recruit histone deacetylases (HDAC) that catalyzes hydrolysis of acetylated *lysine*, restoring the positive charge and the link with DNA. Classical examples of silencing by corepressors are the *s*ilencing *m*ediators of *r*etinoid and *t*hyroid (SMRT) receptors, and the nuclear receptor corepressor, N-COR. These mechanisms of gene silencing have been described in the MAD (Myc associated factor X, also known as MAX) proteins, the retinoic acid receptor (RAR), and the unliganded thyroid hormone receptor (THR), in which the transcription factors bind Sin3A, a transcriptional regulatory protein that interacts with HDAC to form repressive chromatin.

5.5.1 Changes in Chromatin Structure: CBP and CREB/CREM/ICER Family of Transcription Factors

An example of chromatin decondensation is the regulation of gene transcription by CREB, CBP, and p300

(E1A-binding protein p300, also termed EP300). CBP and p300 are structurally related coactivators that interact with several transcription factors to activate transcription. Both coactivators contain five protein interaction domains: nuclear receptor interaction domain (RID), CREB interaction domain, cysteine/histidine domains CH1 and CH3, HAT domain, and bromodomain. The bromodomain recognizes and binds acetylated lysine residues on the amino terminal tails of histones, facilitating protein-histone association and chromatin remodeling. Thus, CBP and p300 activate gene expression by at least three mechanisms: relaxing the chromatin through their HAT activity, recruiting the basal transcriptional machinery to the promoter, and acting as an adaptor.

CRE-binding protein was first recognized as the *trans*-acting factor that binds the *cis*-acting element CRE (TGACGTCA) and confers cAMP-inducible transcription to the somatostatin gene. Six alternatively spliced transcripts are described from the 11 exons of the CREB gene, including CREBΔ, CREBα, CREBγ, CREBαγ, CREBΩ, and CREBψ. The predominant isoforms expressed in most tissues are CREBΔ produced by exclusion of exons 3, 5, and 6, and CREBα produced by exclusion of exons 3 and 6. Both transcripts encode a peptide of about 43 kDa with a basic domain, a bZip motif, and two *glutamine*-rich domains, Q1 and Q2, essential for transcriptional activation. In addition, a PKA consensus sequence is located between Q1 and Q2, termed kinase inducible domain (KID). The rest of the CREB isoforms are truncated peptides that lack KID, Q2, and the bZip motif. cAMP-stimulated PKA activates CREB by phosphorylation at *serine*-133 within the KID domain. Activated CREB forms homodimers through their bZip motifs and binds via its basic regions to CRE in the promoter of target genes. Activated CREB also binds CBP, which recruits the basal transcription complex into the promoter. The two *glutamine*-rich regions in CREB and the *glutamine*-rich region in CBP are required for interaction with the transcription machinery.

cAMP-responsive element modulator is a transcription factor highly homologous to CREB, of which several alternatively spliced isoforms are characterized. CREMα, CREMβ, and CREMγ isoforms are transcribed from the same promoter and contain a bZip motif and a KID region, but lack the *glutamine*-rich domains Q1 and Q2 required for transcriptional activation. These isoforms bind CRE with the same efficiency and specificity as CREB, but regulate gene expression suppressing cAMP-induced transcription. The CREM gene expresses two other groups of isoforms that contain the *glutamine*-rich regions, which function as activators. The first group includes the CREMτ isoforms (τ1, τ2, and τ3) that originate from promoter-1. The second group, termed CREMθ1 and CREMθ2, are transcribed from two newly identified exons that utilize promoters 3 and 4. Promoters 3 and 4 contain several CREs that are responsive to cAMP and therefore are activated by CREB and CREM induction, which allows for positive auto-feedback regulation of the CREMθ isoforms.

The inducible cAMP early repressor (ICER) is a powerful repressor of cAMP-responsive genes also encoded by the CREM gene. ICER originates from an alternative intronic promoter, promoter-2, located close to the 3′ end of the CREB gene that directs the expression of a truncated protein containing the bZip motif only. ICER has four CRE elements in its promoter so its transcription can be activated by CREB and CREM. ICER binds to the CRE of cAMP responsive genes, including its own CRE, and represses transcription. Thus, the CREB/CREM system can be negatively auto-regulated. Overall, the multiple alternatively spliced isoforms of CREB and CREM, some of which function as activators, whereas others are repressors, and the use of alternative promoters exert spatial and temporal specificity on cAMP-regulated genes.

The CREB/CREM family of transcription factors plays a major role in regulating complex processes, including the cAMP-mediated signal transduction during the spermatogenic cycle. In Sertoli cells, CREB levels fluctuate in a cyclical manner along the spermatogenic wave. The FSH-activated cAMP signaling pathway positively autoregulates expression of CREB, which binds to a CRE-like element in its own promoter and activates transcription of genes involved in germ cell differentiation. ICER, which is also activated by CREB, down regulates CREB expression together with its own expression, resetting CREB to basal level enabling a new spermatogenic wave.

In addition to binding CREB, CBP contains distinct interaction domains for various classes of transcription factors, mediating actions of other transcription factors and coactivators as well. When combined with the steroid regulatory coactivator, SRC, CBP enhances nuclear receptor transcription and therefore is considered an *integrator of transcription*. The nuclear receptor coactivator-1, also termed *s*teroid *r*eceptor *c*oactivator-1 (SRC-1), which possesses intrinsic HAT activity and several nuclear receptor interacting domains, function as a transcriptional coregulator. Upon ligand binding, nuclear receptors recruit SRC-1 to the promoter/enhancer region of a gene assisting nuclear receptors in transcription upregulation.

5.5.2 Post-translational Modifications of Transcription Factors

Intracellular kinases including PKA, PKC, mitogen-activated protein (MAP) kinase, Janus kinase (JAK), and Jun N-terminal kinase (JNK) can also regulate gene transcription by post-translational modification of transcription factors. PKA activates C/EBP by phosphorylation, which is consistent with the role of C/EBP in a variety of cAMP regulated genes. The reverse reaction, dephosphorylation, can also induce transcription factor activity. PKC-mediated dephosphorylation of Jun activates AP-1 to bind TRE. PKA and PKC can also activate AP-2, which binds to CRE and is involved in regulation of the proenkephalin gene that responds to a variety of signals including PKA, PKC, and Ca^{2+}-dependent pathways. AP-2 also binds to a variety of other GC sequences with no similarities to CRE, such as $GCCN_3GGC$, and acts as a basal transcription enhancer in the gene promoter of metallothionein, estrogen receptors, cholesterol side-chain cleavage cytochrome P450, and the α- and β-subunits of the human chorionic gonadotropin. Another example of post-translational modification includes the mechanism of interferon-regulated transcription. Interferon binding to the cytokine receptor is followed by the activation of the cytokine receptor-associated kinase, JAK. Activated JAK phosphorylates the cytokine receptors creating sites for interaction with proteins that contain the phosphotyrosine-binding domain SH2, which then recruits the *s*ignal *t*ransducers and *a*ctivators of *t*ranscription (STAT). The SH2 domain of STAT act as docking sites for the SH2 domains of other STATs and bind to the interferon-*g*amma *a*ctivated *s*equence promoter element (GAS) of cytokine-regulated genes. The JAK pathway is negatively regulated by at least three mechanisms:

1. tyrosine phosphatases that remove phosphates from cytokine receptors and activated STATs;
2. *s*uppressors *o*f *c*ytokine *s*ignaling (SOCS) that inhibit STAT phosphorylation by binding and inhibiting JAKs, or competing with STATs for phosphotyrosine binding sites on cytokine receptors; and
3. *p*rotein *i*nhibitors *o*f activated *S*TATs (PIAS), which bind to STAT and block access to the DNA sequences of the cytokines-regulated genes.

Overall, the mechanisms of transcription regulation are not mutually exclusive and some transcription factors may participate in more than one type of regulation. Therefore, the response to intracellular signaling pathways combined with the interaction of multiple transcription factors specific for every tissue can be unique to a particular gene.

5.5.3 Tissue-Specific Gene Expression

Every cell type expresses a limited number of genes, such as the chorionic gonadotropins in the placenta of primates and equidae, growth hormone in the pituitary somatotroph cells, insulin in the β-cells of the pancreatic islet, and β-globin in erythroid cells. Several mechanisms defining cell-specific gene expression have been described, with most of them determined by the interaction of multiple regulatory promoter elements and by their cognate transcription factors. The simplest mechanism of tissue specific transcription appears to be regulation of transcription factor abundance (e.g., CREM). By using alternative promoters and translation initiation sites the CREM gene encodes several different isoforms, which are found in varying amounts in different tissues, some of which operate as activators and others as repressors by transcription factor displacement.

Tissue-specific expression of some genes requires the combinatorial effect of different transcription factors in a particular tissue. For example, expression of the chorionic gonadotropin α-subunit gene is influenced by a tissue-specific regulatory element (TSE) located adjacent to the CRE that binds a protein expressed specifically in the placenta. CRE and TSE form a composite enhancer that imparts the unique property of placental cells to express the chorionic gonadotropin α-subunit gene. Thus, although CRE is present in the promoter of many genes and can bind several transcription factors, the combination of CRE and TSE is unique to the placental cells.

In the pituitary gland the somatotroph cells express a member of the POU-homeodomain family termed Pit-1 that binds to multiple Pit-1 sites in the GH gene promoter. Pit-1 is effective only in the presence of adjacent cell-specific and ubiquitous DNA binding sites for transcription factors expressed exclusively in the pituitary gland. Thus, Pit-1 requires the participation of other enhancer binding factors to fully activate transcription in the pituitary.

The thyroglobulin and thyroid peroxidase genes are expressed specifically in the thyroid gland. Several transcription factors that bind promoters of thyroglobulin and thyroid peroxidase genes have been identified, including the thyroid transcription factor-1 (TTF-1), which is mainly expressed in the thyroid gland, but is also expressed in other tissues, such as the lung and brain, and the paired box member, PAX-8, which is expressed in the thyroid and the kidney. The combined expression of TTF-1 and PAX-8 is unique to the thyroid gland therefore they activate expression of the thyroid-specific thyroglobulin and thyroid peroxidase genes.

The mechanisms of cell-specific expression appear to respond to a combinatorial code that allows unique patterns of transcription for each tissue. Dimerization of transcription factors with the same structural motif allows combinations and interactions of related factors that generate transcriptional switches in which some combinations activate and others inhibit transcription. Overall, specific patterns of transcription appear to be regulated by dimerization and expression of limited transcription factors in each cell type.

5.5.4 Epigenetic Regulation of Gene Expression

Gene expression is also regulated by modifications in chromatin and DNA that are stable and transmitted from generation to generation, but do not involve changes in the nucleotide sequence of the DNA. These modifications, termed *epigenetics*, occur during cell differentiation, allowing cells to acquire different characteristics without changes in the genome. The main epigenetic mechanism of gene regulation is DNA methylation, which involves addition of a methyl group to the fifth carbon of the *cytosine* nucleotides. Common targets of methylation are densely clustered CpG dinucleotide genomic sequences, termed CpG islands ("p" refers to the phosphodiester bond between C and G.) CpG islands are approximately 300–3,000 bp located at or near the Tx of many vertebrate genes.

Methylation is catalyzed by three specific methyltransferases: DNA methyltransferase-1, -3α, and -3β (DNMT1, DNMT3α, DNMT3β). It is believed that DNMT3α and DNMT3β establish a pattern of DNA methylation early in development and that DNMT1 copies the DNA pattern of methylation during DNA replication. Methylation changes DNA binding sites in a gene promoter. Therefore, cognate transcription factors and proteins that would normally bind to the promoter are not recognized, causing silencing of the gene. Methylation is involved in the mechanisms of parental imprinting, where a gene of a contributing parent determines expression of the same gene in the offspring. An example of methylation imprinting is expression of the insulin-like growth factor-II (IGF-II) gene, in which only the paternal allele is active. The mechanism involves the participation of a *methylated insulator* between the IGF-II promoter and enhancer, in a way that the CTC-binding factor can no longer

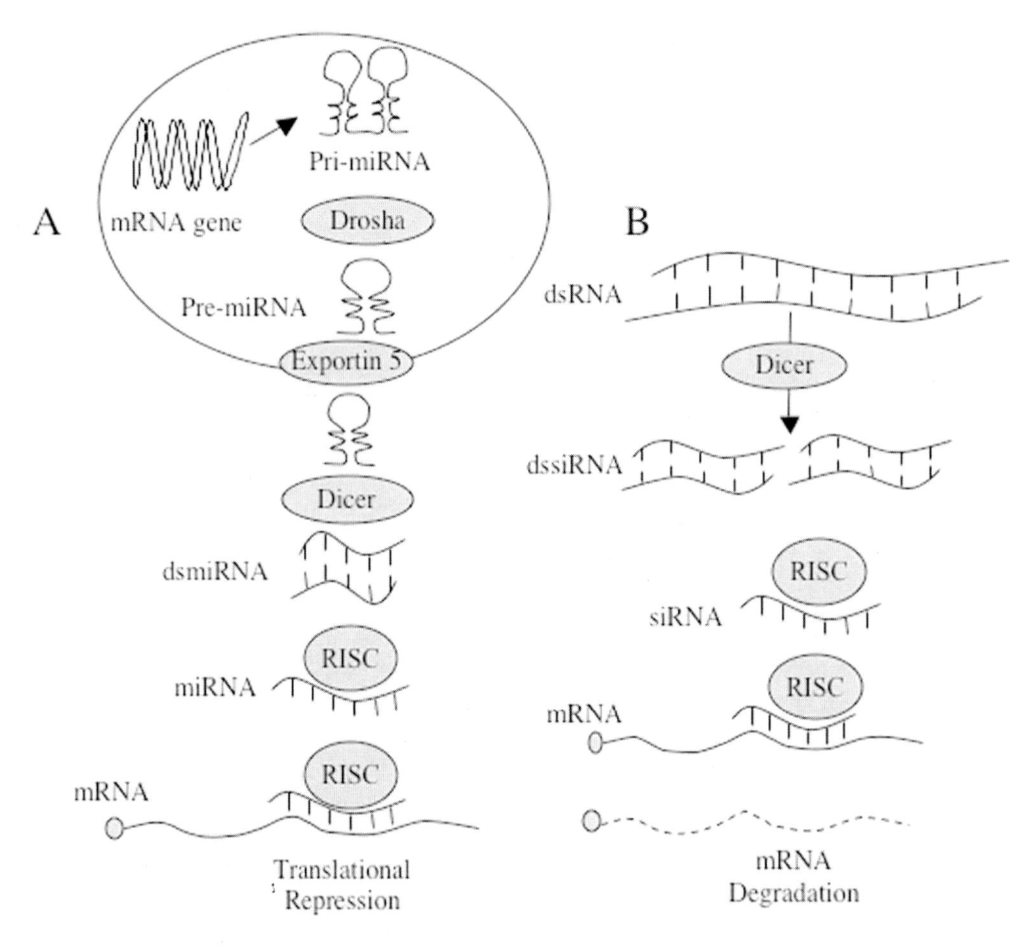

Fig. 5.5 Regulation of gene expression by RNA interference. (**A**) miRNA pathway. A gene expressing miRNA is transcribed into a Pri-miRNA, which is cleaved by Drosha into Pre-miRNA. Pre-miRNA is transported to the cytoplasm by Exportin 5 and cleaved by Dicer to form a short dsmiRNA. dsmiRNA separates into miRNA that binds to mRNA repressing translation. (**B**) siRNA pathway. dsDNA enters the cell and is cleaved by Dicer to form a short dssiRNA that separates into siRNA and complexes with RISC. The RISC-siRNA complex binds to mRNA causing degradation

bind to the insulator. Thus, the enhancer is free to turn on the paternal IGF-II gene promoter.

5.6 Regulation of Gene Expression by RNA Interference

Relatively recent discoveries have lead to increased interest in RNA interference (RNAi), which is a natural regulatory mechanism conserved among many organisms. RNAi regulates gene expression by directing post-transcriptional silencing through small RNA molecules of ~22 bp, such as micro RNAs (miRNAs) and short interfering RNAs (si RNAs). RNAi targets genes in a sequence-specific manner and either decreases mRNA translation or cleaves and degrades mRNAs. Although miRNAs and siRNAs are closely related, sharing some common features and functions, their origin is different and, therefore, their pathways and mechanism of action are different too.

miRNAs originate from non-translated genes, some of which are found in intronic regions and can be transcribed as part of the primary transcript for the corresponding gene (Fig. 5.5A). Others originate from intergenic regions or are in antisense orientation to known genes. miRNA genes are transcribed by a DNA-dependent RNA polymerase-II into primary transcripts termed pri-miRNAs, which are processed into pre-miRNA by a member of the RNase-III family of enzymes, termed *drosha,* in conjunction with a dsRNA-binding protein, termed *pasha* in *Drosophila* and DGCR8 in humans. The pre-miRNAs are imperfect base-paired hairpin loops of approximately 70–100 bp that are transported to the cytoplasm by a protein termed *exportin-5*. In the cytoplasm, another RNase-III enzyme termed *dicer* cleaves pre-miRNA to generate two mature miRNA molecules that are complementary. Dicer also assembles the miRNAs into the *RNA-induced silencing complex* (RISC), which carries an endonuclease termed *argonaute* that degrades the anti-guide (passenger) strand, while the guide strand base pair with the target mRNA. If miRNA is fully complementary to its target, the mRNA is cleaved and degraded. However, if base pairing is partially complementary, translation is repressed. Thus, activation of the miRNA-RNAi pathway results in the down-regulation of gene expression through either degradation of mRNA or repression of translation.

The origin of siRNAs has not yet been clearly defined, however, it is believed that they are synthesized by RNA-dependent RNA polymerases that convert single-stranded RNA (ssRNA) into double-stranded RNA (dsRNA) (Fig. 5.5B). Alternatively, DNA-dependent RNA polymerases produce dsRNA by transcribing inverted DNA repeats. The dsRNAs are transported into the cytoplasm, presumably by a mechanism similar to the miRNAs involving exportin-5. Cytoplasmic dicer recognizes and cleaves both strands of long dsRNA at specific distances from the helical end into numerous double stranded siRNA molecules that are loaded into RISC, where they are separated into two single strands. The antisense strands of siRNA act as guides for RISC-argonaut that targets complementary or partial complementary mRNA leading it to cleavage and degradation, which results in down-regulation of gene expression.

Direct post-transcriptional silencing through RNAi has been described in genes involved in a wide variety of cellular functions, including development, cell proliferation, differentiation and apoptosis, and fat metabolism. The siRNA pathway can also be artificially activated using synthetic silencing reagents designed to degrade target mRNAs. Delivery of exogenous siRNA into the cell is currently utilized as a tool for identification of novel genes and gene functional analysis. As more is learned about RNAi and the target genes of miRNAs and siRNAs, it will be possible too more accurately assess gene expression. RNAi is a significant molecular tool used to study gene silencing, with potential applications in the health science field, such as selected gene silencing to treat diseases and development of disease management therapies.

5.7 Summary

Genes have unique patterns of temporal and spatial expression that ultimately characterize the genetic make-up of all living organisms. Gene expression can be regulated at several levels, which include transcription, post-transcriptional, post-translational, and through epigenetic mechanisms. Genes contain the information necessary to regulate their own expression in the form of specific DNA nucleotide sequences, called *cis*-acting regulatory elements, which can be classified into five main groups: promoters, enhancers, silencers, boundary elements, and response elements. Regulatory elements are recognized by specific transcription factors that are the link between the intracellular signaling pathway and gene transcription. Transcription factors can be classified according to common secondary structural features, or motifs that include helix-turn-helix motif, homeodomain proteins, high mobility proteins, zinc-containing transcription factors, leucine zipper motif, and HLH motif. Transcription regulation involves changes in chromatin structure by histone acetylation, post-translational modifications of transcription factors, and epigenetic modifications such as methylation. Gene expression can also be regulated by RNAi, also known as post-transcriptional gene down-regulation or silencing, through small RNA molecules that either repress gene translation or degrade mRNA. Overall the mechanisms described are not mutually exclusive and more than one mechanism may participate in the regulation of a particular gene.

5.8 Glossary of Terms and Acronyms

AP-1: activating protein-1, Jun-Fos heterodimer

AP-2: activating protein-2

AR: androgen receptor

ARE: androgens response element

bHLH: basic helix-loop-helix

BRE: TFIIB recognition element

bZip: DNA binding domain of the leucine zippers family of heterodimeric proteins

C/EBP: CAAT box/enhancer binding protein

CaMK: Ca^{2+}-mediated kinase

CAP: catabolic activator protein

CBF: CAAT box-binding factor

CBP: CREB binding protein

c-*fos*: cellular protooncogene of the transforming gene of the Finkel-Biskis-Jinkins murine osteosarcoma viruses

c-*jun*: cellular protooncogene of the transforming gene of avian sarcoma virus

COUP-TF: chicken ovoalbumin upstream promoter transcription factor

CRE: cAMP response element

CREB: CRE-response element binding protein

CREM: cAMP-responsive gene modulator

***Cro* repressor:** dimeric protein composed of identical subunits

CTCF: CTC-binding factor

CTF: CAAT binding transcription factor

DNMT: DNA methyltransferase

DPE: downstream promoter element

dsRNA: double-stranded RNA

E-box: sequence CACGTG that binds members of the basic helix-loop-helix

ER: estrogen receptor

ERE: estrogens response element

Fos: transcription factor expressed by the c-*fos* oncogene

FSH: follicle stimulating hormone

GAS: interferon-gamma activated sequence promoter

GATA: a family of transcription factors that contain two zinc finger motif and binds to the DNA sequence (A/T)GATA(A/G)

GH: growth hormone

GR: glucocorticoid receptor

GRE: glucocorticoids response element

GTFs: general transcription factors

HAT: histone acetyl transferase

hCG: chorionic gonadotropin hormone

HDAC: histone deacetylase

HMG: high mobility group of proteins

HMG-box: homologous DNA binding domain of the HMG proteins

HNF: hepatocyte nuclear factor

HRE: hormone response element

HSP: heat shock proteins

HSP70: heat shock protein-70

HTH: helix-turn-helix

ICER: inducible cAMP early repressor

IGF-II: insulin like growth factor-II

Inr: Initiator sequence

IPF: insulin promoter factor

JAK: Janus kinases

JNK: Jun N-terminal kinase

Jun: transcription factor expressed by the c-*jun* gene

KID: kinase inducible domain

LH: luteinizing hormone

MAD/MAX: group of proteins of the bHLH family that can form heterodimers with myc and regulate transcription

MAP: mitogen activated protein

MAPK: mitogen-activated protein kinase

miRNA: micro RNAs

MR: mineralocorticoid receptor

MRE: mineralocorticoids response element

MTF-1: metal-responsive transcription factor-1

Myc: transcription factor expressed by a gene originally described in the avian MC29 myelocytomatosis virus (v-*myc*). A homologous gene (c-*myc*) is located in the long arm of the human chromosome 8.

N-COR: nuclear receptor co-repressor

NF-1: nuclear factor-1.

NF-Y: nuclear factor-Y

p300: E1A-binding protein p300, also termed EP300

P450scc: cytochrome P450 cholesterol side chain cleavage

PAX-8: transcription factor expressed by a member of the paired box (PAX) family of genes that encode proteins that contain a paired box domain, an octapeptide, and a paired-type homeodomain. It is expressed in the thyroid and the kidney and binds thyroglobulin and thyroid peroxidase genes promoters

PIAS: protein inhibitors of activated STATs

Pit-1: a transcription factor expressed specifically in the pituitary gland; member of the POU-homeodomain family

PKA: protein kinase-A

PKC: protein kinase-C

POU: acronym derived from the homeodomain proteins Pit-1, Oct and Unc-86

PR: progesterone receptor

PRE: progesterone response element

pre-miRNA: precursor of miRNA

pri-miRNA: primary transcript of an miRNA gene

RAR: retinoic acid receptor

RID: nuclear receptor interaction domain

RISC: RNA-induced silencing complex

RNAi: RNA interference

RRE: retinoic acid response element

SF-1: steroidogenic factor-1

SH2: Src Homology (SH) region 2, a phosphotyrosine-binding domain originally described in proteins of the Rous sarcoma virus (*src*) oncogene family of tyrosine kinases

Sin3: a negative regulator of transcription in yeast also known as SDII

siRNA: short interfering RNA

SMRT: silencing mediator of retinoid and thyroid receptors

SOCS: suppressors of cytokine signaling

SOX: SRY box

Sp1: selective promoter-1

SRC: steroid receptor coactivator

SRE: serum responsive element

SRF: serum response factor

SRY: sex-determining region of the Y chromosome

ssRNA: single-stranded RNA

STATS: signal transducers and activators of transcription

TBP: TATA-binding protein

TCF/LEF: T-cell factor/lymphoid enhancer factor

TF: transcription factor

TFIID: transcription factor IID

THR: thyroid hormone receptor

TPA: 12-*O*-tetradecanoyl-13-acetate

TR: thyroid hormone receptor

TRE: TPA response element

TSE: tissue specific element

TSH: thyrotropin hormone

TTF-1: thyroid transcription factor

Tx: transcription start site

USF-1 and USF-2: upstream stimulating factors 1 and 2, respectively

VDR: vitamin-D receptor

wHTH: winged HTH

Bibliography

1. Alberts B, Johnson A, Lewis J, et al. Molecular Biology of the Cell, fourth edition. New York: Garland Science, 2002.
2. Andrisani OM. CREB-mediated transcriptional control. Crit Rev Eukaryotic Gene Expr 1999; 9:19–32.
3. Behr R, Weinbauer GF. Germ cell-specific cyclic adenosine 3′,5′-monophosphate response element modulator expression in rodent and primate testis is maintained despite gonadotrophin deficiency. Endocrinology 1999; 140:2746–54.
4. Blendy JA, Kaestner KH, Schmid W, et al. Targeting of the CREB gene leads to up-regulation of a novel CREB mRNA isoform. EMBO J 1996; 15:1098–106.

5. Bolander F. Molecular Endocrinology, third edition. San Diego: Academic Press,2004.
6. Brinkman AO. Steroid hormone receptors. In: Fauser BCJM, editor. Reproductive Medicine, Molecular Cellular and Genetic Fundamentals, second edition. New York: The Parthenon Publishing Group, 2003:279–94.
7. Butler JE, Kadonaga JT. The RNA polymerase 11 core promoter: a key component in the regulation of gene expression. Genes Dev 2002; 16:2583–92.
8. Chan HM, La Thangue NB. P300/CBP proteins: HATs for transcriptional bridges and scaffolds. J Cell Sci 2001; 114: 2363–73.
9. Chawla A, Repa JJ, Evans RM, et al. Nuclear receptors and lipid physiology: opening the X-files. Science 2001; 294:1866–70.
10. Corre S, Galibert MD. USF as a key regulatory element of gene expression. Med Sci (Paris) 2006; 22:62–7.
11. Daniel PB, Rohrbich L, Habener JF. Novel cyclic adenosine 3′,5′-monophosphate (cAMP) response element modulator theta isoforms expressed by two newly identified cAMP-responsive promoters active in the testis. Endocrinology 2000; 141: 3923–30.
12. De Cesare D, Fimia GM, Sassone-Corsi P. CREM, a master-switch of the transcriptional cascade in male germ cells. J Endocrinol Invest 2000; 23:592–6.
13. Don J, Stelzer G. The expanding family of CREB/CREM transcription factors that are involved with spermatogenesis. Mol Cell Endocrinol 2002; 18:115–24.
14. Eckert D, Buhl S, Weber S, et al. The AP-2 family of transcription factors. Genome Biol 2005; 6:246.
15. Foulkes NS, Borrelli E, Sassone-Corsi P. CREM gene: use of alternative DNA-binding domains generates multiple antagonists of cAMP-induced transcription. Cell 1991; 64:739–49.
16. Foulkes NS, Mellstrom B, Benusiglio E, et al. Developmental switch of CREM function during spermatogenesis: from antagonist to activator. Nature 1992; 355:80–4.
17. Foulkes NS, Schlotter F, Pevet P, et al. Pituitary hormone FSH directs the CREM functional switch during spermatogenesis. Nature 1993; 362:264–7.
18. Glass CK, Rose DW, Rosenfeld MG. Nuclear receptor coactivators. Curr Opin Cell Biol 1997; 8:222–32.
19. Glover JN, Harrison SC. Crystal structure of the heterodimeric bZIP transcription factor c-Fos-c-Jun bound to DNA. Nature 1995; 373:257–61.
20. Goldman PS, Tran VK, Goodman RH. The multifunctional role of the co-activator CBP in transcriptional regulation. Recent Prog Horm Res 1997; 52:103–19.
21. Goodman RH, Smolik S. CBP/p300 in cell growth, transformation, and development. Genes Dev 2000; 14:1553–77.
22. Griswold MD, Kim JS, Tribley WA. Mechanisms involved in the homologous down-regulation of transcription of the follicle-stimulating hormone receptor gene in Sertoli cells. Mol Cell Endocrinol 2001; 173:95–107.
23. Grumbach MM, Conte FA. Disorders of sex differentiation. In: Wilson JD, Foster DW, Kronenberg HM, Larsen PR, editors. Williams Textbook of Endocrinology, ninth edition. Philadelphia: W.B. Saunders Company, 1998:1303–425.
24. Hess J, Angel P, Schorpp-Kistner M. AP-1 subunits: quarrel and harmony among siblings. J Cell Sci 2004; 117: 5965–73.
25. Hu X, Lazar MA. Transcriptional repression by nuclear hormone receptors. Trends Endocrinol Metab 2000; 11:6–10.
26. Hummler E, Cole TJ, Blendy JA, et al. Targeted mutation of the CREB gene: compensation within the CREB ATF family of transcription factors. Proc Natl Acad Sci 1994; 91:5647–51.
27. Jameson JL. Transcriptional control of gene expression. In: Endocrine Society, editor. Introduction to Molecular & Cellular Research. Bethesda: The Endocrine Society, 1998:21–31.
28. Johnson W, Albanese C, Handwerger S, et al. Regulation of the human chorionic gonadotropin α- and β-subunit promoters by AP-2. J Biol Chem 1997; 272:15405–12.
29. Johnson W, Jameson JL. Transcriptional control of gene expression. In: Jameson JL, editor. Principles of Molecular Medicine. Totowa: Humana Press, 1998:25–41.
30. Johnson W, Jameson JL. AP-2 (activating protein 2) and Sp1 (selective promoter factor 1) regulatory elements play distinct roles in the control of basal activity and cyclic adenosine 3′,5′-monophosphate responsiveness of the human chorionic gonadotropin-b promoter. Mol Endocrinol 1999; 12:1963–75.
31. Kasper LH, Fukuyama T, Biesen MA, et al. Conditional knockout mice reveal distinct functions for the global transcriptional coactivators CBP and p300 in T-cell development. Mol Cell Biol 2006; 26:789–809.
32. Kisseleva T, Bhattacharya S, Braunstein J, et al. Signaling through the JAK/STAT pathway, recent advances and future challenges. Gene 2002; 285:1–24.
33. Latchman DS. Eukaryotic Transcription Factors, second edition. San Diego: Academic Press, 1995.
34. LaVoie HA. The role of GATA in mammalian reproduction. Exp Biol Med 2003; 228:1282–90.
35. LaVoie HA. Epigenetic control of ovarian function: the emerging role of histone modifications. Mol Cell Endocrinol 2005; 243: 12–8.
36. Mayr B, Montminy M. Transcriptional regulation by the phosphorylation-dependent factor CREB. Nat Rev Mol Cell Biol 2001; 2:599–609.
37. Mclenna NJ, O'Malley BW. Minireview: nuclear receptor coactivators-an update. Endocrinology 2002; 143:2461–5.
38. Molina CA, Foulkes NS, Lalli E, et al. Inducibility and negative autoregulation of CREM: an alternative promoter directs the expression of ICER, an early response repressor. Cell 1993; 75:875–86.
39. Ptashne M. A Genetic Switch: Page l and Higher Organisms, second edition. Cambridge: Blackwell Scientific Publications & Cell Press, 1992.
40. Pestell RG, Jameson JL. Transcriptional regulation of endocrine genes by second messenger signaling pathways. In: Weintraub BD, editor. Molecular Endocrinology: Basic Concepts and Clinical Correlations. New York: Raven Press, Ltd., 1995.
41. Privalsky ML. The role of corepressors in transcriptional regulation by nuclear hormone receptors. Annu Rev Physiol 2004; 66:315–60.
42. Ramji DP, Foka P. CCAAT/enhancer-binding proteins: structure, function, and regulation. Biochem J 2002; 365:561–75.
43. Sanborn BM. Increasing the options-new 3′.5′ cyclic adenosine monophosphate (cAMP)-responsive promoters and new exons in the cAMP response element modulator gene. Endocrinology 2000; 141:3921–2.
44. Spiegelman BM, Heinrich R. Biological control through regulated transcriptional coactivators. Cell 2004; 119:157–67.
45. Tjian R, Maniatis T. Transcriptional activation: a complex puzzle with few easy pieces. Cell 1994; 77:5–8.
46. Vo N, Goodman RH. CREB-binding protein and p300 in transcriptional regulation. J Biol Chem 2001; 276:13505–8.
47. Walker WH, Daniel PB, Habener JF. Inducible cAMP early repressor ICER down-regulation of CREB gene expression in Sertoli cells. Mol Cell Endocrinol 1998; 143:167–78.
48. Williams T, Tjian R. Characterization of a dimerization motif in AP-2 and its function in heterologous DNA-binding proteins. Science 1991; 251:1067–71.

49. Wingender E. Gene Regulation in Eukaryotes. New York: VCH, 1995.
50. Wu L, Belasco JG. 2008. Let me count the ways: Mechanisms of gene regulation by miRNAs and si RNAs. Molecular Cell 2008; 29:1–7.
51. Xing W, Sairam MR. Characterization of regulatory elements of ovine follicle-stimulating hormone (FSH) receptor gene: The role of E-Box in the regulation of ovine FSH receptor expression. Biol Reprod 2001; 64:579–89.
52. Yao TP, Oh SP, Fuchs M, et al. Gene dosage-dependent embryonic development and proliferation defects in mice lacking the transcriptional integrator p300. Cell 1998; 93:361–72.
53. Zhang Y, Dufau ML. Dual mechanisms of regulation of transcription of luteinizing hormone receptor gene by nuclear orphan receptors and histone deacetylase complexes. J Steroid Biochem Mol Biol 2003; 85:401–14.

Chapter 6

Molecular Basis of Abnormal Phenotype

Pedro J. Chedrese and Christine Meaden

6.1 Introduction

With the exception of *gametes*, genes occur in homologous pairs located on chromosomes, which also occur in homologous pairs within the cell nucleus. Each gene of the homologous pair is positioned at corresponding fixed locations called the *locus* on each of the matching chromosomes. Natural random modifications within the DNA nucleotide sequences of genes create one or more alternative forms called *alleles*. The alleles constitute the genetic make-up or *genotype* for each gene and determine normal variability essential for genetic diversity. The genotype provides the primary code for the *phenotype*, which refers to observable structure, function, and behavior that result from gene expression. Identical alleles located at the same locus on each of the matching chromosomes are considered *homozygous* for a trait, while two different alleles are *heterozygous*. The allele expressed by phenotype is said to be *dominant,* while the other allele, which is masked by the dominant gene, is described as *recessive*.

Phenotypic variations produced by alleles that can be distinguished from the traits produced by their homologous genes are known as *genetic polymorphisms*. Normally, polymorphic sites are located approximately once every 500 nucleotides or about 10^7 times in every genome. Sequences of polymorphic sites are used for studying genetic linkage and gene mapping. Dramatic changes to phenotype can occur due to modifications within DNA nucleotide sequences of homologous gene pairs in response to environmental factors, the influence of other genes, or errors during replication. *Heritability* is the proportion of phenotypic variation in a population that is attributable to genetic variation. Heritability analysis estimates the differences in genetic and non-genetic factors relative to the total phenotypic variance in a population. Therefore, it must be considered that genotype and phenotype are not always directly correlated; some genes are triggered only by environmental conditions and some phenotypes are the result of multiple genotypes. Overall, modification within DNA nucleotide sequences may represent either naturally occurring polymorphisms or genetic mutations when they produce uncharacteristic variations.

P.J. Chedrese (✉)
Department of Biology University of Saskatchewan College of Arts and Science, Saskatoon, SK, Canada
e-mail: jorge.chedrese@usask.ca

6.2 Genetic Mutations

A genetic mutation is a heritable change in the genetic material, which occurs mainly due to alteration of nucleotide sequences and inversion or translocation of genes. Rate of mutations may be increased by exposure to *mutagens,* which are physical (i.e., radiation from UV rays, X-rays, or extreme heat) or chemical (natural or synthetic) agents that change genetic information. Mutations that cause expression of abnormal proteins, which produce detrimental effects to the individual, are the basis of genetic syndromes or disease.

However, mutations are not always harmful or affect phenotypic change. Darwin's theory of *natural selection* is based on phenotype or the observable characteristics, in which favorable traits become more common in successive generations while unfavorable traits become less common. Meaning that individuals with favorable phenotypes are more likely to survive and reproduce than those with less favorable phenotypes. This theory is validated throughout evolution, in which changes in the genome were essential for adaptation to the internal and/or external environment, making the species fitter and better adapted to environmental change, coining the term *survival of the fittest.* Also, alleles may increase or decrease in frequency within a population over time through *genetic drift*, which is determined statistically using the law of large numbers theory. The statistical theory is applied to determine the effect "chance" has on survival of the alleles or gene variation. For example, not every member of the

P.J. Chedrese (ed.), *Reproductive Endocrinology*,
DOI 10.1007/978-0-387-88186-7_6, © Springer Science+Business Media, LLC 2009

population will become a parent and not every set of parents will produce the same number of offspring. *Neutral mutations* that occur due to genetic drift do not influence the fitness of the species favorably or negatively. A genetic mutation in a population that began with a few individuals, who carried a particular allele that is inherited by many descendants, is known as the *founder effect.* A mutation that was not inherited from either parent is called a *de novo mutation.* Thus, mutations also create variations in the gene pool.

Mutations can be subdivided into *germline* mutations, which can be passed on through generations and are responsible for inherited genetic syndromes or disease, and *somatic* mutations that cannot be transmitted to descendants. Somatic cell mutations can occur any time after oocyte fertilization and are much higher than germinal cell mutations because the majority of cell divisions that generate an organism are of somatic lineage. It is estimated that a single-cell zygote generates $\sim 10^{13}$ cells of somatic lineages. As molecular technology advances, scientists are increasingly able to identify genetic mutations, thus providing information to diagnose genetic syndromes or diseases (i.e., cancer) affecting individuals. Consequently, progress towards goals of developing interventional therapies and, in some instances, cures for genetic disorders are being achieved. Mutations can be classified into three categories: gene mutations, chromosome mutations, and genome mutations.

```
A normal DNA sequence:   GGT-CAC-TGG-CGG-TTC-TTA-ATG-AAA
Transcribe into:         GGU-CAC-UGG-CGG-UUC-UUA-AUG-AAA
Translate into:          gly-his-trp-arg-phe-leu-met-lys
```

Missense mutation by nucleotide substitution

```
A mutated DNA sequence:  GGT-CTC-TGG-CGG-TTC-TTA-ATG-AAA
Transcribe into:         GGU-CUC-UGG-CGG-UUC-UUA-AUG-AAA
Translate into:          gly-leu-trp-arg-phe-leu-met-lys
```

Nonsense mutation by nucleotide substitution

```
A mutated DNA sequence:  GGT-CAC-TAG-CGG-TTC-TTA-ATG-AAA
Transcribe into:         GGU-CAC-UAG-CGG-UUC-UUA-AUG-AAA
Translat e into:         gly-his-Stop
```

Frameshift mutation by nucleotide deletion

```
                             C
                            ↗
A deleted base pair:     GGT-ACT-GGC-GGT-TCT-TAA-TGA-AA
Transcribe into:         GGU-ACU-GGC-GGU-UCU-UAA-UGA-AA
Translate into:          gly-thr-gly-gly-Ser-Stop
```

Fig. 6.1 Examples of mutations in the DNA and subsequent changes in amino acid sequences

6.2.1 Gene Mutations

In most situations, DNA polymerase has the ability to repair errors that occur during replication. They include nucleotide substitutions, deletions, and insertions, which can be repaired by removing abnormal 3′ nucleotides of the growing DNA strand, and using the normal strand as a template to replace the abnormal nucleotide(s). In addition, errors during DNA replication often occur in the non-coding regions of the DNA, or at sites that do not affect normal expression, therefore the phenotype is not influenced. Gene mutations result when errors during normal DNA replication are not repaired (Fig. 6.1).

Point mutations are the substitution of a single nucleotide with another nucleotide. *Transitions* are substitutions of a purine base for another purine or of a pyrimidine base for another pyrimidine. *Transversions* are a purine substituted for a pyrimidine or a pyrimidine substituted for a purine. Transitions occur more frequently than transversions. A nucleotide substitution in the coding region will affect any amino acid to be synthesized from this gene.

Nonsense mutations are point mutations that change an amino acid triplet codon for a stop codon, thus, the ribosome will stop adding amino acids.

Frameshift mutations are deletions or insertions of nucleotides that result in changes in the length of a gene. When nucleotides are deleted, the coding of proteins that uses that particular DNA sequence will be unable to be produced because each successive nucleotide after the deleted nucleotide will be out of place, thus shifting the reading frame. Similarly, when a nucleotide is inserted into a genetic sequence this also shifts the reading frame.

Missense mutations are substitution of nucleotides that result in a different amino acid. When an amino acid is replaced by one of similar properties, size, charge, and tendency to bend, the substitution may not cause much change in the final shape or properties of the protein, for which they are called *silent mutations.* On the other hand, if the new amino acid has different properties, the mutation can result in the synthesis of an abnormal protein. For example, if a single mutation incorporates a *cysteine* (*Cys*) instead of a *tyrosine* (*Tyr*) in the catalytic active site of a cell signaling protein, this would result in a non-functional signaling pathway. *Cys* forms disulfide bonds with other *cys* residues, which creates the three-dimensional structure of a protein. Thus, the loss of a *cys* residue involved in a disulfide bond may result in an inactive or partially active protein, due to changes in its three-dimensional structure, affecting the response to a variety of hormones and growth factors. Another example is individuals suffering from the genetic disease sickle-cell anemia; a point mutation occurs where a single amino acid replacement on the sickle hemoglobin molecule produces an abnormal protein. Analysis of both the α- and the β-chains that make up the $\alpha 2\beta 2$ form of hemoglobin molecules indicates that the sickle hemoglobin β-chain differed from the normal hemoglobin β-chain, through a specific amino acid substitution of *glutamic acid→valine*, at position 6.

6.2.2 Chromosome Mutations

Chromosome abnormalities usually result from an error within the gametes when cells grow and divide or can occur during early fetal development. Chromosome mutations are a result of gene rearrangements leading to structural variations or changes in number of chromosomes. Depending on which genes are involved, chromosomal mutations may not be viable to produce an organism, result in a lethal genetic disorder, or may produce advantageous changes in the genome. The age of the mother and certain environmental factors can increase the risk of fetal chromosomal abnormality. There are several structural rearrangements associated with chromosome mutations, which can be classified as follows.

Microduplication, also termed *amplification*, occurs when there is an increase in the number of genes at a particular locus, leading to multiple copies of the chromosomal region affected.

Microdeletion and intrachromosomal recombination involves deletion of several adjacent genes causing loss of a portion of the chromosome. Loss of a segment of DNA from a single chromosome may create spontaneous intrachromosomal recombination, bringing together previously distant genes.

Chromosomal inversion is the reverse in the orientation of a chromosomal region when connection between genes breaks.

Chromosomal translocation is the interchange of sections of the gene from chromosomes that are not members of the same pair. Translocations occur when information from one of two homologous chromosomes breaks and attaches to another chromosome, thus producing duplications and deletions within different pairs of chromosomes. Translocations may be *balanced* or *unbalanced*. In a *balanced translocation,* part of one chromosome may break off and attach to another chromosome. Since chromosomes occur in pairs, the individual will have a smaller than normal chromosome, while its homologue is normal. The chromosome with the broken piece attached is longer, while its homologue is also normal. Thus, the person has the correct amount of genetic material that is simply rearranged. They appear phenotypically normal with no related health problems. However, their offspring face several possibilities in that they may inherit the following:

- In the best case scenario, normal chromosomal structure.
- A balanced translocation, therefore the individuals are not affected. However, since they carry the balanced translocation their offspring may be affected or also carry the chromosomal mutation.
- A balanced translocation added onto normal chromosomal structure which produces a disproportionate amount of genetic material and the individual will suffer the affects of the genetic disorder related to the particular *unbalanced translocation*. Alternatively, an unbalanced translocation may cause problems incompatible with life, leading to miscarriage of an affected fetus.

Chromosome mutations occur approximately one rearrangement per 1,700 cell divisions. It is estimated that 1/550 individuals may carry a balanced translocation. If three or more miscarriages occur in a family, chromosomal mutation should be suspected, and the family's genetic history investigated and chromosome studies conducted.

6.2.3 Genome Mutations

The *genome* is defined as the total genetic material contained within the chromosomes of an organism. *Disomy* is the normal condition for *diploid* organisms, which indicates that chromosomes are assembled in pairs. The human genome is comprised of 46 chromosomes arranged into 22 homologous pairs of somatic cell chromosomes called *autosomes* and one pair of sex cell chromosomes, non-autosomal *germline cells*, the ova (X) in females and spermatozoa (Y) in males.

Genome mutations occur within the chromosomes of germline cells during either of the two meiotic divisions when the paired chromosomes fail to segregate properly, called *non-disjunction*, which results in errors in chromosome number. Genome mutations that occur after conception may result in *mosaicism*, a condition in which some cells have the mutation and some do not. Examples of chromosome non-disjunction are shown in Table 6.1 [1]. It is estimated that genome mutations occur at a rate of one event per 25–50 meiotic cell divisions. This result is derived from data collected from chromosomally abnormal stillborn fetuses and live-born infants. However, it is quite possible there are other chromosomal genome mutations that are lethal and therefore, spontaneously aborted shortly after conception.

Table 6.1 Examples of chromosome missegregations

Chromosome missagregation	Karyotype	Frequency in live births
Trisomy 21	47,XY,+21 or 47,XX,+21	1 in 700
Trisomy 18	47,XY,+18 or 47XX,+18	1 in 3,000
Klinefelter's syndrome	47,XXY	1 in 1,000 males
Turners syndrome	45,X or variants	One in 2,500 females
Triple X	47,XXX	One in 1,000 males
XYY syndrome	47,XYY	One in 1,000 males
Triploidy	69,XXY or 69,XXX or 69,XYY	Rare

Modified from Cox & Sinclair [1].

When more than the usual two sets of homologous chromosomes occurs the condition is referred to as *polyploidy*. Types of polyploidy are classified according to the number of chromosome sets: three sets is triploidy, four sets is tetraploidy, and five sets is pentaploidy.

Triploidy consists of three sets of chromosomes, therefore, the individual will possess 69 chromosomes. Triploidy may also occur with mosaicism where some cells have a triploid chromosome. Generally, triploidy is incompatible with life and is associated with hydatidiform moles and miscarriage [2].

When the chromosomes are not all duplicated during meiosis or mitosis the condition is called *aneuploidy*, which is the most common genome mutation (Table 6.1) [1]. Aneuploidy is commonly observed in tumor cells, therefore, in many cases, cancer is considered a genetic disease. Types of aneuploidy are based on the number of chromosomes changed and are classified into monosomy, uniparental disomy, trisomy, tetrasomy, and pentasomy.

If an entire chromosome from a homologous pair is absent, the disorder is called *monosomy.* Partial monosomy occurs when only a portion of the chromosome is absent. Disorders caused by monosomy include the following:

- *Turner Syndrome*, which affect females, only one X chromosome (X0) is present instead of the normal (XX);
- *Cri-du-chat Syndrome* is caused by a deletion of the end of the short p arm of chromosome 5;
- *1p36 Deletion Syndrome* is caused by a deletion at the end of the short p arm of chromosome 1.

Uniparental disomy (UPD) refers to individuals who receive two copies of a chromosome or part of a chromosome from one parent and no copies from the other parent. UPD can occur as a random event during the formation of egg or sperm cells or may happen in early fetal development. Disorders include *Prader-Wili syndrome* and *Angelman syndrome*, although both can also be due to errors involving genes on the long arm of chromosome 15.

Trisomy pertains to conditions having one or more sets of three chromosomes. In many cases trisomy is expressed as *mosaicism*, in which the extra chromosome is present in specific tissue. Individuals with mosaic trisomy tend to have mild forms of conditions associated with full trisomy. Complete trisomy is observed in conditions such as the following:

- *Trisomy 21 (Down syndrome)*, which is perhaps the most frequently observed human genetic disease, occurring in 1:660 newborns. Trisomy 21 is due to the presence of a part, or an extra complete copy of chromosome 21.
- *Klinefelter's syndrome* (47,XXY), or XXY syndrome, is a condition in which affected males have an extra sex chromosome because of a non-disjunction event during sex cell division.
- *Trisomy 18 (Edwards Syndrome)* is the most common form of trisomy after Down syndrome. It is caused by an extra chromosome 18, which usually occurs by non-disjunction, making 24 haploid chromosomes rather than 23. Thus, this mutation in either gamete results in a fetus with three copies of chromosome 18;
- *XYY syndrome* is a condition in which the male has an extra copy of the Y chromosome, producing a 47,XYY karyotype. The 47,XYY is a result of a random non-disjunction event that occurs in metaphase II during the formation of sperm cells. Fertilization with one of these sperm cells causes the fetus to have an extra Y chromosome in each cell.
- Mosaic trisomy involving chromosome 8 and 12, respectively, *is associated with chronic lymphocytic leukemia* and *acute myeloid leukemia.*

Tetrasomy and pentasomy are the presence of four and five copies of a chromosome, respectively. These disorders would appear to be rare, however they may be under reported, since the individual may not have phenotypical changes or symptoms may not be consistent between individuals and thus syndromes of tetrasomy or pentasomy are not identified. With respect to the sex chromosomes an individual could be born with the following karyotypes: (Tetrasomy X)—XXXX, XXXY, XXYY, XYYY; (Pentasomy X) XXXXX, XXXXY, XXXYY, XXYYY, and XYYYY.

6.3 Summary

Natural modifications of DNA create one or more alternative forms of the genes called *alleles* that determine normal variability of the genotype and provides the primary code for the *phenotype*, which is the observable structure, function, and behavior that result from gene expression. A genetic mutation is a heritable change in the genetic material, which occurs mainly due to alteration of nucleotide sequences and inversion or translocation of genes. Mutations that cause expression of abnormal proteins that produce detrimental effects to the individual are the basis of genetic syndromes or disease. Mutations can be subdivided into germline mutations, which can be passed on through generations and are responsible for inherited genetic syndromes or disease, and somatic mutations that cannot be transmitted to descendants. Somatic cell mutations can occur any time after oocyte fertilization and have higher incidence than germinal cell mutations, because the majority of cell divisions that generate an organism are of somatic lineage. As molecular technology advances, scientists are increasingly able to identify genetic mutations, thus providing information to diagnose genetic syndromes or diseases affecting individuals. Consequently, progress toward goals of developing interventional therapies and, in some instances, cures for genetic disorders are being achieved.

Bibilography

1. Graham GE, Allanson JE, Gerritsen JA. Sex chromosome abnormalities. In: Rimoin DL, Connor JM, Pyeritz, Reed E, Korf BR, editors. Emery and Rimoin's Principles and Practice of Medical Genetics, fifth edition. Philadelphia: Churchill Livingstone Elsevier, 2007:1038–57.
2. Hawley RS, Mori CA. The Human Genome: A User's Guide. San Diego: Academic Press, 1999.
3. Maki H. Origins of spontaneous mutations: specificity and directionality of base-substitution, frameshift, and sequence-substitution mutagenesis. Annu Rev Genet 2002; 36:279–303.
4. Milunsky JM. Prenatal diagnosis of sex chromosome abnormalities. In: Milunsky, A, editor. Genetic Disorders and the Fetus: Diagnosis, Prevention, and Treatment, fifth edition. Baltimore: The Johns Hopkins University Press, 2004:297–340.
5. Nussbaum RL, McInnes RR, Willard HF. Thompson & Thompson Genetics in Medicine, Revised Reprint sixth edition. Philadelphia: W.B. Saunders Company 2004:172–4.
6. Ratcliffe SG, Read G, Pan H, et al. Prenatal testosterone levels in XXY and XYY males. Horm Res 1994; 42:106–9.
7. Strachan T, Read AP. Human Molecular Genetics 3. Third Edition. New York: Garland Science. 2004.
8. Robinson DO, Jacobs PA. The origin of the extra Y chromosome in males with a 47,XYY karyotype. Hum Mol Genet 1999; 8:2205–9.
9. Thomson MW, McInnes RR, Willard HF. Thomson & Thompson Genetics in Medicine, fifth edition. .Philadelphia: W. B. Saunders Company, 1991.

References

1. Cox TM, Sinclair J. Molecular Biology in Medicine. Oxford: Blackwell Science, 1995.
2. Balakier H, Bouman D, Sojecki A, et al. Morphological and cytogenetic analysis of human giant oocytes and giant embryos. Hum Reprod 2002; 17:2394–401.

Part II
The Tools of Molecular Reproductive Biology

Chapter 7

Recombinant DNA Technology

Pedro J. Chedrese

7.1 Introduction

The terms *recombinant DNA technology*, *DNA cloning*, *molecular cloning*, or *gene cloning* all refer to use of molecular techniques to select a specific sequence or sequences of DNA from an organism and transfer it into another organism to code for or alter specific traits. Thus, recombinant DNA technology provides a powerful molecular tool that enables scientists to engineer sequences of DNA.

Although the chemical structure of the nucleic acids and the genetic code were completely deciphered during the 1950s and early 1960s, the great progress in recombinant DNA technology occurred in the 1970s, when the advances in enzymology of nucleic acids and molecular genetics of bacterial viruses and plasmids made possible cutting, joining, and modification of DNA. The first successful cloning experiment was reported in the 1970s when a specific gene from one bacterium was removed and inserted into another *host* bacterial cell, from which many identical copies of the desired DNA, or *clones*, were recovered. From then on this technique started to be called *recombinant DNA*. By 1977, interspecies genetic recombinations were attained when genes from different organisms were transferred into bacteria to replicate and be expressed, a process called *transgenesis*. In 1980, the process of recombinant DNA technology used to obtain *transgenic* organisms, "process for producing biologically functional molecular chimeras," was patented. Recombinant DNA technology was used to insert DNA carrying the gene that encodes human insulin into the bacterium *Escherichia coli* (*E. coli*), which resulted in transgenic bacteria able to produce insulin. In 1983, human insulin used to treat diabetes mellitus was the first recombinant biopharmaceutical marketed. Knowledge rapidly evolved and recombinant DNA technology made possible the use of a variety of expression systems, in vitro and in vivo, resulting in a universal tool applied in many fields. In healthcare recombinant DNA technology contributes, to name a few, to produce biopharmaceuticals, create treatments that have the potential to regenerate damaged cells and tissue, and develop genetic therapy for the treatment and/or cure for diseases.

7.2 Restriction Enzymes

A fundamental requisite for molecular cloning is that both the target DNA sequence and the cloning vehicle must to be cut into discrete and reproducible fragments. This became possible with the identification of the host-control system of the bacterium *Haemophilus influenzae*, as this microorganism is able to rapidly breakdown intact foreign phage DNA into discrete pieces. The DNA degradation activity of *Haemophilus influenzae* was subsequently observed in cell-free extracts, from which the first site-specific restriction endonuclease was isolated. This enzyme was able to breakdown *E. coli* DNA, whereas it failed to cut up the DNA of *Haemophilus influenzae* itself. Subsequently, many other enzymes of similar features were isolated and identified. Those capable of cutting DNA molecules internally at specific base pair sequences are called *type II restriction endonucleases*, currently known as *restriction enzymes*, to differentiate from the type I restriction endonucleases that only cut DNA externally. Currently, several hundreds of type-II restriction endonucleases have been isolated from a variety of microorganisms. The recognition sequences of some restriction endonucleases are summarized in Table 7.1. Restriction endonucleases are named according to the following nomenclature: the first letter, capitalized and italicized, represents the genus name of the organism from which the enzyme is isolated; the second and third lowercase, italicized letters are usually the initial letters of the species name of the microorganism of origin; a fourth letter, if any, indicates a particular strain of organism; roman numerals are used to designate the order of

P.J. Chedrese (✉)
Department of Biology, University of Saskatchewan College of Arts and Science, Saskatoon, SK, Canada
e-mail: jorge.chedrese@usask.ca

P.J. Chedrese (ed.), *Reproductive Endocrinology*,
DOI 10.1007/978-0-387-88186-7_7, © Springer Science+Business Media, LLC 2009

Table 7.1 Recognition sequences of some restriction endonucleases

Enzyme	Recognition site	Type of end cut
*Bam*HI	G↓G-A-T-C-C C-C-T-A-G↑G	5′-phosphate extension
*Eco*RI	G↓A-A-T-T-C C-T-T-A-A↑G	5′-phosphate extension
*Hae*III	G-G↓C-C C-C↑G-G	Blunt end
*Hind*II	GTPy↓PuAC CAPu↑PyTG	Blunt end
*Hind*III	A↓AGCTT TTCGA↑A	5′-phosphate extension
*Hpa*I	G-T-T↓A-A-C C-A-A↑T-T-G	Blunt end
*Not*I	G↓C-G-C-C-C-G-C C- G-C-C-G-G-C↑G	5′-phosphate extension
*Sau*3AI	↓G-A-T-C C-T-A-G↑	5′-phosphate extension
*Pst*I	C - T-G-C-A↓G G↑A-C-G-T- C	3′-hydroxyl extension
*Pvu*II	C-A-G↓C-T-G G-T-C↑G-A-C	Blunt end

characterization of the different endonucleases isolated from the same organism. For example, *Hind*I and *Hind*II are the first and second type II restriction endonucleases that were isolated from *Haemophilus influenzae.*

The length of the recognition site for different enzymes can be four, five, six, eight, or more nucleotide pairs. Those that cleave within sites of four and six nucleotide pairs are more commonly used in molecular cloning protocols, because combinations of four and six nucleotides exist in fewer numbers in the genome, therefore the enzymes are more specific. The *Haemophilus influenzae* restriction enzyme, called *Hind*II, recognizes and binds the following sequence:

$$(5')\mathrm{G}-\mathrm{T}-\mathrm{Py}\downarrow\mathrm{Pu}-\mathrm{A}-\mathrm{C}(3')$$

$$(3')\mathrm{C}-\mathrm{A}-\mathrm{Pu}\uparrow\mathrm{Py}-\mathrm{T}-\mathrm{G}(5')$$

*Hind*II cleaves the DNA phosphate bond between the oxygen of the 3′ carbon of the sugar of one nucleotide and the phosphate group attached to the 5′ carbon of the sugar of the adjacent nucleotide. This particular enzyme cuts at the center of the recognition site leaving the DNA with two blunt ends. Another type II restriction endonuclease isolated from *E. coli* designated *Eco*RI, binds DNA in a region with a specific palindromic sequence, and cuts between the guanine and adenine residues on each strand as follows:

$$5'\mathrm{G}\downarrow\mathrm{A}-\mathrm{A}-\mathrm{T}-\mathrm{T}-\mathrm{C}3'$$

$$3'\mathrm{C}-\mathrm{T}-\mathrm{T}-\mathrm{A}-\mathrm{A}\uparrow\mathrm{G}5'$$

This cleavage leaves two single-stranded complementary ends (staggered ends), each with extensions of four nucleotides. In this case, each single-stranded extension ends in a 5′-phosphate group, and the 3′-hydroxyl group from the opposite strand is recessed. They are also called "sticky-ends" because the base pairing of two compatible extensions, aligned by the hydrogen bonds, can hold together. Other restriction endonucleases leave 3′-hydroxyl extensions and 5′-phosphate groups from the opposite strand recessed, such as *Pst*I:

$$(5')\mathrm{C}-\mathrm{T}-\mathrm{G}-\mathrm{C}-\mathrm{A}\downarrow\mathrm{G}(3')$$

$$(3')\mathrm{G}\uparrow\mathrm{A}-\mathrm{C}-\mathrm{G}-\mathrm{T}-\mathrm{C}(5')$$

When a DNA sample is treated with a restriction endonuclease, an identical set of fragments is always produced. The frequency with which a given restriction endonuclease cuts DNA depends on the recognition site of the enzyme. The size of the fragments obtained after digestion can be calculated if the size of the restriction site is known. Therefore it is possible to predict the size and number of DNA fragments that would be obtained by cutting a DNA molecule of known size. Concrete maps of DNA can be constructed by treating DNA with two restriction endonucleases separately, called single digestions, and then with a combination of two restriction endonucleases, called double digestion, to generate identical DNA fragments. The DNA fragments are then separated by agarose gel electrophoresis, a process using an electric field to move the DNA through the pores of an agarose gel in which the smaller fragments move faster than the larger ones. Fragment sizes can then be estimated by comparing their position in the gel to the positions of a DNA standard of known size. This technique gives molecular biologists the capability to produce DNA fragments for further manipulations such as gene mapping, sequencing, and cloning.

7.3 Recombination of DNA and Gene Libraries

The basic strategy of molecular cloning is to move desired genes from the large and complex genome to a smaller, simple one that facilitates manipulation. The process of in vitro recombination makes possible to cut different strands of DNA in vitro with a restriction enzyme and join or recombine them via complementary base pairing. Restriction endonuclease-digested DNA molecules can be recovered intact after separation in agarose gel electrophoresis and can be recombined with the use of DNA ligases, which in the presence of adenosine triphosphate (ATP) restore the linkage between the 3′-hydroxyl group and the 5′-phosphate group of DNA. A DNA ligase expressed by the bacteriophage T4, a virus that infects *E. coli,* is commonly used for this purpose. T4 ligase catalyzes the formation of phosphodiester bonds at the ends of DNA strands, which are already held together by the base pairing of two compatible extensions aligned by their hydrogen bonds, or two blunt ends that come in contact when they bind to the enzyme.

Two different fragments of DNA can be recombined using restriction endonucleases and several recombination are possible, including the following:

- Ligation of two different DNA fragments, previously cleaved with a single restriction endonuclease that leave sticky ends, results in new DNA combinations that are the result of base pairing between the extended regions. This construct will be a recombination in which approximately half of the molecules will be oriented in one direction, while the other half in the opposite direction.
- If the digestions are conducted with a restriction endonuclease that leave blunt ends, the recombinant will also have no defined orientation.
- By digesting two pieces of DNA with two different restriction endonucleases, one that leaves sticky ends and the other that leaves blunt ends, a recombinant product with a defined orientation can be generated.

The use of recombination techniques permitted the identification and analysis of entire genomes in the form of gene libraries, which are a collection of DNA molecules, inserted into cloning vectors. There are two main types of gene libraries: complementary DNA (cDNA) libraries and genomic libraries.

A *cDNA library* is a collection of DNA molecules obtained by reverse transcription of mRNA, using the enzyme reverse transcriptase. Thus, a cDNA library represents only the portions of the genome that are transcribed in a particular tissue under a particular physiological, developmental, or environmental condition. Because cDNA is a copy of the processed mRNA, a cDNA library contains neither introns nor the non-transcribed sections of the gene.

A *genomic library* is a complete collection of all DNA sequences from the entire genome of an organism. Genomic libraries are prepared by digestion of DNA with a restriction endonuclease, a procedure that results in a collection of DNA fragments that are inserted into a cloning vector (usually λ-phage), which is a cosmid or an artificial chromosome (described in Section 7.4.2). In a genomic library all regions of the genome are represented, including exons, introns, and the non-transcribed regions.

7.4 Biological Systems Used in Molecular Biology

A broad variety of organisms are used by molecular biologists to clone and express genes. They include microorganisms, such as the bacterium *E. coli* and the unicellular yeast *Sacharomyces cerevisiae* (*S. cerevisiae*), cloning vectors based on engineered bacterium's parasites and viruses, tissues and cells in culture, and multicellular organisms such as plants, insects, fish, laboratory rodents, and domestic animals.

7.4.1 Microorganisms

E. coli is a gram-negative, non-pathogenic, short (<1 mm in length), motile, rod-shaped bacterium naturally found in the intestines of humans and animals. *E. coli* is the main host cell used for manipulating DNA because it is easy to culture and multiplies by binary fission on a simple medium, consisting of ions (Na^+, K^+, Mg^{2+}, Ca^{2+}, NH_4^{2+},Cl^-, $HP0_4^{2-}$, and $S0_4^{2-}$), trace elements, and a carbon source such as glucose. For DNA cloning, restriction enzyme deficient strains of *E. coli* are used, which are unable to digest cloned DNA. By introducing cloned DNA into *E. coli* investigators can produce unlimited identical copies of the molecule. However, for protein synthesis, *E. coli* is a limited system, mainly because they lack post-translational modification capability and do not possess the biosynthetic ability to perform the complex protein modifications that occurs in eukaryotes. Therefore, for the production of eukaryotic protein, other systems are used, such as *S. cerevisiae,* mammalian and plant cells in culture, and transgenic animals.

S. cerevisiae is a non-pathogenic, single-cell microorganism ~5 mm diameter. One of the best-characterized features of *S. cerevisiae* is its ability to convert sugars to ethanol and carbon dioxide. This property has been used since ancient times for the production of alcoholic beverages and bread. Thus, *S. cerevisiae* can be considered one of the oldest biotechnological tools in history. *S. cerevisiae* reproduces by the *budding off* of a sibling cell from a parent cell and proliferates in the same kind of culture medium used for *E. coli*. *S. cerevisiae* was the first eukaryotic organism to have its complete set of chromosomes entirely sequenced and is a very useful and versatile host cell. A genetic system of *S. cerevisiae* is the yeast artificial chromosome, which has been used for studies of the physical organization of human DNA.

7.4.2 Cloning Vectors

Fragments of DNA can be easily manipulated taking advantage of a variety of natural replicable units derived from plasmids and phages. Engineered versions of these replicable units, termed *cloning vectors*, are able to host foreign DNA and replicate it in restriction enzyme deficient strains of *E. coli.* A common feature of cloning vectors is that one segment of the molecule provides the biological information for maintenance in the host cell, while another segment contains a structure to carry the cloned DNA, also termed passenger DNA.

The classical recombinant experiment involves extraction of DNA from a donor organism, cleaved with a restriction enzyme, and joined with a DNA ligase to the cloning vehicle. The newly recombined DNA, or DNA construct, carries the DNA from the donor organism as an insert. The DNA construct is then transferred into, and maintained within, the host *E. coli* by a process called transformation. The hosts that take up the DNA constructs are identified and selected from those that do not carry the desired DNA insert. Thus, by culturing the transformed *E. coli*, the insert can be reproduced as needed by the investigator, mapped, and its nucleotide sequence analyzed. When the insert is a gene, or its cDNA, it can be expressed in an expression system in which the gene products, its mRNA and/or the protein, can be further investigated. The following are some of the basic types of cloning vectors used: plasmids, phagemids, lytic bacteriophages derived from the phage lamda, cosmids, and artificial chromosome vectors.

Plasmids are autonomous self-replicating, double-stranded, circular DNA molecules that virtually become parasites of most bacteria. They do not integrate into the bacteria genome but replicate independently of the host chromosome. Some plasmids have very specific functions, such as carrying information for their own transfer from one cell to another, defining the origin of their DNA replication in a host cell, and encoding for resistance to antibiotics. Insertion of plasmids into host cells that replicate is called transformation and this produces cells called *transformed* cells. The most commonly used procedure for transformation of *E. coli* involves the heat shock (2 min at 42°C) of permeabilized bacteria, also called competent cells, in the presence of the plasmid DNA, and in a culture media containing calcium chloride ($CaCl_2$). Although the process of transformation is very inefficient and typically yields one transformed cell per 1,000, it is useful for selection as the host cell carrying the recombinant DNA can easily be identified. In the process of transformation, only the recombined plasmid DNA can be expressed in the host cells because any extra chromosomal DNA that lacks an origin of replication cannot be maintained within a bacterial cell. Thus, non-plasmid DNA, such as recombined DNA fragments, does not have any biological effect on the host cell. Plasmids are engineered to be cut with restriction enzymes, recombined with foreign DNA and religated without altering their fundamental biological functions. They also have to be of small size for efficient transfer of foreign DNA into *E. coli*. Transfer efficiency decreases significantly with plasmids larger than 15 kb long. Plasmid vectors for cloning must contain the following:

- A dominant selectable marker, usually a gene conferring resistant to an antibiotic to the host cell; commonly ampicillin, tetracycline, chloramphenicol, or kanamycin is used. Selection and identification of host cells carrying the desired clone rely in the ability of growing in the presence of one of these antibiotics.
- An origin of replication that allows the plasmid to replicate as an extrachromosomal circle, independent of the host genome.
- A cloning site with a unique restriction endonuclease recognition sequence, into which DNA can be inserted. A plasmid-cloned DNA construct can not be digested by the host's restriction endonucleases, therefore, the plasmids are maintained, replicated, and recovered intact.

One of the first plasmid cloning vectors was pBR322, developed in the 1980s by two Mexican scientists, Francisco Bolivar and Raymond Rodriguez. pBR322 is a precursor of a large variety of modern cloning vectors carrying multiple unique cloning sites (MCS), also called *polylinkers*. The modern cloning vectors are more versatile and are designed to facilitate cloning and selection. The first series of these vectors called pUC plasmids were created at the University of California. The pUC vectors, from which many versions are derived, are engineered to have a short fragment of DNA with a *polylinker* for many different restriction endonucleases, located where gene function is not affected, unless exogenous DNA is inserted. The use of *polylinker* has extended the range of enzymes that can be used to generate a restriction fragment. By combining several contiguous restriction sites in the short section of DNA of the polylinker, any two sites can be cleaved simultaneously without affecting vector sequences. The pUC-based plasmids are small (~2,700 bp) and contain the ampicillin resistance gene (amp^R), the pBR322 origin of replication, and a polylinker for multiple endonucleases. The polylinker is located between a segment of the β-galactosidase gene (*lacZ′*) of the *E. coli* lactose operon that expresses a regulatory protein that inhibits transcription of the *lacZ′* gene. Fragments of DNA can be cloned using one or two of the restriction sites of the pUC plasmids and introduced into *E.coli* that are grown in agar plates containing isopropyl-β-D thiogalactopyranoside (IPTG), X-Gal, and the antibiotic ampicillin. IPTG is a lactose analogue inducer of the lactose operon (binds to the product of *lacI* and depress activity of the *lacZ′* gene) and X-Gal is a chromogenic substrate of β-galactosidase that releases a blue indolil derivative. Plasmids carrying the desired cloned DNA have an inactive *lacZ′* gene, therefore, the host cell does not produce β-galactosidase rendering white colonies. This feature is called blue/white selection system that facilitates identification of fragment carrying plasmids.

Phagemids are engineered cloning vectors with *two origin of replication,* one derived from the plasmid Col E1 and the other from the single stranded DNA phage f1. The Col E1 origin of replication is used for replication in the same way as pBR322 and pUC vectors and the phage f1 *origin of replication* is used for induction of synthesis of single stranded DNA. Thus, phagemid transformed *E. coli* produces single stranded DNA. In addition, phagemids have the promoters of the phages T7 and T3, located at both sides of the polylinker, which enables expression of the cloned insert to be obtained regardless of its orientation. Thus, complementary RNA to either strand of DNA can be obtained by incubating phagemids with either T7 or T3 RNA polymerase in the presence of ribonucleotides. Like the pUC vectors, the new generations of phagemids have operon lactose genes inserted in the polylinker, allowing for blue/white color selection.

Bacteriophages are viruses that normally infect bacteria, which have been engineered for DNA cloning. The most commonly used for the construction of gene libraries is the *E. coli* bacteriophage lambda (λ), also called λ-phage. This phage is constituted by tubular shaped proteins that form a capsid head containing 50 kb of DNA, and by a tail of fibers in the form of "pins" that attach to the cell surface of *E. coli* and penetrate the membrane injecting DNA into the bacterium. Once inside the bacterium, the λ-phage can reproduce in the *lytic* and/or the *lysogenic* cycles. In the lytic cycle an infective λ-phage replicates rapidly and the toxins produced lyse the host cell and release phage particles. In the *lysogenic* cycle the λ-phage DNA integrates into the *E. coli* chromosome as a *prophage,* without disrupting the bacterial cell, where it can be maintained as benign guest more or less indefinitely. However under certain conditions, such as nutritional or environmental stress, changes in the host cell can cause excision of the prophage from the chromosome, a process called prophage induction. After excision, the λ-phage enters into the lytic cycle and replication begins.

An infective λ-phage consists of a linear double strand DNA molecule with single-stranded extensions of 12 bp, called cohesive (cos) ends, at the end of each 5′ strand, which are complementary to each other. The genes for the *integration-excision* processes are carried in a middle fragment of the phage of ~20 kb, termed *I/E region*, while the left arm, or *L region*, contains the genetic information for the production of heads and tails and the right arm, or *R region*, carries the genes for DNA replication and cell lysis. As a cloning vector the λ-phage has been engineered carrying two *Bam*HI restriction sites that flank the I/E region. Thus, by cutting with *Bam*HI, a fragment of DNA can be inserted replacing the *I/E region,* after which the phage multiplies only through the lytic cycle. Infective particles can be produced in vitro using a set of reagents containing purified empty heads and tails, which mimic the natural conditions, and a recombined λ-phage DNA of up to 50 kb. If the recombined phage is larger than 52 kb of DNA does not fit into a head, as the enzymatic component of the reaction recognizes the double-stranded cos sequence and cuts it at this site. Thus the location of the cos sequences, which are ~50 kb apart, determines the correct amount of DNA to be packed into the head. In contrast, if less than 38 kb DNA is packed, a non-infective λ-phage particle is generated.

Recombinant λ-phages are perpetuated in a strain of *E. coli* that does not allow reconstituted λ-phage with intact *I/E region* to grow. Colonies containing the λ-phage form zones of lysis, termed plaques, which can be screened by using either a labeled DNA probe or a labeled antibody against the protein encoded by the inserted gene and synthesized during the lytic cycle. In both cases, transfer of the colonies to a solid matrix, usually a membrane of nitrocellulose or nylon, is required. Membranes can be hybridized with the labeled probe or incubated with the antibody. Positive colonies are identified by autoradiography of the membrane, in which the corresponding plaques on the original plate are identified based on the location of the signal. Subcultures of the individual plaques provide limitless source of *E. coli* carrying the recombinant λ-phage.

Cosmids are engineered vectors that combine the properties of plasmids and bacteriophages. The advantages of cosmids are that can carry large DNA inserts, up to 40 kb, and that can be maintained as plasmids in *E. coli*. The commonly used cosmid pLFR-5 is a small molecule of approximately 6 kb that carry the two *cos* sites of the λ-phage separated by a *Sca*I restriction site. It contain six unique restriction enzymes sequences in its multiple cloning site: *Hind*III, *Pst*I, *Sal*I, *Bam*HI, *Sma*I, and *Eco*RI and carries an origin of DNA replication and the tetracycline resistance gene. Cloning into the tpLFR-5 requires purification of DNA of ~40 kb in length partially digested with *Bam*HI, which is inserted into the *Sca*I and *Bam*HI of the vector. The new molecules will be of about 50 kb long, separated by cos sequences located 50 kb apart, which can be packaged into λ-phage heads and maintained as a plasmid-like insert in *E. coli* in media containing tetracyclin.

Artificial chromosome vectors include yeast (YAC), bacterial (BAC), and human (HAC) artificial chromosomes. YAC are artificially designed vectors that contain sequences of telomeres, centromeres, and the autonomously replicating sequence (ARS) that recruits replication proteins and replicate in *S. cerevisiae*. YAC vectors permits cloning of larger fragments of DNA, between 100 and 3,000 kb. BAC is a vector designed based on the fertility plasmid, or F-plasmid, that contains genes that facilitate proper distribution of plasmids after bacterial cell division. BAC vectors contain all the sequences to replicate *E. coli*, like the replication origins, and also the antibiotic resistance genes and sites for DNA cloning, which accept fragments of up to 350 kb. The use of BAC vector facilitated sequencing of large genomes, like the human genome, by reducing the number of clones required covering all sequences.

HAC vectors were designed as an artificial microchromosome of ~6 to 10 megabases, which can operate as an additional chromosome in human cells. Thus, a human cell carrying HAC would have 47 chromosomes instead of 46. HAC can be introduced into HT1080 cell, in which this additional chromosome can be stable for up to 6 months. HACs are used as gene transfer vectors for studying gene expression and for determining human chromosome functions. HACs have been used in gene therapy approaches to complement gene deficiencies in human cultured cells by transfer of large genomic fragments containing the regulatory elements for normal expression. The use of HAC also make possible to express large human transgenes in animals and to develop laboratory models of human genetic diseases.

7.4.3 Tissues and Cells in Culture

Tissues and cells from a variety of plants and animals can be cultured. In many cases, cells in culture retain some of their original physiologic capabilities and serve as ideal models of study. Primary cell cultures can be established by removing tissue from an organism and the use of proteolytic enzymes to release the cells from the extra-cellular matrix. Freed cells are then placed into a nutrient medium that contains amino acids, antibiotics, vitamins, salt, glucose, and a source of protein and growth factors, such as fetal calf serum, and cultured onto a solid surface, generally a plastic culture plate or bottle. The cells are incubated at an appropriate temperature in a humid environment and in the presence of CO_2 where they grow and divide until the solid surface is covered by a monolayer of cells, at which point the cells cease to divide, called confluence. Cells, at or before confluence, can be subcultured or passed by enzymatically removing them from the culture, diluting, and transferring them into fresh medium and placing them onto a new culture surface for continuing growth. Cells are maintained under these conditions for a definite period, usually 50–100 cell generations depending on the type of tissue, before they lose the ability to divide, enter into a senescence period, and begin to die.

A disadvantage of primary cell cultures is that new tissue is required each time cultures need to be established. Although cells in primary cultures may retain some of their physiologic properties for several generations, they sometimes lose many of their original functions after the first few passages, which render them ineffectual for molecular studies. On occasion, during passages of primary cell cultures, some of the cells undergo genetic modifications that allow them to pass the senescence period and survive indefinitely in culture. This process is called spontaneous immortalization and the cells are referred to as a stable cell line. Stable cell lines can also be obtained by culturing tumoral tissue and by in vitro transformation of primary cultured cells with oncogenic genes, such as the SV-40 large T antigen, H-*ras*, and many others. Cells expressing oncogenes are easily identified in culture, since they grow indefinitely and lose their growth factor (serum) dependency for growing.

Some established cell lines retain a few biochemical characteristics of the original cells and may have significant chromosome changes that include extra versions of some chromosomes, and loss of others. These changes often enhance features useful for the study of molecular mechanisms in vitro. Another advantage of stable cell lines is that the pattern of protein glycosylation may be very similar, or identical, to the cell the protein originated from. Stable cell lines are used for maintaining viruses, determining the protein that is encoded by a cloned DNA sequence, studying hormone regulation of gene expression, mapping the regulatory region of hormonally regulated genes, and commercial production of recombinant proteins, such as hormones, growth factors, and vaccines. Therefore established stable cell lines are very valuable study tools and the pursuit of developing and establishing stable cell lines is one of the main goals of the molecular biologist.

7.4.4 Multicellular Organisms

Fish, laboratory rodents, and domestic animals can be used for the expression of cloned genes in vivo in a procedure called animal transgenesis. A transgenic animal carries a foreign gene that has been artificially introduced into its genome. Transgenic animals can be obtained by introducing the gene of interest by injection into the pronucleous of a fertilized mouse egg or by transforming embryonic stem cells (ES cells) in culture. The foreign gene is constructed using a DNA cloning vector carrying the sequences of interest and sequences that facilitates incorporation into the genome and direct the recombinant to be expressed correctly in the targeted tissue of the host. The first transgenic vector was designed to overexpress the growth hormone (GH) gene in mice. This vector was constructed using the structural GH gene driven by the methalothionein gene promoter, which is strong mice promoter induced by heavy metals. This construct targeted the somatotroph cells of the pituitary gland, where the GH is normally produced, and overexpressed the GH gene when the offspring were given water with a heavy metal to drink. The offspring overexpressing the GH gene showed elevated levels of circulating GH and larger body size. Since then, transgenic mice have provided a valuable tool for investigating a variety of biological questions and became the ultimate system for studying gene function. In addition, by using a similar approach, genes can be expressed in the mammary gland of goats, sheep and cows, seminal gland of pigs, and in chicken eggs. Thus, transgenic animals become valuable sources of recombinant proteins for a variety of therapies, for which is a source of pharmaceuticals. This industry is currently recognized as "animal pharming."

Transgenesis not only permit overexpression of a gene but also offers the possibility of gene ablation, also termed gene knockout. By using the ES cells method, cells that are harvested from the inner cell mass (ICM) of mouse blastocysts can be grown and transformed with a modified gene in culture. The modified gene incorporates into the genome by homologous recombination, while the transformed ES cells retain their full potential to produce all the cells of the mature animal including its gametes. Transformed ES cells are re-injected into the ICM of a mouse blastocyst that is transferred to a receptive uterus of a pseudopregnant mouse. Blastocysts carrying the transformed ES cells can implant and develop into pups expressing the modified genes.

7.5 Summary

Recombinant DNA technology is the methodology for transferring genetic information from one organism to another, which provides a powerful molecular tool that enables scientists to engineer sequences of DNA. A classical recombinant DNA experiment consists of taking DNA from a donor organism and joining it to another DNA entity called a cloning vector that is transferred into a replicable biological unit or host cell, usually *E. coli.* Cells are selected from those that do not carry the desired DNA and recultured in a way that the recombinant DNA can be reproduced as required. Molecular cloning became possible using bacterial restriction endonucleases that can cut DNA at specific base pair sequences. Molecular biotechnologists use a number of different biological systems for genetic manipulation. Types of cloning vectors commonly used for manipulation of DNA are plasmids, phagemids, lytic bacteriophages, cosmids, and artificial chromosomes. Genes can also be expressed, overexpressed, or ablated in transgenic plants and animals. Transgenic animals are presently used for the production of mammalian proteins to be used in therapy in human and animals. Recombinant DNA technology contributes to create treatments that have the potential to regenerate damaged cells and tissue, and develop genetic therapy for the treatment and/or cure for diseases.

7.6 Glossary of Terms and Acronyms

amp^r: ampicillin resistance gene

ATP: adenosine triphosphate

BAC: bacterial artificial chromosome

Ca^{2+}: calcium ion

$CaCl_2$: calcium chloride

cDNA: complementary DNA

Cl^-: chlorine ion

CO_2: carbon dioxide

DNA: deoxyribonucleic acid

ES: embryonic stem cells

GH: growth hormone

HAC: human artificial chromosome

HPO_4^{2-}: monohydrogen phosphate ion

ICM: inner cell mass

IPTG: isopropyl-β-D thiogalactopyranoside

K^+: potassium ion

LacZ′: β-galactosidase gene of *E. coli*

MCS: multiple cloning sites also called polylinkers

Mg^{2+}: magnesium ion

mRNA: messenger RNA

Na^+: sodium ion

NH^{4+}: ammonium

RNA: ribonucleic acid

SO_4^{2-}: sulfate ion

YAC: yeast artificial chromosome

Bibilography

1. Ausubel F, Brent R, Kingston RE, et al. Short Protocols in Molecular Biology, fourth edition. New York: Wiley & Sons, 2004.
2. Bolivar F, Rodriguez RL, Greene PJ, et al. Construction and characterization of new cloning vehicles. II. A multipurpose cloning system. Gene 1977; 2:95–113.
3. Butler M. Mammalian Cell Biotechnology: A Practical Approach. New York: IRL Press, 1991.
4. Capecchi MR. Altering the genome by homologous recombination. Science 1989; 244:1288–92.
5. Chedrese PJ, Rodway MR, Swan CL, et al. Establishment of a stable steroidogenic porcine granulosa cell line. J Mol Endocrinol 1998; 20:287–92.
6. Davis L, Kuehl M, Battey J. Basic Methods in Molecular Biology, second edition. Norwalk: Appleton & Lange, 1997.
7. Demain AL, Solomon NA. Biology of Industrial Organisms. Menlo Park: Benjamin/Cummings Publishing Co., 1985.
8. Dujon B. The yeast genome project: what did we learn? Trends Genet 1996; 12:263–70.
9. Glik B, Pasternak J. Molecular Biotechnology, Principles and

Applications of Recombinant DNA. Washington: American Society for Microbiology, 1998.
10. Kaise K, Murray NE. The use of phage lambda replacement vectors in the construction of representative genomic DNA libraries. In: Glober DM, editor. DNA Cloning Volume 1: A Practical Approach. Oxford: IRL Press, 1986.
11. Harrington JJ, Van Bokkelen G, Mays RW, et al. Formation of de novo centromeres and construction of first-generation artificial chromosomes. Nat Genet 1997; 15:345–55.
12. Larin Z, Mejía JE. Advances in human artificial chromosome technology. Trends Genet 2002; 18:313–9.
13. Lodish H, Baltimore D, Berk A, et al. Molecular Cell Biology, third edition. New York: Scientific American Books, 1995.
14. Mortensen RM, Conner DA, Chao S, et al. Production of homozygous mutant ES cells with a single targeting construct. Mol Cell Biol 1992; 12:2301–5.
15. Palmiter RD, Brinster RL, Hammer RE, et al. Dramatic growth of mice that develop from eggs microinjected with metallothionein-growth hormone fusion genes. Nature 1982; 300:611–5.
16. Primrose SB, Twyman RM. Principles of Gene Manipulation and Genomics, seventh edition. Oxford: Blackwell Publishing, 2006.
17. Shizuya H, Birren B, Kim UJ, et al. Cloning and stable maintenance of 300-kilobase-pair fragments of human DNA in *Escherichia coli* using an F-factor-based vector. Proc Natl Acad Sci 1992; 89:8794–7.

Chapter 8

Techniques for DNA Analysis

Javier S. Castresana and Paula Lázcoz

8.1 Introduction

Molecular biology is a dynamic field with techniques and analytical tools continuously being developed. Many of the fundamental DNA analysis techniques were developed more than 30 years ago and have evolved through various modifications, automation, and computerization. This chapter reviews the basic concepts and techniques in order to understand how the procedures progressed into those used today.

8.2 Gel Electrophoresis

The term electrophoresis describes a separation technique based on the migration of charged particles under the influence of an electric field [1]. Thus, gel electrophoresis refers to the technique in which molecules are forced across a span of porous gel, on an electrical field. Activated electrodes at either end of the gel provide the driving force. A molecule's properties including size, shape, and isoelectric point determine how rapidly electricity moves the molecule through a gelatinous medium. Since composition of DNA includes phosphate groups that are negatively charged, techniques were developed to separate DNA fragments using gel electrophoresis. Most commonly used gels are composed of agarose or polyacrylamide (PolyAcrilamide Gel Electrophoresis or PAGE).

Gel electrophoresis is performed for analytical purposes and as a preparative technique to partially purify DNA. During electrophoresis, negatively charged DNA migrates toward the positive electrode. As DNA fragments move, the matrix of the gel slows their migration rate. The result is a continuous separation of DNA according to size, with the smallest fragments moving more easily through the gel and the greatest distance away from the origin, while the large macromolecules are slowed or completely obstructed. A variation of agarose gel electrophoresis, called pulsed-field gel electrophoresis, makes it possible to separate extremely large molecules of DNA. Gel electrophoresis is performed for analytical purposes and as a preparative technique to partially purify DNA.

J.S. Castresana (✉)
Brain Tumor Biology Unit, University of Navarra, Pamplona, Spain
e-mail: jscastresana@unav.es

8.3 Southern Blot Analysis (Southern Blot Hybridization)

Southern blot analysis or Southern hybridization is a method for identifying specific DNA sequences [1]. This method was reported in 1975 by Dr. Edwin M. Southern, as a technique for the initial analysis of the structure of specific genes without prior cloning. Small DNA probes, oligonucleotides that are typically 20 or fewer base pairs, are used for gene detection and to determine gene copy numbers in a genome. Small probes hybridize only one DNA fragment for each gene; therefore, the number of fragments visualized on the blot will correspond to the number of genes.

8.3.1 Methodology

- The DNA to be analyzed is isolated and digested by one or more restriction endonuclease(s).
- The resulting fragments are then separated by agarose gel electrophoresis.
- The DNA is denatured by treating the gel with alkali, usually NaOH.
- The gel is then transferred onto a solid nylon membrane support, by *blotting*, which preserves the positions of the DNA fragments on the gel. The blotting can be driven by capillary action, positive or negative air pressure, or a voltage gradient.

P.J. Chedrese (ed.), *Reproductive Endocrinology*,
DOI 10.1007/978-0-387-88186-7_8, © Springer Science+Business Media, LLC 2009

- The DNA is fixed to the nylon membrane by covalent bonding, using ultraviolet (UV) irradiation.
- The membrane is then hybridized with a probe, which is a short single stranded segment of cloned DNA labeled with a radioactive isotope (i.e., ^{32}P) or a fluorescent tag.
- Depending on the hybridization conditions, the probe will anneal to the complementary fragments of DNA immobilized on the membrane. "Stringency" is the term given to the conditions of annealing, which control the extent of base pair mismatching allowed. If the hybridization and membrane washing is conducted under conditions of "high stringency" such as high temperature or low salt concentration, only identical or very closely related genes will hybridize with the probe (specific hybridization). Conversely, if hybridization and washing are conducted under lower stringency conditions the probe may hybridize to more distantly related genes (non-specific hybridization).
- After hybridization, the membrane is washed and autoradiographed to locate the labeled probe as distinct bands on the X-ray film.

8.4 DNA Sequencing

DNA sequencing, first devised in 1975, allows analysis of genes at the nucleotide level [2–7]. It provides the following information:

- the precise order of nucleotides;
- location of restriction enzyme recognition sites;
- location of introns and exons, from which the amino acid sequence of a peptide can be deduced;
- analysis of the flanking regions of a gene, which provides insight about potential mechanisms that control gene expression.

There are two main techniques for DNA sequencing: *chemical sequencing* or *chemical degradation method*, developed by A. M. Maxam and W. Gilbert, also referred to as Maxam-Gilbert Sequencing, and *enzymatic sequencing* or *chain termination method* developed by F. Sanger and A. R. Coulson, also referred to as Sanger-Coulson Sequencing. Today most DNA sequencing is conducted following the principles of Sanger-Coulson. Various modifications have been introduced and the technique has been automated.

8.4.1 Sanger-Coulson Sequencing Methodology

This method entails two steps, the labeling reaction, in which a labeled source of the DNA to be analyzed is generated, and the termination reaction, where the synthesis of the complementary DNA fragments produced in the first reaction is ended.

During the labeling reaction, a short fragment of complementary DNA is synthesized using the DNA-sequence to be analyzed as a template. The DNA is diluted in a polymerization buffer, divided into four separate reaction tubes and denatured by heat, which causes the H-bonds to break and the DNA to unwind and separate into single strands. The following reagents are added to the denatured DNA in each of the four reaction tubes:

- An oligonucleotide primer of known sequence;
- a mixture of four dNTPs, which include

  a. deoxythymidine triphosphate (dTTP);
  b. deoxycytidine triphosphate (dCTP);
  c. deoxyguanosine triphosphate (dGTP); and
  d. deoxyadenosine triphosphate (dATP), labeled with radioactive isotope (i.e., ^{32}P or ^{35}S) or a fluorescent tag, which permits detection of the DNA fragments by autoradiography; and

- DNA polymerase.

Polymerization begins when the DNA polymerase recognizes the oligonucleotide as a primer and begins synthesizing the complementary strand of DNA in each of the four reaction tubes. The ddNTPs, which include ddTTP, ddCTP, ddGTP, and ddATP, are each added to one of the four labeling reaction tubes to stop the DNA polymerase action. ddNTPs are devoid of an essential –OH group, thereby causing interruption of connection with the next nucleotide. In the automated methods, only one tube is required for the entire reaction instead of four, which simplifies the method. The reaction is terminated using four different colors of fluorescent-labeled ddNTPs. The interruption of polymerization produces various lengths of DNA strands. The newly synthesized DNA strands are run in four adjacent lanes of PAGE in the traditional method, or just in one lane when following the automated method. The gel is then dried onto a filter paper and autoradiographed to locate the labeled dNTP on the X-ray film. The DNA fragments appear as ladders in the autoradiogram, with the labeled dNTP appearing as a rung or band in each of the four lanes. The shortest fragments will migrate the fastest and thus the farthest. Therefore, the bottom band indicates which dideoxynucleotide was added first to the labeled primer. The autoradiogram is read from the bottom to the top, which is the reading of sequences from the 5′ to 3′ of the complementary strand. In the automated method reading is performed using a computerized image analyzer that distinguishes bands of each of the four different colors generated by the fluorescent labeled dNTPs.

8.5 DNA Fingerprinting and Profiling

The chemical structure of DNA in every individual of the same species is identical; on the contrary, the difference among individuals of the same species is due to the sequence of the DNA nucleotides. It was hypothesized that individuals could be identified solely by variability within sequences of their base pairs, and by the late 1970s scientists began searching for methods to identify specific regions in human DNA that were variable. *DNA fingerprinting* was initially based on the use of the Southern blot technique to analyze polymorphic regions of human DNA [8,9]. Although it was thought that the introns do not provide relevant genetic information, repeated sequences of base pairs called *variable number tandem repeats* (VNTRs) were detected. Individuals inherit VNTRs from their parents, half from their mother and half from their father. Using the same restriction enzyme to cut the regions of DNA surrounding the repeated sequences, it was observed that the DNA was not cut at the same sites, which resulted in DNA fragments of varying length, thus demonstrating that restriction sites vary between individuals. By comparing DNA banding patterns between individuals it can be determined whether DNA samples are from the same person, related people or non-related people. The polymerase chain reaction technique (PCR) overcame limitations of DNA fingerprinting; the enhanced technique called *DNA profiling* enables isolation and autoradiography of smaller amounts of DNA. The DNA patterns became simpler to read and interpret and could be stored on a computer database.

8.6 Polymerase Chain Reaction

Polymerase chain reaction is an in vitro technique for enzymatically replicating or amplifying specific regions of DNA [10–13]. The technique allows a small amount of DNA to be amplified exponentially. PCR is easy to execute and has been extensively modified to allow a large number of other applications, including genetic manipulations. PCR is commonly used for detection of hereditary diseases and diagnosis of infectious diseases.

8.6.1 Methodology

To perform a PCR the following materials are required:

- The DNA sample to be amplified.
- Two oligonucleotides of 17–30 bp, designed for the 5′ and 3′ ends of the DNA target region. The oligonucleotides serve as primers that bind to complementary target sequences, thus generating a place for a polymerase to bind and extend the primer by the addition of nucleotides to make a copy of the target sequence.
- A mixture of the four dNTPs: dTTP, dCTP, dGTP, dATP dissolved in a polymerization buffer.
- A thermostable DNA polymerase (i.e., *Taq* polymerase), an enzyme capable of withstanding several cycles of heat required to denature DNA.

The reaction is carried out in a micro centrifuge plastic tube in a thermal reactor, also called thermal cycler, an instrument that can perform programmed, rapid heating and cooling of the reaction. A typical PCR cycle consists of three phases:

- denaturing, in which the DNA sample is heated at 94°C for 90 s to separate the complementary strands;
- annealing, in which the sample is cooled to 55–65°C for 2 min to let the primers bind to the target flanking sequences; and
- polymerization, at 72°C for 3 min, in which *Taq* polymerase extends the primer on the DNA strands.

Amplification of DNA follows a logarithmic progression. Thus, after 25–30 cycles a million-fold increase in the original DNA template can be obtained. Amplified DNA can be visualized on an agarose gel electrophoresis stained with ethidium bromide.

8.7 Real-Time PCR

Real-time PCR also called quantitative real time PCR (qPCR) or kinetic PCR is a technique that simultaneously quantifies and amplifies a specific part of a given DNA molecule [14–19]. It is used to determine whether or not a specific sequence is present in the sample; and if so, the number of copies. The procedure follows the general pattern of PCR, but the DNA is quantified after each round of amplification; this is the "real-time" aspect of it. Two common methods of quantification are the use of fluorescent dyes that intercalate with double-strand DNA and modified DNA oligonucleotide probes that fluoresce when hybridized with a complementary DNA. Frequently, real-time PCR is combined with reverse transcription (RT) PCR to quantify low abundance messenger RNA (mRNA), enabling a researcher to assess relative gene expression at a particular time, or in a particular cell or tissue type. RT-PCR should not be confused with reverse transcription polymerase chain reaction, which may be marketed as qRT-PCR or RT-qPRC.

8.7.1 Real-Time PCR Using Double-Stranded DNA Dyes

As the dye binds to double-stranded DNA (dsDNA) during polylmerization, it causes fluorescence. Therefore, an increase in the amount of DNA leads to an increase in fluorescence intensity, which is measured at each cycle, allowing DNA concentrations to be quantified. However, dsDNA dyes such as SYBR Green will bind to all dsDNA PCR products, including non-specific PCR products called primer dimers. Like other real-time PCR methods, the values obtained do not have absolute units associated with it (i.e., mRNA copies/cell). A comparison of a measured DNA/RNA sample to a standard dilution will only give a fraction or ratio of the sample relative to the standard, allowing only relative comparisons between different tissues or experimental conditions.

8.7.2 Fluorescent Reporter Probe Method

The fluorescent reporter probe qPCR method is the most accurate and reliable, but also the most expensive. qPCR is carried out using an RNA or DNA-based probe that has a fluorescent tag reporter at one end and a quencher of fluorescence at the opposite end. The close proximity of the reporter to the quencher prevents detection of its fluorescence. During the PCR reaction the 5′ to 3′ exonuclease activity of the *Taq* polymerase breaks down the reporter-quencher proximity allowing emission of fluorescence, which can be detected. An increase in the product targeted by the reporter probe at each PCR cycle causes a proportional increase in fluorescence due to the breakdown of the probe and release of the reporter. Only sequence-specific RNA or DNA containing the probe sequence is quantified. Therefore, use of the reporter probe significantly increases specificity, and allows quantification in the presence of non-specific DNA. This method also allows for multiple assays of several genes in the same reaction by using fluorescent reporter probes with different-coloured labels, provided that all genes are amplified with similar efficiency.

8.8 Identification of Gene Mutations

Various techniques detect nucleotide rearrangements in genes, which is critical to understanding the molecular basis of genetic disorders and/or disease.

8.8.1 Restriction Fragment Length Polymorphism (RFLP)

Restriction fragment length polymorphism, which derived from DNA fingerprinting/profiling is aimed at detecting abnormal or mutated DNA sequences, which will create or abolish restriction site(s) [20–23]. RFLP methodology employs the technique of Southern blot analysis to detect mutations that alter recognition sites of specific restriction endonucleases. For example, a copy of one chromosome might possess the nucleotide sequence ATTTCCGG; however, a single point mutation of **T** to **A** on its homologue chromosome changes the nucleotide sequence to ATT**A**CCGG, which is sufficient to abolish a restriction site, thereby yielding different lengths of DNA fragments.

Genetic diseases linked to restriction enzyme polymorphisms have been identified using this approach. The RFLP technique is applied to follow the transmission of an abnormal gene when the exact sequence of the mutated gene is unknown. A gene polymorphism can be detected irrespective of whether the DNA sequence change affects expression of phenotype. The RFLP technique is also used in marker-assisted selection (MAS). RFLP identifies the location of a mutated gene by using marker DNA sequences closely linked to the mutated gene.

8.8.1.1 Methodology

- DNA is extracted, purified and cut into restriction fragments using appropriate endonucleases.
- The restriction fragments are separated according to length by agarose gel electrophoresis.
- The gel is transferred and fixed to a membrane and combined with a hybridization probe.
- The homologous DNA fragments will hybridize, resulting in a restriction map.

Thus, RFLP can be defined as a difference in restriction maps between two individuals. Only two alleles can happen in the population at every given polymorphic point detected by a RFLP probe. Therefore, as a result of the RFLP technique we will obtain a total of three different genotypes in the population: AA, BB, and AB that are produced by combinations of the A and B alleles. In genotype A the enzyme will cut the restriction point, producing a smaller allele/gel band. In genotype B the enzyme will not cut the restriction point, producing a bigger allele/gel band. The AA and BB genotypes correspond to homozygous individuals, e.g., the two alleles containing (or not) the same genetic change at the particular polymorphism, while the AB genotype

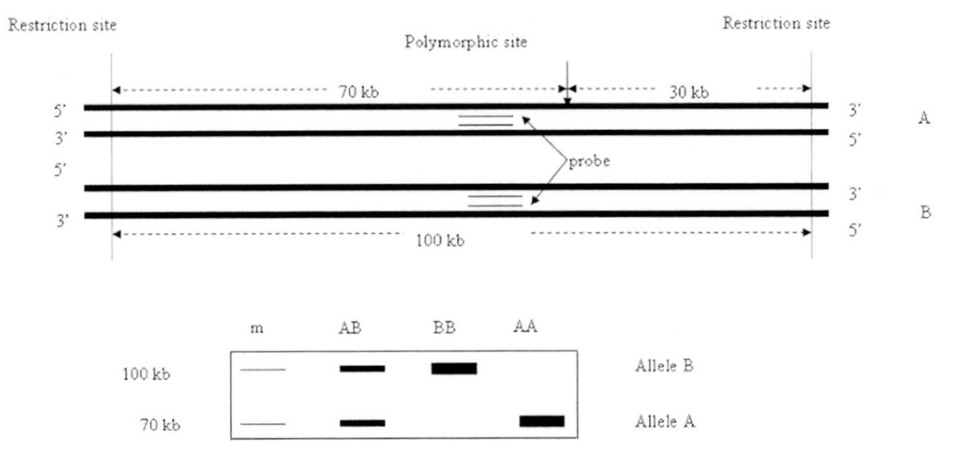

Fig. 8.1 Three different genotypes revealed in three patients of a particular disease when RFLP analysis is performed by Southern blot and the probe used occupies a position as in the picture. Two alleles of 100 and 70 kb are detected by the probe. Genotype AB would be heterozygous, while AA and BB would be homozygous. m: DNA size marker, A: allele A, B: allele B

manifests heterozygosity, e.g., only one of the alleles containing the genetic change.

The RFLP technique can be based on Southern blot (Figs. 8.1, 8.2, and 8.3) or on PCR (Fig. 8.4). Southern-RFLP can produce two different bands in the gels (Figs. 8.1 and 8.2), one for the A allele and one for the B allele, or even three (Fig. 8.3) (when the RFLP probe binds DNA on top of the polymorphic point); in this case two bands associate to the genotype A (smaller) allele and only one to the genotype B (big) allele. So, two bands appearing in a RFLP gel do not always mean heterozygosity. The sizes of the different alleles and how the probe binds the target DNA should be considered into account.

PCR-RFLP will always produce three bands in the gels (Fig. 8.4), similarly to Southern-RFLP with the use of a probe that binds DNA on top of the polymorphic point. A PCR reaction is first produced, being the DNA amplified at a particular region that will be analyzed by restriction enzyme digestion. If the enzyme cuts the two alleles, the genotype will be AA and two gel bands will be produced. If the enzyme does not cut any of the alleles the genotype will be BB (bigger bands in size than those of the AA genotypes) and only one band will be produced. Both cases reflect homozygosity. On the contrary, heterozygosity, or AB genotype, will be revealed as three bands: two will correspond to the allele cut by the enzyme and one, the biggest, will correspond to the uncut allele.

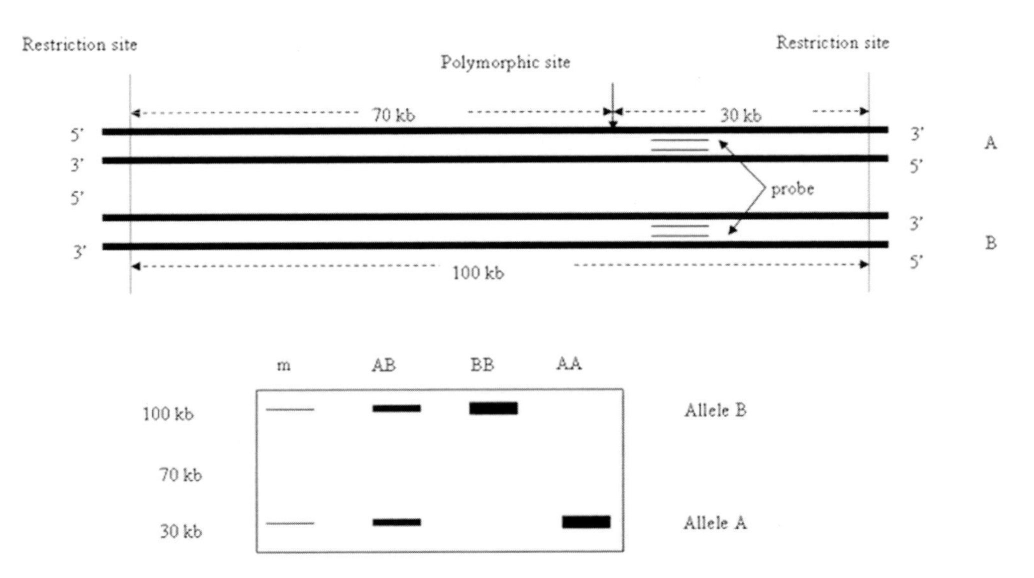

Fig. 8.2 Three different genotypes revealed in three patients of a particular disease when RFLP analysis is performed by Southern blot and the probe used occupies a position as in the picture. Two alleles of 100 and 30 kb are detected by the probe. Genotype AB would be heterozygous, while AA and BB would be homozygous. m: DNA size marker, A: allele A, B: allele B

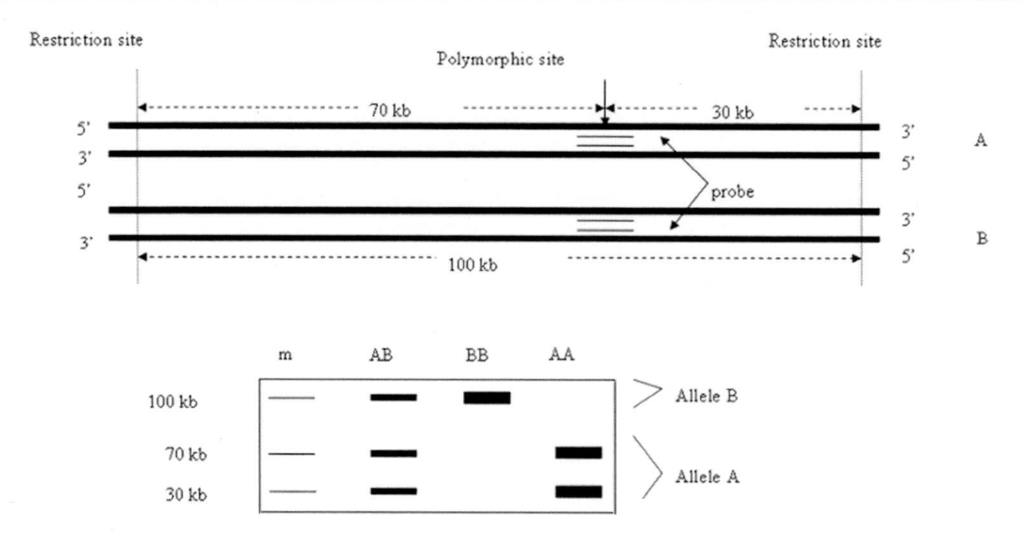

Fig. 8.3 Three different genotypes revealed in three patients of a particular disease when RFLP analysis is performed by Southern blot and the probe used hybridizes at the region of the polymorphic restriction site, as in the picture. Two alleles of 100 and 70+30 kb are detected by the probe (note that although two alleles appear, three bands are produced: 100, 70, and 30 kb). Genotype AB would be heterozygous, while AA and BB would be homozygous. m: DNA size marker, A: allele A, B: allele B

8.8.2 Single-Stranded Conformational Polymorphism (SSCP)

Single-stranded conformational polymorphism is the simplest and most widely used method to detect sequence differences in DNA [24–28]. SSCP involves the electrophoretic separation of single-stranded nucleic acids based on subtle differences in nucleotide sequence, often single basemutations, which result in altered structure (Fig. 8.5). This procedure is based on the changes in mobility of mutated single stranded DNA fragments that migrate as a function of their length and sequence during electrophoresis in non-denaturing gels. Under optimal conditions, approximately 80–90% of the potential base exchanges are detectable by SSCP. Therefore, SSCP is a direct approach, which can be applied to follow transmission of genetic disorders when DNA mutations have already been characterized by deletions or point mutations.

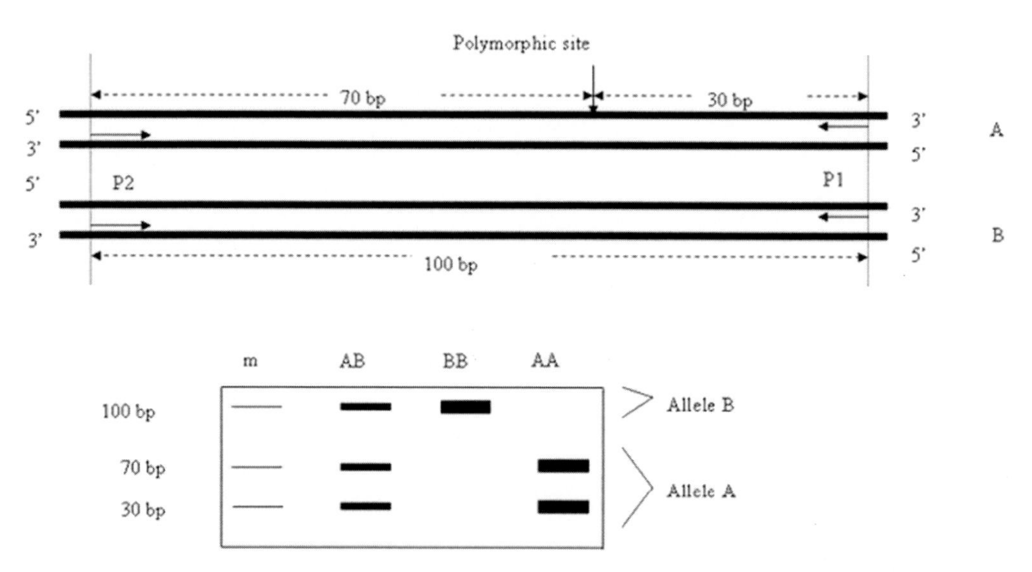

Fig. 8.4 Three different genotypes revealed in three patients of a particular disease when RFLP analysis is performed by PCR on a region containing a polymorphic restriction site as in the picture. No probe is used. Two alleles of 100 and 70+30 kb are revealed by gel electrophoresis (note that although two alleles appear, three bands are produced: 100, 70, and 30 kb). Genotype AB would be heterozygous, while AA and BB would be homozygous. m: DNA size marker, A: allele A, B: allele B

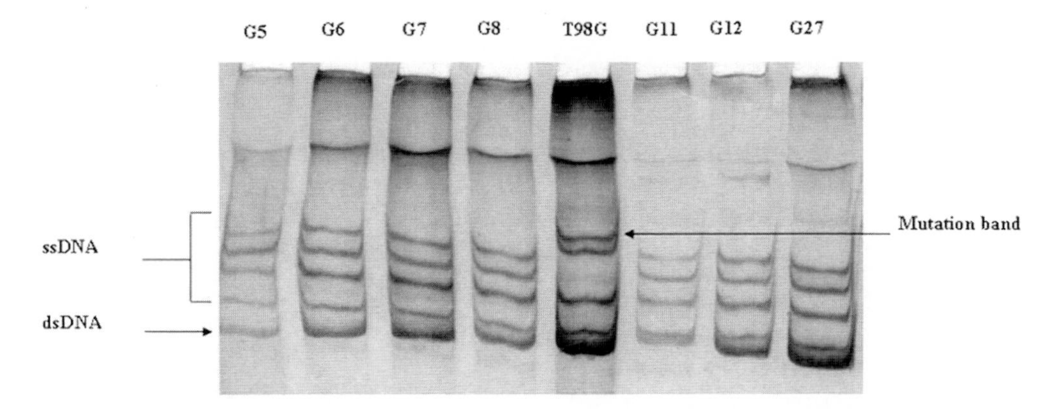

Fig. 8.5 PCR-SSCP non-isotopic mutation analysis at PTEN gene exon 2 in DNAs extracted from seven glioblastomas (G) and one glioblastoma cell line (T98G) that was selected as a positive control. DNA from T98G cell line contains a point mutation at codon 42, exon 2 of the PTEN gene, where CTT changes to CGT, and then Leu changes to Arg in the protein sequence. A silver stained 10% non-denaturing polyacrylamide (19:1 acrylamide:bis-acrylamide) gel is used for electrophoresis. No mutation was detected in the glioblastomas. ssDNA (single stranded DNA); dsDNA (double stranded DNA)

8.8.2.1 Methodology

- The target DNA sequence is amplified by PCR and labeled with a radioactive or fluorescent marker.
- The sample is melted with heat and then rapidly cooled to form single-stranded fragments.
- The sample is run in polyacrylamide gel electrophoresis at a constant temperature.
- Detected variants are confirmed by DNA sequencing.

8.8.3 Denaturing Gradient Gel Electrophoresis (DGGE)

Denaturing gradient gel electrophoresis is a technique for detecting non-RFLP polymorphism single base mutations [29]. Wild type and mutant DNA will migrate at different speeds in a chemical gradient denaturing gel because they have different melting temperatures. In DGGE, DNA is melted by high concentrations of urea salt. DGGE is based on the capability of heteroduplexes, dsDNA molecules, or DNA/RNA hybrids where each strand is from a different source, to separate, or melt, into single-stranded molecules in polyacrylamide gel containing an increasing linear gradient of a chemical denaturant. Initially the dsDNA fragments move according to molecular weight, but as they progress into higher denaturing conditions, each depending on its sequence composition reaches a point where the dsDNA begins to melt. The partial melting markedly changes the shape of the dsDNA and reduces the mobility of the molecule in the gel. Thus, sequence differences, such as a single base mutation in otherwise identical DNA fragments, often cause partial melting at different positions in the gradient and therefore stop mobility at different positions in the gel. By comparing the melting behavior of the polymorphic DNA fragments side by side on denaturing gradient gels, it is possible to detect fragments that have mutations.

In a variation of the DGGE technique, specific mutations may be detected when DNA is run in a gel with a single specific concentration of chemical denaturant rather than an increasing gradient. Once a molecule is partially denatured, its mobility is drastically reduced and heteroduplexes stop at different positions in the gel as compared with normal DNA. Based on the same principles as DGGE, several variations of temperature gradient gel electrophoresis (TGGE) have been designed, in which a gradient of temperature, rather than chemical denaturants, is created within the gel.

8.8.3.1 Methodology

- Target DNA is amplified by PCR and then hybridized to form a heteroduplex with normal DNA.
- The heteroduplexes are run through polyacrylamide gel electrophoresis containing an increasing gradient of chemical denaturant.
- Upon melting, rate of migration decreases dramatically, almost stopping due to the fact that random coil, single-stranded DNA migrates much more slowly than dsDNA.
- When the DNA is visualized on the gel after migration, any alteration in position in comparison with a control will indicate a sequence variation.
- DNA can then be analyzed by DNA sequencing.

8.8.4 Heteroduplex Analysis

Homoduplex (wild type DNA in the two strands) and heteroduplex DNA (wild type DNA in one strand and mutated DNA in the second strand) migrate differently in PAGE.

Therefore, if we can induce the formation of heteroduplexes, we will be able to diagnose mutations through PAGE [30,31].

8.8.4.1 Methodology

- Control DNA (wild type) is denatured and allowed to anneal with denatured sample DNA. Conformational differences of the products will be analyzed on an acrylamide gel.
- The fastest migrating band corresponds to the homoduplex (wild type DNA).
- Heteroduplexes (mutated DNA) will appear high in the gel, as they migrate more slowly due to the mismatches.

8.8.5 *Chemical Mismatch Cleavage (CMC) or Enzyme Mismatch Cleavage (EMC)*

To detect mutations by this method a heteroduplex has to be formed [32]. Heteroduplexes are double stranded molecules of DNA in which one of the strands is wild type, while the other is mutated. Chemical compounds such as osmium tetraoxide or piperidine will break the heteroduplex and produce smaller DNA molecules than wild type DNA. Gel electrophoresis will detect the difference in size of the molecules and define the existence of a mutation in the DNA fragments analyzed. This is the case for CMC, but with EMC a DNA repair enzyme will cut the heteroduplex.

8.8.5.1 Methodology

- Wild type DNA (control) and mutated DNA are amplified separately by some cycles of PCR.
- Wild type and mutated DNA molecules are mixed together.
- PCR continues with more cycles.
- Heteroduplexes are formed.
- CMC or EMC is applied.
- Gel electrophoresis is performed to visualize the small (mutant) DNA fragment.

8.8.6 *Protein Truncation Test (PTT)*

Some abnormal genes present mutations that result in premature termination of translation, which creates a protein of a substantial difference in length than the wild type protein. Those mutations are of different kinds:

- nonsense mutations (that induce a stop codon);
- mutations that change the reading frame and induce a stop codon downstream (small insertions or deletions of a number of bases which is not 3 or a multiple of 3);
- mutations at the splicing donor or acceptor sites.

Missense mutations (change of one amino acid by another one) are not included among the mutations that can be detected by PTT [33,34]. Then, PTT only detects mutations that confer biological significance. On the contrary, neutral polymorphisms are not differentiated from those with phenotypic and pathological problems. PTT was initially developed for the analysis of the Duchenne Muscular Dystrophy gene, but it can be applied to several genes, essentially those that suffer suppressive mutations, like tumor suppressor genes: APC, BRAC1, BRAC2, NF1, NF2, and others.

8.8.6.1 Methodology

The coding region of a gene is screened for the presence of translation terminating mutations using *de novo* protein synthesis from the amplified copy. The technique includes the following steps:

- isolation of genomic DNA and amplification of the target gene coding sequences using PCR or isolation of RNA and amplification of the target sequence using RT-PCR;
- a template for in vitro synthesis of RNA and further in vitro translation of RNA into protein;
- SDS-PAGE analysis of the synthesized protein to determine the length of the protein fragments.

The forward primer used in the PCR amplification step has to contain a T7 promoter sequence at its 5′ end to direct the synthesis of RNA by the T7 RNA polymerase. A consensus initiation sequence for eukaryotic translation has to be incorporated before the template sequence as well.

8.9 Techniques for Detection of Promoter Methylation

DNA methylation can be defined as the covalent addition of a methyl group, catalyzed by the DNA methyltransferases (DNMT), to the 5-carbon of cytosine in a CpG dinucleotide. Methylation plays a critical role in the development and differentiation of mammalian cells, and its deregulation has been implicated in oncogenesis. Generalized hypomethylation in the genome (lack of methyl groups bound to cytosine) and regional hypermethylation (mostly at the CpG dinucleotides placed at the promoters of tumor suppressor genes) have been associated with human neoplasia. Hypermethylation has the potential to inactivate genes, such as tumor suppressor genes and genes that physiologically should not be expressed during differentiation or embryogenesis. Precise mapping of DNA methylation patterns in CpG islands is essential in diverse biological processes such as regulation of imprinted (methylated in one of the parents' chromosomes) genes, X chromosome inactivation, and inactivation of tumor suppressor genes. In general two methods exist to analyse DNA methylation: one by the use of methylation-sensitive enzymes followed by Southern blot or PCR and a second one by bisulfite treatment followed by PCR or sequencing. The latter is the most widely used and will be commented on in this chapter.

8.9.1 Bisulfite Treatment

Cytosines of the DNA chain can react with the bisulfite ion to finally be converted to uracils. This is called DNA modification and it is the principle on which methylation assays rely on. The modification reaction is highly single-strand specific, so DNA should first be incubated with 3 M NaOH at 37°C for 10 min to obtain single-stranded DNA. Salmon sperm DNA, yeast RNA, or tRNA could be used as carriers. There are three processes in the DNA modification by the bisulfite reaction.

- *Sulfonation*: addition of bisulfite to the 5–6 double bond of cytosine and formation of a sulfonated cytosine derivative (cytosine-SO_3). This step is reversible. The reaction is controlled by pH, bisulfite concentration, and temperature. Low pH favors forward reaction.
- *Hydrolytic deamination*: cytosine-SO_3 is hydrolytic deaminated to give a uracilbisulfite derivative (uracil-SO_3). This step is catalyzed by sulfite, bisulfite, and acetate anions. The reaction is irreversible and is favored by pH below 7.0.
- *Alkali desulfonation*: alkali treatment removes the bisulfite adduct to give uracil.

Although 5-metyl-cytosine can also react with the bisulfite, the reaction is too slow and the equilibrium is favored to the 5-methyl-cytosine rather than to the deaminated product.

8.9.2 Methylation Specific PCR (MSP)

The MSP technique can assess the methylation status of a group of CpG sites within a CpG island [35–37]. The treatment of the template with sodium bisulfite converts all unmethylated cytosines to uracils, but methylated cytosines are resistant to the modification. Then, PCR amplification is performed with specific primers for the methylated versus the unmethylated sequence (Fig. 8.6). This technique is very sensitive, and requires 0.1% of methylated alleles of a CpG island. It can be used with paraffin-embedded samples, and it requires small amounts of DNA.

8.9.3 Bisulfite Sequencing PCR (BSP)

DNA is first denatured to create single-stranded DNA and then modified with sodium bisulfite followed by PCR amplification using primers that are specific for the modified DNA but do not contain any CpG sites in their sequence [38,39]. CpG sites are in the DNA sequence amplified between primers. The resulting PCR product is sequenced directly or after cloning.

8.10 Comparative Genomic Hybridization (CGH)

Comparative genomic hybridization is a molecular cytogenetic method that makes possible to screen for copy number aberrations (gains and/or losses) throughout the genome in a single hybridization [40–43]. It is essential in the cytogenetic study of solid tumors because very few metaphase spreads are obtained after cell culture and their quality is often inadequate for recognition of banding patterns. With CGH, it is possible to detect candidate oncogenes or tumor suppressor genes. CGH can detect changes that are present in as little as 30–50% of the specimen cells. The main advantage of CGH compared to traditional cytogenetics is that cell culture is not necessary and this makes it possible to use archived paraffin embedded material for hybridization. However, structural alterations such as balanced translocations, inversions, small deletions, or polyploidization cannot be detected with this technique. DNA copy number changes of less than 10–20 Mb in size are not detectable using CGH, unless they are highly amplified, such as five- to ten-fold amplification of a region of 1 Mb in length. Losses of 10–12 Mb are detectable if they are present in a large proportion of cells.

8.10.1 Methodology

DNA from tumor sample and normal genomic DNA are labeled by nick translation, or random priming using fluorescent green dUTP and fluorescent red dUTP, respectively, and used as a probe. Both DNAs are co-precipitated and resuspended in a buffer that contains formamide.

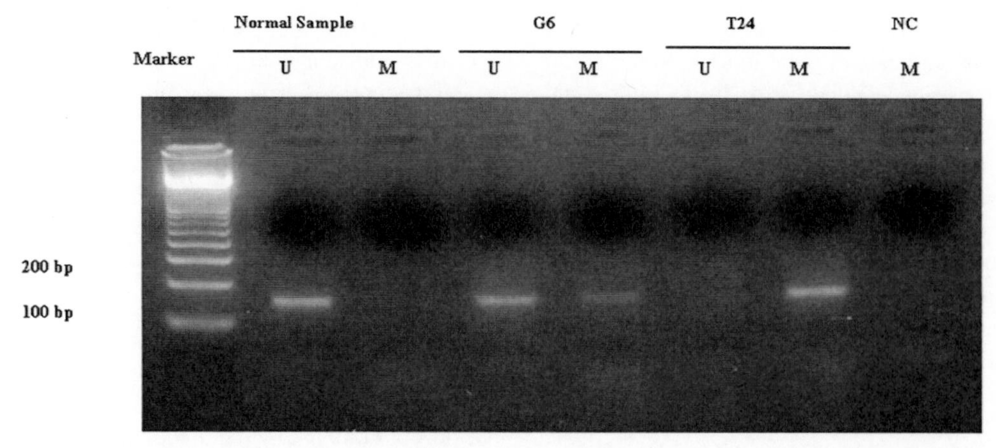

Fig. 8.6 p16 inactivation through promoter hypermethylation in one glioblastoma (G6). An ethidium bromide-stained 2% agarose gel is used for electrophoresis after a PCR assay to detect unmethylated promoter (U) or methylated promoter (M) regions of the p16 gene. Normal Sample: DNA from leukocytes, used as a positive control for the unmethylated p16 promoter. The T24 cell line DNA was used as a positive control of p16 methylation. NC: non-DNA PCR, used as a negative control of the PCR reaction. A 100 base pair ladder is included (first lane) for size determination

The probe is hybridized onto normal lymphocyte denatured metaphase spreads for 3 days at 37°C in a moist dark chamber. Slides are covered with a glass coverslip and sealed with rubber cement. Cover slips are removed and preparations are washed to remove non-specific bounded DNA. After drying the slides, the preparations are counterstained with DAPI/antifade for chromosome identification.

Metaphase images are evaluated by fluorescence microscopy. The three colours image with red, green, and blue are acquired from 10 to 12 metaphases (Fig. 8.7). Metaphase karyotyping is performed using specific software. Chromosomal regions for which the mean green-to-red ratio fell below 0.85 (lower threshold) are considered lost, whereas mean ratios exceeding the value 1.15 (upper threshold) are considered to indicate gains.

8.11 Fluorescence In Situ Hybridization (FISH)

Fluorescence in situ hybridization is a rapid diagnostic test using molecular cytogenetic techniques [44,45]. It has wide applications in many branches of medicine including oncology. The FISH technique supplements conventional cytogenetics and in some cases provides additional information, which is not detected by karyotyping. Using FISH a large number of cells can be studied since interphase nuclei can be analyzed. This helps in the detection of minimal residual disease, assessment of the rate of cytogenetic remission, and detection of disease recurrence.

8.11.1 Methodology

Cells in culture are stimulated with phytohemaglutinin. Then, colchicine is added to stop cell growth in mitosis to obtain a large number of cells in metaphase. Finally, metaphases

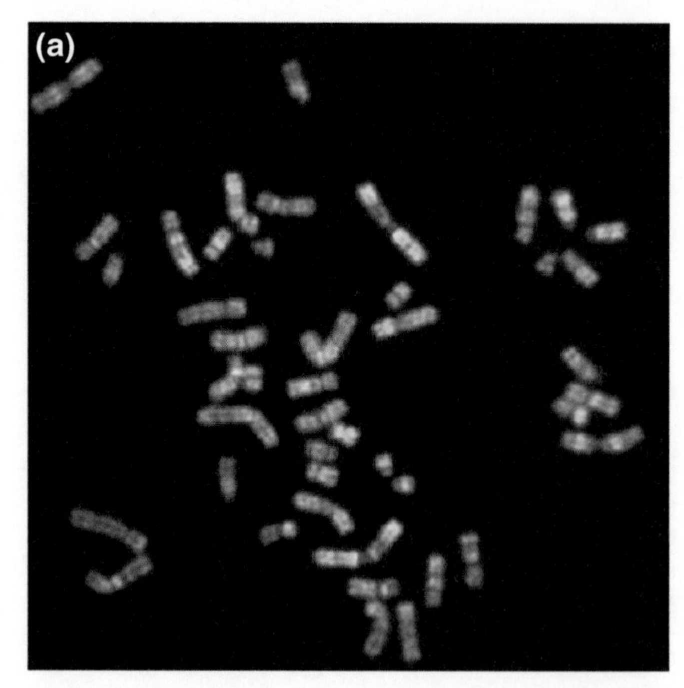

Fig. 8.7 CGH analysis of glioblastoma. (**A**) normal metaphase after hybridization of normal and tumor DNA; (**B**) CGH-karyotype of one glioblastoma; (**C**) CGH-idiogram of one glioblastoma, where gains and losses of genetic material can be detected (usually no less than 10 metaphases—around 20 chromosomes of every of the 22 autosomes—are studied); and (**D**) CGH-idiogram of a collection of glioblastomas, from where we can obtain statistically relevant results of gains and losses of tumor DNA at any particular chromosome in glioblastoma

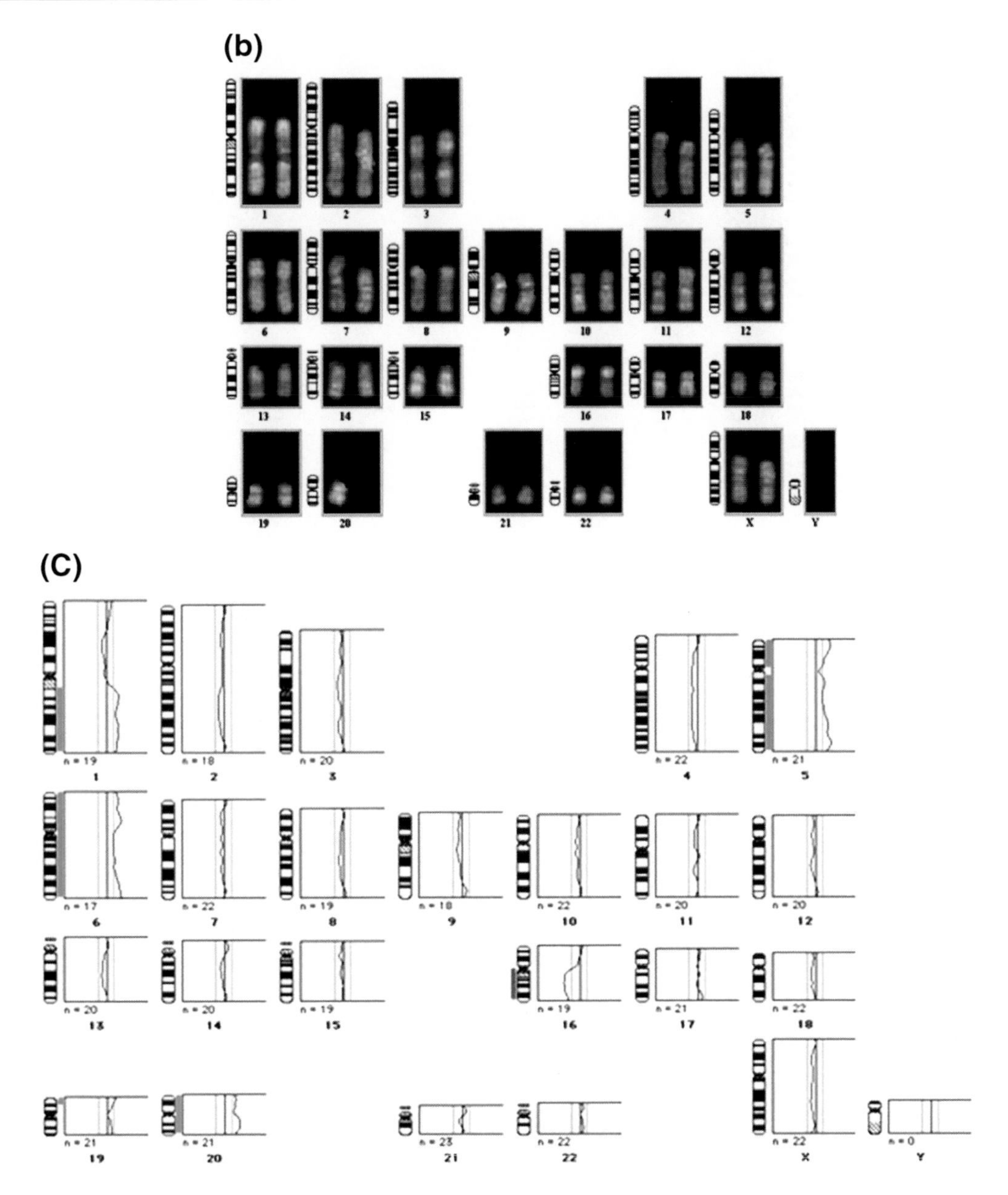

Fig. 8.7 (continued)

are treated with a hypotonic KCl solution and fixed with methanol/acetic acid (3:1) onto slides. Most FISH probes can be obtained labeled and premixed in hybridization buffer ready to use. But sometimes it is necessary to obtain non-commercial probes to analyze specific genes. Non-commercial probes are obtained from bacterial artificial chromosome (BAC) clones. DNA samples from BAC clones are labeled by random priming or nick translation. The probes are hybridized onto cell denatured metaphases spreads for 24 h at 37°C in a moist dark chamber. Slides are covered with a glass cover slip and sealed with rubber cement. Cover slips are removed and preparations are washed to remove non-specific bounded DNA. After drying the slides, the preparations are counterstained with DAPI/antifade for chromosome identification. Metaphase images are assessed by fluorescence microscopy, with which different genetic alterations like amplification, deletion, and translocation can be detected.

8.12 Summary

Molecular biology is a dynamic field with DNA techniques and analytical tools that detect aberrations and facilitate the identification of crucial genes and pathways involved in biological processes and disease. Many of the fundamental

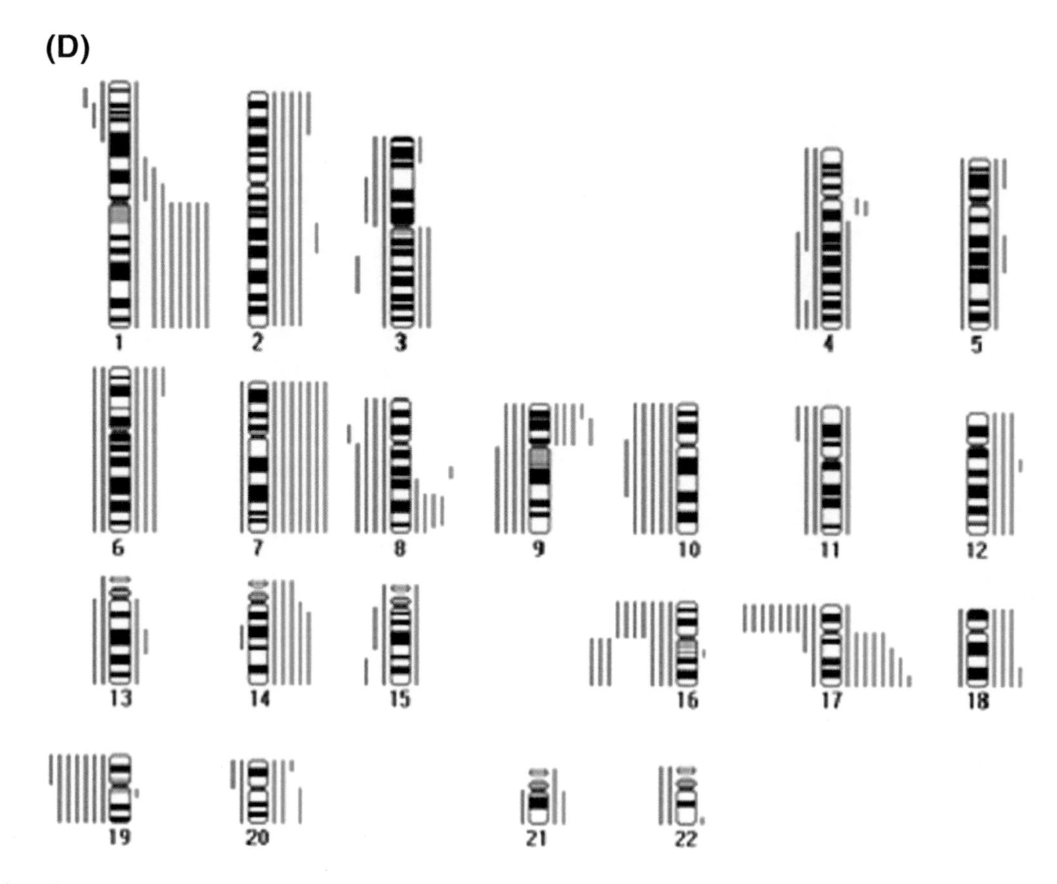

Fig. 8.7 (continued)

DNA analysis techniques discussed in this chapter evolved through various modifications, automation, and computerization. *Gel electrophoresis*, a central technique in molecular biology, allows separation of DNA, RNA, and proteins so that they can be studied individually. *DNA sequencing* methods provide information about the precise order of nucleotides, location of restriction enzyme recognition sites, location of introns and exons, and analysis of the flanking regions of a gene. *DNA fingerprinting* or *DNA profiling* examine the inherited variable lengths of repeat sequences of DNA that differ from individual to individual. *RFLP* is a technique for analyzing the variable lengths of DNA fragments that result from digesting a DNA sample with a restriction endonuclease, which are separated using gel electrophoresis. They are then hybridized with DNA probes that bind to a complementary DNA sequence in the sample. RFLPs can be defined as a difference in restriction maps between two individuals. With the development of newer, more efficient DNA-analysis techniques, RFLP is not used as much as it was because it requires relatively large amounts of DNA. *PCR* is used to make millions of exact copies of DNA from a biological sample. DNA amplification with PCR allows DNA analysis on very small amounts of biological samples and enables even highly degraded samples to be analyzed. Traditional PCR is a qualitative tool for detecting the presence or absence of particular DNA. *Real-time PCR* is a quantitative PCR method. Methods to identify gene mutations include *SSCP* and *RRFP* analysis, *DGGE* or *TGGE, heteroduplex analysis, CMC* or *EMC,* and *PTT.* Precise mapping of *DNA methylation* patterns in CpG islands is essential for regulation of imprinted genes, X chromosome inactivation, and inactivation of tumor suppressor genes. Molecular cytogenetic techniques, *CGH* and *FISH*, encompass aspects of chromosome biology.

8.13 Glossary of Terms and Acronyms

^{32}P: radioactive isotope

5α-DHT: 5α–dihydrotestosterone

BAC: bacterial artificial chromosome

BSP: bisulfite sequencing PCR

CGH: comparative genomic hybridization

CMC: chemical mismatch cleavage

dATP: deoxyadenosine triphosphate

dCTP: deoxycytidine triphosphate

ddATP: dideoxyadenosine triphosphate

ddCTP: dideoxycytidine triphosphate

ddGTP: dideoxyguanosine triphosphate

ddNTP: dideoxynucleoside triphosphate

ddTTP: dideoxythymidine triphosphate

DGGE: denaturing gradient gel electrophoresis

dGTP: deoxyguanosine triphosphate

DNA: deoxyribonucleic acid

DNMT: DNA methyltransferases

dNTP: deoxynucleoside triphosphate

dsDNA: double stranded DNA

dTTP: deoxythymidine triphosphate

dUTP: deoxyuridine triphosphate

EMC: enzyme mismatch cleavage

FISH: fluorescence in-situ hybridization

KCL: potassium chloride

MAS: marker-assisted selection

mRNA: messenger RNA

MSP: methylation specific PCR

NaOH: sodium hydroxide

PAGE: polyacrilamide gel electrophoresis

PCR: polymerase chain reaction

PTT: protein truncation test

qPCR: quantitative real time polymerase chain reaction

RFLP: restriction fragment length polymorphism

RNA: ribonucleic acid

RT-PCR: reverse transcription polymerase chain reaction

SDS-PAGE: sodium dodecyl sulfate polyacrilamide gel electrophoresis

SSCP: single-stranded conformational polymorphism

TGGE: temperature gradient gel electrophoresis

tRNA: transfer RNA

UV: ultra-violet

VNTR: variable number tandem repeats

References

1. Sambrook J, Russell D. Molecular Cloning: A Laboratory Manual, third edition. Cold Spring Harbor: Cold Spring Harbor Laboratory Press, 2001.
2. Gerischer U. Direct sequencing of DNA produced in a polymerase chain reaction. Methods Mol Biol 2001; 167:53–61.
3. Graham CA, Hill AJ. Introduction to DNA sequencing. Methods Mol Biol 2001; 167:1–12.
4. Marziali A, Akeson M. New DNA sequencing methods. Annu Rev Biomed Eng 2001; 3:195–223.
5. Mitnik L, Novotny M, Felten C, et al. Recent advances in DNA sequencing by capillary and microdevice electrophoresis. Electrophoresis 2001; 22:4104–17.
6. Watts D, MacBeath JR. Automated fluorescent DNA sequencing on the ABI PRISM 310 genetic analyzer. Methods Mol Biol 2001; 167:153–70.
7. Zschocke J, Hoffmann GF. Cycle sequencing of polymerase chain reaction-amplified genomic DNA with dye-labeled universal primers. Methods Mol Biol 2001; 167:113–7.
8. Varsha. DNA fingerprinting in the criminal justice system: an overview. DNA Cell Biol 2006; 25:181–8.
9. McClelland M, Welsh J. DNA fingerprinting by arbitrarily primed PCR. PCR Methods Appl 1994; 4:S59–65.
10. Saiki RK, Gelfand DH, Stoffel S, et al. Primer-directed enzymatic amplification of DNA with a thermostable DNA polymerase. Science 1988; 239:487–91.
11. Hagen-Mann K, Mann W. RT-PCR and alternative methods to PCR for in vitro amplification of nucleic acids. Exp Clin Endocrinol Diabetes 1995; 103:150–5.
12. Mullis K, Faloona F, Scharf S, et al. Specific enzymatic amplification of DNA in vitro: the polymerase chain reaction 1986. Biotechnology 1992; 24:17–27.
13. Mullis KB. The unusual origin of the polymerase chain reaction. Sci Am 1990; 262:56–61, 64–5.
14. Arya M, Shergill IS, Williamson M, et al. Basic principles of real-time quantitative PCR. Expert Rev Mol Diagn 2005; 5: 209–19.
15. Rooney PH. Multiplex quantitative real-time PCR of laser microdissected tissue. Methods Mol Biol 2005; 293:27–37.
16. Bustin SA, Benes V, Nolan T, et al. Quantitative real-time RT-PCR – a perspective. J Mol Endocrinol 2005; 34:597–601.
17. Wong ML, Medrano JF. Real-time PCR for mRNA quantitation. Biotechniques 2005; 39:75–85.
18. Valasek MA, Repa JJ. The power of real-time PCR. Adv Physiol Educ 2005; 29:151–9.
19. Kaltenboeck B, Wang C. Advances in real-time PCR: application to clinical laboratory diagnostics. Adv Clin Chem 2005; 40: 219–59.
20. Pourzand C, Cerutti P. Genotypic mutation analysis by RFLP/PCR. Mutat Res 1993; 288:113–21.
21. Woodward SR. RFLP analysis in familial polyposis and Gardner syndrome. Prog Clin Biol Res 1988; 279:305–8.
22. Hammarstrom L, Ghanem N, Smith CI, et al. RFLP of human immunoglobulin genes. Exp Clin Immunogenet 1990; 7:7–19.
23. Permutt MA, Elbein SC. Insulin gene in diabetes. Analysis through RFLP. Diabetes Care 1990; 13:364–74.
24. Hayashi K. PCR-SSCP: a simple and sensitive method for detection of mutations in the genomic DNA. PCR Methods Appl 1991; 1:34–8.
25. Hayashi K. PCR-SSCP: a method for detection of mutations. Genet Anal Tech Appl 1992; 9:73–9.
26. Hayashi K, Yandell DW. How sensitive is PCR-SSCP? Hum Mutat 1993; 2:338–46.

27. Fan E, Levin DB, Glickman BW, et al. Limitations in the use of SSCP analysis. Mutat Res 1993; 288:85–92.
28. Dillon D, Zheng K, Negin B, et al. Detection of Ki-ras and p53 mutations by laser capture microdissection/PCR/SSCP. Methods Mol Biol 2005; 293:57–67.
29. Fodde R, Losekoot M. Mutation detection by denaturing gradient gel electrophoresis (DGGE). Hum Mutat 1994; 3:83–94.
30. Glavac D, Dean M. Applications of heteroduplex analysis for mutation detection in disease genes. Hum Mutat 1995; 6: 281–7.
31. Nataraj AJ, Olivos-Glander I, Kusukawa N, et al. Single-strand conformation polymorphism and heteroduplex analysis for gel-based mutation detection. Electrophoresis 1999; 20: 1177–85.
32. Ellis TP, Humphrey KE, Smith MJ, et al. Chemical cleavage of mismatch: a new look at an established method. Hum Mutat 1998; 11:345–53.
33. Den Dunnen JT, Van Ommen GJ. The protein truncation test: A review. Hum Mutat 1999; 14:95–102.
34. Hauss O, Muller O. The protein truncation test in mutation detection and molecular diagnosis. Methods Mol Biol 2007; 375: 151–64.
35. Derks S, Lentjes MH, Hellebrekers DM, et al. Methylation-specific PCR unraveled. Cell Oncol 2004; 26:291–9.
36. Galm O, Herman JG. Methylation-specific polymerase chain reaction. Methods Mol Med 2005; 113:279–91.
37. Herman JG, Graff JR, Myohanen S, et al. Methylation-specific PCR: a novel PCR assay for methylation status of CpG islands. Proc Natl Acad Sci U S A 1996; 93:9821–6.
38. Taylor KH, Kramer RS, Davis JW, et al. Ultradeep bisulfite sequencing analysis of DNA methylation patterns in multiple gene promoters by 454 sequencing. Cancer Res 2007; 67:8511–8.
39. Warnecke PM, Stirzaker C, Song J, et al. Identification and resolution of artifacts in bisulfite sequencing. Methods 2002; 27:101–7.
40. Shaffer LG, Bejjani BA. Medical applications of array CGH and the transformation of clinical cytogenetics. Cytogenet Genome Res 2006; 115:303–9.
41. Gebhart E. Comparative genomic hybridization (CGH): ten years of substantial progress in human solid tumor molecular cytogenetics. Cytogenet Genome Res 2004; 104:352–8.
42. Davies JJ, Wilson IM, Lam WL. Array CGH technologies and their applications to cancer genomes. Chromosome Res 2005; 13: 237–48.
43. Kallioniemi OP, Kallioniemi A, Sudar D, et al. Comparative genomic hybridization: a rapid new method for detecting and mapping DNA amplification in tumors. Semin Cancer Biol 1993; 4: 41–6.
44. Trask BJ. DNA sequence localization in metaphase and interphase cells by fluorescence in situ hybridization. Methods Cell Biol 1991; 35:3–35.
45. Tibiletti MG. Interphase FISH as a new tool in tumor pathology. Cytogenet Genome Res 2007; 118:229–36.

Chapter 9

Analyzing Gene Expression

Holly A. LaVoie and Pedro J. Chedrese

9.1 Introduction

Changes in activity of regulated genes correlate with the synthesis of the encoded proteins directly affecting the physiological status of the individual. In most of the cases, protein synthesis is directly proportional to the transcriptional activity of its gene, which affects the steady state levels of mRNA encoding the peptide. Thus, measuring mRNA is the obligated first step in investigating how expression of a gene can be regulated.

9.2 Measuring Steady-State Levels of mRNA

Steady-state mRNA levels represent the difference between the amount of mRNA synthesized and the amount degraded. Steady-state levels of mRNA can be measured by a variety of methods, most commonly by Northern blot analysis. More advanced methods such as the nuclease protection assays and the quantitative reverse transcriptase polymerase chain reaction (RT-PCR) are available. Each procedure has its own advantages and disadvantages and no one is necessary the best. Some of the methods may answer a specific question that cannot be addressed by other methods. All the techniques have common basic requirements, such as isolation of RNA from the tissue, internal controls for efficacy of the procedure, and control for specificity of the response.

Most of the methods of mRNA detection and quantification require isolation of total RNA from the tissue under investigation. mRNA is a single-stranded and fragile molecule that is easily degraded by RNAses, which are very abundant in most of the tissues and are a common contaminant in most laboratories. Therefore, the first requirement for isolation of RNA is the immediate inhibition of the tissue RNAse activity. Tissue disruption can be performed using a solution containing guanidinium isothiocyanate (GTC) and 2-mercaptoethanol, which are potent inhibitors of RNAses. Then the disrupted tissue is layered on a solution of CsCl and ultracentrifugated. RNA is the densest molecule within the cell. Therefore, the passage over a cesium chloride cushion efficiently separates rRNA and mRNA from all other cellular components such as DNA, proteins, and lipids. This is an effective method, but requires an ultracentrifuge, that limits the number of samples to be processed. RNA can also be isolated by disrupting tissues with GTC-acidic phenol solutions and chloroform extraction followed by alcohol precipitation from the aqueous phase. This method is very rapid and can be adapted for the processing of large number of samples, but is more likely to have contaminating DNA.

mRNA composes only 2–5% of total tissue RNA. Therefore, detection of very rare message requires an enrichment step, in which mRNA can be separated from total RNA. The adenylated tails (poly-A) present on the 3′ end of the mammalian mRNA facilitates this procedure. mRNA can be isolated from total RNA using a column chromatography of immobilized cellulose beads, containing a short oligonucleotide chain of deoxythimidines (oligo-dT) cellulose. The poly-A tails of the mRNA remain bound to the oligo-dT cellulose, while the rest of the RNA passes through the column (Fig. 9.1). Then, mRNA is eluted from the column with a high salt buffer solution, from which it is precipitated with ethanol. Likewise, oligo-dT coupled magnetic beads are also commonly used in place of columns.

A variety of probes can be used for RNA analysis, including complementary DNA (cDNA), oligonucleotides, and complementary RNA (cRNA). Each type of probe has its own advantages and disadvantages.

cDNA probes—double-stranded cDNA can be easily labeled using short random oligo-nucleotides as primers; in a polymerization reaction that includes the DNA to be labeled, at least one radioactive dNTP, and DNA polymerase, usually the Klenow fragment, a section of the molecule devoid of 5′- to 3′-exonuclease activity. This procedure results in

H.A. LaVoie (✉)
Department of Cell Biology and Anatomy, University of South Carolina School of Medicine, Columbia, SC, USA
e-mail: HLAVOIE@gw.med.sc.edu

P.J. Chedrese (ed.), *Reproductive Endocrinology*,
DOI 10.1007/978-0-387-88186-7_9, © Springer Science+Business Media, LLC 2009

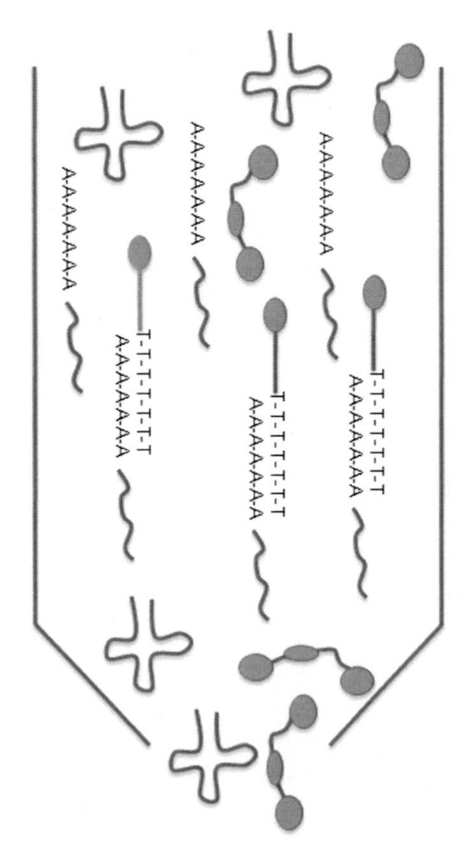

Fig. 9.1 Isolation of mRNA by oligo-dT cellulose column chromatography Samples, diluted in high salt solution buffers, are loaded onto the column. tRNA and rRNA are washed out with a medium salt solution buffer. The mRNA retained by the oligo-dT cellulose beads can be eluted passing a no salt solution buffer through the column

a population of fairly long DNA molecules labeled in both strands, with a very high specific activity that will hybridize to identical complementary DNA sequences or complementary molecules of RNA. Random primer-labeled cDNA probes are useful for mRNA variant detection as well as for detecting rare species of mRNA.

However, since both strands of the DNA are labeled and the labeling procedure results in a heterogeneous population of molecules rather than discrete species, random primed DNA is used only for Northern or slot blots (described below). In addition, the probe must be used at a low concentration (generally less than 10 ng/mL), in the hybridization solution, because of complementary strand reannealing. The use of a cDNA probe from the same species provides a specific tool of mRNA detection. However, sometimes only a heterologous probe is available, which may also be used under less stringent hybridization and washing conditions to allow for some mismatching of bases.

Oligonucleotides probes—if the gene of interest has not been cloned, but a small portion of the RNA or amino acid sequence of the protein is known, oligonucleotides may easily be synthesized for use as a hybridization probe. To design oligonucleotide probes based on the sequence of a protein, we have to take into consideration that some amino acids, such as leucine, are encoded by up to six different codons. Thus, a section of the protein containing amino acid with less redundancy in their codons is preferable. In addition, a combination of various different oligonucleotides with the different possible codons can be used and are referred to as degenerate oligonucleotides.

If the only information available is a DNA sequence from heterologous species, the oligonucleotide should be designed based on the portion of the cDNA encoding a region of the protein with the highest level of homology with the closest species. The use of an oligonucleotide probe is also preferable when a precise sequence within the mRNA is of interest, or to map or distinguish between mRNA isoforms.

Oligonucleotides can be end-labeled using polynucleotide kinase and ATP with ^{32}P in its γ-phosphate group. Under these conditions, the enzyme transfers the ^{32}P to the 5′-hydroxyl group of the oligonucleotide, resulting in a probe labeled at a single base. These probes are of low specific activity, compared with those of cDNA.

cRNA probes—single stranded labeled cRNA probes can be synthesized by in vitro transcription, for which we need to engineer an expression vector, in which the cDNA is inserted next to a specific viral promoter. This vector is first linearized with a restriction nuclease and then incubated with a vector-promoter specific RNA polymerase, in the presence of at least one labeled ribonucleotide, preferably UTP. This method has several advantages, including the following.

- Either a sense or antisense cRNA can be transcribed by cloning the cDNA in different orientation, or by using a different promoter. Commercial expression plasmid vectors are designed with different promoter sites for bacteriophage RNA polymerases such as T7, T3, or SP6 located at each side of the multiple cloning site.
- The cRNA is transcribed with great fidelity and from a defined transcriptional start site.
- The resulting cRNA probe is of high specific activity, as it is internally labeled and can be as long as the inserted cDNA.
- Alternatively, portions of the cDNA can be subcloned into the construct to provide probed targeting precise areas of the mRNA.
- RNA–RNA hybrids have a higher affinity that DNA–RNA hybrids, so hybridization with cRNA probes can be performed at very high temperatures and in formamide-containing buffers, resulting in very specific hybridization signals.
- cRNA probes can be used to make Northern and slot blot analysis more sensitive and suitable to detect rare mRNAs.
- cRNA probes are used in RNase protection assays.

- An alternative to inserting a partial cDNA into a dual-promoter vector is to add the viral polymerase sites to either side of a cDNA by PCR using primers containing the viral promoter sites at their 5′-ends.

The main drawback of this technique is that the use of cRNA makes very difficult to strip the radioactivity from the membranes, which may interfere with subsequent reprobing.

9.2.1 Northern Blot Analysis

The procedure involves electrophoresis of total RNA or poly-A RNA, along with molecular weight markers, to separate molecules by size. RNA samples are dissolved in buffers containing formamide and electrophoresis is conducted in a gel containing formaldehyde. Formaldehyde and formamide are strong denaturing agents that disrupt the secondary structure of the RNA. Thus, linearized RNA molecules can be separated and visualized by staining with fluorescent dyes such as ethidium bromide and the cyanine dye SYBR green. This procedure provides visual information on the length of migration of two distinct rRNA bands, of 28S and 18S, of approximately 5 and 2 kb, respectively (Fig. 9.2), and on their integrity. It also provides information on the amount of RNA applied to the gel. From the gel, RNA is transferred onto a solid matrix, usually nitrocellulose or nylon membrane, by capillary diffusion or electroblotting, and immobilized by cross-linking under ultraviolet light radiation. Membranes are then hybridized with specific labeled nucleic acid probes. After hybridization, the membrane is washed and the hybridized bands can be visualized by auto-radiography (Fig. 9.2).

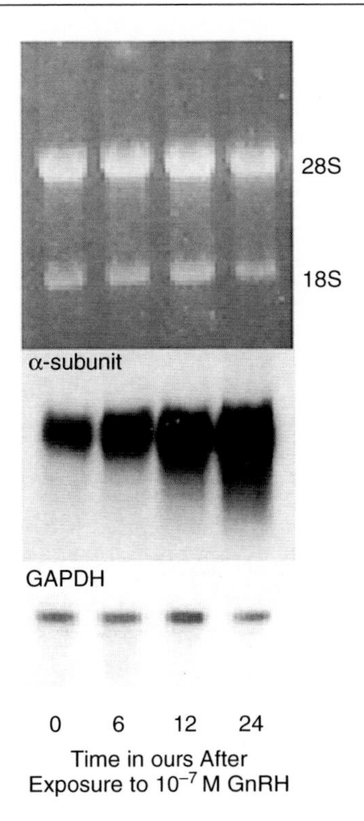

Fig. 9.2 Northern blot analysis of pituitary gonadotrope cultured cells exposed to GnRH. *Top panel*: photograph of a total RNA agarose gel following electrophoresis and stained with ethidium bromide. *Middle panel*: autoradiogram of a Northern blot membrane showing the effect of GnRH on the α-subunit gonadotropin mRNA. *Bottom panel*: autoradiogram of the same membrane hybridized with the glyceraldehyde-3-phosphate-dehydrogenase (GAPDH) cDNA probe

Northern blot analysis is usually the first test performed after characterization of a newly cloned cDNA. It provides information on the entire molecule of mRNA, the number of molecular species transcribed, and their molecular size. To estimate molecular weight, there are available commercial preparations of RNA markers. The main advantage of the Northern analysis is its relative simplicity and that allows the use of a wide variety of molecular probes such as cDNA, cRNA, or oligonucleotides (described below). An additional advantage is that by altering the hybridization stringency, Northern membranes can be probed with heterologous probes, providing information even in the case that a species-specific probe is not available. Nylon membranes are very resistant to many hybridizations and washing. Thus, the same membrane can be re-hybridized with another probe. To determine if the changes in the levels of mRNA are specific, Northern blot membranes can be re-hybridized with a structural or "housekeeping" gene probe, such as glyceraldehyde-3-phosphate-dehydrogenase (GAPDH), shown in Fig. 9.2. In this example, the intensity of the bands hybridizing to the GAPDH probe is the same across treatment groups.

The main limitation of Northern analysis is its low sensitivity. Quantification of Northern blots, although of low accuracy, is a simple procedure using computerized densitometric analysis of film or phosphorimaging autoradiograms. The bands appearing on the autoradiogram are quantified by scanning densitometry, resulting in measured arbitrary densitometric units. Therefore, it is common practice to normalize results by expressing the ratio between the mRNA of interest and the mRNA of a structural gene. This procedure gives only relative values, rather than an accurate number of mRNA molecules. Better estimation of the number of mRNA molecules can be achieved by solution hybridization or quantitative RT-PCR. Finally, probes can hybridize to more than one band in the Northern blot. If the hybridization is conducted at high stringency, the presence of more than one band could suggest the presence of more than one transcript. Several genes encode for multiple transcripts, including the FSH and LH receptors genes.

9.2.2 Slot/Dot Plot Analysis

The slot/dot blotting, or direct blotting, was developed with the idea of designing a better method to detect and quantify mRNA. The methods is based on the Northern blot analysis; but instead of separating the mRNA by size prior to transfer, the RNA is denatured in solution and applied directly to a supporting membrane. For this purpose, a wide variety of simple instruments (manifolds) have been developed that hold the membrane and allow direct application of samples in the form of rectangles (slots) or circles (dots).

The procedure is easy to perform and the apparatus allows the analysis of multiple samples at a time. Several dilutions of a known mass of RNA can be directly blotted into a single membrane, which is then hybridized with any of the previously described probes. The direct blotting technique does not provide information on RNA integrity, specificity of the probe, or number and sizes of hybridizing species. Therefore, Northern blot analysis must be performed first to test specificity of the probe and hybridization (stringency) conditions before the use of direct blotting.

Once these conditions are determined, direct blotting becomes an effective procedure to detect mRNA. The slots provide a discrete signal in the autoradiogram, which facilitates scanning. Quantification can be performed by analyzing several dilutions of the RNA to determine linearity of the response. Standards of mRNA for quantification purposes can be synthesized using transcription vectors and viral RNA polymerases that permit the synthesis of unlabeled sense mRNA. This synthetic mRNA can be used to construct standard curves for the calculation of specific mass.

Specificity of the response can be tested by hybridization of housekeeping genes as was described for Northern blots. In addition, since the slots are located in defined positions, a more precise estimation of the response can be obtained by excising the slots out of the membrane and quantifying them by scintillation counting.

9.2.3 Solution Hybridization Assays

Solution hybridization assays are very sensitive methods for analyzing RNA. They provide information on the RNA structure and its abundance. There are two main methods based on solution hybridization: nuclease protection assay and RNAse protection assay. The two methods differ in the type of probe and the type of enzyme used for digestion, but the principles are the same. Both nuclease and RNase protection assays are performed by hybridizing the RNA in solution to a radioactive, single-stranded DNA or RNA probe. After hybridization, a specific nuclease is added to the solution, which digests the non-hybridized probe and the unprotected sections of the hybrid. The reaction is then resolved by denaturing polyacrylamide gel electrophoresis and exposed to X-ray film or phosphorimager to generate an autoradiogram.

The main advantages of these assays are high sensitivity and the possibility of obtaining structural information on the 5′ or 3′ ends of the RNA, or the detection and quantification of splice variants, which is commonly used for RNA mapping.

S1 nuclease protection assay—this method requires the use of a single-stranded DNA probe, which can be obtained using an M13 viral template internally labeled, or an end-labeled chemically synthesized oligonucleotide. The M13 system can generate high activity cDNA probes, but sometimes they are difficult to purify. The synthesized oligonucleotides, which are only labeled on one residue, result in lower sensitivity of detection.

After hybridization, the reaction is treated with a single-strand specific nuclease S1. This enzyme digests single-stranded DNA and RNA and the non-homologous sections of the probe, leaving intact all the protected, hybridized double-stranded molecules. The hybridized molecules can be resolved by denaturing polyacrylamide gel electrophoresis that separates DNA from RNA and analyzed by autoradiography. A single band in the autoradiogram will represent the amount of DNA fragment remaining. Thus, the intensity of the band will be directly proportional to the amount of mRNA in the sample and can be quantified by densitometric scanning. Standard curves for quantification can be prepared by hybridization with unlabeled sense mRNA synthesized by in vitro transcription.

RNAse protection assay—the RNase protection assay is based on the same principle as the S1 nuclease assay. The difference lies in the use of an internally labeled cRNA as a probe and RNAses instead of the S1 nuclease. Labeled cRNA probes can be prepared as it was described above. RNase A, T1, T7, or a combination of these enzymes is used for digestion of the unhybridized probe. The rest of the procedure is the same. Samples are resolved by denaturing gel electrophoresis and the intensity of the bands are quantified by autoradiography.

9.3 Primer Extension Analysis

Primer extension analysis is a technique aimed at mapping the 5′ end of mRNAs and to look for alternate transcriptional start sites in the gene. It can be used also for quantitative purposes. Primer extension requires the use of an end-labeled DNA oligonucleotide complementary to the section of mRNA to be analyzed. This probe is usually designed to bind within 100 bases of the mRNA 5′ end. In solution, this oligonucleotide is annealed to the mRNA and serves as primer for the enzyme reverse transcriptase. The reverse transcriptase, in the presence of the four dNTPs, forms an RNA/DNA hybrid by extending the DNA to the end of the molecule, using the mRNA as a template. The hybrid is then resolved by denaturing polyacrylamide gel electrophoresis, run along with DNA molecular weight standards or in parallel with a DNA sequencing reaction, and analyzed by autoradiography. The newly synthesized DNA can then be amplified by PCR and sequenced, from which the entire structure of the mRNA molecule can be deduced (mapped). This procedure provides also information on alternatively spliced mRNA. We can determine the structure of the 5′ end of more than one transcript simultaneously when more than one mRNA is transcribed from the same gene, using alternative start sites.

9.4 Polymerase Chain Reaction Amplification of mRNA

The most sensitive technique available to detect and quantify mRNA is RT-PCR. This method is based on amplification of a DNA substrate using a thermal cycler. The procedure can be summarized as follows.

- Total RNA is annealed to a short oligonucleotide composed of multiple dTTP (oligo-dT), which acts as a primer for a viral reverse transcriptase. Under these conditions, only mRNA or RNA containing long regions of A residues are reverse transcribed. However, it is often more efficient to utilize random hexamers as primers for the reaction, which will result in reverse transcription of non-messenger RNAs as well. Alternatively, a sequence specific primer can be used for cDNA synthesis.
- The cDNA is then used in a PCR reaction with sequence-specific primers. The PCR reaction consists of multiple cycles of denaturation, annealing, and extension steps.
- The first PCR step of each cycle is denaturation of the DNA–RNA or DNA–DNA duplexes (94–96°C).
- The next step is the annealing of short, typically 18- to 24-base oligonucleotide primers complementary to each end of the DNA molecule targeted for amplification. The annealing temperature is defined by GC content (40–60%) and the length of the primer. The two primers should have similar melting temperatures. Annealing usually occurs at 50–65°C for 15–30 s.
- The extension step typically occurs at 72°C for *Taq* polymerase. In the presence of the four dNTPs the polymerase synthesizes a new DNA strand starting at the 3′-end of the annealed primer. The extension step is usually 1 min per kilobase amplified.

The rest of the procedure consists of repetition of the denaturation, annealing, and extension steps for 25–35 cycles. A final cycle extension step of 3–10 min at 72°C is usually performed to ensure ends are complete. Amplified DNA can be visualized by electrophoresis using agarose or acrylamide gels stained with ethidium bromide. The main advantage of the RT-PCR is its sensitivity. A particular mRNA can be detected in very small amount of tissue or even from few cells.

For semi-quantitative purposes a labeled oligonucleotide precursor can be incorporated into the reaction, which is terminated during the linear phase of amplification. The amplified DNA is quantified by autoradiography of the gel. The rationale for the quantification is that the amount of DNA amplified will be directly related to the amount of mRNA originally reverse transcribed. Unfortunately this is not always the case, as the amount of product amplified is influenced by a number of variables, including the following:

- efficiency of the reverse transcriptase reaction;
- quantity and concentration of the reagents in the PCR reaction;
- length of DNA to be amplified;
- exact conditions and efficiency of denaturation, primer annealing, and extension steps; and
- rate of temperature change between steps.

Adding an internal control DNA with a mutated internal restriction site or an internal intron circumvents some of these limitations. By using this approach, the primers that amplify the targeted sequence also amplify the internal control. Because both molecules are of different size, they can be distinguished from one another in the autoradiogram. The internal control DNA is added to the amplification mixture in serial dilutions after reverse transcription. The same principle can be applied to measure more than one mRNA simultaneously. Primers can be designed to different mRNAs that result in different cDNAs. Thus, the intensity of the signal of different bands can be used for quantitative analysis.

A more recent yet widely used approach to quantify mRNA is real-time PCR. This technique is based on the same principles of RT-PCR, but the reaction includes fluorescent reporters that allow quantification of the amplified product (amplicon) [1]. Thermal cyclers for real-time PCR include optics for measuring fluorescence emission during each cycle. The fluorescent signal increases in direct proportion to the amount of amplicon. During PCR, a plot of fluorescence versus cycle number yields a sigmoidal curve. The point in the curve where fluorescence begins to increase rapidly can be used to calculate the threshold cycle (Ct) value (Fig. 9.3). Samples with lower Ct values will have higher starting amounts of the cDNA of interest and samples with higher Ct values will have lower starting amounts of the cDNA.

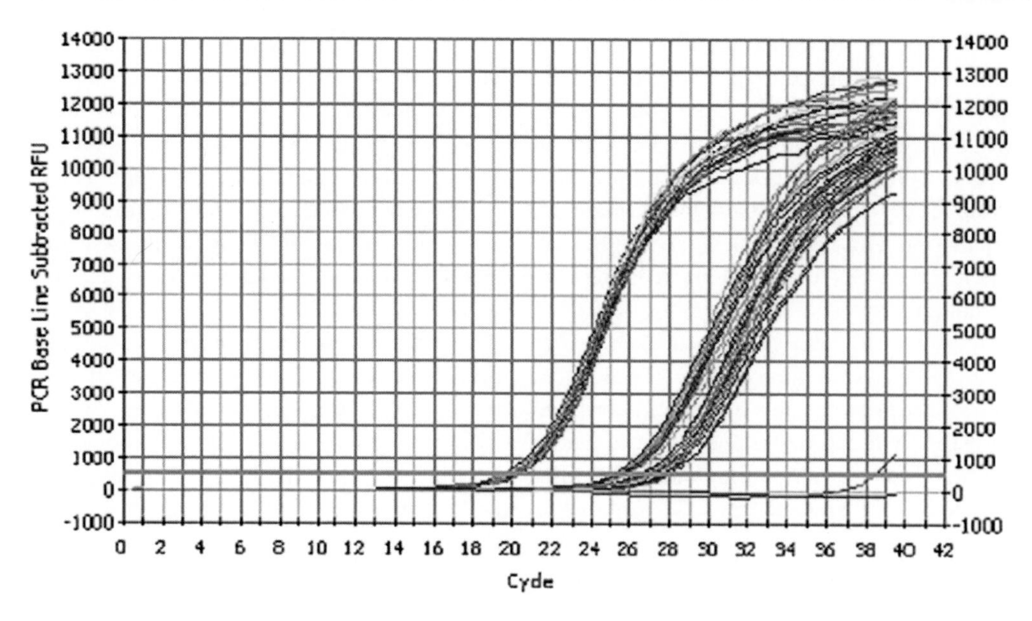

Fig. 9.3 Real-time PCR data. *Curves* represent the relative fluorescence measured in each well during each cycle. The thick *horizontal line* is the threshold set by the user and used to derive the Ct value for each sample well

Several fluorescent reporters are commonly used in real-time PCR reactions including SYBR green, Taqman probes, and molecular beacons [1]:

- SYBR green is a dye that binds to only double-stranded DNA, so its fluorescence increases as the amount of amplicon increases. Steps must be taken to ensure amplification of a single amplicon because the dye will incorporate into any double-stranded DNA product including primer-dimers;
- Taqman probes take advantage of fluorescence resonance energy transfer (FRET). These probes are oligonucleotides that bind an internal region of the amplicon amplified. The oligonucleotide probes have a fluorescent dye on one end and a quenching dye on the other blocking fluorescence emission. The probe is hydrolyzed during the amplification process and fluorescence increases proportional to the amount of the amplicon; and
- Molecular beacon probes take advantage of FRET with a fluorescent dye on one end and a quenching dye on the other end. In their unbound state they assume a hairpin structure that keeps the dyes in close proximity to each other blocking fluorescence emission. Molecular beacons hybridize to the amplicons in each cycle, which separates the dye ends and allows fluorescence emission. The total amount of fluorescence is proportional to the amount of amplicon.

The advantage to Taqman and molecular beacons is that they can be used for multiplexing, amplifying different products in the same well at the same time using different fluorophores.

Quantification of real-time PCR data typically utilizes two methods [2,3]. There is the standard curve method where DNA or cDNA template containing the sequence to be amplified and of known concentration is serially diluted and amplified in parallel with samples. Using the Ct value of the sample, the concentration of the amplicon can be extrapolated from the standard curve. A second method is based on the differences in Ct values between the target gene of interest and an invariant housekeeping or reference gene for both the control and the experimental samples. For example,

$$\Delta\Delta \text{Ct} = [\text{Control } \Delta\text{Ct (target} - \text{reference)}] - [\text{Experimental } \Delta\text{Ct (target} - \text{reference)}].$$

The fold change is $2^{\Delta\Delta \text{Ct}}$.

The above equation assumes primer efficiencies are 100%, which is usually not the case. A modification of this second method which incorporates amplification efficiencies for a primer set into the equation is the method of Pfaffl. Fold changes are derived from the equation:

$$\frac{(\text{E target})^{\Delta\text{Ct target (Control}-\text{Experimental)}}}{(\text{E reference})^{\Delta\text{Ct reference (Control}-\text{Experimental)}}},$$

E represents the PCR efficiency for a primer set. Primer efficiencies can be determined by analyzing the Ct values of a wide range of tenfold serial dilutions and the equation $E = 10^{[-1/\text{slope}]}$.

An advantage to real-time PCR for quantification of mRNA changes over conventional PCR is that after the initial optimization for a primer set, there is no post-reaction processing such as the running of gels or autoradiography.

9.5 In Situ Hybridization

The techniques described above are aimed at analyzing mRNA in tissue homogenates and the information obtained represents levels of expression of all the cells that compose the tissue. This approach may represent a bias in the information obtained, since a heterogeneous tissue may be composed of different cell types that may not express the gene of interest, or are not regulated by the mechanisms being investigated. Exceptions to this generalization are the studies conducted in cell culture, in which a uniform preparation of cells can be maintained for analyzing gene expression.

Thus in situ hybridization techniques were developed and are directed to determine the cells within the tissue that are expressing a particular mRNA and/or if changes in gene expression are occurring in the same fashion in all cell types of a defined tissue. Moreover, by in situ hybridization it is possible to identify specific mRNAs of the same cell type that change its pattern of expression during differentiation and/or at different stages of development.

The procedure includes the following steps:

- tissues frozen and/ or fixed in paraformaldehyde-containing solutions are sliced in thin sections and placed on glass slides;
- the sections of the tissues are used as support to perform hybridization with labeled probes, in a procedure analogous to hybridization on a solid membrane supports;
- after hybridization, the tissue sections are subjected to washing procedures to remove non-specifically bound probes;
- the labeled probes are bound to mRNA within the specific cells in which it was synthesized;
- the slide is covered with a photographic emulsion and subjected to autoradiography;
- the developed autoradiogram reveals patterns of grain images localized on the cells synthesizing the mRNA being investigated.

The probes of choice for in situ hybridization are oligonucleotides or cRNAs labeled with ^{35}S, a low-energy radioisotope that generates a highly localized image. Labeling with other isotopes, fluorescent probes, or colorimetric techniques, such as digoxigenin labeling, are also used. Colocalization studies are conducted with digoxigenin labeling coupled with radiolabeled probes or using probes labeled with different fluorophores.

Quantification of in situ hybridization images is more difficult than in the other techniques described for mRNA analysis. Some approaches include densitometric quantitfication of the gray level variations, which are of directly related to the amount of light transmitted through the autoradiogram image.

9.6 Final Considerations

In the investigation on regulation of gene expression it is generally assumed that extracellular signals affect the steady-state levels of a specific mRNAs, while have no effect on total RNA levels or all mRNAs. However, it must be taken into consideration that immediately after transcription, the first transcript, also called heterogeneous nuclear RNA (hnRNA), is further processed into mature mRNA within the nucleus. This process includes splicing of the introns and joining of the exons at defined base sequences, and by alternative splicing of the hnRNA, a different product of the same gene can be translated into a protein. Thus, splicing represent a second level of regulation of gene expression. In addition, mRNA is a very labile molecule and its stability can be hormonally regulated, for which steady state levels are not a faithful representation of the transcriptional activity of a particular gene and only represent a gross estimation of gene activity at a particular point in time. Finally, after translation, complex proteins suffer modifications that affect their potency, half-life in circulation, and therefore, its biological activity. Therefore, proper assessment of gene expression requires analyzing its activity at the following levels: synthesis, accumulation, and half-life of mRNA; posttranscriptional modifications of mRNA, including splicing, capping, and polyadenylation; translation and protein synthesis; post-translational modifications of the peptides; biological activity of the proteins; and phenotypical and physiological effects. Overall, the biological activity of a protein and its physiological effects are the ultimate manifestation of expression of gene activity.

9.7 Glossary of Terms and Acronyms

ATP: adenosine triphosphate

cDNA: complementary DNA

cRNA: complementary RNA

CsCL: cesium chloride

Ct: threshold cycle

DNA: deoxyribonucleic acid

dNTP: deoxynucleotide triphosphate

FRET: fluorescence resonance energy transfer

GAPDH: glyceral-dehyde-3-phosphate-dehydrogenase

GTC: guanidinium isothiocyanate

hnRNA: heterogeneous nuclear RNA

mRNA: messenger RNA

oligo-dT: oligonucleotide chain of deoxythimidines

PCR: polymerase chain reaction

poly-A: poly-adenylated

RNA: ribonucleic acid

rRNA: ribosomal RNA

RT-PCR: reverse transcriptase polymerase chain reaction

UTP: uridine triphosphate

Bibliography

1. Ausubel F, Brent R, Kingston RE, et al. Short Protocols in Molecular Biology, third edition. New York: Wiley & Sons, 1995.
2. Davis L, Kuehl M, Battey J. Basic Methods in Molecular Biology, second edition. Norwalk: Appleton & Lange, 1997.
3. Farrell RE Jr. RNA Methodologies: A Laboratory Guide for Isolation and Characterization. New York: Academic Press Inc, 1993.
4. Mellon SH, Miller WL. RNA Quantitation and Interpretation. International Symposium on Molecular Biology for Endocrinologists. Washington: Serono Symposia USA, 1996: 53–60.
5. Shupnik MA. Measurement of Gene Transcription and Messenger RNA. In: Weintraub BD, editor. Molecular Endocrinology: Basic Concepts and Clinical Correlations. New York: Raven Press Ltd, 1995: 41–58.
6. Weiss J. Measurement of mRNA. In: The Endocrine Society, editor. Introduction to Molecular and Cellular Research. Bethesda: The Endocrine Society, 1998: 43–9.

References

1. Real-time PCR goes prime time. TechNotes 8:1. Foster City, CA: Applied Biosystems, 2001.
2. SYBR® Green PCR and RT-PCR Reagents Protocol. Foster City, CA: Applied Biosystems, 2001.
3. Pfaffl MW. A new mathematical model for relative quantification in real-time RT-PCR. Nuc Acids Res 2001; 29:2002–7.

Chapter 10

DNA Microarray Analysis

Gheorghe T. Braileanu

10.1 Introduction

The methods described in Chapter 9 are designed to look at the expression of specific selected genes in relatively small numbers. To ascertain mRNA changes that occur on a much larger scale microarray analyses can be performed. A microarray, also known as biochip by analogy with the computer industry terminology, can be described as the sum of a multitude of similar miniaturized assays that take place simultaneously on a small surface. Microarrays types include those for DNA, tissue, protein, and bacteria. DNA microarrays include those covering the whole genome, intergenic regions, and the coding regions of genes. Within this chapter we will focus on DNA microarrays with emphasis on arrays for analyzing gene expression.

The main component of DNA microarrays for assessing gene expression is a set of DNA probes immobilized on a solid matrix, such as a glass slide or a nylon membrane. The solid matrix can accept thousands of probes that may represent the entire genome. Each probe is positioned at a precise coordinate on the solid matrix that permits identification. The interaction between each probe and the complimentary components of the sample after hybridization is visualized (based on the detection of a fluorescent dye, isotope, or biotin) and analyzed. The hybridization signal that is proportional with the amount of mRNA for each corresponding gene in the analyzed sample is compared between treatment and control groups. The results are expressed in relative fold differences. Housekeeping and reference genes are included on each microarray and can be used to normalize the results when comparing different samples. The analysis of the wealth of information on changes induced by a particular experimental condition on a microarray requires the use of extensive bioinformatics tools. Because microarrays analysis can yield false positive signals, the results must be confirmed by other techniques such as Northern blot or real-time PCR.

Each microarray experiment involves several steps including choice of the appropriate matrix, sample preparation, development of biochemical reactions, detection, and visualization of the results. The last step consists of the analysis and interpretation of the data. Many of these steps can be automated.

In the studies of gene expression the use of DNA microarrays offers the unique possibility of detecting changes that occur throughout the entire genome in response to a particular experimental condition in less than two days [1]; whereas, using traditional molecular biology tools, the expression of only one or two genes can be studied during the same amount of time. This striking difference will press toward a large use of this new technology in biology research, including reproduction.

10.2 Types of DNA Microarrays

DNA microarrays (*genome chip*, *DNA chip*, or *gene array)*, can be classified by their purpose, how they were produced, or the number of genes to be analyzed.

10.2.1 By Purpose

DNA microarrays can be used for gene expression profiling, genotyping, or resequencing [2].

Gene expression microarrays—detect differences in mRNA expression between samples for all genes represented on the matrix. They are typically oligonucleotides, specific for each gene, that are hybridized with labeled complementary DNA for all mRNAs in the sample (cRNA). The use of cDNA spotted microarrays was first reported in 1995 [3],

G.T. Braileanu (✉)
Department of Pediatrics, University of Maryland at Baltimore School of Medicine, Baltimore, MD, USA
e-mail: gbrai001@umaryland.edu

P.J. Chedrese (ed.), *Reproductive Endocrinology*,
DOI 10.1007/978-0-387-88186-7_10, © Springer Science+Business Media, LLC 2009

and the first complete eukaryotic genome, *Saccharomyces cerevisiae*, displayed on a microarray was published in 1997 [4]. Under defined physiological or pathological conditions microarrays allow expression analysis of the entire genome. For the interpretation of the results it must be taken in consideration that due to posttranslational modifications the relative amounts of detected mRNAs may or may not be translated proportionally into the corresponding active proteins.

Genotyping microarrays—in genotyping applications, DNA microarrays can be used for rapid identification or measurement of genetic predisposition to a particular disease, the identification of DNA-based drug candidates, or for forensic applications. Examination of genetic variation at specific loci is performed with single nucleotide polymorphism (SNP) microarrays. Polymorphism at the level of a single nucleotide is thought to be responsible for susceptibility to several diseases, such as sepsis, type I diabetes, Crohn's disease, or cancer [5,6]. A 6.0 platform produced by Affimetrix® (Affymetrix, Inc., Santa Clara, CA) with more than 700,000 probes scattered along the entire genome allows initial SNP polymorphism characterization and copy number detection. After detecting a region of interest, a more specific SNP microarray can be used to examine all possible nucleotide combinations found in a population for the corresponding DNA sequence. Genotyping microarrays are also used to profile somatic mutations in cancer, specifically loss of heterozygous events, amplifications, and deletions of different DNA regions. Copy number variation was demonstrated as an important parameter for cancer susceptibility.

Resequencing DNA microarrays—these microarrays can be used to *read* the sequence of a genome. They include genome tiling arrays where overlapping oligonucleotides are designed to cover an entire genomic region of interest. Many companies have successfully designed tiling arrays that entirely cover human chromosomes. These arrays are useful for examining genome variation between different genetic strains of mice and may be used to evaluate germline mutations or somatic mutations in cancer.

10.2.2 By Method of Production

DNA microarrays can be spotted or synthesized *in situ*. These two most used platforms for DNA microarrays production differ in how the probes are attached or synthesized on the substrate surface.

Spotted microarrays (two-channel or two-color microarrays)—in these microarrays, the preformed probes (oligonucleotides, or small fragments of PCR products corresponding to mRNA) are immobilized onto glass slides or membranes using fine-pointed pins. Usually, a specific 50- to 60-mer oligonucleotide corresponding to part of a gene is placed by an ink-jet printing analog process in a precise location on a silica substrate. The oligonucleotides can be also placed on substrate using piezoelectric deposition, a non-contact microdispensing system designed specifically for pipetting sub-nanoliter volumes. This type of array is typically hybridized with cDNA from the two samples to be compared (i.e., treatment and control) each labeled with a different fluorophore, such as Cy3 and Cy5. The scanning of the hybridized product signal intensity on the microarray with two corresponding wave length lasers allows the visualization of the results for each sample in which intensity for each fluorophore is quantified. Relative signal intensity is used to calculate and determine up-regulated and down-regulated gene expression in the experimental sample relative to the control. Generally, a difference greater than twofold it is considered significant. The downside of this approach is that absolute levels of gene expression cannot be observed.

Oligonucleotide synthesis microarrays or single-channel microarrays—contain short oligonucleotide probes of 25–30 nucleotides that are designed to match parts of the known or predicted mRNA sequence. They are produced *in situ* by photolithographic synthesis on a silica substrate using proprietary pre-made masks (Affymetrix, Inc.). They are single-channel microarrays that give estimations of the absolute value of gene expression and therefore the comparison of two conditions requires the use of two separate microarrays. There are commercially available designs that already cover the complete genome. The photo-mediated synthesis chemistry using a Maskless Array Synthesizer (Roche NimbleGen Systems, Madison, WI) allows the building of a large numbers of probes. These arrays can now contain up to 390,000 spots, arranged in a custom design.

10.2.3 By the Number of Genes that can Be Analyzed

DNA microarrays can be high-density microarrays or focused microarrays.

High-density microarrays—are identified by the high number of genes represented and can cover the entire genome. When searching for new genes involved in a biologic process it is advisable to use the high density, whole genome approach.

Focus microarrays—contain smaller numbers of genes and provide answers to specific questions such as the study of a particular cellular pathway, or a limited number of specific biomarker genes for rapid diagnostics.

10.3 Microarray Related Resources

The Electronic Library from TeleChem International, Inc. (http://arrayit.com—Tele Chem e-library) provides free-of-charge to the scientific community a virtual repository that contains citations, abstracts, summary information, and electronic links for papers, patents, and book chapters published on microarrays since the first publication appeared in *Science* magazine in 1995 [5]. The experimental design, the type of microarray, including the number of genes to be analyzed depends on the questions to be answered. Details of microarray technology can be found at different websites, such as www.bio.davidson.edu/courses/genomics/chip/chip.html. Free access depositories for DNA microarray results (like GEO—www.geo.gow) are being established, where each day a large amount of data are stored. The deposit of the extensive microarray results of each experiment in one existing free access repository is now a requirement for publication of microarray data by any major journal. A list databases for DNA microarrays can be found below under "Appendix of Public Databases for Microarray Data."

10.4 Personalized Medicine

Personalized medicine can be defined as the use of genomic information to improve the diagnosis, prevention, and treatment of diseases. Microarrays are considered by US Food and Drug Administration (FDA) as a key technology for advancing new pharmaceutical products and for personalized medicine [7]. In December 2004, the FDA cleared the AmpliChip Cytochrome P450 Genotyping Test (Roche Molecular Systems, Inc., Pleasanton, CA). Physicians can now order this test to gain information on whether the patient has mutations in genes involved in the metabolism of many types of drugs including antidepressants, antipsychotics, beta-blockers, and some chemotherapy drugs. FDA experts found that women with breast cancer who have a slow acting version of the enzyme gene SULT1A1 have lower survival rates on a typical post-surgery regimen of tamoxifen, and may need different doses of the drug [7].

Today, routine genetic tests are used to detect defects in the enzyme thiopurine methyltransferase, which prevents patients from metabolizing the anti-cancer drug 6-mercaptopurine. Based on this test, one patient may need a full dose, whereas another, who has a mutation in the gene, may need less than 10% of that dose. For children with leukemia, for example, getting the wrong dose can make the difference between life and death. Tumors have different genomic variations, and genomic tests may identify cancers that are likely to respond to a particular treatment [8]. Another aspect of pharmacogenomics is testing for drug resistance. For example, the HIV virus genome is always changing, and drug resistance screening will indicate the drug that is best suited for suppression of the virus at a given moment. The TRUGENE HIV-1 Genotyping Kit (Visible Genetics, Inc., Suwanee, GA), cleared by the FDA, detects genetic variations that make the HIV virus resistant to some anti-retroviral drugs. If drug resistance is discovered, another treatment option is considered [7].

The Pharmacogenetics Research Network of the National Institutes of Health is a nationwide collaboration of scientists committed to study the effects of genes in response to a variety of medicines, including antidepressants, chemotherapy, and drugs for asthma and heart disease. However, a patient's genetic information will need to be considered together with other variables, such as family history, medical history, clinical examination, and other non-genomic diagnostic tests when making treatment decisions [7]. More information about progresses in personalized medicine can be found at the following websites—Genomics at the FDA: www.fda.gov/cder/genomics/; Personalized Medicine Coalition: www.personalizedmedicinecoalition.org; the FDA's National Center for Toxicological Research: www.fda.gov/nctr/; the CDC's Office of Genomics and Disease Prevention: www.cdc.gov/genomics/; and NIH Pharmacogenetics Research Network: www.nigms.nih.gov/Initiatives/PGRN/.

10.5 The Use of DNA Microarrays in Reproductive Biology

The use of DNA microarrays technology in reproduction allows comprehensive study of genetic influences on the expression of complex quantitative traits and the results are more precise when this technique is used in concert with laser capture microdissection and linear cRNA amplification [9].

10.5.1 Female Reproduction

The understanding of all aspects of female reproduction such as oocyte fertilization, early embryo development, implantation, and infertility-related disease is beginning to be advanced by the results of microarray studies [9]. By using DNA microarray, pigs selected over the long-term for increased ovulation rate and embryo survival were found to have differential expression of 71 ovarian genes during the follicular phase of the estrous cycle. Many of these genes were not previously associated with reproduction [10, 11]. From these genes, 59 were homologous to genes of known function, 5 had no known matches in GenBank, and 7 were homologous to sequences of unknown function. Among the identified differentially expressed genes are those associated with the transport of cholesterol in the ovarian follicle and synthesis of steroids. Collagen type I receptor (CD36L1, also known as scavenger receptor class B type 1) over expression in higher fertility pigs' ovaries may indicate a greater role for high-density lipoproteins in steroidogenesis. The study also showed how important it is to study the expression of all these genes at different times of the estrous cycle. For instance the steroidogenic acute regulatory protein (StAR) and 3β-hydroxysteroid dehydrogenase (3β-HSD) mRNAs were over expressed in higher producing pig ovaries at day 2 of analysis, while being under expressed at day 3. These results identified other novel genes suggested mechanisms involved in the process of ovulatory follicle selection and maturation [10] and demonstrated changes in follicular gene expression as the result of long-term selection for enhanced reproduction [11]. In another study in rainbow trout 240 genes were observed to be regulated during maturation and development of the ovary [12]. SNP variations between the identified genes are currently being investigated to further characterize and select them for traits of interest. Oocyte-expressed genes are important for follicular growth and development, as well as for early embryogenesis. By microarray 11 genes were identified in bovine, potentially important for reproductive function, with consistently higher expression in fetal ovaries compared to spleen and liver [13].

Detailed knowledge of the changes that occur during the window of implantation is fundamental for the understanding of human reproduction. Microarray expression profiling studies revealed new candidate genes for human uterine receptivity [14] and identified new candidate markers that may be used in the near future to diagnose unequivocally the receptive endometrium [15]. Analysis of the window of implantation has revealed the genes involved in different processes in this short interval, such as angiogenesis, immune modulation, secretion of unique products, transport of ions and water, production of extracellular matrix proteins or unique cell surface glycoproteins, and a variety of transcription factors [14].

Microarrays also allow the detailed study of hormone action. Ninety-one genes were found up regulated and 68 genes down regulated for more than twofold after FSH treatment of MCV152 ovarian epithelial cells [16]. In another example, the LH surge initiates hormonal and structural changes of preovulatory follicle that lead to ovulation and the formation of corpora lutea. Microarray analyses of sow follicles detected a large number of genes whose expression was decreased (107) or increased (43) during the LH-mediated transition from the preovulatory estrogenic phenotype to luteinized phenotype. Approximately 40% of these described genes had unknown function. It was also shown that preovulatory estrogenic follicles had a gene-expression profile of proliferative and metabolically active cells, and preovulatory luteinized follicles had a gene-expression profile of non-proliferative and migratory cells with angiogenic properties [17].

Microarray analysis allowed the description of gene expression patterns and profiles in mouse [18] and human [19] developing placenta. Other microarray experiments have studied fertilization, early embryo development [20, 21, 22], implantation and infertility-related diseases, such as endometriosis and myoma (reviewed in [9]). The different stages of embryonic development in mice are also starting to be comprehensively characterized [23, 24, 25].

A complex approach that illustrates the actual tendency in research was applied to identify new global biological principles that govern molecular events underlying mammary development during pregnancy, lactation, and involution [26]. It included the use of bioinformatic techniques to analyze and correlate microarray data, cellular localization data, protein–protein interactions, gene ontology analysis, pathway analysis, and network analysis. It was shown, for instance, that nearly one third of the transcriptome fluctuates to build, run, and disassemble the lactation apparatus; around 100 transcripts account for the diversity of the milk proteome; the lactation switch is primarily post-transcriptionally mediated; and new regulatory gene targets were identified. It was shown that less than one fifth of the described transcriptionally regulated nodes of the lactation network were described previously. These results have immediate implications for future research in mammary gland and cancer biology [26]. This complex approach is anticipated to be used in the near future in almost all aspects of reproduction and will lead to new hypotheses and a high number of unexpected results.

10.5.2 Male Reproduction

Male infertility represents a good example of a complex trait with a substantial genetic base. Genetic causes account for at least 10–15% of severe cases of spermatogenic impairment [27]. The result of DNA microarray studies suggest that alternative mRNA splicing plays an important role in testis development and spermatogenesis [28]. More and more genes are identified as being important for spermatogenesis in mice [29] and in humans [30]. In mice the study of gene expression during different developmental stages of the testis with microarray identified the novel gene TSC77 located at the chromosome 2G1, which may play an important role during spermatogenesis [29]. Comparing gene expression in nine patients with normal spermatogenesis to 15 patients with maturation arrest, ten novel sterility-related genes were identified in human. Of the ten novel genes, six genes encode protein with predictable functional domains believed to correlate with spermatogenesis and four do not encompass known functional domains [30]. A time-course analysis of the testis development identified more than 290 unknown gene transcripts of which some were potentially essential for male fertility, since they displayed strong expression throughout meiotic and post meiotic steps of spermatogenesis [31]. In another study more than 100 genes were shown to be differentially expressed at different stages of testis development, from which 42 were new and probably important for male fertility [32]. In mice and rats gene expression during development of fetal and adult Leydig cells was studied by microarray. The expression profiles are distinct for these two categories of cells indicating that they may originate from separate pools of stem cells [33]. DNA microarray experiments regarding the profile of hormone-regulated genes in the testis should allow a better understanding of the factors influencing spermatogenesis [34]. If these new genes are found to predict male infertility, microarray technology may be used as a clinical diagnostic tool and might eventually lead to gene therapy [35].

10.5.3 Use of Microarrays for Discovery of New Biomarkers for Reproductive Traits

Another purpose of microarray experiments in reproduction is to identify biomarkers of reproductive performances or for toxicity [36]. A biomarker is a parameter whose level can be associated with or is indicative of a defined biological process such as development or a disease state. So far, using microarrays, a large panel of biomarkers were defined in reproductive development and health, starting with pubertal development [37], adult reproductive health (altered spermatogenesis in male or altered endocrine function in female), and finishing with pregnancy success and outcome. Biomarkers could be established for oocyte quality, embryo competence, and capacity to sustain a successful pregnancy. Microarray screenings of around 30,000 genes on U133P Affymetrix gene chips identified new potential regulators and marker genes, such as BARD1, RBL2, RBBP7, BUB3, or BUB1B, which are involved in oocyte maturation [38]. Recently it was found that fetal mRNA measurement in maternal plasma may be a useful tool for non-invasive prenatal placental gene expressing profiling [39].

10.5.4 Reproductive Diseases

The further development of gene array technology will bring forth the ability to classify disease states at the molecular level and elucidate unique subsets of genes associated with a specific disease [8, 40]. Disease-specific genomic analysis (DSGA) measures the extent to which the disease deviates from a continuous range of normal phenotypes and isolates the aberrant component of data [40]. For instance, there are strong arguments that support a genetic component of preterm birth in human, the leading cause of neonatal morbidity and mortality. Microarrays will lead to the eventual identification and characterization of the genetic etiology of this entity [21].

In 1996 the National Cancer Institute (NCI) initiated the Cancer Genome Anatomy Project (http:/www.ncbi.nlm.nih.gov/projects/CGAP/), an interdisciplinary program with the aim to generate information and techniques needed to decipher the molecular anatomy of the cancer cell. The use of DNA microarrays for different tumors is part of this effort. Cancer of the endometrium represents 13% of all cancers in women. Each year in the US 37,000 new cases are diagnosed with reported 6,000 deaths. Microarray research is identifying new molecular markers to differentiate between grade I and II endometrial tumors from grade III, and would provide new targets for therapy, such as genes involved in suppression of angiogenesis or metastasis [41].

Ovarian cancer is the fifth cause of death from cancer in American women. Most cases are detected at late stages when the survival rate is about 28% at 5 years. However, when the disease is detected early the 5-year survival rate is 95%. Microarray results will contribute to the understanding of the pathogenesis of ovarian carcinogenesis as well as to the identification of new tumor markers for early detection and guidance of the clinical management of the disease [42].

Microarray results indicate an important role of androgen receptors and suggested their use as a therapeutic target for androgen-insensitive prostate cancers [43]. Similar studies have described the patterns of normal prostate development that may be useful for understanding prostate cancer development [44].

An example of a preventive approach is when a genetic test predicts what diseases an individual is likely to develop. For instance, people who have certain mutations in the BRCA1 gene have a high risk of developing breast, ovarian, and possibly prostate, and colon cancers, according to the NCI. Alterations in the BRCA2 gene have been associated with breast, pancreatic, gallbladder, and stomach cancers [7].

10.6 Future Directions

Microarrays are not yet applied on a large scale in a clinical setting due to cost, the difficulty in obtaining high quality RNA, lack of comprehensive databases, potential posttranscriptional modifications, the relative lack of reproducibility in some experiments, and lack of easy analysis and interpretation. However, all these problems are presently being resolved at different paces. To use the growing amount of microarray results deposited each day in data banks, it necessity to be able to compare the results obtained between different microarray platforms and from different laboratories. The Microarray Control Consortium composed of 137 researchers belonging to 51 organizations at more than 40 test sites started to address this important aspect of microarray technology [45] by testing two samples of RNA on more than 20 array platforms including seven well-known platforms, such as Affymetrix, Agilent, Applied Biosystems, Eppendorf, GE Healthcare, Illumina, and NCI_Operon. So far, over 1,300 microarrays have been used in this study that had 12,091 genes in common resulting in the addition of 4.3 GB in public databases. The project was initiated in February 2005 and the first results were published in September 2006. It was reported that "The MAQC project observed intraplatform reproducibility across test sites as well as high interplatform concordance in terms of genes identified as differentially expressed." The committee started the second part of the project and in December 2007 released the results as "Guidance on microarray quality control and data analysis" that will be published in early 2009 [45]. The purpose of the MAQC project is to provide quality control tools to the microarray technology community to avoid procedural failures and to develop uniform guidelines for microarray data analysis. Overall, microarray technology is well on the way to becoming a common diagnostic tool that in the future will be as routine as blood tests. There are few current problems with microarray mainly due to its complexity that will make its common practice in medicine more complicated. However, taking into account the advantages, the incorporation of pharmacogenomics into everyday medicine is only a question of time.

10.7 Summary and Concluding Remarks

The broad use of microarray technology will enrich further our current knowledge in male and female reproduction in humans as well as in animals. The unprecedented volume of new data introduced in the field by the use of this technology in such a short time has already produced significant results. However, the biggest discoveries still lie ahead. Taking in account the present high rate of discoveries we can confidently foresee the time when reproductive diseases could be easily predicted, diagnosed, and will have an individual tailored treatment. Based on microarray results, systems biology advances, which will also include epigenetic influences, in the not so near future will offer the possibility to grow embryos *in vitro*, to select the sex, as well as to influence many other qualitative traits of the offspring, including muscular mass, or even life span. The practical consequences of this rapid development of the reproduction research field will bring into discussion ethical [46] and moral considerations.

10.8 Appendix of Public Databases for Microarray Data

Stanford Microarray database – http://genome-www5.stanford.edu/

Yale Microarray Database – http://www.med.yale.edu/microarray/

UNC Microarray database – https://genome.unc.edu/

MUSC database – http://proteogenomics.musc.edu/ma/musc_madb.php?page=home&act=manage

Gene Expression Omnibus – NCBI – http://www.ncbi.nlm.nih.gov/geo/

ArrayExpress – EBI – http://www.ebi.ac.uk/microarray-as/aer/#ae-main[0]

10.9 Glossary of Terms and Acronyms

Fluxomics: identification of the dynamic changes of molecules within a cell over time.

Genomics: the analysis of gene expression and regulation in cell, tissue or organs under given conditions [2,47] including how the genes interact with each other and with the environment. It has the potential to revolutionize the practice of medicine.

Glycomics: identification of all carbohydrates in a cell or tissues.

Glycoproteomics: a branch of proteomics that identifies, catalogs, and characterizes glycoproteins.

Interactomics: identification of protein–protein interactions but also include interactions between all molecules within a cell.

Metabolomics: identification and measurement of all small metabolites in a cell or tissue.

Pharmacogenomics: the combination of pharmacology and genomics that deals with analysis of the genome and its products (RNA and proteins) related to drug responses.

Phosphoproteomics: a branch of proteomics that identifies, catalogs, and characterizes phosphorylated proteins.

Piezoelectricity: the ability of some materials mainly crystals and certain ceramics to generate an electric charge in response to a mechanical stress. If the material is not shortcircuited, the applied charge induces a voltage across the material.

Protein microarray: a substrate of glass or silicon on which different molecules of protein have been affixed at separate locations in an ordered manner thus forming a microscopic array. These are used to identify protein–protein interactions, substrates of protein kinases, or the targets of biologically active small molecules. The main use of the protein microarrays, also termed protein chip, is to determine the presence and/or the amount of proteins in biological samples, e.g. blood. One common protein microarray is the antibody microarray, where antibodies (most frequently monoclonal) are spotted onto the protein chip and are used as capture-molecules to detect proteins from cell lysates. There are several types of protein chips; however, the most common are glass slide chips and nano-well arrays.

Proteomics: complete identification of proteins and protein expression patterns of a cell or tissue through two-dimensional gel electrophoresis or other multi-dimensional protein separation techniques and mass spectrometry. It is a large-scale study of proteins, particularly of their structure and function under given conditions [2]. This term was created to make an analogy with genomics [47]. Proteomics is much more complicated than genomics: while the genome is a rather constant entity, the proteome differs from cell to cell and is constantly changing through its biochemical interactions with the genome and the environment. One organism has radically different protein expression in different parts of its body, during different stages of its life cycle and under different environmental conditions [47].

SAGE: Serial Analysis of Gene Expression.

System biology: the strategy of pursuing integration of complex data about biological interactions from diverse experimental sources using interdisciplinary tools and high-throughput experiments and bioinformatics.

Transcriptomic: measurement of gene expression in whole cells or tissue by DNA microarray or SAGE.

References

1. Nekrutenko A. Reconciling the numbers: ESTs versus protein-coding genes. Mol Biol Evol 2004; 21:1278–82.
2. Bakalova R, Ewis A, Baba Y. Microarray-based technology: basic principles, advantages and limitations. In: Meyers RA, editor. Encyclopedia of molecular cell biology and molecular medicine. Weinheim: Wiley-VCH Verlag GmbH&Co, 2005: 263–87.
3. Schena M, Shalon D, Davis RW, et al. Quantitative monitoring of gene expression patterns with a complementary DNA microarray. Science 1995; 270:467–70.
4. DeRisi JL, Iyer VR, Brown PO. Exploring the metabolic and genetic control of gene expression on a genomic scale. Science 1997; 278:680–6.
5. Beaudet AL, Belmont JW. Array-based DNA diagnostics: Let the revolution begin. Annu Rev Med 2008; 59:113–29.
6. Baier RJ, Loggins J, Yanamandra K. IL-10, IL-6 and CD14 polymorphisms and sepsis outcome in ventilated very low birth weight infants. BMC Med 2006; 4:10.
7. http://www.fda.gov/fdac/features/2005/605˙genomics.html.
8. Copland JA, Davies PJ, Shipley GL, et al. The use of DNA microarrays to assess clinical samples: the transition from bedside to bench to bedside. Recent Prog Horm Res 2003; 58:25–53.
9. Chen HW , Tzeng CR. Applications of microarray in reproductive medicine. Chang Gung Med J 2006; 29:15–24.
10. Caetano AR, Johnson RK, Ford JJ, et al. Microarray profiling for differential gene expression in ovaries and ovarian follicles of pigs selected for increased ovulation rate. Genetics 2004; 168: 1529–37.
11. Gladney CD, Bertani GR, Johnson RK, et al. Evaluation of gene expression in pigs selected for enhanced reproduction using differential display PCR and human microarrays: I. Ovarian follicles. J Anim Sci 2004; 82:17–31.
12. von Schalburg KR, Rise ML, Brown GD, et al. A comprehensive survey of the genes involved in maturation and development of the rainbow trout ovary. Biol Reprod 2005; 72:687–99.
13. Yao J, Ren X, Ireland JJ, et al. Generation of a bovine oocyte cDNA library and microarray: Resources for identification of genes important for follicular development and early embryogenesis. Physiol Genomics 2004; 19:84–92.
14. Giudice LC. Microarray expression profiling reveals candidate genes for human uterine receptivity. Am J Pharmacogenomics 2004; 4:299–312.
15. Horcajadas JA, Riesewijk A, Martin J, et al. Global gene expression profiling of human endometrial receptivity. J Reprod Immunol 2004; 63:41–9.
16. Ji Q, Liu PI, Chen PK, et al. Follicle stimulating hormone-induced growth promotion and gene expression profiles on ovarian surface epithelial cells. Int J Cancer 2004; 112:803–14.
17. Agca C, Ries JE, Kolath SJ, et al. Luteinization of porcine preovulatory follicles leads to systematic changes in follicular gene expression. Reproduction 2006; 132:133–45.
18. Gheorghe C, Mohan S, Longo LD. Gene expression patterns in the developing murine placenta. J Soc Gynecol Investig 2006; 13: 256–62.
19. Sood R, Zehnder JL, Druzin ML, et al. Gene expression patterns in human placenta. Proc Natl Acad Sci U S A 2006; 103:5478–83.
20. Aiba K, Carter MG, Matoba R, et al. Genomic approaches to early embryogenesis and stem cell biology. Semin Reprod Med 2006; 24:330–9.
21. Esplin MS. Preterm birth: a review of genetic factors and future directions for genetic study. Obstet Gynecol Surv 2006; 61: 800–6.
22. Shuster E. Microarray genetic screening: a prenatal roadblock for life? Lancet 2007; 369:526–9.
23. Tanaka TS , Ko MS. A global view of gene expression in the preimplantation mouse embryo: morula versus blastocyst. Eur J Obstet Gynecol Reprod Biol 2004; 115 Suppl 1:S85–91.
24. Hamatani T, Carter MG, Sharov AA, et al. Dynamics of global gene expression changes during mouse preimplantation development. Dev Cell 2004; 6:117–31.
25. Hamatani T, Daikoku T, Wang H, et al. Global gene expression analysis identifies molecular pathways distinguishing blastocyst dormancy and activation. Proc Natl Acad Sci U S A 2004; 101:10326–31.
26. Lemay DG, Neville MC, Rudolph MC, et al. Gene regulatory networks in lactation: identification of global principles using bioinformatics. BMC Syst Biol 2007; 1:56.
27. Ferlin A, Raicu F, Gatta V, et al. Male infertility: role of genetic background. Reprod Biomed Online 2007; 14:734–45.

28. Huang X, Li J, Lu L, et al. Novel development-related alternative splices in human testis identified by cDNA microarrays. J Androl 2005; 26:189–96.
29. Tang A, Yu Z, Gui Y, et al. Identification of a novel testis-specific gene in mice and its potential roles in spermatogenesis. Croat Med J 2007; 48:43–50.
30. Lin YH, Lin YM, Teng YN, et al. Identification of ten novel genes involved in human spermatogenesis by microarray analysis of testicular tissue. Fertil Steril 2006; 86:1650–8.
31. Schlecht U, Demougin P, Koch R, et al. Expression profiling of mammalian male meiosis and gametogenesis identifies novel candidate genes for roles in the regulation of fertility. Mol Biol Cell 2004; 15:1031–43.
32. Sha J, Zhou Z, Li J, et al. Identification of testis development and spermatogenesis-related genes in human and mouse testes using cDNA arrays. Mol Hum Reprod 2002; 8:511–7.
33. Dong L, Jelinsky SA, Finger JN, et al. Gene expression during development of fetal and adult Leydig cells. Ann N Y Acad Sci 2007; 1120:16–35.
34. Zhou Q, Shima JE, Nie R, et al. Androgen-regulated transcripts in the neonatal mouse testis as determined through microarray analysis. Biol Reprod 2005; 72:1010–9.
35. He Z, Chan WY, Dym M. Microarray technology offers a novel tool for the diagnosis and identification of therapeutic targets for male infertility. Reproduction 2006; 132:11–9.
36. Rockett JC, Kim SJ. Biomarkers of reproductive toxicity. Cancer Biomark 2005; 1:93–108.
37. Rockett JC, Lynch CD, Buck GM. Biomarkers for assessing reproductive development and health: Part 1–Pubertal development. Environ Health Perspect 2004; 112:105–12.
38. Gasca S, Pellestor F, Assou S, et al. Identifying new human oocyte marker genes: a microarray approach. Reprod Biomed Online 2007; 14:175–83.
39. Tsui NB, Dennis Lo YM. Placental RNA in maternal plasma: Toward noninvasive fetal gene expression profiling. Ann N Y Acad Sci 2006; 1075:96–102.
40. Nicolau M, Tibshirani R, Borresen-Dale AL, et al. Disease-specific genomic analysis: Identifying the signature of pathologic biology. Bioinformatics 2007; 23:957–65.
41. Du F, Mahadevappa M, Warrington JA. Gene expression changes in endometrial cancer. In: Warrington JA, Todd R, Wong D, et al., editors. Microarrays and Cancer Research. Westborough: Eaton Publishing Company/Biotechniques Books, 2002: 113–25.
42. Wong KK, Cheng RS, Berkowitz RS. Gene expression analysis of ovarian cancer cells by cDNA microarrays. In: Warrington JA, Todd R, Wong D, et al., editors. Microarrays and Cancer Research. Westborough: Eaton Publishing Company/Biotechniques Books, 2002: 127–38.
43. Li TH, Zhao H, Peng Y, et al. A promoting role of androgen receptor in androgen-sensitive and -insensitive prostate cancer cells. Nucleic Acids Res 2007; 35: 2767–76.
44. Dhanasekaran SM, Dash A, Yu J, et al. Molecular profiling of human prostate tissues: insights into gene expression patterns of prostate development during puberty. Faseb J 2005; 19: 243–5.
45. Shi L. The MicroArray Quality Control (MAQC) Project. Toward consensus on the generation, analysis, and application of microarray data in the discovery, development, and revew of FDA-regulated products [Internet] http://edkb.fda.gov/MAQC/MainStudy/upload/Summary˙MAQC˙DataSets.pdf.
46. Grody WW. Ethical issues raised by genetic testing with oligonucleotide microarrays. Mol Biotechnol 2003; 23:127–38.

Chapter 11

Computer Assisted Analysis of Genes

M. Victoria Fachal and Michael Furlan

11.1 Introduction

The overwhelming volume of molecular data generated over the last 30 years and the need to understand the messages coded by DNA has lead to the development of global bioinformatics resources. The scientific community has collaborated worldwide toward developing and linking computer database DNA and protein repositories and providing computer-based analysis tools that are publicly available on the Internet. Central to these collaborations are institutions such as the National Institutes of Health (NIH) in the United States, which created the National Center for Biotechnology Information (NCBI) and the European Molecular Biology Laboratory (EMBL) located in England that created the European Bioinformatics Institute (EBI). This chapter is a general introduction to using Internet based computing resources to support research in molecular biology. An overview of four popular biological data retrieval systems and five commonly used computer-based molecular analysis tools are discussed.

11.2 Data Retrieval Systems

The information obtained by the Internet from database retrieval systems allows scientists to compare and interpret data, as well as design new experiments. This section provides information of data retrieval systems available in Internet without any cost: *Entrez*, *GenBank*, *Universal Protein Resources*, and the *Ovarian Kaleidoscope database*.

11.2.1 Entrez

The NCBI has available for researchers *Entrez*, a search and database retrieval system that collects and connects information from several biological databases as well as bibliographic data for a variety of health sciences. Within *Entrez*, databases are grouped and classified by subject into 21 data domains including "Protein," "Structure," "Taxonomy," etc. For example, in the "Nucleotide" data domain, nucleotide sequences and biological annotations from several databases such as *GenBank* and *RefSeq*[2] can be found. *Entrez* search page provides interface and access to all the data domains simultaneously where the user can narrow the search by choosing specific subset databases. *Entrez* is also designed to compare and relate records stored within the different databases. Therefore, any protein sequence has direct links to the nucleotide sequence that encodes for it, the three-dimensional structure, if elucidated, as well as related bibliography. In addition, the amino acid sequences can be compared with other sequences within the NCBI's Conserved Domain Database (CDD), and with all other protein databases in *Entrez* making possible to identify conserved domains and similar sequences. NCBI via *Entrez* offers researchers the opportunity to search and compare a wide range of information and computer assisted methods that facilitate the access to biological data. Researchers can access *Entrez* at www.ncbi.nlm.nih.gov.

11.2.2 GenBank

GeneBank, also created by NCBI, is the most popular collection of publicly available genetic sequence database. It contains more than 76 millions of sequences provided by individual labs, large scale sequencing projects, the US Office of Patents and Trademarks, and by the daily exchange of data with the EMBL Nucleotide Sequence Database (EMBL-Bank) in Europe and the DNA Data Bank

M.V. Fachal (✉)
Department of Biology, University of Saskatchewan College of Art and Sciences, Saskatoon, SK, Canada
e-mail: mvf259@mail.usask.ca

P.J. Chedrese (ed.), *Reproductive Endocrinology*,
DOI 10.1007/978-0-387-88186-7_11,

of Japan (DDBJ). Database files are classified according to taxonomic groups or sequencing strategies. The information available in each *GenBank* file includes a nucleotide sequence and its protein translation, descriptions of the regions of biological significance within the sequence, bibliographic references including reported sequence discrepancies and cross-references to the RNA sequences within the journal *Nucleic Acid Research*, and the taxonomic data of the source organism. Researchers can access *GenBank* through *Entrez* retrieval system at the NCBI web site (www.ncbi.nlm.nih.gov).

11.2.3 Universal Protein Resource (UniProt)

UniProt is an efficient protein catalog where biological data of diverse sources can be analyzed and integrated. *UniProt* has been developed by the collaboration of the EBI, the Swiss Institute of Bioinformatics (SIB), and the Protein Information Resource (PIR) from the US. The information within *UniProt* is divided in four databases, each designed for different purposes: *UniProt* Knowledgebase (*UniProtKB*), *UniProt* archive (*UniParc*), *UniProt* Metagenomic and Enviromental sequence database (*UniMES*), and *UniProt* Reference Clusters (*UniRef*).

UniProtKB is a curated protein database that stores reviewed records of protein sequences with their related annotations and cross-references to more than a hundred external databases. Important features of the *UniProtKB* are the annotations performed by biologists and the low level of redundancy on the retrieved records.

UniParc constitutes the main protein sequence repository of *UniProt*, containing sequences from more than 15 databases. *UniParc* avoids redundancy on its records merging on a same database file identical protein sequences in spite of source organisms. The information found in *UniParc* files includes sequences, source databases, accession numbers, version numbers, and the identifier numbers provided by *UniParc*. However, *UniParc* files do not provide biological annotations as *UniProtKB* files do.

UniMES contains protein sequences predicted from nucleotide sequences, which are provided by the genomic analysis of microbes recovered from environmental samples. The data obtained from the Global Ocean Sampling Expedition (GOS) is present within this database.

UniRef is comprised of protein sequences clustered according to the degree of similarity between them at resolutions of 100%, 90%, and 50% in data sets named UniRef100, UniRef90, and UniRef50, respectively. UniRef100 is constituted by data from *UniProtKB* and selected data from *UniParc*. UniRef90 is composed by clusters of UniRef100, and UniRef50 is built from clusters of UniRef90. This organization of the sequences is translated in a database size reduction of 10% for UniRef100, 40% for UniRef90, and 70% for UniRef50, which speeds up the search of similar protein sequences. The record obtained from each cluster contains a protein sequence representative of the group and shows the protein name, taxonomy of the source organisms, and source database of each sequence within the group. This database is usually used in phylogenetic analysis and protein family classifications. Researchers can access *UniProt* through its web site (www.pir.uniprot.org).

11.2.4 The Ovarian Kaleidoscope Database

The *Ovarian Kaleidoscope database (OKdb)*, developed by C. P. Leo and A. Hsueh in 1999 at University of Stanford, is an organ-specific database that provides information about genes expressed in the ovary. The organ-specific perspective of the *OKdb* allows an integrated view of ovarian expressed genes and its interactions, which enlightens their role in development, regulation, and functions of the ovary. Within the *OKdb*, the users can find information related to ovarian genes, regulatory elements, chromosomal localization, and mutations. Searches can be performed based on various parameters, including ovarian cell types, biological process, chemical nature, mutant phenotypes, etc. *OKdb* links to other bioinformatics and bibliographic databases such as *Endometrium Database Resources*, *GenBank*, and *PubMed*. *OKdb* is access through its web site (http://ovary.stanford.edu/).

11.3 Computer-Based Molecular Analysis Tools

Computer-based molecular analysis tools allow expeditious analyses of raw data within databases, such as organizing, comparing, and manipulating nucleotide and protein sequences. Free interface software available on the Internet has led the trend away from commercial software packages. Many Internet bioinformatics database administrators provide a "tool" link to user-friendly computer-based analysis tools and software packages that integrate a range of current tools for sequence analysis such as The European Molecular Biology Open Software Suite (EMBOSS) (http://emboss.sourceforge.net/download/) and EMBOSS-Lite (http://helixweb.nih.gov/emboss_lite/). In the following section, some of the available analysis tools on the Internet are briefly described. All the programs mentioned are free of cost.

11.3.1 Sequence Formatting

Many of the molecular biology softwares are only compatible with specific sequence formats. Thus, programs able to modify sequence formats are necessary to avoid doing this task manually. The EBI has a "tool" link in its web site (www.ebi.ac.uk) that directs the users to *Readseq*. *Readseq* is a program that automatically converts input nucleotide or amino acid sequences into specified formats. This program offers users a wide range of output formats and options to remove or extract part of a sequence and to modify sequence character case.

11.3.2 Sequence Conversion

Sequence conversion programs translate nucleotide sequences into proteins sequences, and protein sequences into nucleotide sequences.

Transeq is a program available at EBI web site (www.ebi.ac.uk) that translates all six reading frames of a nucleotide sequence into the corresponding protein sequences. Standard (universal) genetic code or a selection of non-standard codes can be used with *Transeq*. *Transeq* is used to identify proteins coded by nucleotide sequences.

Reverse Translate is a program designed to translate protein sequences into nucleotide sequences. *Reverse Translate* accepts an amino acid sequence as input and generates a DNA sequence that represents the most likely non-degenerate coding sequence. A consensus sequence derived from all the possible codons for each amino acid is also returned. This program is useful to design oligonucleotide primers and to find protein-coding regions in genomic DNA. *Reverse Translate* is one of many molecular analysis tools accessible through ExPASy web site (www.expasy.org).

11.3.3 Analysis of Restriction Enzyme Recognition Sites

Restriction enzymes recognize and cut specific sections of a nucleotide sequence, making them useful for cloning and manipulation of sequences. There are many programs designed to identify restriction enzyme sites in nucleotide sequences; in this section, *WatCut* (http://watcut.uwaterloo.ca/watcut/watcut/template.php) and *Webcutter* (http://rna.lundberg.gu.se/cutter2/) are described. These programs analyze restriction maps of nucleotide sequences indicating sites within the sequence where a silent mutation can be introduce in order to create a novel restriction site. They also detect single nucleotide polymorphisms (SNPs) if any of the mutated bases generates a cleavage site for a restriction enzyme. *Webcutter* has options to analyze the sequence like circular or linear, restrict the search to a specific group of restriction enzymes, and also a link to a restriction enzyme database.

11.3.4 Design Oligonucleotide Primers

The polymerase chain reaction (PCR) is a popular technique originally developed to amplify DNA segments that is also used in cloning, sequencing, and mutagenesis protocols. The protocol to perform this technique includes materials, such as the four DNA nucleotides, magnesium ions in molar excess to the DNA nucleotides, thermo stable DNA polymerase, and two oligonucleotide primers. All the mentioned elements are commercially available. However, oligonucleotide primers must first be designed using computer programs to ensure efficiently hybridization.

Primer3: www primer tool developed by the University of Massachusetts Medical School (http://biotools.umassmed.edu/bioapps/primer3_www.cgi) is a program used to design PCR and sequencing primers and hybridization probes. To design an oligonucleotide primer a nucleotide sequence and several parameters are input into the program. To obtain an efficient primer the parameters must be carefully chosen, including primer length, melting temperature (Tm), %GC content, and primer annealing temperature. The program then retrieves the most efficient primers for the specified sequence. These designed primers can be tested with others programs like *The Oligonucleotide Property Calculator* accessible through the Northwestern University Medical School of Chicago web site (www.basic.northwestern.edu/biotools/oligocalc.html). This program calculates oligonucleotides primers features, such as primer sequence length, %GC content, Tm, potential hairpin formation, potential self-annealing, and others properties.

11.3.5 Sequence Alignment and Comparison

BLAST, acronym for Basic Local Alignment Search Tool, is a software package created by NCBI that now continues its development at several institutes. The program compares nucleic acid and protein sequences with sequences stored in databases via sequence alignment. *BLAST* performs local alignments, comparing sections of the sequences until the best alignment between them is found. *BLAST* is able to find regions of similarity between sequences, which provide information about its structure, function, evolution, and phylogeny. The software package is subdivided in programs; this section describes basic *BLAST* programs available at the NCBI web site (www.ncbi.nlm.nih.gov), although other programs with different applications are included in the package.

BLASTN compares a nucleotide query sequence to a nucleotide sequence database. It is mostly used to find oligonucleotides or cDNA sequences within a genome, to analyze non-spliced DNA sequence within a genome, to determine the intron–exon structure of genes, and to perform cross-species sequence comparison.

BLASTP compares a protein query sequence to a protein sequence database. This program is useful to identify the query sequence, to make phylogenetic inferences, and to detect conserved regions within sequences.

BLASTX translates nucleotide query sequence in all six reading frames and compares each one against a protein sequence database. *BLASTX* is used to determine if the query sequence corresponds to a known protein or to look for protein coding regions in genomic DNA.

TBLASTN compares a protein query sequence against all six reading frames translations of a nucleotide database which allows the user to find homologous protein coding regions in unannotated nucleotide sequences, such as ESTs, and to find the position of transcripts in genomic DNA.

TBLASTX compares the translation products of a nucleotide query sequence to all six reading frame translations of a nucleotide database. It is a useful tool to identify novel genes and proteins encoded by ESTs.

MEGABLAST is the BLAST program that can accept multiple nucleotide sequence queries. However, there is not *MEGABLAST* program for protein searches. *MEGABLAST* is specifically designed to efficiently find long alignments between more than two nucleotide sequences. Among other uses, *MEGABLAST* determines if an unknown nucleotide query sequence already exists in a public database. The *discontiguous-MEGABLAST* program is designed specifically for comparison of diverged nucleotide sequences and is used to identify cross-species nucleotide sequences, which have alignments with low degree of identity but may encode similar proteins.

Since each alignment is considered an experiment, it is important to know which *BLAST* program is better to use, which database contains the right sequences to compare to, what is meaning of the search parameters, and how to use them to obtain the best records. All this information is available on the NCBI web page under the name of The BLAST Program Selection Guide (www.ncbi.nlm.nih.gov/blast/producttable.shtml#blast).

CLUSTALW2 is a program that performs multiple alignments of DNA or protein sequence queries. ClustalW2 can be used for several purposes such as to identify regions of similarity between sequences, to predict the function and structure of proteins, and to produce cladograms and phylograms to look for evolutionary relationships. *CLUSTALW2* can be used or downloaded through the EBI web site (www.ebi.ac.uk).

11.4 Summary

The scientific community has collaborated worldwide toward linking bioinformatic databases, developing global computer database repositories, and providing computer-based analysis tools that are publicly available on the Internet. A cornerstone to the advances in molecular biology is the ability to rapidly analyze molecular data using sophisticated computer software. By the use of the softwares exposed in these pages, researches can find a DNA sequence in a database, translate it into a protein sequence, change its format, search for restriction enzyme to manipulate it, design primers, compare the sequence obtained with other sequences in databases, find regions of similarity between sequences, and set evolutionary relationships.

Glossary of Terms and Acronyms

%GC: percentage of guanine and cytosine

BLAST: basic local alignment search tools

CDD: conserved domain database

cDNA: complementary DNA

DDBJ: DNA database of Japan

EBI: European Bioinformatics Institute

EMBL: European Molecular Biology Laboratory

EMBOSS: The European Molecular Biology Open Software Suite

ESTs: expressed sequence tag

ExPASy: expert protein analysis system

GOS: global ocean sampling expedition

NCBI: National Center for Biotechnology Information

NIH: National Institutes of Health

OKdb: Ovarian Kaleidoscope database

PCR: polymerase chain reaction

PIR: protein information resources

SIB: Swiss Institute of Bioinformatics

SNP: single nucleotide polymorphism

Tm: melting temperature

UniMES: UniProt metagenomic and environmental sequence database

UniParc: UniProt archive

UniProt: universal protein resource

UniProtKB: UniProt knowledgebase

UniRef: UniProt reference clusters

Bibliography

1. Geer RC, Sayers EW. Entrez making use of its power. Brief Bioinform 2003; 4:179–84.
2. Ostell J. 'The Entrez search retrieval system' in 'The NCBI Handbook' [Internet], National Library of Medicine US 2002, chapter 14.
3. Benson DA, Karsch-Mizrachi I, Lipman DJ, et al. GenBank. ucleic Acids Res 2008; 36:25–30.
4. Bairoch A, Apweiler R, Wu CH, et al. The Universal Protein Resource (UniProt) Nucleic Acids Res 2005; 33:154–9.
5. Ben-Shlomo I, Vitt UA, Hsueh AJW. Perspective: The ovarian kaleidoscope database-II. Functional genomic analysis of an organ-specific database. Endocrinology 2002; 143:2041–4.
6. Leo CP, Vitt UA, Hsueh AJW. The ovarian kaleidoscope database: An online resource for the ovarian research community. Endocrinology 2000; 141:3052–4.
7. Altschul, SF, Madden TL, Schaeffer AA, et al. Gapped BLAST and PSI-BLAST: a new generation of protein database search programs. Nucleic Acids Res 1997; 25:3389–402.
8. Korf I, Yandell M, Bedell J. BLAST, first edition. Sebastopol: O'Reilsly & Associates Inc, 2003.
9. Chenna R, Sugawara H, Koike T, et al. Multiple sequence alignment with the Clustal series of programs. Nucleic Acids Res 2003; 31:3497–500.

Chapter 12

Introduction to Gene Therapy

Ayman Al-Hendy and Salama A. Salama

12.1 Introduction

Gene therapy is broadly defined as the delivery or *transfer* of genetic material to target cells for therapeutic purposes. Elucidation of the molecular bases of inherited diseases as well as acquired diseases, such as cancer, allows for the possibility of therapeutic interventions at the molecular level. The therapeutic benefits may be achieved by, interfering with gene function, restoring lost function, or initiating a new function in the target cells. Current methods of gene transfer include the use of viral and non-viral vectors [1–5]. Viral vectors accomplish gene transfer directly by viral-mediated infection. Non-viral gene transfer, or *transfection,* can be attained through chemical or physical treatment of target cells. Criteria required to fulfill successful gene transfer includes the following:

- a vehicle to deliver the therapeutic gene to the appropriate target cells;
- an appropriate level of expression of the therapeutic gene in the target tissue; and
- most importantly, the transfer and expression of the therapeutic gene must not be deleterious for the patient or the environment.

12.2 Methods for Gene Transfer

Availability of an appropriate method for gene transfer is the major challenge for successful gene therapy. Although tremendous advances in the technology have been achieved with viral and non-viral vectors, an ideal method for gene transfer is not yet available. The ideal vector for gene transfer would fulfill the following criteria:

- amenable to simple methods of production;
- available in high titers to allow delivery of unlimited amounts of genetic material in small volumes;
- able to express precisely regulated transgenes over sustained periods;
- immunologically inert, thus allowing repeated administration;
- site-specific integration into chromosomes of target cells or resides in the nucleus as an episome;
- able to transfect both dividing and non-dividing cells.

12.2.1 *Non-Viral Vectors*

Purified, histone-free DNA, referred to as *naked DNA*, injected into local tissues or systemic circulation is the simplest and safest physical/mechanical approach for gene therapy. Direct injection sites include skeletal muscle, liver, thyroid, heart muscle, urological organs, skin, and tumors [6]. The use of this approach is limited by its comparatively low efficiency [7]. To overcome this hurdle, attempts have been made to use nuclear targeting motifs to direct the passage of the DNA plasmid across the nuclear membrane into the nucleus [8]. Once a well-defined nuclear targeting signal is conjugated to the DNA of interest, the DNA can cross into the nucleus and be expressed at high rate. In addition, other technical problems have been encountered, such as rapid degradation by nucleases and clearance by the mononuclear phagocyte system [9,10]. Consequently, efforts have been directed toward physical manipulation to improve the efficiency of gene transfer that includes microinjection, particle bombardment, and electroporation.

12.2.2 *Microinjection*

Microinjection is the simplest gene transfer method. Microinjection of naked DNA directly into the nucleus was shown to

A. Al-Hendy (✉)
Department of Obstetrics and Gynecology, Meharry Medical College Center for Women Health Research, Nashville, TN, USA
e-mail: ahendy@MMC.edu

P.J. Chedrese (ed.), *Reproductive Endocrinology*,
DOI 10.1007/978-0-387-88186-7_12, © Springer Science+Business Media, LLC 2009

bypass cytoplasmic degradation, resulting in a much higher level of expression than intracytoplasmic injection. However, this technique is a laborious procedure in which only one cell at a time can be injected. Therefore, microinjection, at its current technological level, is impractical for most, if not all, in vivo gene transfer applications. It is most useful for in vitro applications that typically allow a few hundred cells to be transfected per experiment [11].

12.2.3 Particle Bombardment

The concept of gene-mediated particle bombardment is to move naked DNA plasmids into target cells on an accelerated microparticle carrier with a device called the gene gun. Metal particles (i.e., gold, tungsten) may be coated with biologically active DNA or RNA by precipitation in the presence of polyvinypyrrolidone, the polycation spermidine, and $CaCl_2$. The gene gun accelerates the microparticles, typically 1–3 μm in diameter, into the cytoplasm and nucleus by force (i.e., helium pressure), penetrating the cell membrane and bypassing the endosomal compartment [6]. The DNA then dissociates from the gold particles and is expressed intracellularly [12].

Bombardment of many different tissues including the skin, liver, pancreas, kidney, and muscle result in readily detectable transgene activities [13,12,14,15]. However, the major application of the gene gun is genetic immunization targeting the skin. The gene gun primarily propels DNA-coated beads into the epidermal layer [16] where the most efficient DNA immunization is achieved [17]. This approach has been used for genetic vaccination, immunomodulation, and suicide gene therapy to treat cancer [18]. Advantages of particle-mediated bombardment are

- ease of microparticle preparation;
- speed of the delivery vehicle;
- stability of the genetic material;
- absence of viral antigens;
- in vivo transfer able to target different cell types and tissues, tissue depths, and areas;
- rapid shedding of both particles and DNA targeted to the epidermis;
- ability to deliver large DNA molecules.

The limitations of gene bombardment are comparable with those of other non-viral transfection methods, including moderate cellular uptake, transient expression of the delivered gene, and low frequency of stable gene integration.

12.2.4 Electroporation

Electroporation involves the use of electric pulses to transiently permeabilize the cell membrane and is the most efficient method for in vivo transfer of naked DNA [19]. DNA is transferred into cells through the combined effects of electric pulses, which cause permeabilization of the cell membrane and an electrophoretic effect, permitting DNA to move toward or across the cell membrane [20]. Electro gene transfer can be highly efficient, with low variability both in vitro and in vivo, and allows the transfer of genes into tissues without using a virus [14,20]. The only limitation of electroporation is the restriction imposed by the area where the electrodes need to be placed [21–23].

12.2.5 Synthetic Vectors

Synthetic vectors are cationic lipids and polymers complexed with DNA that condenses the genetic material into small particles from ten to several hundred nanometres in diameter [24]. These complexes known as lipoplexes and polyplexes mediate DNA cellular entry and protect the delivered genes by blocking the access of nucleolytic enzymes [25]. The advantages of the synthetic vectors are as follows:

- relatively straightforward and easily produced in large-scale amounts [26];
- capacity to complex larger amounts of DNA;
- versatile and therefore can be used with any size or type of DNA/RNA;
- capable of stable transfection to non-dividing cells;
- cellular transfection does not require specific receptors [27,26];
- minimal toxic or immunological reactions, which allows for repeated administration [28].

Existing forms of synthetic vectors do not have the capacity for cell-specific targeting. However, the chemistry of the synthetic vectors allows for the attachment of targeting moieties that facilitate both increased cell uptake and cell specificity [24]. Ligand-directed tumor targeting of lipoplexes using folate, transferrin, or anti-transferrin receptor scFv are promising approaches for targeted gene transfer and also systemic gene therapy in human breast, prostate, head, and neck cancers [29].

12.2.6 Viral Vectors

The basic concept of viral vectors is to utilize the innate ability of the virus to deliver genetic material into specific cell types. Viral vectors differ in immunogenicity, vector tropism, maximum size of gene that can be packaged, and integration into the host genome leading to persistence of transgene expression in dividing cells. To produce the viral vectors the viruses' non-essential genes for replication are replaced with the therapeutic genes. Thus, by either integrating into the genome or non-integrated on a plasmid of the RNA or DNA, the recombinant viral vectors can transduce the cell type it would otherwise, normally infect. Although a number of viruses have been developed, interest has centered on the following:

- retroviruses (including lentiviruses);
- adenoviruses;
- adeno-associated viruses (AAV).

Retrovirus vector—retroviruses are one of the early vectors developed to introduce DNA into cells that remain among the most commonly used vectors for clinical trials [30]. The properties that make them ideal vectors for stable correction of genetic defects includes the relative simplicity of their genomes, ease of use, and long-term transgene expression in transduced cells or their progeny. Retroviral vectors are also suitable for ex vivo gene therapy, such as transducing CD34+ bone marrow hematopoietic stem cells or peripheral blood lymphocytes [31]. However, in spite of these advantages there exists concern about the safety of retrovirus-based gene therapy, since it has been reported linked to two cases of leukemia in children [32]. Other limitations of retroviral vectors in clinical applications include the following:

- low vector titer concentrations (10^7 infection particles/mL) resulting in low transfection efficiency;
- Inability to transduce non-dividing post-mitotic cells [33], thus restricting in vivo applications of gene transfer to actively dividing cells such as stem cells and cancer cells [34].

Attempts to circumvent problems associated with retroviral vectors has resulted, for example, in re-engineering so that the retroviral insertion takes place only at the desired specific sites of the host cell chromosome.

Adenovirus—replication-incompetent adenoviral vectors have emerged as a very popular and safe vector for human gene therapy and it is expected that soon they will be the most commonly used vehicle in clinical trials [35]. Adenoviral vectors have many advantages, including the following:

- stable and efficient in vivo gene transfer to both dividing and non-dividing cells;
- high levels of gene targeting to the nucleus [35];
- non-oncogenic expression as episomes, ensuring less disruption of host cellular genes;
- can be produced in large-scale amounts and high titer production (up to10^{13} particles/mL) [36];
- transgenes of up to 4.7–4.9 kb can be incorporated, and the cloning capacity of adenovirus can be further increased to 8.3 kb by deleting additional dispensable sequences from its genome; and
- easy manipulation of tropism by inserting receptor-binding ligands into the major capsid proteins, permitting cell-specific transfection.

The utility of adenoviruses in tumor gene therapy is substantiated by the concept of transductional targeting, which decreases the tropism of adenovirus toward the normal cells and enhances the ability of the virus to infect tumor cells [37]. Transcriptional targeting can be achieved by placing the transgene under the control of tissue- or tumor-specific promoters. In addition, the development of conditionally replicative virus vectors holds promises for gene therapy of cancer. Conditionally replicative adenoviruses (CRAds) represent a novel class of anticancer agents designed to selectively replicate in tumor cells and to destroy these cells by inducing lysis. CRAds are engineered to replicate only in the target tissue by modifying the viral genome. These viral genome modifications involve the substitution of the viral genes that alter viral metabolism in the host cell. For instance, crippling the viral replication machinery in a way that can be function only in tumor cells [38]. Therefore, the virus will replicate within the tumor and not in normal tissues. Intratumoral viral replication might induce or augment the development of antitumoral immunity, and this could occur by attracting immune cells into the tumor and increasing the amount of tumor antigens available in an immunostimulatory environment. Several CRAds have been developed for diseases such as cervical and prostate cancer [39–41].

Several challenges face large-scale application of adenoviral vectors for gene transfer. The capacity of the adenoviral vector to transfect target cells depends on the level of expression of *integrin* and the coxsackie-adenovirus receptor (CAR). It has been observed that most primary tumor cells express low levels of CAR compared to normal tissues, therefore, most tumor cells are poorly transfected with adenovirus [42]. Therefore, cells expressing CAR and integrins below the threshold level, such as smooth muscle, endothelial cells, skeletal muscles, fibroblasts, and hematopoietic cells are refractory to adenovirus infection [43,42].

To circumvent limitations of adenovirus vectors new serotypes of adenovirus vectors (i.e., Ad5) have been developed that transduce the target cells utilizing CD46 as a cellular receptor rather than CAR. Thus, these new serotypes of adenovirus vectors are able to efficiently infect cell types expressing low or no levels of CAR [44,45]. An alternative approach to overcome the problem of low transfectability is to modify the adenovirus vector tropism by altering the fiber component of the viral capsid, allowing CAR-independent gene transfer. The initial step of infection involves an interaction between the adenovirus fiber and CAR. However, this interaction is not essential, per se, for infection and only serves as a means of bringing the viral particle into intimate contact with the cellular surface [43]. Other tumor-specific receptors could serve the same purpose. Thus, modification of the adenovirus vector to ablate CAR binding and redirect binding to other tumor-specific receptors would allow for a CAR-independent, targeted vector. It has been reported that by using a bifunctional conjugate consisting antibody agonist adenovirus fiber (anti-knob monoclonal antibody) cross-linked to a ligand of tumor-specific receptor, increased the gene transduction of tumor cells by 7.7- to 44-fold [46,47]. In addition to enhancing the transfectability of tumor cells, this approach has been shown to decrease the dose of adenovirus needed to mediate infection of a given number of tumor cells [48].

Despite the widespread application of the adenovirus, particularly serotype Ad5, potential broad therapeutic applications are limited in humans by the development of vector-neutralizing antibodies that reduce the ability of Ad5-based vectors to transduce important therapeutic target cell types [49]. Adenovirus causes an initial non-specific host response followed by a specific response of cytotoxic T lymphocytes directed against virus-infected cells. In addition, there is activation of B cells and the necessary CD4-positive T cells, leading to a humoral response. Serologic surveys found antibodies against adenovirus serotypes 1, 2, and 5 in 40–60% of children. Cellular immunity eliminates the possibility of transduced cells; whereas, humoral immunity precludes the repeated administration of the adenovirus vector [50–52]. Thus, immune response of the host against adenoviral proteins is the major hurdle to the efficient and safe use of adenoviral vectors.

Several attempts have been made to minimize the immunological reaction to adenovirus. One approach has been to use vectors derived from non-human adenovirus. For example, a simian adenoviral vector has been developed that circumvents the preexisting immunity found in humans [53]. Another approach to avoid antibody-mediated neutralizing immunity and the clearance mechanism, as well, has been to coat the vector with a polymer such as polyethylene glycol or poly-[N-(2-hydroxypropyl) methylyacrylamide].

Incorporation of targeting ligands on the polymer-coated virus produces ligand-mediated, CAR-independent binding and uptake into cells bearing appropriate receptors. This retargeted virus is resistant to antibody neutralization and can selectively infect receptor-positive target cells [54,55]. In an alternative approach to the derivation of cancer cell-specific vectors, adenovirus vectors have been targeted at the level of transcription by placing the therapeutic gene under the control of transcriptional regulatory sequences that are activated in tumor cells but not in normal cells; therefore, expression is selectively targeted to the tumor cell. A higher degree of specificity for cancer cells can be achieved by combining the complementary approaches of transductional and transcriptional targeting [56,57]. Worldwide, over 600 clinical trials using gene therapy have been conducted or are underway, and adenovirus vectors are the most commonly used vectors in these trials [34].

Adeno-associated virus—AAV are single-stranded DNA human parvoviruses that are dependant on a helper virus, usually an adenovirus, to proliferate. They are capable of infecting dividing and non-dividing cells. In contrast to adenoviral vectors, AAV vectors do not cause a significant host inflammatory response and no human disease has been associated with AAV infection [58]. Because of their prolonged transgenic expression and non-pathogenic nature, AAV have become increasingly attractive as vectors for gene transfer [59]. Also, AAV are unique among eukaryotic DNA viruses in their ability to integrate at a specific site within the human chromosome (19q13.3–qter). However, a significant limitation of AAV as vectors is that the viral genome can only accept DNA inserts up to 4.7 kb in length [60]. Integration of AAV into the genome of the trasduced cells is a double-edged sword, as they not only maintain the expression of the therapeutic gene in progeny cells, but also may increase the risk of mutations that are deleterious to the host.

Interestingly, different serotype capsids of AAV can be combined to generate novel tropisms, which permit AAV to transfect tissues that have not been easily infected with the current AAV serotypes. Consequently, the serotypes of AAV developed with varying tropism to different tissues may allow them to become further amenable to AAV-based gene transfer in the future. Many other viral systems are at various stages of development including HSV [61], measles [62], and parvovirus [63]. Each of these vectors has unique features, advantages, and disadvantages, and they can be used accordingly for specific applications.

12.3 Tumor Gene Therapy Strategies

Advances in molecular and tumor biology have contributed greatly to our understanding of the genetic alterations associated with tumor transformation. Thus, gene therapy strategies have been proposed that target genetic alterations specific to tumor cells and tumor pathophysiology. These treatment strategies include mutation compensation, molecular chemotherapy, immunopotentiation, and alteration of drug resistance.

12.3.1 Mutation Compensation

The genetic lesions associated with transformation and progression of neoplasms may be treated by mutation compensation, a correctional gene therapy approach [64]. Mutation compensation strategies focus on functional ablation of expression of dysregulated oncogenes, replacement or augmentation of the expression of tumor suppressor genes, or interference with signaling pathways of growth factors or other biochemical processes that contribute to the initiation or the progression of the tumor [65–68].

Mutation compensation gene therapy relies on several approaches to restore normal function of tumor suppressor genes or block oncogene activity, including antisense oligonucleotide, catalytic ribozymes and small oligonucleotides, dominant negative gene mutation, and, most recently, small interfering RNA (siRNA) technology.

12.3.2 Molecular Chemotherapy

Molecular therapy, which is known also as cytoreductive gene therapy or suicide gene therapy, refers to the delivery of genes that kill cells by direct or indirect effects. Strategies include gene-directed enzyme prodrug therapy, proapoptotic, antiangiogenic, and gene-directed radioisotopic therapy. Gene-directed enzyme prodrug therapy involves delivering a gene that encodes an enzyme, which then converts an innocuous prodrug to a toxic agent within tumor cells. The two most widely used gene-directed enzyme prodrug systems are herpes simplex virus thymidine kinase (HSV-tk) plus ganciclovir (GCV) and *Escherichia coli* bacterial cytosine deaminase plus 5-fluorocytosine (CD/5-FCyt) [69]. The HSV-tk system has many attributes that make it the front-runner of cytoreductive gene therapy strategies. Most importantly, the active ganciclovair metabolite diffuses to adjacent untransduced tumor cells to mediate a bystander effect.

12.3.3 Immunopotentiation

Modulation of immune response is particularly attractive as a modality for cancer gene therapy. A key focus of tumor gene therapy is the enhancement of the immune system's ability to destroy tumor cells. Passive immunopotentiation involves boosting the natural immune response to make it more effective. Active immunopotentiation requires the initiation of an immune response against a previously unrecognized tumor. The immunopotentiation gene therapy capitalizes on strategies such as expression of cytokine genes, which may enhance the activity of antigen-presenting cells and T cells, which are lymphocytes produced mainly by the thymus [70]; expression of co-stimulatory molecules, such as B-1 and B7-2, expressed on the surface of antigen presenting cells that provide signals for T-cell activation, thereby facilitating the recognition and killing of tumor cells [71]; or delivery of exogenous immunogens, which generate local inflammatory reaction that increases the ability of antigen-presenting cells to recognize tumor-associated antigens [72].

12.3.4 Alteration of Drug Resistance

A major hurdle in cancer chemotherapy is drug-related toxicity, mainly myelosuppression that requires discontinuation of chemotherapy, which may result in failure of cancer treatment. The multidrug resistance gene (MDR1) confers resistance toward several anticancer drugs. Gene therapy approaches involving retroviral transfer of MDR1 gene into hematopoietic stem cells and progenitor cells prior to autologous transplantion has been used in the treatment of human germ cell tumors [73] and metastatic breast cancer [74]. For cytotoxic drugs that are not substrate for MDR1, such as cyclophosphamide and methotrexate, retroviral transfer of aldehyde dehydrogenase class 1 gene (*ALDH-1*) and mutated dihydrofolate reductase gene to hematopoietic progenitor cells protects against myelosuppression induced by these drugs [75].

12.4 Gene Therapy of Uterine Leiomyoma

Uterine leiomyomas, known as fibroids, are the most common pelvic tumors in the United States and occur in 20–25% of premenopausal women [76,77]. The development of uterine fibroids involves complex interactions among genes and the environment. They commonly cause severe symptoms such as heavy, irregular, and prolonged menstrual bleeding and anemia. Fibroids have also been associated with infertility and recurrent spontaneous abortion [78,79]. Generally, uterine fibroids are slow-growing tumors, however, these tumors tend to grow rapidly during pregnancy and can cause obstructed labor and/or fetal malpresentation necessitating cesarean section, as well as fetal anomalies, and postpartum hemorrhage secondary to uterine atony.

Fortunately, the localized nature of leiomyoma allows localized, tumor-targeting features of gene therapy, a promising treatment strategy among management options for women with uterine fibroids, either alone or as an adjuvant to existing treatment modalities, while, at the same time, preserving fertility. Thus, generating improved gene therapy delivery systems and refining the mechanisms that control gene expression will be key to the development of safe clinical protocols for treatment of leiomyomas.

The leiomyoma confined in the uterus by fibrous capsules allows ultrasound or endoscopic-guided intratumor targeting of gene therapy vectors (i.e., adenoviral vectors). Directly injecting the viral load into the leiomyoma minimizes systemic toxicity and avoids development of an immunological reaction. Alternatively, the viral vector can be delivered by transarterial injection through a uterine artery catheter threaded through an artery feeding the fibroid. This method would ensure homogenous distribution of the injected virus and may possibly improve efficiency of cell transfection. Subsequently, transductional targeting permits prolonged expression of transgenes minimizes the amount of vector required to reduce tumor size, thus allowing, if necessary, for repeated administration of the viral vector.

For transcriptional targeting, the biological uniqueness of leiomyoma, such as leiomyoma-specific transcription factors or proteins need to be identified in order to design specific promoter/vectors driving therapeutic gene expression. Several biological pathways in leiomyoma represent potential targets for gene therapy application. Angiogenesis, which is essential for leiomyoma growth, is driven by different growth factors such as basic fibroblast growth factor (bFGF), vascular endothelial growth factor (VEGF), and platelet-derived endothelial growth factor (PDGF) [80–83]. Each of these angiogenic factors represents a potential gene therapy target. This general approach may prove to be clinically efficacious in spite of genetic etiology and without the need to completely replace a single, missing gene product. For instance, dominant negative estrogen receptor (ER), delivered via an adenovirus, as well as Ad-HSV-tk/GCV provide two promising approaches to induce apoptosis and eventually shrink leiomyoma.

Evidence has also suggested that apoptotic mechanisms are downregulated in leiomyoma cells compared with normal myometrium [84]. Thus, gene therapy of leiomyoma can be conducted by switching on the apoptosis mechanism through both extrinsic and intrinsic pathways by delivering apoptosis-inducing ligands, such as tumor necroses factor (TNF), TNF apoptosis-inducing ligand (TRAIL), and FasL; or by delivering pro-apoptotic members of the Bcl-2 family, such as Bax or active caspase molecules.

Unlike cancer gene therapy, gene transfer does not have to be achieved in every leiomyoma cell to attain clinical improvement. This is a great advantage because with current gene therapy vectors, it is virtually impossible to achieve 100% gene transfer in vivo. The first attempt to apply the gene therapy approach to uterine leiomyoma was reported by Niu [85]. Christman [86] described a non-viral-mediated transfer of the suicide gene for thymidine kinase into human leiomyoma cells and ELT3, a cell line derived from Eker rat leiomyoma. In this study, a thymidine kinase-ganciclovir-mediated "bystander effect" was demonstrated, with 48.6% (human) and 65.6% (rat) cell death occurring when 5% of the leiomyoma cells were transfected with the HSV-tk expressing thymidine kinase incubated with ganciclovir (pNGVL1-tk) vector; 0.84 and 1.9% of the cells were expected to express thymidine kinase based on the 16.7 and 39.8% transfection efficiency determined by the reporter gene assay in human and rat leiomyoma cells, respectively. Estradiol promoted cell growth and enhanced the bystander effect in rat leiomyoma cells. It has been suggested that the increased estrogen promotes ELT3 cell growth. It is possible that estrogen accelerates the bystander effect in these cells by increasing the number of dividing cells that are sensitive to phosphorylated ganciclovir [85].

The exact mechanisms underlying the bystander effect are not clear. It has been reported, however, that the transfer of phosphorylated ganciclovir through direct cell–cell contact (gap junctions) or through apoptotic vesicles between thymidine kinase-expressing and non-expressing tumor cells is involved [87]. This efficient bystander effect manifested in leiomyomas cells was substantiated by a recent finding indicating that the expression of connexin 43, proteins that form gap junctions between cells in various mammalian tissues, are highly expressed in leiomyoma tissues compared with the adjacent normal myometrium [88]; similar findings were also previously reported by Andersen [89]. These observations suggest that the bystander effect will likely be operational when TK-GCV is applied in vivo and would greatly enhance the efficacy of suicidal gene therapy approach. Additionally, the efficient intercellular gap junction communication in leiomyoma cells compared with normal myometrial cells would limit the transfer of the toxic GCV-triphosphate to the normal myometrium, which exhibits lower gap-junction expression and is physically separated from the tumor tissue by a well-developed avascular capsule, thus, providing a safety margin for the value of the suicidal gene therapy strategy. The recent application of the HSV-TK/GCV to human and rat leiomyoma cells in vitro using an adenoviral delivery system confirmed the above finding [88].

Steroid hormones, such as estrogen, and progesterone play a pivotal role in the development and progression of leiomyoma. Therefore, it is conceivable that the signaling pathway of these hormones may represent a potential target for gene therapy of leiomyoma. Thus, the use a dominant negative ER to intercept the estrogen-signaling pathway was proposed as a gene therapy target [90]. These mutants form heterodimers with wild-type ERs, which make them unable to bind the estrogen-responsive element (ERE) and therefore unable to activate transcription [91]. Ad-LacZ, an adenovirus expressing the marker gene β-lactamase, is capable of infecting both human and rat leiomyoma cells. LM-15, a human leiomyoma cell line, was successfully transduced with Ad-LacZ with maximal (100%) transduction achieved at a multiplicity of infection (MOI) of 10 PFU/cell [90]. Rat leiomyoma cells reached optimal transduction at MOI of 100 plaque forming units/cell. The ability of adenovirus to infect uterine leiomyoma tissue removed directly from hysterectomy specimens was also assessed. The 2–3 mm disks of tissue were incubated with an adenovirus expressing the Ad-LacZ, followed by X-gal staining. A clear nuclear blue color (indicating virus transduction) was noticed in the smooth muscles of the leiomyoma tumor. To disrupt the estrogen signaling pathway, adenoviral vectors expressing a dominant negative ER mutant (ER1-536) under the cytomegalovirus (CMV) promoter (Ad-ER-DN) was used [90]. In an in vitro system using Ad-ER-DN to transduce both human and rat leiomyoma cell lines, Ad-ER-DN induced an increase in both caspase-3 levels and the BAX/Bcl-2 ratio with evident apoptosis in the TdT-mediated dUTP nick-end labeling assay.

In nude mice, ex vivo rat leiomyoma cells transduced with Ad-ER-DN formed significantly smaller tumors compared with Ad-LacZ–treated cells, 5 weeks after implantation. In nude mice with preexisting lesions, treatment by direct intratumor injection of Ad-ER-DN led to immediate arrest of tumor growth. In addition, the Ad-ER-DN-treated tumors demonstrated severely inhibited cell proliferation (BrdU index) and a marked increase in the number of apoptotic cells (TdT-mediated dUTP nick-end labeling index). This study demonstrated that dominant-negative ER gene therapy might provide a non-surgical treatment option for women with symptomatic uterine fibroids who want to preserve their uterus.

In vivo animal data obtained thus far has been generated in the immune-deficient nude mouse model. Although leiomyoma nodules developed in this model resemble the human disease in many features [92,93], the lack of cell-mediated immunity might present an disadvantageous environment for studying the final efficacy of adenovirus transfection and longevity of gene expression [94]. However, two studies reporting on fibroid gene therapy were conducted comparing human leiomyomas cells and, the only immune-competent animal model for uterine leiomyoma [95], ELT3 rat leiomyoma cell line demonstrated quantitative and qualitative differences between the two cell lines [90,85]. Furthermore, experiments are currently underway in our laboratory to extend experiments in the Eker rat, since the fibroid lesions in these rats are in the uterine horns or cervico-uterine junction and share various anatomical, histological, and biological features with human leiomyoma [96,97]. Data obtained from these experiments will provide valuable preclinical data, germane to future studies relevant to human leiomyomas.

Glossary of Terms and Acronyms

AAV: adeno-associated viruses

Ad5: type of adenovirus

Ad-ER-DN: adenoviral vector that expresses a dominant negative ER mutant

Ad-LacZ: adenovirus expressing the marker gene β-lactamase

ALDH-1: aldehyde dehydrogenase class 1 gene

bFGF: basic fibroblast growth factor

$CaCL_2$: calcium chloride

CAR: coxsackie-adenovirus receptor

CD/5-FCyt: cytosine deaminase plus 5-fluorocytosine

CMV: cytomegalovirus

CRAds: conditionally replicative adenoviruses

DNA: deoxyribonucleic acid

ELT3: cell line derived from Eker rat leiomyoma

ER: estrogen receptor

ERE: estrogen receptor element

ERI-536: estrogen receptor mutant

GCV: ganciclovir

HSV: herpes simplex virus

HSV-tk: herpes simplex virus-thymidine kinase

LM-15: human leiomyoma cell line

MDR1: multidrug resistance gene

MOI: multiplicity of infection

PDGF: platelet-derived endothelial growth factor

RNA: ribonucleic acid

siRNA: small interfering RNA

TK-GCV: thymidine kinase – ganciclovir

TNF: tumor necroses factor

TRAIL: TNF apoptosis-inducing ligand

VEGF: vascular endothelial growth factor

References

1. Mullen CA, Blaese RM. Gene therapy of cancer. Canc Chemother Biol Response Modif 1994; 15:176–89.

2. Crystal RG. Transfer of genes to humans: early lessons and obstacles to success. Science 1995; 270:404–10.
3. Blaese M, Blankenstein T, Brenner M, et al. Vectors in cancer therapy: how will they deliver? Cancer Gene Ther 1995; 2:291–7.
4. Ali M, Lemoine NR, Ring CJ. The use of DNA viruses as vectors for gene therapy. Gene Ther 1994; 1:367–84.
5. Schofield JP, Caskey CT. Non-viral approaches to gene therapy. BMJ 51; 56–71.
6. Nishikawa M, Huang L. Nonviral vectors in the new millennium: delivery barriers in gene transfer. Hum Gene Ther 2001; 12: 861–70.
7. Munkonge FM, Dean DA, Hillery E, et al. Emerging significance of plasmid DNA nuclear import in gene therapy. Adv Drug Deliv Rev 2003; 55:749–60.
8. Mattaj IW, Englmeier L. Nucleocytoplasmic transport: the soluble phase. Annu Rev Biochem 1998; 67:265–06.
9. Gallo-Penn AM, Shirley PS, Andrews JL, et al. Systemic delivery of an adenoviral vector encoding canine factor VIII results in short-term phenotypic correction, inhibitor development, and biphasic liver toxicity in hemophilia A dogs. Blood 2001; 97:107–13.
10. Lechardeur D, Sohn KJ, Haardt M, et al. Metabolic instability of plasmid DNA in the cytosol: a potential barrier to gene transfer. Gene Ther 1999; 6:482–97.
11. Hendrie PC, Russell DW. Gene targeting with viral vectors. Mol Ther 2005; 12:9–17.
12. Benn SI, Whitsitt JS, Broadley KN, et al. Particle-mediated gene transfer with transforming growth factor-beta1 cDNAs enhances wound repair in rat skin. J Clin Invest 1996; 98:2894–02.
13. Mahvi DM, Burkholder JK, Turner J, et al. Particle-mediated gene transfer of granulocyte-macrophage colony-stimulating factor cDNA to tumor cells: implications for a clinically relevant tumor vaccine. Hum Gene Ther 1996; 7:1535–43.
14. Andree C, Swain WF, Page CP, et al. *In vivo* transfer and expression of a human epidermal growth factor gene accelerates wound repair. Proc Natl Acad Sci U S A 1994; 91:12188–92.
15. Sun L, Xu L, Chang H, et al. Transfection with aFGF cDNA improves wound healing. J Invest Dermatol 1997; 108:313–18.
16. Larregina AT, Watkins SC, Erdos G, et al. Direct transfection and activation of human cutaneous dendritic cells. Gene Ther 2001; 8:608–17.
17. Loehr BI, Willson P, Babiuk LA, et al. Gene gun-mediated DNA immunization primes development of mucosal immunity against bovine herpesvirus 1 in cattle. J Virol 2000; 74: 6077–86.
18. Davidson JM, Krieg T, Eming SA. Particle-mediated gene therapy of wounds. Wound Repair Regen 2000; 8:452–9.
19. Andre F, Mir LM. DNA electrotransfer: its principles and an updated review of its therapeutic applications. Gene Ther 2004; 11:33–42.
20. Mir LM, Moller PH, Andre F, et al. Electric pulse-mediated gene transfer to various animal tissues. Adv Genet 2005; 54:83–114.
21. Belehradek M, Domenge C, Luboinski B, et al. Electrochemotherapy, a new antitumor treatment. First clinical phase I-II trial. Cancer 1993; 72:3694–700.
22. Mir LM, Glass LF, Sersa G, et al. Effective treatment of cutaneous and subcutaneous malignant tumours by electrochemotherapy. Br J Cancer 1998; 77:2336–42.
23. Gothelf A, Mir LM, Gehl J. Electrochemotherapy: results of cancer treatment using enhanced delivery of bleomycin by electroporation. Cancer Treat Rev 2003; 29:371–87.
24. Pack DW, Hoffman AS, Pun S, et al. Design and development of polymers for gene transfer. Nat Rev Drug Discov 2005; 4:581–93.
25. Abdelhady HG, Allen S, Davies MC, et al. Direct real-time molecular scale visualisation of the degradation of condensed DNA complexes exposed to DNase I. Nucleic Acids Res 2003; 31:4001–5.
26. Worgall S. A realistic chance for gene therapy in the near future. Pediatr Nephrol 2005; 20:118–24.
27. Huang L, Viroonchatapan E. Introduction. In Huang L, Hung MC, Wagner E, editors. Non-viral Vectors for Gene Therapy. San Diego: Academic Press, 1999:3–22.
28. Barron LG, Szoka FC. The preplexing delivery mechanism of lipoplexes. In Huang L, Hung MC, Wagner E, editors. Non-viral Vectors for Gene Therapy. San Diego: Academic Press, 1999: 229–66.
29. Xu L, Pirollo KF, Chang EH. Tumor-targeted p53-gene therapy enhances the efficacy of conventional chemo/radiotherapy. J Control Release 2001; 74:115–28.
30. Culver KW, Vickers TM, Lamsam JL, et al. Gene therapy for solid tumors. BMJ 1995; 51:192–204.
31. Hacein-Bey-Abina S, Le Deist F, Carlier F, et al. Sustained correction of X-linked severe combined immunodeficiency by ex vivo gene therapy. N Engl J Med 2002; 346:1185–93.
32. Check E. Second cancer case halts gene-therapy trials. Nature 2003; 421:305.
33. Barquinero J, Eixarch H, Perez-Melgosa M. Retroviral vectors: new applications for an old tool. Gene Ther 2004; 11:3–9.
34. Verma IM, Weitzman MD. Gene therapy: twenty-first century medicine. Annu Rev Biochem 2005; 74:711–38.
35. Hallenbeck PL, Stevenson SC. Targetable gene transfer vectors. Adv Exp Med Biol 2000; 465:37–46.
36. Kozarsky KF, Wilson JM. Gene therapy: adenovirus vectors. Curr Opin Genet Dev 1993; 3:499–03.
37. Kanerva A, Hemminki A. Adenoviruses for treatment of cancer. Ann Med 2005; 37:33–43.
38. Gomez-Navarro J, Curiel DT. Conditionally replicative adenoviral vectors for cancer gene therapy. Lancet Oncol 2000; 1:148–58.
39. Alemany R, Balague C, Curiel DT. Replicative adenoviruses for cancer therapy. Nat Biotechnol 2000; 18:723–7.
40. Hermiston T. Gene transfer from replication-selective viruses: arming guided missiles in the war against cancer. J Clin Invest 2000; 105:1169–72.
41. Heise C, Kirn DH. Replication-selective adenoviruses as oncolytic agents. J Clin Invest 2000; 105:847–51.
42. Kim M, Zinn KR, Barnett BG, et al. The therapeutic efficacy of adenoviral vectors for cancer gene therapy is limited by a low level of primary adenovirus receptors on tumour cells. Eur J Cancer 2002; 38:1917–26.
43. Barnett BG, Crews CJ, Douglas JT. Targeted adenoviral vectors. Biochim Biophys Acta 2002; 1575:1–14.
44. Stone D, Furthmann A, Sandig V, et al. The complete nucleotide sequence, genome organization, and origin of human adenovirus type 11. Virology 2003; 309:152–65.
45. Vogels R, Zuijdgeest D, van Rijnsoever R, et al. Replication-deficient human adenovirus type 35 vectors for gene transfer and vaccination: efficient human cell infection and bypass of preexisting adenovirus immunity. J Virol 2003; 77:8263–71.
46. Goldman CK, Soroceanu L, Smith N, et al. *In vitro* and *in vivo* gene transfer mediated by a synthetic polycationic amino polymer. Nat Biotechnol 1997b; 15:462–6.
47. Goldman CK, Rogers BE, Douglas JT, et al. Targeted gene transfer to Kaposi's sarcoma cells via the fibroblast growth factor receptor. Cancer Res 1997a; 57:1447–51.
48. Rancourt C, Rogers BE, Sosnowski BA, et al. Basic fibroblast growth factor enhancement of adenovirus-mediated delivery of the herpes simplex virus thymidine kinase gene results in augmented therapeutic benefit in a murine model of ovarian cancer. Clin Cancer Res 1998; 4:2455–61.
49. Stone D, Ni S, Li ZY, et al. Development and assessment of human adenovirus type 11 as a gene transfer vector. J Virol 2005; 79:5090–104.
50. Dai Y, Schwarz EM, Gu D, et al. Cellular and humoral immune responses to adenoviral vectors containing factor IX gene: tolerization of factor IX and vector antigens allows for

long-term expression. Proc Natl Acad Sci U S A 1995; 92: 1401–5.
51. Kafri T, Morgan D, Krahl T, et al. Cellular immune response to adenoviral vector infected cells does not require de novo viral gene expression: implications for gene therapy. Proc Natl Acad Sci U S A 1998; 95:11377–82.
52. Tripathy SK, Black HB, Goldwasser E, et al. Immune responses to transgene-encoded proteins limit the stability of gene expression after injection of replication-defective adenovirus vectors. Nat Med 1996; 2:545–50.
53. Xiang Z, Gao G, Reyes-Sandoval A, et al. Novel, chimpanzee serotype 68-based adenoviral vaccine carrier for induction of antibodies to a transgene product. J Virol 2002; 76:2667–75.
54. Croyle MA, Yu QC, Wilson JM. Development of a rapid method for the PEGylation of adenoviruses with enhanced transduction and improved stability under harsh storage conditions. Hum Gene Ther 2000; 11:1713–22.
55. Fisher KD, Stallwood Y, Green NK, et al. Polymer-coated adenovirus permits efficient retargeting and evades neutralising antibodies. Gene Ther 2001; 8:341–8.
56. Lu B, Makhija SK, Nettelbeck DM, et al. Evaluation of tumor-specific promoter activities in melanoma. Gene Ther 2005; 12: 330–8.
57. Glasgow JN, Bauerschmitz GJ, Curiel DT, et al. Transductional and transcriptional targeting of adenovirus for clinical applications. Curr Gene Ther 2004; 4:1–14.
58. Zaiss AK, Liu Q, Bowen GP, et al. Differential activation of innate immune responses by adenovirus and adeno-associated virus vectors. J Virol 2002; 76:4580–90.
59. Rutledge EA, Russell DW. Adeno-associated virus vector integration junctions. J Virol 1997; 71:8429–36.
60. Smith AE. Viral vectors in gene therapy. Annu Rev Microbiol 1995; 49:807–38.
61. Bowers WJ, Olschowka JA, Federoff HJ. Immune responses to replication-defective HSV-1 type vectors within the CNS: implications for gene therapy. Gene Ther 2003; 10:941–5.
62. Fielding AK. Measles as a potential oncolytic virus. Rev Med Virol 2005; 15:135–42.
63. Ponnazhagan S. Parvovirus vectors for cancer gene therapy. Expert Opin Biol Ther 2004; 4:53–64.
64. Rosenfeld M, Curiel DT. Gene therapy strategies for novel cancer therapeutics. Curr Opin Oncol 1996; 8:39–47.
65. Watanabe M, Nasu Y, Kashiwakura Y, et al. Adeno-associated virus 2-mediated intramural prostate cancer gene therapy: long-term maspin expression efficiently suppresses tumor growth. Hum Gene Ther 2005; 16:699–710.
66. Seth P. Vector-mediated cancer gene therapy: an overview. Cancer Biol Ther 2005; 4:512–17.
67. Elledge RM, Allred DC. The p53 tumor suppressor gene in breast cancer. Breast Cancer Res Treat 1994; 32:39–47.
68. Barnes MN, Deshane JS, Rosenfeld M, et al. Gene therapy and ovarian cancer: a review. Obstet Gynecol 1997; 89:145–55.
69. Greco O, Dachs GU. Gene directed enzyme/pro-drug therapy of cancer: historical appraisal and future prospectives. J Cell Physiol 2001; 187:22–36.
70. Tepper RI, Mule JJ. Experimental and clinical studies of cytokine gene-modified tumor cells. Hum Gene Ther 1994; 5:153–64.
71. Dohring C, Angman L, Spagnoli G, et al. T-helper- and accessory-cell-independent cytotoxic responses to human tumor cells transfected with a B7 retroviral vector. Int J Cancer 1994; 57:754–9.
72. Dalgleish A. The case for therapeutic vaccines. Melanoma Res 1996; 6:5–10.
73. Abonour R, Williams DA, Einhorn L, et al. Efficient retrovirus-mediated transfer of the multidrug resistance 1 gene into autologous human long-term repopulating hematopoietic stem cells. Nat Med 2000; 6:652–8.
74. Cowan KH, Moscow JA, Huang H, et al. Paclitaxel chemotherapy after autologous stem-cell transplantation and engraftment of hematopoietic cells transduced with a retrovirus containing the multidrug resistance complementary DNA (MDR1) in metastatic breast cancer patients. Clin Cancer Res 1999; 5:1619–28.
75. Takebe N, Zhao SC, Adhikari D, et al. Generation of dual resistance to 4-hydroperoxycyclophosphamide and methotrexate by retroviral transfer of the human aldehyde dehydrogenase class 1 gene and a mutated dihydrofolate reductase gene. Mol Ther 2001; 3:88–96.
76. Walker CL, Burroughs KD, Davis B, et al. Preclinical evidence for therapeutic efficacy of selective estrogen receptor modulators for uterine leiomyoma. J Soc Gynecol Investig 2000; 7:249–56.
77. Stewart EA. Uterine fibroids. Lancet 2001; 357:293–8.
78. Surrey ES, Lietz AK, Schoolcraft WB. Impact of intramural leiomyomata in patients with a normal endometrial cavity on *in vitro* fertilization-embryo transfer cycle outcome. Fertil Steril 2001; 75:405–10.
79. Hart R, Khalaf Y, Yeong CT, et al. A prospective controlled study of the effect of intramural uterine fibroids on the outcome of assisted conception. Hum Reprod 2001; 16:2411–7.
80. Di Lieto A, De Falco M, Pollio F, et al. Clinical response, vascular change, and angiogenesis in gonadotropin-releasing hormone analogue-treated women with uterine myomas. J Soc Gynecol Investig 2005; 12:123–8.
81. Gentry CC, Okolo SO, Fong LF, et al. Quantification of vascular endothelial growth factor-A in leiomyomas and adjacent myometrium. Clin Sci 2001; 101:691–5.
82. Hong T, Shimada Y, Uchida S, et al. Expression of angiogenic factors and apoptotic factors in leiomyosarcoma and leiomyoma. Int J Mol Med 2001; 8:141–8.
83. Hyder SM, Huang JC, Nawaz Z, et al. Regulation of vascular endothelial growth factor expression by estrogens and progestins. Environ Health Perspect 2000; 108:785–90.
84. Hoffman PJ, Milliken DB, Gregg LC, et al. Molecular characterization of uterine fibroids and its implication for underlying mechanisms of pathogenesis. Fertil Steril 2004;82:639–49.
85. Niu H, Simari RD, Zimmermann EM, et al. Nonviral vector-mediated thymidine kinase gene transfer and ganciclovir treatment in leiomyoma cells. Obstet Gynecol 1998; 91:735–40.
86. Christman GM, McCarthy JD. Gene therapy and uterine leiomyomas. Clin Obstet Gynecol 2001; 2:425–35.
87. Freeman SM, Abboud CN, Whartenby KA, et al. The "bystander effect": tumor regression when a fraction of the tumor mass is genetically modified. Cancer Res 1993; 53:5274–83.
88. Marwa K, Salama A, Christman GM, et al. Uterine fibroids gene therapy: adenovirus-mediated herpes simplex virus thymidine kinase/ganciclovir treatment inhibits growth of human and rat leiomyoma cells [Abstract]. J Soc Gynecol Investig 2005; 12:94A.
89. Andersen J. Comparing regulation of the connexin43 gene by estrogen in uterine leiomyoma and pregnancy myometrium. Environ Health Perspect 2000; 108:811–5.
90. Al-Hendy A, Lee EJ, Wang HQ, et al. Gene therapy of uterine leiomyomas: Adenovirus-mediated expression of dominant negative estrogen receptor inhibits tumor growth in nude mice. Am J Obstet Gynecol 2004; 191:1621–31.
91. Ince BA, Zhuang Y, Wrenn CK, et al. Powerful dominant negative mutants of the human estrogen receptor. J Biol Chem 1993; 268:14026–32.
92. Howe SR, Gottardis MM, Everitt JI, et al. Rodent model of reproductive tract leiomyomata. Establishment and characterization of tumor-derived cell lines. Am J Pathol 1995; 146: 1568–79.
93. Hodges LC, Hunter DS, Bergerson JS, et al. An *in vivo/in vitro* model to assess endocrine disrupting activity of xenoestrogens in uterine leiomyoma. Ann N Y Acad Sci 2001; 948: 100–11.

94. Al-Hendy A, Magliocco AM, Al Tweigeri T, et al. Ovarian cancer gene therapy: repeated treatment with thymidine kinase in an adenovirus vector and ganciclovir improves survival in a novel immunocompetent murine model. Am J Obstet Gynecol 2000; 182:553–9.
95. Everitt JI, Wolf DC, Howe SR, et al. Rodent model of reproductive tract leiomyomata. Clinical and pathological features. Am J Pathol 1995; 146:1556–67.
96. Cook JD, Walker CL. The Eker rat: establishing a genetic paradigm linking renal cell carcinoma and uterine leiomyoma. Curr Mol Med 2004; 4:813–24.
97. Houston KD, Hunter DS, Hodges LC, et al. Uterine leiomyomas: mechanisms of tumorigenesis. Toxicol Pathol 2001; 29: 100–4.

Part III
Molecular Regulation of Reproductive Hormones

Chapter 13

GnRH-GnRH-Receptor System in the Mammalian Female Reproductive Tract

Indrajit Chowdhury[1] and Rajagopala Sridaran[2]

13.1 Introduction

Reproduction in mammals is controlled by interactions between the hypothalamus, anterior pituitary, and gonads. The hypothalamus secretes synchronized pulses of gonadotropin-releasing hormone (GnRH) as the central initiator of the reproductive hormonal cascade from the diffusely arranged network of neuronal nerve endings of about 800 (rodents) to 1,500–2,000 (human) into the hypophyseal portal system every 30–120 min. GnRH stimulates the biosynthesis and secretion of the gonadotropic hormones, LH, and FSH by the anterior pituitary that in turn regulate the production of gametes and gonadal hormones as a key regulator of the reproductive functions. In the 1970 s, GnRH was first isolated from hypothalamus of pigs and sheep, and the subsequent realization that the decapeptide sequence was conserved across all mammals. This form of GnRH is referred as GnRH-I or type I mammalian GnRH (mGnRH). In 1977, Andrew Schally, Roger Guillemin, and co-workers shared the Nobel Prize in the field of Medicine for their discovery. In the early 1980 s, a second GnRH isoform from chickens was isolated (chicken GnRH; GnRH-II) [1] and a third isoform was identified in fish (salmon GnRH) [2]. Currently, a total of 23 different isoforms of chordate GnRH have been isolated [3–5]. All of these isoforms are decapeptides that share a high degree of sequence identity at both NH_2- and COOH-terminals. This overview describes the recent literature regarding the GnRH primary structure, tissue distribution in female reproductive system, synthesis, secretion, and signaling pathways with current understanding on their cognate receptors and functional significance in relation to female reproductive system.

I. Chowdhury (✉)
[1]Department of Obstetrics and Gynecology;
[2]Department of Physiology, Morehouse School of Medicine, Atlanta, GA, USA

13.2 Primary Structures of GnRH

The identification of mammalian GnRH gene and the translated precursor peptide of GnRH were reported by Seeburg and Adelman [6]. Using techniques, such as immunohistochemistry, in situ hybridization, polymerase chain reaction (PCR), and HPLC, the expression of the diverse forms of GnRH gene and the peptide products were detected and identified in the brains of a wide variety of species across the phylogenetic scale, from tunicate to human [5,7]. GnRH preprohormone mRNA was identified in the diagonal band of Broca (dbB) and the preoptic (POA) area of rat brain. The isolated mammalian hypothalamic GnRH has a unique structure, which is formed by a classical ten amino acid peptide (Fig. 13.1) with a pyro-glutamyl modified NH_2-terminus, an amidated COOH-terminus, and conserved amino acids in positions 1, 2, 4, 9, and 10. The most widely recognized and common structural variation among the different forms of GnRH resides in fifth and eighth amino acids of the peptide sequence. The NH_2-terminus (pGlu-His-Trp-Ser) and COOH-terminus (Pro-Gly-NH_2) sequences are conserved, indicating that these features are critically important for receptor binding and activation. Position 8 is the most variable, followed by positions 6, 5, and 7.

All the 23 different structural isoforms of GnRH peptides consists of ten amino acids and have a similar structure, with at least 50% identity in their sequence. Presently, there are about 14 structurally variant forms of known hypothalamic GnRH. In most vertebrates there are at least two, and usually three forms of GnRH occur in anatomically and distinct neuronal populations as well as different embryonic origins [5,8]. As more forms of GnRH are isolated, they are named after the animals in which they are first found, leading to a

GnRH I: pGlu^1-His^2-Trp^3-Ser^4-**Tyr^5**-Gly^6-**Leu^7**-**Arg^8**-Pro^9-Gly^{10}-NH_2

GnRH II: pGlu^1-His^2-Trp^3-Ser^4-**His^5**-Gly^6-**Trp^7**-**Tyr^8**-Pro^9-Gly^{10}-NH2

Fig. 13.1 A comparative mammalian GnRH primary structure

P.J. Chedrese (ed.), *Reproductive Endocrinology*,
DOI 10.1007/978-0-387-88186-7_13, © Springer Science+Business Media, LLC 2009

rather confusing nomenclature. This general classification of GnRH is based on GnRH encoding regions and a model of two GnRH neuronal systems, namely, the terminal nerve-septo-preoptic system and the midbrain system. Cells of the terminal nerve-septo-preoptic system are originated from olfactory placode, migrate during development into the brain, colonizing the ventral forebrain, and are mostly restricted to the ventral telencephalon, the preoptic area, the basal hypothalamus, and the pituitary gland. These neurons express different GnRH variants according to their respective species and are responsible for the regulation of pituitary function. This form of GnRH is called hypophysiotropic/hypothalamic form of GnRH and designated as GnRH-I. The amino acid sequence of the hypophysiotropic GnRH is identical across mammals, with the exception of the guinea pig in which there are substitutions of amino acids 2 and 7.

The most ubiquitous is chicken GnRH-II (cGnRH-II) system, which was first isolated from chicken brain [1]. cGnRH II is also designated as Type II GnRH/midbrain GnRH or GnRH-II (Fig. 13.1). GnRH-II system has an embryological origin from a non-placodal structure, which gives rise to the posterior system. The posterior neurons express GnRH-II cells that are clustered in a distinct nucleus at the fusion site of the anterior midbrain and the posterior diencephalon. The conserved GnRH-II, expressed by mid-brain neurons, may play a role as a neurotransmitter and/or neuromodulator. Since the GnRH-II structure is totally conserved from teleost fish to human, this is probably the earliest evolved form and has critical functions. In many vertebrate species a third form of GnRH (salmon GnRH) is present and is designated as GnRH-III or the nervus terminalis-telencephalic form. The expression of GnRH-III in human is not yet confirmed.

13.3 Localization and Distribution Pattern of GnRH in the Female Reproductive System

Various techniques, such as immunogold electron microscopy, in-situ hybridization, real time-PCR, and immunocyto-/histo-chemistry confirmed the transcription and translation of GnRH in the peripheral tissues [7,9–17]. The presence of immunoreactive GnRH-I has been demonstrated, apart from hypothalamus, in the ovary, oviduct, placenta, endometrium, trophoblast, fallopian tubes during the luteal phase of the reproductive cycle, in pre-implantation embryos, uterine endometrial cells of human, and specifically in other mammalian reproductive tissues/cell lines. GnRH-II is also widely distributed in the central neural system (pre-optic, mediobasal hypothalamic area, hippocampus, caudate nucleus, amygdale of brain) and peripheral tissues (kidney, bone marrow, endometrium, ovary, trophoblast, placenta, prostate) and ovarian cancer cells. The immunoreactive GnRH-II is present in the mononucleate villus and extravillous cytotrophoblast of placenta. In human, GnRH-II is expressed at significantly higher levels in the peripheral tissues. The expression level of GnRH-II varies from 4- to 30-fold. GnRH or GnRH-like molecules have been detected in human and cow follicular fluid. In situ hybridization studies revealed the localization of GnRH-I and -II mRNA in granulosa cells of primary, secondary, and tertiary follicles in the ovary, ovarian surface epithelial cells, and ovarian cancer cells. Furthermore, immunolocalization studies have shown that both GnRH-I and -II are absent from primordial to the early antral stage, however, both forms are present in the granulosa cell layer of preovulatory follicles, granulosa luteal cells (GLCs), theca luteal cells (TLCs), immortalized ovarian surface epithelial cells (OSE), and in ovarian cancer cells.

13.4 Synthesis and Secretions of GnRH

Synthesis, amplitude, and frequency of secretions of GnRH involve complex processes with many feed-back loops. Like many other proteins and peptides, GnRH is synthesized as precursor and enzymatically processed. Transcription rate, mRNA stability, and post-transcriptional processing potentially influence the rate of production of mature GnRH. Most studies have involved measurement of steady-state levels of mRNA, using techniques such as in situ hybridization and RT-PCR. However, there have been few studies on GnRH synthesis and secretion in female reproductive tissues/cells [7,18].

The gene of GnRH-I is located on the eighth chromosome, 8p11.2—-p21, of the human genome [19]. The gene encoding prepro-GnRH-I have been cloned in a number of mammalian species and is approximately 4,300 bp with four relatively short exons separated by three large introns. Exon 1 encodes the 5′-untranslated region. Exon 2 encodes the signal peptide, GnRH-I decapeptide, and the first 11 amino acids of the GnRH-associated peptide (GAP) region. Exon 3 encodes amino acids 12–43 of GAP and exon 4 encodes the remainder of GAP and the 3′-untranslated region. Gene sequence homology has demonstrated that GnRH-I is conserved throughout evolution and shares a 60% identity between mammals and tunicates. Compared to GnRH-I, the GnRH-II is highly conserved with 100% identity between birds and mammals and ubiquitous in the vertebrates. In human, the gene encoding GnRH-II is present on the 20th chromosome, 20p13. Similar to GnRH-I, GnRH-II gene is organized with 4 exons and 3 introns, but significantly smaller in length.

Most of the studies of transcription of GnRH are restricted to preoptic area of anterior hypothalamus (POA-AH) of rat and human. All three pro-GnRH-I introns are spliced out of

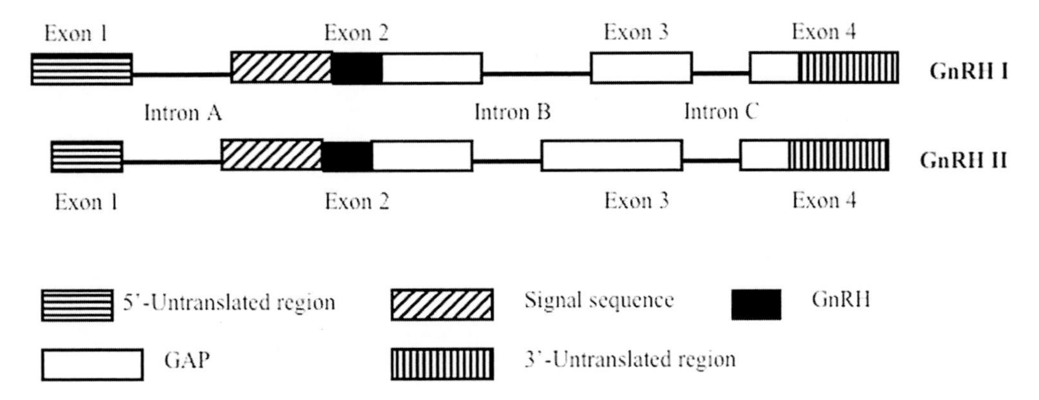

Fig. 13.2 A comparative structure of mammalian GnRH gene

the primary gene transcript resulting in a mature mRNA of about 560 bases, excluding the poly (A) tail (Fig. 13.2). The first processing step of prepro-GnRH I in the POA-AH of rats or in mice cell line are splicing out of intron B and subsequently introns A and C are spliced out in no apparent order. The splicing of introns B and C occurs with greater efficiency than that of intron A. The precise and efficient excision of intron A and the joining of adjacent exons is the most critical regulatory step for the post-transcriptional regulation of GnRH. In human, the pre-pro-GnRH is 92 amino acid long including 56 amino acids in the GAP region.

Both isoforms of GnRH are characterized by post-translational modifications by a *prohormone convertase*. The basic amino acids are removed by a carboxypeptidase, the amino terminus is modified by adding pyro-glutamic acid through the action of *glutaminyl cyclase* and the carboxy-terminus is modified by addition of *glycine* through the action of *peptidylglycine α-amidating mono-oxygenase*. These enzymatic processes produce mature GnRH and GAP. A physiologic role for GAP has not been determined, although it has been postulated to act as a prolactin-inhibitory factor.

Translational studies have demonstrated that both GnRH-I and II are expressed during the follicular development [9,10] in human granulosa-luteal (hGL) and ovarian epithelial (OE) cells, and in both normal and cancer cells. Recent studies using real time PCR and immuno-cytochemical localization have demonstrated expression patterns of GnRH-I and II in rat ovaries throughout the estrous cycle [20] and relatively differential expression patterns in oviduct during post-implantation period of pregnancy [16]. Also, during the menstrual cycle, corpus luteum of monkey has a differential expression pattern of GnRH-I and II [12].

13.5 GnRH Receptor

GnRH acts on target cells/tissues via $G_{q/11}$-coupled seven trans-membrane (7TM) receptor (GnRHR), a typical member of the type I rhodopsin-like receptor family, which consists of seven hydrophobic trans-membrane chains connected to each other with extracellular and intracellular loops [21,22]. The trans-membrane (TM) domains are connected with three extracellular loop (EL) and three intracellular loop (IL) domains (Fig. 13.3). The EL domains are connected with the binding peptides, the TM domains are believed to participate in receptor activation and the transmission of signals, and the IL domains are involved in the interaction with G proteins and also other proteins participating in the intracellular signal transmission.

The GnRH receptors gene were cloned and characterized in mouse, rat, pig, sheep, and humans [21,22]. Since then, it has been identified in other peripheral tissues using various molecular biology techniques. In human, the conventional GnRH receptor (GnRHR-I) gene is located on chromosome

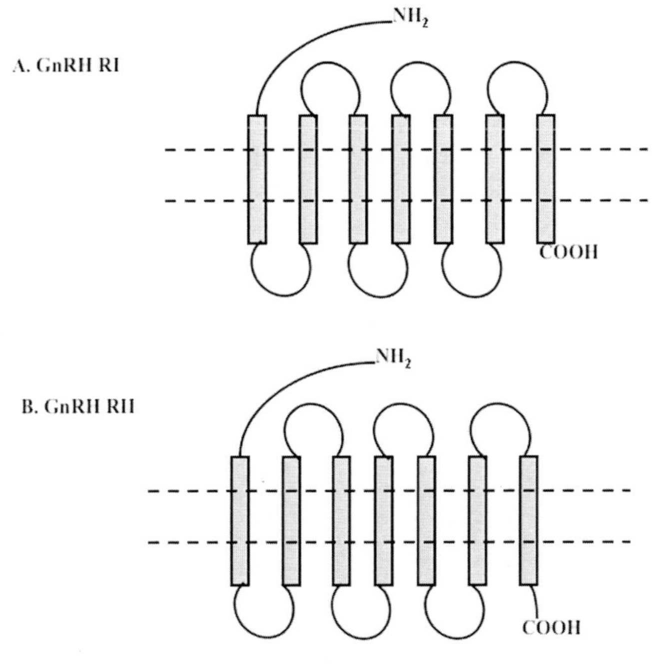

Fig. 13.3 Seven transmembrane structure of GnRHR-I and II

4q13.1-q21.1 with many transcription initiation sites that encodes three exons and two introns and translates for 327–328 amino acids long peptide. Similar to humans, in all mammalian species GnRHR-I gene exist as a single copy, and have a high degree of sequence homology in the coding regions with three exons separated by two introns (Fig. 13.4). The exon–intron boundaries are conserved between species, but the gene differs with regard to the size of the introns, as well as the sequence and the length of the 5′- and 3′-untranslated region (UTR). The tissue specific expressions of GnRHR-I gene are mediated by differential use of various promoter regions. Recently, the ovarian and placental specific promoters were identified. Exon 1 encodes the N-terminal tail as well as TM helices 1, 2, 3, and part of TM 4 of the GnRHRI protein. Exon 2 encodes the rest of TM 4 and the whole of TM 5, whereas exon 3 encodes TM 6 and 7. A common feature of most cloned mammalian GnRHR-I is the absence of a carboxy terminal tail that exists in all other receptors of the GPCR. The absence of the carboxy terminal tail leads to a slow internalization and failure of a quick desensitization of the receptor [11,21,22], which mediates the central control of reproduction.

A putative GnRHR type II gene is present on two different loci in the human genome. The first is located on chromosome 1q12 and overlaps in antisense orientation with the gene encoding the RNA-binding motif protein-8A (RBM-8A) [23]. The GnRHR-II gene has the same exon–intron structure as GnRHRI except that exon 3 includes a cytoplasmic C-terminal tail, which is absent in GnRHR-I. Cytoplasmic C-terminal is present in all non-mammalian GnRHR-II and in recently cloned primate GnRHR-II. In the open reading frame of this gene is disrupted by a frame shift with a premature stop codon (UAA), which is located inframe within exon 2 in the human gene, similar to chimpanzee, marmoset monkey, rhesus monkey, African green monkey, cow, sheep, horse, and rat that makes the gene silent, suggesting that the gene products of GnRHR-II are non-functional. A second human locus containing a pseudogene for GnRHR-II and RBM-8A is on chromosome 14. GnRHR-II gene has also been detected in other mammalian genomes. In human, GnRHR-II exons have 40% sequence identity with the GnRHR-I. Thus, both the GnRH-I and II binds to GnRHR-I only. In contrast, GnRHR-II genomic sequences for some non-primate mammals have also been identified and potentially encode functional proteins as in pigs and

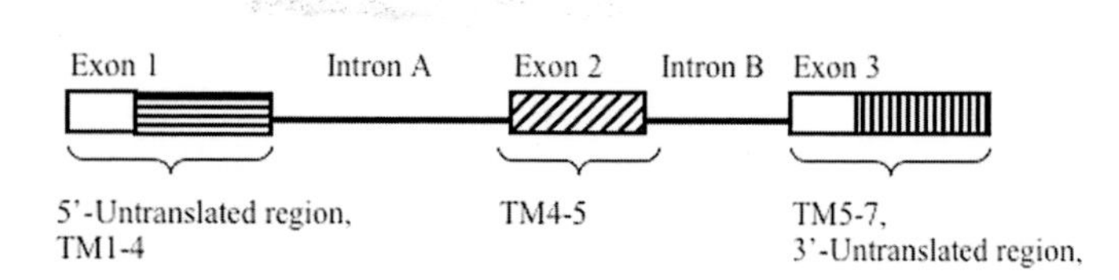

Fig. 13.4 GnRH receptor gene structure. TM: transmembrane domain

dogs. The GnRHR-II gene is completely deleted from mouse genome.

In mammals, besides pituitary, the GnRHR-I and GnRHR-II transcripts are detected in peripheral tissues and cell lines such as endometrium, myometrium, placenta, breast, ovary, oviduct, uterus, prostate, liver, heart, skeletal muscle, kidney, peripheral mononuclear cells, and cancer cells [7]. The density of GnRHRs in the reproductive tissues varies with the state and stage of the reproductive cycle. Immunolocalization studies have shown that both GnRHRs are absent from primordial to early antral stage of ovarian follicles. However, GnRHR-I is present in granulosa cell layer of preovulatory follicles, GLCs, TLCs, and cancer OSE cells [9,10]. The GnRHR is expressed in hGL cells aspirated from preovulatory follicles of women undergoing ovarian stimulation for in vitro fertilization (IVF). The GnRHR mRNA level in the ovary is almost 200-fold lower than in the pituitary. In situ hybridization studies demonstrated a weak expression of the GnRHR mRNA in the corpus luteum. The GnRHR is also present in the ovarian compartments other than follicular and luteal structure.

Recent studies using real time PCR, Western blot, and immuno-cytochemical localization demonstrated expression patterns of GnRHR in rat ovaries throughout the estrous cycle [20] and relatively differential expression pattern in oviduct during post-implantation period of pregnancy [24]. A differential expression pattern of GnRHR-I and -II has been demonstrated in the corpus luteum of monkey during the menstrual cycle [12].

13.6 GnRH Signaling

Upon GnRH binding to its receptors (GnRHRs), a wide range of intracellular signaling pathways are activated in the pituitary and peripheral tissues/cells, which ultimately regulate the synthesis and secretions of gonadotropins, steroids, gametogenesis, and other physiological functions [20,25]. Ligand binding to GnRHRs activates multiple G-proteins (Gαq, Gs, Gi) based on cell context dependent manner and form different signaling complexes with different conformations in target cells, the inositol phosphate pathway, extracellular signal regulated kinases (ERK), jun-N-terminal kinase (JNK), p38-mitogen activated protein kinase (p38MAPK) kinetic pathways, and arrestin-dependent pathways via both protein kinase-A (PKA-) and protein kinase-C (PKC)-dependent and -independent pathways [26–28] (Figs. 13.5 and 13.6). In pituitary gonadotropes GnRHR stimulates preferentially Gq, whereas in tumors, Gi. The C-terminus of both types of GnRH receptors are phosphorylated in response to GnRH and leading to receptor desensitization through the activation of GnRHR coupled Gq/G11 proteins with activation of phospholipase Cβ (PLCβ) that transmits its signal to diacylglycerol (DAG) and inositol 1,4,5-triphosphate (IP3), which are apparently required for conventional intracellular PKC pathway activation and intracellular calcium mobilization. After a short lag, GnRH also activates phospholipase A_2 (PLA_2) and phospholipase D (PLD) apparently to provide late DAG and arachidonic acid (AA) for novel PKCs isoforms activation. AA formation stimulates the formation of 5-lipoxygenase products, however, mechanisms are not yet clarified. Based on functional studies by measuring inositol phosphate production, it has demonstrated that primate GnRHR-I has an approximate 48-fold selectivity for GnRH-I versus GnRH-II, whereas GnRHR-II has a 421-fold preference for GnRH-II versus GnRH-I. PKC activation in response to GnRH also leads to increase activation of MAPK pathways including ERK1, ERK2, ERK5, p38MAPK, and JNK in pituitary cells, gonadotrope cell lines, and in other cells. In gonadotrophs, GnRH activated ERK pathways differentially regulates the transcription, translation, and secretions of α and β subunits of LH and FSH. All these kinetic pathways are activated simultaneously, sequentially or in isolation, and determine by spatiotemporally.

In addition to the direct effects of the GnRHR in activating intracellular signaling, the endogenous mouse GnRHRs mediate PKC-dependent trans-activation of epidermal growth factor (EGF) receptors, with consequent activation of ERK in GT1-7 neurons but cause a large PKC-dependency and EGF receptor-independent activation of ERK in LβT2 gonadotroph lineage cells. The results of recent work suggest that a cross talk with the EGFR may occur at the level of the pituitary. EGFR can be activated via GnRHR by proteolytic release of local EGF-like ligands from transmembrane precursors, which finally activates matrix metalloproteinase (MMP) 2 and 9. GnRH stimulation of the cells induced Src, Ras, and ERK are dependent on the action of the MMPs, whereas activation of c-Jun N-terminal kinase and p38MAPK by GnRH are unaffected by the inhibition of EGFR or MMPs. Together, these studies support direct and indirect effects of GnRH/GnRHR action on diverse intracellular signaling pathways. Thus, the signal transduction pathways of GnRHR has coupled to different Gα proteins in various ovarian cell types, and modulation

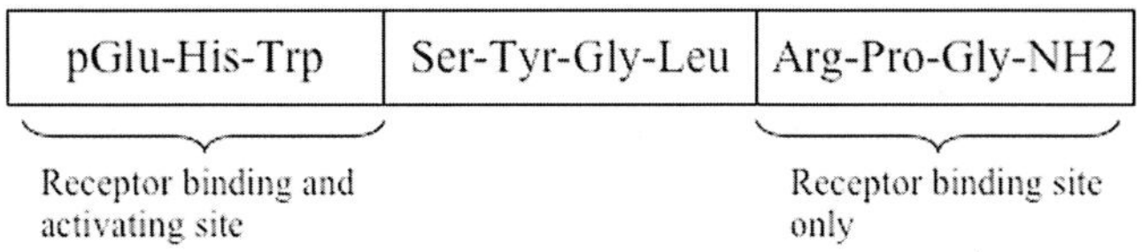

pGlu-His-Trp	Ser-Tyr-Gly-Leu	Arg-Pro-Gly-NH2
Receptor binding and activating site		Receptor binding site only

Fig. 13.5 GnRH I—GnRH receptor interaction sites

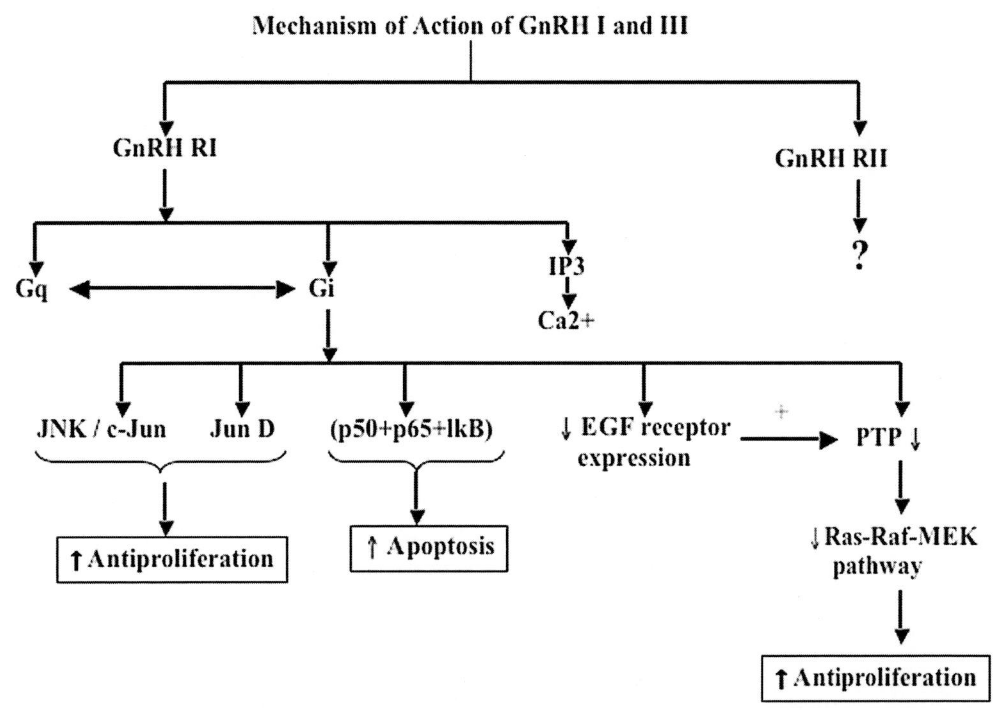

Fig. 13.6 GnRH I and GnRH II signaling in female reproductive system. Upon binding of GnRH I with GnRHR-I activate a phosphotyrosine phosphate (PTP) via GTP-binding proteins which inhibit the mitogenic signal transduction of growth factor receptors, resulting in down-regulation of cell proliferation or by down-regulation of epidermal growth factor (EGF) receptor mRNA expression or activates c-Jun N-terminal kinase (JNK), induces JunD-DNA binding, and stimulates activator protein (AP-1) activity, which reduces proliferation by increasing $G_{0/1}$ phase of cell cycle and decrease DNA synthesis. GnRH I also activates nuclear factor κB (NFκB) through GnRHR-I. Activated NFκB binds to specific DNA site and induces expression of anti-apoptotic mechanisms. The signal transduction pathway for GnRHR-II is unknown

of the activity of MAP kinases (MAPK) appeared to be an important step in mediating GnRH I responses in the ovary.

13.7 Regulation of GnRH and GnRHR Gene Expressions in the Female Reproductive System

The major regulations of GnRHs and GnRHRs have been shown to occur at the transcriptional, translational, and post-translational levels. Most of the studies are restricted to cell lines and few are on female reproductive tissues [20,29,30] (Fig. 13.7).

Studies on transcriptional controls of GnRH have been done largely in GnRH-secreting cell lines, such as GT1-7, GN10, and GN11. The 5′ flanking region of GnRH gene is highly homologous between species. It comprises two major regions for transcription, an enhancer and a promoter. A conserved 173-bp promoter is found just upstream of the transcriptional start site and deoxyribonuclease I footprint analysis of the rat GnRH promoter revealed at least seven regulatory regions located between −173 and 112. The promoter contains binding sites for transcription factors, such as POU-homeodomain Oct-1 and the neuron-restricted homodomain protein, Otx2. The proximal promoter also includes *cis*-acting elements involved in hormonal regulation by glucocorticoids, estradiol and progesterone, and 12-O-tetradecanoylphorbol 13-acetate. A second important region for transcription of the rat GnRH promoter is a 300-bp enhancer that is −1,863 to −1,571 relative to the start site; this confers a 50- to 100-fold transcriptional activation over the promoter alone, at least in GT1-7 cells. Within the enhancer, specific binding sites for a number of transcription factors have been identified such as the zinc finger protein, GATA-4, and pbx-related protein.

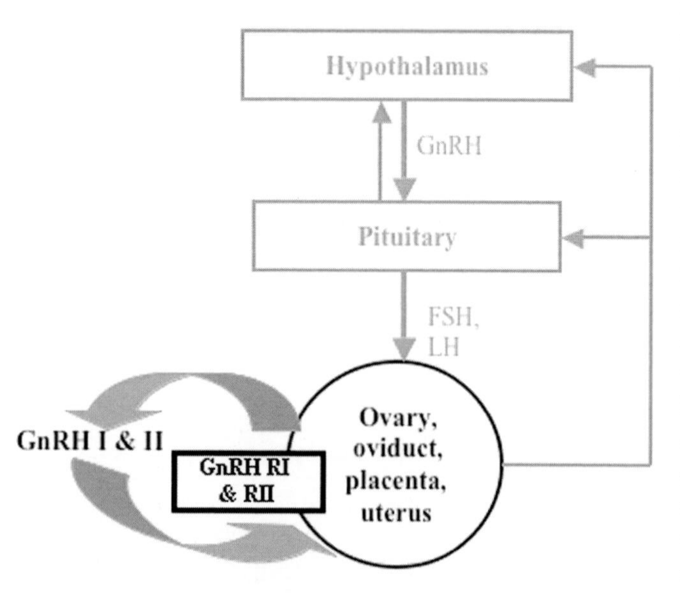

Fig. 13.7 Hypothalamus-pituitary-gonadal axis with autocrine-paracrine loop of GnRH I and II in the female reproductive system

Similar to GnRH gene expression, the regulation of GnRHR numbers has been shown at the transcriptional, translational, and post-translational level. A well-known mechanism for ligand mediated post-translational down-regulation of GPCR numbers on the cell surface involves desensitization, internalization, and degradation. The mammalian GnRHR-I has been shown to internalize slowly due to the lack of a C-terminal tail, whereas GnRHR-II in the marmoset monkey internalizes rapidly, which has a C-terminal cytoplasmic domain. However, very little research has been reported on post-transcriptional regulation of GnRHR-I gene expression. Based on in vivo and in vitro studies using mouse, rat, or human pituitary primary cells or secondary cell lines as an emerging understanding of the complex transcriptional regulatory pathways by which mammals regulate the gene transcription of GnRHR-I gene. An extensive characterization of the human GnRHR-I gene promoter has been performed in human reproductive tissue derived cell lines. However, cell lines differ from mature primary gonadotrophs in that they do not express or secrete the gonadotrophin β-subunits, LHβ and FSHβ.

The 5′ flanking region of GnRHR-I gene has been characterized in the mouse, rat, human, and sheep. The mouse and rat promoters share >80% homology over 1.9 kb; the rat promoter shares 55% homology with the human promoter over 2.2 kb and 63% homology with the sheep promoter over 0.9 kb. There are several highly homologous regions within the proximal 500 bp of the mouse, rat, human, and sheep promoters. A number of *cis*-elements have been conserved, in sequence as well as position, supporting their role as important functional elements. No functional characterization of mammalian GnRHR-II promoters has yet been known.

The mouse GnRHR-I proximal promoter was the first to be isolated and characterized. The major transcription start site in primary pituitary tissue and αT3-1 cells is located at −62 (all numbering is relative to the translation start site) and is not associated with a consensus TATA box. In addition to this site, other pituitary transcription start sites at −90 and −200 bp are identified in αT3-1 cells. Gonadotroph-specific activity of the mouse promoter in αT3-1 cells is conferred by a tripartite basal enhancer, which includes binding sites for steroidogenic factor-1 (SF-1) at −244/−236, and activator protein-1 (AP-1) at −336/−330, respectively, as well as an element originally termed GnRHR activating sequence (GRAS) at −391/−380. The pan pituitary homeobox transcription factor Pitx-1 has been shown by chromatin immunoprecipitation assay to interact with AP-1 in intact LbT2 cells, and functional evidence in other cell types indicate that this interaction might be important for

GnRHR-I gonadotroph-specific, basal promoter activity. In addition, the promoter region around −360, shown to bind LHX3 homeodomain protein in vitro and in intact cells, are recently demonstrated to be important for mouse GnRHR-I basal promoter activity in αT3-1 cells.

Experiments with transgenic mice suggest tissue-specific promoter usage for the mouse GnRHR-I gene because 1,900 bp of mouse GnRHR-I 5′ flanking sequence can drive reporter expression in pituitary, brain, and testis, but not in the ovary, indicating an essential requirement for promoter elements located further upstream for ovary-specific expression in vivo. Five major transcription start sites are identified in αT3-1 cells, four of which are clustered around −103 and one situated at −30, along with several minor start sites. Maximal gonadotroph-specific expression of the rat GnRHR-I has conferred by multiple regulatory domains within 1,260 bp of 5′ flanking region. The proximal 183 bp constitutes a self-sufficient, but fairly weak promoter, which confers basal but not gonadotroph-specific activity. A distal GnRHR-specific enhancer (GnSE), located between −1135 and −753, contains binding sites for GATA-related and LIM homeodomain-related factors, and facilitates gonadotroph-specific expression through functional interaction with an SF-1 site at −245.

There are different mechanisms involved in gonadotroph-specific expression of the mouse and rat GnRHR-I, although both involve SF-1 sites. The mouse, rat, and human promoters all contain several SF-1 sites, with at least one site in each promoter occurring in the 5′ untranslated region. For the human promoter, this site is situated at −140/−134 and is primarily responsible for mediating high cell-specific expression in αT3-1 cells, whereas the same function has not been assigned for similar sites in the mouse and rat promoters (situated at −15/−7 in both species). An upstream Oct-1 binding at −1,718 is also required for basal activity of the human promoter in αT3-1 cells.

The regulatory elements involved in expression of the mouse, rat, and sheep GnRHRs have not been characterized in cells other than pituitary cell lines. However, cell-specific *cis-* and *trans*-elements have recently been identified for the human promoter in ovarian, placental, and neuronal medulloblastoma cell lines. Expression of the human GnRHR-I gene in both αT3-1 mouse gonadotroph cells, and OVCAR-3 human ovarian carcinoma cells requires two promoter regions, located between −771/−557 and between −1351/−1022. However, different *trans*-acting factors appear to bind to these regions in the different cell-types, possibly providing a mechanism for cell-selective expression. Two additional upstream promoters are responsible for high expression levels in human placental and ovarian GLCs, respectively. The granulosa cell specific promoter is situated between −1,300 and −1,018 that contains a GATA element and two putative CCAAT/ enhancer binding protein (C/EBP) motifs that were shown to be crucial in regulating GnRHR-I transcription in the human ovarian granulosa-luteal cell lines SVOG-4o and SVOG-4m [30]. GnRHR-I expression in human placental cells requires a distal promoter region, located between −1737/−1346, in combination with a proximal region, between −707 and −167. At least five placental transcription start sites were identified within the distal promoter region. A strong negative regulatory element is located between −1,018 and −771, with a strong positive regulatory region between −771 and −577. The distal placenta-specific promoter also contains an Oct-1 and an AP-1 binding site, required for basal expression in placental cells and other cell-types, as well as a cAMP response element (CRE) and a GATA element, essential for placenta-specific expression.

Taken together, these studies indicate that various reproductive tissues differentially utilize downstream and upstream promoter elements and transcription factor binding sites for tissue-specific transcription of the human GnRHR-I gene. The transcription factor Oct-1 appears to regulate basal GnRHR-I gene expression both positively and negatively, depending on the species and cell-type. As already mentioned, Oct-1 is required for basal expression of the human GnRHR-I gene in several cell types, including placental, ovarian, and gonadotroph cell lines, via an Oct-1 binding site at −1,718. On the other hand, in placental JEG-3 cells, ovarian OVCAR-3 cells, and αT3-1 cells, Oct-1 acts as a potent repressor of the human GnRHR-I promoter via a negative regulatory element (NRE) at position −1,017. Oct-1 is also involved in basal and GnRH-stimulated activity of the mouse GnRHR-I promoter in αT3-1 cells via the SURG-1 (Sequence Underlying Responsiveness to GnRH) element. The mouse CRE has been found to be essential for basal promoter activity in some pituitary cell lines, such as LbT2 gonadotroph cells and GGH3 somatolactotroph cells, but the rat CRE does not appear to be involved in basal promoter activity in αT3-1 cells. A CRE at position −1,650 is required for placenta specific expression of the human GnRHR-I gene. These findings indicate a cell- and/or species-specific contribution of CREs to basal GnRHR-I expression levels.

Other critical regulating factors of GnRHs and GnRHRs gene expressions in female reproductive system are steroids, gonadotropins (FSH and LH), melatonin, hCG, and GnRH itself in an autocrine and paracrine manner.

13.7.1 Regulation by Gonadal Steroids

Gonadal steroids profoundly influence reproductive functions by supporting maturation and development of reproductive organs and are apparently responsible for success of reproduction. Estradiol and progesterone are the main gonadal steroids in female. The effects of estradiol and progesterone on transcription of GnRH and its receptor are studied in the ovary and in ovarian cancer cells, although the exact mechanism remains unclear. The expression of GnRH-I and GnRH-II genes are differentially regulated by 17β-estradiol in a neuronal cell line, TE-671. The gonadal steroid has shown to up-regulate the mRNA levels of GnRH-II but down-regulate GnRH-I. The GnRH-I mRNA levels are down-regulated by 17β-estradiol in human granulosa/luteal cells and ovarian carcinoma OVCAR-3 cells. This effect can be reversed by co-treatment with an estrogen antagonist, tamoxifen, indicating that the effect of estradiol is mediated via the estrogen receptor (ER). These findings suggest that the regulation of human GnRH-I gene expression by 17β-estradiol is mediated at the promoter/transcriptional level when ER alpha (ERα) is over-expressed. On the contrary, the expression of ER mRNA transcripts and proteins in normal human OSE cells appear to be estrogen-insensitive, since treatment with 17β-estradiol do not alter GnRH-I mRNA levels significantly. The estrogen insensitivity in OSE cells may partly due to their lower expression levels of both ERα and ER beta (ERβ), as compared with the ovarian carcinoma that is supported by estrogen responsive ovarian cancer cell lines with high levels of ER expression in response to estrogen treatment. GnRH-I have been shown to suppress the expression of both ERα and ERβ via a PKC-dependent pathway in the granulosa cells, thus, counteracting the inhibitory actions of estrogen on its own expression.

Responsiveness of GnRHR-I transcription to estradiol appears to vary between ovarian cell types. Treatment with estradiol in human primary OSE cells and OVCAR-3 cells caused a significant down-regulation of GnRHR-I mRNA. In hGL cells, short-term oestradiol treatment (6 h) increased GnRHR-I mRNA levels, whereas long-term treatment (48 h) decreased GnRHR-I mRNA levels. Estradiol also represses GnRHR-I transcription in a ERα-dependent and ERβ-independent way, via an AP-1-like motif at −130/−124 in ovarian cancer cells. Repression of GnRHR-I promoter activity by estradiol did not involve direct binding of the ER to the AP-1 site, suggesting that the ER interacts with other proteins bound to this motif, such as c-Jun or c-Fos.

The role of progesterone in the expression of GnRH-I and GnRH-II has been investigated in cultured hGL cells. The treatment of hGL cells with RU486, a progesterone antagonist, does not affect the levels of GnRH-I. In contrast, GnRH-II mRNA were increased by the progesterone antagonist in a dose- and time-dependent manner, suggesting that endogenous progesterone eventually plays an inhibitory role on GnRH-II expression in the ovary. Progesterone has a positive effect on GnRHR-I promoter activity in the JEG-3 placental cell line in contrast to the repression observed in αT3-1 cells. The GRE/PRE at position −535/−521 has shown to mediate progesterone receptor (PR) regulation in both αT3-1 gonadotroph and JEG-3 placental cells. Furthermore, it has been shown that both PR-A and PR-B isoforms bound to the progesterone receptor elements (PRE) in vitro and that the balance between PR-A and PR-B over-expression in the different cell lines can determine the response to progesterone. Whereas PR-A inhibits transcription in both placental and pituitary cells, PR-B activates transcription in placental cells, and inhibits transcription in pituitary cells.

13.7.2 Regulation by Gonadotropins

The expressions of GnRH-I and II have been shown to be regulated differentially by gonadotropins (FSH, LH) and hCG, which bind to their cognate receptors and stimulate intracellular cAMP production. In hGL cells, treatment with FSH increased GnRH-II but decreased GnRH-I mRNA levels in a dose and time-dependent manner. Using Southern blot fluorescence immunocytochemical analysis and radioimmunoassay demonstrated that cAMP treatment up-regulates GnRH-II mRNA and protein levels in the neuronal TE-671 cells. The stimulatory effects of cAMP on GnRH-II gene expression are through promoter mediation, and a putative cAMP-responsive element located between nucleotides 67 and 60 (relative to the translation start site ATG) within the human GnRH-II promoter is responsible for both the basal and cAMP-induced promoter activities.

FSH, LH, and hCG have inhibitory role on GnRHR mRNA expression. hCG down-regulates GnRHR-I mRNA levels in primary ovarian granulosa-luteal cells without changing GnRH-I expression. In contrast, hCG increased GnRH-II mRNA levels.

13.7.3 Regulation by Melatonin

Melatonin is a pineal hormone that regulates the dynamic physiological adaptations to changes in day length in seasonally breeding mammals. Until recently, regulation of reproductive function by melatonin was assumed to be restricted at the level of the pituitary and the hypothalamus. However, the presence of melatonin in the follicular fluid and of melatonin binding sites in the ovary suggests a role for this hormone in the ovary. By RT-PCR and Southern blot hybridization, RNA transcripts encoding two melatonin receptor subtypes, MT1-R and MT2-R, have been shown to be expressed in hGL cells. These studies suggest that melatonin can also exert effects on reproductive axis by directly binding to granulosa cells in the ovary. In human granulosa cells, both types (MT1 and MT2) of melatonin-receptors are present and melatonin up-regulates LH mRNA-receptor too. LH is essential for the initiation of leuteinization. Melatonin treatment enhances the human chorionic gonadotropin (hCG) stimulated progesterone secretion with an inhibition of GnRH and GnRH receptor expression. In support of this, melatonin reduces both GnRH-I and GnRHR-I mRNA levels in human primary granulosa-luteal cells in a dose-dependent manner. Since GnRH-I has been implicated as a luteolytic factor in the granulosa cells, this melatonin induced down-regulation of GnRH-I and its receptor expressions may play a role in interfering with the demise of corpus luteum during the mid to late luteal phase, suggesting that melatonin directly regulates the ovarian function. GnRH in the ovary act as an important paracrine and/or autocrine regulator and may be involved in the regression of corpus luteum [31].

13.7.4 Autocrine/Homologous Regulation of GnRH and GnRHR

Levels of GnRH-I and GnRH-II mRNA are differentially regulated by GnRH. Treatment of human granulosa/luteal cells and OSE cells with GnRH-I produced a biphasic effect on GnRH-I mRNA levels. The higher concentrations of GnRH-I decreased GnRH-I mRNA levels, whereas low concentrations resulted in up-regulation of GnRH-I gene expression. In contrast, treatment with different concentrations of GnRH-II resulted in homologous down-regulation of its own mRNA levels.

In rat primary granulosa cells the homologous regulation of GnRHR-I by GnRH-I has not been demonstrated consistently. The steady-state mRNA levels of GnRHR-I are differentially regulated in the granulosa/luteal cells by GnRH-I and GnRH-II. Treatment with GnRH-I produced a biphasic change on GnRH-I receptor mRNA levels, whereas treatment with GnRH-II down-regulated GnRH-I receptor expression at various concentrations. Homologous regulation of translation efficiency from GnRHR mRNA has been shown in αT3-1 cells.

Treatment of human granulosa cell lines (SVOG-4o and SVOG-4m) with high and low doses of GnRH-II induced a significant decrease in GnRHR-I mRNA levels, whereas GnRH-I induced a down-regulation at high doses and an up-regulation at low doses, showing that the two ligands regulate GnRHR-I transcription differentially. In the human choriocarcinoma JEG-3 and the immortalized extravillous trophoblast IEVT placental cell lines, the human GnRHR-I mRNA is up-regulated after 24 h of continuous stimulation with GnRH-I. This may be a tissue-specific mechanism through the PKC pathway, as shown in αT3-1 cells, and/or the PKA pathway to help to maintain GnRH-I stimulated hCG secretion throughout pregnancy. This may involve, because the human GnRHR gene is up-regulated by activators of the PKA pathway, via binding of CREB to two AP-1/CRE elements [11].

13.8 Functional Roles of GnRH in Female Reproductive System

Besides the well-established role for GnRH-I and GnRHR-I in gonadotropin regulation in the pituitary, the detection of both forms of the hormones and receptors in multiple mammalian non-pituitary tissues and cells suggests numerous and diverse autocrine, paracrine, and endocrine extra-pituitary roles for GnRHs and GnRHRs [11,22]. It has been demonstrated that both forms of GnRH function as local paracrine and autocrine factors in the mammalian ovary/ovarian follicular development via regulating steroidogenesis, cell proliferation, and apoptosis [32–36]. However, detailed functional studies on female reproductive tract are lacking.

13.8.1 Antigonadotropic Effects of GnRH

GnRH-I possesses antigonadotropic effects in the ovary by down-regulating the expression of FSH and LH receptors, inhibiting gonadotropins-stimulated cAMP production and steroidogenic enzymes. Treatment with GnRH-I agonist rapidly stimulated the phosphorylation of the extracellular signal-regulated kinase (ERK) and caused a significant increase in Elk-1 phosphorylation and c-*fos* messenger RNA expression in human granulosa/luteal cells. Activation of ERK in the granulosa cells is dependent on protein kinase C (PKC) and involvement of the Gαq protein. Pretreatment with MEK 1 inhibitor, PD98059, completely reversed the GnRH-I-inhibited progesterone production. These observations suggest that GnRH-I down-regulates steroidogenesis in the granulosa cells via a PKC-dependent and ERK signaling cascades. Other studies also demonstrated that GnRH-II possesses antigonadotropic effects in the ovary, since treatment with GnRH-II inhibited gonadotropin receptors expression, and basal and hCG-stimulated progesterone secretion. However, in contrast to GnRH-I, GnRH-II treatment does not affect basal and hCG-stimulated intracellular cAMP accumulation, suggesting that these hormones exert their antigonadotropic effects at the receptor level, but not at the cAMP level in human granulosa/luteal cells.

13.8.2 Antiproliferative Effects of GnRH

GnRH-I is a negative autocrine regulator of proliferation in ovarian cancer cells. Treatment with the GnRH-I antagonist, Cetrorelix, suppressed the growth of ovarian cancer, and GnRH-I agonist treatment inhibited the growth of normal human OSE cells in a time- and dose-dependent manner. These antiproliferative effects can be reversed by treatment with GnRH antagonist Antide. Antide is a third generation GnRH antagonist, obtained by several modifications of native amino acids of GnRH (Antide, [N-AC-D-Nal(2)^1,pCl-D-Phe^2,$_D$-Pal(3)^3,Lys(Nic)^5,$_D$-Lys(Nic)^6,Lys(iPR)^8,$_D$-Ala^{10}] GnRH) with long-lasting inhibitory effects on gonadotropin secretion. 17β-estradiol can antagonize the growth inhibitory effects of GnRH-I on an ovarian cancer cell line, OVCAR-3, but not in normal human OSE cells. These evidences provide a potential cross-talk between estradiol/ER and GnRH-I/GnRHR-I systems, which may be an important factor in controlling the growth of normal and neoplastic OSE cells. The growth-suppressive effect of GnRH-I on ovarian cancer has been mediated via the regulation of the MAPK signaling cascade. Other studies have demonstrated that the inhibitory effects of GnRH-I on ovarian cancer cell line are mediated via the ERK signaling cascade and require the Gβγ subunit. In contrast, few studies demonstrated that treatment with GnRH-I agonist antagonized growth factor induced mitogenic signaling in ovarian cancer, possibly via down-regulation of growth factor receptor expression or growth factor-induced tyrosine kinase activity, or both. Taken together, results from these studies indicate that GnRH-I exert its antiproliferative effects in ovarian cancer cells by modulating the activity of mitogenic signaling via the *pertusis toxin*-sensitive $G_{i\alpha}$ protein. By RT-PCR and Southern blot analysis it was demonstrated that GnRH-II also function as a negative autocrine growth factor in human ovary, normal OSE, immortalized OSE, primary cultures of ovarian tumors, and ovarian cancer cell lines. Treatment with GnRH-II produced a dose-dependent reduction of [3H-thymidine] incorporation in both non-tumorigenic and tumorigenic cells. Also, anti-proliferative effects of GnRH-II on ovarian cancer cell lines (EFO-21, OVCAR-3, and SK-OV-3) are dose- and time-dependent manner. These antiproliferative effects are found to be significantly more potent than that produced by equimolar doses of GnRH-I agonist. Treatment with GnRH-I agonist has no effects on GnRHR-I-negative SK-OV-3 cells proliferation. This antiproliferative effects of GnRH-II on ovarian cancer cells are not likely due to cross interaction with the GnRHR-I and, therefore, suggests that the GnRH-II/GnRH-II receptor system represents an additional autocrine growth regulatory system in ovarian cancer.

13.8.3 Role of GnRH in Apoptosis

A few studies suggested a physiologic role of the GnRH system in the control of follicular atresia and luteolytic factor by increasing the number of apoptotic luteinized granulosa cells. In hGL cells isolated from IVF patients the incidence of apoptosis was demonstrated to be stimulated by GnRH-I agonists. GnRH-I agonist treatment caused a time- and dose-dependent increase in DNA fragmentation in rat granulosa cells isolated from preantral and antral follicles. However, the signaling mechanisms mediating the proapoptotic effects of GnRH-I in the granulosa cells are largely unknown. Thus, GnRH can inhibit DNA synthesis, induce apoptosis, and activate genes important for follicular rupture and oocyte maturation such as plasminogen activator, prostaglandin endoperoxide synthase type 2, and PR, and regulators of matrix remodeling such as the MMPs. GnRH-I and II are expressed in human placenta and are key regulator of urokinase-type plasminogen activator (μPA) and its inhibitor, plasminogen activator inhibitor (PAI-I). In endometrial stromal cultures from first trimester decidual tissues, GnRH-I increased expression of μPA mRNA and protein. In contrast GnRH-II decreased PAI-1 mRNA and protein expression.

Direct evidences for the pro-apoptotic effect of GnRH-I in ovarian cancer cells are lacking. The treatment with GnRH-I agonist stimulated the expression of both Fas ligand messenger RNA and immunoreactive Fas ligand, and the stimulated Fas ligand expression could be abolished by antide, indicating that this effect is receptor-mediated in ovarian carcinomas SK-OV-3 and Caov-3 cells, and in cells isolated from GnRHR-1-bearing ovarian tumors. However, in contrast, GnRH-I agonist reduced cytotoxin-induced apoptosis in ovarian cancer cell lines, EFO-21 and EFO-27, via activation of nuclear factor-kappa B (NF-kB) signaling. NF-kB is a ubiquitous, heterodimeric transcription factor that is sequestered in the cytoplasm by inhibitor of kappa B (IkB) family proteins. Phosphorylation of IkB by inhibitor of kappa B kinase A and B targets it for ubiquitination and proteasome mediated degradation. Degradation of IkB frees NF-kB, allowing its translocation into the nucleus and subsequent activation of its target genes. Genes implicated in cell survival that are activated by NF-kB includes the pro-survival Bcl-2 family member Bfl-1/A1 and the caspase inhibitors c-IAP1 and c-IAP2. Although the activation of NF-kB by GnRH-I in the ovarian cancer cells is presumably mediated by Gia, the signal transduction pathway leading to NF-kB activation seems to be independent on the interaction between GnRH-I and the growth factor-mediated mitogenic signaling as mentioned earlier, since treatment with phosphatase inhibitor had no effects on GnRH-I induced NF-kB activation. GnRH-I possess two opposing activities (antiproliferative versus antiapoptotic) in ovarian cancer cells via two distinct signal transduction pathways mediated by Giα protein.

13.9 Summary

The functional physiologic role of GnRH and its receptor in female reproductive tissues/cells are under active investigation. The growing evidence of the presence of endogenous GnRH or GnRH-like molecules and GnRHR systems in female reproductive system supports the concept of an autocrine or paracrine regulator of female reproductive system. However, the biological functions of GnRH-II and its receptors in mammals are largely unclear. Current problems in the field include the detection of GnRH and GnRHR mRNA often with RT-PCR techniques without functional assays of protein expression and ligand-binding assays. Studies of potential physiologic signaling via GnRHR in extrapituitary tissues are flawed by the sole use of GnRH agonists and/or antagonists with long half-lives, often at pharmacologic levels. These agents may trigger signaling via local GnRHR that is different from potential physiologic paracrine signaling from local GnRH/GnRHR activity. Cell systems serve as models for GnRHR signaling. However, many are not physiologic models. This may explain the divergent pathways detected and confuse the complexities of functional pathways in vivo. Finally, few cutting-edge techniques have been used to date to prove that the effects of GnRH agonists or antagonists in peripheral target tissues are via the GnRHR, such as the use of siRNA, antisense technology, or tissue-specific knockouts. Further research is needed to differentiate between GnRH/GnRHR signaling in normal female reproductive tract/sites and show how it differs from that of pituitary or in cancer cells.

13.10 Glossary of Terms and Acronyms

AP-1: activator protein-1

Bcl-2: proto-oncogene

Bfl-1/A1: Bcl-2 family member

Caov-3: ovarian cell line

C/EBP: CCAAT/enhancer binding protein

c-Fos: product of the proto-oncogene c-*fos* that dimerize with c-Jun (c-Fos/c-Jun heterodimer) to form the transcription factor AP-1

c-IAP1: caspase inhibitor

c-IAP2: caspase inhibitor

c-Jun: product of the proto-oncogene c-*jun* that dimerize with c-Fos to form the transcription factor AP-1.

CRE: cAMP response element

DAG: diacylglycerol

dbB: diagonal band of Broca

EFO-21: ovarian cancer cell line

EFO-27: ovarian cancer cell line

EGF: epidermal growth factor

EGFR: epidermal growth factor receptor

EL: extracellular loop

ER: estrogen receptor

ERK: extracellular signal-regulated kinase

FSH: follicle stimulating hormone

GAP: GnRH-associated peptide region

GGH3: somatolactotroph cells

GLCs: granulosa luteal cells

GnRH: gonadotropin-releasing hormone

GnRHR: gonadotropin-releasing hormone receptor

GnSE: GnRHR-specific enhancer

GPCR: G protein-coupled receptors

GRAS: GnRHR activating sequence

GRE/PRE: glucocorticoid responsive element/ progesterone responsive element

GRKs: G protein-coupled receptor kinases

hGL: human granulosa-luteal cells

HPLC: high-performance liquid chromatography

IEVT: immortalized extravillous trophoblast

IkB: inhibitor of kappa B

IL: intracellular loop

IVF: in vitro fertilization

JEG-3: placental cell line

JNK: Jun N-terminal kinase

LbT2: gonadotroph cell line

LH: luteinizing hormone

MAPK: mitogen activated protein kinases

MEK 1: threonine and tyrosine recognition kinase

MMP: matrix metalloproteinase

MT1-R: melatonin receptor subtype

MT2-R: melatonin receptor subtype

NF-kB: nuclear factor-kappa B

NRE: negative regulatory element

OE: ovarian epithelial cells

OSE: ovarian surface epithelial cells

OVCAR-3: ovarian cell line

PAI-I: plasminogen activator inhibitor

PCR: polymerase chain reaction

Pitx-1: pan pituitary homeobox transcription factor

PKC: protein kinase C

POA: preoptic area of the brain

POA-AH: preoptic area of anterior hypothalamus

PR: progesterone receptor

PRE: progesterone receptor elements

Ras: oncogene of the Harvey (*ras*H) and Kristen (*ras*K) rat sarcoma viruses. These genes, which are frequently activated in human tumors, encode a 21 kD G-protein.

RBM-8A: RNA-binding motif protein-8A

RT-PCR: reverse transcriptase polymerase chain reaction

SF-1: steroidogenic factor-1

SK-OV-3: ovarian cell line

Src: name of the first described retroviral oncogene (*v-src*), from the chicken Rous sarcoma retrovirus and its precursor (*c-src*), which encode a membrane-associated protein kinase.

SURG-1: sequence underlying responsiveness to GnRH element

SVOG-4m: human ovarian granulosa-luteal cell line

TLCs: theca luteal cells

SVOG-4o: human ovarian granulosa-luteal cell line

TE-671: neuronal cell line

TM: trans-membrane

UTR: untranslated region

IP3: inositol 1,4,5-triphosphate

αT3-1: ovarian cell line

References

1. Miyamoto K, Hasegawa Y, Nomura M, et al. Identification of a second gonadotropin releasing hormone in chicken hypothalamus: evidence that gonadotropin secretion is probably controlled by two distinct gonadotropin-releasing hormones in avian species. Proc Natl Acad Sci USA 1984; 81:3874–8.
2. Sherwood NM, Harvey B, Brownstein MJ, et al. Gonadotropin-releasing hormone (Gn-RH) in striped mullet (Mugil cephalus), milkfish (Chanos chanos), and rainbow trout (Salmo gairdneri): comparison with salmon GnRH. Gen Comp Endocrinol 1984; 55:174–81.
3. Guilgur LG, Moncaut NP, Canário AV, et al. Evolution of GnRH ligands and receptors in gnathostomata. Comp Biochem Physiol A Mol Integr Physiol 2006; 144:272–83.
4. Morgan K, Millar RP. Evolution of GnRH ligand precursors and GnRH receptors in protochordate and vertebrate species. Gen Comp Endocrinol 2004; 139:191–7.
5. Kah O, Lethimonier C, Somoza G, et al. GnRH and GnRH receptors in metazoa: a historical, comparative, and evolutive perspective. Gen Comp Endocrinol 2007; 153:346–64.
6. Seeburg PH, Adelman JP. Characterization of cDNA for precursor of human luteinizing hormone releasing hormone. Nature 1984; 311:666–8.
7. Millar RP. GnRHs and GnRH receptors. Anim Reprod Sci 2005; 88:5–28.
8. Rispoli LA, Nett TM. Pituitary gonadotropin-releasing hormone (GnRH) receptor: structure, distribution and regulation of expression. Anim Reprod Sci 2005; 88:57–74.
9. Choi JH, Gilks CB, Auersperg N, et al. Immunolocalization of gonadotropin-releasing hormone (GnRH)-I, GnRH-II, and type I GnRH receptor during follicular development in the human ovary. J Clin Endocrinol Metab 2006; 91:4562–70.
10. Choi JH, Choi KC, Auersperg N, et al. Differential regulation of two forms of gonadotropin-releasing hormone messenger ribonucleic acid by gonadotropins in human immortalized ovarian surface epithelium and ovarian cancer cells. Endocrinol Relat Cancer 2006; 13:641–51.
11. Cheng CK, Leung PC. Molecular biology of gonadotropin-releasing hormone (GnRH)-I, GnRH-II, and their receptors in humans. Endocr Rev 2005; 26:283–306.
12. Chakrabarti N, Subbarao T, Sengupta A, et al. Expression of mRNA and proteins for GnRH I and II and their receptors in primate corpus luteum during menstrual cycle. Mol Reprod Dev 2008; 75:1567–1577.
13. Leung PC, Cheng CK, Zhu XM. Multi-factorial role of GnRH-I and GnRH-II in the human ovary. Mol Cell Endocrinol 2003; 202:145–53.
14. Metallinou C, Asimakopoulos B, Schröer A, et al. Gonadotropin-releasing hormone in the ovary. Reprod Sci 2007; 14:737–49.
15. Ramakrishnappa N, Rajamahendran R, Lin YM, et al. GnRH in non-hypothalamic reproductive tissues. Anim Reprod Sci 2005; 88:95–113.
16. Sengupta A, Baker T, Chakrabarti N, et al. Localization of immunoreactive gonadotropin-releasing hormone and relative expression of its mRNA in the oviduct during pregnancy in rats. J Histochem Cytochem 2007; 55:525–34.
17. Sengupta A, Chakrabarti N, Sridaran R. Presence of immunoreactive gonadotropin releasing hormone (GnRH) and its receptor (GnRHR) in rat ovary during pregnancy. Mol Reprod Dev 2008; 75:1031–44.
18. Clarke IJ, Pompolo S. Synthesis and secretion of GnRH. Animal Reprod Sci 2005; 88:29–55.
19. Yang-Feng TL, Seeburg PH, Francke U. Human luteinizing hormone-releasing hormone gene (LHRH) is located on short arm of chromosome 8 (region 8p11.2—p21). Somat Cell Mol Genet 1986; 12:95–100.
20. Schirman-Hildesheim TD, Bar T, Ben-Aroya N, et al. Differential gonadotropin-releasing hormone (GnRH) and GnRH receptor messenger ribonucleic acid expression patterns in different tissues of the female rat across the estrous cycle. Endocrinology 2005; 146:3401–8.
21. Hapgood JP, Sadie H, van Biljon W, et al. Regulation of expression of mammalian gonadotrophin-releasing hormone receptor genes. J Neuroendocrinol 2005; 17:619–38.
22. Millar RP, Lu ZL, Pawson AJ, et al. Gonadotropin-releasing hormone receptors. Endocr Rev 2004; 25:235–75.
23. Faurholm B, Millar RP, Katz AA. The genes encoding the type II gonadotropin-releasing hormone receptor and the ribonucleoprotein RBM8A in humans overlap in two genomic loci. Genomics 2001; 78:15–8.

24. Sengupta A, Sridaran R. Expression and localization of gonadotropin-releasing hormone receptor in the rat oviduct during pregnancy. J Histochem Cytochem 2008; 56:25–31.
25. Ruf F, Fink MY, Sealfon SC. Structure of the GnRH receptor-stimulated signaling network: insights from genomics. Front Neuroendocrinol 2003; 24:181–99.
26. Caunt CJ, Finch AR, Sedgley KR, et al. GnRH receptor signalling to ERK: kinetics and compartmentalization. Trends Endocrinol Metab 2006; 17:308–13.
27. Motola S, Cao X, Ashkenazi H, et al. GnRH actions on rat preovulatory follicles are mediated by paracrine EGF-like factors. Mol Reprod Dev 2006; 73:1271–6.
28. Pawson AJ, McNeilly AS. The pituitary effects of GnRH. Anim Reprod Sci 2005; 88:75–94.
29. Chen A, Laskar-Levy O, Ben-Aroya N, et al. Transcriptional regulation of the human GnRH II gene is mediated by a putative cAMP response element. Endocrinology 2001; 142: 3483–92.
30. Chen A, Ziv K, Laskar-Levy O, et al. The transcription of the hGnRH-I and hGnRH-II genes in human neuronal cells is differentially regulated by estrogen. J Mol Neurosci 2002; 18: 67–76.
31. Stocco C, Telleria C, Gibori G. The molecular control of corpus luteum formation, function, and regression. Endocr Rev 2007; 28:117–49.
32. Lareu RR, Lacher MD, Bradley CK, et al. Regulated expression of inhibitor of apoptosis protein 3 in the rat corpus luteum. Biol Reprod 2003; 68:2232–40.
33. Papadopoulos V, Dharmarajan AM, Li H, et al. Mitochondrial peripheral-type benzodiazepine receptor expression. Correlation with gonadotropin-releasing hormone (GnRH) agonist-induced apoptosis in the corpus luteum. Biochem Pharmacol 1999; 58:1389–93.
34. Sridaran R, Hisheh S, Dharmarajan AM. Induction of apoptosis by a gonadotropin-releasing hormone agonist during early pregnancy in the rat. Apoptosis 1998; 3:51–7.
35. Sridaran R, Philip GH, Li H, et al. GnRH agonist treatment decreases progesterone synthesis, luteal peripheral benzodiazepine receptor mRNA, ligand binding and steroidogenic acute regulatory protein expression during pregnancy. J Mol Endocrinol 1999; 22:45–54.
36. Sridaran R, Lee MA, Haynes L, et al. GnRH action on luteal steroidogenesis during pregnancy. Steroids 1999; 64: 618–23.

Chapter 14

FSH: One Hormone with Multiple Forms, or a Family of Multiple Hormones

Tim G. Rozell and Rena J. Okrainetz

14.1 Introduction

Follicle stimulating hormone (FSH) is produced within pituitary gonadotroph cells and is a critical component of the reproductive process. It has a wide variety of activities in both male and female reproductive tissues including growth, division, and differentiation of Sertoli and granulosa cells. As a result, FSH is directly involved in the production of both male and female gametes, as well as production of hormones (estradiol and inhibin) that feed back to influence secretion of FSH from the pituitary.

Follicle stimulating hormone is a member of the glycoprotein hormone family, which also includes luteinizing hormone (LH), thyroid stimulating hormone (TSH), and human chorionic gonadotropin (hCG). Both LH and FSH are produced in the same cell type (the gonadotroph), while TSH is produced within thyrotroph cells and hCG is produced by placental trophoblast cells [1]. Each member of this family has two subunits, α and β. The α subunit has the same amino acid sequence for each of the glycoprotein hormones, while the β subunit confers a unique immunological and biological conformation to the hormone. Interestingly, in spite of their differences in amino acid composition and in spite of the uniqueness of the β subunit among the different glycoprotein hormones, the α and β subunits have been found to have similar three-dimensional structures in both hCG [2] and FSH [3], as determined by X-ray crystallography.

This chapter will discuss hormonal control of synthesis and secretion of FSH subunits from gonadotrophs, including the roles of inhibin, activin, follistatin, gonadotropin releasing hormone (GnRH), and gonadal and non-gonadal steroids. The glycoprotein hormones are called as such because they contain carbohydrates attached to both subunits. O-linked oligosaccharides have been found to be involved in both biological activity and half-life in the blood, but to have little role in receptor binding or assembly and secretion of the glycoprotein hormones. In contrast N-linked oligosaccharides have been found to be critical for subunit assembly and secretion and are likely to be involved in chaperone-assisted folding as well [4]. The specific features and biological roles of O- and N-linked carbohydrates will be discussed further in Sections 14.4 and 14.5. In addition, the three-dimensional structure, clinical uses and naturally occurring mutations of FSH will be discussed in this chapter.

14.2 Pituitary Gonadotrophs

The anterior pituitary arises from oral ectoderm during embryonic development, eventually coming to rest in a small pocket within the sphenoid bone called the sella tursica. At the same developmental time, the posterior pituitary arises from neuroectoderm. The anterior pituitary contains a number of cell types, each of which are under primary control by the hypothalamus, while the posterior pituitary contains nerve endings from neuroendocrine cells that arise within the hypothalamus. Five distinct endocrine cell types arise within the anterior pituitary: *somatotrophs* that produce growth hormone (somatotropin), *lactotrophs* that produce prolactin, *thyrotropes* that produce TSH, *corticotrophs* that produce adrenocorticotropic hormone (ACTH), and *gonadotrophs* that produce both LH and FSH.

The gonadotroph is therefore unique among pituitary cell types in producing two distinct hormones that are related, but have somewhat different biological roles during the reproductive process. Originally, it was thought that LH and FSH were controlled separately by distinct hypothalamic releasing hormones. However, no apparent candidate for a separate FSH releasing hormone has emerged and GnRH (sometimes called LHRH or luteinizing hormone releasing hormone) is found to cause release of both LH and FSH under a wide variety of conditions.

T.G. Rozell (✉)
Department of Animal Sciences & Industry, Kansas State University, Manhattan, KS, USA
e-mail: trozell@ksu.edu

P.J. Chedrese (ed.), *Reproductive Endocrinology*,
DOI 10.1007/978-0-387-88186-7_14, © Springer Science+Business Media, LLC 2009

An intriguing possibility for how a single cell type within the pituitary could be stimulated to produce more LH than FSH or vice versa at different times during the reproductive cycle comes from the finding that several cell types within the anterior pituitary are capable of responding to different releasing hormones [5] and that certain pituitary cells, including gonadotrophs, may contain different amounts of receptors for hypothalamic releasing hormones at certain stages of life [6]. Thus, it may be a combination of releasing factors that signals the gonadotroph cell to produce and/or release more or less LH and/or FSH.

An alternative explanation comes from the finding that pulse frequency of the hypothalamic releasing factor GnRH determines which gonadotropin is predominately produced by cells [7]. The common α-subunit gene is transcribed to mRNA at the highest rates while GnRH pulses occur at high frequency, but α-subunit mRNA is made in lesser amounts at lower frequencies of GnRH pulses. Therefore, the common α-subunit is made at all times, albeit at a faster rate when GnRH pulse frequencies are highest. At intermediate GnRH pulse frequencies, transcription of the β-subunit for LH is favored, while at lower pulse frequencies of GnRH, transcription of the FSH β-subunit occurs at the highest rate [8–10]. As a result, it can be hypothesized that low pulse frequency of GnRH favors FSH and therefore sets up follicular development during the reproductive cycle. Then, increased pulse frequency of GnRH results in greater production of LH at a time when granulosa cells gain LH receptors and the follicle becomes dominant. Increased synthesis of LH would also be required for the preovulatory surge of LH that results in ovulation, and therefore support for this hypothesis would require a finding of increased GnRH pulse frequency as follicular development progresses. In sheep, the GnRH pulse frequency was found to be highest just before the preovulatory surge of LH, lending support to this hypothesis [11]. However, amplitude of GnRH pulses may also play a role, as GnRH pulse frequency was found to increase during the transition from the luteal to the follicular phase, while the amplitude of GnRH pulses decreased [11].

Still another possible explanation for differential secretion of LH and FSH by pituitary gonadotrophs comes from studies on autocrine and paracrine control of gonadotropin secretion within the pituitary. Autocrine and paracrine factors and their influence on gonadotrophs will be discussed in Section 14.4, specifically in the Section 14.4.1.

14.3 Structure and Characteristics of FSH Genes

As previously discussed, production of a functional FSH protein requires transcription and translation of two different genes within pituitary gonadotrophs.

14.3.1 FSH β-Subunit Gene

The gene encoding the unique β-subunit has been cloned and characterized from a number of species, including humans [12], rats [13], mice [14], cows [15], pigs [16], sheep [17], chickens [18], and ducks [19] to name a few. For the human, the gene encoding the β-subunit specifies a 120-amino acid protein, the first 19 of which serve as a signal sequence [12] that directs co-translational insertion into the endoplasmic reticulum for folding and processing.

The human FSH β-subunit gene also contains three exons and two introns, and this complex gene structure allows alternate splicing to produce multiple forms of the mRNA transcripts [14]. An alternate splicing donor site within exon 1 causes transcripts to have different lengths of 5′ untranslated regions (5′ UTR), with approximately 65% of transcripts containing 63 bases in the 5′ UTR and approximately 35% of transcripts containing 33 bases in the 5′ UTR [14]. These different lengths may affect binding of the ribosome and subsequent rates of protein synthesis.

There are also differences in the size of the poly-A tail (multiple adenosine residues linked to each other) that is added at the 3′ end of the FSHβ transcript. About 80% of all transcripts contain a very long poly-A tail, while the other approximately 20% of transcripts have little to no poly-A tail [14]. The length of the poly-A tail of mRNA transcripts has been linked to the stability of the transcripts within cells, and thus the FSH β-subunit transcripts containing the longest poly-A tails would be expected to have much greater stability within gonadotroph cells. As a result of splice variation and different polyadenylation signals, four unique transcripts are produced for FSH β within gonadotrophs. A similar distribution of poly-A tail lengths is found for transcripts containing the long or short 5′ UTRs, suggesting that alternative splicing and polyadenylation are regulated in an independent fashion [14].

14.3.2 FSH α-Subunit Gene

A single gene containing four exons and three introns encodes the α-subunits of FSH, LH, CG, and TSH (and thus the glycoprotein hormones have identical amino acid sequences within their α-subunits). The common glycoprotein alpha (CGA) subunit gene was characterized from mouse pituitary thyrotropic tumor cells [20,21] and found to produce a precursor form of the protein consisting of 96 amino acids linked to a 24-amino acid leader sequence that is subsequently cleaved from the protein. The mature protein consists of 92 amino acids after post-translational processing [22]. The CGA from the rat is very similar to that of the mouse, with only six conserved amino acid substitution differences between the two [23]. The amino acid sequence of the mouse CGA is similar to that of the cow (93% homology), sheep (91% homology), and pig (98% homology) [20]. The homology of the mouse CGA is slightly lower in horses (82% homology) and humans (75% homology), although the changed amino acids represent conservative substitutions, and thus are similar in hydrophobicity or hydrophilicity [20].

14.4 FSH Synthesis

Synthesis of FSH subunits is the rate-limiting step for production of significant quantities of FSH in gonadotroph cells [24,25], and possibly different combinations of hypothalamic releasing factors or slow pulse frequencies of GnRH are required to increase FSH synthesis. Additionally, hormonal conditions at the time of post-translational processing of FSH may influence the biological characteristics of the secreted hormone by altering carbohydrate composition. Thus, transcriptional control of FSH subunit gene expression may play the largest role in controlling the quantity of FSH available for secretion, while hormonal control of post-translational processing may control the specific activity of the FSH that is subsequently released. Additionally, autocrine or paracrine factors appear to have significant involvement in the synthesis, assembly, and subsequent secretion of FSH subunits.

14.4.1 *Hormonal Control of FSH Gene Expression*

Successful reproductive activity in all mammalian species requires increased release of pituitary FSH in order to stimulate development of small follicles on the ovary, or spermatocytes within the testis. Species which ovulate a single follicle (cows, humans, a number of other primate species and other large ruminants) are required to have elevated FSH for a period of time to support follicular growth to the point where receptors for LH begin to appear within granulosa cells of the largest follicle. At this point in the reproductive cycle, it is important for FSH secretion to decline so that only follicles which can respond to LH can continue to receive enough gonadotropin support for ongoing development. Otherwise, all large follicles would likely ovulate, and there are typically enough sperm present at conception to cause all ovulated oocytes to become fertilized. For many of these species, the energy demands required to raise multiple offspring would be overwhelming, and would limit their ability to survive an often harsh climate. Therefore, it is equally important that negative control be exerted over FSH secretion from the pituitary.

Synthesis of FSH occurs within gonadotroph cells of the pituitary. As previously discussed (see Section 14.2), GnRH is released from the hypothalamus in a pulsatile manner that is controlled by a wide variety of hormones and other factors, and the pattern of pulsatility appears to dictate expression of FSH subunit genes (with slow pulses favoring FSH β-subunit gene transcription). In addition, synthesis of GnRH receptors within gonadotrophs plays an important role in sensitivity to GnRH. Hormones or factors that are known to affect circulating concentrations of FSH have also been found to alter synthesis or at least cell-surface numbers of GnRH receptors in the pituitary [26,27]. GnRH is a decapeptide that is produced within neurons that originate primarily in the preoptic area of the hypothalamus, as well as in other hypothalamic regions, and which terminate on capillaries within the median eminence. These capillaries lead to portal vessels that terminate in capillaries within the anterior pituitary.

The GnRH receptor is a member of the G protein-coupled superfamily of receptors. Upon binding GnRH, the receptor may activate a number of second messenger pathways which have a complex pattern of interaction. Pathways known to be activated by GnRH include phospholipase C, calcium release from intracellular stores, inositol 1,4,5-triphosphate (IP_3), protein kinase C (PKC), and mitogen-activated protein kinase (MAPK) [28,7]. Activation of these pathways may then cause increased transcription of FSH subunit genes by causing increased synthesis or activation of transcription factors, or by modifying chromatin to allow accessibility of transcription factors to FSH subunit gene promoter sequences. Regardless of the precise mechanism, GnRH appears to regulate FSH subunit production primarily by controlling transcription [8]. Additionally, GnRH may have direct or indirect effects on stability of FSH subunit mRNA [29,30].

Not only does GnRH intracellular signaling involve a complex interaction of second messenger pathways, but there are several other hormones that appear to modulate transcription of FSH subunit genes. These hormones include activin, inhibin, follistatin estrogen, progesterone, testosterone, and glucocorticoids [31,32]. Many of these hormones work in a synergistic fashion, either to enhance or suppress production of FSH. Some come from gonadal tissues and thus provide feedback to the pituitary and hypothalamus regarding the reproductive state of the animal, while others come from within the pituitary itself.

Inhibin is a heterodimeric protein that is produced within the gonads, brain, pituitary, and adrenal glands. Although inhibin production does occur locally within the pituitary, its primary activity in decreasing FSH synthesis and secretion occurs through endocrine secretion of inhibin from the gonads. For example, neutralization of circulating inhibin with antibodies results in increased circulating FSH, while injection of recombinant inhibin caused reduced FSH secretion [33].

The inhibin heterodimer is composed of an inhibin/activin β-subunit (either β_A or β_B) and a unique α-subunit. It is a member of the TGF-β family of growth factors and serves as an activin antagonist by interacting with the type-III TGF-β receptor (betaglycan), which binds to either activin or inhibin with high affinity. Upon binding inhibin, the type III TGF-β receptor causes the recruitment and sequestration of the activin receptor [34]. As a result, a primary mode of action for inhibin in suppression of FSH production is through removal of the stimulatory effects of activin (via preventing activin from binding to its receptor). TGF-β may then reduce the activin-suppressing effects of inhibin by interfering with inhibin binding to the type-III TGF-β receptor.

Activin is also a member of the TGF-β superfamily (along with inhibin) and is composed of sulfhydryl-linked dimers of inhibin/activin-β subunits (either β_A or β_B) [33]. Thus it is a homodimer rather than a heterodimer like inhibin, even though three forms of activin may be produced: Activin A (β_A and β_A), activin AB (β_A and β_B), or activin B (β_B and β_B). Activin is produced within the pituitary and gonads, but is known to act locally within the pituitary to regulate gonadotropin secretion from gonadotrophs [35]. Activin has been shown to work in concert with GnRH within gonadotrophs to increase production of FSH β-subunit mRNA [35]. Downstream signaling molecules induced by activin can bind to promoter regions of both the FSH β-subunit and GnRH receptor genes within gonadotrophs, and therefore, activin may increase FSH β-subunit synthesis at the same time it stimulates increased sensitivity to GnRH.

Activin receptors within pituitary cells are members of the TGF-β receptor family, and consist of type I and type II single-transmembrane domain receptors. The type II activin receptor may exist as either ActRIIA (sometimes referred to simply as ActRII) or ActRIIB. Binding of activin to the type II receptor then activates the type I receptor, ALK4 (activin receptor-like kinase 4) [35,36]. Activated ALK4 then activates either Smad2 or Smad3 (some suggestions are that Smad3 is preferentially activated by activin [37]), which in turn activates and binds a co-mediator known as Smad4. The Smad3/Smad4 complex moves into the nucleus where it interacts with a pituitary-specific co-activator known as Pitx2. The entire complex binds to a consensus binding element within the promoter region of the FSH β-subunit gene and within the promoter of the GnRH receptor gene, and causes increased transcriptional activity and production of either FSH-β or GnRH receptor mRNA (see Fig. 14.1).

In addition to causing increased transcriptional activity of FSH-β and GnRH receptor genes, activin has also been found to stimulate production of follistatin, and to do so independently of inhibin in sheep [38]. In turn, endocrine secretion of inhibin and local production of follistatin are key negative regulators of activin production. An additional negative regulator of activin occurs downstream from its intracellular

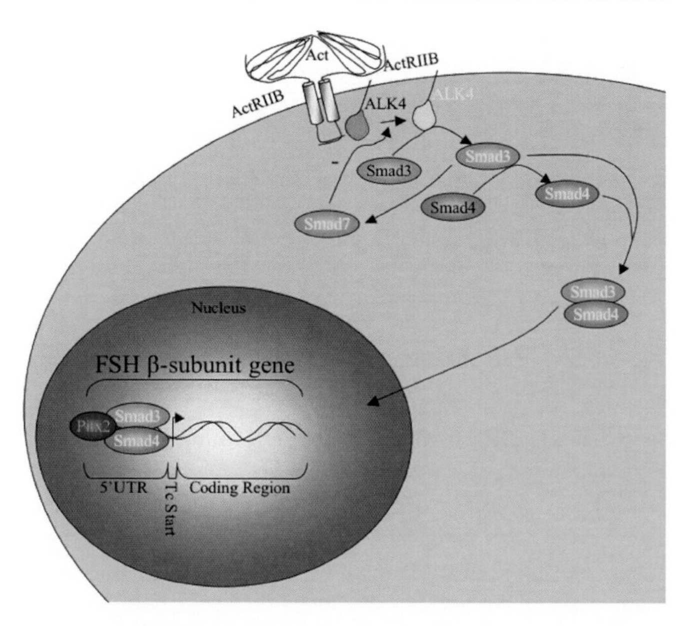

Fig. 14.1 Activin signaling pathway in pituitary gonadotrophs. Activin binds to its receptor, either ActRIIA or ActRIIB (shown here), which then dimerizes. Binding of activin causes activation of the type I activin receptor, activin receptor-like kinase 4 (ALK4), which then activates either Smad2 or Smad3 (Smad3 seems to be preferentially activated within gonadotrophs). Activated Smad3 causes activation of Smad4 and these two activated signaling molecules then associate and translocate to the nucleus. The Smad3/Smad4 complex interacts with Pitx2, a pituitary-specific transcription factor, and the entire complex binds to promoter elements within the upstream portion of the FSHβ gene. Negative feedback occurs within the activin signaling pathway via Smad3 activation of Smad7, which associates with ALK4 and prevents its activation

signaling pathway. Activation of Smad2/3 by ALK4 (as activated by binding of activin to ActRIIA/B) also causes activation of Smad7, which subsequently feeds back to associate with and inhibit the activity of ALK4 (Fig. 14.2). As a result, there appears to be a complex yet elegant system of negative feedback pathways within the pituitary, which is likely to be necessary given that activin is constitutively expressed throughout the reproductive cycle [36].

Follistatin is a glycoprotein monomer that is produced within both gonadotrophs and folliculostellate cells of the anterior pituitary [39]. Folliculostellate cells are cells that surround hormone producing cells and form an interconnected network [40]. Thus, follistatin acts as a paracrine factor to regulate activin activity in gonadotrophs through a powerful and widespread series of protein–protein interactions [36]. The follistatin protein has a unique conformation that allows it to bind specifically to the location on activin where activin would normally interact with ActRIIA/B, and as a result, follistatin prevents interaction of activin with its receptor (Fig. 14.2). Furthermore, the N-terminus of follistatin has the ability to bind with the type I receptor binding domain on activin, effectively preventing it from activating

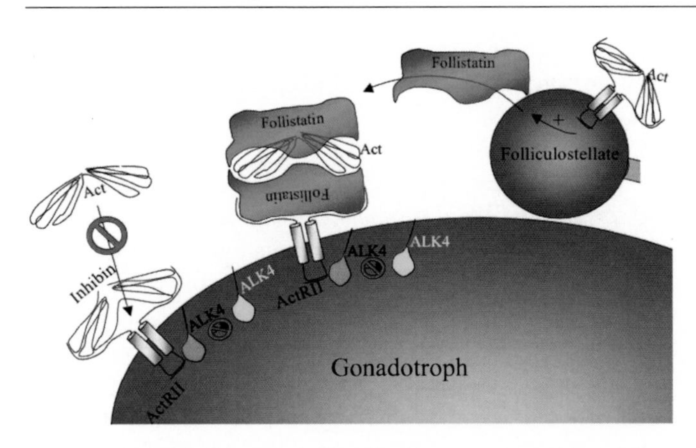

Fig. 14.2 Inhibition of the activin signaling pathway by follistatin or inhibin. Follistatin is produced by folluculostellate cells that form a matrix outside gonadotroph cells within the pituitary. Follistatin production may be increased by activin or by pituitary adenylate cyclase-activating polypeptide (PACAP; not shown), where it then associates with the receptor-binding site on activin. The N-terminal region of follistatin also has affinity for the type II activin receptor (ActRII), such that it covers the binding region on the activin protein as well as covering the activin-binding site within the activin receptor. Inhibin (primarily from the gonads) serves as an activin antagonist by binding to the type III TGFβ receptor (betaglycan) and sequestering activin receptors within the gonadotroph

this receptor and the subsequent Smad2/3 pathway. Thus, a single follistatin protein will mask two receptor binding sites on activin and completely prevent activation of the activin signaling pathway. Two follistatin proteins will eventually cover each activin dimer, and the interaction between these two follistatin molecules may occur in a cooperative manner (i.e., binding of follistatin to activin causes a conformation that promotes binding of a second follistatin to activin, which further promotes binding of the first follistatin to activin). Not surprisingly, then, follistatin overexpression results in infertility [41].

Production of follistatin within folliculostellate cells is controlled by a variety of hormones. Follistatin gene expression is increased by activin [42,43], GnRH, and pituitary adenylate cyclase-activating polypeptide (PACAP) [44], while follistatin gene expression is suppressed by testosterone [43], inhibin [43,45], and by follistatin [43,46]. For the latter two, it seems likely that binding of inhibin or follistatin to activin shuts off the activin-induced increase in follistatin gene expression, and thus suppression of follistatin by inhibin or follistatin may be somewhat indirect.

Interestingly, PACAP also preferentially increases CGA transcription and causes extension of the Poly-A tail on the LHβ mRNA, resulting in increased half-life within gonadotrophs [44]. At the same time, PACAP caused reduced production of FSH β-subunit mRNA within cultured rat pituitary cells [44]. Both hormone producing and folliculostellate cells within the pituitary are found to express both the follistatin gene and PACAP receptors; however, PACAP has been found to increase follistatin production specifically within gonadotrophs and folliculostellate cells [47]. PACAP may thus represent an activin-independent pathway for upregulation of follistatin in order to ensure reduced secretion of FSH at the appropriate time during the reproductive cycle.

Pulse frequency of GnRH release from hypothalamic neurons is known to play a role in preferential production of FSH β-subunit mRNA from gonadotrophs, with slow pulses causing the greatest increase in FSHβ transcription. An additional interaction among hormones within the pituitary (as though additional interactions were needed) involves pulse frequency of GnRH release and its effect on activin production. Pulses of GnRH at a low frequency in rats cause increased transcription of FSHβ and activin β subunit within gonadotrophs, while pulses of GnRH at high frequency preferentially increased transcription of follistatin over that of activin β subunits [46]. Therefore, GnRH pulse frequency may have direct impacts on activin production at the same time as activin has direct impacts on production of GnRH receptors. As GnRH pulse frequency becomes more rapid closer to ovulation, stimulation of follistatin is favored over that of activin, and transcription of LHβ is favored over that of FSHβ (see Section 14.2).

Estrogen is a steroid hormone produced when FSH binds to its receptors within granulosa cells of the developing follicle on the ovary. Estrogen is primarily secreted as estradiol-17β (estradiol) in mammalian species, and is produced when FSH stimulates production of P450 aromatase, an enzyme within granulosa cells that causes the conversion of androstenedione (produced by theca cells of the developing follicle and stimulated by LH) to estradiol. Because FSH has an important physiological impact on estradiol production, it is not surprising that estradiol may feed back and alter either GnRH pulsatility, GnRH receptors, or directly alter FSH synthesis within the pituitary. In ovariectomized ewes, for example, infusion of estradiol in a manner that mimicked the preovulatory rise in endogenous estradiol caused a reduction in FSH secretion [48].

Estradiol appears to have impacts on both the hypothalamus and pituitary to alter FSH synthesis and/or secretion. Cannulation of the portal vessels of sheep revealed decreased GnRH release during the anestrous season in response to estradiol [49]. Thus, it appears that estradiol has a dramatic effect directly on the hypothalamus. However, treatment of cultured pituitary cells with estradiol resulted in an increase in GnRH receptors on those cells [50], while estradiol treatment of ewes in which the effects of GnRH on GnRH receptor production were removed also had increased receptors for GnRH within the pituitary [51]. Estradiol was also found to directly inhibit synthesis and secretion of FSH from cultured gonadotrophs taken from anestrous ewes, and in this same study estradiol also reduced mRNA for activin βB subunit in a dose-dependent manner within cultured pituitary cells

[52]. However, these effects of estradiol may be different in other species or in animals not in their anestrous season. For example, in rats, estradiol had no effect on FSH β-subunit mRNA, while increasing mRNA for activin βB and transiently suppressing follistatin mRNA. In this case, then, the overall effect of estradiol would be positive toward synthesis and secretion of FSH.

Progesterone and testosterone are steroids produced within the gonads, while glucocorticoids are steroids produced within the adrenal cortex. However, these steroids all have similar molecular structures and act through related receptors within cells. Testosterone has been found to increase FSH β-subunit mRNA upon supplementation of castrated rats [53–55]. Progesterone caused a similar increase in FSH β-subunit mRNA when supplemented in rats [56], and anti-progestins have been found to specifically block this response [57]. Further, experiments with reporter genes linked to the FSH β-subunit gene promoter revealed that progesterone increased reporter gene activity in both sheep [58] and rats [59], and further studies found the presence of six progesterone response elements (PREs) within the FSHβ gene promoter in sheep [58] and three PREs within the FSHβ promoter region of rats [59]. Therefore, it appears that progesterone has a direct effect on FSH β-subunit expression by binding to its receptor within gonadotrophs and directly enhancing gene expression by interacting with FSHβ upstream promoter elements.

Both testosterone and progesterone have recently been found to interact with the activin signaling pathway in a cooperative fashion. In these studies, direct protein–protein interactions occurred with either the testosterone or progesterone receptors and Smad proteins that are activated within the activin signaling pathway [31]. Further, both Smad proteins and progesterone or testosterone receptors were required to bind their respective promoter elements within the FSH β-subunit gene to achieve maximal cooperative stimulation of expression [31].

However, some species-specific differences may exist in the mechanism whereby testosterone influences FSH synthesis. Serum FSH levels were found to decrease more dramatically if testosterone was added in combination with a GnRH antagonist in men [60]. A possible explanation for this finding is that testosterone caused increased mRNA for follistatin in a pituitary cell line from primates [61], and thus it is possible that testosterone works by activating a paracrine inhibitor of FSH in humans.

Glucocorticoids have also been found to enhance FSH β-subunit expression within pituitary gonadotrophs [32]. FSH β-subunit mRNA increased in animals in which glucocorticoids were administered, as well as in primary cultures of pituitary cells, indicating that glucocorticoids have direct effects on the pituitary and are not simply enhancing some other stimulatory pathway. As with testosterone and progesterone, glucocorticoids appear to be able to enhance transcription rather than increasing FSH β-subunit mRNA stability, as treatment with glucocorticoids does not change FSHβ mRNA half-life [62].

Obviously it can be difficult to interpret or even to keep track of all the different possible combinations of hormone interactions that regulate FSH production. Perhaps a reasonable summary could be expressed as follows: stimulatory factors like activin and steroidal hormones may work cooperatively, such that activin increases the sensitivity of gonadotrophs to GnRH and steroid hormones and ultimately enhances binding of the steroid receptors to promoter elements within the upstream portion of the FSH β-subunit gene. Steroid hormones may increase production of activin subunits within the pituitary and enhance binding of Smad proteins produced within the activin signaling pathway to promoter elements of the FSH β-subunit gene. Inhibitory factors may reduce availability of GnRH or activin, or may stimulate increased production of follistatin. Follistatin and inhibin then decrease the activity of activin by preventing activin from binding to its receptor. Follistatin does this by interacting with the activin protein, while inhibin does this by interacting with the activin receptor.

14.4.2 Post-translational Processing

After increasing synthesis of mRNA for FSH subunits, these messages must then be translated into protein. Synthesis of FSH subunit proteins occurs on ribosomes on the rough endoplasmic reticulum (RER) within gonadotrophs. As the nascent chains of amino acids are extended, a signal peptide at the N-terminus of each subunit causes translocation of the chains into the lumen of the RER. Shortly thereafter, N-terminal signal peptides are cleaved by a signal peptidase located on the luminal surface of the RER, and thus the signal peptide cleavage occurs co-translationally, while glycosylation occurs both co- and post-translationally. Formation of disulfide bonds and α- and β-subunit dimerization is also initiated in the RER.

Glycosylation of secreted proteins such as FSH is an important process that may be involved in folding, subunit association, secretion from gonadotrophs, and protection from proteases within serum. For example, production of a chimeric form of FSH that contained the highly glycosylated C-terminal extension from hCG within the β-subunit caused the resulting heterodimer to be secreted more readily, and to have a dramatically increased half-life [63–65]. Carbohydrates make up more than 30% of the mass of the FSH heterodimer [66], and their large size may physically block access by proteases, therefore causing increased half-life in serum.

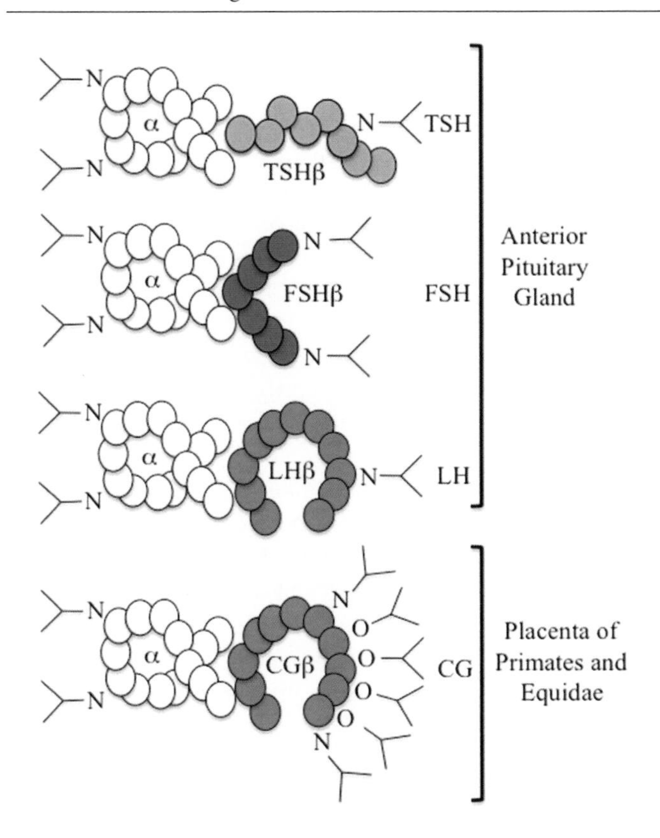

Fig. 14.3 Pattern of glycosylation of the glycoprotein hormone subunits. The alpha subunits of each glycoprotein hormone are identical in amino acid sequence, and contain two sites of N-linked glycosylation. The β-subunits confer hormone specificity and contain variable amounts of N-linked glycosylation. Both TSH and LH β-subunits contain a single site of N-linked glycosylation, while FSH and CG β-subunits contain two sites of N-linked glycosylation. In addition, hCG has an extended C-terminus that contains four sites of O-linked glycosylation

There are two basic types of carbohydrates commonly added to glycoprotein hormones (Fig. 14.3). One type involves linkage of a N-acetylgalactosamine (GalNAc) through the hydroxyl group of a serine or a threonine on the polypeptide. This type of glycosylation is called O-linked. The other type of glycosylation is when N-acetylglucosamine (GlcNAc) is linked through the amide group of asparagine and is called N-linked. The human FSH contains two N-linked oligosaccharides on the α-subunit and two N-linked carbohydrates on the β-subunit. Human chorionic gonadotropin contains a 29-amino acid extension on the β-subunit that contains four sites of O-linked oligosaccharides. This region is the only major difference between hLH and hCG, and yet the half-life of hCG in serum is much longer than that of LH. The β-subunit of FSH contains two sites of N-linked glycosylation, while the β-subunit of LH contains one site of N-linked glycosylation, and FSH has a much longer half-life in serum than does LH. Thus, it appears that both O-linked and N-linked glycosylation protects protein hormones from degradation.

Human FSH, as a glycoprotein hormone, has carbohydrate moieties attached to both the 92-amino acid α-subunit and the 111-amino acid β-subunit (Fig. 14.3) [4,66]. The α-subunit carries N-linked oligosaccharides added to the asparagine (Asn) residues Asn52 and Asn78. When the α-subunit is not combined to a β-subunit it can also contain an O-linked oligosaccharide on Thr43. The FSH β-subunit also contains two glycosylation sites: Asn7 and Asn24 [67]. This results in FSH having four or five sites of glycosylation, although the functional heterodimer would only have four sites. The extent of modification and variation of these glycosylations leads to FSH being secreted as a pool of isohormones varying in their carbohydrate structures (see Section 14.5). Production of either subunit of FSH without carbohydrates allows secretion of the heterodimer from cells, while removal of carbohydrates from both subunits completely prevents secretion [68].

The primary event involved in processing N-linked carbohydrates involves the transfer of an oligosaccharide core consisting of three glucose, nine mannose, and two N-acetylglucosamine residues from a dolichol phosphate donor to specific asparagine residues of the nascent polypeptide chains [69]. The asparagine residues recognized for glycosylation are present in the recognition sequence N-X-S/T (asparagine, any amino acid, except proline, then serine or threonine). This transfer process occurs co-translationally and is mediated by the enzyme oligosacharyl transferase. While the glycoprotein is still in the RER, the three peripheral glucose resides are cleaved. The enzyme α-glucosidase I cleaves the terminal glucose and α-glucosidase II removes the remaining two glucose residues. Varying numbers of mannose residues, 0, 1, or 3 may be removed in the RER by specific glycosidases.

The α- and β-subunits of FSH assemble into heterodimeric form within the RER (Section 14.6) and the hormone is then packaged into transfer vesicles and transferred from the RER to the Golgi. It is within the *cis*-Golgi, medial-Golgi, and *trans*-Golgi compartments that vital post-translation modification of the oligosaccharides takes place to determine the type of isohormone produced. In the *cis*-Golgi, α-1, 2-mannosidase may cleave additional mannose residues yielding the five mannose, two GlcNAc intermediate. This intermediate serves as a substrate for the addition of GlcNAc by GlcNAc transferase I in the medial-Golgi. The resulting intermediate may be converted to varying final products, which can be grouped as neutral, sulfated, sialylated, or sulfated/sialylated oligosaccharides.

14.5 Isoforms of FSH

The oligosaccharides that are attached to Asn56, Asn82 on the α-subunit, or to Asn7 and Asn24 on the β-subunit may have quite variable oligosaccharide chains at each site.

Microheterogeneity of FSH is due not only to variations in the sulfate or sialic acid content of the hormone but also due to differences in the internal structure of the carbohydrate chains [70]. The question is why are different forms of this hormone synthesized and do they serve different biological functions? The different isohormones of FSH may have different physiological roles, as different isoforms are secreted at different rates depending on the endocrine status of the animal [66].

When pituitary, serum, urine, or recombinantly produced FSH are subjected to isoelectric focusing, chromatofocusing, or zone electrophoresis in agarose suspension, multiple variant forms of FSH molecules are found. The difference between these variants of FSH can be correlated to differences in overall isoelectric points (pI) ranging from <3.8 to >7.1 (the pH at which the mature protein is electrically neutral) and also indirectly related to the degree of branching of side chains of the carbohydrates [71]. Thus, variants of FSH that have different isoelectric points are termed, "isohormones" or simply "isoforms."

Different isohormones may have physiological relevance as revealed by in vitro and in vivo studies employed to study these FSH variants. A naturally occurring basically-charged human FSH isoform moderately increased cyclic adenosine monophosphate (cAMP) production by cultured rat granulosa cells while inhibiting estrogen production and tissue-type plasminogen activator enzyme activity [72]. This finding suggests that the pituitary gland may produce FSH variants that can have agonistic or antagonistic properties. When FSH isoforms were separated into seven groups and tested for cAMP production, estrogen, and tissue-type plasminogen activator activity, the more basic isoforms proved more potent than the acidic isoforms [72]. In addition, when the FSH isohormone profiles were investigated by chromatofocusing among subjects of differing physiological status, correlations were found between sex, reproductive status, steroid levels, and phase of estrous cycle with certain FSH isohormone profiles [66].

One physiological factor that may alter the FSH isoform profiles is the transition to puberty. During the pubertal transition in children it was observed that there was a shift to more acidic FSH isoforms, and the median charge was observed to be more basic for girls as compared to boys [73]. Investigators studying experimental induction of puberty in female lambs using GnRH found that serum FSH isoforms of the pubertal lamb eluted predominantly in fractions with less acidic pH [74]. These researches were unable to find differences in pituitary distribution of FSH isoforms between prepubertal and induced pubertal lambs, but they did observe a difference of which isoforms were present in circulation. This may suggest that during different physiological states the pituitary selectively secretes one group of isoforms over another. In support of this idea, Baenziger and Green [69] have suggested that the oligosaccharides present on LH or FSH may act as "recognition-markers" to allow the gonaotrophs to segregate LH and FSH into separate secretory granules. The presence of predominantly sialic acid residues on FSH and sulfate residues on LH may result in the targeting of these hormones to separate secretory granules. The molecular basis for LH bearing predominantly sulfate residues and FSH bearing predominantly sialic acid residues is the recognition sequence for GalNAc-transferase being present on the LH β-subunit but not on the FSH β-subunit [75]. Since sulfate is always added to GalNAc and sialic acid is always added to galactose, this accounts for the difference in glycoform structure of these two hormones produced by the same cell.

One mechanism for altered secretion patterns of FSH isoforms during times of different endocrine profiles could be that gonadotrophs package different isoforms separately. For example, when children were given exogenous GnRH while undergoing pubertal development they were found to have a higher proportion of more basic LH and FSH isoforms in their serum [76]. These same investigators observed that 90 min after GnRH administration LH isoforms returned to the pre-trial profile even though the LH concentrations remained elevated. This observation would support the notion that the gonadotrophs can selectively secrete certain isoforms during GnRH stimulation.

There is a change in FSH isoform distribution between normal menstruating women and post-menopausal women. Oligosaccharides on FSH during the follicular phase are found to have a higher degree of branching and a more complete carbohydrate chain than oligosaccharides on FSH secreted peri-menopausally [77]. The distribution of charge isoforms of FSH also changes concurrently with the phases of the menstrual cycle. In humans, the highest proportion of more basic FSH isoforms was found at midcycle as compared with the follicular or luteal phases [78]. This result may have a physiological relevance as the effects of FSH on granulosa cells are different in the early follicular phase as compared to midcycle. At ovulation, the rise in FSH serves to free the oocyte from follicular attachments and stimulates deposition of a hyaluronic acid matrix. FSH stimulates the production of plasminogen activator, which converts plasminogen to the proteolytic enzyme plasmin. The particular isoforms of FSH secreted at midcycle may be essential for these special effects observed at the time of ovulation. When heifers were evaluated for FSH isoform profiles during follicular growth and the preovulatory gonadotropin surge, it was demonstrated that FSH isoform patterns did not change during the first or second follicular waves, but there was an increase in the proportion of more basic FSH isoforms during the gonadotropin surge, in association with increased estrogen concentrations [79].

Sex steroids appear to affect FSH isoform profile as women and men have different FSH isoform profiles. Men have a greater proportion of acidic FSH isoforms, while women have more basic FSH isoforms. Since FSH profiles change between pre- and post-menopausal women, and between the different phases of the estrous cycle concurrent with estrogen levels, estrogen may modulate which FSH isoforms are predominant. Estrogen modulates expression of the glycosyltransferases that synthesize sulfated oligosaccharides on LH [80]. Within LH GalNAc is added to oligosaccharide acceptors by a GalNAc-transferase that recognizes a tripeptide motif: P-X-R/K (Proline-any amino acid-arginine or lysine) located 6–9 residues upstream from the Asn glycosylation site. Following ovariectomy, there is a three- to fourfold increase in GalNAc-transferase activity. Treatment with estradiol returns the GalNAc-transferase to basal levels. Sulfotransferase activity also increases in response to ovariectomy; however, the increase is less pronounced and occurs over a longer period of time. No change in the levels of pituitary β1,4-galactosyltransferase occurs following ovariectomy in either the presence or absence of estradiol.

Another way in which estrogen regulates FSH isoform production includes regulating an enzyme involved in incorporating sialic acid residues onto the FSH molecule [81]. When estrogen levels are low there is an increase in acidic/sialylated FSH isoforms. The levels of mRNA for the pituitary enzyme Galβ1, 3[4]GlcNAc α2, 3-sialyltransferase were inversely correlated with the serum levels of estrogen [81]. This may be one mechanism in which steroids regulate FSH isoform profiles.

Testosterone also regulates FSH isoform profiles as it was observed that males and hypertestosterone women with polycystic ovary syndrome have FSH isoform profiles that include a greater proportion of acidic FSH isoforms [82]. When female neonatal rats were androgenized by administration of 100 μg of testosterone propionate at 5, 10, 18, 21, 30, 60, and 90 days of age, the control animals 30 days or older exhibited a higher proportion of more basic FSH isoforms as revealed by chromatofocusing [82]. This shift was attenuated in experimentally androgenized animals. When rat pituitary cells in culture are given GnRH they secrete more basic forms of FSH, whereas in the presence of progesterone this effect is prevented [83]. Anovulatory women suffering from polycystic ovary syndrome demonstrated a predominantly acidic pattern of FSH isoform distribution as compared to normally cycling women, and treatment of these women with estrogen shifted the distribution to a more basic FSH isoform profile [84]. Women suffering from polycystic ovary syndrome have elevated levels of testosterone. Since androgens regulate the incorporation of sugar residues into the carbohydrate chains of pituitary FSH, this would favor the synthesis of FSH isoforms with the increased chance of having terminal sialic acid added to galactose residues, thus synthesizing a greater proportion of acidic FSH isoforms [85].

Women affected with galactosemia (excess galactose-1-phosphate in the blood) suffer from ovarian dysfunction and have higher than normal amounts of FSH isoforms with a pI close to neutral (6.4–7.0) [86]. Since the enzyme galactose-1-phosphate uridyl transferase is expressed in the gonadotroph cells, the lack of this enzyme in galactosemia patients may result in loss of galactose incorporation into oligosaccharides on FSH, and therefore no terminal sialic acid residues are incorporated and FSH isoforms expressed are mostly neutral.

Recombinant FSH (rFSH)—rFSH has been synthesized for commercial purposes using various cultured cell lines. Different cell lines used to produce rFSH preparations do not contain the full spectrum of FSH isoforms found in circulation [87]. Researchers have also noted a marked reduction in the pI distribution of recombinant human FSH produced by Chinese hamster ovary (CHO) cells after storage at 4°C, suggestive of an endogenous CHO cell neuraminidase (sialidase) present in the culture supernatant. This suggestion appears to be most likely, as treatment with the neuraminidase inhibitor 2,3-dehydro-2-deoxy-N-acetylneuraminic acid halted the change in the pI profile [88]. Alternatively, rFSH produced in human embryonic kidney cells (HEK-293) was immunologically indistinguishable from a pituitary FSH standard [87]. However, rFSH produced in HEK-293 had a biological potency of three- to sixfold higher when compared to two different pituitary FSH standards. The isoform profile of human FSH produced by HEK-293 cells demonstrated a greater number of basic isoforms than that of pituitary FSH standard [87]. GalNAc-transferase and sulfotransferase are both expressed in the pituitary, salivary gland, and kidney and therefore the co-expression of these enzymes may allow HEK-293 to express the full spectrum of FSH isoforms [80].

Biological relevance of FSH isoforms—different isoforms of FSH appear to have different biological activity. Less acidic/sialyated glycoforms (elution pH 6.6–4.6) induced higher cAMP release, estrogen production, and tissue-type plasminogen activator enzyme activity as well as cytochrome P450 aromatase mRNA expression in cultured rat granulosa cells than the more acidic analogs (pH less than 4.76) [89]. By contrast the more acidic glycoforms induced higher α-inhibin subunit mRNA expression than their less acidic counterparts [89]. Less acidic FSH isoforms (pI range 5–5.6) specifically induced the most rapid growth of mouse follicles during preantral development over a 5 day period and at lower doses than acid isoforms (pI range 3.6–4.6) or mid acidic isoforms (pI range 4.5–5) [90]. When an acidic mix of FSH was administered to prepubertal lambs it cleared more slowly and was a better facilitator of follicular development and maturation that a more basic mix of FSH [91]. In support of the results of Barrios-De-Tomasi [89], treatment of cultured preantral follicles with more acidic fractions of FSH was found to be mostly ineffective in promoting antrum formation, although some positive effects were observed on oocyte quality and subsequent embryo developmental capacity if cultures were maintained for extended periods of time [92]. In contrast, the less acidic fraction of FSH caused much earlier acquisition of developmental capacity of oocytes, when subsequently matured in vitro and fertilized [92]. The results described herein support the idea that less acidic isoforms of FSH have increased biological activity, while more acidic fractions have reduced biological activity. However, when less acidic fractions of FSH were used in these follicular culture experiments, there was a clear maximum dose beyond which oocyte quality and subsequent embryonic development were reduced. Thus, it seems that a mixture of FSH isoforms is somewhat protective against overdose, especially as increasing content of acidic isoforms may suppress overall FSH production and release due to increased production of inhibin.

Receptor affinity of FSH Isoforms—receptor affinity of the different isoforms of FSH may account for the differences observed in biological activity among the different isoforms. More acidic forms of recombinant human FSH (pI 3.5) showed significantly lower affinity for rat testicular FSH receptors than did more basic forms (pI 4.8) [88], and these results correlated well with biological activity. For example, acidic fractions of FSH resulted in reduced estradiol production, as measured by estrogen production by rat granulosa cells. More acidic forms of FSH also exhibit reduced apparent affinity for the human FSH receptor. When nine different FSH isoforms were assayed for induction of cAMP production by HEK-293 cells transfected with the human FSH receptor, the acidic forms exhibited lower potency than the more basic isoforms [93], suggesting reduced receptor binding affinity. However, an alternative explanation might be that the FSH receptor could have promiscuous G protein coupling to G_i or G_o when bound to more acidic forms of FSH [94], due to a slightly different alignment of FSH containing different carbohydrate residues within its binding domain on the FSH receptor. Another possibility could be that FSH receptors dimerize either more or less slowly when bound to more acidic fractions of FSH, resulting in increased rates of down-regulation.

FSH deglycosylation and half-life in circulation—enzymatic removal of N-linked carbohydrates on FSH has been found to cause reduced activation of adenylate cyclase in spite of normal affinity for the FSH receptor [95–97]. Another study actually found that deglycosylated FSH had higher affinity for its receptor in cultured rat granulosa cells [98], indicating that, like hCG, deglycosylated FSH could perhaps serve as a clinical antagonist in certain cases.

The type of glycosylation of FSH plays an important in half life in circulation. The removal of sialic acid by neuraminidase treatment decreases the in vivo half-life as this molecule is rapidly eliminated from circulation by the hepatic asialoglycoprotein receptor [99]. The removal of the terminal sialic acid does not decrease the biological response. When acidic (pI < 4.8) and more basic (pI > 4.8) FSH isoform fractions were compared for clearance rate from rat circulation it was observed that the more basic isoforms had a faster clearance rate [99]. The sialic acid content is one determining factor of the overall charge of the FSH isoform. The sialic acid content and particularly the number of exposed terminal galactose residues of a glycoprotein determines its clearance rate from plasma through a mechanism that involves hepatocyte receptors for those galactose-terminal complex molecules.

14.6 Three-Dimensional Structure of FSH Protein

The crystal structure of human FSH has been described [3], as well as that of human FSH bound to its receptor [100]. Each of these three-dimensional structures reveals several interesting features of the mature FSH heterodimer. The folding and association patterns of the α- and β-subunits of FSH reveal it to be a member of the cysteine-knot family of growth factors. Other members of this family include transforming growth factor β (TGF-β), activin, nerve growth factor, and platelet-derived growth factor (PDGF) [101]. In addition, the overall folding pattern of the FSH heterodimer is similar to that of hCG [3]. The two subunits of FSH are aligned head-to-tail, and associate in what is commonly described as a "hand clasp." In addition, a "seatbelt" structure has been found to form within the heterodimer. A portion of the β-subunit surrounds a loop within the α-subunit, and a disulfide bond forms in order to "latch" the seatbelt structure. Formation of this structure occurs within the endoplasmic reticulum and requires the threading of the α-subunit loop 2 and its attached oligosaccharides through a hole in the β-subunit that forms after the latching of the seatbelt is already accomplished [102]. Because the α-subunit (the CGA) must bind equally well to FSHβ, LHβ, TSHβ, and CGβ, it must be quite flexible in its ability to attach to each β-subunit and subsequently form the seatbelt structure. It is unknown at present how this is accomplished.

Interestingly, amino acids within the common alpha subunit of FSH are involved in receptor binding. Important regions for receptor contact are the C-terminal segments of both α- and β-subunits, as well as α loop 2 and β loop 2 [103]. These findings are interesting in light of the fact that the α-subunit of FSH is identical to that of LH and hCG, yet LH and hCG typically have very low affinity for the FSH receptor. However, receptor contact sites within the FSH α-subunit may help to explain certain promiscuous mutations with the FSH receptor that allow it to bind and become activated by circulating LH or hCG [104,105], hormones that would contain identical α-subunits. The finding that amino acids from both subunits influence binding of FSH to its receptor also helps to explain the complete lack of activity of individual, unassembled FSH subunits.

Crystallography results combined with results from laboratory mutagenesis studies have allowed researchers to conclude that all glycoprotein hormones share similar structural motifs, as well as similar mechanisms of interaction with their receptors [103]. Binding specificity, then, of FSH to its receptor requires the unique amino acids that form the seat-belt structure from within the β-subunit [103], and these results are consistent with those from mutagenesis.

14.7 Clinical Importance of FSH

During the menstrual cycle, a transient rise in FSH occurs at the end of the luteal phase and causes development of a pool of follicles on the ovary. Developmental changes include proliferation and differentiation of granulosa cells such that they begin to produce large amounts of estradiol. Follicles may fail to develop on the ovary for a number of reasons, many of which are indirectly related to secretion of FSH. An example of this is Kallmann syndrome. Kallmann syndrome [106] appears to be due to a genetic deficiency that alters descent of GnRH neurons and prevents release of GnRH onto capillary beds within the median eminence leading to the anterior pituitary. As a result, reduced GnRH stimulation occurs at the gonadotrophs and secretion of both FSH and LH are greatly reduced. Reduced gonadotropin secretion then leads to reduced secretion of gonadal steroids and many of the clinical signs of this syndrome are a result of decreased steroid production. Examples include micropenis in males and primary amenorrhea and lack of breast development in females.

In examples like Kallmann syndrome and many other causes of infertility, normal follicular development (or even development of secondary sex characteristics) may be attained by supplementation of FSH (and in many cases LH). Until fairly recently, FSH used for clinical therapy was obtained by purification from the urine of postmenopausal women, which contains large amounts of both FSH and LH [107,108]. This I.M. formulation is termed human menopausal gonadotropin or hMG. Advances in purification techniques have led to highly purified human urinary FSH, which contains only very small amounts of LH and other urinary proteins. More recently, recombinant FSH (rFSH) has become commercially available, and this formulation obviously contains no LH. In fact, studies on the Old World macaque have indicated that supplementation of primates with both rFSH and rLH may result in higher quality oocytes collected for in vitro fertilization [107].

14.8 Naturally Occurring Mutations within FSH

Genetic alterations in the structure of FSH subunits may also result in reproductive problems. Obviously, because the CGA subunit is shared by TSH, LH, and CG, mutations that significantly alter secondary or tertiary structure of the resulting protein could have lethal impacts, especially if within hCG, which is necessary for maternal recognition of pregnancy. Thus, finding a variety of naturally-occurring mutations in the α-subunit could be quite difficult. However, one mutant form of the α-subunit has been reported from an ectopically secreted α-subunit from carcinoma cells [109]. This mutant contained a single amino acid substitution; however, it appeared to be much larger than the normal α-subunit and was incapable of associating with at least the β-subunit from LH. Possible explanations for this failure would include altered folding and continued association with chaperone proteins (although secretion of the peptide would argue against this), formation of homodimers, or altered glycosylation patterns. Any of these changes could reduce or eliminate the ability of the altered α-subunit to associate with the β-subunit. The physiological relevance of this mutation could certainly be argued, given that it is an ectopically secreted protein and apparently does not affect pituitary secretion of glycoproteins or fertility.

Naturally occurring mutations within the β-subunit of FSH have been reported in at least seven cases, all of which resulted in infertility [110–112]. In the first case, two nucleotides were deleted in codon 61, and this deletion resulted in altered amino acid sequence from codons 61 through 86, and premature stoppage of transcription [113]. The resulting mutant apoprotein was unable to associate with the CGA, and the person in which the mutation was found exhibited primary amenorrhea that was corrected with exogenous FSH.

A second inactivating mutation has been described in FSHβ, in which the mutation from the first case was repeated but with an additional alteration of the FSHβ gene such that Cys at position 51 in the protein was replaced with Gly [114]. The loss of this cysteine residue would prevent the formation of a cys–cys covalent linkage within the β-subunit, dramatically altering the tertiary structure of the resulting protein. This patient also exhibited primary amenorrhea. Interestingly, relatives that were heterozygous for the mutation did not exhibit symptoms, indicating that a single normal copy of the FSHβ gene is all that is necessary to produce biologically active FSH.

In a third female patient, the 2-nucleotide deletion described above was originally misdiagnosed as hypogonadotropic hypergonadism due to high levels of circulating FSH antibodies. A review of the case, however, found that antibody production was due to injections of urinary gonadotropins. In all three female cases described herein, the patients were homozygous for the 2-nucleotide deletion, again indicating that a single wild-type allele is adequate for normal FSH production.

Two additional female patients have been described and found to have a nonsense mutation in codon 76 that replaces a tyrosine residue with no amino acid [112,115]. In these cases, patients were found to have delayed puberty, primary amenorrhea, and partial breast development. Both patients were also homozygous for the mutation.

Two male patients have also been found to have FSHβ mutations [111]. In the first case, the patient presented with normal puberty but azoospermia (absence of spermatozoa in the semen) accompanied by absence of circulating FSH. In this case, the mutation was found to cause a change from cysteine at position 82 to arginine, with the likely result that a cysteine–cysteine bond would be unable to form. A second male patient was found to have the same 2-nucleotide deletion as described for the female patients. In this case, the patient had slightly delayed puberty, small testes, azoospermia, and low circulating FSH [116].

The small number of reported alterations in the FSH β-subunit compared to the LH β-subunit is somewhat surprising, but correlates well with the much smaller number of inactivating mutations found in the FSH receptor as compared to the LH receptor [117]. In support of this, a comprehensive survey of individuals from two populations found only a small number of silent polymorphisms within the FSHβ gene from each population. A likely explanation for the paucity of loss of function mutations in both the FSH β-subunit and within the FSH receptor is that such mutations would affect fertility in a negative manner, and thus would not be passed along within a population. In the case of the FSH receptor, however, there is some suggestion that the protein itself is more resistant to mutagenesis [117], and this may also be the case with the FSH β-subunit.

14.9 Summary

Follicle stimulating hormone consists of a unique β-subunit that is linked to a common α-subunit (identical to that found within other glycoprotein hormones) within gonadotroph cells of the pituitary. Key positive regulators of FSH secretion and/or synthesis are GnRH from the hypothalamus, activin from pituitary folliculostellate cells, and possibly TGF-β produced within the pituitary. Key negative regulators of FSH secretion and/or synthesis are inhibin produced within the gonads (as well as some inhibin produced within the pituitary) and follistatin produced within folliculostellate and gonadotroph cells within the pituitary. Gonadal steroids may have either inhibitory or stimulatory effects on FSH production depending on the stage of the reproductive cycle. The secreted form of FSH contains a number of carbohydrates covalently linked to the protein and these carbohydrates may serve to increase the stability of the protein within the blood and also to change the biological activity of FSH at certain points during the reproductive cycle. Changes in biological activity of different FSH isoforms may be due to changes in receptor binding affinity or alterations in serum half-life. Alterations in FSH production or secretion have been implicated in a number of cases of infertility, and the fairly recent introduction of recombinant FSH has had important clinical benefits. The three-dimensional structure of the fully mature form of FSH has been characterized and several important regions, most notably the seat-belt that fastens the two subunits together, have been identified.

14.10 Glossary of Terms and Acronyms

ACTH: adrenocorticotropic hormone

ALK4: activin receptor-like kinase 4

Asn: asparagine

cAMP: cyclic adenosine monophosphate

CGA: common glycoprotein alpha

CHO cells: Chinese hamster ovary cells

FSH: follicle stimulating hormone

GalNAc: N-acetylgalactosamine

GlcNAc: N-acetylglucosamine

GnRH: gonadotropin releasing hormone, or LHRH – luteinizing hormone releasing hormone

hCG: human chorionic gonadotropin

HEK-293 cells: human embryonic kidney cells

hMG: human menopausal gonadotropin

IP3: inositol 1,4,5-triphosphate

LH: luteinizing hormone

MAPK: mitogen-activated protein kinase

PACAP: pituitary adenylate cyclase-activating polypeptide

PDGF: platelet-derived growth factor

PKC: protein kinase C

PRE: progesterone response element

RER: rough endoplasmic reticulum

rFSH: recombinant follicle stimulating hormone

TGF-β: transforming growth factor-β

TSH: thyroid stimulating hormone

UTR: untranslated region

References

1. Phifer RF, Midgley AR, Spicer SS. Immunohistologic and histologic evidence that follicle-stimulating hormone and luteinizing hormone are present in the same cell type in the human pars distalis. J Clin Endocrinol Metab 1973; 36:125–41.
2. Lapthorn AJ, Harris DC, Littlejohn A, et al. Crystal structure of human chorionic gonadotropin. Nature 1994; 369:455–61.
3. Fox KM, Dias JA, Van Roey P. Three-dimensional structure of human follicle-stimulating hormone. Mol Endocrinol 2001; 15:378–89.
4. Fares F. The role of O-linked and N-linked oligosaccharides on the structure-function of glycoprotein hormones: development of agonists and antagonists. Biochim Biophys Acta 2006; 1760:560–7.
5. Villalobos C, Nunez L, Frawley LS, et al. Multi-responsiveness of single anterior pituitary cells to hypothalamic-releasing hormones: a cellular basis for paradoxical secretion. Proc Natl Acad Sci U S A 1997; 94:14132–7.
6. Senovilla L, Garcia-Sancho J, Villalobos C. Changes in expression of hypothalamic releasing hormone receptors in individual rat anterior pituitary cells during maturation, puberty and senescence. Endocrinology 2005; 146:4627–34.
7. Ferris HA, Shupnik MA. Mechanisms for pulsatile regulation of the gonadotropin subunit genes by GNRH1. Biol Reprod 2006; 74:993–8.
8. Burger LL, Dalkin AC, Aylor KW, et al. GnRH pulse frequency modulation of gonadotropin subunit gene transcription in normal gonadotropes-assessment by primary transcript assay provides evidence for roles of GnRH and follistatin. Endocrinology 2002; 143:3243–9.
9. Shupnik MA. Effects of gonadotropin-releasing hormone on rat gonadotropin gene transcription in vitro: requirement for pulsatile administration for luteinizing hormone-beta gene stimulation. Mol Endocrinol 1990; 4:1444–50.
10. Haisenleder DJ, Dalkin AC, Ortolano GA, et al. A pulsatile gonadotropin-releasing hormone stimulus is required to increase transcription of the gonadotropin subunit genes: evidence for differential regulation of transcription by pulse frequency in vivo. Endocrinology 1991; 128:509–17.
11. Clarke IJ, Thomas GB, Yao B, et al. GnRH secretion throughout the ovine estrous cycle. Neuroendocrinology 1987; 46:82–8.
12. Jameson JL, Becker CB, Lindell CM, et al. Human follicle-stimulating hormone beta-subunit gene encodes multiple messenger ribonucleic acids. Mol Endocrinol 1988; 2:806–15.
13. Gharib SD, Roy A, Wierman ME, et al. Isolation and characterization of the gene encoding the beta-subunit of rat follicle-stimulating hormone. DNA 1989; 8:339–49.
14. Kumar TR, Kelly M, Mortrud M, et al. Cloning of the mouse gonadotropin beta-subunit-encoding genes, I. Structure of the follicle-stimulating hormone beta-subunit-encoding gene. Gene 1995; 166:333–4.
15. Kim KE, Gordon DF, Maurer RA. Nucleotide sequence of the bovine gene for follicle-stimulating hormone beta-subunit. DNA 1988; 7:227–33.
16. Hirai T, Takikawa H, Kato Y. The gene for the beta subunit of porcine FSH: absence of consensus oestrogen-responsive element and presence of retroposons. J Mol Endocrinol 1990; 5:147–58.
17. Guzman K, Miller CD, Phillips CL, et al. The gene encoding ovine follicle-stimulating hormone beta: isolation, characterization, and comparison to a related ovine genomic sequence. DNA Cell Biol 1991; 10:593–601.
18. Shen ST, Yu JY. Cloning and gene expression of a cDNA for the chicken follicle-stimulating hormone (FSH)-beta-subunit. Gen Comp Endocrinol 2002; 125:375–86.
19. Shen ST, Cheng YS, Shen TY, et al. Molecular cloning of follicle-stimulating hormone (FSH)-beta subunit cDNA from duck pituitary. Gen Comp Endocrinol 2006; 148:388–94.
20. Chin WW, Kronenberg HM, Dee PC, et al. Nucleotide sequence of the mRNA encoding the pre-alpha-subunit of mouse thyrotropin. Proc Natl Acad Sci U S A 1981; 78:5329–33.
21. Chin WW, Habener JF, Kieffer JD, et al. Cell-free translation of the messenger RNA coding for the alpha subunit of thyroid-stimulating hormone. J Biol Chem 1978; 253:7985–8.
22. Bousfield GR, Perry WM, Ward DN. Gonadotropins. Chemistry and biosynthesis. In: Knobil E, Neill, JD, editors. The Physiology of Reproduction, second edition 1994:1749–92.
23. Godine JE, Chin WW, Habener JF. alpha Subunit of rat pituitary glycoprotein hormones. Primary structure of the precursor determined from the nucleotide sequence of cloned cDNAs. J Biol Chem 1982; 257:8368–71.
24. Kaiser UB, Jakubowiak A, Steinberger A, et al. Differential effects of gonadotropin-releasing hormone (GnRH) pulse frequency on gonadotropin subunit and GnRH receptor messenger ribonucleic acid levels in vitro. Endocrinology 1997; 138:1224–31.
25. Papavasiliou SS, Zmeili S, Khoury S, et al. Gonadotropin-releasing hormone differentially regulates expression of the genes for luteinizing hormone alpha and beta subunits in male rats. Proc Natl Acad Sci U S A 1986; 83:4026–9.
26. Gregg DW, Schwall RH, Nett TM. Regulation of gonadotropin secretion and number of gonadotropin-releasing hormone receptors by inhibin, activin-A, and estradiol. Biol Reprod 1991; 44:725–32.
27. Braden TD, Conn PM. Activin-A stimulates the synthesis of gonadotropin-releasing hormone receptors. Endocrinology 1992; 130:2101–5.
28. Ruf F, Sealfon SC. Genomics view of gonadotrope signaling circuits. Trends Endocrinol Metab 2004; 15:331–8.
29. Salton SR, Blum M, Jonassen JA, et al. Stimulation of pituitary luteinizing hormone secretion by gonadotropin-releasing hormone is not coupled to beta-luteinizing hormone gene transcription. Mol Endocrinol 1988; 2:1033–42.
30. Weiss J, Crowley WF, Jr., Jameson JL. Pulsatile gonadotropin-releasing hormone modifies polyadenylation of gonadotropin subunit messenger ribonucleic acids. Endocrinology 1992; 130:415–20.
31. Thackray VG, Mellon PL. Synergistic induction of follicle-stimulating hormone beta-subunit gene expression by gonadal steroid hormone receptors and Smad proteins. Endocrinology 2008; 149:1091–102.
32. Thackray VG, McGillivray SM, Mellon PL. Androgens, progestins, and glucocorticoids induce follicle-stimulating hormone beta-subunit gene expression at the level of the gonadotrope. Mol Endocrinol 2006; 20:2062–79.
33. Vale W, Hsueh AJ, Rivier C, et al. The inhibin/activin family of growth factors. In: Sporn MA, Roberts AB, editors. Peptide Growth Factors and Their Receptors, Handbook of Experimental Pharmacology. Heidelberg: Springer-Verlag 1990:211–48.
34. Bilezikjian LM, Blount AL, Donaldson CJ, et al. Pituitary actions of ligands of the TGF-beta family: activins and inhibins. Reproduction 2006; 132:207–15.

35. Bilezikjian LM, Blount AL, Leal AM, et al. Autocrine/paracrine regulation of pituitary function by activin, inhibin and follistatin. Mol Cell Endocrinol 2004; 225:29–36.
36. Lerch TF, Xu M, Jardetzky TS, et al. The structures that underlie normal reproductive function. Mol Cell Endocrinol 2007; 267:1–5.
37. Shimizu A, Kato M, Nakao A, et al. Identification of receptors and Smad proteins involved in activin signalling in a human epidermal keratinocyte cell line. Genes Cells 1998; 3:125–34.
38. Farnworth PG, Thean E, Robertson DM, et al. Ovine anterior pituitary production of follistatin in vitro. Endocrinology 1995; 136:4397–406.
39. Kaiser UB, Lee BL, Carroll RS, et al. Follistatin gene expression in the pituitary: localization in gonadotropes and folliculostellate cells in diestrous rats. Endocrinology 1992; 130:3048–56.
40. Winters SJ, Moore JP. Paracrine control of gonadotrophs. Semin Reprod Med 2007; 25:379–87.
41. Guo Q, Kumar TR, Woodruff T, et al. Overexpression of mouse follistatin causes reproductive defects in transgenic mice. Mol Endocrinol 1998; 12:96–106.
42. Dalkin AC, Haisenleder DJ, Yasin M, et al. Pituitary activin receptor subtypes and follistatin gene expression in female rats: differential regulation by activin and follistatin. Endocrinology 1996; 137:548–54.
43. Bilezikjian LM, Corrigan AZ, Blount AL, et al. Pituitary follistatin and inhibin subunit messenger ribonucleic acid levels are differentially regulated by local and hormonal factors. Endocrinology 1996; 137:4277–84.
44. Winters SJ, Dalkin AC, Tsujii T. Evidence that pituitary adenylate cyclase activating polypeptide suppresses follicle-stimulating hormone-beta messenger ribonucleic acid levels by stimulating follistatin gene transcription. Endocrinology 1997; 138:4324–9.
45. Dalkin AC, Haisenleder DJ, Gilrain JT, et al. Regulation of pituitary follistatin and inhibin/activin subunit messenger ribonucleic acids (mRNAs) in male and female rats: evidence for inhibin regulation of follistatin mRNA in females. Endocrinology 1998; 139:2818–23.
46. Dalkin AC, Haisenleder DJ, Gilrain JT, et al. Gonadotropin-releasing hormone regulation of gonadotropin subunit gene expression in female rats: actions on follicle-stimulating hormone beta messenger ribonucleic acid (mRNA) involve differential expression of pituitary activin (beta-B) and follistatin mRNAs. Endocrinology 1999; 140:903–8.
47. Fujii Y, Okada Y, Moore JP, Jr., et al. Evidence that PACAP and GnRH down-regulate follicle-stimulating hormone-beta mRNA levels by stimulating follistatin gene expression: effects on folliculostellate cells, gonadotrophs and LbetaT2 gonadotroph cells. Mol Cell Endocrinol 2002; 192:55–64.
48. Rozell TG, Keisler DH. Effects of oestradiol on LH, FSH and prolactin in ovariectomized ewes. J Reprod Fertil 1990; 88:645–53.
49. Karsch FJ, Dahl GE, Evans NP, et al. Seasonal changes in gonadotropin-releasing hormone secretion in the ewe: alteration in response to the negative feedback action of estradiol. Biol Reprod 1993; 49:1377–83.
50. Gregg DW, Allen MC, Nett TM. Estradiol-induced increase in number of gonadotropin-releasing hormone receptors in cultured ovine pituitary cells. Biol Reprod 1990; 43:1032–6.
51. Gregg DW, Nett TM. Direct effects of estradiol-17 beta on the number of gonadotropin-releasing hormone receptors in the ovine pituitary. Biol Reprod 1989; 40:288–93.
52. Baratta M, West LA, Turzillo AM, et al. Activin modulates differential effects of estradiol on synthesis and secretion of follicle-stimulating hormone in ovine pituitary cells. Biol Reprod 2001; 64:714–9.
53. Paul SJ, Ortolano GA, Haisenleder DJ, et al. Gonadotropin subunit messenger RNA concentrations after blockade of gonadotropin-releasing hormone action: testosterone selectively increases follicle-stimulating hormone beta-subunit messenger RNA by posttranscriptional mechanisms. Mol Endocrinol 1990; 4:1943–55.
54. Dalkin AC, Paul SJ, Haisenleder DJ, et al. Gonadal steroids effect similar regulation of gonadotrophin subunit mRNA expression in both male and female rats. J Endocrinol 1992; 132:39–45.
55. Burger LL, Haisenleder DJ, Aylor KW, et al. Regulation of luteinizing hormone-beta and follicle-stimulating hormone (FSH)-beta gene transcription by androgens: testosterone directly stimulates FSH-beta transcription independent from its role on follistatin gene expression. Endocrinology 2004; 145:71–8.
56. Attardi B, Fitzgerald T. Effects of progesterone on the estradiol-induced follicle-stimulating hormone (FSH) surge and FSH beta messenger ribonucleic acid in the rat. Endocrinology 1990; 126:2281–7.
57. Ringstrom SJ, Szabo M, Kilen SM, et al. The antiprogestins RU486 and ZK98299 affect follicle-stimulating hormone secretion differentially on estrus, but not on proestrus. Endocrinology 1997; 138:2286–90.
58. Webster JC, Pedersen NR, Edwards DP, et al. The 5′-flanking region of the ovine follicle-stimulating hormone-beta gene contains six progesterone response elements: three proximal elements are sufficient to increase transcription in the presence of progesterone. Endocrinology 1995; 136:1049–58.
59. O'Conner JL, Wade MF, Prendergast P, et al. A 361 base pair region of the rat FSH-beta promoter contains multiple progesterone receptor-binding sequences and confers progesterone responsiveness. Mol Cell Endocrinol 1997; 136:67–78.
60. Tom L, Bhasin S, Salameh W, et al. Induction of azoospermia in normal men with combined Nal-Glu gonadotropin-releasing hormone antagonist and testosterone enanthate. J Clin Endocrinol Metab 1992; 75:476–83.
61. Kawakami S, Fujii Y, Okada Y, et al. Paracrine regulation of FSH by follistatin in folliculostellate cell-enriched primate pituitary cell cultures. Endocrinology 2002; 143:2250–8.
62. Kilen SM, Szabo M, Strasser GA, et al. Corticosterone selectively increases follicle-stimulating hormone beta-subunit messenger ribonucleic acid in primary anterior pituitary cell culture without affecting its half-life. Endocrinology 1996; 137: 3802–7.
63. Fares FA, Suganuma N, Nishimori K, et al. Design of a long-acting follitropin agonist by fusing the C-terminal sequence of the chorionic gonadotropin beta subunit to the follitropin beta subunit. Proc Natl Acad Sci U S A 1992; 89:4304–8.
64. LaPolt PS, Nishimori K, Fares FA, et al. Enhanced stimulation of follicle maturation and ovulatory potential by long acting follicle-stimulating hormone agonists with extended carboxyl-terminal peptides. Endocrinology 1992; 131:2514–20.
65. Klein J, Lobel L, Pollak S, et al. Development and characterization of a long-acting recombinant hFSH agonist. Hum Reprod 2003; 18:50–6.
66. Ulloa-Aguirre A, Timossi C, Barrios-de-Tomasi J, et al. Impact of carbohydrate heterogeneity in function of follicle-stimulating hormone: Studies derived from in vitro and in vivo models. Biol Reprod 2003; 69:379–89.
67. Rathnam P, Saxena BB. Primary amino acid sequence of follicle-stimulating hormone from human pituitary glands. I. alpha subunit. J Biol Chem 1975; 250:6735–46.
68. Bishop LA, Robertson DM, Cahir N, et al. Specific roles for the asparagine-linked carbohydrate residues of recombinant human follicle stimulating hormone in receptor binding and signal transduction. Mol Endocrinol 1994; 8:722–31.
69. Baenziger JU, Green ED. Pituitary glycoprotein hormone oligosaccharides: structure, synthesis and function of the asparagine-linked oligosaccharides on lutropin, follitropin and thyrotropin. Biochim Biophys Acta 1988; 947:287–306.

70. Creus S, Chaia Z, Pellizzari EH, et al. Human FSH isoforms: carbohydrate complexity as determinant of in-vitro bioactivity. Mol Cell Endocrinol 2001; 174:41–9.
71. Ulloa-Aguirre A, Midgley AR, Jr., Beitins IZ, et al. Follicle-stimulating isohormones: characterization and physiological relevance. Endocr Rev 1995; 16:765–87.
72. Timossi CM, Barrios de Tomasi J, Zambrano E, et al. A naturally occurring basically charged human follicle-stimulating hormone (FSH) variant inhibits FSH-induced androgen aromatization and tissue-type plasminogen activator enzyme activity in vitro. Neuroendocrinology 1998; 67:153–63.
73. Phillips DJ, Albertsson-Wikland K, Eriksson K, et al. Changes in the isoforms of luteinizing hormone and follicle-stimulating hormone during puberty in normal children. J Clin Endocrinol Metab 1997; 82:3103–6.
74. Padmanabhan V, Mieher CD, Borondy M, et al. Circulating bioactive follicle-stimulating hormone and less acidic follicle-stimulating hormone isoforms increase during experimental induction of puberty in the female lamb. Endocrinology 1992; 131:213–20.
75. Smith PL, Baenziger JU. Molecular basis of recognition by the glycoprotein hormone-specific N-acetylgalactosamine-transferase. Proc Natl Acad Sci U S A 1992; 89:329–33.
76. Phillips DJ, Wide L. Serum gonadotropin isoforms become more basic after an exogenous challenge of gonadotropin-releasing hormone in children undergoing pubertal development. J Clin Endocrinol Metab 1994; 79:814–9.
77. Creus S, Pellizzari E, Cigorraga SB, et al. FSH isoforms: bio and immuno-activities in post-menopausal and normal menstruating women. Clin Endocrinol (Oxf) 1996; 44:181–9.
78. Wide L, Bakos O. More basic forms of both human follicle-stimulating hormone and luteinizing hormone in serum at midcycle compared with the follicular or luteal phase. J Clin Endocrinol Metab 1993; 76:885–9.
79. Cooke DJ, Crowe MA, Roche JF. Circulating FSH isoform patterns during recurrent increases in FSH throughout the oestrous cycle of heifers. J Reprod Fertil 1997; 110:339–45.
80. Dharmesh SM, Baenziger JU. Estrogen modulates expression of the glycosyltransferases that synthesize sulfated oligosaccharides on lutropin. Proc Natl Acad Sci U S A 1993; 90:11127–31.
81. Damian-Matsumura P, Zaga V, Maldonado A, et al. Oestrogens regulate pituitary alpha2,3-sialyltransferase messenger ribonucleic acid levels in the female rat. J Mol Endocrinol 1999; 23:153–65.
82. Ulloa-Aguirre A, Damian-Matsumura P, Espinoza R, et al. Effects of neonatal androgenization on the chromatofocusing pattern of anterior pituitary FSH in the female rat. J Endocrinol 1990; 126:323–32.
83. Ulloa-Aguirre A, Schwall R, Cravioto A, et al. Effects of gonadotrophin-releasing hormone, recombinant human activin-A and sex steroid hormones upon the follicle-stimulating isohormones secreted by rat anterior pituitary cells in culture. J Endocrinol 1992; 134:97–106.
84. Padmanabhan V, Christman GM, Randolph JF, et al. Dynamics of bioactive follicle-stimulating hormone secretion in women with polycystic ovary syndrome: effects of estradiol and progesterone. Fertil Steril 2001; 75:881–8.
85. Rulli SB, Creus S, Pellizzari E, et al. Immunological and biological activities of pituitary FSH isoforms in prepubertal male rats: effect of antiandrogens. Neuroendocrinology 1996; 63: 514–21.
86. Prestoz LL, Couto AS, Shin YS, et al. Altered follicle stimulating hormone isoforms in female galactosemia patients. Eur J Pediatr 1997; 156:116–20.
87. Flack MR, Bennet AP, Froehlich J, et al. Increased biological activity due to basic isoforms in recombinant human follicle-stimulating hormone produced in a human cell line. J Clin Endocrinol Metab 1994; 79:756–60.
88. Cerpa-Poljak A, Bishop LA, Hort YJ, et al. Isoelectric charge of recombinant human follicle-stimulating hormone isoforms determines receptor affinity and in vitro bioactivity. Endocrinology 1993; 132:351–6.
89. Barrios-De-Tomasi J, Timossi C, Merchant H, et al. Assessment of the in vitro and in vivo biological activities of the human follicle-stimulating isohormones. Mol Cell Endocrinol 2002; 186:189–98.
90. Vitt UA, Kloosterboer HJ, Rose UM, et al. Isoforms of human recombinant follicle-stimulating hormone: comparison of effects on murine follicle development in vitro. Biol Reprod 1998; 59:854–61.
91. West CR, Carlson NE, Lee JS, et al. Acidic mix of FSH isoforms are better facilitators of ovarian follicular maturation and E2 production than the less acidic. Endocrinology 2002; 143: 107–16.
92. Vitt UA, Nayudu PL, Rose UM, et al. Embryonic development after follicle culture is influenced by follicle-stimulating hormone isoelectric point range. Biol Reprod 2001; 65:1542–7.
93. Zambrano E, Zarinan T, Olivares A, et al. Receptor binding activity and in vitro biological activity of the human FSH charge isoforms as disclosed by heterologous and homologous assay systems – implications for the structure-function relationship of the FSH variants. Endocrine 1999; 10:113–21.
94. Arey BJ, Stevis PE, Deecher DC, et al. Induction of promiscuous G protein coupling of the follicle-stimulating hormone (FSH) receptor: A novel mechanism for transducing pleiotropic actions of FSH isoforms. Mol Endocrinol 1997; 11: 517–26.
95. Sairam MR. Role of carbohydrates in glycoprotein hormone signal transduction. FASEB J 1989; 3:1915–26.
96. Sairam MR, Bhargavi GN. A role for glycosylation of the alpha subunit in transduction of biological signal in glycoprotein hormones. Science 1985; 229:65–7.
97. Padmanabhan V, Sairam MR, Hassing JM, et al. Follicle-stimulating hormone signal transduction: role of carbohydrate in aromatase induction in immature rat Sertoli cells. Mol Cell Endocrinol 1991; 79:119–28.
98. Keene JL, Nishimori K, Galway AB, et al. Recombinant deglycosylated human FSH is an antagonist of human FSH action in cultured rat granulosa cells. Endocr J 1994; 2:175–8.
99. D'Antonio M, Borrelli F, Datola A, et al. Biological characterization of recombinant human follicle stimulating hormone isoforms. Hum Reprod 1999; 14:1160–7.
100. Fan QR, Hendrickson WA. Structure of human follicle-stimulating hormone in complex with its receptor. Nature 2005; 433:269–77.
101. Sato A, Perlas E, Ben-Menahem D, et al. Cystine knot of the gonadotropin alpha subunit is critical for intracellular behavior but not for in vitro biological activity. J Biol Chem 1997; 272:18098–103.
102. Xing Y, Myers RV, Cao D, et al. Glycoprotein hormone assembly in the endoplasmic reticulum: III. The seatbelt and its latch site determine the assembly pathway. J Biol Chem 2004; 279: 35449–57.
103. Fan QR, Hendrickson WA. Assembly and structural characterization of an authentic complex between human follicle stimulating hormone and a hormone-binding ectodomain of its receptor. Mol Cell Endocrinol 2007; 260–262:73–82.
104. Vassart G, Pardo L, Costagliola S. A molecular dissection of the glycoprotein hormone receptors. Trends Biochem Sci 2004; 29:119–26.
105. Vasseur C, Rodien P, Beau I, et al. A chorionic gonadotropin-sensitive mutation in the follicle-stimulating hormone receptor as

a cause of familial gestational spontaneous ovarian hyperstimulation syndrome. N Engl J Med 2003; 349:753–9.
106. Cariboni A, Maggi R. Kallmann's syndrome, a neuronal migration defect. Cell Mol Life Sci 2006; 63:2512–26.
107. Stouffer RL, Zelinski-Wooten MB. Overriding follicle selection in controlled ovarian stimulation protocols: quality vs quantity. Reprod Biol Endocrinol 2004; 2:32.
108. Gleicher N, Vietzke M, Vidali A. Bye-bye urinary gonadotrophins? Recombinant FSH: a real progress in ovulation induction and IVF? Hum Reprod 2003; 18:476–82.
109. Nishimura R, Shin J, Ji I, et al. A single amino acid substitution in an ectopic alpha subunit of a human carcinoma choriogonadotropin. J Biol Chem 1986; 261:10475–7.
110. Meduri G, Bachelot A, Cocca MP, et al. Molecular pathology of the FSH receptor: new insights into FSH physiology. Mol Cell Endocrinol 2008; 282:130–42.
111. Themmen APN, Huhtaniemi IT. Mutations of gonadotropins and gonadotropin receptors: elucidating the physiology and pathophysiology of pituitary-gonadal function. Endocr Rev 2000; 21:551–83.
112. Latronico AC, Costa EM, Domenice S, et al. Clinical and molecular analysis of human reproductive disorders in Brazilian patients. Braz J Med Biol Res 2004; 37:137–44.
113. Matthews CH, Borgato S, Beck-Peccoz P, et al. Primary amenorrhoea and infertility due to a mutation in the beta-subunit of follicle-stimulating hormone. Nat Genet 1993; 5:83–6.
114. Layman LC, Lee EJ, Peak DB, et al. Delayed puberty and hypogonadism caused by mutations in the follicle-stimulating hormone beta-subunit gene. N Engl J Med 1997; 337:607–11.
115. Layman LC, Porto AL, Xie J, et al. FSH beta gene mutations in a female with partial breast development and a male sibling with normal puberty and azoospermia. J Clin Endocrinol Metab 2002; 87:3702–7.
116. Phillip M, Arbelle JE, Segev Y, et al. Male hypogonadism due to a mutation in the gene for the beta-subunit of follicle-stimulating hormone. N Engl J Med 1998; 338:1729–32.
117. Zhang M, Tao YX, Ryan GL, et al. Intrinsic differences in the response of the human lutropin receptor versus the human follitropin receptor to activating mutations. J Biol Chem 2007; 282:25527–39.

Chapter 15

The FSH Receptor: One Receptor with Multiple Forms or a Family of Receptors

Tim G. Rozell, Yonghai Li and Lisa C. Freeman

15.1 Introduction

The follicle stimulating hormone (FSH) receptor (FSHR) is expressed within reproductive tissues and primarily allows for control of reproductive activity in response to secretion of FSH from the pituitary gland. Specific reproductive tissues within both males and females express the FSHR, and thus FSH has important roles in the reproductive process in both genders. This chapter will describe the location and expression of the FSHR gene within the gonads, post-transcriptional and post-translational modifications of FSHR protein, mechanisms of FSH binding and activation of the FSHR, and signaling events that occur within cells following FSH binding.

General physiology of the FSHR—in both males and females, the FSHR appears to be expressed primarily at specific times during the reproductive process. In females, the FSHR is found on granulosa cells within developing follicles and responds to FSH by increasing the division rates of granulosa cells, as well as stimulating them to produce certain steroid hormones (cellular differentiation). The FSHR can be found in the largest amounts on granulosa cells within rapidly growing follicles and quickly declines within follicles that undergo atresia [1]. Both LH and FSH work together in the follicle to cause production of estradiol, with LH stimulating production of androgenic precursors within theca cells found just outside the basement membrane of follicles and FSH stimulating the conversion of these androgenic precursors into estrogen within granulosa cells. Thus, from about the time of antrum formation, FSH receptors are found on granulosa cells and the ability to bind and respond to FSH is the rate-limiting step in production of the aromatase enzyme and subsequent conversion of androgens to estrogens by granulosa cells. Both theca and granulosa cells appear to be specific in their ability to respond to either LH or FSH, respectively, and not to both. However, as the follicle develops further, it gains the ability to express receptors for LH within granulosa cells. These developmental changes allow for one follicle to ultimately become "dominant" and to ovulate, as ability to respond to LH is critical due to declining pituitary secretion of FSH at this time.

In males, the FSHR is found exclusively on Sertoli cells within the seminiferous tubules of the testis, where FSH stimulation causes increased cell division and thus increases the number of Sertoli cells. Sertoli cells directly interact with developing germ cells and thus play an important role in the spermatogenic process. Because the ultimate number of Sertoli cells in adulthood may have a profound impact on the number of germ cells, FSH and its receptor thus play a critical role in spermatogenesis. However, animals in which either the FSH-β subunit or the FSHR was knocked out remain fertile, but with reduced numbers and quality of germ cells. These results have led to the idea that FSH is an important regulator of Sertoli cell function, although it is clearly not the only factor involved in spermatogenesis.

History of the FSHR—the FSHR was first cloned from the rat testis (meaning that the genomic DNA encoding the receptor was excised and put into a plasmid and the sequence of nucleotides making up its coding region was determined) [2]. The sequence of nucleotides making up the LH receptor had already been determined at this time [3] and because of the structural similarities between the ligands (LH and FSH), probes corresponding to the LH receptor were used to search a rat Sertoli cell library that contained most, if not all, genes expressed within those cells. A gene was discovered in this manner that was related to but distinct from the LH receptor and this gene was subsequently determined to be the FSH receptor. Early attempts to identify this distinct but related gene involved cloning the cDNA for the gene into an expression vector, then transfecting human embryonic kidney 293 cells with the vector containing the cDNA. When the cells were subsequently treated with FSH, they were found to produce cAMP, the second messenger generated

T.G. Rozell (✉)
Department of Animal Sciences & Industry, Kansas State University, Manhattan, KS, USA
e-mail: trozell@ksu.edu

P.J. Chedrese (ed.), *Reproductive Endocrinology*,
DOI 10.1007/978-0-387-88186-7_15, © Springer Science+Business Media, LLC 2009

by stimulated G protein-coupled receptors. However, the cells did not respond to either LH or TSH, indicating that the cloned gene was indeed the FSH receptor. The cDNA was subsequently sequenced and found to encode a protein that shared many structural features of the large G protein-coupled receptor superfamily [2,4]. Members of this family include the β-adrenergic receptors, the rhodopsin receptor, the LH and TSH receptors, as well as many others, and respond to stimuli as diverse as chemicals in food, light, catecholamines, metabolic signaling molecules, and many others. Because of its classification as a G protein-coupled receptor, many structural and functional features of the FSHR may be deduced from findings on other members of the receptor superfamily.

15.2 Structure and Function of the FSHR

As for all G protein-coupled receptors, the FSH receptor appears to contain seven transmembrane domains as a major structural feature (Fig. 15.1). These transmembrane domains consist of 20–25 hydrophobic amino acids within a structural α-helix that extends through the cell membrane. The seven transmembrane domains are each connected by short (10–22 amino acids) loops that extend slightly into either

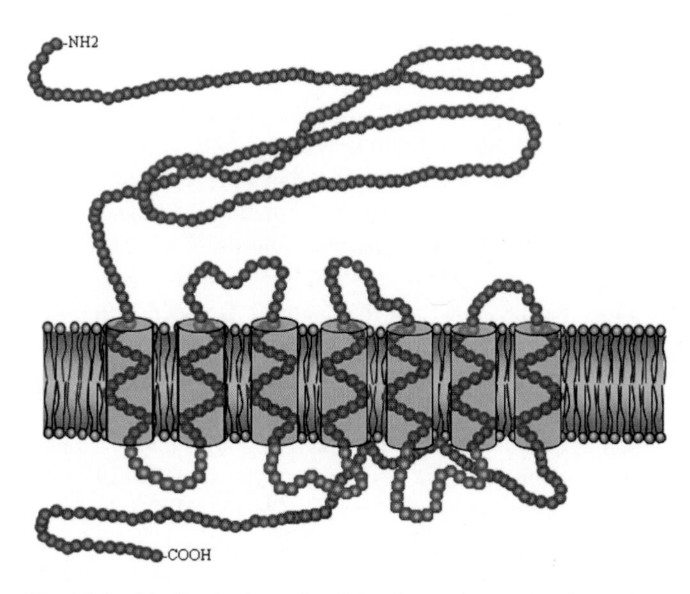

Fig. 15.1 Idealized schematic of the G protein-coupled form of the FSHR, as originally determined for the rat (rFSHR). The rFSHR is related to the rhodopsin family of receptors in which there are seven membrane spanning segments linked by short extracellular and intracellular loop peptides. The rFSHR also has a large intracytoplasmic C-terminal domain composed of 61 amino acids, which is likely involved in association with the G protein and in receptor internalization. The rFSHR also contains a 347 amino acid extracellular N-terminal domain, which contains a series of leucine-rich repeats that are found to be critical for hormone binding affinity and specificity

the extracellular fluid or cytoplasm. Based on X-ray crystallography of the rhodopsin receptor [4], the seven membrane spanning regions of the FSHR likely line up within the cell membrane to form a barrel-like structure, often referred to as the serpentine region. However, in contrast to the rhodopsin receptor, the catecholamine receptors, and many other members of the G protein-coupled receptor superfamily that are known to bind their relatively small ligands within the transmembrane domains, the FSH receptor binds to its very large ligand, on the extracellular domain rather than within the transmembrane domains. In this respect the FSHR exhibits structural similarity with other glycoprotein hormone receptors for LH/hCG (referred to hereafter as the LHR) and TSH (the TSHR).

The glycoprotein hormone receptors all bind large ligands and as a result, all have large, glycosylated extracellular domains attached to the first transmembrane domain. The large extracellular domains of the glycoprotein hormone receptor subclass are what set them apart from other members of the G protein-coupled superfamily. Not surprisingly, then, the extracellular domain has been found to be the region that binds hormone with high affinity, although it may be possible that interaction of the hormone with contact points on the extracellular loops is important for signaling for LH/CG, TSH, and FSH [5].

The FSHR has three potential sites for N-linked glycosylation within its extracellular domain, but only two of these are linked to carbohydrate [6]. The function of glycosylation of the FSHR extracellular domain is unknown, although loss of carbohydrate at either site results in receptor proteins that are trapped intracellularly rather than placed in the outer cell membrane. When these carbohydrates are removed enzymatically after the FSHR is inserted into the outer cellular membrane, high affinity binding to FSH is still observed. However, when the carbohydrates are removed during post-translational processing (actually when no addition of carbohydrates is allowed), the receptor proteins are found to be misfolded and trapped intracellularly [6]. Thus, carbohydrates are likely play an important role in post-translational processing and ultimately in the receptor proteins trafficking correctly to the outer cell membrane. The role of the extracellular domain and extracellular loops in hormone binding will be discussed more thoroughly under Section 15.6. The role of carbohydrates in post-translational processing and receptor trafficking will be discussed more thoroughly under Section 15.5.

15.3 Expression of the FSH Receptor

The FSH receptor gene, as originally cloned and sequenced from the rat, contains 10 exons, with the first nine exons encoding the extracellular domain, and the tenth exon

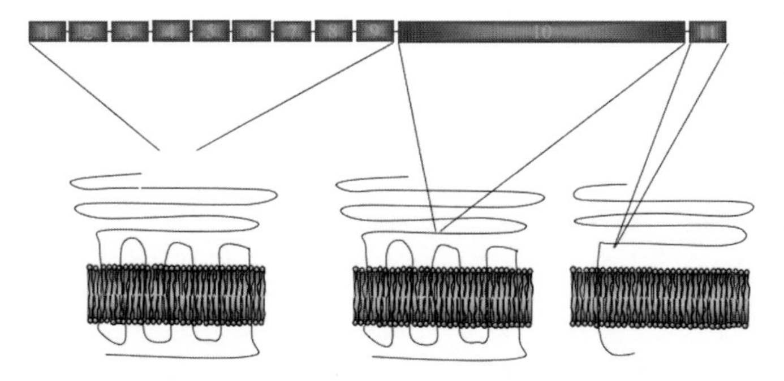

Fig. 15.2 Intron–exon structure of the FSHR gene. The first nine exons encode the extracellular domain, while the tenth exon encodes the transmembrane domain. The novel 11th exon encodes a single transmembrane domain that, when spliced to exons 1–8, result in a splice-variant (FSHR-3) that couples directly to the MAPK signaling pathway

encoding the transmembrane and cytoplasmic domains (Fig. 15.2). More recently, a putative 11th exon has been identified and is implicated in splice-variant forms of the FSHR that may allow it to couple to different signaling pathways (reviewed by [7]). Although the FSHR gene appears to lack traditional hormone response elements within its upstream promoter region, there are clearly differences in FSHR expression at different times during the reproductive cycle. These differences would imply some type of transcriptional control by transcription factors, perhaps induced by hormonal stimulation, binding to regulatory elements within the promoter region of the gene. For example, granulosa cells have been found to increase synthesis of the FSHR when treated with either IGF-1 [8] or FSH [9] in culture. Results such as these indicate that hormones do have some control over transcription of the FSHR gene. In the case of IGF-1, however, it does appear that control of FSHR expression may involve, at least in part, control of mRNA stability [8].

The upstream region of the FSHR gene has been found to be important in regulation of transcription. Many genes have regulatory elements just upstream from the transcription start site, and these regulatory elements specifically bind to transcription factors that often promote or stabilize binding of the RNA polymerase enzyme. For the FSH receptor, it appears that most necessary regulatory elements are located within approximately 100 bp from the transcriptional start site. This was determined by linking portions of the upstream region of the FSHR gene to a reporter gene such that subsequent transcriptional activity could be easily monitored by examining activity of the reporter gene. For example, when the gene for firefly luciferase was linked to different regions of the FSHR promoter, activity of the gene (indicating that transcription and translation were occurring in the transfected cells) was highest when the first 100 bases of the FSHR promoter region were used [10]. The element that appears to be important in this region is an E-box (5′-CACGTG-3′), which binds to upstream stimulatory factors 1 and 2 (Usf-1 and Usf-2). Subsequent studies have revealed that FSHR expression is enhanced by regions that are several hundred bp from the transcriptional start site [11]. Another regulatory factor that appears to bind to the E-box within the FSHR gene promoter is steroidogenic factor 1 (SF-1). Co-transfection of a plasmid expressing the FSHR under control of its promoter along with a plasmid expressing SF-1 was found to cause increased transcriptional activity over that of transfection with the FSHR expressing plasmid alone (reviewed by [12]).

It is not known at the present time how these promoter elements enable cells to produce FSHR protein in different amounts when under different hormonal stimulation. It is possible that Usf-1/2 and/or SF-1 are up or down-regulated by specific hormones. An alternate explanation is that FSHR mRNA is made in a steady-state fashion, and control of receptor protein occurs by regulating stability of this mRNA within the cells. For example, a hormonally induced protein has been found that appears to increase stability of mRNA for the LH receptor at certain times during the reproductive cycle [13], and a similar mechanism could be in place for the FSHR. In further support of this idea is the finding that IGF-1 treatment of cultured granulosa cells increased FSHR mRNA, but did not increase transcriptional activity of the gene, as assessed by nuclear run-on assay [8]. Thus, it does appear likely that hormonal control of FSHR expression occurs via changes in mRNA stability within cells, at least in part.

Expression of the FSHR gene appears to be controlled differently in reproductive tissues from males versus those from females. For example, loss of Usf (via mutation or other engineered conditions) causes different impacts on expression of FSHR in granulosa cells than it does in Sertoli cells [12]. In addition, FSH has been found to upregulate FSHR mRNA in granulosa while downregulating FSHR mRNA in Sertoli cells [9,12]. Thus, expression patterns will be examined separately for each gender.

Expression in females—in the female, expression of at least the G protein-coupled form of the FSHR is primarily if not exclusively within granulosa cells of the follicle. Granulosa cells appear to increase production of the FSHR under control of various hormones, including FSH, estradiol (with FSH + estradiol appearing to be more potent than either hormone alone), TGF-β [14], IGF-1 [8], and others. Interestingly, one study found that estrogen increased FSH receptor expression to a greater extent in cultured rat granulosa if the ooctye was present in the culture medium [15]. These investigators found that response to oocyte addition was independent of cAMP, indicating that some other signaling pathways were involved. A proposed splice-variant form of the FSHR has been reported to couple directly to MAPK rather than activating adenylate cyclase, and thus this form of the receptor may be involved in cAMP independent effects of FSH on granulosa cells. More research will be necessary to further define the role of the oocyte on expression of the FSHR. Hormonal control may also be exerted in a negative fashion on expression of the FSHR. Bone morphogenetic protein-15 (BMP-15) was found to decrease the effects of FSH in granulosa cells by causing a dramatic decrease in expression of the FSHR [16]. Stem cell factor (SCF) was also found to decrease mRNA for FSHR when added to cultured ovaries from neonatal rats [17].

One hormone that does have dramatic, although somewhat controversial, effects on expression of the FSHR is FSH. Both in cultured rat granulosa cells and when injected into rats, FSH has been shown to increase the amount of FSHR mRNA [18–20]. However, porcine granulosa cells that were treated with FSH had increased expression of FSHR mRNA, but decreased cell surface expression of the FSHR protein [9]. These results are intriguing because they argue for post-transcriptional or post-translational control of the FSHR and thus for granulosa cell sensitivity to FSH. Clearly, though, expression of FSHR protein at the cell surface must occur before FSH can possibly have an impact on FSHR expression (FSH must first have a receptor present in order to have any influence on cells). Even a brief scan of the literature in this area reveals that control of FSHR production seems to be regulated in a complex fashion. A possible model is that perhaps granulosa cells gain the ability to constitutively transcribe the FSHR gene at a certain stage of differentiation, then FSHR protein production is regulated by hormones that influence subsequent mRNA stability and posttranslational processing.

Expression in males—as in females, FSH and/or steroid hormones may be important factors in controlling of FSHR gene expression within Sertoli cells of the testis (reviewed by [12]). However, unlike in females, FSH has been found to cause suppression of production of FSHR mRNA rather than stimulation [21,22]. This result was duplicated with other activators of cAMP, indicating that the G protein-coupled form of the receptor was binding to FSH, increasing cAMP, and subsequently reducing its own expression [23]. Activation of the cAMP pathway has been implicated in reduced binding of SF-1 to the E-box within the FSHR promoter [23].

The downregulation of FSHR mRNA caused by FSH does not appear to involve changes in stability of the mRNA within Sertoli cells. When actinomycin D, an antibiotic that blocks transcription, was added to cultured Sertoli cells, the rates of degradation of FSHR mRNA were approximately the same as for cells treated with FSH [22], suggesting that the activity of FSH on suppression of FSHR is at the level of transcription. These results were similar to those obtained in cultured Sertoli cells stably transfected with plasmids expressing human FSHR [24]. In contrast, when a granulosa cell line was transfected with a plasmid expressing FSHR, treatment with FSH was found to increase the stability of FSHR mRNA by up to 50% [25]. These results point out the apparent basic difference between mechanisms associated with FSH activity on FSHR expression in male versus female reproductive tissue.

The expression of the FSHR in the testis appears to be coupled to the cycle of the seminiferous epithelium. Investigators working independently have found FSHR expression to be as much as threefold higher at stages VIII-II [26,27]. These stages are associated with maturation of Sertoli cells and completion of the first generation of spermatocytes developmentally, suggesting that the germ cells may be involved in expression of the FSH receptor. Studies on the FSHR knockout mouse (FORKO) have found that the animals are still capable of producing viable spermatocytes, although at decreased numbers. The FORKO mouse also has reduced numbers of Sertoli cells, which may account for most if not all of the difference in spermatocyte numbers [28].

15.4 Splice Variants of the FSHR

In mammalian systems, tens of thousands of genes are responsible for encoding millions of distinct functional proteins. This is possible because of the occurrence of split genes and RNA splicing. It has been estimated that more than 60% of human genes can undergo alternative splicing in order to create multiple proteins from a single gene. For example, investigation of protein expression in human airway using transcript arrays that allowed description of large numbers of receptor proteins revealed that the 192 G protein-coupled receptors present underwent extensive splicing such that each had an average of five different forms that were expressed. These splice variants were the result of events as diverse as alternative splice donors and acceptors, novel introns, intron retentions, exon(s) skips, and novel exons. The predominant causes of splice variants were exon skips (the final processed form of the mRNA was missing one or more exons) and novel exons [29].

The FSHR is expressed from a ~250 kb gene that has multiple introns and exons. Multiple mRNAs varying in size between 1.2 and 7 kb have been detected in animals of both genders. Alternative splicing of the FSHR gene is responsible for producing at least four different subtypes of the FSHR protein, termed FSHR-1, FSHR-2, FSHR-3, and FSHR-4. These different variants of the FSHR share a common N-terminus and the ability to bind FSH. Variations in the caboxy-termini confer different topologies and functional significance to the four receptor variants. Their molecular bases and biological activities are discussed below.

Figures 15.2 and 15.3 [7] illustrate how four receptor subtypes can be derived from a single FSHR gene. Exons 1–9 of the FSHR gene encode the full-length extracellular domain

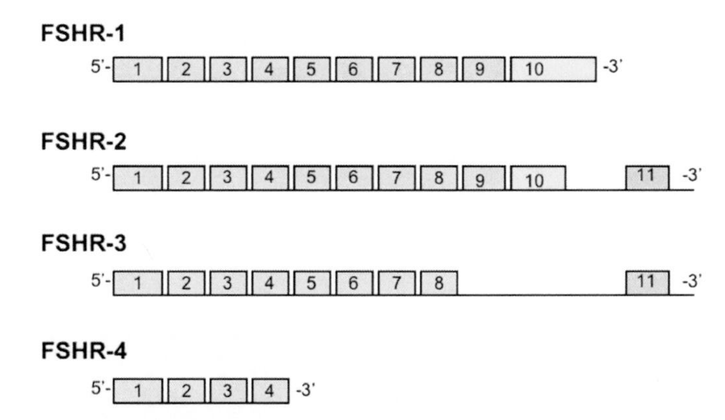

Fig. 15.3 Four FSHR variants are generated from a single gene by alternative splicing. The splicing events and receptor motifs have been best characterized in the sheep (reviewed by [7]). FSHR-1, the full length G-protein coupled receptor is encoded by exons 1–10 of the ovine FSHR gene. The dominant negative receptor FSHR-2 originates from splicing at a site in exon 10 (nucleotide 1925 of sheep FSHR-1). The growth factor type FSHR-3 is generated by a splicing event that joins 392 bp from exon 11 at exon 8. FSHR-4 is encoded by exons 1–4

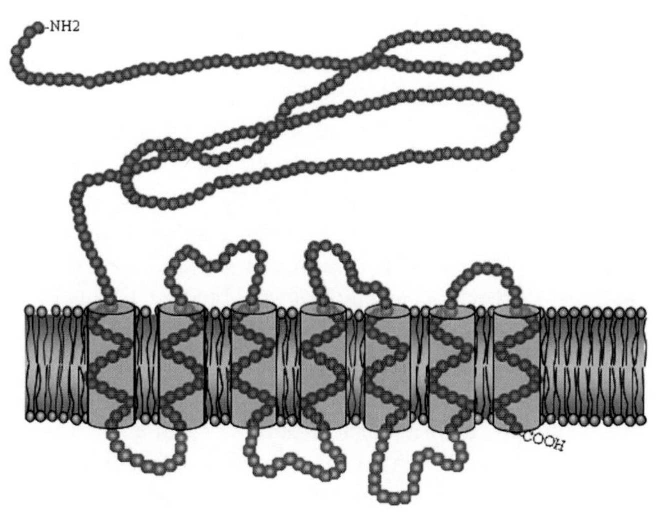

Fig. 15.4 Schematic representation of FSHR-2. This variant of the FSHR is ultimately produced from splicing of exons 1–9 and part of exon 10 to the novel exon, exon 11. The resulting protein contains the full-length extracellular domain, the seven membrane spanning segments, and a greatly truncated intracytoplasmic tail. Amino acids within the seventh membrane spanning segment are encoded by exon 11. The FSHR-2 variant has been found to prevent FSH-induced production of cAMP, even when co-transfected into cells along with the G protein-coupled form of the receptor

present in FSHR-1 and FSHR-2, while exons 1–8 encode the extracellular domain of FSHR-3. This portion (encoded by exons 1–8) of the FSHR protein has been shown to bind FSH with high affinity and its presence defines the FSHR variants expressed on cell surface membranes. The tenth exon encodes the seven membrane-spanning segments, the extra and intracellular loops, and the intracytoplasmic domain of FSHR-1 and the majority of FSHR-2. The tenth exon is slightly truncated in FSHR-2, resulting in a few amino acid differences within the seventh membrane spanning segment. The 11th exon encodes a single membrane-spanning region and is associated with FSHR-3 and the extreme C-terminus of FSHR-2. Exons 1–4, which encode FSHR-4 account for about 40% of the extracellular domain common to the other FSH-R subtypes, and a recombinant version of this protein has been shown to be capable of binding FSH [30].

FSHR-1—the FSHR-1 variant is the well-characterized G protein-coupled form of the FSHR. This FSHR variant is also referred to as the full-length form of the FSHR. The FSHR-1 protein contains 675 amino acids in the sheep and 695 amino acids in the human and exhibits the typical topology of a GPCR, with seven α-helical membrane spanning segments connected by alternating extracellular and intracellular loops (Fig. 15.1). It also contains a large extracellular N-terminal domain (necessary for binding to FSH) and an intracellular C-terminal segment. The intracellular domains of FSHR-1 are responsible for G-protein coupling, activation of specific signal transduction cascades, and termination of the FSH signal. Mutagenesis studies conducted in a number of different laboratories have identified specific amino acid residues involved in various aspects of FSHR function, as summarized in Fig. 15.4 (reviewed by [31]). It is well established that FSH-binding to FSHR-1 triggers activation of the cAMP-adenylate cyclase-protein kinase A cascade. Recently, it has become apparent that FSHR-1 is also associated with the activation of other signal transduction pathways. These signaling events are discussed in more detail below under Section 15.7.

FSHR-2—the proposed FSHR-2 variant of the FSHR contains all the extracellular domain, part of the transmembrane domain, and almost none of the intracellular domain as would be found in the FSHR-1 (Fig. 15.4). This structure is produced from truncation of exon 10, followed by linking to the novel exon, exon 11. In the sheep, the resulting FSHR-2 protein contains 653 amino acids as compared to the 675 amino acids in the full length, G protein-coupled FSHR-1. As expected, FSHR-2 retains full ability to bind FSH with high affinity because it retains the exact hormone-binding domain in the large extracellular portion of the receptor. However, because FSHR-2 lacks the full-length carboxyterminal tail present in FSHR-1, the receptor is apparently incapable of activating the G protein upon binding to FSH.

The precise mechanism by which this truncation interferes with G-protein activation is unknown given that most amino acids that activate G protein have been localized to the third intracellular loop and second membrane spanning segment rather than to the intracytoplasmic portion (reviewed by [32]). A likely scenario, however, is that the amino acid residues of the intracytoplasmic domain either initially make contact with the G protein or somehow stabilize the interaction between the intracellular face of the receptor with the G protein within the cell membrane. Regardless of the mechanism, the FSHR-2 variant has been termed the "dominant negative" form of the receptor due to its inability to activate G proteins. Further, when plasmids expressing this form of the receptor were co-transfected into cells along with plasmids expressing the G protein-coupled form of the receptor, stimulation of the transfected cells with FSH resulted in almost no accumulation of cAMP in spite of cell-surface expression of both types of receptors and high affinity binding to FSH [33]. One possible explanation of these results is that the FSHR-2 variant couples to inhibitory (G_i) rather than stimulatory (G_s) G proteins. The change in the intracytoplasmic domain also reduces the potential number of phosphorylation sites from 12 to 4, a change which could affect downregulation of the receptor following hormone binding (and thus desensitization of the cells expressing the receptor).

The possibility of a dominant negative form of the FSH receptor is intriguing, especially in light of the reduced sensitivity of follicles to FSH at certain times during the reproductive cycle. It is also intriguing to speculate that FSHR-2 might be involved in certain cases of infertility, during which the dominant negative form of the FSHR is somehow preferentially expressed. The novel exon 11 does appear to be expressed at greater amounts during certain stages of the bovine estrous cycle, a preliminary finding implying that preferential expression of certain FSHR variants is a regulated process. Much more research on this variant will be required before its true physiological significance is determined.

FSHR-3—the FSHR-3 variant form of the FSHR arises when the first 8 exons of the receptor gene are linked to the novel exon 11. Because exon 10 which encodes the seven transmembrane domains is missing completely, this variant receptor likely has a much different structural conformation (Fig. 15.5). Interestingly, analysis of hydrophobicity of amino acids produced by the addition of exon 11 suggests that these residues form a single transmembrane domain and that the overall topology of the FSHR-3 is consistent with that of a growth factor type I receptor. Receptors in this family typically activate the MAPK signaling pathway (see Section 15.7). Furthermore, unlike FSHR-1 and FSHR-2, the carboxyterminal portion of FSHR-3 also contains the consensus sequence (PVILSP) for phosphorylation by MAPK. In fact, FSHR-3 has been found to couple to the MAPK pathway in heterologous expression systems and in an ovarian cancer cell line [7,34].

Because the FSHR-3 variant is missing exon 9 and exon 9 encodes a portion of the extracellular domain (the hormone binding portion of the receptor), it is reasonable to question whether it has full ability to bind FSH with high affinity. One study did find hormone binding to the FSHR-3 variant at the cell surface, and several studies have now found phosphorylation of MAPK within a few minutes of FSH treatment in

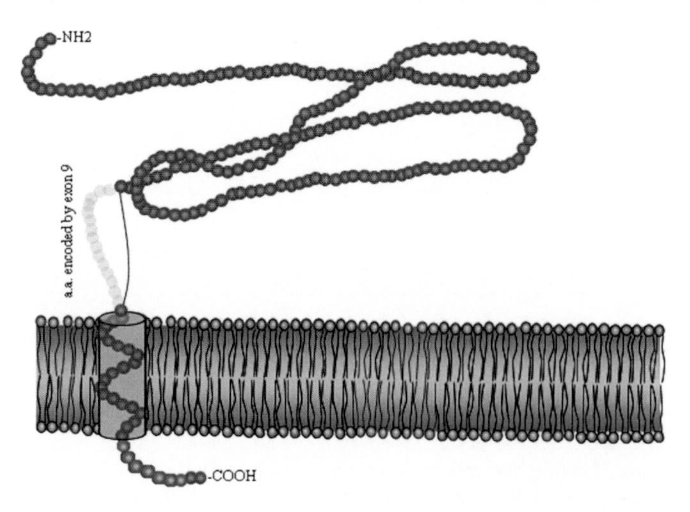

Fig. 15.5 Schematic representation of FSHR-3. This variant of the FSHR is ultimately produced from splicing of exons 1–8 linked to the novel exon, exon 11. The resulting protein contains the majority of the extracellular domain, and has been found to bind FSH with high affinity. However, the loss of exon 10-encoded amino acids results in a receptor protein with only a single membrane-spanning segment. The FSHR-3 variant has been shown to bind FSH and activate the MAPK signaling pathway

cells that only express FSHR-3. These results would imply that normal high affinity binding of FSH occurs. In addition, X-ray crystallography of the extracellular domain of the FSHR bound to FSH has been generated and from the apparent three-dimensional structure of the hormone binding region, the FSHR-3 variant contains all amino acids necessary for high-affinity interaction with FSH [35]. Further, the region of the FSHR hormone-binding domain that remains within FSHR-3 contains all the leucine-rich repeats (LRRs) that have been found to be necessary for formation of the proper hormone binding conformation.

The possible presence of a MAPK-coupled, growth factor type isoform of the FSHR is exciting, especially in view of the fact that there are several cellular events that are apparently stimulated by FSH in the absence of increased cAMP production. In granulosa cells, these events generally occur early in follicular development in less differentiated cells. These FSH-stimulated, cAMP-independent events include DNA synthesis and cell proliferation [36]. Furthermore, in transformed ovarian epithelial cells that express FSHR-3 but not FSHR-1, FSH stimulates activation of MAPK and cell proliferation [34]. Taken together, these findings suggest that, like other growth factor type I receptors, FSHR-3 is important for promoting mitotic activity and cell growth. Additional physiological and pathophysiological implications for this subtype of the FSHR will be discussed in more detail below under Section 15.7.

FSHR-4—FSHR-4 is transcribed from exons 1–4 only. This truncated variant has been described as the predominant form of the FSHR through development in the sheep testis

[37]. Because it lacks any transmembrane domains, FSHR-4 has also been referred to as the soluble FSHR [7]. At this time the functional significance of FSHR-4 is unknown, although a number of paradigms have been proposed. For example, it is possible that FSHR-4 binds to FSH within the extracellular matrix and either stabilizes the hormone or prevents it from binding to the FSHR. This hypothesis assigns FSHR-4 a function similar to the function of IGF-1 binding proteins. It has also been suggested that FSHR-4 is a prohormone for an active signaling molecule because of the presence in its extreme C-terminus of a potential cleavage site for the pro-protein convertase enzyme PC5. PC5 is the same enzyme that generates anti-Mullerian hormone from its precursor in Sertoli cells [38,39].

15.5 Post-translational Modification and Receptor Trafficking

Because all potential variants of the FSHR are glycosylated and inserted into the outer cell membrane (with the probable exception of FSHR-4), they are presumably produced within cells using similar mechanisms and processing pathways. As a result, the following discussion of post-translational modification will focus on research results from FSHR-1, with the presumption that at least FSHR-2 and FSHR-3 follow similar intracellular pathways. Following splicing and production of the final form of the mRNA for the FSHR, translation occurs on the rough ER. The FSHR contains a signal peptide that directs co-translational movement of the nascent protein into the endoplasmic reticulum for further processing. Events that are known to occur for proteins processed within the ER are initial glycosylation, association with chaperone proteins, and initial folding events [40]. The FSHR appears to undergo each of these events within the ER.

Many proteins are found to begin folding into the appropriate secondary structure immediately upon entering the lumen of the ER (thus, folding of the N-terminus may occur even as the C-terminal portion of the protein is being synthesized). In addition to attainment of secondary structure, a very early event may also be initial glycosylation [40]. For all variant forms of the FSHR, early glycosylation would involve attachment of a large oligosaccharide complex consisting of N-acetylglucosamine, several mannose residues, and glucose. These mannose-rich large oligosaccharides are added to the two consensus sequences for glycosylation (N-X-S/T; Asparagine (N) followed by any amino acid except proline, followed by either a serine (S) or threonine (T)). For N-linked glycosylation, N-acetylglucosamine is attached covalently to the asparagine of the consensus sequence. Thus, as the nascent protein enters the lumen of the ER it becomes glycosylated, then the glucose and mannose residues may be trimmed off via endoglycosidases within the ER. The proteins are then transported via special secretory vesicles to the cis-Golgi, where further endoglycosidase activity may trim more mannose residues from the oligosaccharide complex. These endoglycosidases are only found within the ER and Golgi, and thus any proteins that remain sensitive to them after extraction of total cellular proteins via detergent are incompletely processed (immature). The proteins are then transported to the Golgi stack where final processing of carbohydrates occurs, followed by transport to the trans-Golgi for packaging into secretory vesicles.

For the FSHR, the transmembrane regions are placed within the secretory vesicle membranes of the trans-Golgi such that the large N-terminal extracellular domain faces the lumen of the vesicle and the C-terminal cytoplasmic domain faces outside the vesicle (in other words, it is inside-out compared to its final placement in the cell membrane). As the vesicle moves to the outer cell membrane, the vesicle membrane merges with the cell membrane such that the lumen contents of the vesicle are placed outside the cell. In this manner, the transmembrane regions anchor the FSHR in the cell membrane and the extracellular domains extend into the extracellular fluid where they may come in contact with ligand (FSH).

The FSHR actually has three consensus sequences for glycosylation within its extracellular domain, but through mutagenesis studies, only two of these are found to be glycosylated [6]. The role of carbohydrates is unknown for the FSHR, as high affinity binding is found for FSH after carbohydrates are removed from the mature protein via treatment with enzymes (PNGaseF) [6]. It is possible that the large carbohydrate moieties serve a protective function for the extracellular domain of the receptor, as secreted proteins are known to have longer half-lives in serum (i.e., they are protected from proteases) when they have more sites of glycosylation.

The initial three-dimensional folding process (beyond that of simple attainment of secondary structure) of the FSHR within the ER has been found to involve chaperone proteins [41,42]. An exhaustive analysis of specific chaperones that may be involved has not been conducted; however, there are a few chaperone proteins that have been found to at least interact with the FSHR while it is in an immature form. Two chaperones seem to be extensively involved in folding the FSHR into a hormone binding conformation: calnexin and calreticulin [42]. It is likely that these chaperones bind to and stabilize otherwise unstable "folding intermediates," and thus help the receptor protein achieve a final hormone binding conformation. An early study found that calnexin associates with the FSHR when it is in a conformation that is unable to bind hormone [41], lending credence to this idea. A third chaperone protein, BiP (Binding protein; also known as glucose-regulated protein-78 or GRP78), was found to associate with certain misfolded mutant forms of the closely related hLHR, indicating that BiP is likely involved in a quality control role within cells by helping to remove misfolded proteins [42]. Based on other results with the hLHR and preliminary results with the hFSHR, many naturally occurring mutants of the gonadotropin receptors appear to be misfolded and retained intracellularly and thus this area of research will be import for understanding and designing future clinical approaches to certain types of infertility. Naturally occurring mutations of the hFSHR will be discussed more thoroughly below.

15.6 Binding to FSH and Activation of the FSHR

The FSHR has a large extracellular domain consisting of approximately half of the entire protein. The FSHR, like the receptors for LH/CG and TSH, binds to a large ligand consisting of several hundred amino acids. This makes the FSHR, LH/CGR, and TSHR somewhat different from other G protein-coupled receptors, which bind to small ligands, usually within the transmembrane domain. In the case of G protein-coupled receptors that bind small ligands, binding and activation of the receptor (which subsequently activates the G protein) seems to occur in one step due to the location of ligand binding. For the FSHR, LHR, and TSHR, the glycoprotein hormone receptors, the physical size of the ligand (FSH, LH/CG, or TSH) precludes binding within the transmembrane region of the receptor, and thus binding and activation are considered to be sequential rather than simultaneous events. Obviously, then, binding of hormone must somehow signal a change to the receptor such that it can bind to and activate the G protein, which it does not do at times when it is not bound to hormone (except in the cases of some naturally occurring mutations see Section 15.8). Some investigators have suggested that the extracellular domain may somehow inhibit activation of the G protein by the receptor. This idea is supported by the finding that expression of mutant forms of the TSHR containing no hormone-binding domain were constitutively active (stimulated increased cAMP all the time, and in the absence of hormone).

Thus, a model for activation of the receptor by hormone binding has been proposed for the TSHR and can be extended to the FSHR. The extracellular domain, in the absence of hormone, inhibits the transmembrane region from interacting with and/or activating the G protein, most likely because of some configuration assumed by the extracellular domain. Once hormone binds, however, the extracellular domain changes configuration and allows the transmembrane region to activate G proteins.

An alternative model is that the hormone, after binding with high affinity to the extracellular domain, may interact with lower affinity with amino acids in the extracellular loops between the membrane spanning regions of the transmembrane domain. Such interaction then changes the conformation at the internal face of the receptor, allowing it to interact with and activate G proteins. This alternative model has not been well supported by extensive analysis of both naturally occurring and laboratory-generated mutations that activate the receptor in the absence of hormone binding. For example, activating mutations that were duplicated within the transmembrane regions of the TSHR were found to enhance the constitutive activation of the truncated form of the receptor (no extracellular domain), while duplication of activating mutations within the extracellular loops had no effect. This finding suggests that activating mutations at the extracellular face of the receptor are not due to "simulation" of hormone binding, and thus, that the extracellular loops are not required to interact with hormone during the activation step.

The glycoprotein hormone receptors all contain structural features within the large extracellular domain that contribute to hormone binding. Because this chapter is focused on the FSHR, discussion will be limited to that receptor and particularly to findings from studies on FSHR-1. However, unless otherwise noted, it is reasonable to assume that hormone binding and activation will be similar for all glycoprotein hormone receptors and presumably for splice variants of the FSHR as well (although FSHR-3 could be quite different given the single transmembrane segment and altered signaling pathway).

The three-dimensional structure within the extracellular domain of the FSHR most directly involved in making specific contact with FSH is a series of LRRs within the extracellular domain of the FSHR. Each LRR is a sequence of 20–25 amino acids, of which, as the name suggests, a large percentage are leucine residues. Several proteins containing LRRs have been crystallized and their three-dimensional structures have been determined. Each LRR consists of an alpha-helix linked via a short curved amino acid segment to a beta-sheet [43]. Thus, when LRRs are expressed sequentially, the beta-sheet portions form an inner horseshoe-like curved shape, with nine of these beta-sheets located side-by-side and forming a "flat" face in the extracellular domain of the FSHR. Although there is some debate on the issue, it appears that non-leucine amino acids within the short peptide loops that link each alpha-helix and beta-sheet are responsible for specificity of hormone binding. For example, mutagenesis studies have found that substitution of just two of these amino acids from the FSHR with those that would be found in the LHR caused the resulting mutant to bind specifically with LH rather than FSH [43]. These mutants also bound FSH with high affinity, suggesting that other amino acids are also involved with hormone binding specificity.

15.7 FSHR Signaling Pathways

Follicle stimulating hormone actions on target cells in the ovary and testis differ at various stages of follicular and testicular development. There are many gaps in our knowledge about how FSH activation of FSHR can produce diverse responses in a cell context-dependent fashion. For example, it is clear that many, but not all effects of FSH can be attributed to FSH binding to FSHR-1 and subsequent activation of the PKA signaling pathway. However, the mechanisms whereby FSH activates multiple signaling cascades are not fully resolved. In ovarian and testicular cells where multiple subtypes of the FSHR may be expressed, there is no consensus about the contribution of FSHR-2, FSHR-3, and FSHR-4 to the cellular responses associated with FSH stimulation. Furthermore, there are many unanswered questions about crosstalk and convergence among the signaling cascades that are activated when FSH binds to FSHR. The following paragraphs present the current state of knowledge about the distinct signaling cascades activated by FSH in granulosa cells where the effects of FSH have been most extensively investigated [31,44,45].

Stimulation of FSHR-1 by FSH on the surface membrane of granulosa cells causes activation of adenylate cyclase and increases the intracellular production of cAMP. As noted above, FSH binds to the large extracellular N-terminal domain of FSHR-1. Occupancy of FSHR-1 by FSH results in activation of the heteromeric G protein, G_S. The FSHR-1 residues involved in G-protein coupling are highlighted in Fig. 15.6 [31]. The FSHR-1-induced activation of G_S causes the G protein to dissociate into two molecules: the α-subunit

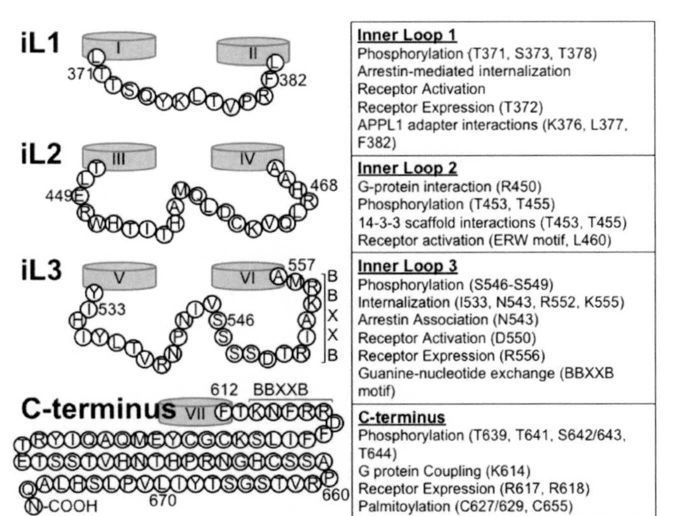

Fig. 15.6 Amino acids in the intracellular loops and C-terminus of the human FSHR are important for FSHR signaling or FSHR interactions with other cellular proteins. These residues have been identified by mutagenesis studies performed in a number of research laboratories [31]

and the β/γ heterodimer. The α-subunit stimulates adenylate cyclase and consequently increases the synthesis of cAMP.

The FSH-induced increase in intracellular cAMP results in activation of PKA, as well as other signaling cascades. Major effects of cAMP in granulosa cells are mediated by PKA. For example, PKA phosphorylates both Histone H3 and the transcription factor CREB (*C*yclic AMP *R*esponse *E*lement *B*inding Protein). Phosphorylation of Histone H3 is associated selectively with the promoters of immediate early and early FSH target genes such as c-FOS and serum glucocorticoid kinase (SGK). The cAMP/PKA/CREB pathway triggers expression of key genes involved in granulosa cell function, including aromatase (*cyp19a*) and inhibin-α (*Inhba*).

There are also cAMP-dependent effects of FSH that are not mediated by PKA. These include FSH actions that occur through exchange factors activated by cAMP (EPAC). In fact, cAMP binding to EPAC is believed to be the primary pathway whereby FSH activates the small GTP binding protein RAP1 and one of the pathways whereby FSH activates the p38 mitogen-activated protein kinase (p38MAPK) [45]. The cAMP/EPAC/RAP1 signaling mechanism is also associated with downstream activation of PI3K and subsequent stimulation of phosphoinositide-dependent kinases (PDK) and protein kinase B (PKB/Akt) [46,47]. The latter kinases impeded the transcription of genes associated with apoptosis, such as FOXO1a and members of the Forkhead family [31,44,45].

The cAMP-dependent, PKA-independent activation of EPAC and its downstream target RAP1 is not the only signaling pathway used by FSH to activate p38MAPK and PKB. FSH stimulates p38MAPK, PKB, and extracellular regulated protein kinases (ERK1/2) through direct activation of the sarcoma oncogene (SRC) family of tyrosine kinases (SFKs) and subsequent activation of RAS. Activation of RAS and ERK1/2 by FSH is also modulated by FSH activation of the epidermal growth factor receptor (EGFR) and cross-talk between the FSH and EGF signal transduction cascades [45]. The importance of FSHR-3 in SFK/RAS/ERK-dependent signaling in granulosa cells is unknown. In summary, FSH provokes a complex pattern of time-dependent changes in granulosa cell gene expression through actions on many different signaling cascades, including those associated with cAMP/PKA, PI3K, MAPKs, ERKs, and SFKs. Figure 15.7 provides a schematic representation of these FSH actions.

15.8 Naturally Occurring Mutations Within the FSHR

Mutations within the FSHR that occur in the human population have been described and, although of obvious clinical interest, are also intriguing because they reveal

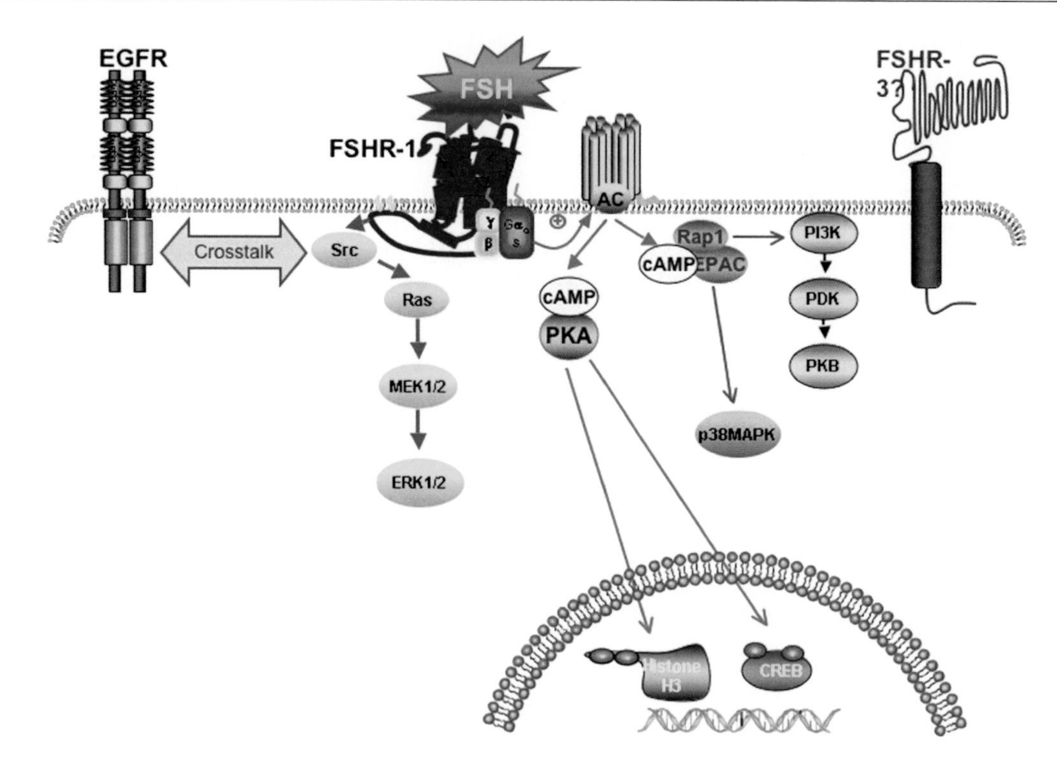

Fig. 15.7 Schematic representation of key granulosa cell signaling cascades activated by FSH (see text for detailed explanation)

further information about the structure/function relationships within the FSHR. These mutations in the DNA sequence encoding the FSHR result in a change in the amino acid sequence of the resulting protein. There are two possibilities for mutations: they either activate the FSHR in the absence of hormonal stimulation (constitutively activating mutations; CAM) or inactivate the receptor and thus result in infertility even at times when hormone concentrations are very high.

Activating mutations—although several naturally occurring mutations have been identified in the LH/CG receptor that cause the mutants to be constitutively active (reviewed by [32,48]), there has only been one reported constitutively activating mutation in the FSHR from the human population [49]. This was from a male patient who had been hypophysectomized and had undetectable gonadotropin secretion and yet exhibited spermatogenesis. The location of the amino acid substitution was within the third intracellular loop at a site that corresponds to a constitutively activating mutation within the LH receptor [50]. However, reconstruction and expression of this mutation in the laboratory proved somewhat inconclusive, as it did not cause higher basal production of cAMP in the absence of FSH stimulation than did the wild-type receptor. After transfection with smaller amounts of cDNA (producing cells which presumably express smaller amounts of the receptor proteins), basal activity was however determined to be approximately 1.5-fold higher for the mutant receptor [49].

An interesting subcategory of activating mutations has also been described. This subcategory includes mutations that cause the FSHR to bind and respond to human chorionic gonadotropin, resulting in hyperstimulation of the ovaries (Ovarian Hyperstimulation Syndrome; OHSS). Mutations that resulted in increased sensitivity to hCG were subsequently found to be constitutively active, although at reduced levels (less cAMP production) as compared to FSH-stimulated wild-type receptors [51]. Expression of an hCG-active FSHR mutant in the laboratory revealed that FSH also bound to the mutant form with the same affinity and subsequently increased cAMP production as it did to the wild-type receptor. Interestingly, however, the mutant bound hCG and caused increased cAMP production in a dose-dependent manner, and with a higher EC50 (a measure of potency) than that observed for hCG binding to and activating the wild-type LH receptor [52].

Surprisingly, there were no apparent pathological effects other than high rates of pregnancy with FSHR mutations that caused activation by hCG [53–56]. Other investigators have examined the likelihood that certain types of granulosa cell tumors or other forms of high FSH sensitivity (such as in families with high rates of non-identical twins) involved constitutively active or promiscuous mutant forms of the FSHR; however, none of these studies were able to find activating mutations within the FSHR [57–59].

Unexpectedly, studies on naturally occurring mutant forms of the FSHR that exhibit hCG sensitivity have resulted in the finding that the mutations are located within the transmembrane domain of the receptor [52,56]. This finding is surprising given that the extracellular domain of the receptor is known to provide hormone specificity and high affinity binding. In fact, other studies with mutations created in the laboratory have found FSHR binding to LH or hCG when certain amino acids within the LRRs of the extracellular domain were changed [43]. Although the precise reasons that increased sensitivity to hCG resulting from amino acid substitutions within the transmembrane region of the FSHR are unknown, it is interesting to note that these mutant forms of the FSHR were also more sensitive to TSH [56], in addition to the previously mentioned constitutive activation. It is still possible, then, that the transmembrane domain of the FSHR is inhibited from becoming active by the extracellular domain because it forms a barrier to activation. Mutations within the transmembrane domain likely lower this barrier to inhibition by the extracellular domain, resulting in a protein that is much easier to activate, even with low affinity interactions with other hormones.

There appear to be structural differences between the FSHR and LHR that may somehow alter the ability of mutant forms of the FSHR to become constitutively active. One study reported that the construction of FSHR mutants that corresponded to constitutively active LHR mutants often resulted in no constitutive activation of the resulting FSHR mutants [60]. Other FSHR mutants did exhibit constitutive activation (although at lower levels than corresponding LHR mutants) and these mutant FSHRs also exhibited promiscuous activation by hCG and TSH. Corresponding LHR mutants exhibited much lower levels of promiscuous activation by hFSH and hTSH. These results indicate that some structural feature of the FSHR prevents it from being as susceptible to activating mutations as the LHR.

Inactivating mutations—inactivating mutations of the FSHR occurring within human populations may be difficult for researchers to find due to their likely result of reducing or eliminating fertility. Thus, the mutant genes are not passed on along familial lines like several of those found within the LHR. However, inactivating mutations of the FSHR have been found in certain clinical cases, particularly in patients presenting with ovarian dysgenesis. Causes of reduced activity or complete loss of function by inactivating mutations within the FSHR are unknown, although they could include complete loss of hormone binding, inability to couple to and/or activate G proteins, or the receptors becoming trapped intracellularly in a misfolded form. In support of the latter possibility, one study found that substitution of an aspartate residue in the rFSHR with the very similar glutamate amino acid caused the resulting mutants to become trapped intracellularly in a form that remained sensitive to endoglycosidase H [61]. However, cell-surface binding of FSH and production of cAMP (at reduced levels) have been observed for some naturally occurring FSHR mutations [52]. See Fig. 15.8 for a summary of activating and inactivating mutations that

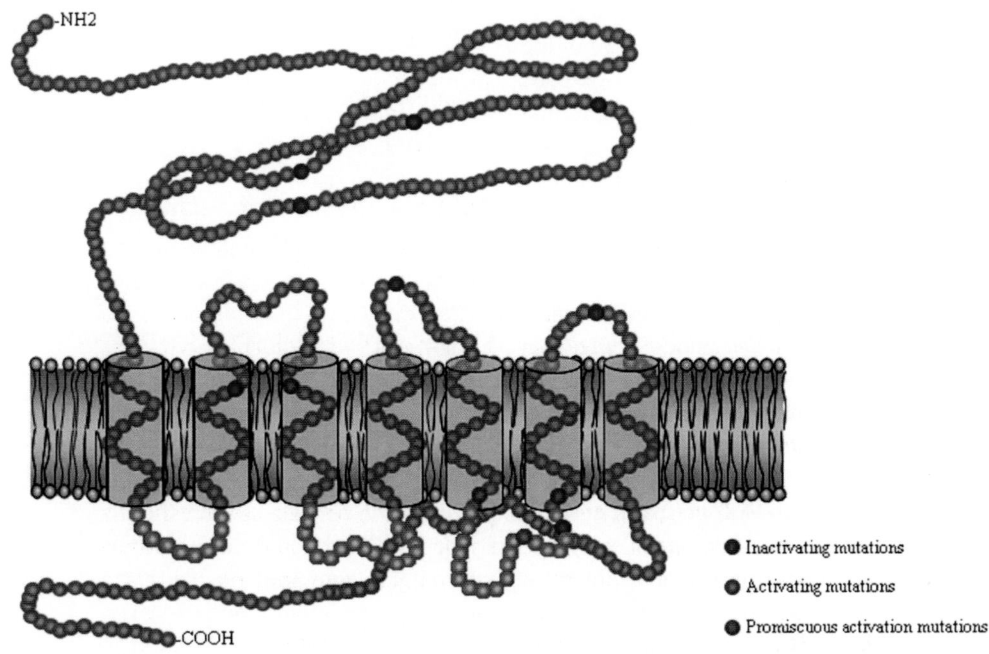

Fig. 15.8 Naturally-occurring mutations of the human FSHR. Amino acid locations shown in *red* are sites of inactivating mutations, the amino acid shown in *green* is the relative location of an activating mutation, and the amino acids shown in *blue* are the relative locations of promiscuous mutations. The latter mutations result in both constitutive activation and activation by other glycoprotein hormones. Locations of amino acids in this figure are not intended to be precise, but instead are designed to illustrate relative locations where mutations alter signaling function by the FSHR

have been found to occur within human populations and their relative locations within the receptor protein.

15.9 FSHR and Ovarian Cancer

Ovarian cancer is one of the leading causes of cancer-related death among women in developed countries. The vast majority of ovarian tumors (>80%), including those that contribute most to mortality, are derived from the ovarian surface epithelium (OSE). Follicle stimulating hormone receptors are expressed by normal OSE, as well as by epithelial ovarian cancer cell and cell lines [62]. Interestingly, expression of FSHR is higher in ovarian epithelial tumors than in normal OSE [62,63]. Moreover, FSH levels are higher in the peritoneal fluid and tumor fluid of women with ovarian cancer than women with non-cancerous cysts or tumors [64,65]. On this basis, it has been postulated that FSHR in the OSE may play an important role in the induction and progression of cancer [66].

A significant amount of experimental data are consistent with this hypothesis. For example, FSH has been shown to stimulate the growth of ovarian cancer cells and to activate the signal transduction pathways associated with cell proliferation and oncogenesis [62,63]. Whereas some researchers have proposed that FSH promotes cancer cell growth through actions on the cAMP/PKA pathway, most have suggested that FSH-stimulation of cancer cell growth is mediated primarily by other signal transduction cascades, particularly PI3K/Akt and ERK [62,63]. The growth factor variant of the FSHR, FSHR-3, has been shown to mediate FSH-induced cell proliferation in at least one ovarian cancer cell line [34]. Treatment with FSH changes gene expression in both normal and malignant OSE, and the actions of FSH differ in these two cellular contexts [67,68]. In ovarian cancer cells, the genes upregulated by FSH were largely related to metabolism, cell proliferation, and oncogenesis, whereas the downregulated genes were associated with cell differentiation, recognition of apoptotic signals, and tumor suppression [67,68].

As described above in the Section 15.8, the ovarian response to FSH is affected by the FSHR genotype [69]. In fact, single nucleotide polymorphisms (SNPs) in the coding region of the FSHR have been associated with susceptibility to ovarian cancer; moreover, histological subtypes of ovarian cancer have been shown to associate differently with FSHR SNPs. Specifically, the two SNPs identified as affecting ovarian cancer risk are Thr307Ala, situated within the extracellular domain and located at nucleotide position 919 of the FSHR coding region (GenBank NM_000145), and Asn608Ser, situated within the C-terminus and located at nucleotide position 2,039 [70]. Individuals who are 307Ala carriers and 680 Ser carriers are significantly more likely to develop epithelial ovarian cancers when compared to non-carriers. It is likely that these SNPs are associated respectively with altered FSH binding and altered coupling to the PKA pathway.

Interestingly, loss of FSHR signaling, as occurs in the FSHR knockout mouse model, is also associated with ovarian epithelial tumorigenesis [71]. Thus, although the FSHR may contribute to the induction and progression of ovarian cancer, expression of the FSHR in the OSA is not an absolute requirement for epithelial carcinogenesis.

15.10 Summary

The receptor for FSH appears in multiple forms and couples to multiple signaling pathways. Variant forms of the receptor result from both splice variation and the addition of a novel exon, and produce receptor proteins that couple to different signaling pathways following stimulation by FSH. Expression of the FSHR genes is hormonally regulated, although it appears to be different in males and females. Regulation may be through transcriptional control, mRNA stability, or control of processes occurring as FSHR proteins are synthesized and trafficked from the ER to the cell membrane. The FSHR interacts with FSH via a series of LRRs within its extracellular domain, and this interaction subsequently results in the extracellular domain switching from an antagonist to an agonist of receptor activation. The FSHR is implicated in a number of pathophysiological conditions, including naturally occurring mutations that affect fertility either positively or negatively, and growth of ovarian tumors.

15.11 Glossary of Terms and Acronyms

cAmp: cyclic adenosine monophosphate

cDNA: complementary deoxyribonucleic acid

ERK: extracellular-regulated kinase

FSH: follicle stimulating hormone

FSHR: follicle stimulating hormone receptor

hCG: human chorionic gonadotropin

LH: luteinizing hormone

LHR: luteinizing hormone receptor

LRR: leucine-rich repeat

MAPK: mitogen-activated protein kinase

PKA: protein kinase A

SF-1: steroidogenic factor-1

SFK: Src-family kinases

TSH: thyroid stimulating hormone

TSHR: thyroid stimulating hormone receptor

USF-1/2: upstream regulatory factor-1/2

References

1. Xu Z, Garverick HA, Smith GW, et al. Expression of follicle stimulating hormone and luteinizing hormone receptor messenger ribonucleic acids in bovine follicles during the first follicular wave. Biol Reprod 1995; 53:951–7.
2. Sprengel R, Braun T, Nikolics K, et al. The testicular receptor for follicle stimulating hormone: structure and functional expression of cloned cDNA. Mol Endocrinol 1990; 4:525–30.
3. McFarland KC, Sprengel R, Phillips HS, et al. Lutropin-choriogonadotropin receptor: an unusual member of the G protein-coupled receptor family. Science 1989; 245:494–9.
4. Palczewski K, Kumasaka T, Hori T, et al. Crystal structure of rhodopsin: A G protein-coupled receptor. Science 2000; 289: 739–45.
5. Moyle WR, Xing Y, Lin W, et al. Model of glycoprotein hormone receptor ligand binding and signaling. J Biol Chem 2004; 279:44442–59.
6. Davis D, Liu X, Segaloff DL. Identification of the sites of N-linked glycosylation on the follicle stimulating hormone (FSH) receptor and assessment of their role in FSH receptor function. Mol Endocrinol 1995; 9:159–70.
7. Sairam MR, Babu PS. The tale of follitropin receptor diversity: a recipe for fine tuning gonadal responses? Mol Cell Endocrinol 2007; 260–262:163–71.
8. Minegishi T, Hirakawa T, Kishi H, et al. A role of insulin-like growth factor I for follicle-stimulating hormone receptor expression in rat granulosa cells. Biol Reprod 2000; 62:325–33.
9. Sites CK, Patterson K, Jamison CS, et al. Follicle-stimulating hormone (FSH) increases FSH receptor messenger ribonucleic acid while decreasing FSH binding in cultured porcine granulosa cells. Endocrinology 1994; 134:411–7.
10. Heckert LL, Daggett MA, Chen J. Multiple promoter elements contribute to activity of the follicle-stimulating hormone receptor (FSHR) gene in testicular Sertoli cells. Mol Endocrinol 1998; 12:1499–512.
11. Hermann BP, Heckert LL. Transcriptional regulation of the FSH receptor: New perspectives. Mol Cell Endocrinol 2007; 260–262:100–8.
12. Heckert LL, Griswold MD. The expression of the follicle-stimulating hormone receptor in spermatogenesis. Recent Prog Horm Res 2002; 57:129–48.
13. Kash JC, Menon KMJ. Identification of a hormonally regulated luteinizing hormone/human chorionic gonadotropin receptor mRNA binding protein. J Biol Chem 1998; 273:10658–64.
14. Dunkel L, Tilly JL, Shikone T, et al. Follicle-stimulating hormone receptor expression in the rat ovary: increases during prepubertal development and regulation by the opposing actions of transforming growth factors beta and alpha. Biol Reprod 1994; 50:940–8.
15. Otsuka F, Moore RK, Wang X, et al. Essential role of the oocyte in estrogen amplification of follicle-stimulating hormone signaling in granulosa cells. Endocrinology 2005; 146:3362–7.
16. Otsuka F, Yamamoto S, Erickson GF, et al. Bone morphogenetic protein-15 inhibits follicle-stimulating hormone (FSH) action by suppressing FSH receptor expression. J Biol Chem 2001; 276:11387–92.
17. Jin X, Han CS, Zhang XS, et al. Stem cell factor modulates the expression of steroidogenesis related proteins and FSHR during ovarian follicular development. Front Biosci 2005; 10:1573–80.
18. Camp TA, Rahal JO, Mayo KE. Cellular localization and hormonal regulation of follicle-stimulating hormone and luteinizing hormone receptor messenger RNAs in the rat ovary. Mol Endocrinol 1991; 5:1405–17.
19. LaPolt PS, Tilly JL, Aihara T, et al. Gonadotropin-induced up- and down-regulation of ovarian follicle-stimulating hormone (FSH) receptor gene expression in immature rats: effects of pregnant mare's serum gonadotropin, human chorionic gonadotropin, and recombinant FSH. Endocrinology 1992; 130:1289–95.
20. Tilly JL, LaPolt PS, Hsueh AJ. Hormonal regulation of follicle-stimulating hormone receptor messenger ribonucleic acid levels in cultured rat granulosa cells. Endocrinology 1992; 130:1296–302.
21. Themmen AP, Blok LJ, Post M, et al. Follitropin receptor down-regulation involves a cAMP-dependent post-transcriptional decrease of receptor mRNA expression. Mol Cell Endocrinol 1991; 78:R7–13.
22. Maguire SM, Tribley WA, Griswold MD. Follicle-stimulating hormone (FSH) regulates the expression of FSH receptor messenger ribonucleic acid in cultured Sertoli cells and in hypophysectomized rat testis. Biol Reprod 1997; 56:1106–11.
23. Griswold MD, Kim JS, Tribley WA. Mechanisms involved in the homologous down-regulation of transcription of the follicle-stimulating hormone receptor gene in Sertoli cells. Mol Cell Endocrinol 2001; 173:95–107.
24. Zhu C, Tian H, Xiong Z, et al. Follicle-stimulating hormone (FSH) induced internalization of porcine FSH receptor in cultured porcine granulosa cells and Chinese hamster ovary cells transfected with recombinant porcine FSH receptor cDNA. J Tongji Med Univ 2001; 21:188–90.
25. Manna PR, Pakarainen P, Rannikko AS, et al. Mechanisms of desensitization of follicle-stimulating hormone (FSH) action in a murine granulosa cell line stably transfected with the human FSH receptor complementary deoxyribonucleic acid. Mol Cell Endocrinol 1998; 146:163–76.
26. Rannikko A, Penttila TL, Zhang FP, et al. Stage-specific expression of the FSH receptor gene in the prepubertal and adult rat seminiferous epithelium. J Endocrinol 1996; 151:29–35.
27. Griswold MD, Heckert L, Linder C. The molecular biology of the FSH receptor. J Steroid Biochem Mol Biol 1995; 53:215–8.
28. Johnston H, Baker PJ, Abel M, et al. Regulation of Sertoli cell number and activity by follicle-stimulating hormone and androgen during postnatal development in the mouse. endocrinology 2004; 145:318–29.
29. Einstein R, Jordan H, Zhou W, et al. Alternative splicing of the G protein-coupled receptor superfamily in human airway smooth muscle diversifies the complement of receptors. Proc Natl Acad Sci U S A 2008; 105:5230–5.
30. Khan H, Jiang LG, Jayashree GN, et al. Recognition of follicle stimulating hormone (alpha-subunit) by a recombinant receptor protein domain coded by an alternately spliced mRNA and expressed in Escherichia coli. J Mol Endocrinol 1997; 19:183–90.
31. Ulloa-Aguirre A, Zarinan T, Pasapera AM, et al. Multiple facets of follicle-stimulating hormone receptor function. Endocrine 2007; 32:251–63.
32. Themmen APN, Huhtaniemi IT. Mutations of gonadotropins and gonadotropin receptors: elucidating the physiology and pathophysiology of pituitary-gonadal function. Endocr Rev 2000; 21:551–83.
33. Sairam MR, Jiang LG, Yarney TA, et al. Follitropin signal transduction: alternative splicing of the FSH receptor gene produces a dominant negative form of receptor which inhibits hormone action. Biochem Biophys Res Commun 1996; 226:717–22.

34. Li Y, Ganta S, Cheng C, et al. FSH stimulates ovarian cancer cell growth by action on growth factor variant receptor. Mol Cell Endocrinol 2007; 267:26–37.
35. Fan QR, Hendrickson WA. Structure of human follicle-stimulating hormone in complex with its receptor. Nature 2005; 433:269–77.
36. Delidow BC, White BA, Peluso JJ. Gonadotropin induction of c-fos and c-myc expression and deoxyribonucleic acid synthesis in rat granulosa cells. Endocrinology 1990; 126:2302–6.
37. Yarney TA, Fahmy MH, Sairam MR, et al. Ontogeny of FSH receptor messenger ribonucleic acid transcripts in relation to FSH secretion and testicular function in sheep. J Mol Endocrinol 1997; 18:113–25.
38. Lusson J, Vieau D, Hamelin J, et al. cDNA structure of the mouse and rat subtilisin/kexin-like PC5: a candidate proprotein convertase expressed in endocrine and nonendocrine cells. Proc Natl Acad Sci U S A 1993; 90:6691–5.
39. Nachtigal MW, Ingraham HA. Bioactivation of Mullerian inhibiting substance during gonadal development by a kex2/subtilisin-like endoprotease. Proc Natl Acad Sci U S A 1996; 93:7711–6.
40. Ellgaard L, Helenius A. Quality control in the endoplasmic reticulum. Nat Rev Mol Cell Biol 2003; 4:181–91.
41. Rozell TG, Davis DP, Chai Y, et al. Association of gonadotropin receptor precursors with the protein folding chaperone calnexin. Endocrinology 1998; 139:1588–93.
42. Mizrachi D, Segaloff DL. Intracellularly located misfolded glycoprotein hormone receptors associate with different chaperone proteins than their cognate wild-type receptors. Mol Endocrinol 2004; 18:1768–77.
43. Smits G, Campillo M, Govaerts C, et al. Glycoprotein hormone receptors: determinants in leucine-rich repeats responsible for ligand specificity. EMBO J 2003; 22:2692–703.
44. Hunzicker-Dunn M, Maizels ET. FSH signaling pathways in immature granulosa cells that regulate target gene expression: branching out from protein kinase A. Cell Signal 2006; 18:1351–9.
45. Wayne CM, Fan HY, Cheng X, et al. Follicle-stimulating hormone induces multiple signaling cascades: evidence that activation of Rous sarcoma oncogene, RAS, and the epidermal growth factor receptor are critical for granulosa cell differentiation. Mol Endocrinol 2007; 21:1940–57.
46. Gonzalez-Robayna IJ, Falender AE, Ochsner S, et al. Follicle-Stimulating hormone (FSH) stimulates phosphorylation and activation of protein kinase B (PKB/Akt) and serum and glucocorticoid-Induced kinase (Sgk): evidence for A kinase-independent signaling by FSH in granulosa cells. Mol Endocrinol 2000; 14:1283–300.
47. Richards JS. New signaling pathways for hormones and cyclic adenosine 3′,5′-monophosphate action in endocrine cells. Mol Endocrinol 2001; 15:209–18.
48. Latronico AC, Segaloff DL. Naturally occurring mutations of the luteinizing-hormone receptor: lessons learned about reproductive physiology and G protein-coupled receptors. Am J Hum Genet 1999; 65:949–58.
49. Gromoll J, Simoni M, Nieschlag E. An activating mutation of the follicle-stimulating hormone receptor autonomously sustains spermatogenesis in a hypophysectomized man. J Clin Endocrinol Metab 1996; 81:1367–70.
50. Laue L, Chan WY, Hsueh AJ, et al. Genetic heterogeneity of constitutively activating mutations of the human luteinizing hormone receptor in familial male-limited precocious puberty. Proc Natl Acad Sci U S A 1995; 92:1906–10.
51. Vassart G, Pardo L, Costagliola S. A molecular dissection of the glycoprotein hormone receptors. Trends Biochem Sci 2004; 29:119–26.
52. Meduri G, Bachelot A, Cocca MP, et al. Molecular pathology of the FSH receptor: new insights into FSH physiology. Mol Cell Endocrinol 2008; 282:130–42.
53. De Leener A, Montanelli L, Van Durme J, et al. Presence and absence of follicle-stimulating hormone receptor mutations provide some insights into spontaneous ovarian hyperstimulation syndrome physiopathology. J Clin Endocrinol Metab 2006; 91: 555–62.
54. Montanelli L, Delbaere A, Di Carlo C, et al. A mutation in the follicle-stimulating hormone receptor as a cause of familial spontaneous ovarian hyperstimulation syndrome. J Clin Endocrinol Metab 2004; 89:1255–8.
55. Smits G, Olatunbosun O, Delbaere A, et al. Ovarian hyperstimulation syndrome due to a mutation in the follicle-stimulating hormone receptor. N Engl J Med 2003; 349:760–6.
56. Vasseur C, Rodien P, Beau I, et al. A chorionic gonadotropin-sensitive mutation in the follicle-stimulating hormone receptor as a cause of familial gestational spontaneous ovarian hyperstimulation syndrome. N Engl J Med 2003; 349:753–9.
57. Fuller PJ, Verity K, Shen Y, et al. No evidence of a role for mutations or polymorphisms of the follicle-stimulating hormone receptor in ovarian granulosa cell tumors. J Clin Endocrinol Metab 1998; 83:274–9.
58. Ligtenberg MJ, Siers M, Themmen AP, et al. Analysis of mutations in genes of the follicle-stimulating hormone receptor signaling pathway in ovarian granulosa cell tumors. J Clin Endocrinol Metab 1999; 84:2233–4.
59. Montgomery GW, Duffy DL, Hall J, et al. Mutations in the follicle-stimulating hormone receptor and familial dizygotic twinning. Lancet 2001; 357:773–4.
60. Zhang M, Tao YX, Ryan GL, et al. Intrinsic differences in the response of the human lutropin receptor versus the human follitropin receptor to activating mutations. J Biol Chem 2007; 282:25527–39.
61. Rozell TG, Wang H, Liu X, et al. Intracellular retention of mutant gonadotropin receptors results in loss of hormone binding activity of the follitropin receptor but not the lutropin/choriogonadotropin receptor. Mol Endocrinol 1995; 9:1727–36.
62. Choi JH, Wong AS, Huang HF, et al. Gonadotropins and ovarian cancer. Endocr Rev 2007; 28:440–61.
63. Leung PC, Choi JH. Endocrine signaling in ovarian surface epithelium and cancer. Hum Reprod Update 2007; 13:143–62.
64. Halperin R, Hadas E, Langer R, et al. Peritoneal fluid gonadotropins and ovarian hormones in patients with ovarian cancer. Int J Gynecol Cancer 1999; 9:502–7.
65. Chudecka-Glaz A, Rzepka-Gorska I. Concentrations of follicle stimulating hormone are increased in ovarian tumor fluid: implications for the management of ovarian cancer. Eur J Gynaecol Oncol 2008; 29:37–42.
66. Bose CK. Follicle stimulating hormone receptor (FSHR) antagonist and epithelial ovarian cancer (EOC). J Exp Ther Oncol 2007; 6:201–4.
67. Ho SM, Lau KM, Mok SC, et al. Profiling follicle stimulating hormone-induced gene expression changes in normal and malignant human ovarian surface epithelial cells. Oncogene 2003; 22:4243–56.
68. Ji Q, Liu PI, Chen PK, et al. Follicle stimulating hormone-induced growth promotion and gene expression profiles on ovarian surface epithelial cells. Int J Cancer 2004; 112:803–14.
69. Gromoll J, Simoni M. Genetic complexity of FSH receptor function. Trends Endocrinol Metab 2005; 16:368–73.
70. Yang CQ, Chan KY, Ngan HY, et al. Single nucleotide polymorphisms of follicle-stimulating hormone receptor are associated with ovarian cancer susceptibility. Carcinogenesis 2006; 27: 1502–6.
71. Chen CL, Cheung LW, Lau MT, et al. Differential role of gonadotropin-releasing hormone on human ovarian epithelial cancer cell invasion. Endocrine 2007; 31:311–20.

Chapter 16

Regulation of the Early Steps in Gonadal Steroidogenesis

Steven R. King and Holly A. LaVoie

16.1 Introduction

De novo synthesis of sex steroid hormones occurs primarily in the Leydig cells of the testes and the theca cells and theca- and granulosa-derived luteal cells of the ovary. Other major sites of de novo steroid biosynthesis include the placenta, the adrenal cortex, and the brain. In steroidogenic cells, cholesterol is taken up in response to trophic stimuli and transported into the mitochondria through the action of the steroidogenic acute regulatory protein (StAR). Inside the mitochondria, the cytochrome P450 cholesterol side-chain cleavage enzyme (P450scc) converts cholesterol to pregnenolone, the steroid from which all sex steroid hormones are derived (Chapter 18 and Fig. 16.1). Hence, the presence of P450scc defines a cell as being steroidogenic. Leydig, theca, luteal, and placental trophoblast cells additionally possess 3β-hydroxysteroid dehydrogenase/Δ4-Δ5 isomerase (3β-HSD), which generate progesterone from pregnenolone. 3β-HSD also converts Δ5-pathway steroids 17α-hydroxypregnenolone, dehydroepiandrosterone (DHEA), or androstenediol to 17α-hydroxyprogesterone, androstenedione, and testosterone, respectively.

Gonadal steroidogenesis is regulated by growth factors and gonadotropins, like luteinizing hormone (LH). Increased steroid production is usually due to increased transcription of steroidogenic enzymes and StAR [1, 2]. The half-lives of some steroidogenic mRNAs are also altered in response to hormonal signals [2, 3]. The activities of select proteins involved in steroidogenesis are affected by post-translational modifications as well [4, 5]. The focus of this chapter is the transcriptional and post-translational regulation of the proteins that mediate the first steps in steroidogenesis in gonadal cells: StAR, P450scc, and 3β-HSD.

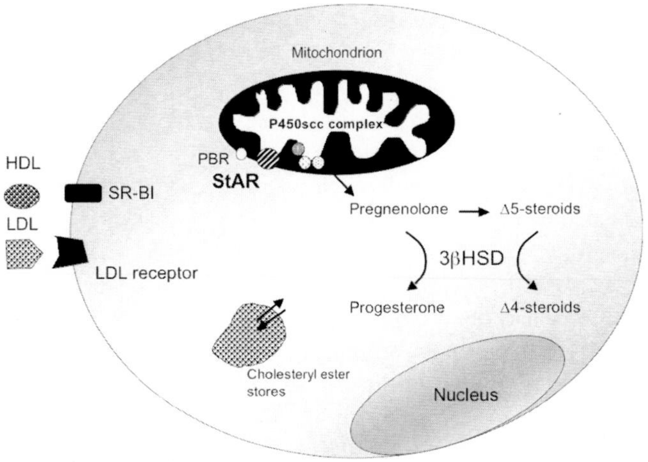

Fig. 16.1 Steps in *de novo* steroidogenesis and progesterone production. Extracellular sources of cholesterol are supplied by high density lipoproteins (HDL) and low density lipoproteins (LDL) and are taken up by their respective receptors, scavenger type receptor, class B, type 1 (SR-BI) or LDL receptor. StAR acts at the mitochondrial membrane to transfer cholesterol to the inner mitochondrial membrane and may functionally interact with PBR. The side-chain of cholesterol is cleaved at the inner mitochondrial membrane by the collective actions of the P450scc complex, which is comprised of adrenodoxin reductase, adrenodoxin, and P450scc, to yield pregnenolone. Pregnenolone can be converted to progesterone by 3β-HSD or go down the Δ5-pathway, catalyzed by P450c17 and be modified subsequently by 3β-HSD

16.2 Transcriptional Control of Steroidogenesis

16.2.1 Hormonal Regulation of Gonadal StAR mRNA Expression

Given the importance of StAR in initiating steroidogenesis, its regulation has become the focus of extensive study. LH or its homologue, human chorionic gonadotropin (hCG), stimulates StAR mRNA expression in many steroidogenic cell types including theca, luteinized granulosa, luteal, and Leydig cells [6, 7, 8, 9, 10, 11, 12, 13, 14].

S.R. King (✉)
Scott Department of Urology Baylor College of Medicine Houston, TX, USA
e-mail: srking100@yahoo.com

P.J. Chedrese (ed.), *Reproductive Endocrinology*,
DOI 10.1007/978-0-387-88186-7_16, © Springer Science+Business Media, LLC 2009

During follicle development in the adult ovary, StAR mRNA and protein are expressed in theca-interstitial cells of most species but not in granulosa cells until the LH surge [15, 16]. One possible exception is the cow, where one study showed that StAR mRNA is expressed in the granulosa cells of larger antral follicles [17]. However, a second bovine study indicated that StAR expression was confined to granulosa cells of atretic antral follicles [18]. With the LH surge, StAR is dramatically upregulated and continues to be strongly expressed in differentiated theca and mural granulosa cells that compose the corpus luteum [11, 15, 16, 19]. Steroidogenic granulosa cells in the cumulus layer also maintain and increase StAR expression after ovulation [20].

The gonadotropin follicle stimulating hormone (FSH) induces StAR mRNA expression in cultured granulosa cells undergoing luteinization, but does not appear to do so in the non-luteinized follicle [15, 16, 21, 22]. In cultured granulosa cells, FSH may therefore mimic the actions of LH on the StAR gene. Adult human and rat testicular Sertoli cells of the also possess low levels of StAR protein that can be stimulated in cultured rat Sertoli cells by FSH [11, 23, 24].

Depending on the species and cell type, select growth factors can either directly affect StAR mRNA levels or modify gonadotropin stimulation of StAR gene transcription. For example, insulin-like growth factor type I (IGF-I) by itself modestly stimulates StAR mRNA expression in porcine granulosa and additively or synergistically augments gonadotropin- and cAMP-stimulation of StAR [9, 21, 25]. IGF-I also augments hCG-driven StAR mRNA accumulation in primary rat Leydig cells and murine Leydig tumor (mLTC-1) cells [7, 26]. On the other hand, epidermal growth factor (EGF) alone does not affect StAR gene expression in porcine granulosa cells; however, it can suppress FSH or cAMP stimulation of StAR mRNA [25, 27]. In contrast, EGF increases StAR transcript levels in mLTC-1 cells and at submaximal concentrations enhances hCG and cAMP stimulation of StAR mRNA [28]. These studies reveal that gonadotropins and growth factors have some common effects on the StAR gene, whereas others are cell-specific. Numerous other hormones and growth factors also positively or negatively regulate StAR mRNA levels in steroidogenic cells in the gonads (Table 16.1).

Table 16.1 Summary of hormones and growth factors shown to regulate StAR mRNA

Hormone/Growth factor	Cell types
Positive Regulators	
LH/hCG	Granulosa[1], theca, luteal, and Leydig cells [8,9,10,170]
FSH	Granulosa cells [21,22]
Estradiol	Luteal cells[2] [171]
Progesterone	Leydig cells [172]]
Thyroid hormone	Leydig cells [173,174]
Prolactin	Luteal[3] and Leydig cells[3,4] [175,176]
Growth Hormone	Luteal cells [10]
CRH	Leydig [177]
Insulin	Granulosa and theca cells [8,9,178]
IGF-I/IGF-II	Granulosa and Leydig cells [9,21,26,179]
EGF/TGF-α	Leydig cells [28,180]
GDF-9	Granulosa cells [181]
Amphiregulin	Cumulus-oocyte-complexes [182]
MIS	Leydig cells [183]
Leptin	Granulosa cells[4] [184]
Negative Regulators	
PGF2α	Luteal cells [10]
EGF	Granulosa cells[3] [25,27]
TGFβ	Thecal cells [185]
TNFα	Luteal cells [175]
Interferon γ	Leydig cells [7]
Leptin	Granulosa[4] and Leydig cells [184,186]
Prolactin	Leydig cells[3,4] [176]
BMP-6	Granulosa cells [187]

[1] Granulosa refers to granulosa cells at various stages of differentiation prior to ovulation.
[2] No mRNA studies, protein was studied.
[3] When combined with gonadotropin/cAMP stimulus.
[4] Low and high doses show different responses in the same cell type.

16.2.2 StAR Promoter Regulation in Gonadal Cells

Studies of the transcriptional regulation of StAR and other steroidogenic genes are mostly based on transfection studies using plasmids harboring their 5′-flanking regions linked to reporter genes. Although reporter construct studies may not perfectly model endogenous gene regulation, they provide a wealth of information about potential regulatory signals. Most studies have addressed steroidogenic gene promoter regulation in clonal cells, while a few have utilized primary cell types. The StAR gene promoter has been extensively evaluated and studies reveal complex regulation of this gene. This section will summarize the *cis* and *trans* elements that regulate StAR gene transcription with emphasis on gonadotropin/cAMP regulation in gonadal cells.

Steroidogenic Factor 1 (SF-1) and Liver Receptor Homologue 1 (LRH-1)—the proximal region of the StAR promoter is highly conserved across multiple species (Fig. 16.2) and exhibits responsiveness to gonadotropins and cAMP. Evaluation of this region has facilitated the identification of functional transcription factor binding sites [29, 30]. Early studies implicated the transcription factor SF-1 (NR5A1) in the activation of the StAR gene. SF-1 is an orphan nuclear receptor that participates in the regulation of the reproductive axis and genes mediating steroidogenesis. Multiple conserved SF-1 sites have been identified with at least two functional sites in the proximal StAR promoter at −105 to −95 and −35 to −42 (position relative to human

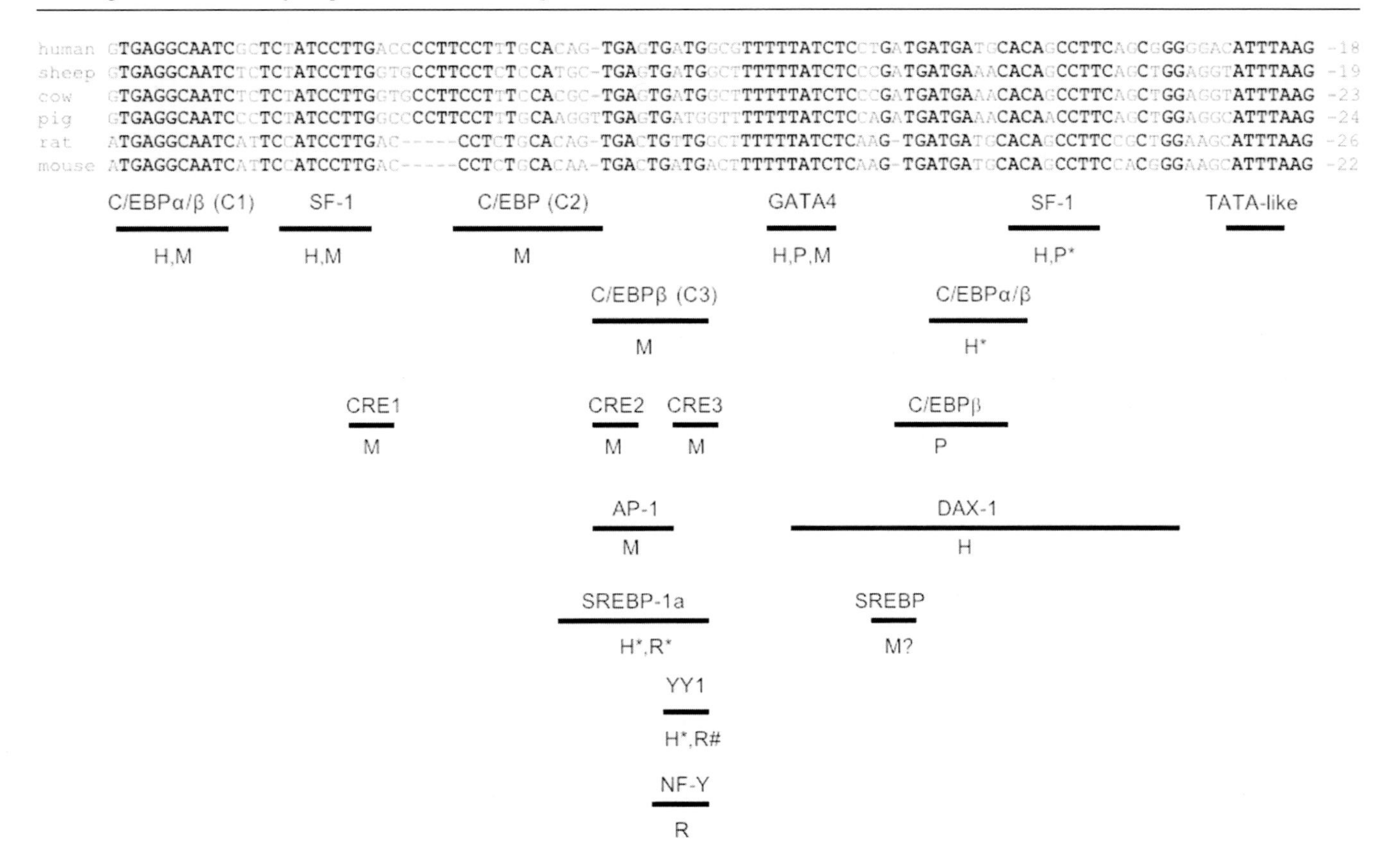

Fig. 16.2 Comparison of the StAR gene proximal promoter region of several mammalian species reveals high homology. Bold text indicates perfectly conserved nucleotides. Bars underneath the sequence correspond to DNA elements participating in StAR gene regulation. * means DNA binding demonstrated by recombinant protein only. H, M, R, and P indicate evidence exists for human, mouse, rat, or pig sequences, respectively. # means recombinant protein did not bind DNA. ? indicates site is functionally important but there is no protein-DNA studies to support binding of the indicated factor. All sequences were obtained from Genbank. See text for references

start site) [31]. One or more of these sites is needed for full basal and/or cAMP responsiveness in Leydig and cultured granulosa cells [31, 32, 33]. Chromatin immunoprecipitation (ChIP) assays confirm increased SF-1 association with the murine StAR promoter in response to a cAMP stimulus in MA-10 mouse Leydig tumor cells and granulosa cells [34]. Two studies in granulosa cells did not support the requirement of the proximal SF-1 site in FSH stimulation of StAR promoter activity [22, 35]. Recently, SF-1 family member LRH-1 (NR5A2) was shown to transactivate the StAR promoter in clonal and human granulosa tumor cells and can bind to the −105 to −95 SF-1 recognition sequence [36, 37]. Taken together, evidence supports an integral role for SF-1 or LRH-1 in cAMP-mediated StAR gene transcription.

GATA4 and GATA6—the transcription factors GATA4 and GATA6 are found in the somatic cells of both the ovary and the testis [38]. A highly conserved GATA site is located at −63 to −58 of the human promoter and is conserved in all mammalian species examined to date. GATA4 is the major factor binding to the GATA site in cultured granulosa cells although some GATA6 binding is detectable [22, 35]. Recombinant GATA6 also efficiently binds the site [35].

GATA4 associates with the endogenous murine StAR promoter in granulosa cells as shown by ChIP assay following an ovulatory hCG stimulus [34]. Interestingly, in overexpression studies, GATA4 and GATA6 both effectively drive porcine StAR promoter activity in homologous granulosa cells [39]. The GATA site in the murine promoter contributes about 20% to the cAMP responsiveness of the promoter in MA-10 Leydig cells [33]. Since GATA4 levels decline in luteal cells following the LH surge *in vivo* [39, 40, 41, 42], it is likely that GATA6 is the family member that stimulates endogenous StAR gene transcription following the surge. In summary, GATA4/6 contributes to maximal cAMP responsiveness of the StAR promoter in gonadal cells.

CCAAT/Enhancer Binding Proteins (C/EBP)—in close proximity to the highly conserved SF-1 sites in the proximal StAR promoter are recognition sequences for C/EBPs. Several C/EBP family members are present in steroidogenic gonadal cells [43, 44]. C/EBPβ is positively regulated by gonadotropins, cAMP, and luteinization [22, 43, 44] and is required for ovulation [45]. One C/EBP site, denoted C1, is conserved in mammalian species and is located at −119 to −110 in the human.

Second and possible third C/EBP sites are more proximal and their locations and sequences vary by species. In humans a more proximal C/EBP site exists at −50 to −41, which partially overlaps the C/EBP site in the porcine promoter at −58 to −49 [35, 46]. The proximal porcine C/EBP site is perfectly conserved in the ovine and bovine promoters. In the human mutation of both C/EBP sites affects basal but not cAMP-stimulated StAR promoter activity in human luteinized granulosa cells [46]. In porcine granulosa cells, the proximal C/EBP site binds C/EBPβ in nuclear extracts and is necessary for full responsiveness to FSH [35].

The mouse StAR promoter has two additional C/EBP sites downstream of the C1 site, named C2 and C3. The more proximal site at −81 to −72 (C3) is necessary for full responsiveness to FSH in rat granulosa cells and binds C/EBPβ in electromobility shift assays [22]. The most proximal C/EBP site in both mouse and porcine promoters show cooperation with the adjacent GATA site to confer maximal responsiveness to FSH in granulosa cells [22, 35]. C/EBP sites C1 and C2 are required for SF-1 activation of the murine promoter and are important for basal promoter activity in MA-10 cells [47]. C/EBPβ and SF-1 also physically interact with each other in the presence of a DNA fragment of the proximal murine StAR promoter containing regions that bind C/EBPβ in MA-10 cell extracts [47]. More recent studies suggest that C/EBPβ may be more important for StAR activity in granulosa cells than in Leydig cells (see below).

Cyclic AMP Response Element Binding Protein (CREB) family—although the StAR promoter responds to cAMP in all steroidogenic gonadal cells examined, the promoter lacks a consensus cAMP response element (CRE) [48]. However, CRE half-sites are present in most species. The −110 to −67 region of the murine StAR promoter contains three, designated CRE1, CRE2, and CRE3. Overexpression of CREB in MA-10 Leydig cells drives murine StAR promoter activity in this region [49]. Further analyses reveal that all three half-sites participate in the cAMP response in MA-10 cells, with the CRE2 site being the most important and the major binding protein being CRE modulator (CREM). Interestingly, the CRE2 site falls within the C3-CEBP site important for cAMP-responsiveness in rodent granulosa cells. These findings were clarified by the groundbreaking work of Hiroi and coworkers who performed ChIP assays with the murine StAR promoter in both MA-10 and mouse periovulatory granulosa cells following cAMP or hCG stimulation [34]. Their studies reveal that in response to the cAMP stimulus, CREB/CREM but not C/EBPβ increased association with the proximal StAR promoter in Leydig cells, whereas C/EBPβ preferentially increased association with the proximal promoter in granulosa cells. These studies demonstrated cell- and sex-specific differences in StAR gene regulation.

Sterol Regulatory Element Binding Protein (SREBP) and Yin Yang 1 (YY1)—sterol regulatory element binding proteins (SREBPs) are implicated in the stimulation of StAR promoter activity. Under conditions of sterol-depletion, the active fragment of SREBP migrates to the nucleus and promotes the transcription of genes involved in sterol availability such as the LDL receptor and HMG-CoA reductase genes [50]. Several putative sterol response elements (SREs) localize to the rat distal StAR promoter region. In the proximal StAR promoter of human and rat, an SRE is located within positions −81 to −70 and −87 to −79, respectively [30, 51]. Overexpression of SREBP-1a in granulosa-derived or clonal cells increases StAR promoter activity [30, 51, 52]. Recombinant SREBP-1a binds the human and rat SRE sequences [30, 51]. Another SREBP site is proposed to exist 4 bp downstream of the GATA site in the mouse promoter. The mutation of the sequence decreases basal promoter activity, but there is no experimental evidence of SREBP binding to this region [28].

The transcription factor Yin Yang 1 (YY1) interferes with the activation of sterol-regulated genes by SREBP [50]. In rat ovaries, YY1 increases in response to a luteolytic dose of PGF2α [51]. Under similar conditions PGF2α downregulates StAR mRNA [53]. YY1 does not bind the proximal SRE site in the rat promoter but does inhibit SREBP-1a and NF-Y binding and activation in clonal cells [51]. In humans, YY1 binds the DNA directly and inhibits SREBP-1a-mediated StAR promoter activation [30].

Although evidence supports a role for SREBP activation of and YY1 inhibition of StAR promoter activity, no studies in gonadal cells have documented the endogenous proteins binding the StAR promoter under physiological conditions. Thus, it remains an open question as to whether SREBPs participate in hormonal regulation of the StAR gene.

Activator Protein 1 (AP-1) family—AP-1 family members both positively and negatively impact StAR promoter activity. A fairly well conserved AP-1 site is located at −81 to −75 of the mouse StAR promoter sequence, overlapping the CRE2 site [54]. In the rat, mutation of the analogous AP-1 site at −85 to −79 increases basal and reduces SF-1-stimulated StAR promoter activity in Y1 adrenocortical cells [55].

c-Fos and Fra-1 inhibit basal StAR promoter activity in mLTC-1 cells [28]. In MA-10 Leydig cells, c-Fos, FosB, c-Jun, JunB, and JunD overexpression enhances and Fra-1 and Fra-2 decreases basal murine StAR promoter activity [54]. In addition, all these AP-1 factors attenuate the response of the promoter to a cAMP analogue.

In MA-10 nuclear extracts, c-Fos, Fra-2, and JunD appear to be the major AP-1 factors binding to the mouse −81 to −75 region. Mammalian two-hybrid assays in HeLa cells showed an interaction of c-Fos and c-Jun with SF-1, GATA4, and C/EBPβ, suggesting these molecules cooperate to facilitate StAR transcription [54]. Further studies of the mouse CRE2/AP-1 site show that c-Fos/c-Jun represses CREB stimulation of the StAR promoter, most likely by competing for the coactivator CREB-binding protein (CBP) [56]. In summary, the AP-1 family represses cAMP-mediated StAR promoter activity through a common DNA element and competition for coactivator molecules.

Other factors regulating StAR transcription—several other transcription factors have been implicated in StAR promoter regulation. Sp1 promotes StAR promoter activity in clonal cells [57]. DAX-1 binds hairpin conformations of the proximal StAR promoter and inhibits basal and forskolin-driven promoter activity in Y1 adrenocortical cells [58]. DAX-1 also interferes with SF-1 activation of the murine StAR promoter in mouse Leydig tumor cells [59]. Forkhead box protein FOXL2 represses basal activity of the human StAR promoter in clonal cells [60].

Conclusions—the StAR promoter has multiple validated and potential regulatory elements that contribute to its activity in gonadal cells. When studies are evaluated for gonadotropin action and by gonadal cell type, two themes emerge. In ovarian cultured granulosa cells, gonadotropin-stimulated StAR promoter activity is primarily regulated by SF-1/LRH-1, GATA4/6, and C/EBPβ. In Leydig cells, cAMP-regulated StAR promoter activity primarily involves SF-1, CREM, GATA4, and modulation by AP-1 factors. However, the regulation of the gene in luteal cells is unknown.

Post-transcriptional regulation of StAR—changes in StAR mRNA stability is another way that cells evolved to more precisely and quickly regulate steroidogenesis. StAR mRNA is generated primarily as approximately 3.5 kb and 1.6 kb long [61]. The disparities in size reflect differences in the 3′ untranslated region. Upon stimulation, the 3.5-kb form predominates as a result of enhanced stability through the activation of cAMP pathways [62]. Cessation of stimulation causes preferential degradation of this form, thus providing an additional mechanism to quickly halt cholesterol transfer to P450scc and steroid synthesis [3].

16.2.3 Hormonal Regulation of Gonadal CYP11A1 mRNA Expression

The CYP11A1 gene encodes for P450scc [63]. Gonadotropins are the primary regulators of CYP11A1 expression in adult gonads. In many respects, its hormonal regulation overlaps that of the StAR gene. For example, the theca cells of developing ovarian follicles express both StAR and CYP11A1 [15, 16, 64, 65]. In Leydig cells, CYP11A1 is expressed at high levels basally unlike StAR and increases more slowly than StAR in response to stimulation by cAMP, LH, or hCG [66, 67, 68, 69].

But there are distinct differences from StAR. While StAR mRNA is virtually undetectable in the granulosa cells of the pre-surge follicle in most species, CYP11A1 mRNA appears after antrum formation and increases with follicular development [16, 18, 65]. The precise timing of its expression is species-dependent. Of note, the cow has unusually high levels of CYP11A1 in granulosa in dominant follicles [18]. CYP11A1 expression is further increased in granulosa cells by the ovulatory gonadotropin surge and is strongly expressed by luteal cells [65].

In cultured granulosa cells CYP11A1 is induced by both FSH and LH/hCG, depending on the state of luteinization [22, 70, 71]. Although there is a fair consensus that LH is needed to initiate the increase in CYP11A1 expression in luteal cells, the requirement for LH to sustain CYP11A1 mRNA during luteal function may be species-dependent [65, 72]. In some cases, luteal CYP11A1 expression is constitutive [65]. In the testes, P450scc is not normally expressed in Sertoli cells, but appears in cultured cells with FSH stimulation [73].

Gonadotropin stimulation of CYP11A1 expression is also modified by growth factors and some growth factors may act independently in a species-specific fashion. For example, in rodents IGF-1 alone has no effect on CYP11A1 expression in cultured granulosa cells, but synergizes with gonadotropin [74]. In the pig, IGF-I does increase CYP11A1 expression in granulosa cells and acts in concert with gonadotropin [75, 76]. The species differences in IGF-I responsiveness can be attributed to the presence of an IGF-I responsive element in the porcine promoter (see below) [77]. Another contrast with StAR gene regulation is the fact that EGF does not inhibit gonadotropin/cAMP-induction of CYP11A1 in porcine granulosa cells [25, 27]. Table 16.2 summarizes the other hormones and growth factors that regulate CYP11A1 gene expression in gonadal cells.

16.2.4 CYP11A1 Promoter Regulation in Gonadal Cells

The CYP11A1 promoter has been extensively studied in adrenal and placental cells with fewer studies in gonadal cells. CYP11A1 promoter activation shares some common *trans* regulatory factors with the StAR gene, but also has its own distinct regulatory factors. As with StAR, gonadotropins and cAMP stimulate CYP11A1 promoter activity in granulosa, theca, luteal, and Leydig cells [78, 79, 80, 81]. Two regions of the human 5′-flanking region of the gene are cAMP responsive. One is located proximally within the first −120 bp and encompasses a fairly well conserved region (Fig. 16.3) [82]. The second is between −1,633 and −1,553

Table 16.2 Summary of hormones and growth factors shown to regulate CYP11A1 mRNA

Hormone/Growth factor	Cell types
Positive Regulators	
LH/hCG	Granulosa[1], theca, luteal, and Leydig cells [65,71,72,103,188,189,190,191]
FSH	Granulosa cells [65,70,189]
Estradiol	Granulosa cells[2] [65]
Progesterone	Granulosa cells [192]
Thyroid hormone	Leydig cells [174]
Growth Hormone	Granulosa cells[3] [193]
Insulin	Granulosa cells [71,178]
IGF-I/IGF-II	Granulosa, theca, and Leydig cells [75,179,190]
EGF/TGF-α	Granulosa and Leydig cells [25,27,180]
Amphiregulin	Cumulus-oocyte-complexes [182]
Negative Regulators	
PGF2α	Granulosa and luteal cells [194,195]
Glucocorticoids	Leydig cells [196]
TGFβ	Leydig cells [197]
TNFα	Granulosa and Leydig cells [198,199,200]
Interferon γ	Leydig cells [201]
Leptin	Leydig cells [186]
IL-1	Leydig cells [202]
GDF-9	Theca cells [203]
BMP-6	Granulosa cells [187]

[1] Granulosa refers to granulosa cells at various stages of differentiation prior to ovulation.
[2] When combined with gonadotropin/cAMP stimulus.
[3] When combined with IGF-I

of the human sequence and is not conserved. In addition to these two cAMP response sequences (CRS), other tissue- and species-specific enhancers and repressors exist as well. The following section will summarize the major *cis* and *trans* elements involved in gonadal CYP11A1 gene regulation.

SF-1 and LRH-1—like StAR, studies implicate SF-1 in the regulation of CYP11A1 promoter activity. In the proximal region of the human CYP11A1 promoter, known as the proximal CRS, resides a highly conserved binding site for SF-1 between −46 and −38. This site binds SF-1 in extracts from granulosa, luteal, and adrenal cells as well as recombinant LRH-1 [78, 80, 83]. Both SF-1 and LRH-1 can drive the human CYP11A1 promoter in granulosa cells [83].

The proximal SF-1 site almost completely overlaps with an AP-2 binding site that participates in promoter activity in the mouse placenta where SF-1 is absent [84]. Forskolin activation of the rat CYP11A1 promoter in granulosa cells is partly localized to the −73 bp region containing the proximal SF-1 element (SCC1) [78]. Deletion of a second SF-1 binding element (SCC2) in the rat in −83 to −64 does not affect forskolin-driven promoter activity in granulosa cells. In MA-10 Leydig cells, mutation of both rat SCC1 and SCC2 reduces basal activity but minimally affects cAMP-stimulated promoter activity [81]. The proximal SF-1 site is needed for both basal and cAMP-mediated activity of the bovine promoter in luteal cells [80].

In the human, mutation of the proximal SF-1 site in the CYP11A1 promoter eliminates expression of a lacZ fusion gene in transgenic mice, suggesting that it is essential for CYP11A1 expression *in vivo*. A second and unconserved SF-1 site resides in the upstream CRS at positions −1,617 to −1,609 of the human promoter. Mutation of this site decreases testicular and adrenal activation of the gene in transgenic mice but does not affect ovarian expression [85]. Further studies using CRE-lox technology to mutate the proximal SF-1 site of the endogenous mouse CYP11A1 promoter show reduced adrenal steroid hormones in transgenic mice [86]. These animals exhibit normal ovarian and placental expression of P450scc, whereas Leydig cells have reduced P450scc levels with no apparent defects in reproduction. These studies suggest that the proximal SF-1 site is not essential for expression in the ovary and enhances but is not required for Leydig cell CYP11A1 expression. Another interpretation of these data is that other elements can substitute for the function of the proximal SF-1 site. In summary, the bulk of the data implicate SF-1 as a contributor to basal and/or cAMP-regulated CYP11A1 promoter activity.

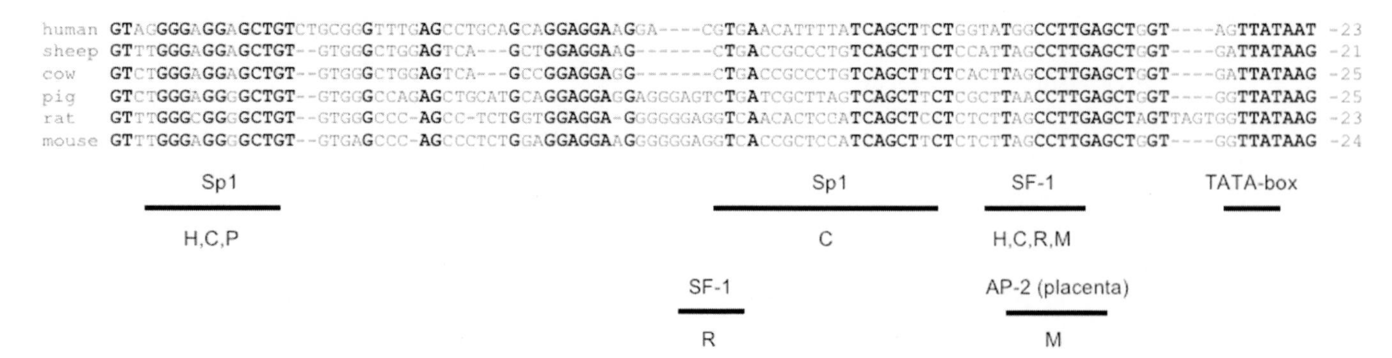

Fig. 16.3 Comparison of the CYP11A1 gene proximal promoter region of several mammalian species shows conserved Sp1 and SF-1 sites. Bold text indicates perfectly conserved nucleotides. Bars underneath the sequence correspond to DNA elements participating in CYP11A1 gene regulation and/or having demonstrated binding of the factor in nuclear extracts. H, C, P, R, and M indicate evidence exists for human, cow, pig, rat, and mouse sequences, respectively. All sequences were obtained from Genbank. See text for references

Sp1—the proximal region of the CYP11A1 promoter is G-rich and one or more Sp1 sites have been identified within it. The −118 to −100 region of the bovine promoter binds Sp1 and/or Sp3 in luteal and Y1 adrenocortical cell extracts [87, 88]. Mutagenesis studies show that this site cooperates with the proximal SF-1 site to drive basal and cAMP-stimulated promoter activity in both bovine luteal and Y1 cells [89]. A second Sp1 binding site at −70 to −32 of the bovine promoter was also confirmed in Y1 cell extracts.

In porcine granulosa cells, forskolin and IGF-I independently activate the porcine CYP11A1 promoter through the Sp1 binding site at −130 to −100 [77]. This region also binds polypyrimidine tract-binding protein-associated splicing factor (PSF), which modulates growth factor responsiveness [90]. Finally, the human −117 to −94 promoter region was also found to bind Sp1 in adrenal cell extracts [91].

GATA and CREB/AP-1—one or more GATA sites exist within the CYP11A1 promoter and their location(s) are species-specific. Overexpression of GATA4 and GATA6 can drive the CYP11A1 promoter in COS-1, HeLa, and/or adrenal cells [92, 93, 94]. Only one study using the mouse promoter has characterized the GATA binding site and its function in a cell-specific context. The site located at −475 to −470 binds GATA4 in nuclear extracts of rat granulosa cells and binds GATA2 in mouse giant trophoblast cells [95]. The GATA site participates in FSH activation of the promoter in granulosa cells and in cAMP-independent promoter activity in trophoblasts.

Just downstream of the mouse GATA site resides a multifunctional CRE half-site at −450 to −447 [95]. The CRE half-site binds CREB-1 in less differentiated granulosa and binds AP-1 family member Fra-2 in more differentiated granulosa cells. In trophoblast cells, the CRE half-site also binds CREB-1, which contributes to basal activity.

Conclusions—studies on the CYP11A1 promoter in gonadal cells have shown that SF-1, LRH-1, Sp1, GATA4, CREB-1, and AP-1 family members participate in basal or cAMP-mediated activity. Additional elements necessary for Leydig cell-specific expression of the mouse CYP11A1 promoter may reside 2.5–5 kb upstream of the transcriptional start site [69].

16.2.5 Hormonal Regulation of Gonadal HSD3B mRNA Expression

Numerous forms of 3β-HSD and their corresponding HSD3B genes have been identified with different family members serving tissue-specific functions [96]. The number of enzyme isoforms varies greatly between species. The human, for instance, expresses two forms while the mouse has six, all independently regulated. The numerical naming of the HSD3B gene family members designates the order of discovery in most species rather than function.

Generally speaking, this enzyme is not unique to steroidogenic cell types, but all steroidogenic cell types in reproductive tissues express 3β-HSD [97]. In humans the major gonadal and adrenal transcript derives from the HSD3B2 gene, whereas in other species the major adrenal/gonadal form derives from the HSD3B1 gene [96]. For the sake of this chapter, these gonadal forms of the gene will be referred to as HSD3B.

The distribution of HSD3B is similar in many respects to CYP11A1 with some species variability. HSD3B mRNA and its protein are expressed in Leydig cells and the theca of developing follicles [69, 98, 99]. In most species 3β-HSD protein is absent in the granulosa cell layer until late in follicle development [98, 99]. An exception is cow, where 3β-HSD protein and its mRNA are present in the granulosa of early dominant follicles [18, 98, 100]. HSD3B mRNA and its protein are further induced by the ovulatory gonadotropin surge and expressed in luteal cells [98, 101].

As with StAR and CYP11A1, gonadotropins and cAMP-signaling play a major role in gonadal HSD3B expression. In culture, LH, hCG, and cAMP-signaling agonists enhance HSD3B mRNA and protein levels in theca, granulosa, and Leydig cells [102, 103, 104]. FSH also stimulates HSD3B mRNA and protein levels in rodent and human cultured granulosa cells [74, 105].

A variety of growth factors directly affect or alter HSD3B mRNA expression. Insulin and IGF-I can increase HSD3B expression in granulosa cells [74, 105]. In human granulosa cells, insulin promotes FSH stimulation of HSD3B mRNA in a more than additive fashion [105]. In contrast, cooperation between IGF-I and FSH was not observed for rodent granulosa cells [74]. EGF also can induce HSD3B expression in rodent granulosa cells and augments stimulation by FSH [106]. Table 16.3 summarizes the known hormones and growth factors shown to regulate HSD3B expression.

16.2.6 HSD3B Promoter Regulation in Gonadal Cells

The HSD3B promoter has been studied much less than the StAR and CYP11A1 gene promoters with few studies in gonadal cells. Part of the reason for the paucity of promoter studies is that the 5′-flanking regions of many gonadal/adrenal type HSD3B genes have not yet been isolated. Moreover, the 5′-flanking regions of mouse and human genes have little homology between them. All HSD3B promoter studies performed to date have used the human promoter (Fig. 16.4).

Table 16.3 Summary of hormones and growth factors shown to regulate HSD3B mRNA

Hormone/Growth Factor	Cell Types
Positive Regulators	
LH/hCG	Granulosa[1], theca, luteal, and Leydig cells [72,102,103,104,204,205,206,207]
FSH	Granulosa cells [74,105]
Progesterone	Granulosa cells[2] [208]
Insulin/IGF-I/IGF-II	Granulosa and theca cells [74,105,209]
EGF	Granulosa[2] and Leydig cells[2] [106,210]
TGFβ	Theca cells[2,3] and Leydig cells[2] [211,212]
FGF	Theca cells[2,3] [211]
Activin A	Leydig cells [212]
Prolactin	Luteal cells[3] [207]
Negative Regulators	
PGF2α	Granulosa and luteal cells [194,213]
Glucocorticoids	Leydig cells[2] [214]
TNFα	Leydig cells [215]
FGF	Leydig cells[2] [216,217]
PDGF	Leydig cells[2] [217]
5α-dihydrotestosterone	Leydig cells[2] [218]
Testosterone	Leydig cells [219]
IL-1	Leydig cells [215]
BMP-2/-6	Granulosa cells[2] [220]
Prolactin	Luteal cells[4] [221]

[1] Granulosa refers to granulosa cells at various stages of differentiation prior to ovulation.
[2] Protein levels or activity evaluated, no mRNA studies.
[3] When combined with gonadotropin/cAMP stimulus.
[4] Luteolytic dose of prolactin.

SF-1 and LRH-1—an SF-1 site resides at positions −64 through −56 in the human HSD3B2 promoter and is responsible for promoter activity in adrenal cells [107]. Two additional SF-1/LRH-1 binding sites are upstream at −906 to −900 and −315 to −309 and bind LRH-1 in human luteinized granulosa cell extracts [108]. While promoter activity in granulosa cells is driven by cAMP independent of LRH-1, LRH-1 in combination with cAMP stimulation additively increases promoter activity. Mutation of either of the two upstream SF-1/LRH-1 sites decreases LRH-1-driven promoter activity in granulosa tumor cells [108]. DAX-1 inhibits LRH-1-stimulated promoter activity in these cells.

GATA4 and GATA6—the 5′-flanking region of the human HSD3B promoter has at least four potential GATA binding sites at approximately −966, −570, −364, and −196. GATA4 and GATA6 can cooperate with SF-1 and LRH-1 to drive HSD3B promoter activity in Leydig and adrenal cell lines [109]. The proximal −196 GATA site by itself is sufficient to confer GATA-responsiveness.

NR4A family and STAT5—an NR4A binding site was identified at −133 to −125 of the human promoter. Overexpression of NR4A family members Nur77, Nurr1, and Nor1 stimulates promoter activity in MA-10 Leydig and adrenal cells [110].

A STAT5 consensus sequence also resides at positions −118 to −110 in the human. Although there are no studies in gonadal or adrenal cells, HeLa cells transfected with the prolactin receptor and STAT5 increase HSD3B2 promoter activity [111]. These studies could have potential significance to the corpus luteum where STAT5 levels are increased with advancing stage of luteal development and steroidogenic capacity [112].

Conclusions—as with the StAR and CYP11A1 gene promoters, SF-1/LRH-1 and GATA4/6 foster human HSD3B2 promoter activity in transfection assays. Therefore, SF-1 and GATA family members may be a common mechanism by which gonadotropins, particularly LH, regulate the StAR, CYP11A1, and HSD3B gene promoters. The regulation of the HSD3B gene is distinguished in part by the activity of NR4A and possibly STAT5. Numerous growth factors can also modify the effects of gonadotropins on HSD3B in a species- and cell-specific manner. Modulation of gene activity by growth factors may explain the differences observed among species and different gonadal cell types in both the male and female. Studies of transactivation of the HSD3B gene are still in their preliminary stages and more progress is expected as the promoter regions of species other than the mouse and human become available for comparison.

16.3 Biochemistry and Post-translational Regulation

16.3.1 The Cholesterol Side-Chain Cleavage Complex (CSCC)

The first and rate-limiting enzymatic step in steroidogenesis is the conversion of cholesterol to pregnenolone by P450scc [113]. The enzyme, formerly known as 20,22-desmolase,

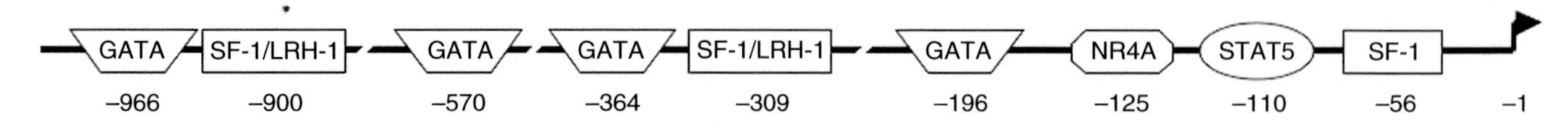

Fig. 16.4 Diagram of the human HSD3B2 gene promoter regions showing sites implicated in transactivation. The −900 and −300 SF-1/LRH-1 sites, the −196 GATA site, and the −125 NR4A site have functional significance in gonadal cells. See text for references

carries out this step on the matrix half of the inner mitochondrial membrane [114, 115]. Pregnenolone is the product of three consecutive monooxygenase reactions catalyzed by P450scc: two hydroxylations of the cholesterol side chain to yield, first, 22R-hydroxycholesterol and then 20α,22R-dihydroxycholesterol, and the final cleavage of the bond between carbons 20 and 22 to liberate pregnenolone and isocaproic acid. Each step requires two reducing equivalents supplied by the flavoprotein NADPH-adrenodoxin reductase/ferredoxin reductase via an electron shuttle, the iron-sulfur protein adrenodoxin/adrenal ferredoxin. Together with P450scc, they comprise the CSCC.

Adrenodoxin reductase uses NADPH to reduce its flavin adenine dinucleotide (FAD) group and binds adrenodoxin. This enzyme then transfers the two electrons from $FADH_2$ to the iron-sulfur cluster of adrenodoxin. This stimulates dissociation of adrenodoxin from the reductase binding site and frees the molecule to interact with P450scc [116].

Binding of cholesterol by P450scc promotes the binding of adrenodoxin and the transfer of electrons from adrenodoxin to its heme ring for the first hydroxylation reaction [116]. Reduced adrenodoxin preferentially binds oxidized P450scc-substrate complexes as well, thus keeping the side-chain cleavage reaction moving forward [117]. Since P450scc is present in excess compared to other constituents of the CSCC, the availability of reduced adrenodoxin and the efficiency of electron transfer determine how swiftly the side-chain cleavage reaction proceeds. Accordingly, adrenodoxin and adrenodoxin reductase are present at higher levels in steroidogenic tissues than in other tissue types in the body [118, 119].

Like P450scc, adrenodoxin levels increase with stimulation of cAMP pathways in JEG-3 cytotrophoblasts, adrenal cortical cells, rhesus monkey fetal ovarian cells, and human fetal adrenal and testicular cells [118–120]. Supporting levels of both adrenodoxin and adrenodoxin reductase rise simultaneously with P450scc in cultured rat granulosa cells and adrenodoxin and P450scc increase with luteal phase in the bovine ovary, ensuring adequate supplies of reducing equivalents [115, 121]. Adrenodoxin has also been shown to increase with P450scc in rhesus monkey fetal ovarian cells and human fetal adrenal and testicular cells [118]. By contrast, the level of adrenodoxin reductase in the human placenta is so low as to limit the rate of pregnenolone production [122]. Production of mitochondrial supporting NADPH is also linked to trophic hormone stimulation through increases in calcium [123].

The lipid composition of the inner membrane additionally influences P450scc activity. In particular, the phospholipid cardiolipin has been implicated in enhancing the binding of cholesterol by P450scc [124, 125] and in the formation of contact sites relevant to cholesterol supply to the enzyme [126].

Interestingly, casein kinase II phosphorylation of Thr71 in mature bovine adrenodoxin can increase the efficiency of electron transfer to P450scc and the rate of catalysis; its relevance *in vivo* is under investigation [127].

Steroid synthesis itself is a closely regulated process. In the absence of stimulation, cells may produce little or no steroid despite the presence of P450scc. Addition of hydroxylated forms of cholesterol like 22R-hydroxycholesterol, which freely diffuse into the mitochondria, are still readily converted to steroid though [128]. This shows that the CSCC is constitutively active and is only constrained by the supply of cholesterol substrate. This is the true rate-limiting step in steroid production and it is carried out by StAR [129].

16.3.2 StAR

Upon trophic hormone stimulation, StAR is rapidly synthesized and goes to the mitochondria, where it mediates the delivery of cholesterol from the outer mitochondrial membrane to the cholesterol-poor inner membrane and P450scc. Exogenous expression of this protein in Leydig tumor cells results in steroid production in the absence of stimulation [6]. If cells are stimulated to produce steroid in the presence of a protein synthesis inhibitor, StAR is not produced [130, 131], cholesterol is not transferred to the inner membrane, and steroidogenesis is blocked [132, 133, 134].

Thus, the acute phase of steroidogenesis depends on StAR-mediated intermembrane cholesterol transfer. Prolonged chronic stimulation, however, results in abundant supply of cholesterol in the inner membrane. At this point, the rate of steroidogenesis now depends upon the capacity of the system, i.e., the level of P450scc and availability of reducing equivalents. Most studies on the mechanism of StAR action have been conducted using adrenocortical or Leydig cell lines as model systems. StAR expression results from the stimulation of pathways related to cAMP, protein kinase C, chloride, and calcium [135]. Production of StAR due to cAMP stimulation is critically supported by arachidonic acid release.

Mechanism of StAR-mediated intermembrane cholesterol transfer—while the function of StAR is fairly straightforward, the mechanism of its action remains elusive. StAR is produced as a 285-amino acid precursor with an apparent molecular weight of 37 kDa [12–14, 130, 131, 136–140]. Its N-terminal signal sequence targets StAR into the mitochondria. During import, the signal sequence is cleaved, giving rise to 30-kDa mature forms of StAR [6, 141, 142]. Interestingly, once inside the mitochondria, StAR no longer supports cholesterol transfer. In fact, blocking the import of StAR into the mitochondria by, for instance, deleting the N-terminal 62 amino acids does not prevent StAR from stimulating steroidogenesis [143, 144]. Therefore, StAR must promote cholesterol transfer to the inner mitochondrial membrane and P450scc while outside the mitochondria. Import then switches off this activity. As a result, continual synthesis of steroid requires the continual synthesis of StAR [139, 140].

Several models exist to explain how StAR may cause cholesterol transfer. One model suggests that StAR acts with cofactors like the peripheral benzodiazepine receptor (PBR) and voltage-dependent anion channel (VDAC), while in vitro studies indicate that StAR may transport cholesterol itself [145, 146, 147, 148].

What is known is that the C-terminus of StAR is essential for its function. This region contains the so-called StAR-related lipid transfer (START) domain, a motif that is conserved across a large family of proteins [149]. All inactivating mutations of StAR, also referred to as START domain 1 (StarD1), affect this region. The protein has proven difficult to crystallize, but X-ray crystallographic analysis of a homolog, MLN64, indicates that the START domain forms a helix-grip fold structure that provides a binding pocket for cholesterol [150, 151]. Thus, StAR may bind cholesterol in the outer membrane and release it to the inner through an unknown mechanism at sites of contact between the membranes [139, 152].

The promotion of intermembrane cholesterol transfer by StAR requires ATP hydrolysis and an intact electrochemical potential across the inner mitochondrial membrane [153, 154]. This activity is doubled with protein kinase A (PKA) phosphorylation of StAR on a conserved residue within the START domain (Ser195 in human) [5, 130]. Treatment of steroidogenic cells with phorbol ester stimulates StAR production via protein kinase C activation, but StAR is not phosphorylated and steroidogenesis remains low [155]. Concomitant addition of low levels of a cAMP analogue, however, causes the newly made StAR to be phosphorylated and dramatically stimulates steroidogenesis in the absence of further increases in StAR protein levels. The mechanism by which phosphorylation alters StAR activity is unknown, but one study suggests that it increases StAR protein stability [4].

The modification likely occurs prior to import. The PKA responsible is thought to be moored to the outer mitochondrial membrane through binding of its R2A regulatory subunit with A-kinase anchor protein 121 (AKAP121) [156]. This PKA may then help bring nascent StAR into contact with the mitochondria through its own association with the outer membrane.

There is a second, uncharacterized modification of StAR whose importance is unclear. Both phosphorylation and this second modification reaction are inefficient. A significant portion of StAR remains unmodified or carrying one of the changes [141].

16.3.3 Congenital Lipoid Adrenal Hyperplasia (Lipoid CAH)

Causes—mutations in StAR and P450scc cause the fatal disorder, lipoid CAH (reviewed in [157]). In this condition, all steroid production is compromised or lost. As a result, patients must be sustained by glucocorticoid supplementation.

Since all steroid production is compromised, it means that progesterone production by fetally derived trophoblasts is compromised too. However, rabbit and mice with mutations in StAR or P450scc still come to term, since pregnancy relies on progesterone from corpora lutea. Patients with mutations in StAR survive to birth, since placental progesterone production is independent of StAR.

Why human fetuses with catastrophic mutations in P450scc survive is not as easily explained, since after week 9, the primary source of progesterone for pregnancy is the placenta. One possibility is that the roles of progesterone at this time are not essential for the maintenance of pregnancy. Alternatively, progesterone filtering in from the maternal circulation could be sufficient to compensate for the loss in placental output.

Effects on reproductive tissues—male patients and animal models of lipoid CAH exhibit phenotypic sex reversal with intra-abdominal or inguinal testes, accumulation of lipid in Leydig cells, underdeveloped vas deferentia and epididymides, testes with disorganized seminiferous tubules, and defects in spermatogenesis [157]. There is a further diminution or loss of the bulbourethral gland, prostate, and seminal vesicles.

In post-pubertal females, the ovaries generally exhibit lipid accumulation in theca cells associated with maturing follicles, become enlarged, and eventually undergo primary ovarian failure. Ovarian cysts and irregular menses are also described in human patients. Interestingly, estrogen synthesis persists in cases of lipoid CAH that originate from catastrophic mutations in StAR [158, 159]. This observation reflects the existence of StAR-independent pathways for steroid production.

16.3.4 StAR-Independent Steroidogenesis

Steroidogenic cells contain other typically minor pathways to initiate steroid synthesis independent of StAR. These pathways are generally regarded as not subject to acute regulation and relevant for basal or constitutive steroid synthesis. In certain cases, these pathways are essential, such as in the human placenta, which produces progesterone in the absence of StAR expression. Various candidates have been proposed to mediate intramitochondrial cholesterol transfer in place of StAR such as the ubiquitous StAR homolog MLN64 [129, 150, 160, 161]. However, definitive proof to demonstrate such a role for any candidate has not yet emerged.

Hydroxylated cholesterol molecules, also called oxysterols, likely account for some StAR-independent steroidogenesis. As mentioned before, oxysterols are more hydrophilic than cholesterol and thus, can move more freely across both mitochondrial membranes for direct conversion to steroid. And since many cholesterol hydroxylases localize to the endoplasmic reticulum, their own generation does not require StAR.

The production of steroid from oxysterols may have physiological relevance, such as in the testes [162]. There, resident macrophages produce 25-hydroxycholesterol, which is then converted into testosterone by neighboring Leydig cells. This is hypothesized to help direct testicular development early in life. The brain also generates high levels of 24-hydroxycholesterol. This oxysterol may be relevant in neurosteroid production, although after modification the hydroxyl group is usually sulfated [163, 164, 165].

16.3.5 3β-HSD

Mechanism of action—after pregnenolone is synthesized, it can be further metabolized by either 3β-HSD or cytochrome P450 17α-hydroxylase/17,20-lyase (P450c17), depending on cell type and species. Both enzymes are found on the endoplasmic reticulum, but 3β-HSD also localizes to the matrix half of the mitochondrial inner membrane. There it associates with the CSCC, thereby facilitating progesterone production [166, 167].

3β-HSD generally converts Δ5-3β-hydroxysteroids into hormonally active Δ4-3-ketosteroids (Fig. 16.1). This NAD^+-dependent process occurs in two steps, catalyzing dehydrogenation and isomerization of a double bond in the steroid molecule, with the first dehydrogenase step being rate-limiting [97]. Like most steroidogenic enzymes, it is constitutively active.

However, not all isoforms possess the same activity. In the rodent unlike the human, certain isoforms expressed primarily in non-steroidogenic tissues are only NADPH-dependent 3-ketosteroid reductases that inactivate steroid hormones [97, 168]. This makes the investigation of the regulation and role of this enzyme family more complicated.

CAH due to 3β-HSD deficiency—inactivating mutations in 3β-HSD II, which is the predominant isoform in the gonads of the human, also cause a rare and sometimes fatal form of CAH [169]. In regards to sexual differentiation, females appear virilized due to elevated DHEA production. Males are more severely affected as the loss of testosterone can cause phenotypic sex reversal and abdominal testes.

16.4 Summary and Conclusions

Steroidogenesis in reproductive tissues is a critical process and thus, highly regulated at the level of transcription, post-transcription, and post-translation. Future challenges will include putting together a model as to how different transcriptional factors come together to temporally coordinate the induction of all three genes. Further exploration of the mechanism by which StAR acts will also lead to a better model of how cholesterol transfer for pregnenolone production is modulated, perhaps uncovering other regulated proteins that assist in this step.

16.5 Glossary of Terms and Acronyms

3β-HSD: 3β-hydroxysteroid dehydrogenase/Δ4–Δ5 isomerase

AKAP121: A-kinase anchor protein

AP-1: Activator Protein 1

C/EBP: transcription factor CCAAT/Enhancer Binding Protein

c-Fos: transcription factor product of the cellular protooncogene homologue of v-FOS identified as the transforming gene of the Finkel-Biskis-Jinkins murine osteosarcoma virus; member of the AP-1 family

ChIP: chromatin immunoprecipitation

c-Jun: transcription factor product of the cellular protooncogene homologue of v-Jun identified as the transforming gene of avian sarcoma virus 17; member of the AP-1 family of transcription factors

CRE: consensus cAMP response element

CREB: transcription factor cyclic AMP response element binding protein

CREM: CRE modulator

CRS: cAMP response sequences

CYP11A1: gene coding for P450scc

DAX-1: dosage-sensitive sex reversal, adrenal hypoplasia critical region, on chromosome X, gene 1

DHEA: dehydroepiandrosterone

EGF: epidermal growth factor

FAD: flavin adenine dinucleotide

$FADH_2$: reduced FAD

FOXL2: transcription factor Forkhead box protein L2

Fra: transcription factor Fos-related antigen

GATA4 and GATA6: members of a family of transcription factors that contain a two-zinc-finger motif and bind to the DNA sequence (A/T)GATA(A/G)

hCG: human chorionic gonadotropin

HDL: high density lipoprotein

HMG-CoA reductase: 3-hydroxy-3-methyl-glutaryl-CoA reductase

HSD3B: gene(s) coding for 3β-HSD

IGF-I: insulin-like growth factor 1

JunB: transcription factor related to c-Jun

JunD: transcription factor related to c-Jun

LDL: low density lipoprotein

Lipoid CAH: congenital Lipoid Adrenal Hyperplasia

LRH-1: transcription factor Liver Receptor Homologue-1

MA-10: stable cell line originated from a Leydig cell tumor

MLN64: StAR homologue

mLTC-1: murine Leydig tumor cells

NADP: nicotinamide adenine dinucleotide phosphate

NADPH: reduced NADP

NR4A: transcription factor nuclear receptor 4A subgroup

NR5A1: transcription factor SF-1

NR5A2: SF-1 family member LRH-1

Nur77, Nurr1 and Nor1: family members of NR4A

P450c17: cytochrome P450 17α-hydroxylase/17,20-lyase

P450scc: cytochrome P450 cholesterol side-chain cleavage enzyme

PBR: peripheral benzodiazepine receptor

PKA: protein kinase A

R2A: PKA regulatory subunit

SCC1: SF-1 binding element

SCC2: SF-1 binding element

SF-1: transcription factor steroidogenic factor 1, also termed NR5A1

Sp1: transcription factor stimulatory protein 1

SR-BI: scavenger type receptor, class B, type 1

SRE: putative sterol response element

SREBP: transcription factor sterol regulatory element binding protein

StAR: steroidogenic acute regulatory protein

StarD1: START domain-containing protein 1 (StAR)

START: StAR-related lipid transfer domain

STAT: signal transducers and activators of transcription

VDAC: voltage-dependent anion channel

YY1: transcription factor Yin Yang 1

References

1. Clark BJ, Combs R, Hales KH et al. Inhibition of transcription affects synthesis of steroidogenic acute regulatory protein and steroidogenesis in MA-10 mouse Leydig tumor cells. Endocrinology 1997; 138:4893–901.
2. Brentano ST, Miller WL. Regulation of human cytochrome P450scc and adrenodoxin messenger ribonucleic acids in JEG-3 cytotrophoblast cells. Endocrinology 1992; 131:3010–8.
3. Zhao D, Duan H, Kim YC et al. Rodent StAR mRNA is substantially regulated by control of mRNA stability through sites in the 3′-untranslated region and through coupling to ongoing transcription. J Steroid Biochem Mol Biol 2005; 96:155–73.
4. Clark BJ, Ranganathan V, Combs R. Steroidogenic acute regulatory protein expression is dependent upon post-translational effects of cAMP-dependent protein kinase A. Mol Cell Endocrinol 2001; 173:183–92.
5. Arakane F, King SR, Du Y et al. Phosphorylation of steroidogenic acute regulatory protein (StAR) modulates its steroidogenic activity. J Biol Chem 1997; 272:32656–62.
6. Clark BJ, Wells J, King SR et al. The purification, cloning, and expression of a novel luteinizing hormone-induced mitochondrial protein in MA-10 mouse Leydig tumor cells. Characterization of the steroidogenic acute regulatory protein (StAR). J Biol Chem 1994; 269:28314–22.
7. Lin T, Hu J, Wang D et al. Interferon-gamma inhibits the steroidogenic acute regulatory protein messenger ribonucleic acid expression and protein levels in primary cultures of rat Leydig cells. Endocrinology 1998; 139:2217–22.
8. Zhang G, Garmey JC, Veldhuis JD. Interactive stimulation by luteinizing hormone and insulin of the steroidogenic acute regulatory (StAR) protein and 17alpha-hydroxylase/17,20-lyase (CYP17) genes in porcine theca cells. Endocrinology 2000; 141:2735–42.
9. Sekar N, LaVoie HA, Veldhuis JD. Concerted regulation of steroidogenic acute regulatory gene expression by luteinizing hormone and insulin (or insulin-like growth factor I) in primary cultures of porcine granulosa-luteal cells. Endocrinology 2000; 141:3983–92.
10. Juengel JL, Meberg BM, Turzillo AM et al. Hormonal regulation of messenger ribonucleic acid encoding steroidogenic acute regulatory protein in ovine corpora lutea. Endocrinology 1995; 136:54239.

11. Pollack SE, Furth EE, Kallen CB et al. Localization of the steroidogenic acute regulatory protein in human tissues. J Clin Endo Metab 1997; 82:4243–51.
12. Stocco DM, Chaudhary LR. Evidence for the functional coupling of cyclic AMP in MA-10 mouse Leydig tumour cells. Cell Signal 1990; 2:161–70.
13. Stocco DM, Chen W. Presence of identical mitochondrial proteins in unstimulated constitutive steroid-producing R2C rat Leydig tumor and stimulated nonconstitutive steroid-producing MA-10 mouse Leydig tumor cells. Endocrinology 1991; 128:1918–26.
14. Sugawara T, Holt JA, Driscoll D et al. Human steroidogenic acute regulatory protein: functional activity in COS-1 cells, tissue-specific expression, and mapping of the structural gene to 8p11.2 and a pseudogene to chromosome 13. Proc Natl Acad Sci U S A 1995; 92:4778–82.
15. Kiriakidou M, McAllister JM, Sugawara T et al. Expression of steroidogenic acute regulatory protein (StAR) in the human ovary. J Clin Endocrinol Metab 1996; 81:4122–8.
16. Ronen-Fuhrmann T, Timberg R, King SR et al. Spatio-temporal expression patterns of steroidogenic acute regulatory protein (StAR) during follicular development in the rat ovary. Endocrinology 1998; 139:303–15.
17. Pescador N, Soumano K, Stocco DM et al. Steroidogenic acute regulatory protein in bovine corpora lutea. Biol Reprod 1996; 55:485–91.
18. Bao B, Garverick HA. Expression of steroidogenic enzyme and gonadotropin receptor genes in bovine follicles during ovarian follicular waves: a review. J Anim Sci 1998; 76:1903–21.
19. LaVoie HA, Benoit AM, Garmey JC et al. Coordinate developmental expression of genes regulating sterol economy and cholesterol side-chain cleavage in the porcine ovary. Biol Reprod 1997; 57:402–7.
20. Feuerstein P, Cadoret V, bies-Tran R et al. Gene expression in human cumulus cells: one approach to oocyte competence. Hum Reprod 2007; 22:3069–77.
21. Balasubramanian K, LaVoie HA, Garmey JC et al. Regulation of porcine granulosa cell steroidogenic acute regulatory protein (StAR) by insulin-like growth factor I: synergism with follicle-stimulating hormone or protein kinase A agonist. Endocrinology 1997; 138:433–9.
22. Silverman E, Eimerl S, Orly J. CCAAT enhancer-binding protein β and GATA-4 binding regions within the promoter of the steroidogenic acute regulatory protein (StAR) gene are required for transcription in rat ovarian cells. J Biol Chem 1999; 274:17987–96.
23. Gregory CW, DePhilip RM. Detection of steroidogenic acute regulatory protein (stAR) in mitochondria of cultured rat Sertoli cells incubated with follicle-stimulating hormone. Biol Reprod 1998; 58:470–4.
24. Ishikawa T, Hwang K, Lazzarino D et al. Sertoli cell expression of steroidogenic acute regulatory protein-related lipid transfer 1 and 5 domain-containing proteins and sterol regulatory element binding protein-1 are interleukin-1beta regulated by activation of c-Jun N-terminal kinase and cyclooxygenase-2 and cytokine induction. Endocrinology 2005; 146:5100–11.
25. Pescador N, Stocco DM, Murphy BD. Growth factor modulation of steroidogenic acute regulatory protein and luteinization in the pig ovary. Biol Reprod 1999; 60:1453–61.
26. Manna PR, Chandrala SP, Jo Y et al. cAMP-independent signaling regulates steroidogenesis in mouse Leydig cells in the absence of StAR phosphorylation. J Mol Endocrinol 2006; 37:81–95.
27. Rusovici R, Hui YY, LaVoie HA. Epidermal growth factor-mediated inhibition of follicle-stimulating hormone-stimulated StAR gene expression in porcine granulosa cells is associated with reduced histone H3 acetylation. Biol Reprod 2005; 72:862–71.
28. Manna PR, Huhtaniemi IT, Wang XJ et al. Mechanisms of epidermal growth factor signaling: regulation of steroid biosynthesis and the steroidogenic acute regulatory protein in mouse Leydig tumor cells. Biol Reprod 2002; 67:1393–404.
29. Reinhart AJ, Williams SC, Stocco DM. Transcriptional regulation of the StAR gene. Mol Cell Endocrinol 1999; 151:161–9.
30. Christenson LK, Osborne TF, McAllister JM et al. Conditional response of the human steroidogenic acute regulatory protein gene promoter to sterol regulatory element binding protein-1a. Endocrinology 2001; 142:28–36.
31. Sugawara T, Kiriakidou M, McAllister JM et al. Multiple steroidogenic factor 1 binding elements in the human steroidogenic acute regulatory protein gene 5′-flanking region are required for maximal promoter activity and cyclic AMP responsiveness. Biochemistry 1997; 36:7249–55.
32. Caron KM, Ikeda Y, Soo SC et al. Characterization of the promoter region of the mouse gene encoding the steroidogenic acute regulatory protein. Mol Endocrinol 1997; 11:138–47.
33. Wooton-Kee CR, Clark BJ. Steroidogenic factor-1 influences protein-deoxyribonucleic acid interactions within the cyclic adenosine 3,5-monophosphate-responsive regions of the murine steroidogenic acute regulatory protein gene. Endocrinology 2000; 141:1345–55.
34. Hiroi H, Christenson LK, Chang L et al. Temporal and spatial changes in transcription factor binding and histone modifications at the steroidogenic acute regulatory protein (StAR) locus associated with StAR transcription. Mol Endocrinol 2004; 18:791–806.
35. LaVoie HA, Singh D, Hui YY. Concerted regulation of the porcine steroidogenic acute regulatory protein gene promoter activity by follicle-stimulating hormone and insulin-like growth factor I in granulosa cells involves GATA-4 and CCAAT/enhancer binding protein beta. Endocrinology 2004; 145:3122–34.
36. Sirianni R, Seely JB, Attia G et al. Liver receptor homologue-1 is expressed in human steroidogenic tissues and activates transcription of genes encoding steroidogenic enzymes. J Endocrinol 2002; 174:R13–R7.
37. Kim JW, Peng N, Rainey WE et al. Liver receptor homolog-1 regulates the expression of steroidogenic acute regulatory protein in human granulosa cells. J Clin Endocrinol Metab 2004; 89:3042–7.
38. LaVoie HA. The role of GATA in mammalian reproduction. Exp Biol Med 2003; 228:1282–90.
39. Gillio-Meina C, Hui YY, LaVoie HA. GATA-4 and GATA-6 transcription factors: expression, immunohistochemical localization, and possible function in the porcine ovary. Biol Reprod 2003; 68:412–22.
40. Anttonen M, Ketola I, Parviainen H et al. FOG-2 and GATA-4 are coexpressed in the mouse ovary and can modulate Müllerian-inhibiting substance expression. Biol Reprod 2003; 68:1333–40.
41. LaVoie HA, McCoy GL, Blake CA. Expression of the GATA-4 and GATA-6 transcription factors in the fetal rat gonad and in the ovary during postnatal development and pregnancy. Mol Cell Endocrinol 2004; 227:31–40.
42. Silverman E, Yivgi-Ohana N, Sher N et al. Transcriptional activation of the steroidogenic acute regulatory protein (StAR) gene: GATA-4 and CCAAT/enhancer-binding protein beta confer synergistic responsiveness in hormone-treated rat granulosa and HEK293 cell models. Mol Cell Endocrinol 2006; 252:92–101.
43. Nalbant D, Williams SC, Stocco DM et al. Luteinizing hormone-dependent gene regulation in Leydig cells may be mediated by CCAAT/enhancer-binding protein-beta. Endocrinology 1998; 139:272–9.
44. Sirois J, Richards JS. Transcriptional regulation of the rat prostaglandin endoperoxide synthase 2 gene in granulosa cells. Evidence for the role of a cis-acting C/EBP beta promoter element. J Biol Chem 1993; 268:21931–8.

45. Sterneck E, Tessarollo L, Johnson PF. An essential role for C/EBPbeta in female reproduction. Genes Dev 1997; 11:2153–62.
46. Christenson LK, Johnson PF, McAllister JM et al. CCAAT/enhancer-binding proteins regulate expression of the human steroidogenic acute regulatory protein (StAR) gene. J Biol Chem 1999; 274:26591–8.
47. Reinhart AJ, Williams SC, Clark BJ et al. SF-1 (steroidogenic factor-1) and C/EBP beta (CCAAT/enhancer binding protein-beta) cooperate to regulate the murine StAR (steroidogenic acute regulatory) promoter. Mol Endocrinol 1999; 13:729–41.
48. Manna PR, Wang XJ, Stocco DM. Involvement of multiple transcription factors in the regulation of steroidogenic acute regulatory protein gene expression. Steroids 2003; 68:1125–34.
49. Manna PR, Dyson MT, Eubank DW et al. Regulation of steroidogenesis and the steroidogenic acute regulatory protein by a member of the cAMP response-element binding protein family. Mol Endocrinol 2002; 16:184–99.
50. Ericsson J, Usheva A, Edwards PA. YY1 is a negative regulator of transcription of three sterol regulatory element-binding protein-responsive genes. J Biol Chem 1999; 274:14508–13.
51. Nackley AC, Shea-Eaton W, Lopez D et al. Repression of the steroidogenic acute regulatory gene by the multifunctional transcription factor Yin Yang 1. Endocrinology 2002; 143:1085–96.
52. Ruiz-Cortes ZT, Martel-Kennes Y, Gevry NY et al. Biphasic effects of leptin in porcine granulosa cells. Biol Reprod 2003; 68:789–96.
53. Sandhoff TW, McLean MP. Repression of the rat steroidogenic acute regulatory (StAR) protein gene by PGF2alpha is modulated by the negative transcription factor DAX-1. Endocrine 1999; 10:83–91.
54. Manna PR, Eubank DW, Stocco DM. Assessment of the role of activator protein-1 on transcription of the mouse steroidogenic acute regulatory protein gene. Mol Endocrinol 2004; 18:558–73.
55. Shea-Eaton W, Sandhoff TW, Lopez D et al. Transcriptional repression of the rat steroidogenic acute regulatory (StAR) protein gene by the AP-1 family member c-Fos. Mol Cell Endocrinol 2002; 188:161–70.
56. Manna PR, Stocco DM. Crosstalk of CREB and Fos/Jun on a single cis-element: transcriptional repression of the steroidogenic acute regulatory protein gene. J Mol Endocrinol 2007; 39:261–77.
57. Sugawara T, Saito M, Fujimoto S. Sp1 and SF-1 interact and cooperate in the regulation of human steroidogenic acute regulatory protein gene expression. Endocrinology 2000; 141:2895–903.
58. Zazopoulos E, Lalli E, Stocco DM et al. DNA binding and transcriptional repression by DAX-1 blocks steroidogenesis. Nature 1997; 390:311–5.
59. Manna PR, Roy P, Clark BJ et al. Interaction of thyroid hormone and steroidogenic acute regulatory (StAR) protein in the regulation of murine Leydig cell steroidogenesis. J Steroid Biochem Mol Biol 2001; 76:167–77.
60. Pisarska MD, Bae J, Klein C et al. Forkhead l2 is expressed in the ovary and represses the promoter activity of the steroidogenic acute regulatory gene. Endocrinology 2004; 145:3424–33.
61. Clark BJ, Soo SC, Caron KM et al. Hormonal and developmental regulation of the steroidogenic acute regulatory protein. Mol Endocrinol 1995; 9:1346–55.
62. Duan H, Jefcoate CR. The predominant cAMP-stimulated 3×5 kb StAR mRNA contains specific sequence elements in the extended 3′UTR that confer high basal instability. J Mol Endocrinol 2007; 38:159–79.
63. Simpson ER. Cholesterol side-chain cleavage, cytochrome P450, and the control of steroidogenesis. Mol Cell Endocrinol 1979; 13:213–27.
64. Garmey JC, Guthrie HD, Garrett WM et al. Localization and expression of low-density lipoprotein receptor, steroidogenic acute regulatory protein, cytochrome P450 side-chain cleavage and P450 17-alpha-hydroxylase/C17-20 lyase in developing swine follicles: in situ molecular hybridization and immunocytochemical studies. Mol Cell Endocrinol 2000; 170:57–65.
65. Goldring NB, Durica JM, Lifka J et al. Cholesterol side-chain cleavage P450 messenger ribonucleic acid: evidence for hormonal regulation in rat ovarian follicles and constitutive expression in corpora lutea. Endocrinology 1987; 120:1942–50.
66. Payne AH. Hormonal regulation of cytochrome P450 enzymes, cholesterol side-chain cleavage and 17 alpha-hydroxylase/C17-20 lyase in Leydig cells. Biol Reprod 1990; 42:399–404.
67. Lejeune H, Sanchez P, Chuzel F et al. Time-course effects of human recombinant luteinizing hormone on porcine Leydig cell specific differentiated functions. Mol Cell Endocrinol 1998; 144:59–69.
68. Clark BJ, Combs R, Hales KH et al. Inhibition of transcription affects synthesis of steroidogenic acute regulatory protein and steroidogenesis in MA-10 mouse Leydig tumor cells. Endocrinology 1997; 138:4893–901.
69. Payne AH, Youngblood GL. Regulation of expression of steroidogenic enzymes in Leydig cells. Biol Reprod 1995; 52:217–25.
70. Urban RJ, Garmey JC, Shupnik MA et al. Follicle-stimulating hormone increases concentrations of messenger ribonucleic acid encoding cytochrome P450 cholesterol side-chain cleavage enzyme in primary cultures of porcine granulosa cells. Endocrinology 1991; 128:2000–7.
71. Sekar N, Garmey JC, Veldhuis JD. Mechanisms underlying the steroidogenic synergy of insulin and luteinizing hormone in porcine granulosa cells: joint amplification of pivotal sterol-regulatory genes encoding low-density lipoprotein (LDL) receptor, steroidogenic acute regulatory (StAR) protein and cytochrome P450 side-chain cleavage (P450scc) enzyme. Mol Cell Endocrinol 2000; 159:25–35.
72. Ravindranath N, Little-Ihrig L, Benyo DF et al. Role of luteinizing hormone in the expression of cholesterol side-chain cleavage cytochrome P450 and 3 beta-hydroxysteroid dehydrogenase, delta 5-4 isomerase messenger ribonucleic acids in the primate corpus luteum. Endocrinology 1992; 131:2065–70.
73. Ford SL, Reinhart AJ, Lukyanenko Y et al. Pregnenolone synthesis in immature rat Sertoli cells. Mol Cell Endocrinol 1999; 157:87–94.
74. Eimerl S, Orly J. Regulation of steroidogenic genes by insulin-like growth factor-1 and follicle-stimulating hormone: differential responses of cytochrome p450 side-chain cleavage, steroidogenic acute regulatory protein, and 3beta-hydroxysteroid dehydrogenase/isomerase in rat granulosa cells. Biol Reprod 2002; 67:900–10.
75. Urban RJ, Garmey JC, Shupnik MA et al. Insulin-like growth factor type I increases concentrations of messenger ribonucleic acid encoding cytochrome P450 cholesterol side-chain cleavage enzyme in primary cultures of porcine granulosa cells. Endocrinology 1990; 127:2481–8.
76. Winters TA, Hanten JA, Veldhuis JD. In situ amplification of the cytochrome P-450 cholesterol side-chain cleavage enzyme mRNA in single porcine granulosa cells by IGF-1 and FSH acting alone or in concert. Endocrine 1998; 9:57–63.
77. Urban RJ, Shupnik MA, Bodenburg YH. Insulin-like growth factor-I increases expression of the porcine P-450 cholesterol side chain cleavage gene through a GC-rich domain. J Biol Chem 1994; 269:25761–9.
78. Clemens JW, Lala DS, Parker KL et al. Steroidogenic factor-1 binding and transcriptional activity of the cholesterol side-chain cleavage promoter in rat granulosa cells. Endocrinology 1994; 134:1499–508.

79. Nelson-DeGrave VL, Wickenheisser JK, Cockrell JE et al. Valproate potentiates androgen biosynthesis in human ovarian theca cells. Endocrinology 2004; 145:799–808.
80. Liu Z, Simpson ER. Molecular mechanism for cooperation between Sp1 and steroidogenic factor-1 (SF-1) to regulate bovine CYP11A gene expression. Mol Cell Endocrinol 1999; 153:183–96.
81. Chau YM, Crawford PA, Woodson KG et al. Role of steroidogenic-factor 1 in basal and 3′,5′-cyclic adenosine monophosphate-mediated regulation of cytochrome P450 side-chain cleavage enzyme in the mouse. Biol Reprod 1997; 57:765–71.
82. Guo IC, Shih MC, Lan HC et al. Transcriptional regulation of human CYP11A1 in gonads and adrenals. J Biomed Sci 2007; 14:509–15.
83. Kim JW, Havelock JC, Carr BR et al. The orphan nuclear receptor, liver receptor homolog-1, regulates cholesterol side-chain cleavage cytochrome p450 enzyme in human granulosa cells. J Clin Endocrinol Metab 2005; 90:1678–85.
84. Ben Zimra M, Koler M, Orly J. Transcription of cholesterol side-chain cleavage cytochrome P450 in the placenta: activating protein-2 assumes the role of steroidogenic factor-1 by binding to an overlapping promoter element. Mol Endocrinol 2002; 16:1864–80.
85. Hu MC, Hsu NC, Pai CI et al. Functions of the upstream and proximal steroidogenic factor 1 (SF-1)-binding sites in the CYP11A1 promoter in basal transcription and hormonal response. Mol Endocrinol 2001; 15:812–18.
86. Shih MC, Hsu NC, Huang CC et al. Mutation of mouse Cyp11a1 promoter caused tissue-specific reduction of gene expression and blunted stress response without affecting reproduction. Mol Endocrinol 2008; 22:915–23.
87. Ahlgren R, Suske G, Waterman MR et al. Role of Sp1 in cAMP-dependent transcriptional regulation of the bovine CYP11A gene. J Biol Chem 1999; 274:19422–8.
88. Begeot M, Shetty U, Kilgore M et al. Regulation of expression of the CYP11A (P450scc) gene in bovine ovarian luteal cells by forskolin and phorbol esters. J Biol Chem 1993; 268:17317–25.
89. Liu Z, Simpson ER. Steroidogenic factor 1 (SF-1) and SP1 are required for regulation of bovine CYP11A gene expression in bovine luteal cells and adrenal Y1 cells. Mol Endocrinol 1997; 11:127–37.
90. Urban RJ, Bodenburg Y. PTB-associated splicing factor regulates growth factor-stimulated gene expression in mammalian cells. Am J Physiol Endocrinol Metab 2002; 283:E794–E8.
91. Guo IC, Tsai HM, Chung BC. Actions of two different cAMP-responsive sequences and an enhancer of the human CYP11A1 (P450scc) gene in adrenal Y1 and placental JEG-3 cells. J Biol Chem 1994; 269:6362–9.
92. Ho CK, Strauss JF, III. Activation of the control reporter plasmids pRL-TK and pRL-SV40 by multiple GATA transcription factors can lead to aberrant normalization of transfection efficiency. BMC Biotechnol 2004; 4:10.
93. Wood JR, Nelson VL, Ho C et al. The molecular phenotype of polycystic ovary syndrome (PCOS) theca cells and new candidate PCOS genes defined by microarray analysis. J Biol Chem 2003; 278:26380–90.
94. Jimenez P, Saner K, Mayhew B et al. GATA-6 is expressed in the human adrenal and regulates transcription of genes required for adrenal androgen biosynthesis. Endocrinology 2003; 144:4285–8.
95. Sher N, Yivgi-Ohana N, Orly J. Transcriptional regulation of the cholesterol side chain cleavage cytochrome P450 gene (CYP11A1) revisited: binding of GATA, cyclic adenosine 3′,5′-monophosphate response element-binding protein and activating protein (AP)-1 proteins to a distal novel cluster of cis-regulatory elements potentiates AP-2 and steroidogenic factor-1-dependent gene expression in the rodent placenta and ovary. Mol Endocrinol 2007; 21:948–62.
96. Simard J, Ricketts ML, Gingras S et al. Molecular biology of the 3beta-hydroxysteroid dehydrogenase/delta5-delta4 isomerase gene family. Endocr Rev 2005; 26:525–82.
97. Payne AH, Abbaszade IG, Clarke TR et al. The multiple murine 3 beta-hydroxysteroid dehydrogenase isoforms: structure, function, and tissue- and developmentally specific expression. Steroids 1997; 62:169–75.
98. Conley AJ, Kaminski MA, Dubowsky SA et al. Immunohistochemical localization of 3 beta-hydroxysteroid dehydrogenase and P450 17 alpha-hydroxylase during follicular and luteal development in pigs, sheep, and cows. Biol Reprod 1995; 52:1081–94.
99. Teerds KJ, Dorrington JH. Immunohistochemical localization of 3 beta-hydroxysteroid dehydrogenase in the rat ovary during follicular development and atresia. Biol Reprod 1993; 49:989–96.
100. Tian XC, Berndtson AK, Fortune JE. Differentiation of bovine preovulatory follicles during the follicular phase is associated with increases in messenger ribonucleic acid for cytochrome P450 side-chain cleavage, 3 beta- hydroxysteroid dehydrogenase, and P450 17 alpha-hydroxylase, but not P450 aromatase. Endocrinology 1995; 136:5102–10.
101. Yuan W, Lucy MC. Messenger ribonucleic acid expression for growth hormone receptor, luteinizing hormone receptor, and steroidogenic enzymes during the estrous cycle and pregnancy in porcine and bovine corpora lutea. Domest Anim Endocrinol 1996; 13:431–44.
102. Chedrese PJ, Zhang D, Luu T, V et al. Regulation of mRNA expression of 3 beta-hydroxy-5-ene steroid dehydrogenase in porcine granulosa cells in culture: a role for the protein kinase-C pathway. Mol Endocrinol 1990; 4:1532–8.
103. Voss AK, Fortune JE. Levels of messenger ribonucleic acid for cholesterol side-chain cleavage cytochrome P-450 and 3 beta-hydroxysteroid dehydrogenase in bovine preovulatory follicles decrease after the luteinizing hormone surge. Endocrinology 1993; 132:888–94.
104. Keeney DS, Mason JI. Expression of testicular 3 beta-hydroxysteroid dehydrogenase/delta 5-4-isomerase: regulation by luteinizing hormone and forskolin in Leydig cells of adult rats. Endocrinology 1992; 130:2007–15.
105. Mcgee E, Sawetawan C, Bird I et al. The effects of insulin on 3 beta-hydroxysteroid dehydrogenase expression in human luteinized granulosa cells. J Soc Gynecol Investig 1995; 2:535–41.
106. Bendell JJ, Dorrington JH. Epidermal growth factor influences growth and differentiation of rat granulosa cells. Endocrinology 1990; 127:533–40.
107. Leers-Sucheta S, Morohashi K, Mason JI et al. Synergistic activation of the human type II 3beta-hydroxysteroid dehydrogenase/delta5-delta4 isomerase promoter by the transcription factor steroidogenic factor-1/adrenal 4-binding protein and phorbol ester. J Biol Chem 1997; 272:7960–7.
108. Peng N, Kim JW, Rainey WE et al. The role of the orphan nuclear receptor, liver receptor homologue-1, in the regulation of human corpus luteum 3beta-hydroxysteroid dehydrogenase type II. J Clin Endocrinol Metab 2003; 88:6020–8.
109. Martin LJ, Taniguchi H, Robert NM et al. GATA factors and the nuclear receptors, steroidogenic factor 1/liver receptor homolog 1, are key mutual partners in the regulation of the human 3beta-hydroxysteroid dehydrogenase type 2 promoter. Mol Endocrinol 2005; 19:2358–70.
110. Martin LJ, Tremblay JJ. The human 3beta-hydroxysteroid dehydrogenase/Delta5-Delta4 isomerase type 2 promoter is a novel target for the immediate early orphan nuclear receptor Nur77 in steroidogenic cells. Endocrinology 2005; 146:861–9.

111. Feltus FA, Groner B, Melner MH. Stat5-mediated regulation of the human type II 3beta-hydroxysteroid dehydrogenase/delta5-delta4 isomerase gene: activation by prolactin. Mol Endocrinol 1999; 13:1084–93.
112. Soloff MS, Gal S, Hoare S et al. Cloning, characterization, and expression of the rat relaxin gene. Gene 2003; 323:149–55.
113. Miller WL. Minireview: regulation of steroidogenesis by electron transfer. Endocrinology 2005; 146:2544–50.
114. Farkash Y, Timberg R, Orly J. Preparation of antiserum to rat cytochrome P-450 cholesterol side chain cleavage, and its use for ultrastructural localization of the immunoreactive enzyme by protein A-gold technique. Endocrinology 1986; 118:1353–65.
115. Hanukoglu I, Suh BS, Himmelhoch S et al. Induction and mitochondrial localization of cytochrome P450scc system enzymes in normal and transformed ovarian granulosa cells. J Cell Biol 1990; 111:1373–81.
116. Lambeth JD, Seybert DW, Lancaster JR, Jr. et al. Steroidogenic electron transport in adrenal cortex mitochondria. Mol Cell Biochem 1982; 45:13–31.
117. Lambeth JD, Pember SO. Cytochrome P-450scc-adrenodoxin complex. Reduction properties of the substrate-associated cytochrome and relation of the reduction states of heme and iron-sulfur centers to association of the proteins. J Biol Chem 1983; 258:5596–602.
118. Voutilainen R, Picado-Leonard J, DiBlasio AM et al. Hormonal and developmental regulation of adrenodoxin messenger ribonucleic acid in steroidogenic tissues. J Clin Endocrinol Metab 1988; 66:383–8.
119. Brentano ST, Black SM, Lin D et al. cAMP post-transcriptionally diminishes the abundance of adrenodoxin reductase mRNA. Proc Natl Acad Sci U S A 1992; 89:4099–103.
120. Hanukoglu I, Feuchtwanger R, Hanukoglu A. Mechanism of corticotropin and cAMP induction of mitochondrial cytochrome P450 system enzymes in adrenal cortex cells. J Biol Chem 1990; 265:20602–8.
121. Rodgers RJ, Waterman MR, Simpson ER. Cytochromes P-450scc, P-450(17)alpha, adrenodoxin, and reduced nicotinamide adenine dinucleotide phosphate-cytochrome P-450 reductase in bovine follicles and corpora lutea. Changes in specific contents during the ovarian cycle. Endocrinology 1986; 118:1366–74.
122. Tuckey RC, Sadleir J. The concentration of adrenodoxin reductase limits cytochrome p450scc activity in the human placenta. Eur J Biochem 1999; 263:319–25.
123. Hanukoglu I, Rapoport R. Routes and regulation of NADPH production in steroidogenic mitochondria. Endocr Res 1995; 21:231–41.
124. Lambeth JD. Cytochrome P-450scc. Cardiolipin as an effector of activity of a mitochondrial cytochrome P-450. J Biol Chem 1981; 256:4757–62.
125. Schwarz D, Kisselev P, Wessel R et al. Alpha-branched 1,2-diacyl phosphatidylcholines as effectors of activity of cytochrome P450SCC (CYP11A1). Modeling the structure of the fatty acyl chain region of cardiolipin. J Biol Chem 1996; 271:12840–6.
126. Gasnier F, Rey C, Hellio Le Graverand MP et al. Hormone-induced changes in cardiolipin from Leydig cells: possible involvement in intramitochondrial cholesterol translocation. Biochem Mol Biol Int 1998; 45:93–100.
127. Bureik M, Zollner A, Schuster N et al. Phosphorylation of bovine adrenodoxin by protein kinase CK2 affects the interaction with its redox partner cytochrome P450scc (CYP11A1). Biochemistry 2005; 44:3821–30.
128. Chaudhuri AC, Harada Y, Shimuzu K et al. Biosynthesis of pregnenolone from 22-hydroxycholesterol. J Biol Chem 1962; 237:703–4.
129. Stocco DM, Clark BJ. Regulation of the acute production of steroids in steroidogenic cells. Endo Rev 1996; 17:221–44.
130. Krueger RJ, Orme-Johnson NR. Acute adrenocorticotropic hormone stimulation of adrenal corticosteroidogenesis. Discovery of a rapidly induced protein. J Biol Chem 1983; 258: 10159–67.
131. Stocco DM, Kilgore MW. Induction of mitochondrial proteins in MA-10 Leydig tumour cells with human choriogonadotropin. Biochem J 1988; 249:95–103.
132. Ferguson JJ, Jr. Protein synthesis and adrenocorticotropin responsiveness. J Biol Chem 1963; 238:2754–9.
133. Garren LD, Ney RL, Davis WW. Studies on the role of protein synthesis in the regulation of corticosterone production by adrenocorticotropic hormone in vivo. Proc Natl Acad Sci U S A 1965; 53:1443–50.
134. Privalle CT, Crivello JF, Jefcoate CR. Regulation of intramitochondrial cholesterol transfer to side-chain cleavage cytochrome P-450 in rat adrenal gland. Proc Natl Acad Sci U S A 1983; 80:702–6.
135. Stocco DM, Wang X, Jo Y et al. Multiple signaling pathways regulating steroidogenesis and steroidogenic acute regulatory protein expression: more complicated than we thought. Mol Endocrinol 2005; 19:2647–59.
136. Pon LA, Orme-Johnson NR. Acute stimulation of steroidogenesis in corpus luteum and adrenal cortex by peptide hormones. Rapid induction of a similar protein in both tissues. J Biol Chem 1986; 261:6594–9.
137. Pon LA, Epstein LF, Orme-Johnson NR. Acute cAMP stimulation in Leydig cells: rapid accumulation of a protein similar to that detected in adrenal cortex and corpus luteum. Endocr Res 1986; 12:429–46.
138. Pon LA, Orme-Johnson NR. Acute stimulation of corpus luteum cells by gonadotrophin or adenosine 3′,5′-monophosphate causes accumulation of a phosphoprotein concurrent with acceleration of steroid synthesis. Endocrinology 1988; 123:1942–8.
139. Epstein LF, Orme-Johnson NR. Regulation of steroid hormone biosynthesis. Identification of precursors of a phosphoprotein targeted to the mitochondrion in stimulated rat adrenal cortex cells. J Biol Chem 1991; 266:19739–45.
140. Stocco DM, Sodeman TC. The 30-kDa mitochondrial proteins induced by hormone stimulation in MA-10 mouse Leydig tumor cells are processed from larger precursors. J Biol Chem 1991; 266:19731–8.
141. King SR, Ronen-Fuhrmann T, Timberg R et al. Steroid production after in vitro transcription, translation, and mitochondrial processing of protein products of complementary deoxyribonucleic acid for steroidogenic acute regulatory protein. Endocrinology 1995; 136:5165–76.
142. Yamazaki T, Matsuoka C, Gendou M et al. Mitochondrial processing of bovine adrenal steroidogenic acute regulatory protein. Biochim Biophys Acta 2006; 1764:1561–7.
143. Arakane F, Sugawara T, Nishino H et al. Steroidogenic acute regulatory protein (StAR) retains activity in the absence of its mitochondrial import sequence: implications for the mechanism of StAR action. Proc Natl Acad Sci USA 1996; 93:13731–6.
144. Sasaki G, Ishii T, Jeyasuria P et al. Complex role of the mitochondrial targeting signal in the function of steroidogenic acute regulatory protein revealed by bacterial artificial chromosome transgenesis in vivo. Mol Endocrinol 2008; 22:951–64.
145. Bose M, Whittal RM, Gairola CG et al. Molecular mechanism of reduction in pregnenolone synthesis by cigarette smoke. Toxicol Appl Pharmacol 2008; 229:56–64.
146. Papadopoulos V. In search of the function of the peripheral-type benzodiazepine receptor. Endocr Res 2004; 30:677–84.
147. Petrescu AD, Gallegos AM, Okamura Y et al. Steroidogenic acute regulatory protein binds cholesterol and modulates mitochondrial membrane sterol domain dynamics. J Biol Chem 2001; 276:36970–82.

148. Kallen CB, Billheimer JT, Summers SA et al. Steroidogenic acute regulatory protein (StAR) is a sterol transfer protein. J Biol Chem 1998; 273:26285–8.
149. Ponting CP, Aravind L. START: a lipid-binding domain in StAR, HD-ZIP and signalling proteins. Trends Biochem Sci 1999; 24:130–2.
150. Alpy F, Stoeckel ME, Dierich A et al. The steroidogenic acute regulatory protein homolog MLN64, a late endosomal cholesterol-binding protein. J Biol Chem 2001; 276:4261–9.
151. Tsujishita Y, Hurley JH. Structure and lipid transport mechanism of a StAR-related domain. Nat Struct Biol 2000; 7:408–14.
152. Stocco DM. Tracking the role of a star in the sky of the new millennium. Mol Endocrinol 2001; 15:1245–54.
153. King SR, Stocco DM. ATP and a mitochondrial electrochemical gradient are required for functional activity of the steroidogenic acute regulatory (StAR) protein in isolated mitochondria. Endocrine Res 1996; 22:505–14.
154. King SR, Liu Z, Soh J et al. Effects of disruption of the mitochondrial electrochemical gradient on steroidogenesis and the Steroidogenic Acute Regulatory (StAR) protein. J Steroid Biochem Mol Biol 1999; 69:143–54.
155. Jo Y, King SR, Khan SA et al. Involvement of protein kinase C and cyclic adenosine 3′,5′-monophosphate-dependent kinase in steroidogenic acute regulatory protein expression and steroid biosynthesis in Leydig cells. Biol Reprod 2005; 73:244–55.
156. Dyson MT, Jones JK, Kowalewski MP et al. Mitochondrial A-kinase anchoring protein 121 binds type II protein kinase A and enhances steroidogenic acute regulatory protein-mediated steroidogenesis in MA-10 mouse leydig tumor cells. Biol Reprod 2008; 78:267–77.
157. Bhangoo A, Anhalt H, Ten S et al. Phenotypic variations in lipoid congenital adrenal hyperplasia. Pediatr Endocrinol Rev 2006; 3:258–71.
158. Lin D, Sugawara T, Strauss JF, III et al. Role of steroidogenic acute regulatory protein in adrenal and gonadal steroidogenesis. Science 1995; 267:1828–31.
159. Ogata T, Matsuo N, Saito M et al. The testicular lesion and sexual differentiation in congenital lipoid adrenal hyperplasia. Helv Paediatr Acta 1989; 43:531–8.
160. Tomasetto C, Regnier C, Moog-Lutz C et al. Identification of four novel human genes amplified and overexpressed in breast carcinoma and localized to the q11-q21.3 region of chromosome 17. Genomics 1995; 28:367–76.
161. Watari H, Arakane F, Moog-Lutz C et al. MLN64 contains a domain with homology to the steroidogenic acute regulatory protein (StAR) that stimulates steroidogenesis. Proc Natl Acad Sci U S A 1997; 94:8462–7.
162. Chen JJ, Lukyanenko Y, Hutson JC. 25-hydroxycholesterol is produced by testicular macrophages during the early postnatal period and influences differentiation of Leydig cells in vitro. Biol Reprod 2002; 66:1336–41.
163. Prasad VV, Ponticorvo L, Lieberman S. Identification of 24-hydroxycholesterol in bovine adrenals in both free and esterified forms and in bovine brains as its sulfate ester. J Steroid Biochem 1984; 21:733–6.
164. Bjorkhem I, Lutjohann D, Breuer O et al. Importance of a novel oxidative mechanism for elimination of brain cholesterol. Turnover of cholesterol and 24(S)-hydroxycholesterol in rat brain as measured with 18O2 techniques in vivo and in vitro. J Biol Chem 1997; 272:30178–84.
165. King SR, Manna PR, Ishii T et al. An essential component in steroid synthesis, the steroidogenic acute regulatory protein, is expressed in discrete regions of the brain. J Neurosci 2002; 22:10613–20.
166. Cherradi N, Rossier MF, Vallotton MB et al. Submitochondrial distribution of three key steroidogenic proteins (steroidogenic acute regulatory protein and cytochrome p450scc and 3beta-hydroxysteroid dehydrogenase isomerase enzymes) upon stimulation by intracellular calcium in adrenal glomerulosa cells. J Biol Chem 1997; 272:7899–907.
167. Pelletier G, Li S, Luu-The V et al. Immunoelectron microscopic localization of three key steroidogenic enzymes (cytochrome P450(scc), 3 beta-hydroxysteroid dehydrogenase and cytochrome P450(c17)) in rat adrenal cortex and gonads. J Endocrinol 2001; 171:373–83.
168. Mason JI, Howe BE, Howie AF et al. Promiscuous 3beta-hydroxysteroid dehydrogenases: testosterone 17beta-hydroxysteroid dehydrogenase activities of mouse type I and VI 3beta-hydroxysteroid dehydrogenases. Endocr Res 2004; 30:709–14.
169. Simard J, Moisan AM, Morel Y. Congenital adrenal hyperplasia due to 3beta-hydroxysteroid dehydrogenase/Delta(5)-Delta(4) isomerase deficiency. Semin Reprod Med 2002; 20:255–76.
170. Stocco DM, King S, Clark BJ. Differential effects of dimethylsulfoxide on steroidogenesis in mouse MA-10 and rat R2C Leydig tumor cells. Endocrinology 1995; 136:2993–9.
171. Townson DH, Wang XJ, Keyes PL et al. Expression of the steroidogenic acute regulatory protein in the corpus luteum of the rabbit: dependence upon the luteotropic hormone, estradiol-17 beta. Biol Reprod 1996; 55:868–74.
172. Schwarzenbach H, Manna PR, Stocco DM et al. Stimulatory effect of progesterone on the expression of steroidogenic acute regulatory protein in MA-10 Leydig cells. Biol Reprod 2003; 68:1054–63.
173. Manna PR, Tena-Sempere M, Huhtaniemi IT. Molecular mechanisms of thyroid hormone-stimulated steroidogenesis in mouse leydig tumor cells. Involvement of the steroidogenic acute regulatory (StAR) protein. J Biol Chem 1999; 274:5909–18.
174. Manna PR, Kero J, Tena-Sempere M et al. Assessment of mechanisms of thyroid hormone action in mouse Leydig cells: regulation of the steroidogenic acute regulatory protein, steroidogenesis, and luteinizing hormone receptor function. Endocrinology 2001; 142:319–31.
175. Chen YJ, Feng Q, Liu YX. Expression of the steroidogenic acute regulatory protein and luteinizing hormone receptor and their regulation by tumor necrosis factor alpha in rat corpora lutea. Biol Reprod 1999; 60:419–27.
176. Manna PR, El-Hefnawy T, Kero J et al. Biphasic action of prolactin in the regulation of murine Leydig tumor cell functions. Endocrinology 2001; 142:308–18.
177. Huang BM, Stocco DM, Li PH et al. Corticotropin-releasing hormone stimulates the expression of the steroidogenic acute regulatory protein in MA-10 mouse cells. Biol Reprod 1997; 57: 547–51.
178. Mamluk R, Greber Y, Meidan R. Hormonal regulation of messenger ribonucleic acid expression for steroidogenic factor-1, steroidogenic acute regulatory protein, and cytochrome P450 side-chain cleavage in bovine luteal cells. Biol Reprod 1999; 60:628–34.
179. Lin T, Wang D, Hu J et al. Upregulation of human chorionic gonadotrophin-induced steroidogenic acute regulatory protein by insulin-like growth factor-I in rat Leydig cells. Endocrine 1998; 8:73–8.
180. Millena AC, Reddy SC, Bowling GH et al. Autocrine regulation of steroidogenic function of Leydig cells by transforming growth factor-alpha. Mol Cell Endocrinol 2004; 224:29–39.
181. Elvin JA, Clark AT, Wang P et al. Paracrine actions of growth differentiation factor-9 in the mammalian ovary. Mol Endocrinol 1999; 13:1035–48.
182. Shimada M, Hernandez-Gonzalez I, Gonzalez-Robayna I et al. Paracrine and autocrine regulation of epidermal growth factor-like factors in cumulus oocyte complexes and granulosa cells: key

roles for prostaglandin synthase 2 and progesterone receptor. Mol Endocrinol 2006; 20:1352–65.
183. Houk CP, Pearson EJ, Martinelle N et al. Feedback inhibition of steroidogenic acute regulatory protein expression in vitro and in vivo by androgens. Endocrinology 2004; 145:1269–75.
184. Ruiz-Cortes ZT, Martel-Kennes Y, Gevry NY et al. Biphasic effects of leptin in porcine granulosa cells. Biol Reprod 2003; 68:789–96.
185. Attia GR, Dooley CA, Rainey WE et al. Transforming growth factor beta inhibits steroidogenic acute regulatory (StAR) protein expression in human ovarian thecal cells. Mol Cell Endocrinol 2000; 170:123–9.
186. Tena-Sempere M, Manna PR, Zhang FP et al. Molecular mechanisms of leptin action in adult rat testis: potential targets for leptin-induced inhibition of steroidogenesis and pattern of leptin receptor messenger ribonucleic acid expression. J Endocrinol 2001; 170:413–23.
187. Otsuka F, Moore RK, Shimasaki S. Biological function and cellular mechanism of bone morphogenetic protein-6 in the ovary. J Biol Chem 2001; 276:32889–95.
188. Oonk RB, Krasnow JS, Beattie WG et al. Cyclic AMP-dependent and -independent regulation of cholesterol side chain cleavage cytochrome P-450 (P-450scc) in rat ovarian granulosa cells and corpora lutea. cDNA and deduced amino acid sequence of rat P-450scc. J Biol Chem 1989; 264:21934–42.
189. Voutilainen R, Tapanainen J, Chung BC et al. Hormonal regulation of P450scc (20,22-desmolase) and P450c17 (17 alpha-hydroxylase/17,20-lyase) in cultured human granulosa cells. J Clin Endocrinol Metab 1986; 63:202–7.
190. Magoffin DA, Weitsman SR. Effect of insulin-like growth factor-I on cholesterol side-chain cleavage cytochrome P450 messenger ribonucleic acid expression in ovarian theca-interstitial cells stimulated to differentiate in vitro. Mol Cell Endocrinol 1993; 96: 45–51.
191. Mellon SH, Vaisse C. cAMP regulates P450scc gene expression by a cycloheximide-insensitive mechanism in cultured mouse Leydig MA-10 cells. Proc Natl Acad Sci U S A 1989; 86:7775–9.
192. Swan CL, Agostini MC, Bartlewski PM et al. Effects of progestins on progesterone synthesis in a stable porcine granulosa cell line: control of transcriptional activity of the cytochrome p450 side-chain cleavage gene. Biol Reprod 2002; 66:959–65.
193. Xu YP, Chedrese PJ, Thacker PA. Growth hormone amplifies insulin-like growth factor I induced progesterone accumulation and P450scc mRNA expression. Mol Cell Endocrinol 1995; 111:199–206.
194. Li XM, Juorio AV, Murphy BD. Prostaglandins alter the abundance of messenger ribonucleic acid for steroidogenic enzymes in cultured porcine granulosa cells. Biol Reprod 1993; 48: 1360–6.
195. Neuvians TP, Schams D, Berisha B et al. Involvement of pro-inflammatory cytokines, mediators of inflammation, and basic fibroblast growth factor in prostaglandin F2alpha-induced luteolysis in bovine corpus luteum. Biol Reprod 2004; 70:473–80.
196. Hales DB, Payne AH. Glucocorticoid-mediated repression of P450scc mRNA and de novo synthesis in cultured Leydig cells. Endocrinology 1989; 124:2099–104.
197. Gautier C, Levacher C, Saez JM et al. Transforming growth factor beta1 inhibits steroidogenesis in dispersed fetal testicular cells in culture. Mol Cell Endocrinol 1997; 131:21–30.
198. Veldhuis JD, Garmey JC, Urban RJ et al. Ovarian actions of tumor necrosis factor-alpha (TNF alpha): pleiotropic effects of TNF alpha on differentiated functions of untransformed swine granulosa cells. Endocrinology 1991; 129:641–8.
199. Xiong Y, Hales DB. The role of tumor necrosis factor-alpha in the regulation of mouse Leydig cell steroidogenesis. Endocrinology 1993; 132:2438–44.
200. Lin T, Wang D, Nagpal ML et al. Recombinant murine tumor necrosis factor-alpha inhibits cholesterol side-chain cleavage cytochrome P450 and insulin-like growth factor-I gene expression in rat Leydig cells. Mol Cell Endocrinol 1994; 101:111–9.
201. Orava M, Voutilainen R, Vihko R. Interferon-gamma inhibits steroidogenesis and accumulation of mRNA of the steroidogenic enzymes P450scc and P450c17 in cultured porcine Leydig cells. Mol Endocrinol 1989; 3:887–94.
202. Lin T, Wang TL, Nagpal ML et al. Interleukin-1 inhibits cholesterol side-chain cleavage cytochrome P450 expression in primary cultures of Leydig cells. Endocrinology 1991; 129:1305–11.
203. Spicer LJ, Aad PY, Allen DT et al. Growth differentiation factor 9 (GDF9) stimulates proliferation and inhibits steroidogenesis by bovine theca cells: influence of follicle size on responses to GDF9. Biol Reprod 2008; 78:243–53.
204. Kaynard AH, Periman LM, Simard J et al. Ovarian 3 beta-hydroxysteroid dehydrogenase and sulfated glycoprotein-2 gene expression are differentially regulated by the induction of ovulation, pseudopregnancy, and luteolysis in the immature rat. Endocrinology 1992; 130:2192–200.
205. Chaffin CL, Dissen GA, Stouffer RL. Hormonal regulation of steroidogenic enzyme expression in granulosa cells during the peri-ovulatory interval in monkeys. Mol Hum Reprod 2000; 6:11–8.
206. Magoffin DA, Weitsman SR. Insulin-like growth factor-I stimulates the expression of 3 beta-hydroxysteroid dehydrogenase messenger ribonucleic acid in ovarian theca-interstitial cells. Biol Reprod 1993; 48:1166–73.
207. Martel C, Labrie C, Couet J et al. Effects of human chorionic gonadotropin (hCG) and prolactin (PRL) on 3 beta-hydroxy-5-ene-steroid dehydrogenase/delta 5-delta 4 isomerase (3 beta-HSD) expression and activity in the rat ovary. Mol Cell Endocrinol 1990; 72:R7–13.
208. Tanaka N, Iwamasa J, Matsuura K et al. Effects of progesterone and anti-progesterone RU486 on ovarian 3 beta-hydroxysteroid dehydrogenase activity during ovulation in the gonadotrophin-primed immature rat. J Reprod Fertil 1993; 97:167–72.
209. McGee EA, Sawetawan C, Bird I et al. The effect of insulin and insulin-like growth factors on the expression of steroidogenic enzymes in a human ovarian thecal-like tumor cell model. Fertil Steril 1996; 65:87–93.
210. Sordoillet C, Chauvin MA, de PE et al. Epidermal growth factor directly stimulates steroidogenesis in primary cultures of porcine Leydig cells: actions and sites of action. Endocrinology 1991; 128:2160–8.
211. McAllister JM, Byrd W, Simpson ER. The effects of growth factors and phorbol esters on steroid biosynthesis in isolated human theca interna and granulosa-lutein cells in long term culture. J Clin Endo Metab 1994; 79:106–12.
212. Mauduit C, Chauvin MA, de PE et al. Effect of activin A on dehydroepiandrosterone and testosterone secretion by primary immature porcine Leydig cells. Biol Reprod 1991; 45:101–9.
213. Juengel JL, Meberg BM, McIntush EW et al. Concentration of mRNA encoding 3 beta-hydroxysteroid dehydrogenase/delta 5,delta 4 isomerase (3 beta-HSD) and 3 beta-HSD enzyme activity following treatment of ewes with prostaglandin F2 alpha. Endocrine 1998; 8:45–50.
214. Agular BM, Vind C. Effects of dexamethasone on steroidogenesis in Leydig cells from rats of different ages. J Steroid Biochem Mol Biol 1995; 54:75–81.
215. Xiong Y, Hales DB. Differential effects of tumor necrosis factor-alpha and interleukin-1 on 3 beta-hydroxysteroid dehydrogenase/delta 5–>delta 4 isomerase expression in mouse Leydig cells. Endocrine 1997; 7:295–301.
216. Murono EP, Washburn AL, Goforth DP et al. Effects of acidic fibroblast growth factor on 5-ene-3 beta-hydroxysteroid

dehydrogenase-isomerase and 5 alpha-reductase activities and [125I] human chorionic gonadotrophin binding in cultured immature Leydig cells. J Steroid Biochem Mol Biol 1993; 45: 477–83.

217. Murono EP, Washburn AL. Platelet derived growth factor inhibits 5 alpha-reductase and delta 5-3 beta-hydroxysteroid dehydrogenase activities in cultured immature Leydig cells. Biochem Biophys Res Commun 1990; 169:1229–34.
218. Fanjul LF, Quintana J, Gonzalez J et al. Testicular 3 beta-hydroxysteroid dehydrogenase/delta 5-4 isomerase in the hypophysectomized rat: effect of treatment with 5 alpha-dihydrotestosterone. J Endocrinol 1992; 133: 237–43.
219. Heggland SJ, Signs SA, Stalvey JR. Testosterone decreases 3beta-hydroxysteroid dehydrogenase-isomerase messenger ribonucleic acid in cultured mouse Leydig cells by a strain-specific mechanism. J Androl 1997; 18:646–55.
220. Brankin V, Quinn RL, Webb R et al. Evidence for a functional bone morphogenetic protein (BMP) system in the porcine ovary. Domest Anim Endocrinol 2005; 28:367–79.
221. Martel C, Labrie C, Dupont E et al. Regulation of 3 beta-hydroxysteroid dehydrogenase/delta 5-delta 4 isomerase expression and activity in the hypophysectomized rat ovary: interactions between the stimulatory effect of human chorionic gonadotropin and the luteolytic effect of prolactin. Endocrinology 1990; 127:2726–37.

Chapter 17

Prostaglandins and Their Mechanisms of Action in the Cyclic Ovary

Jorge A. Flores and Christy Barlund

17.1 Introduction

Prostaglandins (PG) belong to a diverse family of cell signaling molecules known as eicosanoids. Other members in this family include prostacyclins (PGI), thromboxanes (TX), and leukotrienes (LT). All eicosanoids (from eicosa-, Greek for "twenty") are synthesized by oxygenation of 20-carbon essential fatty acids (EFA). There are three parallel pathways by which eicosanoids are produced depending on the starting EFA. In all these parallel pathways, eicosanoid production is mediated by the activity of two families of enzymes: cyclooxygenases and lypooxygenases. Cyclooxygenases, or COX, generate prostanoids, the collective term used for PG, PGI, and TX, while lypooxigenases generate the LT. Eicosanoids are found in all living cells; in mammals eicosanoids derived from arachidonic acid are primarily involved in inflammation, immunity, and in central nervous system functions. Eicosanoids derived from the other two EFA, Dihomo-gamma-linoleic acid (DGLA) and eicosapentenoic acid (PA) are inactive or even anti-inflammatory [1].

The term prostaglandin was derived from the "prostate" gland because when it was first isolated from seminal fluid, it was believed to be a prostatic secretion [2]. Eventually, PG was found to be a secretory product of the seminal vesicles and many other tissues for various functions. The diverse functions associated with PG are more than can be mentioned here, but some include smooth muscle contraction/relaxation, platelet aggregation, inhibition of gastric acid secretion, stimulation of gastric mucus secretion, inhibition of lipolysis, increase of autonomic neurotransmission, ion transport, and thermal regulation. PG are key to reproductive events including implantation, decidualization, parturition, ovulation, and luteolysis. This discussion will be limited to the PG involved in two reproductive processes, ovulation and luteolysis.

J.A. Flores (✉)
Department of Biology, West Virginia University, Morgantown, WV, USA
e-mail: jflores@wvu.edu

17.2 Biosynthesis of Prostaglandins

Eicosanoid synthesis begins with the release of free EFA from biological membranes by the enzymatic activity of phospholipase A_2 (PLA2, EC3.1.1.4) and phospholipase C (PLC, EC 3.1.4.11). Diacylglycerol formed by the activity of PLC can be metabolized into arachidonic acid by the enzyme diglyceride lipase. Phospholipase A_2 includes several unrelated proteins sharing common enzymatic activity. Important members within these include the calcium-independent PLA_2 and the secreted and cytosolic PLA_2. Activation of receptors linked to the enzyme PLA cause cytoplasmic calcium to rise above threshold cellular levels, activating cytosolic PLA_2 with an increase in arachidonic acid production. After prolonged and intense PLA-linked receptor activation, the isoenzyme secreted PLA_2 will amplify the loop of arachidonic acid synthesis, maximizing the available substrates for the cyclooxygenase pathway. Arachidonic acid formed in this manner by enzymatic activity of phospholipases can be directed into one of either the cyclooxygenase pathway or the lipoxygenase pathway to form either prostaglandin and thromboxane or leukotriene. The cyclooxygenase pathway produces thromboxane, prostacyclin, and prostaglandin D, E, and F. The lipoxygenase pathway is active in leukocytes and in macrophages and synthesizes leukotrienes (Fig. 17.1).

Prostaglandins are ubiquitous in all tissues, where they commonly are autocrine or paracrine lipid mediators that act upon their target cells by interacting with specific plasma membranes, and in some instances, nuclear receptors. They are synthesized in the cell from three EFA: gamma-linolenic acid, arachidonic acid, and eicosapentaenoic acid. Depending on which of these EFA substrates is used, prostanoids series -1, -2, and -3 are respectively synthesized.

P.J. Chedrese (ed.), *Reproductive Endocrinology*,
DOI 10.1007/978-0-387-88186-7_17,

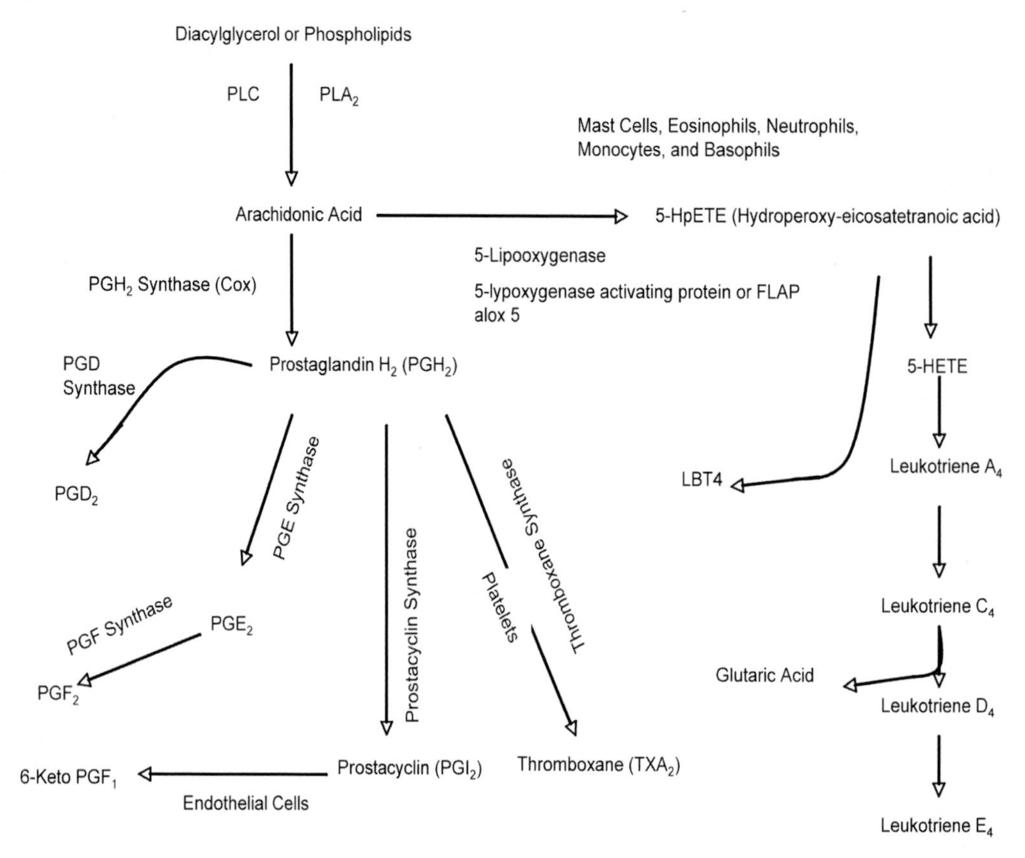

Fig. 17.1 Metabolic pathway for the synthesis of eicosanoids. Arachidonic acid formed by activation of Phospholipase C- (PLC) and A_2-linked receptors is metabolized by the activity of two families of enzymes: cyclooxygenases (PGH_2 or Cox) and lypooxygenases. Cyclooxygenases, or Cox, generate prostanoids (PG, PGI, and TX), while lypooxigenases generate the Leukotrienes (LT)

The enzyme cyclooxygenase or COX converts arachidonic acid (AA, a ω-6 polyunsaturated fatty acid) to prostaglandin H_2 (PGH_2), the precursor of the series-2 prostanoids (Fig. 17.1). COX contains two active sites: a heme with peroxidase activity, responsible for the reduction of PGG_2 to PGH_2, and a cyclooxygenase site, where arachidonic acid is converted into the hydroperoxy endoperoxide prostaglandin G_2 (PGG_2). Two O_2 molecules then react with the arachidonic acid radical, yielding PGG_2.

Terminal enzymes act subsequent to PGH_2 and ultimately lead to the tissue-specific production of individual PG (Fig. 17.1). The most pertinent PG in the female reproductive system are prostaglandin E_2 and prostaglandin $F_{2\alpha}$. Prostaglandin H_2 is metabolized to PGE_2 by prostaglandin E synthases, and several isozymes have been described. A prostaglandin E synthase associated with microsomal membranes, prostaglandin E synthase-1, is a key enzyme in the formation of PGE_2. In the central nervous system, prostaglandin D_2 is formed from PGH_2 by a glutathione - independent prostaglandin D synthase. In non-nervous tissue this reaction is carried out by a glutathione-dependent prostaglandin D synthase. Similarly, prostaglandin $F_{2\alpha}$ is synthesized from prostaglandin D_2 by the enzyme prostaglandin F synthase (PGFS), which has a liver-type and lung-type isoforms. An alternate pathway exists for the production of prostaglandin $F_{2\alpha}$ from PGE_2 utilizing the enzyme 9-keto-PGE_2-reductase (9 K-PGR).

17.3 Prostaglandins Release from Cells and Subsequent Transport

Prostaglandins are lipophyllic in nature but predominate as charged anions; consequently they diffuse poorly through membranes [3]. Prostaglandin transporter (PGT), a member of the Solute Carrier Organic Anion, mediates both the efflux and influx of PG. This PGT will uptake PGE_2 and $PGF_{2\alpha}$ more readily than other PG and arachidonic acid. In ruminants, transport of $PGF_{2\alpha}$ is critical for regulation of the estrous cycle and establishment of pregnancy. Prostaglandin $F_{2\alpha}$ is released from endometrial epithelial cells, effluxed into the venous vessels, and then transported into the ovarian artery via the utero-ovarian plexus [4, 5]. Expression of PGT in bovine and ovine endometrium during the estrous cycle and pregnancy has been elegantly demonstrated [6, 7, 8]. The bovine PGT consists of 1,935 nucleotides and encodes

644 amino acids. Similarily, the ovine PGT cDNA consists of 1,935 nucleotides encoding 644 amino acids. The predicted amino acid sequence of ovine PGT shared 92, 85, 83, 81, and 79% identity with bovine, canine, human, and rat homolog, respectively. Furthermore, the functional role for PGT in regulation of $PGF_{2\alpha}$ efflux and influx in ovine endometrial cells that influence luteolytic mechanisms has been recently reported [8]. Although PG more often participate in paracrine and autocrine cell signaling, in this particular instance, they participate in an endocrine regulation of the ruminant CL. The importance of this selective transport of PG is critical, given that the lung must be bypassed, to prevent PG clearance. In cattle, one pass through the lungs will metabolize 65% of PG in blood, while sheep can metabolize up to 99%.

17.4 Prostaglandin Receptors

The human thromboxane receptor was the first prostanoid receptor to be cloned in 1991 [9]. It is a protein composed of 343 amino acids and is a G protein-coupled receptor with seven putative transmembrane domains. Homology screening based on its sequence was performed in various species, and all of the eight types and subtypes of the prostanoid receptors previously defined pharmacologically were identified. These include the PGD receptor DP, the PGE receptors with four subtypes, $EP_{1\text{-}4}$, the PGF receptor FP, the PGI receptor IP, and the thromboxane receptor TxA. Four additional isoforms (A–D) have been characterized for the EP_3 receptor subtype [10].

Signal transduction pathways of prostanoid receptors have been characterized by examining agonist-induced changes in cAMP, calcium, inositol phosphates, and by identifying G protein coupling by various receptors. On the basis of their signal transduction and action, the eight types of prostanoid receptors have been grouped into three categories: the relaxant receptors, the contractile receptors, and the inhibitory receptors. The relaxant receptors consist of the IP, DP, EP_2, and EP_4 receptors; and they typically mediate increases in cAMP and induce smooth muscle relaxation. The contractile receptors consist of the TP, FP, and EP_1 receptors; they mediate calcium mobilization and induce smooth muscle contraction. The EP_3 receptor is an inhibitory receptor that mediates decreases in cAMP and inhibits smooth muscle relaxation [10]. Because the most pertinent PG in the female reproductive system are prostaglandin E_2 and prostaglandin $F_{2\alpha}$, the EP_2 and the FP receptors are also the most relevant receptors in reproductive tissues. The EP_2 receptors are coupled to G_s and increases cAMP. The FP receptors are typically coupled via G_q with increases in inositol phosphates and calcium [10].

17.5 Role of Prostaglandins in Ovulation

The relationship between prostaglandin biosynthesis and ovulation first emerged in the early 1970s. It was also around this period when one of the underlying action mechanisms of aspirin was demonstrated to be the inhibition of COX-1 and COX-2 and, consequently, prostaglandin synthesis [11]. With the availability of other non-steroidal anti-inflammatory drugs or NSAIDs, indomethacin-dependent inhibition of ovulation was soon reported in numerous species [12, 13]. Although there were studies questioning the obligatory role of PG during ovulation [14], this issue was eventually resolved by genetic studies in mice, and now the role of PGE_2 acting through EP_2 in ovulation is well established.

The preovulatory surge of luteinizing hormone (LH) results in a transient, specific, and developmental stage-specific marked induction of COX-2 expression in granulosa cells of ovulatory follicles prior to ovulation. Selective COX-2 inhibitors further underscored the role of COX-2 in ovulation. Genetic inactivation of COX-2 has provided compelling evidence for the importance of COX-2 in the ovulatory process. Mice carrying a null mutation for COX-1 were fertile, but mice deficient in COX-2 proved to be infertile. The transient induction of COX-2 results in an increased concentrations of PGE_2 and $PGF_{2\alpha}$ within ovarian follicles of several ruminants, rodents, and primates. As mentioned earlier, the granulosa cell layer of the preovulatory follicles are the primary sites of prostaglandin synthesis in the ovary. The most significant prostaglandin produced is in the E_2 series, and genetic inactivation of prostanoids EP_2 receptor has provided convincing support to this interpretation that PGE_2 has a dominant role in ovulation. Constructing mice deficient in prostanoid receptors has demonstrated that only those deficient in EP_2 receptors have impaired ovulation and fertilization [15]. In primate periovulatory granulosa cells, expression and function of three distinct PGE_2 receptors have been reported [16]. Exposure of granulosa cells to an ovulatory dose of human chorionic gonadotropin increased levels of EP_2, EP_3, and, possibly, EP_1 receptor proteins; granulosa cells responded to stimulation of each of these PGE_2 receptors with increased intracellular signal generation only just before ovulation. These PGE_2 receptors may regulate different intracellular events, suggesting that PGE_2 may use multiple pathways to stimulate periovulatory events in primate granulosa cells. Both PGE_2 levels and PGE_2 receptor responses peak just before ovulation, ensuring the maximal ovulatory effect of PGE_2 [16].

Interestingly, length of the ovulatory process (a species-specific aspect) appears to determine the time of COX-2 induction after the preovulatory surge of LH, with the interval of time from COX-2 induction to ovulation being highly conserved across species [17]. Therefore, it has been suggested that COX-2 could serve as a molecular determinant that sets the alarm of the mammalian ovulatory clock [18].

In addition to LH, there are other stimulatory signaling molecules for prostaglandin synthesis in the ovulatory follicle. These include GnRH, Transforming Growth factor α, interleukin-1 (probably mediated by PG), GDF9, and any agent capable of increasing cAMP. In cumulus granulosa cells, the pharmacologically cAMP/PKA activated COX-2 induction is sensitive to inhibition of p38-MAPk or Erk1/2, indicating that Erk-mediated transcriptional induction, distinct from or in conjunction with classical cAMP pathway, is required for COX-2 expression [19]. Within the COX-2 promoter region, the proximal 200 base pairs are the most integral for gonadotropin induction of transcription. *Cis*-acting regulatory elements that might be mediators for diverse regulation of COX-2 transcription are interferon response element (IRE), cAMP responsive element (CRE), CAAT enhancing binding protein (C/EBP), nuclear factor kappa B (NF kB), or E-box. In bovine granulosa cells, induced regulation of COX-2 occurs primarily by binding of transcription factors to the E-box. The E-box is a cognate binding site for the basic helix-loop-helix (bHLH) family of transcription factors situated at –50 base pair of the PGHS-2 promoter. Upstream stimulatory factor (USF) is a basic helix–loop–helix/leucine-zipper binding protein that initiates transcription at the E-box. The E-box is the principle regulatory element for COX-2 transcription [17].

17.6 Role of Prostaglandins in Luteolysis

If fertilization fails, luteal progesterone secretion declines within a few hours (functional luteolysis). Subsequently, within 12 hours, there is a dramatic increase in luteal tissue remodeling and in the number of luteal cells undergoing apoptosis (structural luteolysis) that culminate with the demise of the CL and allowing the beginning of a new ovarian cycle. The entire process of luteal demise is known as luteolysis, but as mentioned above, functional and structural components are now recognized in it.

Lifespan regulation of the mammalian CL displays pronounced interspecies variation. For instance, since the initial observation by Loeb [20] in the guinea pig that the uterus controls the length of the estrous cycle, $PGF_{2\alpha}$ of uterine origin has been demonstrated to be the physiological luteolysin that causes CL regression in cycling sheep and cow [21], horse and pig [22], rat [23], rabbit [24], and hamster [25]. In some species however, including the dog [26], ferret [27], squirrel [28], mouse [29], monkey [30], opossum [31], marsupials [32], and human [33], removal of the uterus exerts no control of luteal lifespan. In some of these species where hysterectomy does not alter cyclic ovarian activity, it has been suggested that luteolytic $PGF_{2\alpha}$ is synthesized within the CL itself. In ruminants, although the initial source of $PGF_{2\alpha}$ is the uterus, evidence for the idea that the CL contributes to the synthesis of $PGF_{2\alpha}$ that influences or determines its own lifespan is currently expanding. The observation that corpora lutea of domestic animals can secrete PG dates to the early-mid 1970 s [34], but studies to systematically address the possibility of autocrine and paracrine actions of luteal PGs in CL function are more recent [7, 35]. In literature pertaining to the human CL, the role of $PGF_{2\alpha}$ in luteolysis is less clear, although luteolytic effects in vitro and implications for an in vivo effect of $PGF_{2\alpha}$ synthesized in the CL as an active luteolysin do exist [36].

The interaction of neurohypophysial oxytocin (OT) with endogenous OT receptors is thought to evoke the secretion of luteolytic pulses of uterine $PGF_{2\alpha}$. McCracken [37] have postulated the concept of a central OT pulse generator that functions as a pacemaker for luteolysis. It is postulated that the uterus transduces hypothalamic signals in the form of episodic OT secretion into luteolytic pulses of uterine $PGF_{2\alpha}$ secretion. In ruminants, uterine $PGF_{2\alpha}$ pulses induce luteal OT released so that a positive feedback loop is established that amplifies neural OT. The onset of this feedback loop and episodic $PGF_{2\alpha}$ secretion is thought to be controlled by the appearance of OT receptors in the endometrium.

As mentioned above, it is now clear that mammalian corpora lutea can secrete PG. Interestingly, once ovulation occurs and granulosa cells differentiate into large steroidogenic luteal cells, transcriptional regulation of COX-2 appears to switch from predominantly PKA-dependent pathway in pre-ovulatory follicles to a predominantly PKC-dependent pathway in the corpus luteum [38]. It has been suggested that the early CL favors responsiveness and synthesis of PGE_2 over $PGF_{2\alpha}$. Prostaglandin E_2 is known to be a luteotropic factor of the early CL in several species [34, 39].

Although the ability of $PGF_{2\alpha}$ to regress the corpus luteum CL is well documented, if $PGF_{2\alpha}$ is administered outside of a specific developmental window, it can, in a species-dependent manner, have no effect, stimulate, or inhibit luteal progesterone production [40, 41]. This wide range of effects of $PGF_{2\alpha}$ on luteal progesterone secretion in vivo has been observed also in vitro [42, 43]. However, the literature contains many statements asserting that observations in vitro are in contrast with the "known inhibitory" actions of $PGF_{2\alpha}$ in vivo [36]. This apparent discrepancy however might be resolved by the interesting biological observation that cell responsiveness to the same agonist can be altered during luteal development [42, 43, 44].

Although $PGF_{2\alpha}$ has been well established as the most important luteolytic hormone for more than 30 years, the cellular mechanisms by which $PGF_{2\alpha}$ exerts its luteolytic effects remain incompletely understood. For instance, the control mechanisms for functional and structural luteolysis might be separately regulated with the participation of endo-, para-, and auto-crine interactions. In this context, whether $PGF_{2\alpha}$ induces its inhibitory effects on progesterone production by direct effects on luteal steroidogenic cells, or through indirect actions via other luteal cells continues to be debated. A variety of paracrine factors have been studied as mediator of $PGF_{2\alpha}$-induced luteal regression, among these, the luteal endothelin system has received a great deal of attention.

17.6.1 Luteal Endothelin System

Endothelin-1 (EDN1) is a 21-amino acid peptide produced by endothelial cells, and is a member of a family of structurally related peptides that includes Endothelin-2 (EDN2) and Endothelin-3 (EDN3) [45]. Two classes of receptors named type A (EDNRA) and B (EDNRB) mediate the actions of the different members of the endothelin family [46]. While EDNRB shows equal affinity for all three endothelin peptides, EDNRA shows greatest affinity for EDN1.

17.6.2 Role of Endothelin in Mediating Luteolytic Actions of $PGF_{2\alpha}$

While known to be a potent vasoconstrictor, there is evidence that luteal EDN1 plays an important role in the anti-steroidogenic actions of $PGF_{2\alpha}$ [47]. Studies have demonstrated anti-steroidogenic actions of EDN1 in vitro. It reduced both basal and LH-stimulated progesterone in dispersed ovine and bovine luteal cells [48, 49]. This effect of EDN1 on ovine luteal cells was reduced by preincubation with a selective EDNRA antagonist [48]. A mediatory role for EDN1 during $PGF_{2\alpha}$ induced luteolysis was supported by (1) gene expression of *EDN1* and its receptor, *EDNRA*, are greater when corpora lutea are responsive to $PGF_{2\alpha}$ (approximately day 6 and beyond during the estrous cycle) [50, 51], and (2) regulation of gene expression of the luteal endothelin system acquires responsiveness to $PGF_{2\alpha}$ during the late luteal phase [52, 53]. However, during the early phase of the bovine CL, before day 10, the endothelin system is steadily up regulated in a $PGF_{2\alpha}$ independent manner [54]. Furthermore, cells isolated from the early CL are responsive to endogenous and exogenous EDN1 [42]. Therefore, these observations raise the possibility that in the early CL, EDN1 might have a different function from its intermediary role of $PGF_{2\alpha}$ action in the late luteal phase.

If luteal EDN1 indeed plays a mediatory role in the anti-steroidogenic actions of $PGF_{2\alpha}$, blocking the luteal endothelin system should then abolish the inhibitory actions of $PGF_{2\alpha}$. However, in vivo experiments designed to abolish the luteolytic effects of $PGF_{2\alpha}$ by blocking the endothelin system have had only limited success [48] or have altogether failed [55]. For example, direct infusion of an EDNRA antagonist into the bovine CL failed to mitigate the anti-steroidogenic actions of $PGF_{2\alpha}$ [55]. Consequently, the manner in which EDN1 and $PGF_{2\alpha}$ interact during luteolysis remains controversial. A recent study in which an endothelin receptor antagonist was infused within the CL by means of a mini-osmotic pump implanted 2 days before inducing luteal regression by an intramuscular injection of $PGF_{2\alpha}$, strongly indicated the involvement of EDN1 as mediator of $PGF_{2\alpha}$ in the process of luteal regression [56]. A second and perhaps more important contribution revealed in that study was that there were early luteolytic actions of $PGF_{2\alpha}$ (up to 12 h after the $PGF_{2\alpha}$injection) that were independent of mediation by EDN1; however, 12 h after $PGF_{2\alpha}$administration, the anti-steroidogenic actions were reversed effectively by an EDNRA antagonist. These observations could mean that during the early part of luteal regression, $PGF_{2\alpha}$ acts directly on luteal steroidogenic cells, or via some paracrine or autocrine factor other than EDN1 [56]. There is consensus that both EDNRA and EDNRB are expressed in the mammalian CL. However there are species differences regarding the receptor type that participates in the mediatory actions of EDN1 on $PGF_{2\alpha}$-induced luteal regression. In rodents, EDNRB appears to be most important, whereas EDNRA appears of greater importance in ruminant species and in the human CL. The cellular distribution of these receptors is not well characterized but certainly, steroidogenic, endothelial, and smooth muscle luteal cells are target for EDN1 actions. In the ovine CL, there is an indication that the small steroidogenic cells might be the luteal cells expressing the EDNRA [56].

In the CL of the ewe, there are endocrine, paracrine, and possible autocrine interactions between $PGF_{2\alpha}$, EDN1, oxytocin, and progesterone during luteal regression (Fig. 17.2). Endometrial epithelial cells are the initial source of $PGF_{2\alpha}$, which in an endocrine fashion acts upon large luteal steroidogenic cells (LLC) and luteal endothelial cells [42, 57]. On the LLC, $PGF_{2\alpha}$ induces oxytocin release, which acts as a paracrine factor on the small luteal steroidogenic cells (SLC) to decrease their response to PKA-stimulated (by LH for instance) progesterone secretion. Another effect of $PGF_{2\alpha}$ on LLC appears to be to induce a decrease in their basal progesterone secretion. It is not clear if this is a direct $PGF_{2\alpha}$ action or if it is mediated through another paracrine factor. What is clear is that this early antisteroidogenic effect of $PGF_{2\alpha}$ is independent of EDN1, and that it might account for the initial decline in luteal progesterone during functional regression [56]. This initial decline in progesterone might be critical, as there is evidence that a decline in progesterone modifies the response of SLC to oxytocin. Specifically, a decline in progesterone increases the calcium responses initiated by OT on SLC, which presumably would

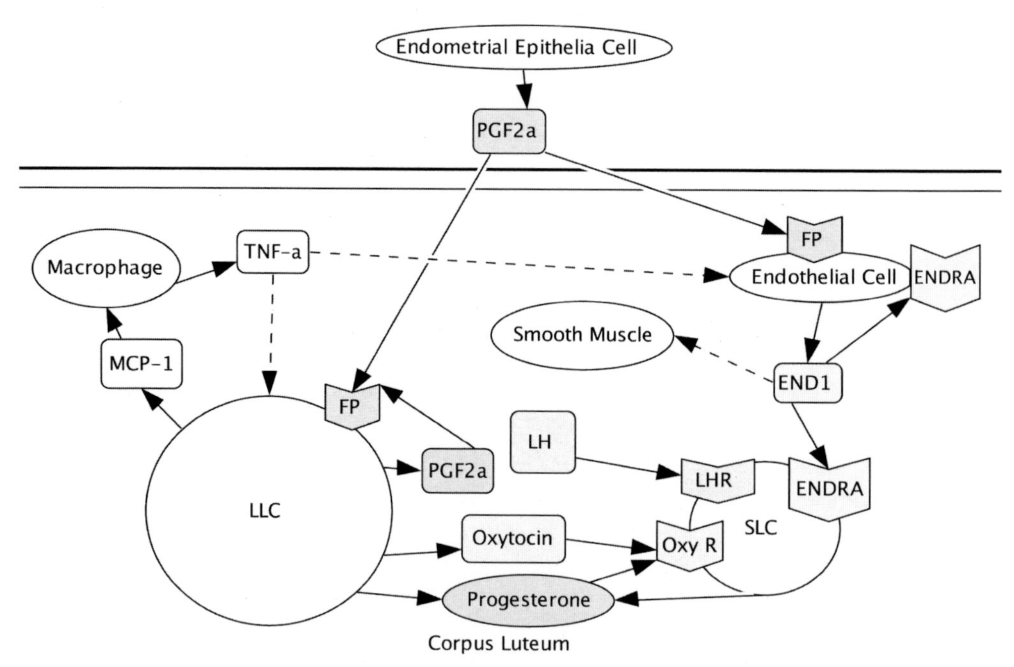

Fig. 17.2 Endocrine, paracrine, and possibly autocrine interactions between $PGF_{2\alpha}$, endothelin-1 (EDN1), oxytocin, and progesterone during luteal regression of the ovine CL. Endometrial epithelial cells are the initial source of $PGF_{2\alpha}$, which in an endocrine fashion acts upon large luteal steroidogenic cells (LLC) and luteal endothelial cells by interacting with plasma membrane specific $PGF_{2\alpha}$ receptors (FP). Prostaglandin $F_{2\alpha}$elicited responses by LLC include oxytocin and possible monocyte chemoattractant protein-1 (MCP-1) secretion, and reduction of basal progesterone secretion. Oxytocin secreted by LLC acts in a paracrine fashion on SLC by activating its specific plasma membrane receptors on these cells (Oxy R). The main cell response elicited by oxytocin on SLC is reducing luteinizing hormone (LH)-stimulated progesterone secretion. Endocrine and possibly paracrine action of $PGF_{2\alpha}$ on luteal endothelial induce secretion of endothelin-1 (EDN1). Endothelin-1 activating its specific receptors on SLC (ENDRA) induces further reduction on progesterone secretion and initiates structural regression. MCP-1 attracts monocytes, which differentiate into macrophages. Macrophages secreting factors such as tumor necrosis factor alpha (TNFα) can have additional paracrine actions on steroidogenic and endothelial cells

mediate an increase in the number of cells undergoing apoptosis [58]. An EDN1-dependent action of $PGF_{2\alpha}$ on SLC appears to be exerted with the participation of luteal endothelial cells. Prostaglandin $F_{2\alpha}$, which at this stage is most likely of luteal and uterine origin, induces EDN1 release from luteal endothelial cells. EDN1 acting as a paracrine factor on SLC might then be responsible for the further decline in progesterone observed during the onset of structural luteal regression [56]. Prostaglandin $F_{2\alpha}$ has been shown to induce COX-2 in large luteal cells and therefore at this stage $PGF_{2\alpha}$ might be acting also as an autocrine/paracrine factor on LLC and SLC, respectively [58]. Additional interactions between LLC and macrophages are likely, but they are at this time still poorly defined. There is an increase in the number of resident macrophages in the corpus luteum during luteal regression. Although limited, there is some information regarding regulation of monocyte chemoattractant protein-1 (MCP-1) during luteal regression [59]. Potentially, luteal, steroidogenic, and endothelial cells are putative sources of MCP-1 [60, 61, 62]. Whether secretion of MCP-1 by endothelial cells is regulated by $PGF_{2\alpha}$ is not settled. There is debate as to whether or not PF receptors are expressed in luteal endothelial cells and $PGF_{2\alpha}$ has been reported not to regulate MCP-1 secretion in a commercially available bovine endothelial cell line [57, 62]. Upon taking residence in the CL, monocytes would differentiate into macrophages and could be a source of tumor necrosis-alpha (TNF-a), which could potentially mediate further interactions directly on steroidogenic cells and on endothelial cells (Fig. 17.2).

17.6.3 Luteolytic Capacity of Prostaglandins in the Young Corpus Luteum

During the ovarian cycle of most species, including humans, the transition from early to mid-luteal phase is associated with changes in susceptibility to the luteolysin $PGF_{2\alpha}$. For instance, in cows, the CL is resistant to exogenous $PGF_{2\alpha}$ prior to day 5 of the estrous cycle [63]. The cellular basis controlling luteal function during these physiological transitions, although studied intensely, is incompletely understood. Several ideas have been suggested to explain the lack of progesterone-inhibitory response to $PGF_{2\alpha}$ in the early bovine CL. For instance, it has been proposed that the luteolytic resistance of the early CL might be due to alterations in components of the signal transduction associated with the receptor by locally produced hormones. This is supported by the observed increased expression in $PGF_{2\alpha}$ catabolizing enzyme, 15-hydroxyprostaglandin dehydrogenase (PGDH), in the early CL [64]. Similarly, the inability of $PGF_{2\alpha}$ to induce COX-2 expression and intraluteal $PGF_{2\alpha}$ synthesis in the early bovine CL has been reported to be implicated in luteolytic insensitivity to $PGF_{2\alpha}$ [35].

There might be species differences regarding the steroidogenic cells expressing FP receptors, but it is accepted that in luteal steroidogenic cells, FP plasma membrane G-protein-coupled receptors activate the membrane-bound phosphoinositide-specific phospholipase C (PLC), yielding inositol 1,4,5-trisphosphate (IP_3), diacylglycerol [65], and resulting in mobilization of intracellular Ca^{2+} [66]. Accordingly, calcium and protein kinase C (PKC) have been shown to be the intracellular mediators of PGF_{2a} actions in luteal steroidogenic cells [67]. The regulatory effects of intracellular calcium concentration ($[Ca^{2+}]i$) on progesterone might be biphasic as there is also evidence for a calcium requirement to support progesterone synthesis by bovine luteal cells and LH, a luteotropic hormone, increases IP_3, and $[Ca^{2+}]i$ in bovine luteal cells and in porcine granulosa cells [68, 69, 70]. There is a possibility that there might exist thresholds of $[Ca^{2+}]i$ that support or inhibit progesterone synthesis.

It has been suggested that the lack of progesterone-inhibitory response to $PGF_{2\alpha}$ in the early bovine CL is not due to low expression of FP receptor number, but rather by alterations in components of the signal transduction associated with the receptor, and/or (as mentioned earlier) by modifications due to locally produced hormones [71]. For instance, in ewes, an increase in mRNAs encoding PKC inhibitors, and the associated increase in the corresponding proteins, might participate in the resistance of the CL to inhibitory actions of exogenous $PGF_{2\alpha}$during the early part of the estrous cycle [72]. Studies testing the ability of increasing concentrations of $PGF_{2\alpha}$ to increase the $[Ca^{2+}]i$ in large and small bovine luteal cells as a function of development have indicated that the lower efficacy of $PGF_{2\alpha}$ in the early CL was likely related to signal transduction differences associated with the FP receptor at those two developmental stages [42].

The array of PKC isozymes expressed in the whole bovine CL includes α, βI, βII, ε, and μ [73, 74, 75], and it has been demonstrated that the amount of PKCε expressed in the day-10 CL is greater than in the day-4 CL [75]. The latter observation led Sen et al [44] to propose that differential expression of PKCε as a function of development could play a role in the observed transitional resistance/susceptibility to $PGF_{2\alpha}$-induced luteal regression. Furthermore, it is hypothesized that regulation of $[Ca^{2+}]i$ is a cellular mechanism through which PKCε could mediate inhibitory actions of $PGF_{2\alpha}$ on progesterone secretion [44]. There is additional evidence indicating that when bovine follicular theca cells are isolated and their luteinization is induced under in vitro tissue culture conditions, they express PKCδ [76]. As PKCδ has been reported to play an important role in other species such as in rabbits and rodents [77, 78], this PKC isozyme might also be important for the physiology of the CL of some mammals.

The roles of specific PKC isozymes in luteal physiology have received little attention to date. Studies using PKC isozyme-specific inhibitors or specific down-regulation of PKCε by siRNA technology have provided interesting evidence about the function of this PKC isozyme in luteal physiology. This isozyme appears to regulate quantitatively the intracellular calcium signal initiated by $PGF_{2\alpha}$ on luteal steroidogenic cells and this in turn might have consequences (at least in part) in the ability of $PGF_{2\alpha}$ to inhibit progesterone secretion [44]. Interestingly, if the FP receptor and its associated signal transduction is bypassed with a pharmacological agent to increase the $[Ca^{2+}]i$, the LH-stimulated progesterone secretion in day-4 steroidogenic cells is eliminated, an action that cannot be induced by $PGF_{2\alpha}$ at this developmental stage. Conversely, if the increase in $[Ca^{2+}]i$ typically induced by $PGF_{2\alpha}$ on day-10 steroidogenic luteal cells is buffered by a pharmacological agent, then the ability of $PGF_{2\alpha}$ to inhibit the LH-stimulated progesterone secretion is abrogated. The overall data support the hypothesis that luteal resistance to the luteolytic actions of $PGF_{2\alpha}$ is associated with a compromised ability of $PGF_{2\alpha}$ to induce a rise in $[Ca^{2+}]i$ [79]. Details of the cellular mechanisms utilized by PKCε to influence the ability of $PGF_{2\alpha}$ to induce a rise in $[Ca^{2+}]i$ remain unknown.

17.7 Summary

Prostaglandins belong to a family of cell-signaling molecules, termed prostanoids, involved in key reproductive events, such as ovulation, corpus luteum function, and luteolysis. In animals where the uterus controls the length of the estrous cycle, $PGF_{2\alpha}$ of uterine origin has been demonstrated to be the physiological luteolysin that causes CL regression. However, in species where the uterus exerts no control of luteal lifespan the luteolytic $PGF_{2\alpha}$ is synthesized within the CL itself. It has been postulated that a central hypothalamic oxytocin pulse generator functions as a pacemaker for luteolysis and that the uterus transduces these signals into luteolytic pulses $PGF_{2\alpha}$. Endometrial $PGF_{2\alpha}$ acts upon large luteal steroidogenic cells and luteal endothelial cells to induce a decrease in progesterone secretion ending the functional life of the corpus luteum. This initial decline in progesterone might be critical, as it modifies the response of SLC to oxytocin. Specifically, a decline in progesterone increases the calcium responses initiated by oxytocin on small luteal cells undergoing apoptosis. EDN1 act as a paracrine regulator on small luteal cells and is responsible for the further decline in progesterone observed during the onset of structural luteal regression.

17.8 Glossary of Terms and Acronyms

[Ca2+]i: intracellular calcium concentration

9K-PGR: 9-keto-PGE2-reductase

AA: arachidonic acid

bHLH: basic helix-loop-helix

C/EBP: CAAT enhancing binding protein

CL: corpus luteum

COX: cyclooxygenases

CRE: cAMP responsive element

DGLA: dihomo-gamma-linoleic acid

DP: PGD receptor

EDN1: endothelin-1

EDN2: endothelin-2

EDN3: endothelin-3

EDNRA: type A endothelin receptors

EDNRB: type B endothelin receptors

EFA: essential fatty acids

EP1-4: PGE receptors with four subtypes

FP: PGF receptor

IP: PGI receptor

IP3: inositol 1,4,5-trisphosphate

IRE: interferon response element

LH: luteinizing hormone

LLC: large luteal steroidogenic cells

LT: leukotrienes

MCP-1: monocyte chemoattractant protein-1

NF kB: or E-box nuclear factor kappa B

NSAID: non-steroidal anti-inflammatory drugs

OT: oxytocin

P4: progesterone

PA: eicosapentenoic acid

PG: prostaglandins

PGDH: 15-hydroxyprostaglandin dehydrogenase

PGFS: prostaglandin F synthase

PGG2: hydroperoxy endoperoxide prostaglandin G_2

PGH2: prostaglandin H_2

PGI: prostacyclins

PGT: prostaglandin transporter

PKC: Protein Kinase C

PLA2: phospholipase A_2

PLC: phospholipase C

siRNA: silencing RNA

SLC: small luteal steroidogenic cells

TNF-α: tumor necrosis-alpha

TX: thromboxanes

TxA: thromboxane receptor

USF: Upstream stimulatory factor

References

1. Tilley SL, Coffman TM, Koller BH. Mixed messages: modulation of inflammation and immune responses by prostaglandins and thromboxanes. J Clin Invest 2001; 108:15–23.
2. Goldblatt MW. Properties of human seminal plasma. J Physiol 1935; 84:208–18.
3. Banu SK, et al. Molecular cloning and spatio-temporal expression of the prostaglandin transporter: a basis for the action of prostaglandins in the bovine reproductive system. Proc Natl Acad Sci U S A 2003; 100:11747–52.
4. Stefanczyk-Krzymowska S, et al. Local transfer of prostaglandin E_2 into the ovary and its retrograde transfer into the uterus in early pregnant sows. Exp Physiol 2005; 90:807–14.
5. Einer-Jensen N, Hunter RHF. Counter-current transfer in reproductive biology. Reproduction 2005; 129:9–18.
6. Arosh JA, et al. Molecular cloning and characterization of bovine prostaglandin E2 receptors EP2 and EP4: expression and regulation in endometrium and myometrium during the estrous cycle and early pregnancy. Endocrinology 2003; 144:3076–91.
7. Arosh JA, et al. Prostaglandin biosynthesis, transport, and signaling in corpus luteum: a basis for autoregulation of luteal function. Endocrinology 2004; 145:2551–60.
8. Banu SK, et al. Molecular cloning and characterization of Prostaglandin (PG) transporter in Ovine Endometrium: Role for multiple cell signaling pathways in transport of PGF2{alpha}. Endocrinology 2008; 149:219–31.
9. Hirata M, et al. Cloning and expression of cDNA for a human thromboxane A_2 receptor. Nature 1991; 349:617–29.
10. Narumiya S, Sugimoto Y, Ushikubi F. Prostanoid receptors: Structures, properties, and functions. Physiol Rev 1999; 79:1193–226.
11. Vane JR. Inhibition of prostaglandin synthesis as a mechanism of action for aspirin-like drugs. Nat New Biol 1971; 231:232–5.
12. Armstrong DT, Moon YS, Zamecnik J. Evidence for a role of ovarian prostaglandins in ovulation. In: Moudgal NR, editor. Gonadotropins and Gonadal Function. New York: Academic Press, 1974:345–56.
13. Armstrong DT. Involvement of prostaglandins in ovarian regulation. In: James VH, editor. Endocrinology. Proceedings of the 5th International Congress of Endocrinology. Amsterdan: Excerpta Medica, 1977:391–5.

14. Murdoch WJ, Hansen TR, McPherson LA. A review—role of eicosanoids in vertebrate ovulation. Prostaglandins 1993; 46: 85–115.
15. Kennedy CRJ, et al. Salt-sensitive hypertension and reduced fertility in mice lacking the prostaglandin EP_2 receptor. Nature 1999; 5:217–20.
16. Markosyan N, et al. Primate Granulosa cell response via Prostaglandin E2 receptors increases late in the Periovulatory interval. Biol Reprod 2006; 75:868–76.
17. Sirois J, et al. Cyclooxygenase-2 and its role in ovulation: a 2004 account. Hum Reprod Update 2004; 10:373–85.
18. Richards J. Sounding the alarm—does induction of prostaglandin endoperoxide synthase-2 control the mammalian ovulatory clock? Endocrinology 1997; 138:4047–8.
19. Russell DL, Robker RL. Molecular mechanisms of ovulation: co-ordination through the cummulus complex. Hum Reprod Update 2007; 13:289–312.
20. Loeb L. The effect of extirpation of the uterus on the life and function of the corpus luteum in the guinea pig. Proc Sot Exptl Biol Med 1923; 20:441–64.
21. Wiltbank JN, Casida LE. Alteration of ovarian activity by hysterectomy. L Animal Sci 1956; 15:134–40.
22. Spies HG, et al. Influence of hysterectomy and exogenous progesterone on size and progesterone content of the corpora lutea in gilts. J Animal Sci 1958; 17:1234–958.
23. Bradbury JT. Prolongation of the life of the corpus luteum by hysterectomy in the rat. Anat Record Suppl 1937; 1:51.
24. Asdell SA, Hammond L. The effects of prolonging the life of the corpus luteum in the rabbit byhysterectomy. Am J Physiol 1933; 103:600–9.
25. Caldwell BV, Mazer RS, Wright PA. Luteolysis as affected by uterine transplantation in the syrian hamster. Endocrinology 1967; 80:477–82.
26. Cheval M. Ovarian and uterine grafts. Proc Roy Soc Med 1934; 27:1395–406.
27. Deanesly R, Parker AS. The effect of hysterectomy on the oestrous cycle in the ferret. J Physiol London 1933; 78:80–4.
28. Dreys D. Hysterectomy in squirrel has no effect. Am J Anat 1919; 25:117.
29. Dewar AD. Effects of hysterectomy on corpus luteum activity in the cyclic, pseudopregnant and pregnant mouse. J Reprod Fertility 1973; 33:77–90.
30. Burford TH, Diddle AW. Effect of total hysterectomy upon the ovary of the Macacus rhesus monkey. Surg Gynecol Obstec 1936; 62:600–9.
31. Hartman CG. Hysterectomy and the estrous cycle in the opossum. Am J Anat 1925; 35: 25.
32. Clark MJ, Sharman GB. Failure of hysterectomy to affect the ovarian cycle of the marsupial Trichosurus vulpecula. J Reprod Fertility 1965; 10:459–61.
33. Jones GES, Elinde RW. The metabolism of progesterone in the hysterectomized woman. Am J Obstet Gynecol 1961; 41:682–7.
34. Shemesh M, Hansel W. Stimulation of prostaglandin synthesis in bovine ovarian tissues by arachidonic acid and luteinizing hormone. Biol Reprod 1975; 13: 448–52.
35. Tsai SJ, Wiltbank MC. Prostaglandin F2α induces expression of prostaglandin G/H synthase-2 in the ovine corpus luteum; a potential positive feedback loop during luteolysis. Biol Reprod 1997; 57:1016–22.
36. Bennegard B, et al. Local luteolytic effect of prostaglandin F2 alpha in the human corpus luteum. Fertil Steril 1991; 56:1070–6.
37. McCracken JA, Custer EE, Lamsa JC. Luteolysis: aneuroendocrine-mediated event. Physiol Rev 1999; 79: 263–323.
38. Wu YL, Wiltbank MC. Transcriptional regulation of the cyclooxygenase-2 gene changes from protein kinase (PK) A- to PKC-dependence after luteinization of granulosa cells. Biol Reprod 2002; 66:1505–14.
39. Houmard B, Ottobre J. Progesterone and prostaglandin production by primate luteal cells collected at various stages of the luteal phase: modulation by calcium ionophore. Biol Reprod 1989; 41:401–6.
40. Wuttke W, et al. Synergistic effects of prostaglandin F2α and tumor necrosis factor to induce luteolysis in the pig. Biol Reprod 1998; 58:1310–5.
41. Gastal EL, et al. Responsiveness of the early corpus luteum to PGF2α and resulting progesterone, LH, and FSH interrelationships in mares. Anim Reprod 2005; 2:240–9.
42. Choudhary E, et al. Developmental sensitivity of the bovine corpus luteum to prostaglandin F2α (PGF2α) and endothelin-1 (ET-1): Is ET-1 a mediator of the luteolytic actions of PGF2α or a tonic inhibitor of progesterone secretion? Biol Reprod 2005; 72: 633–42.
43. Denning-Kendall PA, Wathes DC. Acute effects of prostaglandin F_{2alpha}, luteinizing hormone, and estradiol on second messenger system and on the secretion of oxytocin and progesterone from granulosa and early luteal cells of the ewe. Biol Reprod 1994; 40:765–73.
44. Sen A, et al. Effects of selective protein kinase C isozymes in prostaglandin F2α -induced Ca2+ signaling and luteinizing hormone-induced progesterone accumulation in the mid-phase bovine corpus luteum. Biol Reprod 2005; 71:976–80.
45. Inoue A, et al. The human endothelin family: three structurally and pharmacologically distinct isopeptides predicted by three separate genes. Proc Natl Acad Sci U S A 1989; 86:2863–7.
46. Arai H, et al. Cloning and expression of a cDNA encoding an endothelin receptor. Nature 1990; 348:730–2.
47. Girsh E, Greber Y, Meidan R. Luteotrophic and luteolytic interactions between bovine small and large luteal-like cells and endothelial cells. Biol Reprod 1995; 52:954–62.
48. Hinckley ST, Milvae RA. Endothelin-1 mediates prostaglandin F2α-induced luteal regression in the ewe. Biol Reprod 2001; 64:1619–23.
49. Girsh E, et al. Effect of endothelin-1 on bovine luteal cell function: role in prostaglandin F2α-induced anti-steroidogenic action. Endocrinology 1996; 137:1306–12.
50. Nussdorfer GG. Autocrine-paracrine endothelin system in the physiology and pathology of steroid-secreting tissues. Pharmacol Rev 1999; 51:403–35.
51. Levy N, et al. Administration of PGF2α during the early bovine luteal phase does not alter the expression of ET-1 and of its type A receptor: a possible cause for corpus luteum refractoriness. Biol Reprod 2000; 63:377–82.
52. Girsh E, et al. Regulation of endothelin-1 expression in the bovine corpus luteum: elevation by prostaglandin F2α. Endocrinology 1996; 137:5191–6.
53. Ohtani M, Kobayashi S, Miyamoto A, et al. Real-time relationships between intraluteal and plasma concentrations of endothelin, oxytocin, and progesterone during prostaglandin F2α-induced luteolysis in the cow. Biol Reprod 1998; 58:103–8.
54. Choudhary E, et al. Prostaglandin F2α independent and dependent regulation of the bovine luteal endothelin system. Domest Anim Endocrinol 2004; 27:63–79.
55. Watanabe S, et al. Effect of intraluteal injection of endothelin type A receptor antagonist on PGF2α-induced luteolysis in the cow. J Reprod Dev 2006; 52:551–9.
56. Doerr MD, Goravanahally MP, Rhinehart DJ, et al. Effects of endothelin receptor type-A and type-B antagonists on prostaglandin F2alpha-induced luteolysis of the sheep corpus luteum. Biol Reprod 2008; 78:688–96.
57. Meidan R, Levy N. The ovarian endothelin network: an evolving story. Trends Endocrinol Metab 2008; 18:379–85.

58. Niswender GD, et al. Judge, jury and executioner: the autoregulation of luteal function. Soc Reprod Fertil Suppl 2007; 64:191–206.
59. Senturk LM, et al. Monocyte chemotactic protein-1 expression in human corpus luteum. Mol Hum Reprod 1999; 5:697–702.
60. Hosa K, et al. Porcine luteal cells express monocyte cheoattractant protein-2 (MCP-2): analysis vy cDNA cloning and northern analysis. Biochem Biophys Res Commun 1994; 2005:148–53.
61. Nagaosa K, Shiratsuchi A, Nakanishi Y. Determination of cell type specificity and Estrous cycle dependency of monocyte chemoattractant protein-1 expression in Corpora Lutea of normally cycling rats in relation to Apoptosis and Monocyte/Macrophage accumulation. Biol Reprod 2002; 67:1502–8.
62. Cavicchio VA, et al. Secretion of monocyte chemoattractant protein-1 by endothelial cells of the bovine corpus luteum: regulaion by cytokines but not prostaglandin F2α. Endocrinology 2002; 143:3552–9.
63. Copelin JP, et al. Responsiveness of bovine corpora lutea to prostaglandin F2α: comparison of corpora lutea anticipated to have short or nomal lifespans. J Anim Sci 1988; 66: 1236–45.
64. Silva PJ, et al. Prostaglandin metabolism in the ovine corpus luteum: catabolism of prostaglandin F(2alpha) (PGF(2alpha)) coincides with resistance of the corpus luteum to PGF(2alpha). Biol Reprod 2000; 63:1229–36.
65. Davis JS, et al. Acute effects of prostaglandin F 2α on inositol phospholipid hydrolysis in the large and small cells of bovine corpus luteum. Mol Cell Endo 1988; 58:43–50.
66. Davis JS, et al. Prostaglandin F2 alpha stimulates phosphatidylinositol 4,5-bisphosphate hydrolysis and mobilizes intracellular calcium in bovine luteal cells. Proc Natl Acad Sci USA 1987; 84:3728–32.
67. Wiltbank MC, Diskin MG, Niswender GD. Differential actions of second messenger system in the corpus luteum. J Reprod Fertil Suppl 1991; 43:65–75.
68. Davis JS, et al. Luteinizing hormone increases inositol trisphosphate and cytosolic free Ca2+ in isolated bovine luteal cells. J Biol Chem 1987; 262:8515–21.
69. Flores JA, et al. Luteinizing Hormone (LH) stimulates both intracellular Calcium Ion ($[Ca^{2+}]_i$) mobilization and transmembrane cation influx in single ovarian (Granulosa) cells: Recruitment as a cellular mechanism of LH-$[Ca^{2+}]_i$ dose response. Endocrinol 1998; 139:3606–12.
70. Alila HW, et al. Differential effects of calcium on progesterone production in small and large bovine luteal cells. J Steroid Biochem 1990; 36:387–93.
71. Skarzynski DJ, Okuda K. Sensitivity of bovine corpora lutea to prostaglandin $F_{2\alpha}$ is dependent on progesterone, oxitocyn and prostaglandis. Biol Reprod 1999; 60:1292–8.
72. Juengel JL, et al. Steady-state concentrations of mRNA encoding two inhibitors of protein kinase C in ovine luteal tissue. J Reprod Fertil 1998; 113:299–305.
73. Orwig KE, et al. Immunochemical characterization and cellular distribution of protein kinase C isozymes in the bovine corpus luteum. Comp Biochem Physiol 1994; 108B:53–7.
74. Davis JS, May JV, Keel BA. Mechanisms of hormone and growth factor action in the bovine corpus luteum. Theriogenology 1996; 45:1351–80.
75. Sen A, et al. Expression and activation of protein kinase C isozymes by prostaglandin F (PGF2α) in the early- and mid-luteal phase bovine corpus luteum. Biol Reprod 2004; 70: 379–84.
76. Budnik LT, Mukhopadhyay AK. Lysophosphatidic acid-induced nuclear localization of protein kinase C δ in bovine theca cells stimulated with luteinizing hormone. Biol Reprod 2002; 67: 935–44.
77. Maizels ET, et al. Hormonal regulation of PKC-delta protein and mRNA levels in thje rabbit corpus luteum. Mol Cell Endocrinol 1996; 122:213–21.
78. Peters CA, et al. Induction of relaxin messenger RNA expression in response to prolactin receptor activation requires protein kinase C-delta signaling. Mol Endocrinol 2002; 14: 576–90.
79. Goravanahally MP, et al. PKCepsilon and an increase in intracellular calcium concentration are necessary for PGF2alpha to inhibit LH-stimulated progesterone secretion in cultured bovine steroidogenic cells. Reprod Biol Endocrinol 2007; 5:37.

Chapter 18

Androgens—Molecular Basis and Related Disorders

Christine Meaden and Pedro J. Chedrese

18.1 Introduction

The term *androgen* refers to any natural or synthetic compounds that stimulate or control development and maintenance of masculine characteristics. Most commonly, androgens refer to endogenous steroid sex hormones responsible for virilizing the accessory male sex organs and secondary sex characteristics. Androgens are mainly synthesized by the testes, although females also produce small amounts, which are important for positive protein balance, maintaining strong muscles and bones, and contribute to libido. There are two major androgens secreted by the testes: *testosterone* and *5α-dihydrotestosterone* (5α–DHT). Two weaker androgens primarily synthesized in the adrenal cortex and in smaller amounts by the testes and ovaries are *dehydroepiandrosterone* (DHEA) and *androstenedione,* which are converted metabolically to testosterone and other androgens. There is one common androgen receptor (AR) that all androgens bind to, although their target genomic responses are distinctly different. Testosterone is the most abundant androgen with 4–10 mg secreted daily in adult men. 5α-DHT binds the AR with higher affinity than all other androgens, making 5α-DHT the most potent androgen. This chapter focuses on the fundamental molecular mechanisms of the effects of androgens, androgen metabolism in males and defects in the AR that is meant to communicate the complexity of intersex disorders.

18.2 Androgen Biosynthesis and Metabolism in Males

Cholesterol, a lipid found in cell membranes and circulating in blood plasma, is required for all steroid hormone synthesis. Androgens are synthesized by the Leydig cells of the testes from a cholesterol substrate, incorporated from the circulation in the form of low-density lipoproteins (LDL), and from internal sources of acetate [1]. Lutenizing hormone (LH) stimulates testosterone biosynthesis by increasing mobilization and transport of cholesterol into the steroidogenic pathway. Cholesterol is transported from the outer to the inner mitochondrial membrane by the steroidogenic acute regulatory protein (StAR), the rate-limiting step in steroid biosynthesis [2]. In the mitochondria, cholesterol is converted into *pregnenolone*, the precursor steroid required for synthesis of all steroid hormones. The 27-carbon cholesterol is then cleaved twice to reduce its size to the 19-carbon testosterone in a process that requires five enzymatic steps (Fig. 18.1):

- Reduction of the side chain of cholesterol by the 20,22 desmolase activity of cytochrome P450 side chain cleavage (P450scc), to form 21-carbon pregnenolone. This step is regulated by LH, which controls the overall rate of testosterone synthesis.
- Hydroxylation/isomerization of the steroids is catalyzed by 3β-hydroxysteroid dehydrogenase/Δ4-Δ5 isomerase (3β-HSD). Because of its isomerase activity, 3β-HSD drives the steroid synthesis from the Δ5-pathway to the Δ4-pathway. Thus, by using four different substrates, pregnenolone, 17α-pregnenolone, DHEA, and androstenediol, 3β-HSD can synthesize progesterone, 17α-progesterone, androstenedione, and testosterone, respectively. The Δ4-pathway is predominant in human testes, therefore, pregnenolone is mainly converted into progesterone.
- Progesterone is hydroxylated to 17α-hydroxyprogesterone by 17α -hydroxylase.
- 17α-hydroxylase, which is also a 17,20-lyase, converts the 21-carbon 17α-hydroxyprogesterone into the 19-carbon androstenedione.
- Finally, 17β-hydroxysteroid dehydrogenase (17β-HSD) converts androstenedione into testosterone.

Testosterone enters circulation bound to serum albumin and to steroid hormone-binding globulin (SHBG), a glycosylated

C. Meaden (✉)
Department of Biology, University of Saskatchewan College of Art and Sciences, Saskatoon, SK, Canada
e-mail: cmeaden@shaw.ca

P.J. Chedrese (ed.), *Reproductive Endocrinology*,
DOI 10.1007/978-0-387-88186-7_18, © Springer Science+Business Media, LLC 2009

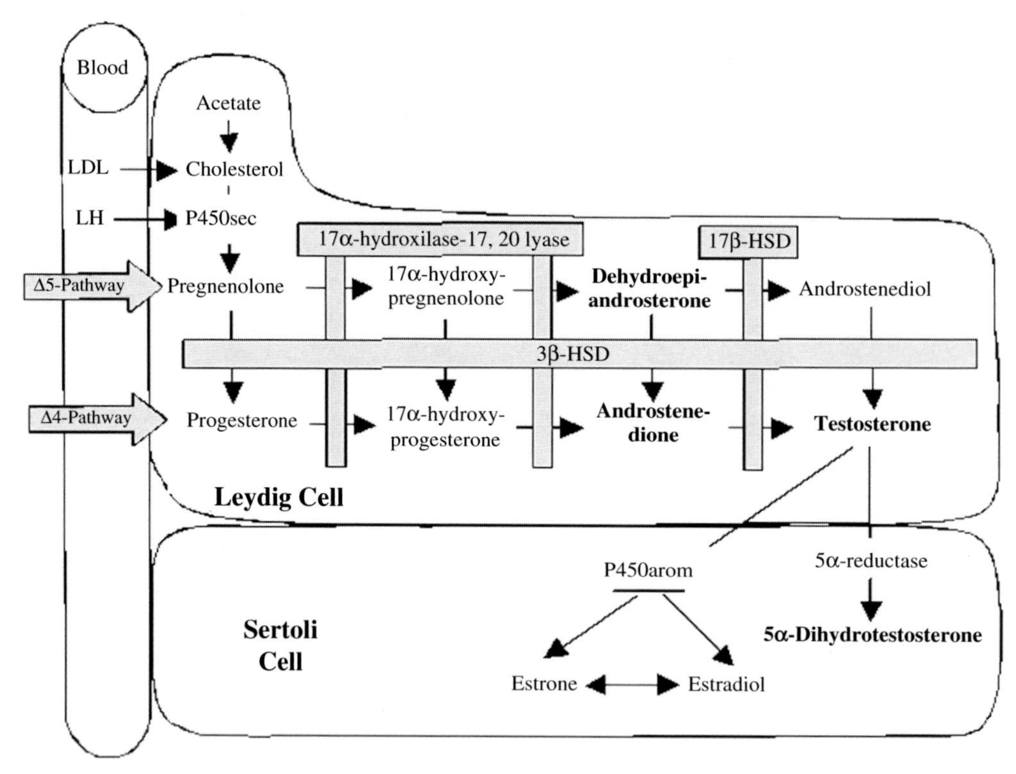

Fig. 18.1 Steroidogenic pathways in the testis. In the Leydig cells 3β-HSD drives the steroid synthesis from the Δ5-pathway to the Δ4-pathway. 17α-hydroxylase-17,20-lyase catalyzes the conversion of progesterone into androstenedione, which is converted into testosterone by 17β-HSD. In the Sertoli cells, testosterone is reduced to 5α-dehydrotestosterone by 5α-reductase. The main sex steroids secreted are in the bold. 3β-HSD: 3β-hydroxysteroid dehydrogenase/Δ4-Δ5 isomerase; 17β-HSD: 17β-hydroxysteroid dehydrogenase; LDL: low density lipoprotein; LH: luteinizing hormone; P450scc: cytochrome P450 side chain cleavage; P450arom: cytochrome P450 aromatase

dimeric protein expressed mainly in the liver. Approximately 2% dissociates from the proteins and becomes available to directly enter the cells by diffusion. Testosterone is converted into the more potent androgen 5α-DHT by two different microsomal isoenzymes: 5α-reductase-1 and 5α-reductase-2. 5α-reductase-1 is encoded by the SRD5A1 gene located on chromosome 5 (5p15) and is widely expressed in peripheral tissues, including the skin, liver, and specific regions of the brain. 5α-reductase-1 converts circulating testosterone into 5α-DHT required for androgen-mediated tissue growth. 5α-reductase-2 is encoded by the five exon structural SRD5A2 gene located on chromosome 2 (2p23) and is limited to expression in androgen-dependent tissues such as the genitals. 5α-reductase-2 is responsible for the local conversion of gonadal testosterone into 5α-DHT. Therefore, expression of the SRD5A2 gene is required for the normal development of the male external genitalia and urogenital sinus formation during fetal differentiation.

Testosterone is also aromatized to estradiol-17β by the cytochrome P450 aromatase (P450arom) in peripheral tissue. Estradiol-17β is a major female steroid sex hormone that is also required in smaller amounts by males to maintain healthy bones, neuro- and cardiovascular health, spermatogenesis, and healthy libido. Overall, approximately 80% of the circulating 5α-DHT and estradiol-17β found in males is derived from peripheral conversion. Table 18.1 summarizes the contributions of the testes, adrenals, and peripheral tissues to the circulating levels of steroid sex hormones in adult men [3].

Table 18.1 Relative contribution (%) of the androgen-secreting tissues to the circulatory levels of androgens in men. 5α-DHT: 5α-dehydrotestosterone, DHEA: dehydroepiandrosterone

	Testes	Adrenals	Peripheral tissues
Testosterone	95	<1	<5
5α-DHT	20	<1	80
Estradiol-17β	20	<1	80
Estrone	2	<1	98
DHEA	<10	90	–

The liver converts circulating androgens into various metabolites, which include *androsterone* and *etiocholanolone*. After conjugation with glucuronic or sulfuric acid, androgen metabolites are excreted in the urine as 17-ketosteroids, which represent 20–30% of the total urinary

17-ketosteroids. The remaining 17-ketosteroids found in urine originate from the metabolism of adrenal steroids.

18.3 Biological Effects of Androgens

Expression of the AR gene is ubiquitous and found in a wide array of genital and non-genital tissues in both males and females. Therefore, the sum of the effects of testosterone, 5α-DHT, DHEA, and androstenedione are varied and tissue-specific.

18.3.1 Role of Androgens in Fetal Sexual Differentiation

Androgens are essential for proper differentiation of the internal and external male genital system throughout fetal development [4]. During the first 6 weeks of gestation the genital anatomy of the male (46,XY karyotype) and female (46,XX karyotype) fetuses are indistinguishable. They consist of undeveloped tissues, the genital tubercle, which will eventually become a penis and scrotum or a clitoris and labia, respectively. In the normal male fetus, gene expression of the *sex-determining region of the Y chromosome* (SRY) induces testes to form in the fetal abdomen within a few weeks of conception. By the seventh week the fetal testes begin to produce testosterone, which is the androgenic signal that modulates secretion of LH by the hypothalamic-pituitary axis. Testststosterone stimulates, development of the Wolffian's duct system, the precursor of the male genital tract, and causes regression of the Müllerian duct system, the precursor of the female genital tract [5].

Fetal testosterone metabolized into 5α-DHT functions as an *intracrine* regulator inducing formation of the urogenital sinus, differentiation of the prostate, and virilization of the external genitalia. 5α-DHT also inhibits growth of the vesicovaginal septum, thereby preventing the development of a vagina in the male fetus (Fig. 18.2). Thus, absence or deficiency of androgens in the male fetus inhibits formation of external genitalia, urogenital sinus, and causes hypoplasia of the prostate and feminization. Complete male differentiation of the external genitalia and the urogenital sinus occurs only if the androgenic stimulus is received during the critical 7–12 week period of fetal development; if not complete by the 13th week, no amount of testosterone later will change the location of the urethral opening or close a vaginal opening [6]. In the absence of the Y chromosome the undifferentiated gonadal tissue develops into ovaries within 7–8 weeks of conception. However, it remains unclear if this process occurs by a default pathway or is controlled by specific gene(s). The

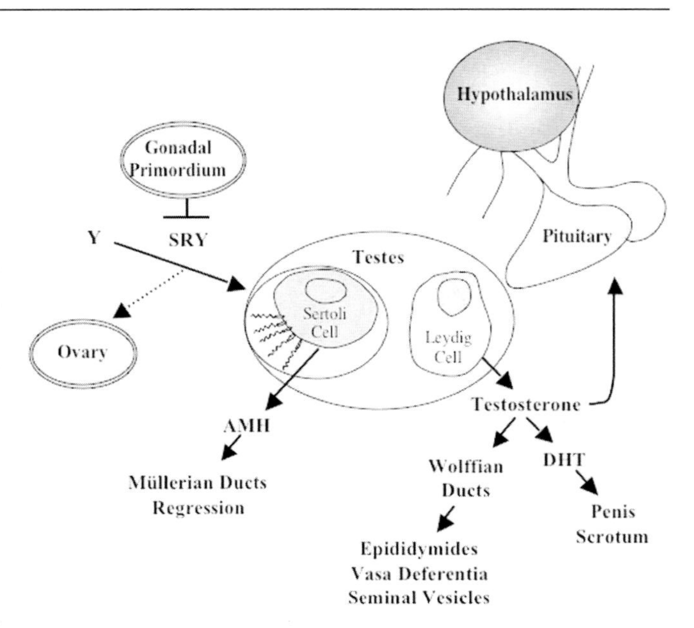

Fig. 18.2 Genetic mechanism of male sex differentiation. SRY induces formation of the testes. By the 7th week testes begin to produce testosterone, which stimulates development of the Wolffian's duct system precursor of the male genital tract and cause regression of the Müllerian duct system. AMH: Anti-Müllerian Hormone; DHT: 5α-dehydrotestosterone; SRY: sex determining region of the Y chromosome gene

Müllerian duct system differentiates spontaneously without hormonal stimulus to form the upper.

18.3.2 Effects of Androgens at Puberty

During puberty, under control of the gonadotropins, LH, and follicle stimulating hormone (FSH), androgens induce secondary sexual characteristics in males and to a lesser extent in females. Androgens in conjunction with estradiol-17β and inhibin participate in a negative feedback mechanism that regulates secretion of the pituitary gonadotropins. LH stimulates the Leydig cells of the testes and theca cells in the ovaries to produce testosterone, while FSH stimulates the spermatogenic tissue in males and the granulosa cells of the ovarian follicles in females. The biological effects of the androgens at puberty can be summarized as follows:

- support spermatogenesis;
- influence libido and aggressive behavior;
- promote protein anabolism resulting in skeletal muscle growth and strength;
- play a crucial role in neuromuscular maintenance and normal muscle function and coordination;
- promote ossification of epiphyseal cartilaginous plates accounting for pubertal growth spurt;
- stimulate erythropoiesis, the process of red blood cell formation;

- stimulate sebaceous gland activity, which can result in acne, and hair growth, including pubic, axillary, facial, chest, abdomen, and back.

18.4 Androgen Receptor

The AR belongs to the family of ligand-activated Zn-finger nuclear receptor transcription factors [7]. The gene encoding the AR (NR3C4) is located on the X chromosome (qX11-12) and contains eight coding exons and seven introns. Two isoforms of the AR have been characterized, AR-A and AR-B [8]. AR-A is a truncated isoform that lacks the first 187 amino acids of the amino terminus with a molecular weight of 87 kDa. The AR-B is the full-length isoform, a single subunit phosphoprotein of 918 amino acids with a molecular weight of 110 kDa. The AR is organized into four main domains: the *N-terminal regulatory domain*, the *DNA-binding domain* (DBD), the *hinge region,* and the *ligand-binding domain* (LBD). The amino terminal regulatory domain contains *activation function-1* (AF-1), a recognition site required for ligand activated transcriptional activity, located between residues 101 and 370 [9, 10], and *activation function-5* (AF-5), required for constitutive activity of the receptor located between residues 360 and 485 [11]. The hinge region connects the DBD with the LBD and contains a ligand-dependent nuclear localization signal that drives the protein into the nucleus. The LBD contains *activation function-2* (AF-2), which is involved in the recognition of coactivator proteins carrying the LXXLL or FXXFL motifs, essential for the interaction with other nuclear transcription factors and for agonist induced activity [12, 13].

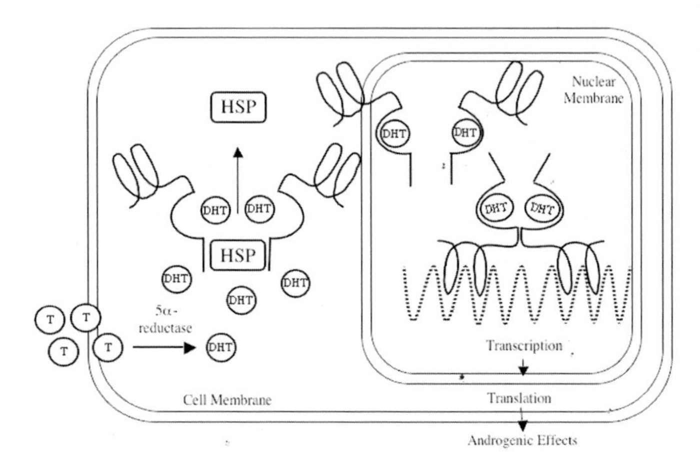

Fig. 18.3 Mechanism of androgen action. Androgen binding causes dissociation of the receptor from the HSP followed by activation of the receptor and translocation to the nucleus. The ligand-bound receptor dimerizes and binds the palindromic *cis*-acting element ARE. ARE: androgen responsive element; DHT: 5α-dihydrotestosterone; HSP: heat shock protein; T: testosterone

In the absence of ligand, the AR is complexed to cytoplasmic chaperone heat shock protein (HSP) (Fig. 18.3). Androgen binding causes phosphorylation of *serine* residues at the AF-1, causing dissociation of the receptor from the HSP and a conformational change, which is followed by activation of the receptor and translocation to the nucleus (Fig. 18.3). The ligand-bound receptor dimerizes and binds palindromic steroid responsive *cis*-acting elements termed androgen response elements (ARE), (GGTACAnnnTGTTCT), located upstream of the GAAT and TATA boxes of androgen-target genes [12]. The AR recruits RNA polymerase and other transcription factors and co-activators, including the androgen receptor-associated protein (ARA) that initiates transcription of target genes [10].

18.5 Androgen Receptor Gene Mutations

Any essential step missed between hormone binding and gene expression results in an aberrant hormone response, which can have negligible to mild or severe pathological consequences. A registry maintained at McGill University, Montreal, Canada, lists over 200 mutations and single nucleotide polymorphisms of the AR gene (http://www.androgendb.mcgill.ca/).

Point mutations introduce either premature stop codons, which result in the synthesis of an incomplete and therefore inactive protein; or an amino acid substitutions, termed *missense* mutations, which may result in either a non-functional protein or only a slightly structurally altered protein, which may not have impaired function. Missense mutations have been reported in the five exons encoding the LBD and the two exons encoding the DBD.

Frameshift mutations occur when a nucleotide is deleted or added to the coding region of a gene. Frameshift mutations disarrange the encoded information, which then becomes translated into a defective protein. Deletion of either exon-3 or exon-4 of the AR gene, which result in a non-functional protein lacking either the second zinc cluster or the hinge region and the N-terminal part of the LBD has been reported.

Splice-site mutations or ***intronic mutations*** occur within the splicing region of the intron, which causes interference with the transcription of pre-mRNA. These mutations are rare and include both splice donor (5′ end of the introns) and acceptor (3′ end of the intron) sites [14]. Defective splicing results in one or more exons being spliced out or the use of a cryptic splice donor site within the preceding exon. For all splice donor sites in the AR gene, the consensus splice donor site sequence GTAAG/A is present. The reported mutations in donor splice sites are all substitutions either at position +1 (G to A or G to T), position +3 (A to T), position + 4 (A to T), or position + 5 (G to A) [15,16]. The corresponding protein is defective in DNA-binding because the insertion has occurred between the first and second zinc cluster [17].

In general 70% of the AR gene mutations are transmitted in an X-linked recessive manner and approximately 30% of the mutations arise de novo. When de novo mutations of the AR occur after the zygotic stage, they result in somatic mosaicisms, which have been associated with metastatic prostate cancer [18].

18.6 Androgen Disorders

Defects have been characterized at the molecular level in androgen biosynthesis, the AR, the ARE, and other transcription factors. The most commonly studied defects are those affecting androgen biosynthesis pathways and the AR and corresponding genes.

Literature describing intersex disorders date back throughout human history and many cultures. During the nineteenth century, *hermaphrodite* was the term used to define the phenomena of a child born with ambiguous external genitalia who develop both male and female secondary sexual characteristics. *Pseudohermaphrodite* described people whose secondary sex characteristics were not as expected based on phenotype at birth. In 1947, E. F. Reifenstein described a "familial male hypogonadism" disorder, which was named *Reifenstein Sydrome*, as a form of male pseudohermaphroditism characterized by severe hypospadias, gynecomastia, and infertility. During the 1950s it was thought that intersex disorders related to virilization failure were due to resistance to testosterone. At that time the terms *androgen resistance* and *testicular feminization* were applied to describe this category of intersex disorders. By the 1960s, it became clear that the severity of this condition varied and that genotype/phenotype mismatches were based on level of response to androgens rather than level of resistance. However, hormone levels were not able to be measured, which, furthermore, remains a complicated subject in present day medicine.

Since the 1980s, molecular biology tools have been available for the analysis of the molecular mechanisms involved in hormone binding and gene expression. Thus, molecular biology has greatly expanded understanding of the mechanisms and etiology of the clinical features related to intersex disorders. This information has contributed significantly to health care professionals' understanding, leading to revised clinical practice and multidisciplinary approaches to gender identity, emphasizing the value of molecular genetic information for affected individuals and their families.

18.6.1 Androgen Insensitivity Syndrome

The cloning of the AR cDNA in 1988 and the subsequent elucidation of the genomic organization of the gene revealed that a defective AR renders genetic males unable to respond to testosterone. However, the cells remain capable of responding to estrogen produced by the aromatization of testosterone, resulting in the development of female primary and secondary sex characteristics [19, 20]. Sexual differentiation disorders associated with AR gene mutations are collectively referred to as *Androgen Insensitivity Syndrome* (AIS). There are many phenotypic variations associated with AIS defects. It is believed that missense mutations in the AR gene may explain the partial phenotypes and the different degrees of virilization observed in AIS. Thus, AIS represents a continuum rather than separate disorders. Many previously described disorders such as *androgen resistance, testicular feminization, and Reifenstein syndrome* and its variants are considered forms of AIS. The clinical manifestations of AIS are diverse and categorized into *complete AIS (CAIS), partial AIS (PAIS), and mild AIS (MAIS).*

18.6.1.1 Genetic Molecular Pathophysiology

AIS is an X-linked recessive trait that affects male offspring. Females with a mutated AR gene are "carriers" of AIS and their male offspring will have 50% chance of being affected by AIS, while their female offspring may also be carriers of the mutated AR gene. AR gene mutations that have been detected in individuals with AIS include point mutations, complete or partial deletions, intronic mutations, and somatic mosaicisms [21]. In addition, a single case of complete androgen insensitivity has been attributed to an abnormality of the co-activator that binds AF-1, rather than a defective AR [22]. Therefore, it was proposed that the CAIS resulted from a defect of the AF-1-specific co-activator crucial to the activation of the AR [23]. The discovery of a defective AR coregulator in association with AIS supports the importance of the AF-1 region in the AR-mediated transcription. In addition, from this observation was proposed a new clinical concept, a *coactivator disease*, in which a defective coactivator is responsible for the testosterone resistance observed in CAIS [23].

18.6.1.2 Clinical Manifestations, Diagnosis, and Management

The basic clinical manifestations affecting genetic males with AIS are described in Table 18.2. Genetic females with a mutated AR gene are minimally affected and may display

Table 18.2 Comparative clinical manifestations of the androgen insensitivity syndrome (AIS) in males (46,XY karyotype). The clinical manifestations of AIS have been divided into phenotypes, which correspond to the mutation in the androgen receptor. AIS represents a continuum rather than separate disorders. Some individuals affected by AIS will fall between the phenotypes described. Reifenstein syndrome and other familial male intersex and hypogonadal conditions, which were considered variants of the Reifenstein syndrome, are now *all* considered forms of PAIS. AMH: anti-Müllerian hormone; CAIS: complete androgen insensitivity syndrome; PAIS: partial androgen insensitivity syndrome

CAIS	PAIS	MAIS
Mutation(s) in the AR Gene		
Severe mutation that cause complete insensitivity to testosterone	Mutations are not sufficient to cause complete insensitivity to testosterone	Mutations minimally affect responsiveness to testosterone causing only mild undervirilization
Phenotype		
External genitalia appears female at birth	Degree of feminization is less compared to CAIS with a wide variety of phenotypes, from male appearing to female appearing	Male external genitalia present at birth
Gonads and Genitalia		
Testes in abdomen or inguinal canals;	Testes in abdomen or inguinal canals;	Descended gonadal testes;
No ovaries;	No ovaries;	No ovaries;
Absence of Wolffian duct structures, epididymis, vasa deferens, seminal vesicles, and ejaculatory ducts. Absence of prostate and external male genitalia;	Wolffian duct structures may vary from fully formed to undeveloped traces.	Underdeveloped Wolffian duct structures.
Testicular AMH suppresses development of the Müllerian duct female structures: upper vagina, uterus, and fallopian tubes;	AMH: Same as CAIS	AMH: Same as CAIS
Shallow blindly ending perineal pouch termed a pseudovagina.	Variable severity of ambiguous external genitalia at birth, including Virilized female external genitalia, labioscrotal folds almost fused midline forming pseudovagina, clitoral enlargement Normal male external genitalia at birth that will remain abnormally small as the person grows, hypospadia.	Under virilized male: penis and scrotum do not grow normally.
Puberty		
Sparse axillary and pubic hair growth, gynecomastia, lack of menses, infertile.	If present, penis and scrotum abnormally small; sparse to normal androgenic axillary and pubic hair growth; gynecomastia, lack of menses, infertile.	Small penis and scrotum; normal androgenic axillary and pubic hair growth; sperm count and fertility may be normal, reduced, or infertile, gynecomastia.

minor traits such as sparse axillary and pubic hair; they are unlikely to suffer acne during puberty and may have delayed menarche.

Diagnosis is based on thorough endocrine examination and DNA analysis. Among the endocrinological data collected are serum hormone levels that usually reveal normal or elevated testosterone levels, elevated LH levels indicating androgen resistance at the hypothalamic-pituitary level, elevated levels of estrogen due to unopposed aromatization of testosterone, and elevated androgen sensitivity index (ASI), which is the serum testosterone levels (nmol/liter^2) multiplied by LH levels (IU) to determine the product value. DNA sequencing of each AR exon and the flanking intron sequences can be analyzed to differentiate AIS from other genetic intersex disorders presenting with similar phenotypes.

The affected person and their family require an individualized health care plan and a multidisciplinary health care team approach. Education and information about AIS and community support agencies, groups, and networks, as well as psychological counseling, is essential. Referral for genetic counseling to explain the implications of X-linked inheritance is strongly recommended to families affected by AIS.

Often the first major decision in the management of AIS, particularly those with ambiguous genitalia, is *gender assignment*, which has been amended over the years. In the beginning, female assignment with reconstructive surgery was usually chosen by physicians and parents. In the mid 1990s, male assignment with reconstructive surgery increased in popularity. However, a third choice of course has evolved which is delaying gender assignment until gender identity and sexual orientation declare themselves and the affected person can make informed choices with regard to surgery. Individuals with CAIS often identify and chose to live as women, while those affected by MAIS are not severely affected and they identify and live as men. However, individuals with PAIS may encounter more complex problems related to gender identity and role and may have limited choices with regard to reconstructive surgery depending on severity of the AIS.

18.6.2 X-linked Spinal and Bulbar Muscular Atrophy

Androgen receptors are expressed in clusters located in the spinal bulbar muscles and spinal and cranial motor neurons. They play a crucial role in maintaining normal muscle coordination and strength. Spinal and bulbar muscular atrophy (SBMA), also known as *Kennedy Disease,* is a rare X-linked recessive neuromuscular disorder resulting from defects in the AR. Dr. W. R. Kennedy was among the first to identify the disease as separate from other neurological conditions [24]. In 1986, K. H. Fischbeck reported the trinucleotide (CAG) repeat mutation of the AR gene that affects the spinal bulbar muscles and spinal and cranial motor neurons as the cause of the symptoms of SBMA [25].

18.6.2.1 Genetic Molecular Pathophysiology

Spinal and bulbar muscular atrophy is linked to a CAG repeat expansion in the first exon of the AR gene encoding a *polyglutamine* tract. In normal alleles of the AR gene there is a specific CAG triplet in the first exon that repeats 12–30 times. However, in individuals with SMBA this triplet is amplified anywhere from 40 to 62 times. The amplification of CAG encodes a protein that alters the function of the AR gene, making it harmful to motor neurons, therefore, the bulbar spinal muscles eventually weaken and deteriorate. However, the protein can still function relatively normally for sexual development and therefore symptoms of AIS are usually mild. It has been observed that the greater the number of abnormal CAG repeats, the earlier the onset and the more severe the symptoms.

18.6.2.2 Clinical Manifestations, Diagnosis, and Management

Spinal and bulbar muscular atrophy, which affects only males, is characterized by muscle degeneration and bulbar dysfunction. Ages of onset and severity of manifestations vary from adolescence to old age, but most commonly diagnosed during middle age. Retrospective history may reveal earlier symptoms such as abnormally premature muscle exhaustion during adolescence. Also, many years of muscle cramps may precede muscle weakness resulting in gastrocnemius (calf) muscle hypertrophy. Other signs and symptoms include oromandibular and oropharyngeal musculature weakness leading to difficulty in speaking and swallowing, intention tremors, and involuntary muscle twitches. Patients experience progressive motor-sensory neuropathy resulting in proximal limb and back muscle weakness causing chronic back pain and atrophy, and weakness of the distal limb muscles in the hands, legs, and feet. Since manifestations of SMBA are similar to many other neuromuscular or neurodegenerative diseases, investigations follow a typical course based on clinical presentation, including detailed medical and family history, physical and neurological examination, blood and urine tests, electroencephalography (EEG), and psychological tests. Depending on the individual's symptoms, imaging tests such as X-rays, computed topography (CT), or magnetic resonance imaging (MRI) scan might be required. Electromyography (EMG) and nerve conduction studies may be used to determine the function of muscles and nerves. Affected individuals have characteristics of AIS that include primary and secondary sexual characteristics, such as small external genitalia and gynecomastia. They may, or may not, be impotent but most often are infertile. Examination of the endocrine component and a skin biopsy for genetic testing can prevent misdiagnosis, since SBMA has been misdiagnosed for other neurological disorders such as amyotrophic lateral (ALS) [26]. As a consequence of bulbar muscle weakness, control and strength of the laryngeal and pharyngeal muscles may cause chronic malnutrition, prevent adequate cough to clear airway debris, and cause upper airway obstruction. In addition, back pain can become excruciating making adequate chronic pain management essential. There is no cure for SMBA and management is directed toward supportive care and preventative therapies such as physical and occupational therapy, psychosocial support, and genetic counseling.

18.6.3 5α-reductase-2 Deficiency Syndrome

5α-reductase-2 deficiency syndrome was originally referred to as *pseudovaginal perineoscrotal hypospadia*, first identified among a sub-group of populations in the Dominican Republic, New Guinea, and Turkey [27, 28]. 5α-reductase-2 deficiency syndrome is an androgen metabolism disorder characterized by impaired enzymatic activity of 5α-reductase type-2 that results in diminished 5α-DHT synthesis affecting virilization of the external genitalia and formation of the urogenital sinus during crucial stages of male fetal development.

18.6.3.1 Genetic Molecular Pathophysiology

5α-reductase-2 deficiency syndrome is as an autosomal recessive disorder caused by mutations that have been found in all the five exons of the SRD5A2-gene. These mutations include amino acid substitutions, complete gene deletion, deletions resulting in either a premature stop codon or an inframe amino acid deletion, nonsense mutations, and a splice donor site mutation in intron 4. Mutations have been reported worldwide in different ethnic groups and have resulted both *de novo* and through the founder effect, which is directly related to isolated geographical populations with a high degree of consanguinity. The majority of reported cases are homozygous, and heterozygous and compound heterozygous cases have also been reported.

18.6.3.2 Clinical Manifestations, Diagnosis, and Management

5α-reductase-2 deficiency syndrome is characterized by decreased 5α-DHT synthesis. During crucial stages of male fetal development decreased 5α-DHT synthesis causes diminished virilization of the external genitalia and abnormal formation of the urogenital sinus, which results in hypospadia and the absence or underdevelopment of the prostate. Therefore, at birth, genetic males (46,XY) may have variable phenotypes, ranging from appearing completely female to genital ambiguity or minimally masculinized genitalia to normal male, depending on the type of mutation and its effect on enzyme activity. There is evidence that despite complete loss of function of 5α-reductase-2, some noticeable virilization of the external genitalia occurs as a result of the weak direct effects of testosterone during fetal development [29]. Differentiation of the Wolffian duct structures is not dependent on 5α-DHT. Therefore affected individuals have normal formation of the epididymis, vasa deferens, seminal vesicles, and ejaculatory ducts. AMH production by the testes during fetal development ensures there is no development of an upper vagina, uterus, and fallopian tubes. In contrast to AIS, the testosterone and estrogen produced by the testes are within normal limits and counterbalanced, therefore, there is no breast development during puberty.

At puberty, circulating total testosterone rises dramatically, resulting in virilization. Direct action of the elevated levels of circulating testosterone causes the external genitalia to enlarge to form a penis and scrotum and the testes descend but spermatogenesis is absent or impaired. In addition, some circulating testosterone may be converted to 5α-DHT by the action of 5α-reductase-1, which is expressed in the peripheral tissues such as skin, liver, and specific regions of the brain. There is increased muscular growth and deepening of the voice. However, facial and body hair growth is sparse or absent and there is a typical female pattern of pubic hair growth. In addition, male pattern baldness has never been reported in affected individuals. 5α-reductase-2 deficient males are usually infertile due to the absence or underdevelopment of the prostate. However, male carriers of a single mutant allele have normal fertility [30]. Genetic females (46,XX) deficient in 5α-reductase-2 are minimally affected. Pubic and axillary hair may be reduced and menarche delayed, suggesting 5α-reductase-2 plays a role in hair growth, follicular development, and menarche in female physiology.

Diagnosis of 5α-reductase type-2 deficiency in genetic males can be confirmed by the presence of a high ratio of serum testosterone to 5α-DHT. The reference range of testosterone to 5α-DHT ratio is 8–16:1, while individuals with 5α-reductase type-2 deficiency most often have a ratio greater than 35:1. Urinary metabolites of testosterone and 5α-DHT can also be used to establish the diagnosis in a similar fashion. The difference in pH optimal catalytic activity between 5α-reductase type 1 (pH 8.0) and type 2 (pH 5.5) can be used diagnostically for the differential assessment of isoenzyme conversion of testosterone to 5α-DHT in selected tissues. Imperato-McGinley and Saenger demonstrated that decreased conversion of testosterone to 5α-DHT in the genital skin fibroblast cells establishes that 5α-reductase type-2 is deficient, supporting the diagnosis of 5α-reductase deficiency syndrome [31, 32].

When dealing with intersex disorders, it is important to acknowledge the psychosexual issues of *gender identity*, defined as the awareness of oneself as male or female, *gender role* as learned social behavior and *sexual orientation*, characterized by erotic response to one sex or the other. Often, genetic males affected by 5α-DHT-2-deficiency syndrome born with ambiguous genitalia have been assigned a female gender and raised as females. Thus, matters become even more complicated by the unexpected masculinizing influence of the hormones at puberty, lending validity to the line of reasoning that definitive gender assignment in individuals with intersex disorders should be postponed until after puberty.

18.7 Summary

Androgens are the steroid sex hormones mainly synthesized by the testes that masculinize the accessory male sex organs and secondary sex characteristics. However, females also produce small amounts of androgens, which contribute to overall health. Through complex mechanisms, androgenic effects of the endocrine system are balanced within healthy individuals. There are two major physiological androgens, *testosterone* and *5α-dihydrotestosterone,* and two weaker androgens, which are *dehydroepiandrosterone* and *androstenedione.* Androgens are synthesized from a cholesterol substrate that is converted into pregnenolone, the precursor steroid required for synthesis of all steroid hormones, and the rate-limiting step in testosterone biosynthesis. In peripheral tissues, testosterone is converted into the more potent androgen *5α-dihydrotestosterone* by two different microsomal isoenzymes: 5α-reductase type-1 and type-2. All androgens bind to the same AR, which is a zinc finger nuclear transcription factor involved in activation of several androgen-regulated genes. Expression of the AR gene is ubiquous and found in a wide array of genital and non-genital tissues in both males and females. Thus, the sum of the effects of androgens are varied and tissue specific. Any essential step missed between hormone binding and gene expression results in an aberrant hormone response and can have negligible to mild or severe pathological consequences. Mutations in the AR gene or androgen metabolism can cause a variety of genetic abnormalities, collectively classified as intersex disorders, with major effects in males. In general, anomalies may include ambiguous sexual differentiation involving the external genitalia and internal reproductive organs. There are many phenotypic variations in accordance with the structure and sensitivity of the abnormal AR. As the individual grows and reaches puberty abnormalities of secondary sexual characteristics occur. Molecular biology has greatly expanded understanding of the mechanisms and etiology of the clinical features related to AR and metabolism disorders as well as other intersex disorders. This information has contributed significantly to health care professionals' understanding, leading to revised clinical practice and multidisciplinary approaches to gender identity. Referral for genetic counseling to explain the inheritance of intersex disorders is strongly recommended. It is important to acknowledge the psychosexual issues of *gender identity*, defined as the awareness of oneself as male or female, *gender role* as learned social behavior and *sexual orientation*, characterized by erotic response to one sex or the other. Current best practice recommends that gender assignment should be delayed until gender identity and sexual orientation declare themselves and the affected person can make informed choices with regard to surgery.

18.8 Glossary of Terms and Acronyms

17β-HSD: 17β-hydroxsteroid dehydrogenase

3β-HSD: 3β-hydroxysteroid dehydrogenase/Δ4-Δ5 isomerase

5α-DHT: 5α-dihydrotestosterone

ABP: androgen-binding protein, a glycosylated dimeric protein secreted by the Sertoli cells homologous to steroid hormone-binding globulin

AF-1, -2 and -5: activation function-1, -2 and -5.

AIS: androgen insensitivity syndrome

ALS: amyotrophic lateral sclerosis

Amenorrhea: absence of menstruation

AMH: anti-Müllerian hormone

AR: androgen receptor

AR-A and AR-B: androgen receptor isoforms

ARA: AR associated proteins

ARE: androgen response element

ASI: androgen sensitivity index

Bulbar: describe any bulb-shaped organ of the body

CAIS: complete androgen insensitivity syndrome

CT: computed topography

DBD: DNA-binding domain

DHEA: dehydroepiandrosterone

EEG: electroencephalography

EMG: electromyography

FSH: follicle stimulating hormone

Gynecomastia: abnormal overdevelopment of the male breasts

HSP: heat shock protein

Hypospadia: failure of the distal urethra to develop normally, resulting in a ventral urinary meatus

IU: international units

LBD: ligand-binding domain

LDL: low density lipoproteins

LH: luteinizing hormone

MAIS: mild AIS

MIS: Müllerian-inhibiting substance

Mosaicism: the presence of two populations of cells with different genotypes in one individual originated from a single fertilized egg

MRI: magnetic resonance imaging

P450arom: cytochrome P450 aromatase

P450scc: cytochrome P450 side chain cleavage

PAIS: partial AIS

SBMA: X-linked spinal and bulbar muscular atrophy or Kennedy Disease

SHBG: steroid hormone-binding globulin, a glycosylated dimeric protein homologous to the androgen-binding protein secreted by the Sertoli cells of the testes

SRD5A1: 5α-reductase-1 gene

SRD5A2: 5α-reductase-2 gene

SRY: sex determining region of the Y chromosome

StAR: steroidogenic acute regulatory protein

References

1. Gwynne JT, Strauss JF III. The role of lipoproteins in steroidogenesis and cholesterol metabolism in steroidogenic glands. Endoc Rev 1982; 3:299–329.
2. Stocco DM, Clark BJ. Regulation of the acute production of steroids in steroidogenic cells. Endocr Rev 1966; 17: 221–44.
3. Braunstein GD. Testes. In: Greenspan FS, Gardner DG, editors. Basic & Clinical Endocrinology, sixth edition. New York: McGraw-Hill, 2001:422–52.
4. Mooradian AD, Morley JE, Korenman SG. Biological actions of androgens. Endocr Rev 1987; 8:1–28.
5. Davison Sl, Bell R. Androgen physiology. Semin Reprod Med 2006; 24:71–7.
6. Sinisi AA, Pasquali D, Notaro A, et al. Sexual differentiation. Endocrinol Invest 2003; 26:23–8.
7. Trapman J, Klaassen P, Kuiper GG. Cloning, structure and expression of a cDNA encoding the human androgen receptor. Biochem Biophys Res Commun 1988; 153:241–8.
8. Wilson CM. McPhaul MJ. A and B forms of the androgen receptor are present in human genital skin fibroblasts. Proc Natl Acad Sci USA 1994; 91:1234–8.
9. Jenster G, Van der Korput HA, Trapman J. Identification of two transcription activation units in the N-terminal domain of the human androgen receptor. J Biol Chem 1995; 270: 7341–6.
10. Klokk TI, Kurys P, Elbi C. Ligand-specific dynamics of the androgen receptor at its response element in living cells. Mol Cell Biol 2007; 27:1823–43.
11. Schaufele F, Carbonell X, Guerbadot M. The structural basis of androgen receptor activation: intramolecular and intermolecular amino-carboxy interactions. Proc Natl Acad Sci U S A 2005; 102:9802–7.
12. Berrevoets CA, Doesburg P, Steketee K. Functional interactions of the AF-2 activation domain core region of the human androgen receptor with the amino-terminal domain and with the transcriptional coactivator TIF2 (transcriptional intermediary factor2). Mol Endocrinol 1998; 12:1172–83.
13. Dubbink HJ, Hersmus R, Verma CS. Distinct recognition modes of FXXLF and LXXLL motifs by the androgen receptor. Mol Endocrinol 2004; 18:2132–50.
14. Gottlieb B, Beitel LK, Wu JH, et al. The androgen receptor gene mutations database (ARDB): 2004 update. Hum Mutat 2004; 23:527–33.
15. Evans BAJ, Ismail RA, France T, et al. Analysis of the androgen receptor gene structure in a patient with complete androgen insensitivity syndrome. J Endocrinology 1991; 129: Abstr 65.
16. Yong EL, Chua KL, Yang M, et al. Complete androgen insensitivity due to a splice-site mutation in the androgen receptor gene and genetic screening with single-stranded conformation polymorphism. Fertil Steril 1994; 61:856–62.
17. Ris-Stalpers C, Kuiper GG, Faber PW, et al. Aberrant splicing of androgen receptor mRNA results in synthesis of a nonfunctional receptor protein in a patient with androgen insensitivity. Proc Natl Acad Sci U S A 1990; 87:7866–70.
18. Kohler B, Lumbroso S, Leger J, et al. Androgen insensitivity syndrome: somatic mosaicism of the androgen receptor in seven families and consequences for sex assignment and genetic counselling. J Clin Endocrinol Metab 2005; 90:106–11.
19. Kuiper GG, Faber PW, van Rooij HC, et al. Structural organization of the human androgen receptor gene. J Mol Endocrinol 1989; 2:R1–4.
20. Lubahn DB, Brown TR, Simental JA, et al. Sequence of the intron/exon junctions of the coding region of the human androgen receptor gene and identification of a point mutation in a family with complete androgen insensitivity [published erratum appears in Proc Natl Acad Sci U S A 1990; 87:4411]. Proc Natl Acad Sci U S A 1989; 86:9534–8.
21. Holterhus PM, Brüggenwirth HT, Hiort O, et al. Mosaicism due to a somatic mutation of the androgen receptor gene determines phenotype in androgen insensitivity syndrome. J Clin Endocrinol Metab 1997; 82:3584–9.
22. Adachi M, Takayanagi R, Tomura A, et al. Androgen-insensitivity syndrome as a possible coactivator disease. New Eng J Med 2000; 343:856–62.
23. Yanase T, Adachi M, Goto K, et al. Coregulator-related diseases. Intern Med 2004; 43:368–73.
24. Kennedy WR, Alter M, Sung JH. Progressive proximal spinal and bulbar muscular atrophy of late onset: a sex-linked recessive trait. Neurology 1968; 18:671–80.
25. Fischbeck KH, Ionasescu V, Ritter A, et al. Localization of the gene for X-linked spinal muscular atrophy. Neurology 1986; 36:1595–8.
26. Dejager S, Bry-Gauillard H, Bruckert E. A comprehensive endocrine description of Kennedy's disease revealing androgen insensitivity linked to CAG repeat length. J Clin Endocrinol Metab 2002; 87:3893–901.
27. Imperato-McGinley J, Miller M, Wilson JD, et al. A cluster of male pseudohermaphroditer with 5α-reductase deficiency in Papua New Guinea. Clin Endocrinol 1991; 34:293–8.
28. Katz MD, Cai L-Q, Zhu Y-S, et al. The biochemical and phenotypic characterization of female homozygous for 5α-reductase 2 deficiency. J Clin Endocrinol Metab 1995; 80:3160–7.
29. Hiort, Olaf, Schütt, et al. A novel homozygous disruptive mutation in the SRD5A2-gene in a partially virilized patient with 5α-reductase deficiency. Int J Androl 2002; 25:55.
30. Brinkmann AO. Androgen physiology: Receptor and metabolic disorders. In: McLachlan R, editor. Endocrinology of Male Reproduction, [Internet] Endotext.com, 2006, Chapter 3.
31. Imperato-McGinley J, Guerrero L, Gautier T, et al. Steroid 5α-reductase deficiency in man: an inherited form of male pseudohermaphrodism. Science 1974; 186:1213–5.
32. Saenger P, Goldman AS, Levine LS, et al. Prepubertal diagnosis of steroid 5α-reductase deficiency. J Clin Endocrinol Metab 1978; 46:627–34.

Chapter 19

Leptin as a Reproductive Hormone

Michael C. Henson and V. Daniel Castracane

19.1 Introduction

Leptin is an adipokine (i.e., adipocytokine). Adipokines are adipocyte-derived secretory proteins that are linked to physiological mechanisms that influence energy metabolism, inflammation, and an array of pathological conditions that range from infectious disease to diabetes and from preeclampsia to cancer. After only a few years of research on these versatile physiological mediators, their obvious importance served to elevate adipose tissue from its previously accepted role as a simple storage compartment for lipids to that of a powerful endocrine organ [1, 2]. Adipokines, therefore, are now recognized as molecular agents of communication connecting adipose tissue, brain, vasculature, liver, pancreas, muscle, the immune system, and many of the specialized reproductive tissues and organs in both males and females [3].

Although the implications of the adipokines as regulators of growth, development, and reproduction in animals of agricultural significance have been recognized [4], it is the universal acknowledgment of obesity as a human health issue affecting both adults and children in unprecedented numbers that drives current research efforts worldwide [5, 6, 7]. To this end, a better understanding of the regulation of energy balance by these proteins via neural and peripheral pathways is needed to combat the ravages of the metabolic diseases related to enhanced adiposity [8, 9, 10]. Roles for the adipokines are now known to transcend the mechanisms that link accumulated fat stores with satiety and pathological conditions with obesity to also include the basic pathways that signal the hypothalamic-pituitary-gonadal axis to release gonadotropins and regulate spermatogenesis, menstrual cyclicity, and resultant fertility [11, 12]. These regulatory influences extend to life in utero, where adipokines have been recognized as important contributors to the endocrine milieu that orchestrates successful human conceptus development [13] and perhaps, the hormonal interplay of the maternal-fetoplacental unit itself [14–18].

Leptin, the prototypical adipokine, was discovered and first characterized in 1994 [19] and because of its obvious importance as a regulator of metabolism via its physiological roles in obesity, energy expenditure, and diabetes, it rapidly became one of the most intensely studied new hormonal proteins. The *ob/ob* (now Lep^{ob}/Lep^{ob}) mouse is characterized by hyperphagia, obesity, transient hyperglycemia, and elevated plasma insulin concentrations that are associated with an increase in the number and size of pancreatic beta cells. Following the discovery of leptin, some of the first experiments were to treat the Lep^{ob}/Lep^{ob} mouse with the adipokine. Administration resulted in dosage- and time-dependent diminutions of body weight, body fat, food intake, and serum glucose. The effective reversal of such adverse signs of adiposity suggested roles for leptin in preventing obesity and controlling insulin resistance and diabetes and led to the hope that it might be a useful pharmaceutical tool in the medical management of these conditions.

19.2 Leptin as a Neuroendocrine Regulator of Reproduction

Early studies tested the hypothesis that leptin is involved in gonadotropin regulation. McCann and colleagues [20] incubated hemi-anterior pituitaries of adult male rats with increasing concentrations of leptin for 3 hours and observed a dosage-dependent stimulation of follicle stimulating hormone (FSH) and luteinizing hormone (LH) release. Prolactin secretion also increased dosage-dependently, but only at higher leptin concentrations. They also examined the action of leptin on median eminence-arcuate nucleus transplants and found a stimulation of gonadotropin-releasing hormone (GnRH) release at the lowest leptin concentrations and a

M.C. Henson (✉)
Department of Obstetrics and Gynecology, Texas Tech, University Health Sciences Centers, Odessa, TX, USA
e-mail: henson@calumet.purdue.edu

P.J. Chedrese (ed.), *Reproductive Endocrinology*,
DOI 10.1007/978-0-387-88186-7_19, © Springer Science+Business Media, LLC 2009

suppression of GnRH release at higher leptin levels. When leptin was administered through a third ventricle cannula 72 hours after estrogen treatment, the result was a significant increase in plasma LH. Fasting results in declines in serum LH and testosterone concentrations in several species [21–24], with fasting for only 24–48 hours causing a decrease in serum leptin levels [25, 26]. Finn and coworkers [27] demonstrated the cessation of LH pulses in monkeys within 48 hours of fasting and determined that 2 days of leptin infusion would restore LH pulses.

As described by Ahima and Lazar [9], the leptin receptor (LR) is a member of the class 1 cytokine receptor family and is manifested as five alternatively spliced isoforms (LRa, LRb, LRc, LRd, and LRe). LRa, the main "short" LR isoform, lacks the cytoplasmic domain needed for signaling via the Janus tyrosine kinases—signal transduction and activators of transcription (JAK/STAT). It is found throughout the brain, most notably in vascular endothelium and in peripheral tissues, and may be involved in leptin transport and signaling via the mitogen-activated protein kinase (MAPK) and extracellular-signal-regulated kinase (ERK) pathways. The "long" LR isoform (LRb) is most commonly associated with the hypothalamic satiety and neuroendocrine centers and mediates intracellular signaling. As depicted in Fig. 19.1 [9], LRb-bound leptin associates with JAK2, resulting in its autophosphorylation, the phosphorylation of LRb tyrosine residues 985 and 1,138, and STAT3 activation. STAT3 is then translocated to the nucleus and brings about neuropeptide transcription. Phosphorylated Tyr985 of LRb binds Src homology 2 (SH2)-containing tyrosine phosphatase-2, which in turn activates ERK. Tyr985 then binds the suppressor of cytokine signaling (SOCS) 3, which ends LRb signaling. Leptin also stimulates the phosphorylation of Tyr1077 on LRb and activates STAT5 and ribosomal protein S6 kinase. Tyr1138 supports STAT5 phosphorylation and serves with STAT3 to attenuate STAT5-dependent transcription. Leptin also up-regulates protein-tyrosine phosphatase 1B activity and inactivates JAK2 and leptin signaling [9, 28].

With respect to leptin signaling in the hypothalamus, LRb functions in the brain to modulate energy balance, satiety, and glucose metabolism. As illustrated in Fig. 19.2 [9], leptin inhibits hypothalamic neurons in the arcuate nucleus that express neuropeptide Y (NPY) and agouti-related protein (AGRP), but induces proopiomelanocortin (POMC) and cocaine- and amphetamine-regulated transcript (CART). These neurons extend from the arcuate to the paraventricular nucleus and perifornical, dorsomedial, and lateral hypothalamic areas to inhibit feeding, stimulate thermogenesis, and enhance lipid oxidation and insulin sensitivity in somatic tissues via thyrotropin-releasing hormone (TRH) and corticotropin-releasing hormone (CRH). Leptin also indirectly controls expression of melanin-concentrating-hormone (MCH) and orexins (ORX) in the lateral hypothalamus, as well as in mesolimbic dopaminergic circuits. A deficiency of leptin, LRb, and STAT3 in POMC neurons stimulates hyperphagia and impairs thermogenesis, the combination of which can result in morbid obesity.

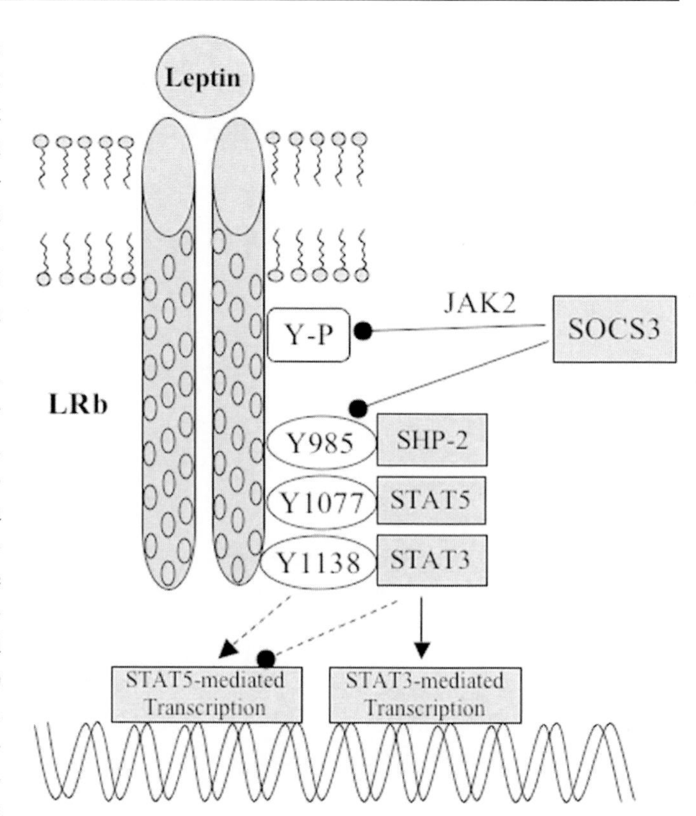

Fig. 19.1 Intracellular signaling pathways regulated by LRb. Information summarized from R.S Ahima and M.A. Lazar [9]

As described by Hill and colleagues [12], leptin is uniquely positioned to signal changes in energy status to the hypothalamic-pituitary-gonadal axis in response to metabolic indicators. This intimate association between energy and reproductive success has often been demonstrated in humans, with anorexia, cachexia, and excessive exercise interrupting reproductive cyclicity and the secretion of gonadal steroids. Related mechanisms impacting obesity and diabetes can also result in decreased fertility. Although the mechanisms that regulate these relationships remain somewhat unclear, GnRH neurons in the preoptic area control pituitary LH release via the pulsatile release of GnRH from terminals in the median eminence into the portal vasculature. The GnRH pulse generator is exquisitely sensitive to energetic stress and is easily inhibited by food restriction and extremes of ambient temperature or exercise, with pulsatility quickly returning when a forced insult to energy balance ends.

As proposed in Fig. 19.3, it is intriguing with respect to the range of mechanisms linking energy balance, adiposity, and diabetes that coordinated leptin and insulin signaling

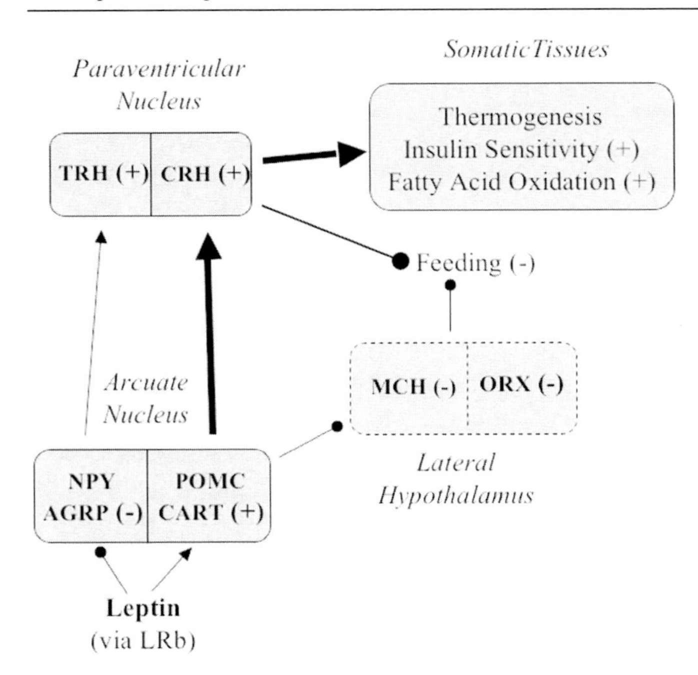

Fig. 19.2 Leptin signaling in the hypothalamus. Information summarized from R.S. Ahima and M.A. Lazar [9]

in POMC and NPY cells could contribute to joint regulation of GnRH release [12]. Further, recent work describing G-protein-coupled receptor-54 (GPR54) and the cognate ligands kisspeptins [29–31] indicates that KiSS-1 is expressed in hypothalamic nuclei, including the anteroventral periventricular (AVPV) and the arcuate nuclei, with administration of kisspeptin stimulating LH and FSH release in mice [32]. Kisspeptin may act directly on GnRH release as influenced by steroid hormones, mediating the negative feedback of steroids on GnRH secretion via neurons in the arcuate nucleus and the positive feedback of sex steroids via AVPV neurons [33]. Although KiSS-1 and LR are both expressed in the arcuate nucleus, the specific hypothalamic sites where leptin acts to stimulate KiSS-1 in this scheme are unknown. However, results suggest that leptin and perhaps other adipokines stimulate KiSS-1 expression, trigger kisspeptin synthesis, and stimulate GnRH release. It has been hypothesized that infertility previously associated with leptin deficiencies could be linked to diminished expression of KiSS-1 and/or KISS-1 receptor [12].

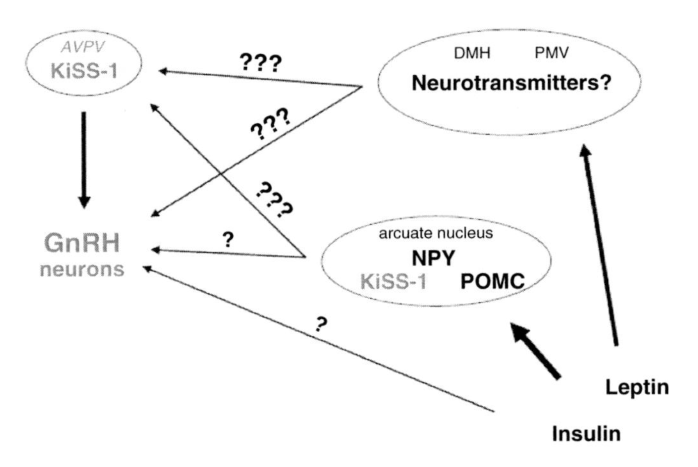

Fig. 19.3 Schematic representation of potential brain pathways mediating leptin and insulin actions in reproduction. Leptin and insulin receptors are distributed in a variety of brain nuclei, but the connections with areas related to reproductive control including the anteroventral periventricular (AVPV) and GnRH neurons are unknown. Moreover, although the arcuate nucleus neurons expressing POMC or NPY have been investigated, their projections to the AVPV are as yet uncharacterized and the innervation of GnRH neurons remains unclear. Reproduced with permission from the American Journal of Physiology [12]

19.3 Leptin as a Direct Regulator of Reproduction

The original classification of leptin as an adipokine was based on its origin in white adipose tissue, but it soon became clear that the polypeptide was expressed, often with its receptors, in many other tissues as well. Therefore, leptin's links to reproduction became evident as studies demonstrated direct roles in fertility, pubertal development, and gender-specific regulatory differences [34]. As discussed previously, leptin was originally characterized in the rodent as the product of the *ob* (now *Lep*) gene and was not present in the Lep^{ob}/Lep^{ob} obese mouse. It had long been known that while food restriction would restore normal body weight in obese females, it would not restore fertility. However, leptin administration served to accomplish both goals [35].

19.3.1 Steroid Interactions

A positive correlation exists between serum leptin levels with body fat mass [36] and although females generally have a greater percentage body fat than males of similar body weight, or BMI (body mass index), leptin levels are greater in females over males of equal fat mass [37]. Although gender-specific differences in relative amounts of subcutaneous versus visceral adipose tissues may be important, other factors include inherent differences in the endogenous hormonal milieu, with estrogens increasing leptin production [38] and androgens suppressing it [39, 40]. Ovariectomy diminished leptin gene expression in white adipose tissue and caused a decline in serum leptin levels in rats [41–43], while administration of estradiol (E_2) reversed all the effects of ovariectomy. Ovariectomy also reduced serum leptin levels in humans [44]. Although leptin and E_2 demonstrate similar profiles during the human menstrual cycle, leptin levels were unaffected by the relatively small increases in estrogen associated with normal menstrual cyclicity, but were upregulated by the large increases that typically result from ovulation induction, effects that may identify estrogen as a dose-dependent regulator [45–48]. Because commensurate administration of E_2 and progesterone to normally cycling women resulted in increased serum leptin concentrations, cooperative mechanisms mediated by the two steroids might also be implied during the luteal phase of the menstrual cycle [49]. However, in late pregnancy when placental progesterone production is at its height, progesterone has been reported to inhibit leptin secretion by human placental cells in culture [50].

Numerous studies have demonstrated the gender difference in neonatal samples (cord blood) with female leptin levels generally greater than male neonates [51–54]. Generally, E_2 and testosterone levels were not different between males and females in term deliveries [53, 54] and an apparent decline in leptin levels was evident by the fifth postnatal day [55]. In all cases, fetal leptin concentrations correlated with fetal weight. Gender differences between boys and girls between 3 and 90 days of life were not observed, despite increasing leptin levels during this time period [56]. During pubertal development, leptin levels in boys and girls are similar until activation of gonadal steroidogenesis. In girls, leptin levels increase with the initial increases in gonadal estrogen production, but conversely, as the testes become active in testosterone production, a clear decline in serum leptin is evident [39, 57]. In normal adults, the classic gender differences described by many investigators are present and are clearly correlated with indices of body fat [45, 58, 59].

19.3.2 Puberty

With the first availability of the leptin protein to investigators, several studies demonstrated that leptin administration to prepubertal mice would shorten the time to puberty [60, 61]. Furthermore, in the leptin-lacking Lep^{ob}/Lep^{ob} mouse that never enters puberty, administration of the polypeptide resulted in a normal pubertal process [62]. In the rat, mild food restriction was found to result in delayed sexual maturation, which could be overcome with leptin administration. When the food restriction was more severe, leptin was not able to overcome this effect. Cheung [63] concluded that leptin has a permissive role, but is not the major metabolic factor that initiates pubertal development in the rat. Leptin administered into the cerebral ventricle of the rat initiated early puberty onset, indicating an action for leptin in the central nervous system [64]. In swine and sheep, an increase in serum leptin occurs prior to pubertal development [65, 66], and administered leptin has even been reported to advance puberty in the domestic hen [67].

Similarly, human serum leptin concentrations rise during the years preceding pubertal development and reach high levels around the time of puberty onset. In females, the levels continue to rise, while in males they decline, probably related to the testosterone effect [57]. Conversely, the role of leptin is also seen in other conditions, for example, an earlier age for puberty has been observed in obese girls, presumably related to the higher leptin levels in this group [68]. Young girls in rigorous gymnastic or ballet programs may evidence delayed pubertal development and serum leptin levels in both of these groups may be decreased [69]. Perhaps the most revealing clinical situation concerning the role of leptin in pubertal development is seen in those children with the rare mutation that results in the absence of leptin. In these children, no sign of pubertal development is typically observed, but when treatment with leptin is initiated the first endocrine changes characteristic of puberty are noted [62]. In the rhesus monkey, several studies report no observable increase in serum leptin prior to the initiation of puberty and suggest that leptin may not be involved in the pubertal process in this species [70–72]. This remains controversial [73, 74], however, and whether or not leptin plays a direct role in rhesus pubertal development, adequate serum levels of leptin may still play a permissive role and cannot be discounted.

19.3.3 Menstrual Cycle

The action of leptin on steroidogenesis is well documented and effects of steroid hormones on leptin synthesis have been described. These are part of the intricate interactions by which leptin exerts its influence on the menstrual cycle. When samples were obtained every 1–2 days throughout the menstrual cycle, plasma leptin was seen to increase from the early follicular phase to peak at the mid-luteal phase, returning to baseline by menses [75]. Despite minor differences, the luteal phase, either mid or late, is reported to be the period of peak serum leptin concentrations, with some studies also noting an increase during the late follicular phase. Several reports have demonstrated the regulation of leptin production and its relationship to progesterone and may account for the generally reported increase in serum leptin during the luteal phase. In one such study, following ovariectomy, women received no treatment, treatment with E_2, or treatment with E_2 plus progesterone. In both untreated and E_2-treated women, a decline in serum leptin concentrations over 4 days was observed, but following treatment with E_2 plus progesterone, leptin levels were significantly increased [76]. In a similar study, cycling women were either untreated or treated with E_2 or E_2 plus progesterone during the early follicular phase. In the untreated and E_2-treated women there was no increase in serum leptin, but those that received E_2

plus progesterone demonstrated enhanced leptin concentrations over the 3 days of treatment, with levels declining after cessation of treatment [77]. This relationship might account for the previously noted luteal phase increase in serum leptin levels and probably represents the action of ovarian steroids on production of the polypeptide in adipose tissue. The increase seen in the late follicular phase of the menstrual cycle may be due to E_2 increases at that phase of the cycle, since there are multiple reports that suggest or demonstrate a stimulatory effect of estrogens on serum leptin levels [37, 38]. The small increase in progesterone, which begins in the late follicular phase, may also contribute to this effect.

19.3.4 Female Infertility

Obesity has a deleterious effect on fertility, affecting many parameters that range from poor ovulation to the cessation of menstrual cycles, and ultimately the development of polycystic ovarian syndrome (PCOS). It has been reported that obese hyperandrogenic, amenorrheic women were less likely to ovulate after clomiphene citrate (CC) medication, which is the first line of treatment for these anovulatory patients. Peripheral leptin levels in these obese women are markedly elevated and leptin concentrations are generally more reliable than BMI or waist to hip ratio to predict which of those patients will remain anovulatory after CC medication. These investigators suggested that leptin is more involved in ovarian dysfunction in these patients than are other endocrine events and that this may be a direct action of leptin on ovarian dysfunction [78]. Several studies have demonstrated that leptin levels are influenced by the functional ovarian state and that obesity-related high leptin levels are associated with reduced ovarian response. The increase in serum leptin following FSH treatment of infertile women may serve a diagnostic role for pregnancy rate, with higher leptin levels being less successful [79–81].

The use of leptin in conditions of low levels of this hormone has been reported to have beneficial effects. A group of women with hypothalamic amenorrhea and low leptin levels were studied by Welt And Colleagues [82] for 1 control month before receiving recombinant human leptin for up to 3 months. A separate control group received no treatment. Controls were essentially unchanged over the course of the study. Conversely, recombinant human leptin increased LH levels and LH pulse frequency after 2 weeks, with an increase in maximum follicular diameter and the number of dominant follicles, ovarian volume, and E_2 levels over 3 months. Ovulatory menstrual cycles were reported in three subjects and two other women had preovulatory follicular development with withdrawal bleeding during treatment. These studies suggest that leptin may be required for normal reproductive and neuroendocrine function and may be of clinical utility in the treatment of leptin deficiency in women with hypothalamic amenorrhea. These studies indicate an association of leptin with the normal physiology of the menstrual cycle and a potential role for the polypeptide in the treatment of women undergoing controlled ovarian hyperstimulation.

19.3.5 Pregnancy and the Maternal-Fetoplacental Unit

The physiological roles and regulatory mechanisms for leptin during mammalian pregnancy have been the subjects of intensive study since the initial determination of its synthesis [83] and the later discernment of its ontogeny [84] in human placental trophoblast [14, 15, 16, 17, 85, 86, 87]. Serum leptin concentrations, which are enhanced in the maternal circulation over those in the nonpregnant state, suggest physiological roles during gestation apart from the previously defined responsibility for appetite suppression. As reviewed by Haguel-de Mouzon and coworkers [17], the increased mobilization of maternal fat stores to allow enhanced transplacental transfer of lipid substrates to the developing fetus are one such proposed function. This role may be reflected by delayed clearance of the polypeptide from the maternal circulation due to the enhanced level of soluble leptin receptor in the bloodstream that results from placental membrane shedding throughout pregnancy. The existence of syncytiotrophoblastic leptin receptors suggests pregnancy-specific autocrine and paracrine roles for the adipokine at the maternal-fetal interface. This is an important consideration in itself, as Lappas and colleagues [88] reported that in human adipose tissue and placental tissues explants, exogenous leptin increased the secretion of tumor necrosis factor (TNF)-α, interleukin (IL)-1β, IL-1β, and IL-6, as had already been reported for human chorionic gonadotropin (hCG) in cytotrophoblasts [89]. Recent data has brought this regulatory pathway full circle by indicating that hCG exerts a proadipogenic effect in human preadipocytes by simulating leptin secretion [90]. Most significantly, the regulation of proinflammatory cytokines produced in the intrauterine environment may suggest roles for leptin in modulating myometrial contractility, cervical ripening, and the rupture of conceptus membranes [88], all of which significantly impact the processes associated with parturition. Coya [91] reported that leptin led to a dosage-dependent decline in E_2 release by human term placental cells in culture. Taken collectively along with a synchronous increase in testosterone release, this suggested an important role for the adipokine in regulating trophoblastic steroidogenesis in late pregnancy. Indeed, leptin may be important for placental maintenance itself, as leptin introduced into choriocarcinoma cell cultures (a model for trophoblastic development) proved to be a significant trophic and mitogenic factor by inhibiting apoptosis and promoting cell proliferation in a dosage- and time-dependent manner [92].

As illustrated in Fig. 19.4, although nomenclature still varies in the literature, "long" and "short" intracellular domain LR isoforms are generated by alternative splicing and signal via JAK/STAT (made possible by the BOX2 motif) and MAPK pathways, respectively, in most tissues. A soluble leptin receptor (solLR) has been characterized in rodents that is also formed via alternative splicing [93], which in humans is probably cleaved from the membrane by a metalloproteinase [94, 95]. The physiological significance of the solLR in pregnancy is great, as its ability to modulate leptin action [96] and prevent passage of the blood-brain barrier [97] might be critical in enhancing resistance to the satiety inducing effect of the adipokine during pregnancy, a period of increased energy demand. It is quite interesting to note the suggestion of recent work that this mechanism could be augmented by direct endocrine influences at the hypothalamic level brought about by high concentrations of placental lactogen, albeit by a yet uncharacterized mechanism [98].

To better understand leptin regulation and function in human pregnancy, we have characterized leptin dynamics in old- and new world nonhuman primates [99] and found the baboon (*Papio* sp.) to be an excellent model [100]. Leptin concentrations in pregnant animals (term ~184 days) are much higher than in either cycling or postpartum baboons [101]. As in humans, leptin mRNA transcripts in placental villous tissue decline between early and late gestation, but maternal serum leptin levels increase almost threefold with pregnancy and are correlated with advancing gestational age. Because the presence of both leptin and its receptor in the placenta, amnion, chorion, and umbilical vasculature suggest important roles in human pregnancy, we assessed these

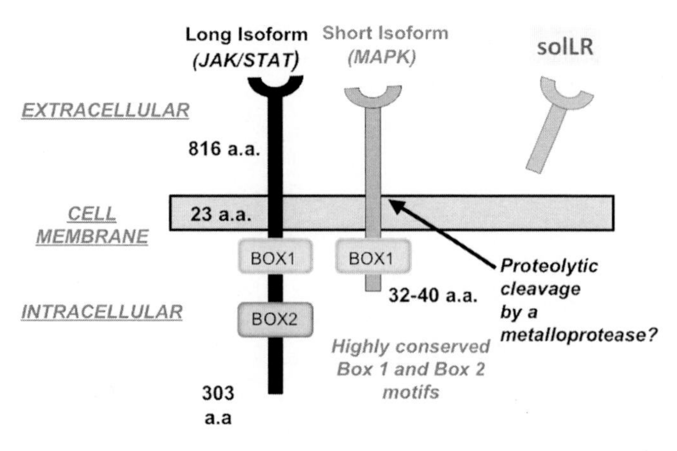

Fig. 19.4 Leptin receptor isoforms. References to "long" and "short" isoforms are derived from the relative lengths of their intracellular signaling domains (a.a., amino acids; solLR, soluble leptin receptor)

tissues, as well as omental and subcutaneous fat at early (day 60), mid (day 100), and late (day 160) baboon pregnancy [102]. A resurgent corpus luteum, decidual tissue, and fetal brain (hypothalamic region) tissues were also collected on day 160. Expression of long and short LR ($LEPR_L$ and $LEPR_S$) mRNA transcripts was detected in all tissues, utilizing human LR primers. As in humans [84], in situ hybridization localized transcripts for leptin and both receptor isoforms in baboon trophoblast. Expression intensity for leptin was greatest in early pregnancy, reflecting the enhanced abundance of *LEP* transcripts at that time [101]. In the human, receptor isoforms are similar in size to those identified in the baboon, with results suggesting a role for enhanced LR concentrations in regulating leptin bioavailability [103].

We have hypothesized that elevated maternal leptin levels may be owed to enhanced transcriptional regulation in maternal adipose tissue and/or placenta, resulting from enhanced estrogen levels during pregnancy. Estrogens may regulate leptin expression by acting on a portion of the estrogen response element in the leptin promoter [104] with leptin production by cultured first trimester human cytotrophoblast cells being dose-responsively potentiated by E_2 [18]. The presence of estrogen receptor (ER) in primate trophoblast [105] suggests that, as in adipose tissue [106], this is an ER-mediated phenomenon. Like the human, the baboon possesses a true maternal-fetoplacental unit, which relies on androgen precursors from the fetal adrenal gland for placental estrogen synthesis [100]. Thus, the surgical removal of the fetus, but not the placenta (fetectomy), at day 100 of gestation inhibits estrogen production by the syncytiotrophoblast and reduces maternal serum estrogen levels to near baseline. We collected placental villous tissue, omental adipose tissue, and subcutaneous adipose tissue from baboons in late (day 160) pregnancy [107]. In another group of pregnant baboons, estrogen production was inhibited by fetectomy and placentae were left in situ until day 160 of gestation when they were surgically retrieved. Maternal adipose tissues were collected on days 100 and 160. Fetectomy elicited an 87% decrease in maternal serum E_2 concentrations, while leptin levels were unaltered by fetectomy. However, in subcutaneous fat *LEP* mRNA transcript abundance declined fivefold as a consequence of fetectomy, while transcripts increased almost threefold in placental villous tissue. In subcutaneous fat, leptin protein levels in fetectomized baboons were about one-half that of controls, while placental levels were threefold higher in fetectomized animals than in those with intact pregnancies. Therefore, although adipose leptin expression declined, increased placental expression suggested a compensatory, tissue-specific regulatory role for estrogen (stimulatory in adipose tissue, inhibitory in placenta).

Many physiological roles have been suggested for leptin in human pregnancy [14, 15, 16, 17, 86]. Although no direct correlation exists between maternal serum leptin concentrations and fetal size at term, assumptions that the polypeptide acts to regulate conceptus growth has been based on often observed correlations between fetal mass and umbilical levels of the cytokine. These observations, usually noted with respect to cases of either large- or small-for-gestational age infants, have been difficult to distinguish as being causative, rather than as being the simple consequence of differences in fetal adiposity. As reviewed [15, 17], in addition to being an index of fetal size, the expression of LR in a variety of fetal tissues suggests a regulatory involvement of leptin in the stimulation of uterine implantation, and fetal erythropoiesis, vasculogenesis, osteogenesis, and lymphopoiesis. The adipokine may also be associated with pulmonary development in utero. In this capacity, insufficient maturation of the fetal lungs can be characterized by inadequate production of pulmonary surfactant by epithelial type II cells. Increasing cortisol at term prompts the differentiation of type II cells and surfactant synthesis, although in preterm infants, insufficient surfactant levels lead to pulmonary deficiency. Torday and colleagues [108] observed that leptin was expressed by fibroblasts and that LR was expressed by fetal rat lung type II cells. Later experiments indicated that leptin plays a direct role in enhancing surfactant production in this species [109].

In this capacity, we reported [110] that in late baboon pregnancy the abundance of $LEPR_L$ mRNA transcripts in fetal lung was approximately tenfold greater and $LEPR_S$ transcript abundance was about eightfold greater than in early pregnancy (Fig. 19.5) [110]. LR protein, undetectable in fetal lungs at early and mid gestation, was detected by immunoblotting in late gestation and localized immunohistochemically in distal pulmonary epithelial cells, including type II cells (Fig. 19.6) [110]. Therefore, because fetal serum leptin concentrations are significant and up-regulation of LR occurs in late gestation, we proposed that leptin could contribute to fetal lung maturation in primates. Interestingly, later work by Kirwin and colleagues [111] demonstrated an increase in surfactant proteins in fetal rat type II alveolar cells as a result of leptin administration both in vivo and in cultured cells, strongly suggesting the adipokine to be a modulator of fetal lung maturity in that species as well. Indeed, the ability for leptin to regulate pulmonary function may also extend to the postnatal period, as it has recently been reported that alveolar size and surface area were enhanced in Lep^{ob}/Lep^{ob} mice administered leptin through 7 weeks of age [112].

Leptin has also been associated with significant gestational pathologies, including diabetes and preeclampsia. In cases of pregnancy-associated diabetes, maternal leptin levels are positively correlated with BMI as in nonpregnant women, although there is a substantial increase in fetoplacental leptin levels [113–115]. Enhanced umbilical leptin levels suggest that this is exacerbated in cases of macrosomia (birth weight $\geq$ 4,000 g) associated with diabetes mellitus [115], although conflicting evidence exists [116]. Hypertension is the hallmark of preeclampsia, as is the potential for placental insufficiency, and maternal serum leptin concentrations are typically elevated in documented cases, or even before clinical signs appear [117]. Enhanced maternal leptin levels originate in the placental trophoblast as a result of increased *LEP* mRNA expression [118], with the increased rate of synthesis probably being driven by hypoxia [119]. Although leptin has been implicated by numerous studies as a regulatory or signaling molecule linking preeclampsia with intrauterine growth restriction (IUGR), a precise role remains undefined and pathways still remain to be elucidated [14, 15, 16, 86].

19.3.6 Male Reproduction

Perhaps of greater significance are the leptin-mediated events within the testes [120]. While all splice variants of the leptin receptor were recognized, the LRb isoform was highest in

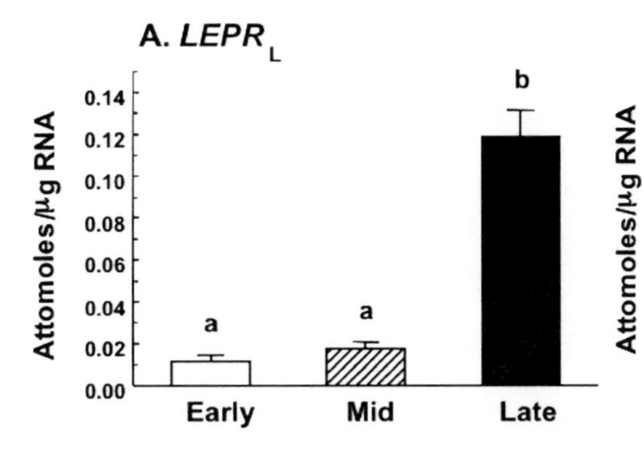

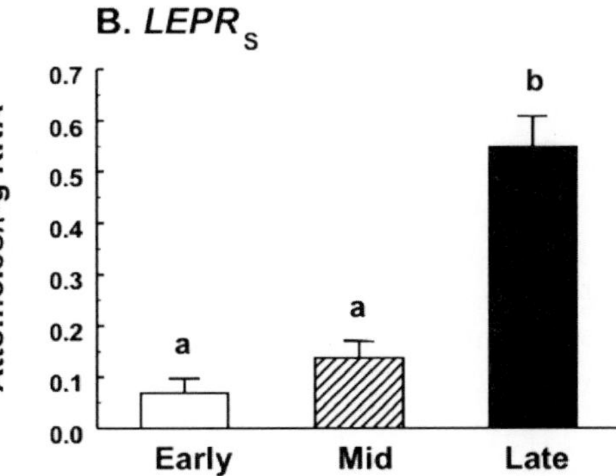

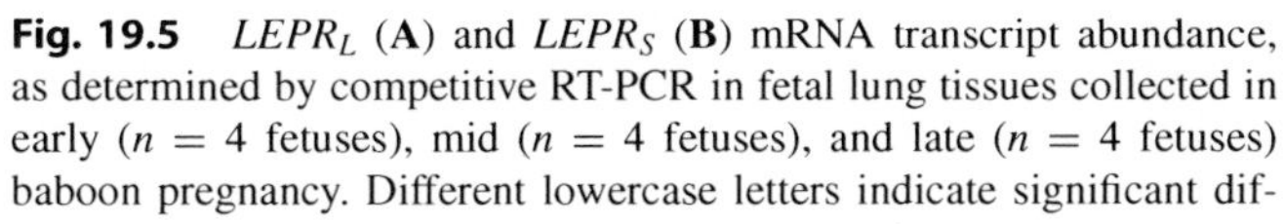

Fig. 19.5 $LEPR_L$ (**A**) and $LEPR_S$ (**B**) mRNA transcript abundance, as determined by competitive RT-PCR in fetal lung tissues collected in early ($n = 4$ fetuses), mid ($n = 4$ fetuses), and late ($n = 4$ fetuses) baboon pregnancy. Different lowercase letters indicate significant differences between means ± SEM (ab, $P < 0.01$). (**C**) Society for Reproduction and Fertility (2004). Reproduced with permission from Reproduction [110]

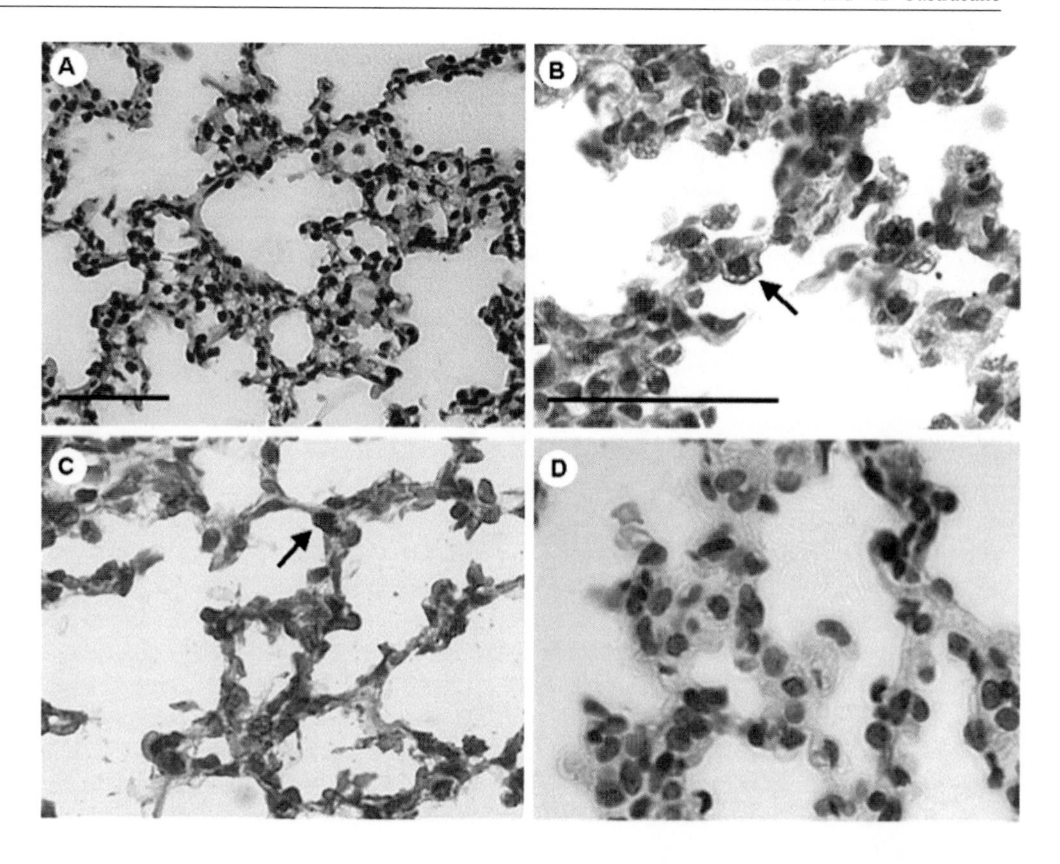

Fig. 19.6 Photomicrographs of lung tissue from fetal baboons in late gestation depicting the results of (**A**) hematoxylin-eosin staining (200X), and immunohistochemical localization of (**B**) surfactant protein A (400X), or (**C**) LEPR (400X) protein in pulmonary epithelial cells. A (**D**) negative (without primary antibody) immunohistochemical control for LEPR (400X) is included. Reference lines in panels A (200X) and B (400X) are 50 μm. *Arrows* denote pulmonary type II cells. (C) Society for Reproduction and Fertility (2004). Reproduced with permission from Reproduction [110]

pubertal testes (15–30 day old rats) and declined in adulthood. Testicular Lepr mRNA expression was sensitive to neonatal endocrine influence, since neonatal treatment with E_2 benzoate (E_2B) resulted in a permanent increase in the relative expression of *Lepr* mRNA. E_2B treatment had a differential effect on the different isoforms of the leptin receptor. These studies were the first to indicate a direct role for leptin in testicular regulation in the rat. Caprio [55] reported a different pattern of leptin receptor in the rat testes. Using an immunohistochemical approach, they demonstrated that Lepr is absent in early embryonic stages (14.5 days) and only appears in late embryonic testes (19.5 days). In postnatal life, leptin receptor immunoreactivity was only evident after sexual maturation (after 35 days) and was absent in testes from sexually immature rats (less than 21 days). RT-PCR analysis would reveal leptin receptor expression in embryonic, prepubertal, and adult rat testes and demonstrates the difference of sensitivity between these two methodologies. Leptin addition to adult rat Leydig cell cultures would inhibit hCG-stimulated testosterone production, but had no effect on the steroidogenic function of prepubertal Leydig cells and suggests that no functional LR are present in the prepubertal testes. Further studies also demonstrated that leptin acts as a direct inhibitory signal for testicular steroidogenesis and that this effect is due to suppression of several upstream factors (SF-1, StAR, and P450scc) in the steroidogenic pathway [121]. The administration of leptin for 5 days to adult male mice was investigated using a variety of techniques. Immunohistochemical testosterone staining revealed more intense staining in leptin treated than in control mice. Testicular weights and seminiferous tubule diameters were also increased by leptin administration. These results in the mouse indicated that leptin administration stimulates testicular function and testosterone synthesis. It is not clear from these studies, whether the leptin effect is directly on the testes or through a hypothalamic-pituitary effect of leptin [122].

In men, there was no significant difference in histochemical staining of the testes between infertile and normal control males. There was leptin staining in seminiferous tubules and Leydig cells. These results suggest a leptin action as a central neuroendocrine effect rather than a direct effect on testicular tissue [123]. Glander [124] examined leptin and testes physiology in infertile male patients and individuals following vasectomy. The concentration of leptin in seminal plasma was significantly lower in normal semen samples than in that from infertile patients and showed a negative correlation with the percentage of motile sperm. Leptin concentrations in serum showed no relationship to any sperm parameters and in seminal plasma was unchanged following vasectomy. Aquila and colleagues [125] demonstrated that human sperm expresses leptin. There was a large difference in leptin secretion between uncapacitated and capacitated sperm. Greater leptin release from capacitated sperm suggests a functional role for sperm produced leptin in capacitation.

The role of leptin in the male may extend beyond reproductive involvements. Mice lacking the androgen receptor (ARKO) were used to study the relationship between the androgen receptor (AR) and insulin resistance. In ARKO mice, a progressive reduced insulin sensitivity and impaired glucose tolerance was observed with advancing age. These mice also had an accelerated weight gain, hyperinsulinemia, and hyperglycemia, as well as higher leptin levels. These studies demonstrate that the action of androgen at the AR has a role in the development of insulin resistance, which may contribute to the development of type 2 diabetes. Moderately obese men have a decreased androgen profile with suppressed serum levels of total and free testosterone. In massively obese men, there is a consistently low level of free testosterone [126]. These investigators speculate that these results represent an action of leptin on LH pulse amplitude and serum LH levels, as well as possible negative actions of excess circulating leptin on testicular steroidogenesis. Semen leptin concentrations are inversely correlated with serum testosterone levels and directly with serum leptin levels [127].

19.4 Summary

Leptin's identity as a reproductive hormone seems assured and its practically unparalleled pleiotropism has served to dramatically expand our original understanding of its interactions that were once limited to those regulating adiposity and metabolism. To this end, relationships with the gonadotropins demonstrate hypothalamic roles for the polypeptide in controlling menstrual cyclicity and ovarian function, while as a modulator of energy homeostasis it may permissively affect the advent of puberty and enhance fertility [3, 11, 34, 87]. Clinical reports, in vitro experiments using human tissues, and in vivo experiments in rodents and nonhuman primates have demonstrated an array of receptor-mediated roles for the adipokine with respect to implantation, placental function, and conceptus development. Further associations with preeclampsia, pregnancy-associated diabetes, macrosomia, and IUGR have combined to place leptin in the vanguard of today's research in reproductive endocrinology [15, 17, 87].

Glossary of Terms and Acronyms

AGRP: agouti-related protein

AR: androgen receptor

ARKO: androgen receptor knockout mouse

AVPV: anteroventral periventricular

BMI: body mass index

BOX2: amino acid sequence [V/L]E[V/L]L present in the single chain cytokine receptors required in some cases for full activation of Jak2

CART: cocaine- and amphetamine-regulated transcript

CRH: corticotropin-releasing hormone

E_2: estradiol

E_2B: estradiol benzoate

ER: estrogen receptor

ERK: extracellular-signal-regulated kinase (ERK) pathways

GnRH: Gonadotripin-releasing hormone

GPR54: G-protein-coupled receptor-54

hCG: human chorionic gonadotropin

IL: interleukin

IUGR: intrauterine growth restriction

JAK/STAT: Janus tyrosine kinases/signal transduction and activators of transcription

KiSS-1: cognate ligands of the GPR54

LEP: human/primate leptin

Lep: rodent leptin

***Lep^{ob}/Lep^{ob}*:** obese, genetically leptin deficient, mouse

$LEPR_L$: long isoform, primate leptin receptor

$LEPR_S$: short isoform, primate leptin receptor

LR: leptin receptor

LRa: a short LR isoform

LRb: long LR isoform

MAPK: mitogen-activated protein kinase

MCH: melanin-concentrating hormone

NPY: neuropeptide Y

***ob*:** original name for *Lep* gene (*obese*)

ORX: orexins

P450scc: cytochrome P450 side chain cleavage

POMC: proopiomelanocortin

SF-1: steroidogenic factor-1

SH2: Src homology 2

solLR: soluble leptin receptor

StAR: sterodogenic acute regulatory protein

TNF-α: tumor necrosis factor-α

TRH: thyrotropin-releasing hormone

References

1. Scherer PE. Adipose tissue: from lipid storage compartment to endocrine organ. Diabetes 2006; 55:1537–45.
2. Tilg H, Moschen AR. Adipocytokines: mediators linking adipose tissue, inflammation and immunity. Nat Rev Immunol 2006; 6:772–83.
3. Budak E, Fernandez Sanchez M, Bellver J, et al. Interactions of the hormones leptin, ghrelin, adiponectin, resistin, and PYY3-36 with the reproductive system. Fertil Steril 2006; 85: 1563–81.
4. Miner JL. The adipocyte as an endocrine cell. J Anim Sci 2004; 82:935–41.
5. Farooqi S, O'Rahilly S. Genetics of obesity in humans. Endocr Rev 2006; 27:710–18.
6. Crowley V. Overview of human obesity and central mechanisms regulating energy homeostasis. Ann Clin Biochem 2008; 45: 245–55.
7. Pischon T, Nothlings U, Boeing H. Obesity and cancer. Proc Nutr Soc 2008; 67:128–45.
8. Bulcao C, Ferreira SR, Giuffrida FM, et al. The new adipose tissue and adipocytokines. Curr Diabetes Rev 2006; 2:19–28.
9. Ahima RS, Lazar MA. Adipokines and the Peripheral and Neural Control of Energy Balance. Mol Endocrinol 2008; 22:1023–31.
10. de Ferranti S, Mozaffarian D. The Perfect Storm: Obesity, Adipocyte Dysfunction, and Metabolic Consequences. Clin Chem 2008; 54:945–55.
11. Mitchell M, Armstrong DT, Robker RL, et al. Adipokines: implications for female fertility and obesity. Reproduction 2005; 130:583–97.
12. Hill J, Elmquist JK, Elias CF. Hypothalamic Pathways Linking Energy Balance and Reproduction. Am J Physiol Endocrinol Metab 2008; 294:E827-E32.
13. Kiess W, Petzold S, Topfer M, et al. Adipocytes and adipose tissue. Best Pract Res Clin Endocrinol Metab 2008; 22:135–53.
14. Henson MC, Castracane VD. Leptin in pregnancy. Biol Reprod 2000; 63:1219–28.
15. Henson MC, Castracane VD. Leptin in pregnancy: an update. Biol Reprod 2006; 74:218–29.
16. Henson MC, Castracane VD. Leptin in primate pregnancy. In: Henson MC, Castracane VD, editors. Leptin and Reproduction. New York: Kluwer Academic/Plenum, 2003:239–63.
17. Hauguel-de Mouzon S, Lepercq J, Catalano P. The known and unknown of leptin in pregnancy. Am J Obstet Gynecol 2006; 194:1537–45.
18. Islami D, Bischof P. Leptin in the placenta. In: Henson MC, Castracane VD, editors. Leptin and Reproduction. New York: Kluwer Academic/Plenum, 2003:675–2.
19. Zhang Y, Proenca R, Maffei M, et al. Positional cloning of the mouse obese gene and its human homologue. Nature 1994; 372:425–32.
20. Yu WH, Kimura M, Walczewska A, et al. Role of leptin in hypothalamic-pituitary function. Proc Natl Acad Sci U S A 1997; 94:1023–8.
21. Cameron JL, Nosbisch C. Suppression of pulsatile luteinizing hormone and testosterone secretion during short term food restriction in the adult male rhesus monkey (Macaca mulatta). Endocrinology 1991; 128:1532–40.
22. Lujan ME, Krzemien AA, Reid RL, et al. Caloric restriction inhibits steroid-induced gonadotropin surges in ovariectomized rhesus monkeys. Endocrine 2005; 27:25–31.
23. Maciel MN, Zieba DA, Amstalden M, et al. Leptin prevents fasting-mediated reductions in pulsatile secretion of luteinizing hormone and enhances its gonadotropin-releasing hormone-mediated release in heifers. Biol Reprod 2004; 70:229–35.
24. Steiner J, LaPaglia N, Kirsteins L, et al. The response of the hypothalamic-pituitary-gonadal axis to fasting is modulated by leptin. Endocr Res 2003; 29:107–17.
25. Boden G, Chen X, Mozzoli M, et al. Effect of fasting on serum leptin in normal human subjects. J Clin Endocrinol Metab 1996; 81:3419–23.
26. Hardie LJ, Rayner DV, Holmes S, et al. Circulating leptin levels are modulated by fasting, cold exposure and insulin administration in lean but not Zucker (fa/fa) rats as measured by ELISA. Biochem Biophys Res Commun 1996; 223:660–5.
27. Finn PD, Cunningham MJ, Pau KY, et al. The stimulatory effect of leptin on the neuroendocrine reproductive axis of the monkey. Endocrinology 1998; 139:4652–62.
28. Arora A, Arora S. Leptin and its metabolic interactions – an update. Diabetes Obes Metab 2008, 10:973–993.
29. Fernandez-Fernandez R, Martini AC, Navarro VM, et al. Novel signals for the integration of energy balance and reproduction. Mol Cell Endocrinol 2006; 254–255:127–32.
30. Smith JT, Clifton DK, Steiner RA. Regulation of the neuroendocrine reproductive axis by kisspeptin-GPR54 signaling. Reproduction 2006; 131:623–30.
31. Kauffman AS, Clifton DK, Steiner RA. Emerging ideas about kisspeptin- GPR54 signaling in the neuroendocrine regulation of reproduction. Trends Neurosci 2007; 30:504–11.
32. Gottsch ML, Cunningham MJ, Smith JT, et al. A role for kisspeptins in the regulation of gonadotropin secretion in the mouse. Endocrinology 2004; 145:4073–7.
33. Smith JT, Acohido BV, Clifton DK, et al. KiSS-1 neurones are direct targets for leptin in the ob/ob mouse. J Neuroendocrinol 2006; 18:298–303.
34. Bluher S, Mantzoros CS. Leptin in reproduction. Curr Opin Endocrinol Diabetes Obes 2007; 14:458–64.
35. Chehab FF, Lim ME, Lu R. Correction of the sterility defect in homozygous obese female mice by treatment with the human recombinant leptin. Nat Genet 1996; 12:318–20.
36. Considine RV, Caro JF. Leptin and the regulation of body weight. Int J Biochem Cell Biol 1997; 29:1255–72.
37. Rosenbaum M, Leibel RL. Clinical review 107: Role of gonadal steroids in the sexual dimorphisms in body composition and circulating concentrations of leptin. J Clin Endocrinol Metab 1999; 84:1784–9.
38. Shimizu H, Shimomura Y, Nakanishi Y, et al. Estrogen increases in vivo leptin production in rats and human subjects. J Endocrinol 1997; 154:285–92.
39. Mantzoros CS, Flier JS, Rogol AD. A longitudinal assessment of hormonal and physical alterations during normal puberty in boys V. Rising leptin levels may signal the onset of puberty. J Clin Endocrinol Metab 1997; 82:1066–70.
40. Horlick MB, Rosenbaum M, Nicolson M, et al. Effect of puberty on the relationship between circulating leptin and body composition. J Clin Endocrinol Metab 2000; 85:2509–18.
41. Chu SC, Chou YC, Liu JY, et al. Fluctuation of serum leptin level in rats after ovariectomy and the influence of estrogen supplement. Life Sci 1999; 64:2299–306.

42. Machinal F, Dieudonne MN, Leneveu MC, et al. In vivo and in vitro ob gene expression and leptin secretion in rat adipocytes: evidence for a regional specific regulation by sex steroid hormones. Endocrinology 1999; 140:1567–74.
43. Yoneda N, Saito S, Kimura M, et al. The influence of ovariectomy on ob gene expression in rats. Horm Metab Res 1998; 30: 263–5.
44. Messinis IE, Milingos SD, Alexandris E, et al. Leptin concentrations in normal women following bilateral ovariectomy. Hum Reprod 1999; 14:913–8.
45. Castracane VD, Kraemer RR, Franken MA, et al. Serum leptin concentration in women: effect of age, obesity, and estrogen administration. Fertil Steril 1998; 70:472–7.
46. Lindheim SR, Sauer MV, Carmina E, et al. Circulating leptin levels during ovulation induction: relation to adiposity and ovarian morphology. Fertil Steril 2000; 73:493–8.
47. Unkila-Kallio L, Andersson S, Koistinen HA, et al. Leptin during assisted reproductive cycles: the effect of ovarian stimulation and of very early pregnancy. Hum Reprod 2001; 16:657–62.
48. Yamada M, Irahara M, Tezuka M, et al. Serum leptin profiles in the normal menstrual cycles and gonadotropin treatment cycles. Gynecol Obstet Invest 2000; 49:119–23.
49. Geisthovel F, Jochmann N, Widjaja A, et al. Serum pattern of circulating free leptin, bound leptin, and soluble leptin receptor in the physiological menstrual cycle. Fertil Steril 2004; 81:398–402.
50. Coya R, Martul P, Algorta J, et al. Progesterone and human placental lactogen inhibit leptin secretion on cultured trophoblast cells from human placentas at term. Gynecol Endocrinol 2005; 21:27–32.
51. Ertl T, Funke S, Sarkany I, et al. Postnatal changes of leptin levels in full-term and preterm neonates: their relation to intrauterine growth, gender and testosterone. Biol Neonate 1999; 75:167–76.
52. Laml T, Hartmann BW, Preyer O, et al. Serum leptin concentration in cord blood: relationship to birth weight and gender in pregnancies complicated by pre-eclampsia. Gynecol Endocrinol 2000; 14:442–7.
53. Maffeis C, Moghetti P, Vettor R, et al. Leptin concentration in newborns′ cord blood: relationship to gender and growth-regulating hormones. Int J Obes Relat Metab Disord 1999; 23:943–7.
54. Matsuda J, Yokota I, Iida M, et al. Serum leptin concentration in cord blood: relationship to birth weight and gender. J Clin Endocrinol Metab 1997; 82:1642–4.
55. Caprio M, Fabbrini E, Ricci G, et al. Ontogenesis of leptin receptor in rat Leydig cells. Biol Reprod 2003; 68:1199–207.
56. Akcurin S, Velipasaoglu S, Akcurin G, et al. Leptin profile in neonatal gonadotropin surge and relationship between leptin and body mass index in early infancy. J Pediatr Endocrinol Metab 2005; 18:189–95.
57. Garcia-Mayor RV, Andrade MA, Rios M, et al. Serum leptin levels in normal children: relationship to age, gender, body mass index, pituitary-gonadal hormones, and pubertal stage. J Clin Endocrinol Metab 1997; 82:2849–55.
58. Considine RV, Sinha MK, Heiman ML, et al. Serum immunoreactive-leptin concentrations in normal-weight and obese humans. N Engl J Med 1996; 334:292–5.
59. Maffei M, Halaas J, Ravussin E, et al. Leptin levels in human and rodent: measurement of plasma leptin and ob RNA in obese and weight-reduced subjects. Nat Med 1995; 1:1155–61.
60. Ahima RS, Dushay J, Flier SN, et al. Leptin accelerates the onset of puberty in normal female mice. J Clin Invest 1997; 99:391–5.
61. Chehab FF, Mounzih K, Lu R, et al. Early onset of reproductive function in normal female mice treated with leptin. Science 1997; 275:88–90.
62. Farooqi IS. Leptin and the onset of puberty: insights from rodent and human genetics. Semin Reprod Med 2002; 20:139–44.
63. Cheung CC, Thornton JE, Kuijper JL, et al. Leptin is a metabolic gate for the onset of puberty in the female rat. Endocrinology 1997; 138:855–8.
64. Gruaz NM, Lalaoui M, Pierroz DD, et al. Chronic administration of leptin into the lateral ventricle induces sexual maturation in severely food-restricted female rats. J Neuroendocrinol 1998; 10:627–33.
65. Barb CR, Hausman GJ, Czaja K. Leptin: a metabolic signal affecting central regulation of reproduction in the pig. Domest Anim Endocrinol 2005; 29:186–92.
66. Foster DL, Nagatani S. Physiological perspectives on leptin as a regulator of reproduction: role in timing puberty. Biol Reprod 1999; 60:205–15.
67. Paczoska-Eliasiewicz HE, Proszkowiec-Weglarz M, Proudman J, et al. Exogenous leptin advances puberty in domestic hen. Domest Anim Endocrinol 2006; 31:211–26.
68. Jaruratanasirikul S, Mo-suwan L, Lebel L. Growth pattern and age at menarche of obese girls in a transitional society. J Pediatr Endocrinol Metab 1997; 10:487–90.
69. Munoz MT, de la Piedra C, Barrios V, et al. Changes in bone density and bone markers in rhythmic gymnasts and ballet dancers: implications for puberty and leptin levels. Eur J Endocrinol 2004; 151:491–6.
70. Mann DR, Akinbami MA, Gould KG, et al. A longitudinal study of leptin during development in the male rhesus monkey: the effect of body composition and season on circulating leptin levels. Biol Reprod 2000; 62:285–91.
71. Plant TM, Durrant AR. Circulating leptin does not appear to provide a signal for triggering the initiation of puberty in the male rhesus monkey (Macaca mulatta). Endocrinology 1997; 138:4505–8.
72. Urbanski HF, Pau KY. A biphasic developmental pattern of circulating leptin in the male rhesus macaque (Macaca mulatta). Endocrinology 1998; 139:2284–6.
73. Plant TM. Leptin, growth hormone, and the onset of primate puberty. J Clin Endocrinol Metab 2001; 86:458–60.
74. Suter KJ, Pohl CR, Wilson ME. Growth signals and puberty. J Clin Endocrinol Metab 2001; 86:460.
75. Riad-Gabriel MG, Jinagouda SD, Sharma R, et al. Changes in plasma leptin during the menstrual cycle. Eur J Endocrinol 1998; 139:528–31.
76. Messinis IE, Kariotis I, Milingos S, et al. Treatment of normal women with oestradiol plus progesterone prevents the decrease of leptin concentrations induced by ovariectomy. Hum Reprod 2000; 15:2383–7.
77. Messinis IE, Papageorgiou I, Milingos S, et al. Oestradiol plus progesterone treatment increases serum leptin concentrations in normal women. Hum Reprod 2001; 16:1827–32.
78. Imani B, Eijkemans MJ, de Jong FH, et al. Free androgen index and leptin are the most prominent endocrine predictors of ovarian response during clomiphene citrate induction of ovulation in normogonadotropic oligoamenorrheic infertility. J Clin Endocrinol Metab 2000; 85:676–82.
79. Brannian JD, Schmidt SM, Kreger DO, et al. Baseline non-fasting serum leptin concentration to body mass index ratio is predictive of IVF outcomes. Hum Reprod 2001; 16:1819–26.
80. Butzow TL, Moilanen JM, Lehtovirta M, et al. Serum and follicular fluid leptin during in vitro fertilization: relationship among leptin increase, body fat mass, and reduced ovarian response. J Clin Endocrinol Metab 1999; 84:3135–9.
81. Zhao Y, Kreger DO, Brannian JD. Serum leptin concentrations in women during gonadotropin stimulation cycles. J Reprod Med 2000; 45:121–5.

82. Welt CK, Chan JL, Bullen J, et al. Recombinant human leptin in women with hypothalamic amenorrhea. N Engl J Med 2004; 351:987–97.
83. Masuzaki H, Ogawa Y, Sagawa N, et al. Nonadipose tissue production of leptin: leptin as a novel placenta-derived hormone in humans. Nat Med 1997; 3:1029–33.
84. Henson MC, Swan KF, O'Neil JS. Expression of placental leptin and leptin receptor transcripts in early pregnancy and at term. Obstet Gynecol 1998; 92:1020–8.
85. Castracane VD, Henson MC. Regulation of leptin and leptin receptor in the human uterus: possible roles in implantation and uterine pathology. In: Henson MC, Castracane Vd, editors. Leptin and Reproduction. New York: Kluwer Academic/Plenum, 2003:111–5.
86. Henson MC, Castracane VD. Leptin: roles and regulation in primate pregnancy. Semin Reprod Med 2002; 20:113–22.
87. Henson MC, Castracane VD. Roles and regulation of leptin in reproduction. In: Castracane VD, Henson MC, editors. Leptin. New York: Springer, 2006:149–82.
88. Lappas M, Yee K, Permezel M, et al. Release and regulation of leptin, resistin and adiponectin from human placenta, fetal membranes, and maternal adipose tissue and skeletal muscle from normal and gestational diabetes mellitus-complicated pregnancies. J Endocrinol 2005; 186:457–65.
89. Chardonnens D, Cameo P, Aubert ML, et al. Modulation of human cytotrophoblastic leptin secretion by interleukin-1alpha and 17beta-oestradiol and its effect on HCG secretion. Mol Hum Reprod 1999; 5:1077–82.
90. Dos Santos E, Dieudonne MN, Leneveu MC, et al. In vitro effects of chorionic gonadotropin hormone on human adipose development. J Endocrinol 2007; 194:313–25.
91. Coya R, Martul P, Algorta J, et al. Effect of leptin on the regulation of placental hormone secretion in cultured human placental cells. Gynecol Endocrinol 2006; 22:620–6.
92. Magarinos MP, Sanchez-Margalet V, Kotler M, et al. Leptin promotes cell proliferation and survival of trophoblastic cells. Biol Reprod 2007; 76:203–10.
93. Lee GH, Proenca R, Montez JM, et al. Abnormal splicing of the leptin receptor in diabetic mice. Nature 1996; 379:632–5.
94. Maamra M, Bidlingmaier M, Postel-Vinay MC, et al. Generation of human soluble leptin receptor by proteolytic cleavage of membrane-anchored receptors. Endocrinology 2001; 142:4389–93.
95. Ge H, Huang L, Pourbahrami T, et al. Generation of soluble leptin receptor by ectodomain shedding of membrane-spanning receptors in vitro and in vivo. J Biol Chem 2002; 277:45898–903.
96. Yang G, Ge H, Boucher A, et al. Modulation of direct leptin signaling by soluble leptin receptor. Mol Endocrinol 2004; 18:1354–62.
97. Tu H, Kastin AJ, Hsuchou H, et al. Soluble receptor inhibits leptin transport. J Cell Physiol 2008; 214:301–5.
98. Augustine RA, Ladyman SR, Grattan DR. From feeding one to feeding many: hormone-induced changes in bodyweight homeostasis during pregnancy. J Physiol 2008; 586:387–97.
99. Castracane VD, Hendrickx AG, Henson MC. Serum leptin in nonpregnant and pregnant women and in old and new world nonhuman primates. Exp Biol Med (Maywood) 2005; 230:251–4.
100. Henson MC. Pregnancy maintenance and the regulation of placental progesterone biosynthesis in the baboon. Hum Reprod Update 1998; 4:389–405.
101. Henson MC, Castracane VD, O'Neil JS, et al. Serum leptin concentrations and expression of leptin transcripts in placental trophoblast with advancing baboon pregnancy. J Clin Endocrinol Metab 1999; 84:2543–9.
102. Green AE, O'Neil JS, Swan KF, et al. Leptin receptor transcripts are constitutively expressed in placenta and adipose tissue with advancing baboon pregnancy. Proc Soc Exp Biol Med 2000; 223:362–6.
103. Edwards DE, Bohm RP, Jr., Purcell J, et al. Two isoforms of the leptin receptor are enhanced in pregnancy-specific tissues and soluble leptin receptor is enhanced in maternal serum with advancing gestation in the baboon. Biol Reprod 2004; 71:1746–52.
104. O'Neil JS, Burow ME, Green AE, et al. Effects of estrogen on leptin gene promoter activation in MCF-7 breast cancer and JEG-3 choriocarcinoma cells: selective regulation via estrogen receptors alpha and beta. Mol Cell Endocrinol 2001; 176:67–75.
105. Albrecht ED, Aberdeen GW, Pepe GJ. The role of estrogen in the maintenance of primate pregnancy. Am J Obstet Gynecol 2000; 182:432–8.
106. Mystkowski P, Schwartz MW. Gonadal steroids and energy homeostasis in the leptin era. Nutrition 2000; 16:937–46.
107. O'Neil JS, Green AE, Edwards DE, et al. Regulation of leptin and leptin receptor in baboon pregnancy: effects of advancing gestation and fetectomy. J Clin Endocrinol Metab 2001; 86:2518–24.
108. Torday JS, Rehan VK. Stretch-stimulated surfactant synthesis is coordinated by the paracrine actions of PTHrP and leptin. Am J Physiol Lung Cell Mol Physiol 2002; 283:L130–5.
109. Torday JS, Sun H, Wang L, et al. Leptin mediates the parathyroid hormone-related protein paracrine stimulation of fetal lung maturation. Am J Physiol Lung Cell Mol Physiol 2002; 282:L405–10.
110. Henson MC, Swan KF, Edwards DE, et al. Leptin receptor expression in fetal lung increases in late gestation in the baboon: a model for human pregnancy. Reproduction 2004; 127:87–94.
111. Kirwin SM, Bhandari V, Dimatteo D, et al. Leptin enhances lung maturity in the fetal rat. Pediatr Res 2006; 60:200–4.
112. Huang K, Rabold R, Abston E, et al. Effects of leptin deficiency on postnatal lung development in mice. J Appl Physiol 2008, 105:249–59.
113. Lea RG, Howe D, Hannah LT, et al. Placental leptin in normal, diabetic and fetal growth-retarded pregnancies. Mol Hum Reprod 2000; 6:763–9.
114. Meller M, Qiu C, Vadachkoria S, et al. Changes in placental adipocytokine gene expression associated with gestational diabetes mellitus. Physiol Res 2006; 55:501–12.
115. Vela-Huerta MM, San Vicente-Santoscoy EU, Guizar-Mendoza JM, et al. Leptin, insulin, and glucose serum levels in large-for-gestational-age infants of diabetic and non-diabetic mothers. J Pediatr Endocrinol Metab 2008; 21:17–22.
116. Ategbo JM, Grissa O, Yessoufou A, et al. Modulation of adipokines and cytokines in gestational diabetes and macrosomia. J Clin Endocrinol Metab 2006; 91:4137–43.
117. Anim-Nyame N, Sooranna SR, Steer PJ, et al. Longitudinal analysis of maternal plasma leptin concentrations during normal pregnancy and pre-eclampsia. Hum Reprod 2000; 15:2033–6.
118. Haugen F, Ranheim T, Harsem NK, et al. Increased plasma levels of adipokines in preeclampsia: relationship to placenta and adipose tissue gene expression. Am J Physiol Endocrinol Metab 2006; 290:E326–33.
119. Mise H, Sagawa N, Matsumoto T, et al. Augmented placental production of leptin in preeclampsia: possible involvement of placental hypoxia. J Clin Endocrinol Metab 1998; 83:3225–9.
120. Tena-Sempere M, Pinilla L, Zhang FP, et al. Developmental and hormonal regulation of leptin receptor (Ob-R) messenger ribonucleic acid expression in rat testis. Biol Reprod 2001; 64:634–43.
121. Tena-Sempere M, Manna PR, Zhang FP, et al. Molecular mechanisms of leptin action in adult rat testis: potential targets for leptin-induced inhibition of steroidogenesis and pattern of leptin receptor messenger ribonucleic acid expression. J Endocrinol 2001; 170:413–23.
122. Kus I, Sarsilmaz M, Canpolat S, et al. Immunohistochemical, histological and ultrastructural evaluation of the effects of leptin on testes in mice. Arch Androl 2005; 51:395–405.

123. Soyupek S, Armagan A, Serel TA, et al. Leptin expression in the testicular tissue of fertile and infertile men. Arch Androl 2005; 51:239–46.
124. Glander HJ, Lammert A, Paasch U, et al. Leptin exists in tubuli seminiferi and in seminal plasma. Andrologia 2002; 34:227–33.
125. Aquila S, Gentile M, Middea E, et al. Leptin secretion by human ejaculated spermatozoa. J Clin Endocrinol Metab 2005; 90:4753–61.
126. Lima N, Cavaliere M, Knobel M, et al. Decreased androgen levels in massively obese men may be associated with impaired function of the gonadostat. J Obes Relat Metab Disord 2003; 24:1433–7.
127. von Sobbe HU, Koebnick C, Jenne L, et al. Leptin concentrations in semen are correlated with serum leptin and elevated in hypergonadotrophic hypogonadism. Andrologia 2003; 35:233–7.

Chapter 20

Neurosteroids and Sexual Behavior and Reproduction

Steven R. King

20.1 Introduction

Gonadal and adrenal steroids strongly determine sexual behavior and reproductive function through their effects on the brain. These "neuroactive steroids" act directly on neural cells or following their conversion to other metabolites locally. For instance, the production of estrogen from testicular testosterone by neural aromatase is essential for masculinization of the brain. However, steroids synthesized de novo within the central and peripheral nervous systems (CNS and PNS), termed "neurosteroids," may also strongly impact sexual differentiation of the brain and sexual function. Specifically, neurosteroids affect sexual and gender-typical behaviors, ovulation, and behaviors that influence sexual interest and motivation like aggression, anxiety, and depression.

20.2 Neurosteroids

Both the developing and adult nervous systems produce neurosteroids as demonstrated in cell culture and gonadectomized (GDX) and adrenalectomized (ADX) animals, which cannot produce steroids from endocrine sources. Select neural cell types express the necessary synthetic enzymes, including the two proteins that initiate steroid production—the steroidogenic acute regulatory (StAR) protein and cytochrome P450scc (Chapters 16 and 23) [1–3]. Steroidogenic enzyme levels are estimated to be 2–5 orders of magnitude lower in the brain than in the adrenal cortex or gonads [4, 5]. Still, local steroid concentrations at neurons and glia can be much greater than in the serum.

Neurosteroidogenesis is an evolutionarily conserved process, discovered thus far in mammals, birds, and amphibians [e.g., 6–10]. In mammals, steroidogenic enzymes and neurosteroids are localized in the PNS, the spinal cord, and select regions of the brain including those critical for sexual behavior and function, like the amygdala, olfactory bulb, and hypothalamus [2, 5, 11, 12]. Specific populations of neurons, type 1 astrocytes, and Schwann cells generate a variety of neurosteroids, such as progesterone, dehydroepiandrosterone (DHEA), 17β-estradiol, and 5α-reduced steroids, 3α,5α-tetrahydrodeoxycorticosterone (THDOC), 3α-androstanediol, 3α,5α-tetrahydroprogesterone (allopregnanolone/THP), and 3α,5β-tetrahydroprogesterone (pregnanolone) (Fig. 20.1) [5, 13, 14]. The primate brain may also produce steroid sulfates like DHEA-sulfate (DHEA-S) and pregnenolone sulfate; the rodent brain may generate sulfated and/or lipid-associated steroids [15, 16, 17].

20.3 Receptor Types Targeted by Neurosteroids

Neurosteroids act in a distinct manner from steroids that slowly infiltrate the brain from the bloodstream. Neuroactive steroids typically instigate long-term genomic changes via classical nuclear steroid receptors. Neurosteroids, on the other hand, act in an autocrine or paracrine manner and have more immediate, nongenomic effects that occur within seconds to milliseconds. This is accomplished through their interactions with G protein-coupled and ligand-gated ion channel membrane receptors such as glycine receptors, metabotropic sigma type 1 (σ_1) receptors, and ionotropic glutamate receptors like the *N*-methyl-D-aspartic acid (NMDA) receptor (Table 20.1) [18]. Other receptor targets include plasma membrane-localized estrogen receptors [19], G protein-coupled membrane progesterone receptors [20], α_1-adrenergic [21], neuronal nicotinic acetylcholine (nnAchR) [22, 23], and dopamine type 1 (D_1) receptors [24].

S.R. King (✉)
Scott Department of Urology Baylor College of Medicine, Houston, TX, USA
e-mail: srking100@yahoo.com

P.J. Chedrese (ed.), *Reproductive Endocrinology*,
DOI 10.1007/978-0-387-88186-7_20,

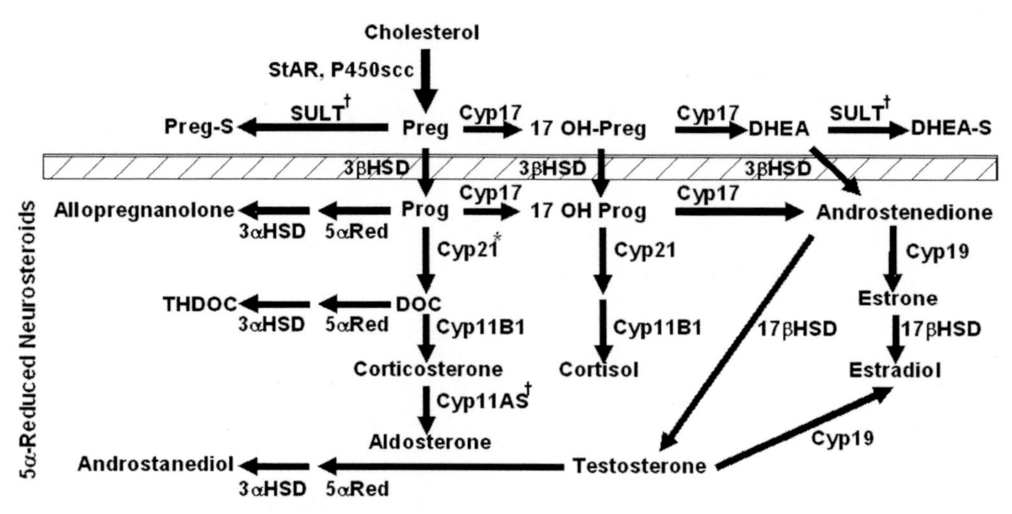

Fig. 20.1 Major pathways for neurosteroidogenesis. Neurosteroids synthesized prior to conversion by 3βHSD (hatched box) generally have different effects than progesterone and its metabolites. The exact identities of the enzymatic isoforms in neural cells are ambiguous for 17βHSD, 3αHSD, and 5α-reductase (5αRed). *Recent data implicate cytochrome P450 2D or trace expression of P450c21 as responsible for DOC synthesis in the CNS [13]. The presence of sulfotransferase (SULT) and aldosterone synthase/Cyp11B2 (Cyp11AS) in the CNS is controversial [14]. Other abbreviations not in the text: Cyp19, aromatase; Prog, progesterone; Preg, pregnenolone

Neurosteroids can also allosterically modulate and directly activate γ-aminobutyric acid type A ($GABA_A$) receptors [25–28]. Through synaptic and extrasynaptic receptors, neurosteroids effect changes in $GABA_A$ channel chloride currents, resulting in phasic and tonic inhibition of neurons, respectively. Nanomolar levels of 5α-reduced androstanediol, allopregnanolone, and THDOC reduce neuronal excitability with a 20- to 200-fold higher efficacy than benzodiazepines and barbiturates [25, 29, 30, 31]. On the other hand, pregnenolone sulfate and DHEA-S antagonize $GABA_A$ channel activity [32, 33, 34]. Thus, locally synthesized neurosteroids can precisely and quickly alter specific neuronal functions like those related to reproduction in a manner unachievable by neuroactive steroids permeating the nervous system from the blood.

Table 20.1 General modulatory effects of specific neurosteroids on nonsteroidal membrane receptors. In certain instances, the effect of a neurosteroid can depend upon dose and the subunit composition of the receptor. σ_1: metabotropic sigma type 1 glycine receptor; AMPA: α-Amino-3-hydroxy-5-methyl-4-isoxazolepropionic acid; $GABA_A$ receptor: γ-aminobutyric acid type A receptor; $GABA_c$ receptor: γ-aminobutyric acid type C receptor; NMDA: *N*-methyl-D-aspartic acid; nnAchR: neuronal nicotinic acetylcholine receptor

Receptor	Neurosteroid	Modulatory effect
$GABA_A$	Allopregnanolone, THDOC, androstanediol	Positive
	Pregnenolone sulfate, DHEA-S	Negative
$GABA_C$	Allopregnanolone, THDOC	Positive
	Pregnanolone	Negative
Glycine	Progesterone, pregnanolone, allopregnanolone, pregnenolone sulfate, DHEA-S	Negative
σ_1	DHEA, pregnenolone sulfate, DHEA-S	Positive
	Progesterone	Negative
AMPA	Pregnenolone sulfate	Negative
Kainate	Progesterone	Positive
	Pregnenolone sulfate	Negative
NMDA	DHEA, pregnenolone sulfate	Positive
	Estradiol	Negative
5-hydroxytryptamine type 3/5-HT_3 (Serotonin)	Progesterone, allopregnanolone, testosterone, estradiol	Negative
nnAchR	Progesterone, allopregnanolone, pregnenolone sulfate	Negative
Muscarinic	Pregnenolone, progesterone	Negative
Oxytocin	Progesterone	Negative
Vanilloid type 1/TRPV1	DHEA, pregnenolone sulfate	Negative

20.4 Neurosteroids in Sexual Function

During folliculogenesis, ovarian estrogen and progesterone regulate follicle-stimulating hormone (FSH) and luteinizing hormone (LH) release directly at the anterior pituitary and indirectly by regulating hypothalamic secretion of gonadotropin-releasing hormone (GnRH) (Chapters 13 and 14). The climactic LH surge that triggers the process leading to ovulation is estrogen-dependent. But the estrogen-induced release of GnRH for this surge may require the local production of progesterone, which then acts on relevant hypothalamic progesterone receptors [35, 36].

Astrocytes in the hypothalamus appear to be the source of this steroid. They synthesize and secrete progesterone in response to the activation of estrogen-sensitive membrane receptors. Administration of supraphysiologic levels of estradiol (50 μg) to ovariectomized (OVX)/ADX rats induces progesterone generation specifically in the medial basal hypothalamus and consequently, the LH surge [37, 38]. Inhibition of hypothalamic progesterone synthesis blocks the surge [37]. Importantly, estrogen only induces hypothalamic progesterone production in post-pubertal not neonatal astrocytes and not in castrate ADX males [39].

Estrogen also induces allopregnanolone synthesis in the anterior pituitary and hypothalamus of OVX animals [40]. In the hypothalamus, allopregnanolone potentiates the activation of post-synaptic $GABA_A$ receptors on GnRH neurons in diestrus female mice, which induces GnRH release [41, 42]. The age-related reduction in the level of this steroid in the hypothalamus may thus be one cause of the changes in ovarian function that occur with age [40].

The ability of allopregnanolone to potentiate GnRH release may be site-specific within the hypothalamus. Allopregnanolone suppresses ovulation when it is delivered by intracerebroventricular (i.c.v.) injection. This method provides access of allopregnanolone to the ventromedial nucleus (VMN) in the hypothalamus, suggesting that activation of hypothalamic $GABA_A$ receptors here inhibits GnRH secretion [43].

Sulfated neurosteroids also regulate GnRH release. DHEA-S inhibits the activation of post-synaptic $GABA_A$ receptors in GnRH neurons of diestrus female mice [41]. On the other hand, pregnenolone sulfate potentiates glutamate or NMDA-stimulated GnRH release from hypothalamic neurons [44]. Allopregnanolone also potentiates $GABA_A$ channel activity and reduces basal GnRH release in male hypothalamic cultures [45]. This suppression is selectively overcome by pregnenolone sulfate.

The allopregnanolone precursor 3α-hydroxy-4-pregnen-20-one (3αHP), produced by the pituitary gland, is also a neurosteroid that regulates the $GABA_A$ receptor [46]. This steroid rapidly inhibits basal and GnRH-stimulated FSH secretion in both sexes [47–49]. This effect involves calcium and protein kinase C signaling pathways utilized by GnRH and is independent of any conversion to allopregnanolone [50, 51]. Interestingly, LH itself can regulate pregnenolone synthesis by neurons [52]. Thus, neurosteroids can regulate gonadotropin release. While it can have dramatic effects on the female, its impact on male reproductive function has not been described.

20.5 Neurosteroids in Sexual Behavior

20.5.1 *Female Sexual Behavior*

Neurosteroids may have vital roles in female sexual behavior [53]. Receptivity in the rodent requires ovarian estrogen, which increases progesterone receptor number in the ventral portion of the VMN [54]. The resulting increased sensitivity to ovarian progesterone facilitates lordosis and appropriate tactile stimulation. However, locally produced allopregnanolone may also be important. Participation in paced mating behavior selectively increases allopregnanolone levels in specific regions of the CNS such as the midbrain and the hippocampus [55].

Infusion of allopregnanolone or THDOC into the ventral tegmental area (VTA) enhances and maintains progesterone-facilitated lordosis in estrogen-primed OVX hamsters and rats [56, 57]. Inhibition of allopregnanolone synthesis by blocking 3βHSD or 5α-reductase activity in the VTA attenuates lordosis, whereas local infusion of allopregnanolone rescues this behavior [58, 59]. The mechanism of allopregnanolone action involves the activation of D_1 and $GABA_A$ receptors and cAMP pathways in the VTA [24, 60, 61, 62].

However, the stimulatory effect of allopregnanolone is, again, site-specific. Allopregnanolone levels specifically in the ventral medial hypothalamus fluctuate with the estrus cycle [43]. Its lowest levels are reached at proestrus, a time when exhibitions of receptive behavior coincide with rising serum estrogen and progesterone. Intracerebroventricular administration of allopregnanolone suppresses lordotic behavior by OVX rats [43, 63]. This implies that circulating allopregnanolone (as opposed to neurosteroid allopregnanolone in the VTA) in intact animals is inhibitory through its access to the VMN. In fact, i.c.v. injection of antiserum against allopregnanolone during proestrus augments lordosis [43]. Thus, female sexual behavior may be a product of facilitation by progesterone and disinhibition by serum-derived and possibly neurosteroid allopregnanolone in the VMN, whereas neurosteroid allopregnanolone augments VMN activity through its actions in the VTA, where the progesterone receptor is absent.

Complicating matters, systemic administration of high doses of allopregnanolone can also induce lordosis in estrogen-primed progesterone receptor-knockout OVX aged females [64]. One interpretation of this finding is that allopregnanolone can rescue age-related losses in sexual receptivity when progesterone receptor levels in the VMN are compromised due to lowered central allopregnanolone levels and when neurons may have higher sensitivity to allopregnanolone and $GABA_A$-channel activity [40]. A separate study found that high doses of estrogen also elicit lordosis independent of progesterone receptor action [65]. This implies that estrogen can stimulate receptive behavior by altering allopregnanolone synthesis in the CNS.

Infusion of another 5α-reduced steroid, androstanediol, has similar paradoxical effects on receptive behavior and promotes aggression in female rats, possibly through inhibition of $GABA_A$ receptors in the medial preoptic area (POA) and VMN [66–68]. Neurosteroids are also implicated in the preference for male odors and pheromone-stimulated release of GnRH and ovulation [69]. Therefore, neurosteroids may act at several levels in the CNS to regulate sexual motivation and receptive behavior.

20.5.2 Male Sexual Behavior

Allopregnanolone also figures prominently in the regulation of male sexual behavior. This steroid potentiates $GABA_A$ channel activity in neurons in the medial POA, a region essential for sexual interest, erection, copulation, and ejaculation [70–72]. Pregnenolone sulfate opposes the action of allopregnanolone, as may progesterone.

Loss of the progesterone receptor or subcutaneous (s.c.) infusion of the antiprogestin RU486 enhances mount and intromission frequency in the absence of changes in testosterone or testicular function [73]. Physiologically, it is unclear whether it is serum-derived progesterone, or local synthesis, or both that may regulate male sexual activity.

Estradiol rapidly induces copulatory behavior through its actions in the medial POA and amygdala [74–76]. Again, it is unclear if this behavior is regulated by estrogen synthesized de novo in the brain, although local aromatization of circulating testosterone is clearly required for sexual behavior and maintenance of sensitivity for sexual stimulation.

Neurosteroids are also generated in components of the vomeronasal pathway, which relays stimulatory pheromone signals to the bed nucleus of the stria terminalis and the POA. Consistent with this observation, odor preference for estrus females by male rodents is enhanced by i.c.v. administration of 3αHP and reduced by pregnenolone sulfate [77, 78]. Male rats also exhibit higher cerebellar levels of StAR, P450scc, and aromatase mRNAs, and by implication, higher levels of neurosteroids [79].

Male behavior is influenced by neurosteroids in other types of animals as well. In the zebra finch, sex-dependent differences in time and level of de novo estrogen synthesis in the brain determine the establishment of a key neuronal circuit for male song independent of the presence of gonadal steroids [80–82]. In the quail, diurnal changes in brain levels of the neurosteroid 7α-hydroxypregnenolone occur in males, but not in females, and correspond to increased locomotor activity [83]. Variations in the level of pregnenolone and progesterone in the brains of male newts coincide with day length and breeding season independent of plasma steroid levels [7]. Interestingly, P450scc and 3βHSD primarily localize to the POA in the newt, indicating that the changes in neurosteroids directly relate to sexual behavior and performance.

20.6 Neurosteroids in Mood, Sexual Interest, and Gender-typycal Behaviors

Anxiety, stress, and overall mood strongly impact sexual desire and behavior. Some of the effects on these emotions previously attributed to the actions of neuroactive steroids are now being recognized as mediated by or in coordination with neurosteroids.

Locally produced 5α-reduced steroids allopregnanolone and THDOC are anxiolytic, analgesic, and sedative largely through their regulation of $GABA_A$ channels [26]. Inhibition of 5α-reductase activity in the amygdala increases anxiety and depression [84, 85]. Declines in anxiety elicited by the drug etifoxine correlate with increased allopregnanolone levels in the CNS of sham-operated and GDX/ADX rodents [84]. Circulating levels of the steroid also rise in intact animals suggesting an additional contribution from peripheral sources. Anxiogenic drugs similarly increase levels of anxiolytic steroids in both the brain and the serum [86, 87].

Other neurosteroids such as pregnenolone and its sulfate conjugate increase anxiety and oppose the anxiolytic and sedative actions of benzodiazepines and alcohol [88–90]. Lower doses (0.1 μg/kg versus 1.0 μg/kg) of pregnenolone sulfate administered i.p. are anxiolytic, however, possibly reflecting in part conversion to allopregnanolone [88].

20.6.1 Anxiety and Female Sexual Interest and Behavior

Libido in women is not clearly reliant on ovarian status, but depends more on psychological factors like depression and societal influences [53, 91]. Therefore, anxiolytic neurosteroids may greatly affect sexual interest and motivation. But as with sexual receptivity, the effect of neurosteroids as observed in rodents is specific to individual structures in the brain. Inhibition of allopregnanolone production through the introduction of the 5α-reductase inhibitor finasteride (Proscar) into the amygdala decreases social interactions of estrus females, while infusion of the inhibitor systemically or into the hippocampus increases these encounters [85, 92].

Conversely, addition of allopregnanolone into the VTA in diestrus increases social interactions in female rats along with paced mating behavior to levels similar to proestrus [93]. The changes observed in social activities associated with sexual behavior in proestrus may therefore involve allopregnanolone generated in the brain in response to ovarian steroids. Beyond their effects on anxiety, neurosteroids are part of a larger role in mood and adaptation to stress. Pregnenolone may improve mood, while anxiolytic DHEA is antidepressant [90]. Postpartum depression and anxiety may involve anxiogenic neurosteroids in addition to changes in neuroactive steroids [94, 95].

20.6.2 Anxiety and Male Sexual Interest and Behavior

For men, depression can be an important factor in erectile dysfunction. Interestingly, how neurosteroids affect depression may be gender-dependent. Postnatal stress models indicate a lower sensitivity of adult male rats to repeated exposures of i.c.v.-administered allopregnanolone than females as measured by anxiety-related behavior, such as reduced grooming [96].

Social isolation stress also reduces CNS levels of 5α-reductase and allopregnanolone in males in an androgen-dependent manner, causing increased contextual fear responses and aggressiveness as assessed by resident-intruder tests [97–99]. The changes in allopregnanolone synthesis appear to be restricted to select neurons with outputs to the amygdala and basolateral amygdala glutamatergic neurons, all of which can alter emotional responses [100]. Decreased allopregnanolone levels in the corticolimbic system due to the use of a 5α-reductase inhibitor mimicked the effects of social isolation as measured for fear [101]. Fluoxetine (Prozac) rescues the changes in allopregnanolone and aggression [102].

Chronic treatment with DHEA also inhibits aggression by GDX male mice toward lactating female intruders and correlates with declines in measured pregnenolone sulfate levels in the brain [103]. DHEA has similar effects on aggressive behavior and pregnenolone sulfate in androgenized females [104]. Thus, through their effects on mood, anxiety, and stress, neurosteroids may have additional functions in regulating sexual interest and gender-typical behaviors.

20.7 The Future of Neurosteroids in the Clinic

Neurosteroids are currently being investigated as therapies for a wide variety of psychological disorders, addictive behaviors, neurological disorders, and, due to their role in neuroprotection, neurodegenerative conditions [90]. For instance, a synthetic version of allopregnanolone, ganaxolone, is in advanced clinical trials as a treatment for seizures [105]. Neurosteroids also appear to be prophylactic against age-related declines in nervous system function.

Treatments to directly target sexual and reproductive dysfunctions do not yet exist, although current gonadal replacement strategies to address problems of libido may act in part through the induction of local neurosteroid synthesis. Moreover, therapies that address psychological conditions such as mood disorders may positively impact sexual desire. Early experimental trials have met with mixed success [90, 106]. Most notably, DHEA supplementation has not proven reliable in improving mood in patients [90].

Treatments that involve systemic administration of neurosteroids may not be the best method to address neurosteroid-related problems. Not only does this method increase steroid levels in the target region of the brain but also throughout the brain and the circulation. The presence of a particular steroid in other regions of the brain may have potentially opposing effects, given the site-specificity of neurosteroids. Whereas intrahippocampal administration of pregnenolone sulfate improves memory in aging rats [107], systemic administration of the steroid to increase hippocampal levels may also increase belligerence.

Changes in the level of steroids in nonneuronal tissues due to systemic administration may have adverse effects as well. This is an ongoing concern with the use of estrogen and testosterone to replace faltering gonadal steroid production to improve sexual desire in aging patients. Moreover, neurosteroids can differentially affect target cells depending on concentration and length of treatment.

Another problem is that if endogenous production of a particular neurosteroid is not compromised, further supplementation may not augment its activity. Addition of estrogen to hippocampal neurons only affects cell survival and proliferation when native estrogen synthesis is blocked [108]. This can be an important consideration such as in clinical depression, which has yet to be conclusively linked to a deficiency in anti-depressive neurosteroids (see below) [106]. Consequently, neurosteroids are eyed as a novel treatment for conditions in which their production is compromised, as in age-related cognitive decline and Alzheimer's and Niemann-Pick type C disease [107, 109, 110, 111, 112]. The application of certain neurosteroids like progesterone can also assist in the remyelination and survival of damaged nerves [113].

An alternative approach is to manipulate neurosteroid concentrations in the brain through the use of other pharmacological agents, recently termed "selective brain steroidogenic stimulants" (SBSS) [101]. Antipsychotic agents clozapine and olanzapine can increase allopregnanolone synthesis in the CNS, which in turn may be a mechanism by which these drugs improve schizophrenia [114]. Similarly, reductions in aggression elicited by fluoxetine may primarily result from its stimulation of brain allopregnanolone levels, since the drug is effective at levels tenfold lower than those required to block serotonin reuptake [102]. This is an important finding, since selective serotonin-reuptake inhibitors (SSRIs) like fluoxetine as they are used now are notorious for their adverse impact on libido and sexual function. While withdrawal from SSRI use generally restores sexual interest and function, rare patients do experience persistent genital arousal disorder while on SSRIs [115]. Speculatively, the appearance of this disorder may be hypothesized to be related to SSRI dose and effects on allopregnanolone levels in the brain.

Negative side effects of drugs may also be explained by neurosteroid modulation. Preliminary studies indicate that as in the rodent, finasteride may increase depression in men due to the loss of allopregnanolone [116, 117]. This raises the possibility that changes in neurosteroid levels contribute to the decrease in sexual desire reported by some patients. Treatment strategies that take into account effects on neurosteroids and the development of drugs like SBSSs that singularly target neurosteroid biosynthesis may therefore provide superior therapies for various conditions including those affecting sexual behavior and reproduction.

20.8 Summary

Current research suggests a broad range of functions for neurosteroids in sexual behavior and reproduction. Since select neurosteroids are neuroprotective, they may also indirectly preserve sexual function by promoting neuronal survival in the CNS and PNS. However, a full description of the functions that are truly governed by neurosteroids, not neuroactive steroids, remains a focus of investigation. Future gains in the use of neurosteroids to treat sexual disorders require clear delineation of their functions and how they complement the influences of gonadal and adrenal steroids on the CNS.

References

1. Mellon SH, Deschepper CF. Neurosteroid biosynthesis: genes for adrenal steroidogenic enzymes are expressed in the brain. Brain Res 1993; 629:283–92.
2. King SR, Manna PR, Ishii T, et al. An essential component in steroid synthesis, the steroidogenic acute regulatory protein, is expressed in discrete regions of the brain. J Neurosci 2002; 22:10613–20.
3. King SR, Ginsberg SD, Ishii T, et al. The steroidogenic acute regulatory protein is expressed in steroidogenic cells of the day-old brain. Endocrinology 2004; 145:4775–80.
4. Furukawa A, Miyatake A, Ohnishi T, et al. Steroidogenic acute regulatory protein (StAR) transcripts constitutively expressed in the adult rat central nervous system: colocalization of StAR, cytochrome P-450scc (CYP XIA1), and 3β-hydroxysteroid dehydrogenase in the rat brain. J Neurochem 1998; 71: 2231–8.
5. Compagnone NA, Mellon SH. Neurosteroids: biosynthesis and function of these novel neuromodulators. Front Neuroendocrinol 2000; 21:1–56.
6. Takase M, Ukena K, Yamazaki T, et al. Pregnenolone, pregnenolone sulfate, and cytochrome P450 side-chain cleavage enzyme in the amphibian brain and their seasonal changes. Endocrinology 1999; 140:1936–44.
7. Inai Y, Nagai K, Ukena K, et al. Seasonal changes in neurosteroid concentrations in the amphibian brain and environmental factors regulating their changes. Brain Res 2003; 959: 214–25.
8. Tsutsui K, Matsunaga M, Miyabara H, et al. Neurosteroid biosynthesis in the quail brain: a review. J Exp Zoolog A Comp Exp Biol 2006; 305:733–42.
9. London SE, Monks DA, Wade J, et al. Widespread capacity for steroid synthesis in the avian brain and song system. Endocrinology 2006; 147:5975–87.
10. London SE, Schlinger BA. Steroidogenic enzymes along the ventricular proliferative zone in the developing songbird brain. J Comp Neurol 2007; 502:507–21.
11. Sierra A, Lavaque E, Perez-Martin M, et al. Steroidogenic acute regulatory protein in the rat brain: cellular distribution, developmental regulation and overexpression after injury. Eur J Neurosci 2003; 18:1458–67.
12. Compagnone NA, Bulfone A, Rubenstein JL, et al. Expression of the steroidogenic enzyme P450scc in the central and peripheral nervous systems during rodent embryogenesis. Endocrinology 1995; 136:2689–96.
13. Kishimoto W, Hiroi T, Shiraishi M, et al. Cytochrome P450 2D catalyze steroid 21-hydroxylation in the brain. Endocrinology 2004; 145:699–705.
14. Gomez-Sanchez EP, Ahmad N, Romero DG, et al. Is aldosterone synthesized within the rat brain? Am J Physiol Endocrinol Metab 2005; 288:E342–6.
15. Weill-Engerer S, David JP, Sazdovitch V, et al. Neurosteroid quantification in human brain regions: comparison between Alzheimer's and nondemented patients. J Clin Endocrinol Metab 2002; 87:5138–43.
16. Kriz L, Bicikova M, Hill M, et al. Steroid sulfatase and sulfuryl transferase activity in monkey brain tissue. Steroids 2005; 70:960–9.
17. Ebner MJ, Corol DI, Havlikova H, et al. Identification of neuroactive steroids and their precursors and metabolites in adult male rat brain. Endocrinology 2006; 147:179–90.
18. Rupprecht R, Holsboer F. Neuroactive steroids: mechanisms of action and neuropsychopharmacological perspectives. Trends Neurosci 1999; 22:410–6.

19. Chaban VV, Lakhter AJ, Micevych P. A membrane estrogen receptor mediates intracellular calcium release in astrocytes. Endocrinology 2004; 145:3788–95.
20. Zhu Y, Bond J, Thomas P. Identification, classification, and partial characterization of genes in humans and other vertebrates homologous to a fish membrane progestin receptor. Proc Natl Acad Sci U S A 2003; 100:2237–42.
21. Dong Y, Fu YM, Sun JL, et al. Neurosteroid enhances glutamate release in rat prelimbic cortex via activation of α1-adrenergic and sigma1 receptors. Cell Mol Life Sci 2005; 62:1003–14.
22. Bertrand D, Valera S, Bertrand S, et al. Steroids inhibit nicotinic acetylcholine receptors. Neuroreport 1991; 2:277–80.
23. Bullock AE, Clark AL, Grady SR, et al. Neurosteroids modulate nicotinic receptor function in mouse striatal and thalamic synaptosomes. J Neurochem 1997; 68:2412–23.
24. Frye CA, Walf AA, Sumida K. Progestins' actions in the VTA to facilitate lordosis involve dopamine-like type 1 and 2 receptors. Pharmacol Biochem Behav 2004; 78:405–18.
25. Gee KW, McCauley LD, Lan NC. A putative receptor for neurosteroids on the GABAA receptor complex: the pharmacological properties and therapeutic potential of epalons. Crit Rev Neurobiol 1995; 9:207–27.
26. Lambert JJ, Belelli D, Peden DR, et al. Neurosteroid modulation of GABAA receptors. Prog Neurobiol 2003; 71:67–80.
27. Reddy DS. Role of neurosteroids in catamenial epilepsy. Epilepsy Res. 2004; 62:99–118.
28. Belelli D, Lambert JJ. Neurosteroids: endogenous regulators of the GABA(A) receptor. Nat Rev Neurosci 2005; 6:565–75.
29. Morrow AL, Suzdak PD, Paul SM. Steroid hormone metabolites potentiate GABA receptor-mediated chloride ion flux with nanomolar potency. Eur J Pharmacol 1987; 142:483–5.
30. Brot MD, Akwa Y, Purdy RH, et al. The anxiolytic-like effects of the neurosteroid allopregnanolone: interactions with GABA(A) receptors. Eur J Pharmacol 1997; 325:1–7.
31. Weir CJ, Ling AT, Belelli D, et al. The interaction of anaesthetic steroids with recombinant glycine and GABAA receptors. Br J Anaesth 2004; 92:704–11.
32. Majewska MD, Harrison NL, Schwartz RD, et al. Steroid hormone metabolites are barbiturate-like modulators of the GABA receptor. Science 1986; 232:1004–7.
33. Majewska MD, Schwartz RD. Pregnenolone-sulfate: an endogenous antagonist of the gamma-aminobutyric acid receptor complex in brain? Brain Res 1987; 404:355–60.
34. Majewska MD, Demirgoren S, Spivak CE, et al. The neurosteroid dehydroepiandrosterone sulfate is an allosteric antagonist of the GABAA receptor. Brain Res 1990; 526:143–6.
35. Chappell PE, Schneider JS, Kim P, et al. Absence of gonadotropin surges and gonadotropin-releasing hormone self-priming in ovariectomized (OVX), estrogen (E2)-treated, progesterone receptor knockout (PRKO) mice. Endocrinology 1999; 140:3653–8.
36. Chappell PE, Levine JE. Stimulation of gonadotropin-releasing hormone surges by estrogen I. Role of hypothalamic progesterone receptors. Endocrinology 2000; 141:1477–85.
37. Micevych P, Sinchak K, Mills RH, et al. The luteinizing hormone surge is preceded by an estrogen-induced increase of hypothalamic progesterone in ovariectomized and adrenalectomized rats. Neuroendocrinology 2003; 78:29–35.
38. Soma KK, Sinchak K, Lakhter A, et al. Neurosteroids and female reproduction: estrogen increases 3β-HSD mRNA and activity in rat hypothalamus. Endocrinology 2005; 146:4386–90.
39. Micevych PE, Chaban V, Ogi J, et al. Estradiol stimulates progesterone synthesis in hypothalamic astrocyte cultures. Endocrinology 2007; 148:782–9.
40. Genazzani AR, Stomati M, Bernardi F, et al. Conjugated equine estrogens reverse the effects of aging on central and peripheral allopregnanolone and β-endorphin levels in female rats. Fertil Steril 2004; 81 Suppl 1:757–66.
41. Sullivan SD, Moenter SM. Neurosteroids alter gamma-aminobutyric acid postsynaptic currents in gonadotropin-releasing hormone neurons: a possible mechanism for direct steroidal control. Endocrinology 2003; 144:4366–75.
42. El-Etr M, Akwa Y, Fiddes RJ, et al. A progesterone metabolite stimulates the release of gonadotropin-releasing hormone from GT1-1 hypothalamic neurons via the gamma-aminobutyric acid type A receptor. Proc Natl Acad Sci U S A 1995; 92:3769–73.
43. Genazzani AR, Palumbo MA, de Micheroux AA, et al. Evidence for a role for the neurosteroid allopregnanolone in the modulation of reproductive function in female rats. Eur J Endocrinol 1995; 133:375–80.
44. El-Etr M, Akwa Y, Baulieu EE, et al. The neuroactive steroid pregnenolone sulfate stimulates the release of gonadotropin-releasing hormone from GT1-7 hypothalamic neurons, through N-methyl-D-aspartate receptors. Endocrinology 2006; 147: 2737–43.
45. Calogero AE, Palumbo MA, Bosboom AM, et al. The neuroactive steroid allopregnanolone suppresses hypothalamic gonadotropin-releasing hormone release through a mechanism mediated by the gamma-aminobutyric acidA receptor. J Endocrinol 1998; 158:121–5.
46. Wiebe JP, Boushy D, Wolfe M. Synthesis, metabolism and levels of the neuroactive steroid, 3α-hydroxy-4-pregnen-20-one (3αHP), in rat pituitaries. Brain Res 1997; 764:158–66.
47. Wiebe JP, Wood PH. Selective suppression of follicle-stimulating hormone by 3 α-hydroxy-4-pregnen-20-one, a steroid found in Sertoli cells. Endocrinology 1987; 120:2259–64.
48. Wood PH, Wiebe JP. Selective suppression of follicle-stimulating hormone secretion in anterior pituitary cells by the gonadal steroid 3 α-hydroxy-4-pregnen-20-one. Endocrinology 1989; 125:41–8.
49. Beck CA, Wolfe M, Murphy LD, et al. Acute, nongenomic actions of the neuroactive gonadal steroid, 3 α-hydroxy-4-pregnen-20-one (3 α HP), on FSH release in perifused rat anterior pituitary cells. Endocrine 1997; 6:221–29.
50. Wiebe JP, Dhanvantari S, Watson PH, et al. Suppression in gonadotropes of gonadotropin-releasing hormone-stimulated follicle-stimulating hormone release by the gonadal- and neurosteroid 3 α-hydroxy-4-pregnen-20-one involves cytosolic calcium. Endocrinology 1994; 134:377–82.
51. Dhanvantari S, Wiebe JP. Suppression of follicle-stimulating hormone by the gonadal- and neurosteroid 3 α-hydroxy-4-pregnen-20-one involves actions at the level of the gonadotrope membrane/calcium channel. Endocrinology 1994; 134:371–6.
52. Liu T, Wimalasena J, Bowen RL, et al. Luteinizing hormone receptor mediates neuronal pregnenolone production via up-regulation of steroidogenic acute regulatory protein expression. J Neurochem 2007; 100:1329–39.
53. King SR, Lamb DJ. Why we loose interest in sex: do neurosteroids play a role? Sexuality, Reproduction, and Menopause 2006; 4:20–3.
54. Pfaff D, Schwartz-Giblin S. Cellular mechanisms of female reproductive behaviors. In: Knobil E, Neill J, Ewing L, et al, editors. The Physiology of Reproduction. New York: Raven Press, 1998:1487–568.
55. Frye CA, Paris JJ, Rhodes. Engaging in paced mating, but neither exploratory, anti-anxiety, nor social behavior, increases 5α-reduced progestin concentrations in midbrain, hippocampus, striatum, and cortex. Reproduction 2007; 133:663–74.
56. Frye CA, DeBold JF. 3 α-OH-DHP and 5 α-THDOC implants to the ventral tegmental area facilitate sexual receptivity in hamsters after progesterone priming to the ventral medial hypothalamus. Brain Res 1993; 612:130–7.

57. Frye CA, Gardiner SG. Progestins can have a membrane-mediated action in rat midbrain for facilitation of sexual receptivity. Horm Behav 1996; 30:682–91.
58. Frye CA, Vongher JM. Ventral tegmental area infusions of inhibitors of the biosynthesis and metabolism of 3α,5α-THP attenuate lordosis of hormone-primed and behavioural oestrous rats and hamsters. J Neuroendocrinol 2001; 13:1076–86.
59. Petralia SM, Jahagirdar V, Frye CA. Inhibiting biosynthesis and/or metabolism of progestins in the ventral tegmental area attenuates lordosis of rats in behavioural oestrus. J Neuroendocrinol 2005; 17:545–52.
60. Frye CA. The role of neurosteroids and non-genomic effects of progestins and androgens in mediating sexual receptivity of rodents. Brain Res Brain Res Rev 2001; 37:201–22.
61. Frye CA. The role of neurosteroids and nongenomic effects of progestins in the ventral tegmental area in mediating sexual receptivity of rodents. Horm Behav 2001; 40:226–33.
62. Petralia SM, Frye CA. In the ventral tegmental area, G-proteins and cAMP mediate the neurosteroid 3α,5α-THP's actions at dopamine type 1 receptors for lordosis of rats. Neuroendocrinology 2004; 80:233–43.
63. Laconi MR, Cabrera RJ. Effect of centrally injected allopregnanolone on sexual receptivity, luteinizing hormone release, hypothalamic dopamine turnover, and release in female rats. Endocrine 2002; 17:77–83.
64. Frye CA, Sumida K, Lydon JP, et al. Mid-aged and aged wild-type and progestin receptor knockout (PRKO) mice demonstrate rapid progesterone and 3α,5α-THP-facilitated lordosis. Psychopharmacology (Berl) 2006; 185:423–32.
65. Apostolakis EM, Garai J, Lohmann JE, et al. Epidermal growth factor activates reproductive behavior independent of ovarian steroids in female rodents. Mol Endocrinol 2000; 14:1086–98.
66. Frye CA, Van Keuren KR, Erskine MS. Behavioral effects of 3 α-androstanediol. I: Modulation of sexual receptivity and promotion of GABA-stimulated chloride flux. Behav Brain Res 1996; 79:109–18.
67. Frye CA, Duncan JE, Basham M, et al. Behavioral effects of 3 α-androstanediol. II: Hypothalamic and preoptic area actions via a GABAergic mechanism. Behav Brain Res 1996; 79: 119–30.
68. Frye CA, Van Keuren KR, Rao PN, et al. Progesterone and 3 α-androstanediol conjugated to bovine serum albumin affects estrous behavior when applied to the MBH and POA. Behav Neurosci 1996; 110:603–12.
69. More L. Mouse major urinary proteins trigger ovulation via the vomeronasal organ. Chem Senses 2006; 31:393–401.
70. Haage D, Johansson S. Neurosteroid modulation of synaptic and GABA-evoked currents in neurons from the rat medial preoptic nucleus. J Neurophysiol 1999; 82:143–51.
71. Uchida S, Noda E, Kakazu Y, et al. Allopregnanolone enhancement of GABAergic transmission in rat medial preoptic area neurons. Am J Physiol Endocrinol Metab 2002; 283:E1257–65.
72. Haage D, Backstrom T, Johansson S. Interaction between allopregnanolone and pregnenolone sulfate in modulating GABA-mediated synaptic currents in neurons from the rat medial preoptic nucleus. Brain Res 2005; 1033:58–67.
73. Schneider JS, Burgess C, Sleiter NC, et al. Enhanced sexual behaviors and androgen receptor immunoreactivity in the male progesterone receptor knockout mouse. Endocrinology 2005; 146:4340–8.
74. Huddleston GG, Michael RP, Zumpe D, et al. Estradiol in the male rat amygdala facilitates mounting but not ejaculation. Physiol Behav 2003; 79:239–46.
75. Balthazart J, Baillien M, Cornil CA, et al. Preoptic aromatase modulates male sexual behavior: slow and fast mechanisms of action. Physiol Behav 2004; 83:247–70.
76. Huddleston GG, Paisley JC, Clancy AN. Effects of estrogen in the male rat medial amygdala: infusion of an aromatase inhibitor lowers mating and bovine serum albumin-conjugated estradiol implants do not promote mating. Neuroendocrinology 2006; 83:106–16.
77. Kavaliers M, Wiebe JP, Galea LA. Male preference for the odors of estrous female mice is enhanced by the neurosteroid 3 α-hydroxy-4-pregnen-20-one (3 α HP). Brain Res 1994; 646: 140–4.
78. Kavaliers M, Kinsella DM. Male preference for the odors of estrous female mice is reduced by the neurosteroid pregnenolone sulfate. Brain Res 1995; 682:222–6.
79. Lavaque E, Mayen A, Azcoitia I, et al. Sex differences, developmental changes, response to injury and cAMP regulation of the mRNA levels of steroidogenic acute regulatory protein, cytochrome p450scc, and aromatase in the olivocerebellar system. J Neurobiol 2006; 66:308–18.
80. Wade J, Arnold AP. Functional testicular tissue does not masculinize development of the zebra finch song system. Proc Natl Acad Sci U S A 1996; 93:5264–8.
81. Holloway CC, Clayton DF. Estrogen synthesis in the male brain triggers development of the avian song control pathway in vitro. Nat Neurosci 2001; 4:170–5.
82. Forlano PM, Schlinger BA, Bass AH. Brain aromatase: new lessons from non-mammalian model systems. Front Neuroendocrinol 2006; 27:247–74.
83. Tsutsui K, Inoue K, Miyabara H, et al. 7α-hydroxypregnenolone mediates melatonin action underlying diurnal locomotor rhythms. Neurosci 2008; 28:2158–67.
84. Verleye M, Akwa Y, Liere P, et al. The anxiolytic etifoxine activates the peripheral benzodiazepine receptor and increases the neurosteroid levels in rat brain. Pharmacol Biochem Behav 2005; 82:712–20.
85. Walf AA, Sumida K, Frye CA. Inhibiting 5α-reductase in the amygdala attenuates antianxiety and antidepressive behavior of naturally receptive and hormone-primed ovariectomized rats. Psychopharmacology (Berl) 2006; 186:302–11.
86. Barbaccia ML, Roscetti G, Bolacchi F, et al. Stress-induced increase in brain neuroactive steroids: antagonism by abecarnil. Pharmacol Biochem Behav 1996; 54:205–10.
87. Barbaccia ML, Roscetti G, Trabucchi M, et al. Isoniazid-induced inhibition of GABAergic transmission enhances neurosteroid content in the rat brain. Neuropharmacology 1996; 35: 1299–305.
88. Melchior CL, Ritzmann RF. Pregnenolone and pregnenolone sulfate, alone and with ethanol, in mice on the plus-maze. Pharmacol Biochem Behav 1994; 48:893–7.
89. Meieran SE, Reus VI, Webster R, et al. Chronic pregnenolone effects in normal humans: attenuation of benzodiazepine-induced sedation. Psychoneuroendocrinology 2004; 29:486–500.
90. Strous RD, Maayan R, Weizman A. The relevance of neurosteroids to clinical psychiatry: from the laboratory to the bedside. Eur Neuropsychopharmacol 2006; 16:155–69.
91. Avis NE, Zhao X, Johannes CB, et al. Correlates of sexual function among multi-ethnic middle-aged women: results from the Study of Women's Health Across the Nation (SWAN). Menopause 2005; 12:385–98.
92. Rhodes ME, Frye CA. Inhibiting progesterone metabolism in the hippocampus of rats in behavioral estrus decreases anxiolytic behaviors and enhances exploratory and antinociceptive behaviors. Cogn Affect Behav Neurosci 2001; 1:287–96.
93. Frye CA, Rhodes ME. Infusions of 3α,5α-THP to the VTA enhance exploratory, anti-anxiety, social, and sexual behavior and increase levels of 3α,5α-THP in the midbrain, hippocampus, diencephalon, and cortex of female rats. Behav Brain Res 2008; 187:88–99.

94. Maayan R, bou-Kaud M, Strous RD, et al. The influence of parturition on the level and synthesis of sulfated and free neurosteroids in rats. Neuropsychobiology 2004; 49:17–23.
95. Maayan R, Strous RD, bou-Kaoud M, et al. The effect of 17β estradiol withdrawal on the level of brain and peripheral neurosteroids in ovarectomized rats. Neurosci Lett 2005; 384:156–61.
96. Zimmerberg B, Rackow SH, George-Friedman KP. Sex-dependent behavioral effects of the neurosteroid allopregnanolone (3α,5α-THP) in neonatal and adult rats after postnatal stress. Pharmacol Biochem Behav 1999; 64:717–24.
97. Dong E, Matsumoto K, Uzunova V, et al. Brain 5α-dihydroprogesterone and allopregnanolone synthesis in a mouse model of protracted social isolation. Proc Natl Acad Sci U S A 2001; 98:2849–54.
98. Pinna G, Agís-Balboa RC, Doueiri MS, et al. Brain neurosteroids in gender-related aggression induced by social isolation. Crit Rev Neurobiol 2004; 16:75–82.
99. Pinna G, Costa E, Guidotti A. Changes in brain testosterone and allopregnanolone biosynthesis elicit aggressive behavior. Proc Natl Acad Sci U S A 2005; 102:2135–40.
100. Agís-Balboa RC, Pinna G, Pibiri F, et al. Down-regulation of neurosteroid biosynthesis in corticolimbic circuits mediates social isolation-induced behavior in mice. Proc Natl Acad Sci U S A 2007; 104:18736–41.
101. Pibiri F, Nelson M, Guidotti A, et al. Decreased corticolimbic allopregnanolone expression during social isolation enhances contextual fear: a model relevant for posttraumatic stress disorder. Proc Natl Acad Sci U S A 2008; 105:5567–72.
102. Pinna G, Dong E, Matsumoto K, et al. In socially isolated mice, the reversal of brain allopregnanolone down-regulation mediates the anti-aggressive action of fluoxetine. Proc Natl Acad Sci U S A 2003; 100:2035–40.
103. Young J, Corpechot C, Haug M, et al. Suppressive effects of dehydroepiandrosterone and 3 β-methyl-androst-5-en-17-one on attack towards lactating female intruders by castrated male mice. II. Brain Neurosteroids. Biochem Biophys Res Commun 1991; 174:892–7.
104. Robel P, Young J, Corpechot C, et al. Biosynthesis and assay of neurosteroids in rats and mice: functional correlates. J Steroid Biochem Mol Biol 1995; 53:355–60.
105. Nohria V, Giller E. Ganaxolone. Neurotherapeutics 2007; 4: 102–5.
106. Uzunova V, Sampson L, Uzunov DP. Relevance of endogenous 3α-reduced neurosteroids to depression and antidepressant action. Psychopharmacology (Berl) 2006; 186:351–61.
107. Vallée M, Mayo W, Darnaudéry M, et al. Neurosteroids: deficient cognitive performance in aged rats depends on low pregnenolone sulfate levels in the hippocampus. Proc Natl Acad Sci U S A 1997; 94:14865–70.
108. Fester L, Ribeiro-Gouveia V, Prange-Kiel J, et al. Proliferation and apoptosis of hippocampal granule cells require local oestrogen synthesis. J Neurochem 2006; 97:1136–44.
109. Wang JM, Irwin RW, Liu L, et al. Regeneration in a degenerating brain: potential of allopregnanolone as a neuroregenerative agent. Curr Alzheimer Res 2007; 4:510–7.
110. Griffin LD, Gong W, Verot L, et al. Niemann-Pick type C disease involves disrupted neurosteroidogenesis and responds to allopregnanolone. Nat Med 2004; 10:704–11.
111. Chen G, Li HM, Chen YR, et al. Decreased estradiol release from astrocytes contributes to the neurodegeneration in a mouse model of Niemann-Pick disease type C. Glia 2007; 55:1509–18.
112. Liu B, Li H, Repa JJ, et al. Genetic variations and treatments that affect the lifespan of the NPC1 mouse. J Lipid Res 2008; 49: 663–9.
113. Schumacher M, Guennoun R, Stein DG, et al. Progesterone: therapeutic opportunities for neuroprotection and myelin repair. Pharmacol Ther 2007; 116:77–106.
114. Marx CE, Shampine LJ, Duncan GE, et al. Clozapine markedly elevates pregnenolone in rat hippocampus, cerebral cortex, and serum: candidate mechanism for superior efficacy? Pharmacol Biochem Behav 2006; 84:598–608.
115. Leiblum SR, Goldmeier D. Persistent genital arousal disorder in women: case reports of association with anti-depressant usage and withdrawal. J Sex Marital Ther 2008; 34:150–9.
116. Altomare G, Capella GL. Depression circumstantially related to the administration of finasteride for androgenetic alopecia. J Dermatol 2002; 29:665–9.
117. Rahimi-Ardabili B, Pourandarjani R, Habibollahi P, et al. Finasteride induced depression: a prospective study. BMC Clin Pharmacol 2006; 6:7.

Part IV
Molecular Regulation of the Reproductive Organs and Tissues

Chapter 21

Autocrine and Paracrine Regulation of the Ovary

Marta Tesone, Dalhia Abramovich, Griselda Irusta, and Fernanda Parborell

21.1 Introduction

Ovarian follicular development and regression is a continuous and cyclic process that depends on a number of endocrine, paracrine, and autocrine signals. Although many of the factors involved in follicular growth have been characterized, it remains unknown why one or more, depending on species, preovulatory follicle(s) emerge as dominant and the others regress. It is postulated that the selected dominant follicle(s) possesses a higher sensitivity to FSH due to increased expression of FSH receptors. As a result, increases in estradiol and inhibin trigger a negative feedback mechanism that prevents the other follicles from continuing their development. The ovarian theca and granulosa cells also play an important role, since they produce steroid hormones required for normal follicular growth. In addition, ovarian growth factors, cytokines, and neuropeptides participate as regulators of follicular growth and in the formation of the ovarian cell compartments as well.

21.2 Selection of the Dominant Follicle

In the first days of a woman's menstrual cycle the circulating levels of FSH increase and as a consequence, a group of antral follicles escape the process of apoptosis, that lead them to follicular atresia. Within this group, approximately ten antral follicles begin to grow faster and produce increased amounts of estrogens and inhibin. One follicle is dominant and is selected to ovulate. The selected preovulatory follicle possesses a higher sensitivity to FSH due to increased expression of FSH and/or LH receptors and produces more estrogens than the remaining follicles. Associated with this process, estradiol and local growth factors exert a permissive effect, amplifying the action of FSH in the maturing follicles. Also, the increase in estradiol and inhibin trigger a negative feedback mechanism at the hypothalamus-hypophysis level that causes a decrease in the release of FSH, which further prevents the other follicles from continuing their development [1, 2]. The decrease in FSH causes a decrease in the activity of FSH-dependent aromatase, which limits the availability of estrogens to the less mature follicles. This leads to both a decrease in the proliferation of granulosa cells and an increase in androgens, causing irreversible atresia. The thecal layer becomes highly vascularized, which increases supply of circulating FSH to the dominant follicle resulting in stimulation of granulosa cell proliferation.

In rats treated with estrogens, FSH stimulates expression of LH and prolactin receptors in granulosa cells, reaching maximum levels just before ovulation [3, 4]. Estrogens exert a positive feedback effect on the pituitary gland that triggers the preovulatory discharge of LH. These synchronized processes appear to be the determining factors for the selection of multiple dominant follicles that finally ovulate [1]. In addition, the dominant follicles produce atretogenic factors that cause atresia of neighbouring or subordinate follicles [1, 5].

21.3 The Role of Steroids: The Two-Cell Two-Gonadotropin Concept

The microenvironment of the follicular ovarian antrum facilitates the access of FSH and LH, which amplify the paracrine and autocrine signals produced in the ovary. Therefore, all the cells in one follicle are immersed in the same environment, while contiguous follicles can be at different stages of growth. The presence of estrogens and FSH in the antral fluid is essential for granulosa cell proliferation and follicular growth [6]. High concentrations of estrogens and a lower androgen/estrogen relationship cause higher rates of cell proliferation within antral follicles; therefore, these follicles are

M. Tesone (✉)
Institute of Experimental Biology and Medicine-CONICET, Buenos Aires, Argentina
e-mail: mtesone@dna.uba.ar

P.J. Chedrese (ed.), *Reproductive Endocrinology*,
DOI 10.1007/978-0-387-88186-7_21, © Springer Science+Business Media, LLC 2009

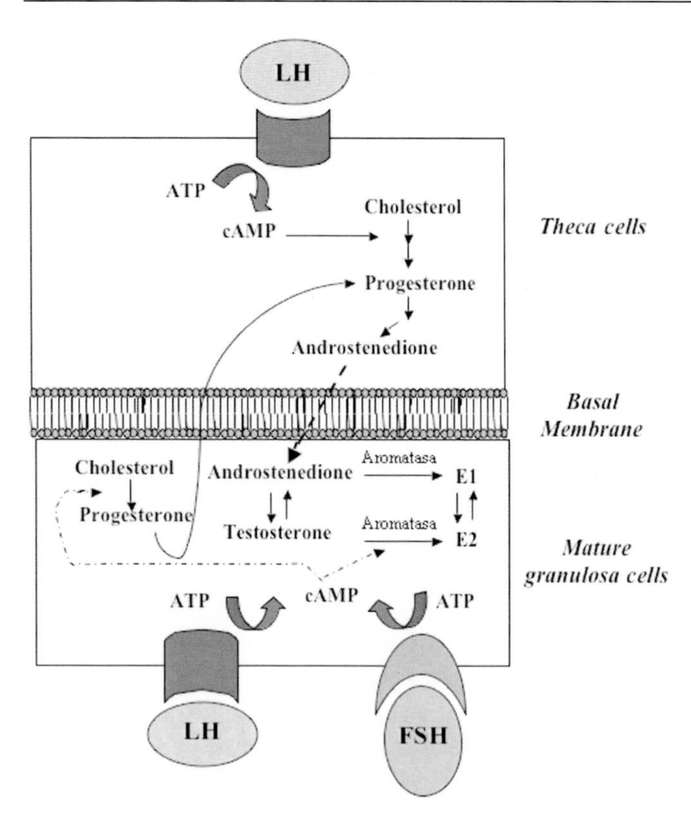

Fig. 21.1 Scheme describing the "two-cell, two gonadotropin" concept. Modified from Fortune and Armstrong [11]

able to maintain a viable oocyte, whereas an androgenic environment promotes oocyte degeneration [7].

Synthesis of the ovarian steroid hormones—androgens, estrogens, and progestins—is compartimentalized within each follicle of the human ovary [6, 8]. Granulosa cells in response to the gonadotropins produce progesterone that diffuses to the thecal cells, which lack FSH receptors, and serves as a substrate for androgen synthesis under the control of LH [9]. The androgens diffuse back through the basal membrane to the granulose cells where they are aromatized into estrogens under control of FSH [10]. The capability of these two cell types to each produce a different steroid hormone leads to the theory of "two-cell, two-gonadotropin" regulation of steroidogenesis in the ovary, a physiological concept currently accepted (Fig. 21.1) [11].

21.4 Intraovarian Regulators

Although the role of gonadotropins and gonadal steroids in ovarian folliculogenesis is unquestionable, the fact that follicles reach different outcomes indicates that intraovarian modulating systems play a role. Growth factors, cytokines, and neuropeptides are among the intraovarian regulators that have been thoroughly studied. These agents are not classical endocrine extracellular signals; they are recognized as intraovarian regulators that modulate growth and function of the ovarian cell compartments.

21.4.1 Epidermal Growth Factor/Transforming Growth Factor-α (EGF/TGF-α)

The mature epidermal growth factor (EGF) consists of a single polypeptide strand of 53 amino acids and three internal disulfide bonds. EGF is able to induce a great variety of physiological responses. EGF stimulates the proliferation of different cell types in culture, such as epidermal cells, fibroblasts, crystalline cells, glial cells, and vascular endothelial cells [12, 13]. The EGF receptor is a monomeric glycoprotein that binds EGF with high affinity and specificity. In the ovary, the EGF receptor is present in thecal cells, granulosa cells, and corpus luteum [14]. EGF binding induces activation of *tyrosine* kinases, causing both autophosphorylation and phosphorylation of other cellular substrates on their *tyrosine* residues. The *tyrosine* kinase inhibitor, *genistein*, blocks the apoptosis suppressive effect of EGF in granulosa cells ECF [15].

Transforming growth factor-α (TGF-α) is a structural analogue of EGF composed of 50 amino acids, capable of binding to a common receptor for both EGF and TGF-α [16]. TGF-α was first isolated from supernatants of tumor cells in culture, but later also found in the pituitary gland, brain, ovary, and macrophages. In vivo, FSH positively regulates expression of the EGF and TGF-α genes [17, 18]. The paracrine effects of the EGF/TGF-α secreted by interstitial thecal cells and the autocrine effects of the *basic fibroblastic growth factor* (bFGF) prevent apoptosis of granulosa cells from follicles selected to ovulate [15].

21.4.2 Insulin-like Growth Factors

Insulin-like growth factor-I (IGF-I) is a ubiquitous 70-amino acid polypeptide that has various functions in different tissues. In the rat ovary, IGF-I mRNA is localized in granulosa cells of developing follicles [19]. These cells possess receptors that bind IGF-I with higher affinity than IGF-II or insulin [20]. IGF-I acts synergistically with gonadotropins, stimulating synthesis of estradiol and progesterone and expression of LH/hCG receptors [20]. This synergistic effect suggests that IGF-I plays an important role in the selection of preovulatory follicles.

Insulin-like growth factor-II (IGF-II) is a 67-amino acid polypeptide, with 62% homology with IGF-I. IGF-II is expressed in theca cells of antral follicles and granulosa cells of dominant follicles [21, 22]. Thus, the switch in the expression of the IGF-II gene from theca to granulosa cells during follicle selection may be considered a distinct functional feature of the dominant follicle. These observations suggest that IGFs support ovarian intercompartmental communications during follicular development.

Insulin-like growth factor activity is modulated by low molecular weight proteins termed IGF binding proteins (IGBP) that bind specifically to IGFs. There are six IGFBPs that regulate IGF expression in the ovary, either by binding IGFs or by exerting a direct effect on steroidogenesis. The regulation of expression is dependant on the presence of FSH [23]. IGFBP-4 and IGFBP-5 found in rat granulosa cells, have been reported to decrease with FSH treatment [20]. In situ analysis detected the presence of IGFBPs in atretic follicles but not healthy follicles [24]. Treatment with IGF-I alone, as well as in combination with FSH and hCG, suppresses spontaneous apoptosis in rat preovulatory follicles [25]. These results suggest that IGF-I acts as a survival factor in ovarian follicular cells.

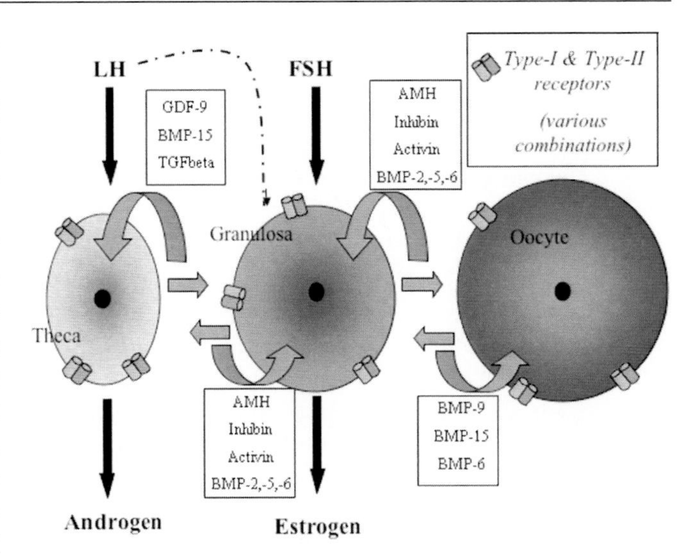

Fig. 21.2 The members of the *TGF-β* superfamily are among the main extracellular ligands involved in the bi-directional communication between theca and granulosa cells and between granulosa cells and the oocyte. They participate both by autocrine (*gray arrows*) and paracrine (*black arrows*) signals, depending on the expression of the adequate receptors on the cell surface

21.4.3 Transforming Growth Factor-β

Transforming growth factor-β (TGF-β), also referred to as tumor growth factor when upregulated in cancer, is part of a superfamily of ubiquitous proteins involved in numerous physiological processes. The TGF-β superfamily has been classified into different subfamilies according to their structural characteristics, which include *bone morphogenetic protein* (BMP), *growth and differentiation factors* (GDF), *inhibins* and *activins*, *neurotrophic factors* (GDNF) derived from glial cells, and *anti-Müllerian hormone* (AMH).

Transforming growth factor-β consists of 25-kDa polypeptides composed of two homodimeric chains. TGF-β is synthesized as an inactive latent form that is regulated by enzymatic conversion to its active form [26]. Three isoforms of TGF-β have been identified: TGF-β1, TGF-β2, and TGF-β3. TGF-β exerts regulatory actions on a variety of tissues, including antiproliferative effects involving control of expression of the *myc* gene, stimulation of chemotaxis in fibroblasts and other cell types, stimulation of tissue repair and bone formation, and increase in survival of neurons. In the ovary, the presence of TGF-β1 and TGF-β2 mRNA has been found in oocytes and granulosa and thecal cells [27, 28, 29]. TGF-β1 and TGF-β2 factors act synergistically, in the ovary, with the gonadotropins to control follicular cell differentiation. TGF-β is involved in the bi-directional communication between granulosa and thecal cells and also between granulosa cells and the oocyte (Fig. 21.2) as well. It was also reported that TGF-β1 alters the proliferation and differentiation of granulosa cells in rats [30, 31]. Inhibins are glycoproteins consisting of two α (18 kDa) and β (12 kDa) subunits. Inhibin subunits form heterodimers, constituted by a common α-subunit, but with different β-subunits, known as βA and βB. Inhibin A (αβA) and inhibin B (αβB) are mainly produced in the gonads. Ovarian inhibins, synthesized in granulosa cells inhibit synthesis of FSH in the pituitary gland. Inhibin B is secreted during the early follicular phase and decreases thereafter. The levels of inhibin A are low in the follicular phase but increase in the luteal phase [32].

Activins are composed of dimers of inhibin β-subunits: βAβB, βAβA, or βBβB. In humans activins are bound mainly to the serum protein, follistatin. During the menstrual cycle, fewer variations of activin A than inhibin are observed [33]. High levels of activins have been observed in the mid-cycle and at the end of the luteal phase. In addition, serum levels of activins are high during pregnancy. Activins synthesized by the granulosa cells exert autocrine and paracrine effects in the follicle, stimulating expression of LH receptors in the granulosa cells and inhibiting LH-induced synthesis of androgens in thecal cells. Lastly, activins possess an important role in the development of the oocyte and cumulus cells, which express the inhibin/activin subunits, α, βA, βB, and follistatin [34].

21.4.4 Fibroblastic Growth Factors

Fibroblastic growth factors (FGFs) are 146-amino acid polypeptides with mitogenic effects in several cells derived from the mesoderm and the neuroectoderm. There are several homologs of the FGFs. Basic fibroblastic growth factor (bFGF) is a truncated homolog that lacks the first 15 residues and has been identified in the ovarian corpus luteum [35]. Both the mRNA and the receptor for bFGF have been identified in human granulosa cells [36]. It has been suggested that bFGF modulates angiogenesis, cell proliferation, progesterone synthesis, and apoptosis in the ovary [37, 38, 39]. Like EGF and TGF-α, FGF acts through a *tyrosine* kinase receptor. Studies carried out in rat granulosa cell cultures and preovulatory follicles have shown that bFGF, EGF, and TGF-α are capable of inhibiting DNA apoptotic fragmentation [15]. These results suggest that these growth factors have important functions in the modulation of programmed cell death in ovarian follicles.

21.4.5 Neuronal Factors

Neurotransmitters, such as noradrenalin and dopamine, reach the ovary by functioning as neuroendocrine signals between oocytes and granulosa cells. However, it has been demonstrated that some neurotransmitters are produced intrinsically in neuronal and non-neuronal cells of the ovary. Dopamine, synthesized in the granulosa cells, is actively transported to the oocyte where it is converted into noradrenalin [40]. Granulosa cells also produce acetylcholine [41] that acts as an autocrine regulator, stimulating granulosa cell proliferation [42] and synthesis of steroidogenic acute regulatory protein (StAR) [43], which increases sensitiviy to LH. The ovary also produces other neurotransmitters such as γ-aminobutyric acid (GABA) [44], somatostatin [45], and cholecystokinin [46], although their functions have not yet been elucidated. In addition, the ovary contains four of the five known neurotrophins, *nerve growth factor* (NGF), *brain-derived neurotrophic factor* (BDNF), *neurotrophin 4/5* (NT-4/5), and *neurotrophin 6* (NT-6) and their respective receptors [47]. These neurotrophins promote the proliferation of granulosa cells regulating early follicular development [48]. NGFs also appear to act during ovulation, emitting a signal that induces loss of intercellular adhesion between the oocyte complex and the follicular wall [49].

21.4.6 Ovarian GnRH Peptides

There is substantial evidence to support the existence of *gonadotropin releasing hormone* (GnRH) peptides and their receptors in the ovary [50]. It has been proposed that GnRHs are synthesized locally and, therefore, play either an autocrine or a paracrine role in gonadal function.

GnRH analogs are synthetic nonapeptides, designed to interact with the GnRH receptor that are used for therapeutic purposes, such as control of ovarian hyperstimulation in assisted reproductive techniques. GnRH agonist (GnRH-a) is a GnRH analog that activates the pituitary GnRH receptor initially increasing secretion of FSH and LH; however, after prolonged administration reduces the endogenous secretion of gonadotropins, FSH and LH, through a mechanism of GnRH receptor desensitization, causing a decrease in the ovarian steroids, particularly estrogens. Antigonadal action of GnRH-a has been described in granulosa cells and the corpus luteum in both rats and humans [50, 51, 52, 53, 54]. In patients undergoing assisted fertilization treatments, the generic drug, leuprolide acetate, which acts as a GnRH-a, is administered together with gonadotropins to inhibit the endogenous peak of LH.

In addition, an inhibitory action of GnRH-a on follicular development due to increased follicular apoptosis has been observed. GnRH-a increases follicular DNA fragmentation and is related to an imbalance in the relation of anti-apoptotic/proapoptotic proteins of the bcl-2 family, with an increase in the expression of the proapoptotic protein of Bclx-s [55, 56, 57]. Administration of a GnRH-a to superovulated rats reduces androgens and estradiol serum levels, mediated by changes in expression of StAR [58] and decreased expression of the CYP 17 hydroxylase. These findings suggest that GnRH acts as an intraovarian factor able to interfere with gonadotropin regulated follicular development by increasing apoptosis and changing expression of steroidogenic enzymes.

21.4.7 Angiogenic Factor

Angiogenesis is not a common physiological phenomenon as the endothelium of most tissues maintains a low mitotic rate of a relatively stable population of cells [59]. Angiogenesis is observed mainly in tissue repair, such as the healing of wounds and fractures. However, the tissues of the female reproductive system, including the ovary, uterus, and placenta, present a high mitogenic rate comparable with tumor growth [60]. However, unlike tumor growth, the growth of these tissues is limited and sustained by the fast development of a highly organized vascular network [60]. The ovary is one of the few organs with active angiogenesis at regular intervals. Therefore, the ovary is used as experimental model to study not only its reproductive functions but also for angiogenesis in general.

Preantral ovarian follicles do not possess their own vasculature they depend on the stromal blood supply [61]. During the formation of the antrum, capillaries develop in the thecal cells forming two vascular networks, one in the theca interna and another in the theca externa that provide an adequate flow of gonadotropins, growth factors, oxygen, steroid precursors, and other substances to the developing follicle. The acquisition of an adequate blood supply is the limiting step in the selection and maturation of the dominant follicle [61]. The degeneration of the capillary bed causes atresia in follicles [2]. Both the ovarian follicles and the corpus luteum produce diverse angiogenic factors. However, it is thought that the *vascular endothelium growth factor* (VEGFA or VEGF) plays an essential role in ovarian angiogenesis [62]. The existence of VEGF and its receptors has been reported in both granulosa and thecal cells [63, 64].

Expression of VEGF in the ovary depends on the size of the follicle. In bovine and porcine follicles expression of VEGF is weak during the early stages of follicle development, but increases during the development of the dominant follicle [63, 64]. Similar results, together with increased expression of VEGF in the zona pellucida, have been described in the rat ovary [65]. VEGF is a potent mitogenic factor that stimulates migration of endothelial cells. It also plays a role in the structural maintenance of the already formed blood vessels increasing capillary permeability [66]. VEGF exerts its action through the binding to two *tyrosine-kinase-like* receptors VEGFR1 and VEGFR2, which are mainly expressed in endothelial cells [67] (Fig. 21.3).

The formation and differentiation of a mature and functional vascular network requires the coordinated action of

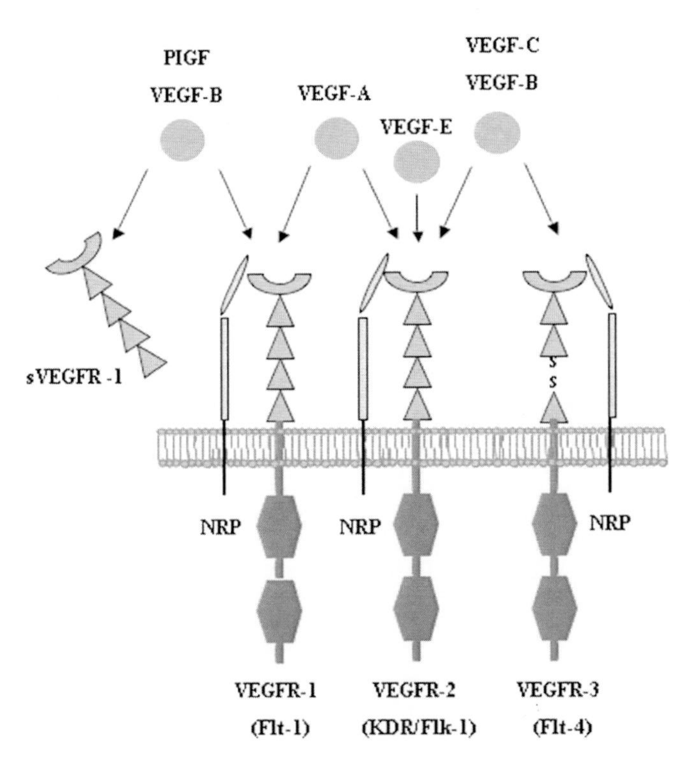

Fig. 21.3 VEGF receptors and ligands

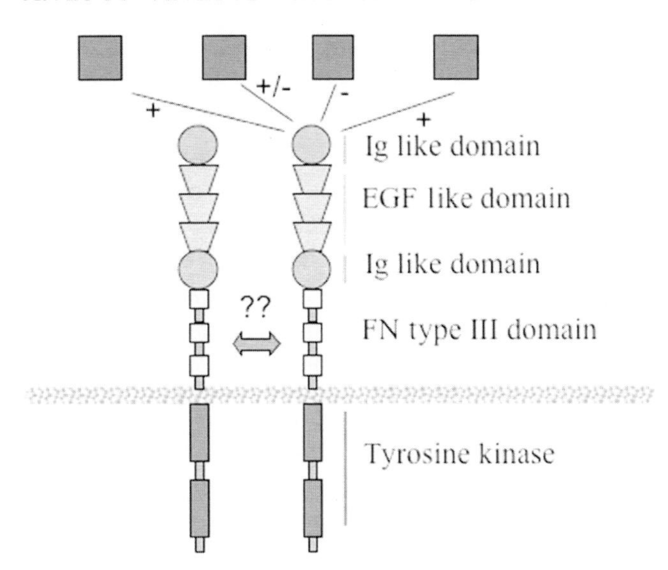

Fig. 21.4 Receptors and ligands of ANGPT

several factors, including *angiopoietin 1* (ANGPT1) and *angiopoietin 2* (ANGPT2), which act through tyrosine kinase Tie-1 and Tie-2 receptors [68] (Fig. 21.4). Unlike VEGF, ANGPT1 is unable to stimulate the proliferation of endothelial cells [69], but is essential in the recruitment of perivascular cells, which allows the maturation and stabilization of the recently formed vessels [70]. ANGPT2 is a natural antagonist of ANGPT1 and promotes the maintenance of a less stable capillary endothelium, which allows endothelial cell migration and neovascularization [68]. Therefore, it is believed that in the presence of VEGF, ANGPT2 can promote the branching of the vessels by blocking the ANGPT1 signal, whereas in the absence of VEGF, induces regression of the vessels [71, 72]. In situ hybridization studies in rat ovaries have shown that ANGPT2 mRNA is not detectable in follicular cells until the preovulatory stage, whereas ANGPT1 mRNA is expressed in thecal cells uniformly throughout follicular development [68]. ANGPT3, which has been identified in mice acts as an agonist of the Tie-2 receptor, whereas ANGPT4 has been identified in humans and acts as a natural antagonist of the Tie-2 receptor [73].

Several in vivo studies have been carried out to inhibit angiogenic factors. The administration of an antibody against VEGFR2 or Flk-1/KDR inhibits gonadotropin-dependent follicular angiogenesis in mice, which, in turn, blocks the development of mature antral follicles [74]. The antagonist known as *Trap* also inhibits VEGF causing a decrease in follicular development and expression of VEGFR1 or Flt-1 and VEGFR2 in non-human primates [75]. Intrafollicular injection of Trap in rhesus monkeys prevents ovulation and the consequent development and functionality of the corpus luteum [76]. *Trap* also delays rat follicular development and increases apoptosis of granulosa cells [77]. It has also been shown that the inhibition of the action of ANGPT1 causes an increase in follicular atresia through an imbalance in the ratio of antiapoptotic to proapoptotic proteins [78]. Other investigators have observed that local administration of either ANGPT1 or ANGPT2 inside the preovulatory follicle of female rhesus monkeys altered the balance between the angiogenic (VEGF, ANGPT1) and angiolytic (ANGPT2) factors, thus preventing ovulation and corpus luteum formation [79]. Defects in ovarian angiogenesis can contribute to a variety of reproductive disorders including anovulation, infertility, miscarriage, ovarian hyperstimulation syndrome, polycystic ovarian syndrome (PCOS), and ovarian neoplasms.

21.5 Summary

Although it has been known for sometime that gonadotropins and gonadal steroids are the main regulators of ovarian folliculogenesis, more recent experimental evidence supports the concept that local regulators, including growth factors, cytokines and neuropeptides, play important roles as non-classical endocrine extracellular signals in the ovary. Intraovarian regulators modulate growth and function of the ovarian cell compartments and participate in regulation of steroidogenesis, follicular growth, and apoptotic follicular regression. These mechanisms are altered in several disorders of the reproductive system, such as polycystic ovary syndrome, precocious puberty or menopause, hirsutism, endometriosis, and infertility caused by failures in follicular development, ovulation, and formation of the corpus luteum. Thus, understanding how the intraovarian regulators in concert with the gonadotropins control ovarian function will lead to better understanding of the normal physiology as well as pathological functioning of the ovary. This knowledge will contribute to the development of new therapeutics in the treatment of reproductive disorders and provide information for the design of new female contraceptive methods.

21.6 Glossary of Terms and Acronyms

AMH: anti-Müllerian hormone

ANGPT1: angiopoietin 1

ANGPT2: angiopoietin 2

ANGPT3: angiopoietin 3

ANGPT4: angiopoietin 4

bFGF: basic fibroblastic growth factor

BMP: bone morphogenetic protein

EGF: epidermal growth factor

FGF: fibroblastic growth factor

FSH: follicle-stimulating hormone

GABA: γ-aminobutyric acid

GDF: growth and differentiation factors

GDNF: neurotrophic factors derived from glial cells

GnRH: gonadotroping releasing hormone

GnRH-a: gonadotroping releasing hormone agonist

hCG: human Chorionic Gonadotropin

IGFBPs: insulin like growth factor binding protein

IGF-I: insulin like growth factor-I

IGF-II: insulin like growth factor-II

LH: luteinizing hormone

NGF: nerve growth factor

NT-4/5: neurotrophin 4/5

StAR: steroidogenic acute regulatory protein

TGF-α: transforming growth factor alpha

TGF: transforming growth factor

Tie1: angiopoietin receptor type 1

Tie2: angiopoietin receptor type 2

Trap: VEGF antagonist

VEGFA or VEGF: vascular endothelial growth factor

VEGFR1 or Flt-1: vascular endothelial growth factor receptor type 1

VEGFR2 or Flk-1/KDR: vascular endothelial growth factor receptor type 2

References

1. McGee EA, Hsueh AJ. Initial and cyclic recruitment of ovarian follicles. Endocr Rev 2000; 21:200–14.
2. Zeleznik AJ, Schuler HM, Reichert LE, Jr. Gonadotropin-binding sites in the rhesus monkey ovary: role of the vasculature in the selective distribution of human chorionic gonadotropin to the preovulatory follicle. Endocrinology 1981; 109:356–62.
3. Zeleznik AJ, Midgley AR, Jr., Reichert LE, Jr. Granulosa cell maturation in the rat: increased binding of human chorionic gonadotropin following treatment with follicle-stimulating hormone in vivo. Endocrinology 1974; 95:818–25.
4. Wang XN, Greenwald GS. Synergistic effects of steroids with FSH on folliculogenesis, steroidogenesis and FSH- and hCG-receptors in hypophysectomized mice. J Reprod Fertil 1993; 99:403–13.

5. Vitale AM, Gonzalez OM, Parborell F, et al. Inhibin a increases apoptosis in early ovarian antral follicles of diethylstilbestrol-treated rats. Biol Reprod 2002; 67:1989:95.
6. McNatty KP, Makris A, DeGrazia C, et al. The production of progesterone, androgens, and estrogens by granulosa cells, thecal tissue, and stromal tissue from human ovaries in vitro. J Clin Endocrinol Metab 1979; 49:687–99.
7. McGee EA. The regulation of apoptosis in preantral ovarian follicles. Biol Signals Recept 2000; 9:81–6.
8. Hillier SG, van den Boogaard AM, Reichert LE, Jr., et al. Alterations in granulosa cell aromatase activity accompanying preovulatory follicular development in the rat ovary with evidence that 5alpha-reduced C19 steroids inhibit the aromatase reaction in vitro. J Endocrinol 1980; 84:409–19.
9. McNatty KP. Regulation of follicle maturation in the human ovary: a role for 5-reduced androgens. In: Cumming IA FJMF, editor. Endocrinology. Camberra: Australian Academy of Sciences, 1980:1–51.
10. Moon YS, Tsang BK, Simpson C, et al. 17 beta-Estradiol biosynthesis in cultured granulosa and thecal cells of human ovarian follicles: stimulation by follicle-stimulating hormone. J Clin Endocrinol Metab 1978; 47:263–7.
11. Fortune JE, Armstrong DT. Androgen production by theca and granulosa isolated from proestrous rat follicles. Endocrinology 1977; 100:1341–7.
12. Ashkenazi H, Cao X, Motola S, et al. Epidermal growth factor family members: endogenous mediators of the ovulatory response. Endocrinology 2005; 146:77–84.
13. De Celis JF, Mari-Beffa M, Garcia-Bellido A. Cell-autonomous role of Notch, an epidermal growth factor homologue, in sensory organ differentiation in Drosophila. Proc Natl Acad Sci U S A 1991; 88:632–6.
14. Maruo T, Ladines-Llave CA, Samoto T, et al. Expression of epidermal growth factor and its receptor in the human ovary during follicular growth and regression. Endocrinology 1993; 132: 924–31.
15. Tilly JL, Billig H, Kowalski KI, et al. Epidermal growth factor and basic fibroblast growth factor suppress the spontaneous onset of apoptosis in cultured rat ovarian granulosa cells and follicles by a tyrosine kinase-dependent mechanism. Mol Endocrinol 1992; 6: 1942–50.
16. Yeh J, Yeh YC. Transforming growth factor-alpha and human cancer. Biomed Pharmacother 1989; 43:651–9.
17. Tsafriri A, Adashi EY. Local nonsteroidal regulators of ovarian function. In: Knovil E, Neill J, editors. The Physiology of Reproduction. New York: Raven Press Ltd, 1994: 817–43.
18. Kudlow JE, Kobrin MS, Purchio AF, et al. Ovarian transforming growth factor-alpha gene expression: immunohistochemical localization to the theca-interstitial cells. Endocrinology 1987; 121:1577–9.
19. Oliver JE, Aitman TJ, Powell JF, et al. Insulin-like growth factor I gene expression in the rat ovary is confined to the granulosa cells of developing follicles. Endocrinology 1989; 124:2671–9.
20. Hsueh AJ, Billig H, Tsafriri A. Ovarian follicle atresia: a hormonally controlled apoptotic process. Endocr Rev 1994; 15:707–24.
21. el Roeiy A, Chen X, Roberts VJ, et al. Expression of the genes encoding the insulin-like growth factors (IGF-I and II), the IGF and insulin receptors, and IGF-binding proteins-1-6 and the localization of their gene products in normal and polycystic ovary syndrome ovaries. J Clin Endocrinol Metab 1994; 78: 1488–96.
22. Geisthovel F, Moretti-Rojas I, Asch RH, et al. Expression of insulin-like growth factor-II (IGF-II) messenger ribonucleic acid (mRNA), but not IGF-I mRNA, in human preovulatory granulosa cells. Hum Reprod 1989; 4:899–02.
23. Iwashita M, Kudo Y, Yoshimura Y, et al. Physiological role of insulin-like-growth-factor-binding protein-4 in human folliculogenesis. Horm Res 1996; 46:31–6.
24. Erickson GF, Nakatani A, Ling N, et al. Localization of insulin-like growth factor-binding protein-5 messenger ribonucleic acid in rat ovaries during the estrous cycle. Endocrinology 1992; 130: 1867–78.
25. Chun SY, Billig H, Tilly JL, et al. Gonadotropin suppression of apoptosis in cultured preovulatory follicles: mediatory role of endogenous insulin-like growth factor I. Endocrinology 1994; 135:1845–53.
26. Armstrong DG, Webb R. Ovarian follicular dominance: the role of intraovarian growth factors and novel proteins. Rev Reprod 1997; 2:139–46.
27. Chegini N, Williams RS. Immunocytochemical localization of transforming growth factors (TGFs) TGF-alpha and TGF-beta in human ovarian tissues. J Clin Endocrinol Metab 1992; 74:973–80.
28. Li S, Maruo T, Ladines-Llave CA, et al. Expression of transforming growth factor-alpha in the human ovary during follicular growth, regression and atresia. Endocr J 1994; 41:693–01.
29. Drummond AE. TGFbeta signalling in the development of ovarian function. Cell Tissue Res 2005; 322:107–15.
30. Magoffin DA, Gancedo B, Erickson GF. Transforming growth factor-beta promotes differentiation of ovarian thecal-interstitial cells but inhibits androgen production. Endocrinology 1989; 125:1951–8.
31. Knight PG, Glister C. TGF-beta superfamily members and ovarian follicle development. Reproduction 2006; 132:191–06.
32. Cook RW, Thompson TB, Jardetzky TS, et al. Molecular biology of inhibin action. Semin Reprod Med 2004; 22:269–76.
33. Drummond AE, Le MT, Ethier JF, et al. Expression and localization of activin receptors, Smads, and beta glycan to the postnatal rat ovary. Endocrinology 2002; 143:1423–33.
34. Silva CC, Groome NP, Knight PG. Immunohistochemical localization of inhibin/activin alpha, betaA and betaB subunits and follistatin in bovine oocytes during in vitro maturation and fertilization. Reproduction 2003; 125:33–42.
35. Dodson WC, Schomberg DW. The effect of transforming growth factor-beta on follicle-stimulating hormone-induced differentiation of cultured rat granulosa cells. Endocrinology 1987; 120: 512–6.
36. Di Blasio AM, Vigano P, Cremonesi L, et al. Expression of the genes encoding basic fibroblast growth factor and its receptor in human granulosa cells. Mol Cell Endocrinol 1993; 96: R7–11.
37. Tapanainen J, Leinonen PJ, Tapanainen P, et al. Regulation of human granulosa-luteal cell progesterone production and proliferation by gonadotropins and growth factors. Fertil Steril 1987; 48:576–80.
38. Chun SY, Eisenhauer KM, Minami S, et al. Growth factors in ovarian follicle atresia. Semin Reprod Endocrinol 1996; 14:197–02.
39. Berisha B, Schams D. Ovarian function in ruminants. Domest Anim Endocrinol 2005; 29:305–17.
40. Mayerhofer A, Smith GD, Danilchik M, et al. Oocytes are a source of catecholamines in the primate ovary: evidence for a cell-cell regulatory loop. Proc Natl Acad Sci U S A 1998; 95:10990–5.
41. Fritz S, Wessler I, Breitling R, et al. Expression of muscarinic receptor types in the primate ovary and evidence for nonneuronal acetylcholine synthesis. J Clin Endocrinol Metab 2001; 86: 349–54.
42. Fritz S, Fohr KJ, Boddien S, et al. Functional and molecular characterization of a muscarinic receptor type and evidence for expression of choline-acetyltransferase and vesicular acetylcholine transporter in human granulosa-luteal cells. J Clin Endocrinol Metab 1999; 84:1744–50.

43. Fritz S, Grunert R, Stocco DM, et al. StAR protein is increased by muscarinic receptor activation in human luteinized granulosa cells. Mol Cell Endocrinol 2001; 171:49–51.
44. Erdo S, Varga B, Horvath E. Effect of local GABA administration on rat ovarian blood flow, and on progesterone and estradiol secretion. Eur J Pharmacol 1985; 111:397–400.
45. McNeill DL, Burden HW. Neuropeptide Y and somatostatin immunoreactive perikarya in preaortic ganglia projecting to the rat ovary. J Reprod Fertil 1986; 78:727–32.
46. McNeill DL, Burden HW. Peripheral pathways for neuropeptide Y- and cholecystokinin-8-immunoreactive nerves innervating the rat ovary. Neurosci Lett 1987; 80:27–32.
47. Dissen GA, Hirshfield AN, Malamed S, et al. Expression of neurotrophins and their receptors in the mammalian ovary is developmentally regulated: changes at the time of folliculogenesis. Endocrinology 1995; 136:4681–92.
48. Matzuk MM, Burns KH, Viveiros MM, et al. Intercellular communication in the mammalian ovary: oocytes carry the conversation. Science 2002; 296:2178–80.
49. Mayerhofer A, Dissen GA, Parrott JA, et al. Involvement of nerve growth factor in the ovulatory cascade: trkA receptor activation inhibits gap junctional communication between thecal cells. Endocrinology 1996; 137:5662–70.
50. Hsueh AJ, Jones PB. Extrapituitary actions of gonadotropin-releasing hormone. Endocr Rev 1981; 2:437–61.
51. Guerrero HE, Stein P, Asch RH, et al. Effect of a gonadotropin-releasing hormone agonist on luteinizing hormone receptors and steroidogenesis in ovarian cells. Fertil Steril 1993; 59:803–88.
52. Kang SK, Tai CJ, Nathwani PS, et al. Differential regulation of two forms of gonadotropin-releasing hormone messenger ribonucleic acid in human granulosa-luteal cells. Endocrinology 2001; 142:182–92.
53. Andreu C, Parborell F, Vanzulli S, et al. Regulation of follicular luteinization by a gonadotropin-releasing hormone agonist: relationship between steroidogenesis and apoptosis. Mol Reprod Dev 1998; 51:287–94.
54. Sridaran R, Philip GH, Li H, et al. GnRH agonist treatment decreases progesterone synthesis, luteal peripheral benzodiazepine receptor mRNA, ligand binding and steroidogenic acute regulatory protein expression during pregnancy. J Mol Endocrinol 1999; 22:45–54.
55. Parborell F, Dain L, Tesone M. Gonadotropin-releasing hormone agonist affects rat ovarian follicle development by interfering with FSH and growth factors on the prevention of apoptosis. Mol Reprod Dev 2001; 60:241–7.
56. Parborell F, Irusta G, Vitale AM, et al. GnRH antagonist Antide inhibits apoptosis of preovulatory follicle cells in rat ovary. Biol Repr 2005; 72:659–66.
57. Parborell F, Pecci A, Gonzalez O, et al. Effects of a gonadotropin-releasing hormone agonist on rat ovarian follicle apoptosis: Regulation by EGF and the expression of Bcl-2-related genes. Biol Reprod 2002; 67:481–6.
58. Irusta G, Parborell F, Peluffo M, et al. Steroidogenic acute regulatory protein in ovarian follicles of gonadotropin-stimulated rats is regulated by a gonadotropin-releasing hormone agonist. Biol Reprod 2003; 68:1577–83.
59. Klagsbrun M, D'Amore PA. Vascular endothelial growth factor and its receptors. Cytokine Growth Factor Rev 1996; 7: 259–70.
60. Reynolds LP, Grazul-Bilska AT, Killilea SD, et al. Mitogenic factors of corpora lutea. Prog Growth Factor Res 1994; 5:159–75.
61. Stouffer RL, Martinez-Chequer JC, Molskness TA, et al. Regulation and action of angiogenic factors in the primate ovary. Arch Med Res 2001; 32:567–75.
62. Tamanini C, De Ambrogi M. Angiogenesis in developing follicle and corpus luteum. Reprod Domest Anim 2004; 39:206–16.
63. Geva E, Jaffe RB. Role of vascular endothelial growth factor in ovarian physiology and pathology. Fertil Steril 2000; 74: 429–38.
64. Berisha B, Schams D, Kosmann M, et al. Expression and localisation of vascular endothelial growth factor and basic fibroblast growth factor during the final growth of bovine ovarian follicles. J Endocrinol 2000; 167:371–82.
65. Celik-Ozenci C, Akkoyunlu G, Kayisli UA, et al. Localization of vascular endothelial growth factor in the zona pellucida of developing ovarian follicles in the rat: a possible role in destiny of follicles. Histochem Cell Biol 2003; 120:383–90.
66. Redmer DA, Doraiswamy V, Bortnem BJ, et al. Evidence for a role of capillary pericytes in vascular growth of the developing ovine corpus luteum. Biol Reprod 2001; 65:879–89.
67. Jakeman LB, Winer J, Bennett GL, et al. Binding sites for vascular endothelial growth factor are localized on endothelial cells in adult rat tissues. J Clin Invest 1992; 89:244–53.
68. Maisonpierre PC, Suri C, Jones PF, et al. Angiopoietin-2, a natural antagonist for Tie2 that disrupts in vivo angiogenesis. Science 1997; 277:55–60.
69. Davis S, Aldrich TH, Jones PF, et al. Isolation of angiopoietin-1, a ligand for the TIE2 receptor, by secretion-trap expression cloning. Cell 1996; 87:1161–9.
70. Suri C, Jones PF, Patan S, et al. Requisite role of angiopoietin-1, a ligand for the TIE2 receptor, during embryonic angiogenesis. Cell 1996; 87:1171–80.
71. Hanahan D. Signaling vascular morphogenesis and maintenance. Science 1997; 277:48–50.
72. Yancopoulos GD, Davis S, Gale NW, et al. Vascular-specific growth factors and blood vessel formation. Nature 2000; 407: 242–8.
73. Valenzuela DM, Griffiths JA, Rojas J, et al. Angiopoietins 3 and 4: diverging gene counterparts in mice and humans. Proc Natl Acad Sci U S A 1999; 96:1904–9.
74. Zimmermann RC, Hartman T, Kavic S, et al. Vascular endothelial growth factor receptor 2-mediated angiogenesis is essential for gonadotropin-dependent follicle development. J Clin Invest 2003; 112:659–69.
75. Wulff C, Wilson H, Wiegand SJ, et al. Prevention of thecal angiogenesis, antral follicular growth, and ovulation in the primate by treatment with vascular endothelial growth factor Trap R1R2. Endocrinology 2002; 143:2797–807.
76. Hazzard TM, Xu F, Stouffer RL. Injection of soluble vascular endothelial growth factor receptor 1 into the preovulatory follicle disrupts ovulation and subsequent luteal function in rhesus monkeys. Biol Reprod 2002; 67:1305–12.
77. Abramovich D, Parborell F, Tesone M. Effect of a vascular endothelial growth factor (VEGF) inhibitory treatment on the folliculogenesis and ovarian apoptosis in gonadotropin-treated prepubertal rats. Biol Reprod 2006; 75:434–41.
78. Parborell F, Abramovich D, Tesone M. Intrabursal Administration of the Antiangiopoietin 1 Antibody Produces a Delay in Rat Follicular Development Associated with an Increase in Ovarian Apoptosis Mediated by Changes in the Expression of BCL2 Related Genes. Biol Reprod 2007; 78: 506–13.
79. Xu F, Hazzard TM, Evans A, et al. Intraovarian actions of anti-angiogenic agents disrupt periovulatory events during the menstrual cycle in monkeys. Contraception 2005; 71: 239–48.

Chapter 22

Ovarian Endocrine Activity: Role of Follistatin, Activin, and Inhibin

Stella Campo, Nazareth Loreti, and Luz Andreone

22.1 Introduction

The first evidence of a testicular non-steroidal factor able to regulate the activity of the pituitary gland was reported by Roy McCullagh in 1932 [1]. He observed that the administration of an aqueous testicular preparation to castrated rats was able to restore the physiological characteristics of pituitary cells that had been altered after castration. This bioactive factor was called "inhibin." When FSH and LH were purified, standards were available and specific radioimmunoassays were developed to determine their levels in biological fluids. The application of these methodologies to patients with gonadal dysfunctions confirmed the existence of a gonadal regulatory mechanism specific for FSH secretion. In adult males with damaged seminiferous tubules epithelium, a marked increase of FSH levels, concomitantly with normal testosterone and LH levels, was reported [2]. Similar observations were described in pre-menopausal women, with elevated serum FSH levels and normal estradiol, during the follicular phase of the menstrual cycle [3].

Inhibins were isolated from bovine and porcine follicular fluids in 1985 [4, 5, 6]. When inhibin was purified to homogeneity it was possible to design oligonucleotide probes that permitted cloning of the gene and the determination of its molecular structure [7]. Inhibins are heterodimeric glycoproteins composed of a common α-subunit linked with one of the highly homologous β-subunits by disulphide bonds. Heterodimerization of the α-chain with either βA- or βB-subunit generates dimeric inhibin A (α-βA) and inhibin B (α-βB), respectively. These dimeric molecules are the only bioactive molecular forms that strongly inhibit pituitary FSH secretion. Both α- and β-subunits are synthesized as high molecular weight precursors: the α-subunit as a preproprotein and the β-subunit as a proprotein (Fig. 22.1) [8, 9]. The proportion of the α-subunit in the tissue where it is synthesized, as well as in circulation, is higher than that of the β-subunit [10]. Although the physiological relevance of this difference is still unknown, it has been proposed that the formation of the heterodimers is contingent upon the relative abundance of the α-subunit. The opposite situation would favor the formation of activins (FSH releasing proteins, FRP) [11]. Activins are formed by homo- or heterodimerization of the β-chains; combinational assembly of the βA and βB can generate activin A (βA-βA), activin B (βB-βB), and activin AB (βA-βB) (Fig. 22.1) [8, 9]. These peptides potently stimulate FSH secretion from the pituitary [12, 13].

Activins are members of the TGF-β superfamily, a group of growth and differentiation factors that includes TGF-β, müllerian inhibiting substance (MIS), bone morphogenic proteins, and the amphibian protein VG-1. The membership in this family help to understand the variety of activin functions, some of which remain still unexplained. Activins are synthesized in a large number of tissues and are considered paracrine rather endocrine factors [14]. The presence of activins in spermatocytes at defined stages, in unfertilized and fertilized ova, and in various organs throughout embryogenesis suggests that they must be important to development [15].

Other fractions isolated from follicular fluid might modulate the secretion of FSH. They are proteins with molecular weights varying from 32 kDa to 43 kDa depending on the degree of glycosylation [16]. These proteins, now termed follistatins, can inhibit the secretion of FSH "in vitro" with less potency than that of inhibin and have subsequently been shown to bind activins with high affinity and neutralize their biological activity [17]. Follistatins are widespread in tissues, suggesting that they are ubiquitous proteins, regulating a wide variety of local activin actions [18]. Gospodarowicz and Lau [19] showed that in the pituitary gland, folliculostellate cells are able to produce follistatin "in vitro." This finding was confirmed by Kaiser [20], who proposed that this glycoprotein might have a paracrine role in the modulation of FSH biosynthesis and secretion. Separate genes code for

S. Campo (✉)
Centro de Investigaciones Endocrinológicas, Hospital de Niños Ricardo Gutiérrez, Buenos Aires, Argentina
e-mail: scampo@cedie.org.ar

P.J. Chedrese (ed.), *Reproductive Endocrinology*,
DOI 10.1007/978-0-387-88186-7_22, © Springer Science+Business Media, LLC 2009

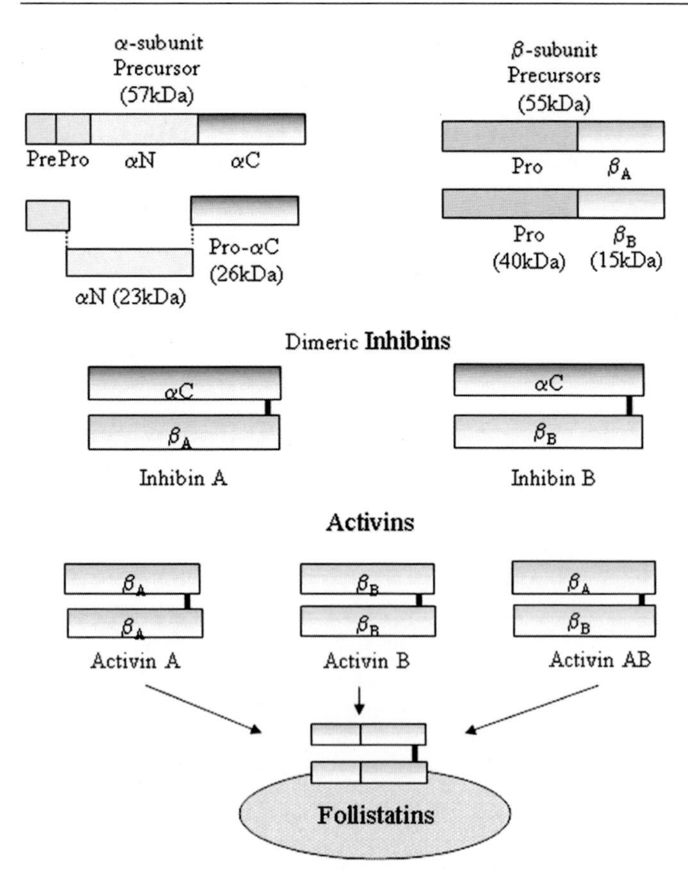

Fig. 22.1 Molecular forms of inhibins, activins, and high molecular weight subunit precursors. Inhibins are present in a wide range of molecular weight as a result of the combination of mature forms and precursors of each subunit. Follistatins bind activins with high affinity and neutralize their biological activity. Information summarized from Burger [8], and Knight and Glister [9]

the α, βA, and βB subunits; the human α and βB genes are located on opposite ends of chromosome 2, whereas the βA gene is present on chromosome 7 [21].

22.2 Biological Effects of Inhibin

The isolation and purification of inhibin made possible to study its direct effect on synthesis and secretion of pituitary FSH. Cultures of dispersed pituitary cells were used to demonstrate that inhibin blocks FSH production under basal conditions and in the presence of GnRH [22] (Fig. 22.2) [23]. In vitro studies carried out on dispersed pituitary cells isolated from sheep and non-human primates showed that inhibins selectively reduce the level of FSH β-subunit mRNA [24, 25]. It has also been proposed that inhibins exert antagonistic effects, presumably through the β-subunit that they share with activins, by competing with activins for the activin receptor-II [26].

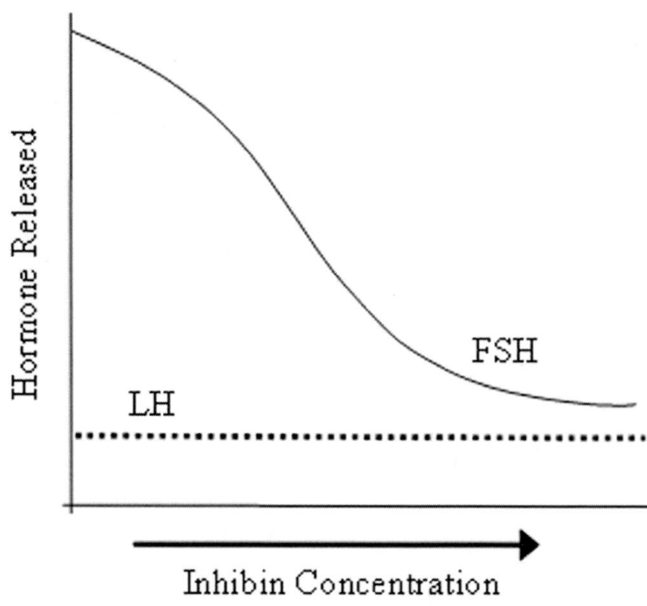

Fig. 22.2 Effect of increasing concentrations of purified ovine inhibin on basal FSH production by pituitary cells in culture. Information summarized from Vale [23]

22.3 Sources of Inhibin

Inhibins are secreted by granulosa cells of the follicle or granulosa-lutein cells of the corpus luteum in the ovary and by Sertoli cells in the testis [27–29]. In addition, α-, βA-, and βB-subunit RNAs have been detected in the placenta, pituitary gland, adrenal glands, bone marrow, kidneys, spinal cord, and brain [30]. The presence of inhibin α- and β-subunits in the placenta and the pituitary gland, two cell types that are regulated by exogenous inhibin, may reflect paracrine and/or autocrine processes in these tissues. Detection of inhibin RNAs in the brain and spinal cord reflect neuroregulatory functions in the central and peripheral nervous systems.

Activins and follistatins also produced by folliculostellate cells are involved in a paracrine regulatory mechanism of FSH synthesis in pituitary gonadotropes [18, 31]. FSH and LH mainly regulate inhibin production in the ovary; both gonadotropins stimulate its production in granulosa cells and in the corpus luteum, respectively [29, 32].

22.4 Physiology of Immunoreactive Inhibin

22.4.1 Pubertal Development

A heterologous radioimmunoassay (RIA) developed by McLachlan at Monash University in Melbourne, Australia, was used to determine the serum profile of inhibin under different physiological and pathological conditions [33]. Burger showed that immunoreactive inhibin levels increase concomitantly with those of FSH, LH, and estradiol, from early to advanced puberty in females [34]. Based on these unexpected findings it was proposed that (a) the increment in FSH levels observed at Tanner stage II induces follicular development; (b) in response to this stimulus, the ovary progressively produces more inhibin; and (c) once serum inhibin rises to adult levels, the negative feedback between FSH and inhibin becomes established, although there is some evidence suggesting that the negative feedback may be functional earlier in life, perhaps with a different threshold.

22.4.2 Development of Specific and More Sensitive Methods for the Determination of Dimeric and Monomeric Inhibin

The radioimmunoassay of inhibin had specificity limitations as bioinactive free α-inhibin subunits, which are present in biological fluids, cross-react in this assay. Thus, the limited availability of purified inhibin for immunization and the weak immunogenicity of this peptide rendered the use of RIA of no clinical value, particularly, to assess seminiferous epithelium function in men. With increasing interest in the physiology of inhibin, the need for specific and highly sensitive serum inhibin assays has become important. The synthetic peptide approach was successfully used to prepare monoclonal antibodies that were able to recognize the N-terminal portion of the 20kDa inhibin α-subunit, the βA and the βB subunits [35]. Using two monoclonal antibodies together, the two-site immunoassays for free inhibin α-subunit and both dimeric forms, inhibin A and B, were set up and became worldwide available [36,37,38].

22.5 Inhibins as Reliable Markers of Ovarian Activity

The measurement of serum inhibin by ELISA is useful as a marker of ovarian function from birth to puberty and throughout the normal menstrual cycle in adult women [39, 40].

22.5.1 Neonatal Period

When the serum profile of dimeric inhibins, A and B, and the monomeric α-inhibin precursor protein, Pro-αC, was determined in the newborn, sexual dimorphism in gonadal inhibin expression was noted. This dimorphism starts during fetal development. At mid-gestation the inhibin α-subunit is not expressed in the fetal ovaries, whereas there is a very strong expression in the fetal testis [41]. Dimorphism remains reflected at birth; the first day of life, dimeric inhibins are undetectable in females' serum; whereas inhibin B levels reach adult values in males.

In the course of the first week of life it is possible to detect all forms of inhibin in the serum of females; concentration of inhibin B largely predominates over that of inhibin A. Serum levels of both dimers increase during the first weeks of life (Figure 22.3) [42]; LH seems to be the driving force behind this production in female newborns [43].

It is believed that during the last months of pregnancy, the fetal ovary may acquire the capacity to synthesize the inhibin α-subunit, since there is no difference in inhibin Pro-αC serum levels between males and females at birth, which remain high after birth when the hCG β-subunit is no longer detectable in circulation. This information suggests that the fetal ovary and not the placenta is the main source of inhibin α-subunit production in newborn females (Fig. 22.4) [43].

Morphologic studies have shown that during the neonatal period, primordial follicles in the ovary coexist with preantral, small antral, and cystic follicles thought to be at early stages of atresia [44]. The contribution of inhibin B by follicles at early stages of development and the production of inhibin A by follicles at more advanced stages of maturation with signs of luteinization may explain the concomitant high serum concentrations of both dimeric inhibins during this period. Serum inhibin B concentrations in newborn females are as high as those found in pubertal females. Serum inhibin A concentrations decrease progressively during the second

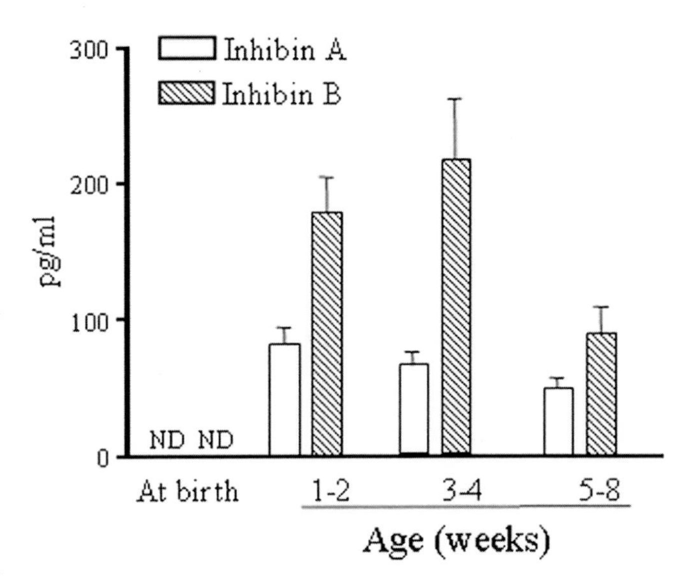

Fig. 22.3 Serum levels of inhibin A and B in normal girls during the first 2 months of life. Information summarized from Bergada [42]

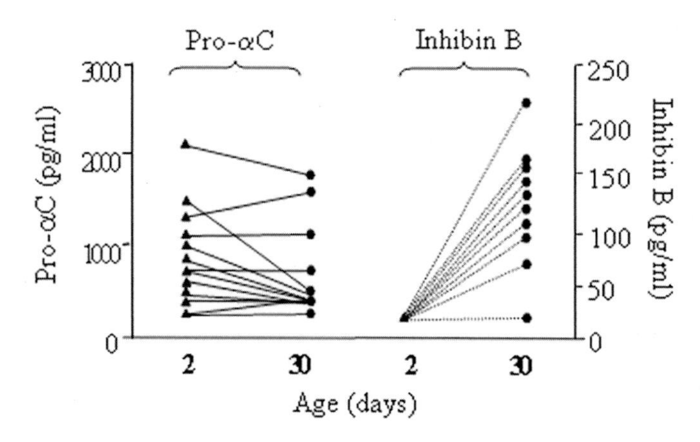

Fig. 22.4 Serum levels of inhibin B and Pro-αC determined in girls within the first 2 days of life and at 30 days of age. Information summarized from Bergada and Milani [43]

month of life. Measurement of serum inhibin A and B during the first months of life may be a valuable tool for the assessment of ovarian function, particularly since highly specific and sensitive estradiol assays are not available. Furthermore, the serum inhibin profile may help to elucidate the ontogeny of human follicles and aid in the investigation of congenital disorders of the female gonad.

22.5.2 Infancy and Childhood

After 6 months of age, serum concentrations of inhibin A are undetectable in normal females and they remain below the detection limit of the assay until the onset of puberty. During infancy, high inhibin B and Pro-αC levels drop to prepubertal values faster in females than in males (Table 22.1) [39]. In females, by 6 months of age inhibin B and Pro-αC reach prepubertal concentration, whereas FSH reaches prepubertal concentration by age four. Interestingly, this FSH-inhibin relationship is the opposite of that found in males during this period. The interpretation of these findings is difficult because follicular development persists beyond the first 6 months of life [45] and low levels of inhibin B and non-detectable inhibin A are concomitant with elevated FSH levels [46]. The presence of measurable serum levels of inhibin B in females during childhood may reflect functionally active gonads. However, the concomitant absence of inhibin A in circulation suggests that the follicles present in the ovary have not reached an advanced stage of development. Close to puberty, the observed increase of inhibin A levels in circulation without any significant increase of inhibin B may indicate re-activation of gonadal activity, which stimulates follicular development.

22.5.3 Pubertal Development

After the onset of puberty, increases in inhibin B levels concomitant with the characteristic increase of gonadotropins and estradiol levels is observed in females (Fig. 22.5) [40]. Levels of inhibin B increase throughout pubertal Tanner stages II and III [47], possibly reflecting its production by small antral follicles in response to gonadotropin stimulation. This could indicate that follicles are recruited as puberty progresses and they reach a more advanced stage of development before undergoing atresia. Inhibin B levels attained at Tanner stage III are higher than those observed in adult women. Based on these observations, it has been proposed that this stage of pubertal development may represent a period of consistently high ovarian follicular activity before the development of the adult menstrual cycle, with ovulation and a luteal phase [40]. Chada [48] found correlation between inhibin B and FSH, LH, and estradiol in early puberty; based on this evidence it was proposed that FSH regulates inhibin B secretion during early pubertal development. The progressive and sustained increment of inhibin B serum levels throughout pubertal development would be the signal that the pituitary gland receives to initiate a full functioning negative feedback mechanism between this peptide and FSH.

22.5.4 Profile of Inhibin in the Normal Menstrual Cycle

Once the cyclic activity of the ovary begins, a specific secretion pattern of inhibin A and inhibin B through the menstrual cycle is observed. Inhibin B is principally secreted in the follicular phase, and inhibin A is principally secreted

Table 22.1 Serum levels of inhibin A, inhibin B, and Pro-αC in normal girls from birth to prepuberty. ND: non-detectable. Information summarized from Bergada [39]

Age	Inhibin A (pg/ml)	SEM	Inhibin B (pg/ml)	SEM	Pro-αC (pg/ml)	SEM
0–6 (months)	29.9	8.7	83	18.3	86.6	37.4
6–24 (months)	ND		17.5	1.6	24.2	5.4
2–4 (years)	ND		38	8.4	49.4	5.9
4–6 (years)	ND		40.2	6.7	50.4	11.6
6–8 (years)	ND		38.3	5.3	36	4.7
8–10 (years)	10.5	4.2	39	8.3	64	11.4

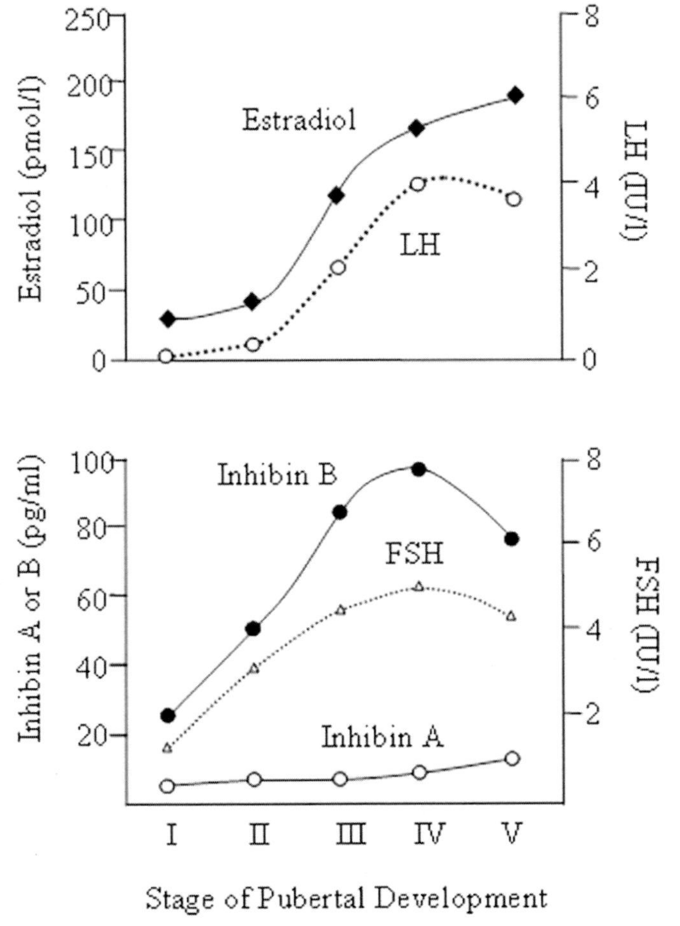

Fig. 22.5 Serum profile of inhibins, gonadotropins, and estradiol in normal girls during pubertal development. Information summarized from Sehested [40]

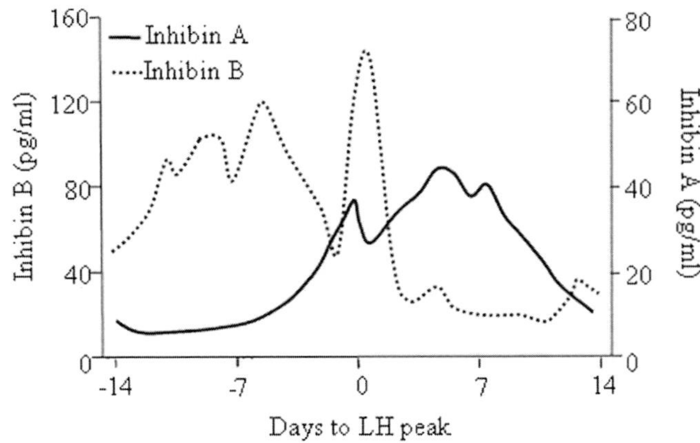

Fig. 22.6 Serum profile of dimeric inhibins throughout the normal menstrual cycle. Information summarized from Sehested [40]

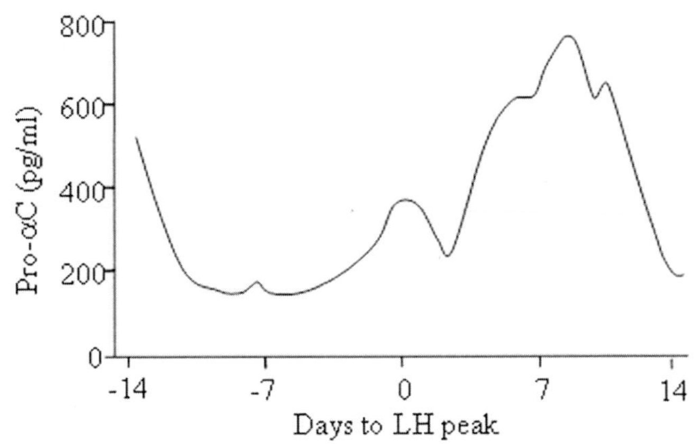

Fig. 22.7 Serum profile of the monomeric form of inhibin throughout the normal menstrual cycle. Modified from Groome [49]

in the luteal phase; its pattern follows very closely that of progesterone (Figure 22.6) [40]. The inhibin profile during the menstrual cycle reflects different stages of development and maturation of the ovarian follicle. Inhibin B is mainly produced during the early stages of follicular development, possibly stimulated by FSH [38]. This is the reason why it cannot be detected in serum during the luteal phase. On the other hand, inhibin A is produced by pre-ovulatory follicles and by the corpus luteum. This is the reason why it cannot be detected in serum during the follicular phase [36]. The high molecular weight precursor of the inhibin α-subunit is present in circulation during the follicular phase as well as during the luteal phase (Figure 22.7) [49].

It has been proposed that the ovary produces inhibin α-subunit in high amounts, compared to those of the β-subunits, to make possible the heterodimerization of subunits to form dimeric inhibins [50]. If this were not the case, the synthesis of activins would be favored. Although this monomeric form of inhibin does not exert any effect on FSH secretion, immunization of sheep with this peptide impairs fertility, by an unknown mechanism [51] and it has been claimed that it is able to modulate the binding of FSH to its receptor [52]. As for its possible clinical utility, due to the high sensitivity of the assay, Pro-αC has been proposed as a reliable marker of gonadal steroidogenesis and it may have an important application in the detection and monitoring of ovarian tumors [49].

22.6 Regulation of Inhibin Production by Granulosa Cells

Regulation of the synthesis of inhibin in granulosa cells by FSH and ovarian local regulators was demonstrated using a heterologous RIA that does not discriminate between the dimeric forms of inhibin and the free α-subunit [28, 53]. Recent studies have confirmed that FSH and estradiol exert their regulatory effect on expression of inhibin subunits α, βA, and βB [54, 55]. However, the measurement of mRNA does not accurately reflect the actual levels of inhibin because β chains can also dimerize to produce activin.

When new methods were applied to study the hormonal regulation of inhibin production by granulosa cells, it was possible to determine which factors may favor the production of each inhibin dimer. In vitro, non-treated granulosa cells obtained from immature, estrogen-treated rats produced the three molecular forms of inhibin; however, they produced ten times more inhibin A than inhibin B. Interestingly, inhibin B was found to be the predominant inhibin dimer in this experimental model in vivo [56]. This discrepancy suggests that intraovarian factors are able to shift inhibin production toward the predominance of inhibin B.

Follicle-stimulating hormone (FSH) stimulates inhibin A and B production in granulosa cells, with a more pronounced effect on inhibin A. The relative sensitivity of granulosa cells to respond to FSH in terms of inhibin production is affected by IGF-I. The addition of IGF-I in the presence of low doses of the gonadotropin produces a marked decrease in the inhibin A/B ratio. Estradiol stimulates inhibin A production with a minor effect on inhibin B.

TGF-β, activin-A, and oocyte-derived factor(s) all induce a decrease in the relative ratio of inhibin A to inhibin B [57]. TGF-β induces the secretion of both inhibin A and B with a clear preferential stimulation of the B dimer; activin A is more potent that TGF-β in stimulating inhibin B production; coculture with meiotically immature oocytes is able to induce a selective stimulation of inhibin B, resembling the effect observed with TGF-β or activin A. Therefore, the inhibin/activin-βB-subunit gene may be a specific target for the TGF-β family of peptides. Considering that inhibin B is produced predominantly by early preantral follicles, whereas inhibin A is preferentially secreted by more differentiated cells from antral follicles variations in the ratio of the inhibin dimers in serum may be a reflection of the change in the concentrations of intrafollicular regulators [55, 12]. Each inhibin dimer may be responding to different environmental signals acting on the granulosa cells. Inhibin A is more sensitive to FSH stimulation during the later stages of follicular growth, whereas inhibin B reflects the action of the members of the TGF-β superfamily in preantral follicles (Fig. 22.8).

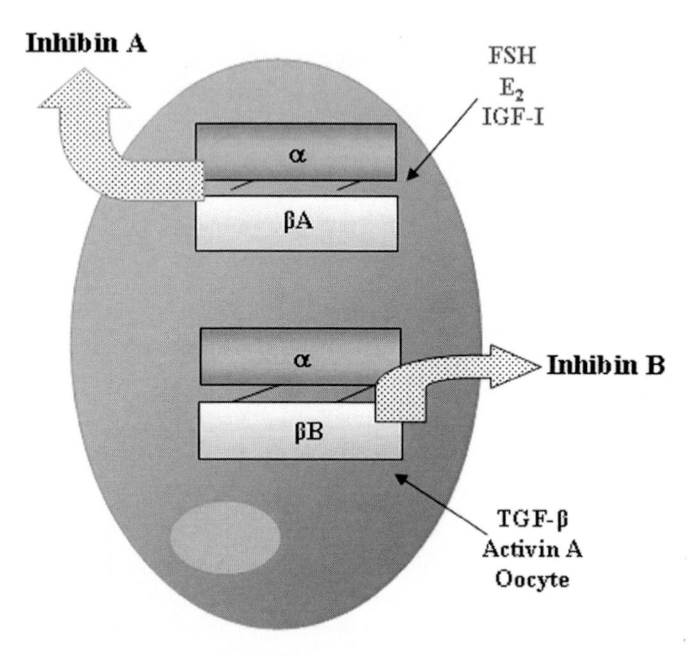

Fig. 22.8 Regulatory factors involved in the regulation of inhibin A and B production by cultured granulosa cells isolated from immature, estrogen-treated rats

22.7 Clinical Uses of Inhibin Assays

Dimeric inhibins are reliable serum markers in assessing ovarian endocrine activity in the newborn as well as in infancy, prepuberty, and during pubertal development. In adult women, they are very useful to evaluate follicular recruitment and development. In particular, determination of inhibin B and AntiMullerian hormone (AMH) levels are widely used to evaluate ovarian reserves. Dimeric inhibins are also useful tools in monitoring the ovarian response to exogenous gonadotropin stimulation in IVF programs.

22.8 Summary

Inhibin, activin, and follistatin are proteins involved in the control of FSH secretion. The development of specific and sensitive assays for the different molecular forms of these peptides permitted the elucidation of their sources, the hormonal regulation of their production, and the possible clinical application of their serum levels in the assessment of gonadal function at different stages of the reproductive life. Activins and follistatins exert their action as paracrine or autocrine growth factors through local production at many sites. Inhibins are mainly produced by the ovarian follicle and released into the circulation to exert their action on the pituitary FSH secretion. FSH and several local factors regulate inhibin A and B production by granulosa cells. A predominant production of inhibin B is sustained by TGF-β, activin A, and oocyte-derived factors in a low estrogenic follicular tone. FSH and estradiol stimulate both dimeric inhibins production, but this effect is more pronounced on inhibin A. Inhibin A and B are absent in serum in girls at birth but their levels increase concomitantly during the first 2 weeks of life. This unique profile characterizes the postnatal period in girls and reflects the presence of granulosa cells in an advanced stage of maturation and differentiation. Inhibin B is the only dimeric form of inhibin produced in the ovary during infancy and childhood and its levels increase concomitantly with those of FSH during pubertal development.

Inhibin A is detectable again in circulation before the clinical onset of puberty. Once the ovarian cyclic activity is established, changes in serum inhibin profile may be a reflection of the change in the concentration of intrafollicular regulators and FSH stimulation. Inhibin B is the only dimer present in the follicular phase and reflects the presence of follicles at early stages of development. Inhibin A appears in circulation in the pre-ovulatory period and reaches its maximal levels in the luteal phase reflecting the endocrine activity of pre-ovulatory follicles and the corpus luteum. At present, dimeric inhibins are considered reliable markers to assess ovarian endocrine activity from birth to adulthood. Inhibin B is a widely used parameter to evaluate ovarian reserve concomitantly with FSH and AMH in infertile women.

22.9 Glossary of Terms and Acronyms

AMH: anti-Müllerian hormone, also termed MIS

FRP: FSH releasing proteins

FSH: follicle stimulating hormone

GnRH: gonadotropin releasing hormone

hCG: human chorionic gonadotropin

IGF-1: insulin like growth factor-1

IVF: in vitro fertilization

LH: luteinizing hormone

MIS: müllerian inhibiting substance

mRNA: messenger RNA

RIA: radioimmunoassay

RNA: ribonucleic acid

TGF: transforming growth factor

References

1. McCullagh D. Dual endocrine activity of the testes. Science 1932; 76:19–20.
2. de Kretser D, Burger H, Fortune D, et al. Hormonal, histological and chromosomal studies in adult males with testicular disorders. J Clin Endocrinol Metab 1972; 35:392–401.
3. Sherman B, West J, Korenman S. The menopausal transition: analysis of LH, FSH, estradiol and progesterone concentrations during menstrual cycles of older women. J Clin Endocrinol Metab 1976; 42:629–36.
4. Robertson D, Foulds L, Leversha L, et al. Isolation of inhibin from bovine follicular fluid. Biochem Biophys Res Commun 1985; 126:220–6.
5. Miyamoto K, Hasegawa Y, Fukuda M et al. Isolation of porcine follicular fluid inhibin of 32 K daltons. Biochem Biophys Res Commun 1985; 129:396–403.
6. Ling N, Ying S, Ueno N, et al. Isolation and partial characterization of a Mr 32,000 protein with inhibin activity from porcine follicular fluid. Proc Natl Acad Sci U S A 1985; 82:7217–21.
7. Mason A, Hayflick J, Ling N, et al. Complementary DNA sequences of ovarian follicular fluid inhibin show precursor structure and homology with transforming growth factor-beta. Nature 1985; 318:659–63.
8. Burger H. Clinical review 46: Clinical utility of inhibin measurements. J Clin Endocrinol Metab 1993; 76:1391–6.
9. Knight P, Glister C. Potential local regulatory functions of inhibins, activins and follistatin in the ovary. Reproduction 2001; 121: 503–12.
10. Knight P, Beard A, Wrathall J, et al. Evidence that the bovine ovary secretes large amounts of monomeric inhibin alpha subunit and its isolation from bovine follicular fluid. J Mol Endocrinol 1989; 2:189–200.
11. Halvorson L, de Cherney A. Inhibin, activin, and follistatin in reproductive medicine. Fertil Steril 1996; 65:459–69.
12. Woodruff T. Regulation of cellular and system function by activin. Biochem Pharmacol 1998; 1:953–63.
13. Pangas S, Woodruff T. Activin signal transduction pathways. Trends Endocrinol Metab 2000; 11:309–14.
14. Welt C, Crowley W. Activin: an endocrine or paracrine agent? Eur J Endocrinol 1998; 139:469–71.
15. Mitrani E, Ziv T, Thomsen G, et al. Activin can induce the formation of axial structures and is expressed in the hypoblast of the chick. Cell 1990; 63:495–501.
16. Ueno N, Ling N, Ying S, et al. Isolation and partial characterization of follistatin: a single-chain Mr 35,000 monomeric protein that inhibits the release of follicle-stimulating hormone. Proc Natl Acad Sci U S A 1987; 84:8282–6.
17. Nakamura T, Takio K, Eto Y, et al. Activin-binding protein from rat ovary is follistatin. Science 1990; 16:836–8.
18. Kogawa K, Ogawa K, Hayashi Y, et al. Immunohistochemical localization of follistatin in rat tissues. Endocrinol Jpn 1991; 38:383–91.
19. Gospodarowicz D, Lau K. Pituitary follicular cells secrete both vascular endothelial growth factor and follistatin. Biochem Biophys Res Commun 1989; 165:292–8.
20. Kaiser U, Lee B, Carroll R, et al. Follistatin gene expression in the pituitary: localization in gonadotropes and folliculostellate cells in diestrous rats. Endocrinology 1992; 130:3048–56.
21. Barton D, Yang-Feng T, Mason A, et al. Mapping of genes for inhibin subunits alpha, beta A, and beta B on human and mouse chromosomes and studies of jsd mice. Genomics 1989; 5:91–9.
22. Farnworth P, Robertson D, de Kretser D, et al. Effects of 31 kDa bovine inhibin on FSH and LH in rat pituitary cells in vitro: antagonism of gonadotrophin-releasing hormone agonists. J Endocrinol 1988; 119:233–41.
23. Vale W, Rivier C, Hsueh A, et al. Chemical and biological characterization of the inhibin family of protein hormones. Recent Prog Horm Res 1988; 44:1–34.
24. Attardi B, Keeping H, Winters S, et al. Effect of inhibin from primate Sertoli cells and GnRH on gonadotropin subunit mRNA in rat pituitary cell cultures. Mol Endocrinol 1989; 3:1236–42.
25. Carroll R, Corrigan A, Gharib S, et al. Inhibin, activin, and follistatin: regulation of follicle-stimulating hormone messenger ribonucleic acid levels. Mol Endocrinol 1989; 3:1969–76.
26. Xu J, McKeehan K, Matsuzaki K, et al. Inhibin antagonizes inhibition of liver cell growth by activin by a dominant-negative mechanism. J Biol Chem 1995; 270:6308–13.
27. Steinberger A. Inhibin production by Sertoli cells in culture. J Reprod Fertil Suppl 1979; 26:31–45.
28. Bicsak T, Tucker E, Cappel S, et al. Hormonal regulation of granulosa cell inhibin biosynthesis. Endocrinology 1986; 119: 2711–9.

29. Zhang Z, Carson R, Herington A, et al. Follicle-stimulating hormone and somatomedin-C stimulate inhibin production by rat granulosa cells in vitro. Endocrinology 1987; 120:1633–8.
30. Meunier H, Rivier C, Evans R, et al. Gonadal and extragonadal expression of inhibin and subunits in various tissues predicts diverse functions. Proc Nat Acad Science USA 1988; 85:247–51.
31. Roberts V, Meunier H, Vaughan J, et al. Production and regulation of inhibin subunits in pituitary gonadotropes. Endocrinology 1989; 124:552–4.
32. Tsonis C, Hillier S, Baird D. Production of inhibin bioactivity by human granulosa-lutein cells: stimulation by LH and testosterone in vitro. J Endocrinol 1987; 112:R11–4.
33. McLachlan R, Robertson D, Burger H, et al. The radioimmunoassay of bovine and human follicular fluid and serum inhibin. Mol Cell Endocrinol 1986; 46:175–85.
34. Burger H, McLachlan R, Bangah M, et al. Serum inhibin concentrations rise throughout normal male and female puberty. J Clin Endocrinol Metab 1988; 67:689–94.
35. Muttukrishna S, Fowler P, Groome N, et al. Serum concentrations of dimeric inhibin during the spontaneous human menstrual cycle and after treatment with exogenous gonadotrophin. Hum Reprod 1994; 9:1634–42.
36. Groome N, Illingworth P, O'Brien M, et al. Detection of dimeric inhibin throughout the human menstrual cycle by two-site enzyme immunoassay. Clin Endocrinol (Oxf) 1994; 40:717–23.
37. Groome N, Illingworth P, O'Brien M, et al. Measurement of dimeric inhibin B troughout the human menstrual cycle. J Clin Endocrinol Metab 1996; 81:1401–5.
38. Groome N, O'Brien M. Immunoassays for inhibin and its subunits. Further applications of the synthetic peptide approach. J Immunol Methods 1993; 165:167–76.
39. Bergada I, Rojas G, Ropelato G, et al. Sexual dimorphism in circulating monomeric and dimeric inhibins in normal boys and girls from birth to puberty. Clin Endocrinol 1999; 51:455–60.
40. Sehested A, Juul A, Andersson A, et al. Serum inhibin A and B in healthy prepubertal, pubertal and adolescent girls and adult women: relation to age, stage of puberty, menstrual cycle, follicle-stimulating hormone, luteinizing hormone and estradiol levels. J Clin Endo Metabol 2000; 85:1634–40.
41. Martins da Silva S, Bayne R, Cambray N, et al. Expression of activin subunits and receptors in the developing human ovary: activin A promotes germ cell survival and proliferation before primordial follicle formation. Dev Biol 2004; 266:334–45.
42. Bergada I, Ballerini G, Ayuso S, et al. High serum concentrations of dimeric inhibins A and B in normal newborn girls. Fertil Steril 2002; 77:363–5.
43. Bergada I, Milani C, Bedecarrás P, et al. Time course of the serum gonadotropin surge, inhibins, and anti-Müllerian hormone in normal newborn males during the first month of life. J Clin Endocrinol Metab 2006; 91:4092–8.
44. Peters H, Byskov A, Grinsted J. Follicular growth in fetal and prepubertal ovaries of humans and other primates. Clin Endocrinol Metab 1978; 7:469–85.
45. Polhemus D. Ovarian maturation and cyst formation in children. Pediatrics 1953; 588–94.
46. Winter J, Faiman C. Pituitary-gonadal relations in female children and adolescents. Pediatr Res 1973; 7:948–53.
47. Marshall W, Tanner J. Variations in pattern of pubertal changes in girls. Arch Dis Child 1969; 44:291–303.
48. Chada M, Průsa R, Bronský J, et al. Inhibin B, follicle stimulating hormone, luteinizing hormone, and estradiol and their relationship to the regulation of follicle development in girls during childhood and puberty. Physiol Res 2003; 52:341–6.
49. Groome N, Illingworth P, O'Brien M, et al. Quantification of Inhibin Pro-αC-containing forms in the human serum by a new ultrasensitive two-site enzyme-linked immunoabsorbent assay. J Clin Endocrinol Metab 1995; 80:2926–32.
50. Findlay J, Russell D, Doughton B, et al. Effect of active immunization against the aminoterminal peptide (alpha-n) of the alpha-43-kda subunit of inhibin (alpha-43) on fertility of ewes. Reprod Fertil Dev 1994; 6:265–7.
51. Schneyer A, Sluss P, Whitcomb R, et al. Precursors of α-inhibin modulate FSH receptor binding and biological activity. Endocrinology 1991; 129:1987–99.
52. Mason A, Schwall R, Renz M, et al. Human inhibin and activin: structure and recombinant expression in mammalian cells. In: Burger H, de Kretser D, Findlay J, editors. Inhibin: non-steroidal regulation and follicle stimulating hormone secretion. New York: Raven Press, 1987:89–103.
53. Suzuki T, Miyamoto K, Hasegawa Y, et al. Regulation of inhibin production by rat granulosa cells. Mol Cell Endocrinol 1987; 54:185–95.
54. LaPolt P, Piquette G, Soto D, et al. Regulation of inhibin subunit messenger ribonucleic acid levels by gonadotropins, growth factors, and gonadotropin-releasing hormone in cultured rat granulosa cells. Endocrinology 1990; 127:823–31.
55. Fahy P, Wilson C, Beard A, et al. Changes in inhibin-A (α-βA dimer) and total α-inhibin in the peripheral circulation and ovaries of rats after gonadotrophin-induced follicular development and during the normal oestrous cycle. J Endocrinol 1995; 147:271–83.
56. Woodruff T, Besecke L, Groome N, et al. Inhibin A and inhibin B are inversely correlated to follicle-stimulating hormone, yet are discordant during the follicular phase of the rat estrous cycle, and inhibin A is expressed in a sexually dimorphic manner. Endocrinology 1996; 137:5463–7.
57. Lanuza G, Groome N, Barañao L, et al. Dimeric inhibin A and B production are differentially regulated by hormones and local factors in rat granulosa cells. Endocrinology 1999; 140: 2549–54.

Chapter 23

Hormonal and Molecular Regulation of the Cytochrome P450 Aromatase Gene Expression in the Ovary

Carlos Stocco

23.1 Introduction

The cytochrome P450 aromatase (aromatase), encoded by the *Cyp19a1* (*Cyp19*) gene, is the enzyme that converts androgens into estrogens in the ovarian granulosa cells. Estradiol-17β (estradiol), the most potent estrogen, is crucial for female and male fertility, as proved by the severe reproductive defects observed when its synthesis [1] or actions [2] are blocked. In the ovary, locally produced estradiol in concert with the pituitary gonadotropins is required for successful folliculogenesis and steroid production. Estradiol modulates the structure and function of female reproductive tissues, such as uterus and oviduct, and is required for the fluctuating patterns of biosynthetic and secretory activity of the gonadotropins in the pituitary gland, to generate the preovulatory surge of luteinizing hormone (LH), and for the cyclical variations in sexual female behavior. Therefore, the coordinated and cell-specific expression of the aromatase gene in the ovary is crucial for the normal progression of the menstrual/estrous cycle. This chapter reviews the molecular, cellular, and normal physiological mechanisms regulating the expression of the aromatase gene in the ovary.

23.2 Dynamics of $P450_{arom}$ Gene Expression in the Ovary

The enzymatic capacity to produce ovarian estradiol is acquired during embryonic life [3]. Aromatase expression in fetal ovaries is extremely low, although it can be increased by treatment with a cAMP analog, but not by FSH [4]. Whether this minute expression of aromatase in fetal ovaries plays any role in the regulation of ovarian development is not clear, since estradiol does not appear to be critical for the normal development of the ovary during fetal life [5]. Aromatase activity and mRNA can also be found after birth in the granulosa cell layer of growing follicles [6], which progressively increases in both preantral and small antral follicles in infantile ovaries. However, in the ovaries of sexually mature animals, aromatase is no longer expressed by growing follicles, but is confined to healthy large antral follicles [6], and in the corpus luteum (Fig. 23.1). In these structures, expression of the aromatase gene is controlled in a cell-specific, temporal, and spatial manner, which limits estradiol production only to mural granulosa cells of healthy large antral follicles and to the corpus luteum.

23.2.1 Expression of Aromatase in Ovarian Follicles

Within the granulosa cell layer of antral follicles, aromatase expression is highest in the mural granulosa cells at the outer edge of the follicle when compared to granulosa cells closest to the antral cavity [7]. Moreover, aromatase is not expressed in cumulus granulosa cells [7]. The marked compartmentalization of aromatase expression within the granulosa cell layer is particularly striking in mature proestrus follicles. The molecular basis and physiological implication of the differential expression of aromatase in mural and cumulus granulosa cells are not clear. Aromatase expression in cumulus granulosa cells is probably silenced by oocyte-derived compounds. Factors produced by the oocytes such as BMP-15 [8] and GDF-9 [9, 10] inhibit the FSH-induced increase aromatase expression. In addition, in GDF-9 deficient mice, which manifest an arrest of preantral follicular growth, there is a premature expression of aromatase in preantral follicles [11]. Recently, it was suggested that the tumor-suppressor gene BRCA1 may contribute to the specific distribution of aromatase in antral follicles. In contrast to aromatase, BRCA1 expression is restricted to the

C. Stocco (✉)
Department of Obstetrics Gynecology and Reproductive Sciences, Yale University School of Medicine, New Haven, CT, USA
e-mail: carlos.stocco@yale.edu

P.J. Chedrese (ed.), *Reproductive Endocrinology*,
DOI 10.1007/978-0-387-88186-7_23, © Springer Science+Business Media, LLC 2009

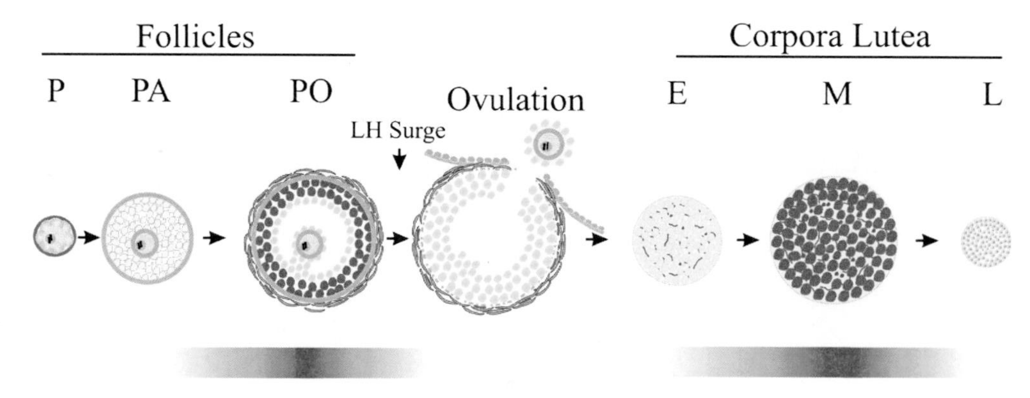

Fig. 23.1 Aromatase expression in the mature ovary. In the ovary of adult animals, aromatase is observed only in preovulatory follicles (PO) but not in primordial (P) preantral (PA) follicles. Moreover, expression of aromatase is not uniform throughout the antral follicles, being highest in the mural granulosa cells at the periphery of the follicle and absent from the granulosa cells surrounding the antrum and within the cumulus. Expression of aromatase is downregulated after the LH surge, which induces ovulation and luteinization. Aromatase is not expressed in newly formed corpora lutea (E), but becomes highly expressed in fully functional corpora lutea during the middle of the gestation (M) in rats. Toward the end of pregnancy, aromatase expression decreases again, becoming undetectable in the regressing corpora lutea (L)

cumulus cells of antral follicles [12]. Ectopic expression of BRCA1 decreases aromatase in a granulosa tumor cell line [13]. Although BRCA1 seems to bind to the aromatase promoter [14], the region where BRCA1 binds and the mechanism by which BRCA1 represses aromatase remains to be elucidated. Nevertheless, it is possible that BRCA1 does not regulate aromatase, and the differential expression of BRCA1 and aromatase in mural and cumulus cells is only a consequence of the different phenotypes of these two cells. Dias [15] recently demonstrated that oocyte-stimulated signaling opposes the action of FSH, whereas FSH stimulates expression of mural marker transcripts and suppresses the expression of cumulus markers. Thus, oocytes and FSH establish opposing gradients of influence that define the cumulus and mural granulosa cell phenotypes.

Factors affecting aromatase expression in granulosa cells—FSH is the major inducer of aromatase activity in granulosa cells. However, the stimulatory effect of FSH is subject to modulations by numerous factors (Fig. 23.2), including estradiol that augments the actions of FSH on granulosa cells. Moreover, both hormones are necessary to establish a fully differentiated and healthy preovulatory follicle. This feed-forward effect of estradiol is believed to play a key role in follicle dominance. In the rat, one of the effects of estradiol is to enhance FSH-induced aromatase activity [16, 17]. Estradiol actions in rodent granulosa cells are mediated by the activation of estrogen receptor β (ERβ) [18]. Accordingly, in immature ERβ knockout mice, the stimulation of aromatase expression by FSH is impaired [19]. Since estradiol alone does not affect aromatase expression [16] and no ER response elements are present in the aromatase promoter, it is not known if ERβ is able to directly affect the activity of the aromatase promoter.

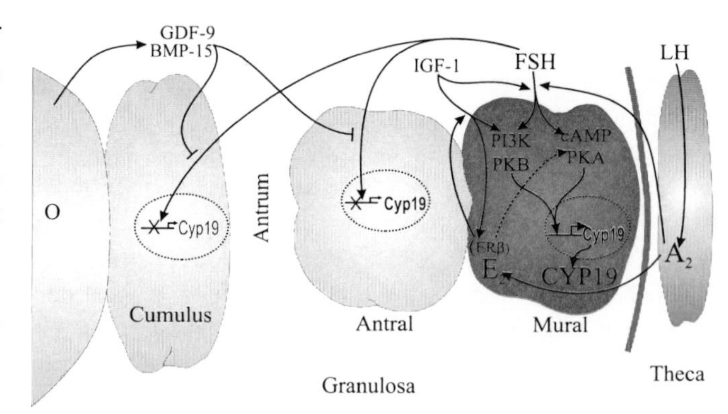

Fig. 23.2 Hormone interactions in the regulation of aromatase expression in follicles. FSH activates the cAMP/PKA and PI3K/PKB signaling pathways, which are known to mediate its stimulatory effect on aromatase expression. Androgens, IGF-1, and estradiol potentiate this effect of FSH. Androgens have direct effects probably mediated by the activation of androgen receptors and indirect effects throughout its conversion to estradiol. IGF-1 stimulates the expression of the FSH receptor and probably synergizes with FSH in the activation of PKB. Estradiol also enhances the stimulatory effect of FSH. The positive feed-back of estradiol may be mediated by a cooperative effect on the activation of the PI3K/PKB and cAMP/PKA pathways. On the other hand, the oocyte (O) produces factors (GDF-9 and BMP-15) that block the stimulatory effect of FSH in cumulus cells and in antral granulosa cells in contact with the follicular fluid, limiting the expression of aromatase to mural granulosa cells. BM: basal membranes

Androgens also enhance the FSH-induced aromatase expression, testosterone being more effective than estradiol, whereas the non-aromatizable androgen, dihydrotestosterone, is as effective as estradiol [16, 20]. This evidence indicates that theca cell androgens act not only as a substrate of estrogen synthesis but also modulate FSH action via activation of androgen receptors [16, 20]. In vitro studies suggest that androgens enhance FSH-stimulated steroidogenesis by increasing cAMP synthesis [21, 22]. In the rat, androgen receptor expression is highest in preantral/early antral follicles and gradually decreases as follicles mature at the time that aromatase expression increases [23]. These findings suggest that androgens enhance FSH action at an early stage of follicular development, but during the final stages they mainly serve as a substrate for estrogen synthesis [20]. Aromatase expression and estradiol levels are normal in the ovaries of androgen receptor knockout mice [24], suggesting that androgens are not crucial for aromatase expression in vivo.

FSH-induced stimulation of aromatase activity is potentiated by insulin-like growth factor-1 (IGF-1) [25, 26, 27], which also amplifies the synergism between FSH and testosterone on aromatase expression [22]. In mice, aromatase mRNA is found only in IGF-I and FSH receptor positive follicles [28]. Further evidence of a functional link between the IGF system and stimulation of aromatase by FSH is provided by the fact that IGF-binding protein (IGFBP-4) is a potent inhibitor of FSH-induced estradiol production by murine [29] and human [30] granulosa cells. IGF-I likely acts primarily by augmenting the capacity of granulosa cells to respond to FSH, since FSH receptor expression is reduced in IGF-1 knockout mice [28]. In human granulosa cells, however, IGF-I alone increases estradiol production to levels comparable to those induced by FSH, although together these hormones also have synergistic effects [31].

Intracellular signaling pathway and aromatase expression—FSH activation of the cAMP-dependent protein kinase A (PKA) is the main activator or aromatase expression [32]. Fitzpatrick and Richards identified two elements in the aromatase promoter that mediate the stimulatory effect of cAMP/PKA on the aromatase promoter [33]. One element is a hexameric DNA sequence recognized by nuclear orphan receptors [33]. The second and more distal element was identified as a cAMP-response element-like sequence (CLS) [34]. FSH activates several other intracellular signaling pathways, including those for the extracellular regulated kinases (ERKs), p38 mitogen-activated protein kinases (MAPKs), and phosphatidylinositol-3 kinase (PI3K) [35]. Of these pathways, PI3K, which activates protein kinase B (PKB or viral proto-oncogene 1; Akt), also participates in the induction of aromatase expression by FSH [36, 37]. FSH-stimulated aromatase is amplified by expression of constitutively activated Akt, whereas treatment with an inhibitor of the PI3K [37] or overexpression of dominant negative Akt prevents aromatase up-regulation [36]. FSH stimulation of Akt seems to relieve aromatase from a repressive effect of forkhead box-O1 (FOXO1), since a constitutively active FOXO1 protein prevents aromatase stimulation by FSH and activin [37]. Consistent with this hypothesis, both FSH and IGF-I, act post-translationally to phosphorylate FOXO1 [38], resulting in cytoplasmic localization and, presumably, loss of FOXO1 repression activity. Moreover, after PMSG treatment of immature rats, FOXO1 expression decreases [38], remaining only in cumulus cells and antral granulosa cells, which as mentioned before do not expressed aromatase [39]. The mechanism by which FOXO1 blocks aromatase expression is not known. However, it has been demonstrated that the DNA binding ability of FOXO1 is crucial to repress aromatase, since overexpression of a constitutively active FOXO1 protein that does not bind DNA has no effect. Since no FOXO1 binding sites have been described in the aromatase proximal promoter, the effect of this transcription factor is probably indirect. It is also possible that the DNA-binding region of FOXO1 is necessary for the interaction of this factor with other regulatory proteins directly at the level of the aromatase promoter; in this case the effect of FOXO1, although independent of DNA binding, would be direct.

Estradiol enhances the expression of components of the IGF-I pathway such as the IGF-1 receptor; in turn, IGF-1 stimulates the expression of ERβ [38]. Moreover, the stimulation of Akt by IGF-1 [40] is potentiated by estradiol in the rat [41]. These results suggest that the synergistic effect of estradiol, IGF-1, and FSH on the induction of aromatase could converge on Akt. On the other hand, synergism also exists between the ERβ and the cAMP/PKA signaling pathways [42]. Therefore, the two signaling pathways activated by FSH (PI3K/Akt and cAMP/PKA) known to be involved in the regulation of aromatase may receive positive inputs from the estradiol/ERβ and the IGF-1/PI3K/Akt pathways (Fig. 23.2).

Kinetics of aromatase expression—ONE intriguing characteristic of the effect of FSH on aromatase expression is its the relative long time (24–48 h) required to elevate aromatase mRNA [16]. Since FSH stimulates cAMP production very rapidly, it has been proposed that an increase in the expression of the regulatory and catalytic units of PKA or proteins synthesized as a consequence of PKA activation may be required for aromatase induction [16]. Interestingly, FSH rapidly stimulates aromatase in the Sertoli cells of immature rats [43]. In these cells, the stimulatory action of FSH on aromatase expression also depends on the cAMP/PKA and PI3K/AKT1 pathways [43]. Studies aimed to compare the regulation of the aromatase gene in Sertoli and granulosa cells could provide information on the mechanisms by which the expression of this gene is delayed in granulosa cells.

23.2.2 Aromatase Expression in Luteal Cells

Aromatase is highly expressed in the corpora lutea of pregnant rats. Luteal cell function is greatly affected by locally produced estradiol, which stimulates both progesterone biosynthesis and luteal cell hypertrophy [44]. As

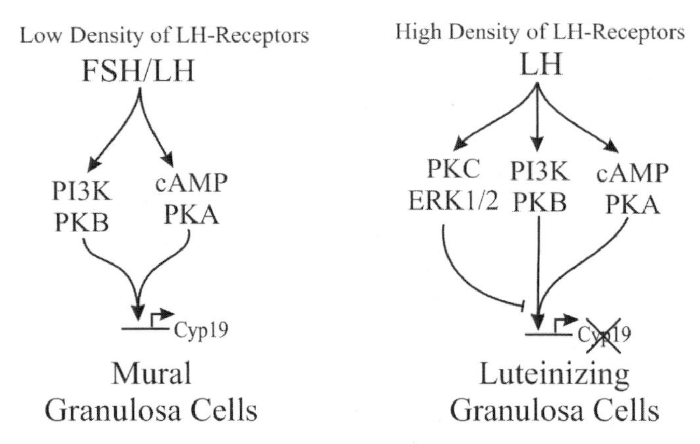

Fig. 23.3 Changes on the intracellular signaling pathways that control aromatase expression of granulosa cells during luteinization. Growing follicles express low level of LH and FSH receptors, which leads to the stimulation of the cAMP/PKA and PI3K/PKB, which in turn stimulate the expression of the aromatase gene. In growing follicles FSH also increase LH-R expression. Thus, in preovulatory follicles, the LH-R is extensively expressed in granulosa cells. The impact of the LH surge on preovulatory granulosa cells expressing high levels of LH-R leads to the activation of not only the cAMP/PKA and PI3K/PKB, but also to the activation of PKC, which in turn activates ERK1/2. ERK1/2 blocks the stimulatory effect of PKB and PKA resulting in the downregulation of aromatase expression

shown in Fig. 23.3, luteal aromatase mRNA content is low on day 4 of pregnancy, increases progressively to reach high levels of expression between days 14 and 19, and decreases from day 20 to reach undetectable levels on day 23, the day of parturition [45]. To facilitate the description of the mechanisms that may control luteal aromatase expression, three periods will be considered: (i) luteinization, (ii) pregnancy, and (iii) before parturition.

Aromatase expression during luteinization—aromatase expression reaches its zenith in preovulatory follicles, which ovulate and luteinize in response to the LH surge. The LH surge also induce a rapid decrease in aromatase mRNA [32]. This inhibitory effect of LH contrasts with the stimulatory action of FSH, since both hormones mediate their effects through the adenylyl cyclase/cAMP. Differences in the magnitude and duration of the cAMP signal that each receptor produces [16] and the generation of specific intracellular signals [36] have been proposed to explain the differential effects of FSH and LH on aromatase expression. In support of the latter mechanism, Ascoli and collaborators have proposed that the differential effects of FSH and LH may depend on the density of their receptors. Thus, the number of LH receptor molecules expressed in granulosa cells dictates the signaling pathway activated by this hormone [46]. By using an adenovirus to direct the expression of LH receptors in primary cultures of immature rat granulosa cells, Donadeu and Ascoli [46] demonstrated that only cells with a high density of receptors respond to LH by activating the inositol phosphate cascade in addition to the cAMP and Akt pathways. In this situation, the cAMP and Akt induction of aromatase is inhibited by activation of the inositol phosphate cascade [46]. At low receptor density, however, LH stimulates aromatase expression. Similar to aromatase, the LH receptor is highly expressed in preovulatory follicles [47], suggesting that only in preovulatory granulosa LH stimulates inositol phosphate production and intracellular calcium release. The intracellular increase of these second messengers leads to the activation of PKC, which inhibit the FSH-induced aromatase [48]. Activation of PKC in rat granulosa cells results in decreased expression of the catalytic subunit of PKA and the transcription factor steroidogenic factor 1(SF-1 or Nr5a1) [48]. Whether or not the decreased expression of the catalytic subunit of PKA and SF-1 are responsible for the inhibitory effect of LH on aromatase expression remains to be determined.

The inhibitory effect of PKC may be mediated by extracellular regulated kinase (ERK1/2 or MAPK) pathway. LH stimulates ERK1/2 phosphorylation in granulosa cells [49, 50, 51]. Remarkably, LH causes a rapid and transient activation of ERK1/2 in granulosa cells expressing low levels of LH receptors; whereas, if high levels of LH receptor are present, a delayed and more sustained activation of ERK1/2 take place [52]. The early increase in ERK1/2 phosphorylation is PKA dependent and PKC independent [50]. In contrast the delayed effect is PKA and PKC dependent and seems to be mediated by an increase in epidermal growth-like factors (EGF-1) [52]. Moreover, an inhibitor of MEK, the upstream kinase of ERK1/2, overcomes the ability of PKC activation to antagonize the induction of aromatase. This results suggest that PKC-mediated phosphorylation of ERK1/2 may be ultimately responsible for the inhibition of aromatase in immature granulosa cells expressing a high density of LH receptors (Fig. 23.3).

Inhibition of protein synthesis partially prevents the loss of aromatase mRNA induced by LH or PKC activation [16, 48], suggesting that the expression of a repressor protein is needed. In fact, the transcription factors C/EBPβ (CCAAT/enhancer-binding protein-β) and ICER (inducible cAMP early repressor), which are rapidly induced by the LH surge [53, 47], seem to participate in silencing the aromatase gene. In C/EBPβ-null mice, the decrease in aromatase expression observed after the LH surge does not occur [54], and overexpression of C/EBPβ in endometriotic stromal cells prevents cAMP stimulation of the aromatase promoter [55]. Moreover, downregulation of C/EBPβ in endometriotic stromal cells has been proposed as the cause of elevated aromatase expression in endometriosis [55].

Inducible cAMP early repressor is an isoform of CREM (cAMP-responsive element modulator) that represses cAMP-induced transcription [47]. ICER has been shown to participate in the repression of the inhibin α-subunit gene that takes place after the surge of LH [56]. Overexpression of ICER decreases aromatase promoter activity induced by forskolin [57] and mediates the inhibition of aromatase expression by tumor necrosis factor-α (TNF-α) [58]. Whether the induction of C/EBPβ and ICER is due to PKC activation by the LH surge remains to be determined, but this seems to be unlikely because these genes do not respond to PKC [59, 60]. Additional studies are needed to bring together the effects of LH/inositol, Cebpb, and ICER on the downregulation of aromatase in granulosa cells.

Aromatase expression in the corpus luteum of pregnancy—in newly formed rat corpus luteum, aromatase remains expressed at low levels, which are not affected by the elevation of intracellular cAMP [32, 61]. Luteal aromatase expression starts to increase between days 6 and 8 of pregnancy [62]. Similarly, in vitro experiments demonstrated that luteinized granulosa cells express very low levels of aromatase; however, aromatase expression is recovered after 5–6 days of incubation [32, 63]. The factors involved in the reactivation of the aromatase gene either in vivo or in vitro are not known.

During the first week of pregnancy, corpus luteum function is maintained by the actions of prolactin (PRL) and LH [64]. PRL is secreted from the pituitary gland until approximately days 9–10 of gestation [65]. Although PRL is essential to maintain luteal function, evidence suggests that this hormone inhibits luteal aromatase. For instance, PRL decreases aromatase expression in luteinized granulosa cells and in the corpora lutea of pseudopregnant rats [62]. In pregnant rats, treatment with a blocker of PRL secretion on day 5 of gestation leads to an increase in aromatase mRNA expression that can be prevented by treatment with PRL (Stocco, unpublished results).

Placental lactogens and estradiol are required to maintain rat luteal function during the second half of gestation [64]. These hormones have also been implicated in the regulation of aromatase expression. Placental lactogens removal after day 10 of pregnancy is followed by low expression and activity of luteal aromatase [66]. Placental lactogens signal through the same receptor that PRL uses [64]. Accordingly, in hysterectomized pregnant rats treated daily with PRL (or PRL plus either testosterone or estrogen), aromatase expression increases, reaching levels similar to those found in intact rats on day 15 [66]. Thus, in contrast to the inhibitory effect of PRL during the first part of gestation, placental lactogens are necessary to sustain luteal aromatase expression during the second part of gestation.

Administration of human chorionic gonadotropin (hCG) does not affect aromatase expression before (days 8–10) or after (days 13–19) midgestation; however, aromatase mRNA and protein increase in intact pregnant rats treated with hCG between days 10 and 12 [66]. In agreement with these results, the induction of aromatase on day 15 of gestation can be blocked by the administration of an anti-LH antibody on day 10 but not when given on day 12 [66].

In summary, luteal aromatase seems to be maintained by PRL at a low level during the first part of pregnancy and modulated by LH at midgestation. During the second half of pregnancy, aromatase becomes highly expressed by the stimulatory action of placental lactogens and testosterone in coordination with the stimulatory effect of ovarian estradiol itself (Fig. 23.4). The molecular and cellular mechanisms by which these hormones control the expression of the aromatase gene in the corpora lutea of pregnant rats are not known. In humans, estrogen levels also undergo a transient decline after the gonadotropin surge and a later increase due to marked induction of aromatase mRNA [67]. In contrast, in cattle, this gene remains expressed at very low levels during pregnancy [68, 69], suggesting that the pattern of luteal estrogen production may differ between species.

Aromatase expression in the regressing corpora lutea—aromatase expression rapidly decreases just before parturition. We have demonstrated that the luteolytic hormone prostaglandin F2α (PGF2α) represses luteal aromatase mRNA and protein levels when administered to rats on day 19 of pregnancy [45]. This inhibitory effect of PGF2α on aromatase expression was confirmed using PGF2α

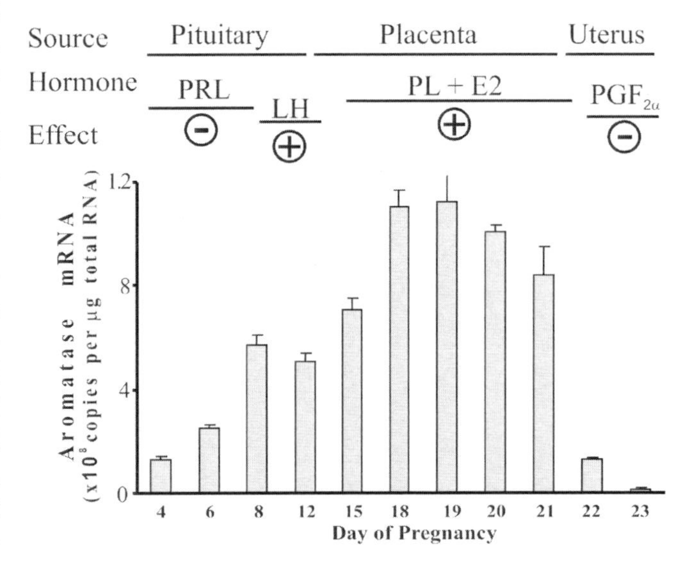

Fig. 23.4 Hormonal regulation of luteal aromatase during pregnancy. Bars represent aromatase mRNA levels in corpora lutea of rats on different days of pregnancy. Aromatase and ribosomal protein L19 mRNA levels were determined by real-time PCR and the results expressed as aromatase mRNA molecules per µg of total RNA was determined by using a standard curve generated from known quantities of aromatase cDNA. Values represent average ± SEM. Luteal aromatase expression seems to be maintained by prolactin (PRL) at a low level during the first part of pregnancy, is modulated by LH at midgestation, and becomes highly expressed by the stimulatory action of placental lactogens (PL) and testosterone in coordination with the stimulatory effect of ovarian estradiol (E2). Prostaglandin F2α (PGF2α) is involved in the rapid downregulation of aromatase toward the end of pregnancy

receptor-knockout mice [45]. In these mutant animals, parturition does not occur and luteal expression of aromatase remains high. In vitro experiments demonstrated that PGF2α inhibits aromatase gene transcription in luteinized granulosa cells [45].

23.3 Regulation of the Ovarian Aromatase Promoter

23.3.1 Structure of the Aromatase Gene

In all species studied thus far, expression of the aromatase gene is under the control of two or more tissue-specific promoters. These promoters are controlled by tissue-specific signaling pathways and produce several alternative forms of exon-I that are then spliced onto a common 3′-splice acceptor site in exon-II [70]. This 3′-splice acceptor is located upstream of the translation start site; therefore, all transcripts contain an identical open reading frame and encode the same protein regardless of the promoter used (Fig. 23.5A). In humans, the following promoters are known: I.1, I.2, and IIa (placenta); I.3 (breast cancer); I.4 (normal adipose cells, skin, and fetal liver); I.5 (fetal liver); I.7 (endothelial cells and breast cancer); I.f (brain); and I.6 (bone). Promoter-II drives transcription in the ovary and lies immediately upstream of the translation start site; therefore, promoter-II

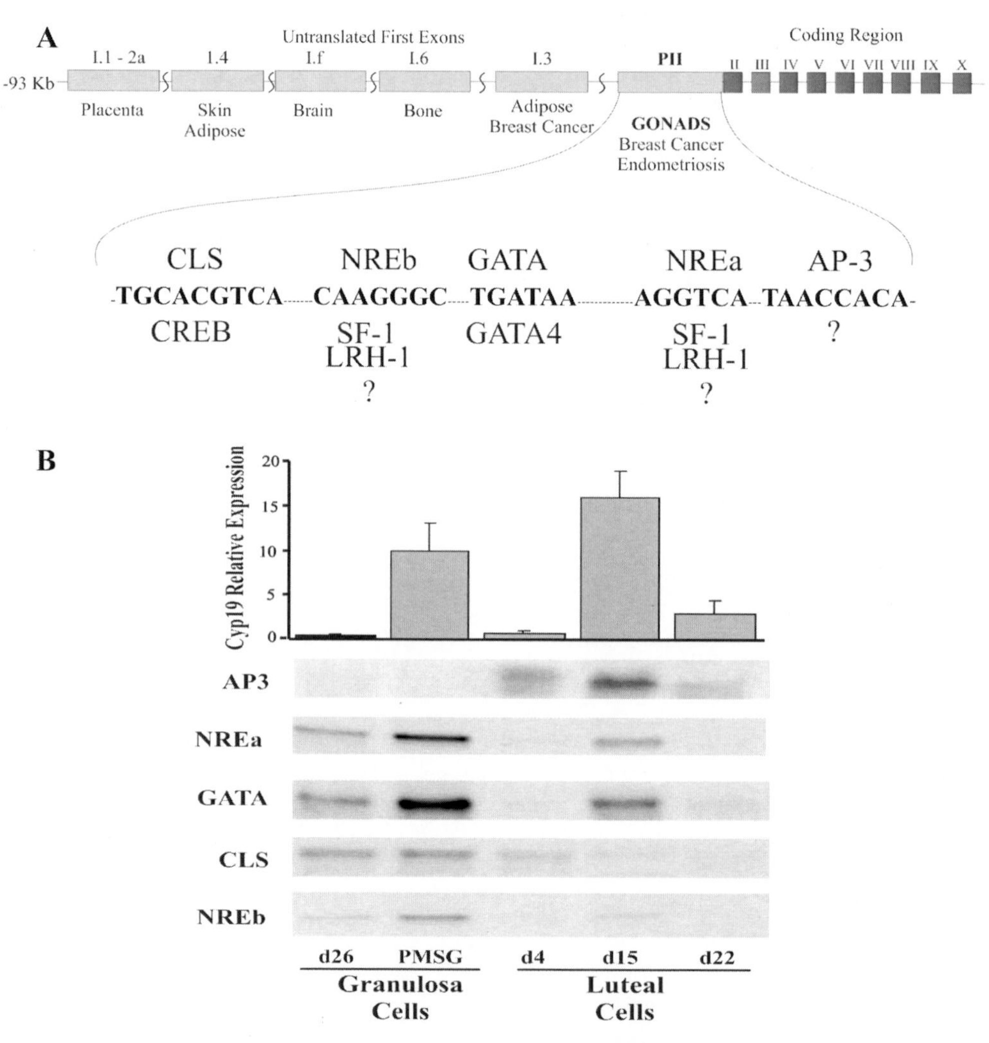

Fig. 23.5 A: Scheme of the human aromatase promoter indicating the different alternative tissue-specific promoters. The PII or proximal promoter is highly conserved between species. This promoter is used to control the aromatase expression in the ovary of humans and rodents. Within this promoter a cAMP-responsive element-like sequence (CLS), two binding sites for members of the nuclear receptors 5A family of transcription factors (NREa and NREb), one GATA response element, and one AP-3 binding site are present. **B**: Dynamics of the activation of the aromatase promoter in ovarian cells. Aromatase mRNA levels and DNA protein binding in 26-day-old immature rats (d26), immature rats treated with PMSG (PMSG), and from corpora lutea of rats on days 4 (d4), 15 (d15), or 23 (d23) of pregnancy. Aromatase mRNA levels were determined as in Fig. 23.3 and expressed as the ratio between aromatase and ribosomal L19 mRNA. DNA protein binding was investigated using gel shift analyses with oligonucleotides spanning the cAMP response element-like sequence (CLS), the nuclear receptor elements a and b (NREa and NREb), the GATA binding sites, or the AP-3 binding site

is also called proximal promoter. The proximal promoter or ovarian promoter is highly conserved between species [45]. In breast adipose tissue of breast cancer patients, aromatase activity and *Cyp19* expression are elevated when compared to normal breast tissue [71]. This increase in *Cyp19* expression is associated with a switch in promoter usage from the normal adipose-specific promoter I.4 to the proximal promoter [72].

In the rat, two aromatase gene promoters have been described: a proximal promoter, which controls aromatase expression in granulosa and luteal cells [45, 48] as well as in Leydig cells [73], and a distal promoter, which drives aromatase expression in the brain [74]. Transient transfection experiments have demonstrated that the first 160 bp of the ovarian promoter are sufficient to mediate cAMP stimulation in rat granulosa cells [33]. Both the brain and the ovarian transcripts are produced in the rat corpus luteum [45], although transcripts produced by the brain promoter are expressed at very low levels when compared to the ovarian transcripts. Throughout pregnancy, only the expression of the ovarian transcript correlates with the changes observed in luteal aromatase protein levels. The low and constant levels of the brain transcript suggest that activation of the brain promoter drives the nominal and cAMP-independent expression of aromatase found at the beginning of pregnancy.

The region necessary for the activation of the proximal promoter was investigated in transgenic mice carrying 2,700, 278, or 43 base pairs (bp) of the region upstream of the translation start site of the human aromatase gene linked to the human growth hormone coding region as reporter [75]. In mice carrying 2,700 or 278 bp, the reporter is specifically expressed in the ovary, whereas animals carrying the 43 bp construct show no expression of the reporter. This in vivo evidence suggests that the –278 to –43 region of the ovarian promoter contains the elements that drive aromatase expression in the ovary. However, the 278 bp reporter construct is active in ovarian stromal cells of adult animals, which do not express aromatase. Moreover, this construct is expressed throughout large antral follicles while aromatase expression is restricted to mural granulosa cells (Fig. 23.1). This indicates that the region 278 does not contain the elements that silence aromatase expression in cumulus granulosa cells and in stromal cells. It is also possible that chromatin modifications are important for the cell-specific expression of aromatase in the ovary. Since the reported construct could have been integrated anywhere in the genome, the activity of these constructs are not controlled by the same chromatin modification as the aromatase gene. As the author noticed [75], this abnormal expression could also be due to the presence of high level of growth hormone that was used as reporter.

23.3.2 Regulation of the Cis-Elements in the Aromatase Gene Ovarian Promoter

The first 300-bp region of the proximal promoter of the aromatase gene contains the element necessary for expression in mural granulosa cells. As shown in Fig. 23.4, within this region a cAMP-responsive element-like sequence (CLS) [32] and two binding sites for members of the nuclear receptors 5A family of transcription factors (NREa and NREb) are present [33, 76]. The proximal promoter also contains one response element for members of the zinc finger family of transcription factors known as GATA and one AP-3 binding site [77]. These binding sites are highly conserved between species.

CLS (TGCACGTCA)—this region differs from a consensus CRE binding site (TGACGTCA) by the insertion of a cytosine between the second and third nucleotides. Reporter experiments suggest that it plays a key role in the activation of the proximal promoter in granulosa cells [34, 78]. However, mutation of this element greatly reduces but does not completely block the induction of promoter activity by cAMP. This region is recognized by the cAMP-responsive element binding (CREB) protein as demonstrated using gel shift assays [34]. CLS is also recognized by C/EBPβ, which has an inhibitory effect on aromatase promoter activity in human endometrial cells [55].

cAMP-responsive element binding is rapidly phosphorylated by treatment of granulosa cells with FSH, but the expression and subcellular localization of CREB does not change during granulosa cells differentiation [34]. Overexpression of a non-phosphorylatable mutant of CREB in primary cultures of rat granulosa cells decreases estradiol production induced by FSH [79], suggesting that CREB activation is required for the expression of the aromatase gene. Noticeably, the FSH-induced increase in aromatase mRNA is detectable only 24 h after treatment [16], but FSH stimulation of CREB phosphorylation occurs within 1 h and declines to basal levels by 6 h. There is no consensus as to whether phosphorylation of CREB is subsequently increased. For instance, no [80], low [61], or very high [81] levels of CREB phosphorylation have been reported 24 and 48 h after FSH treatment. Moreover, in forskolin-treated cells, phospho-CREB remains high in the presence of an inhibitor of phosphodiesterase 4 (PDE4), but no alterations in either the magnitude or the pattern of aromatase expression occur [61]. These results suggest that the effect of CREB on aromatase expression is complex and that CREB may have a direct effect via CLS and indirect effects by inducing the expression of other proteins involved in the activation of the aromatase gene.

The proximal promoter of the bovine aromatase gene has a 1-bp deletion in the CLS element that destroys this element as a site for CREB binding [82]. Consequently, bovine aromatase reporter constructs do not respond to cAMP [82]. However, bovine granulosa cells do express aromatase in response to FSH [83]. Moreover, rat luteal cells, which expressed high levels of aromatase, do not contain proteins that bind to CLS [77] (Fig. 23.5). In addition, in rat luteal cells, CREB resides in the cytoplasm [61], suggesting that neither CREB nor CLS activation is involved in the regulation of aromatase in the rat corpus luteum. This evidence supports the concept that activation of CLS is cell specific and that additional factor(s) and regulatory elements are necessary to activate the aromatase proximal gene promoter in the ovary.

NREa (AGGTCA)—because the aromatase gene promoter does not have a consensus cAMP response element, this nuclear receptor half-site was originally proposed to mediate the cAMP stimulation of aromatase [33]. Later, it was demonstrated that CLS and NREa interact in an additive manner to control aromatase expression in granulosa cells [81]. NREa is recognized by steroidogenic factor-1 (SF-1, also termed Nr5a1 or Ad4BP) and liver receptor homolog-1 (LRH-1, also termed Nr5a2). Although both SF-1 and LRH-1 are derived from different genes, these proteins share a high degree of identity, particularly in their DNA binding domains [84]. In fact, in vitro translated mouse SF-1 and LRH-1 proteins [76] or SF-1 expressed in bacteria [85] bind to oligonucleotides containing the NREa present in the aromatase gene. However, because both factors are expressed in granulosa cells [86, 87], whether SF-1 or LRH-1 or both are important for aromatase expression in these cells remains under debate (describe below), although evidence suggests that SF-1 has a more prominent role. Exogenous expression of LRH-1 in bovine granulosa cells [76], mouse Leydig and Sertoli cell lines [88], or 3T3L1 mouse preadipocytes [89] increases aromatase promoter activity. However, overexpression of LRH-1 in rat granulosa cells does not enhance estrogen production or aromatase expression in either the presence or absence of FSH, although it did increase progesterone production [90]. This suggests that LRH-1 may not be crucial for aromatase expression in granulosa cells.

On the other hand, strong evidence suggests that SF-1 is a main player in the regulation of aromatase in the ovaries of rodents. Ovaries of granulosa-cell-specific SF-1 knockout mice contain hemorrhagic follicles [91] and similar abnormalities were found in the ovaries of the aromatase knockout mice [92]. Moreover, exogenous expression of SF-1 in rat granulosa cells stimulates aromatase expression [93, 94]. Recent evidence further supports a prominent role of SF-1 in aromatase expression in granulosa cells. FSH- and cAMP-dependent regulation of aromatase in granulosa cells is enhanced by β-catenin (Ctnnb1). This stimulatory effect of β-catenin is mediated through an interaction with SF-1 [94]. Using a granulosa cell line, Parakh demonstrated for the first time that SF-1 binds to the endogenous ovarian-specific aromatase promoter and that this binding is stimulated by cAMP [94]. Moreover, overexpression of constitutively active mutant FOXO1, which prevent aromatase stimulation by FSH and activin, also decreases SF-1 expression [37]. Levels of SF-1 protein do not exhibit major changes during follicular growth or in response to low levels of FSH in culture but decrease rapidly after the LH surge along with the downregulation of aromatase [86], further supporting the idea of an important role for SF-1 in the regulation of aromatase expression. A second NRE upstream of CLS (Fig. 23.5) was recently described as able to bind in vitro transcribed SF-1 or LRH1 proteins and that this response element is necessary for the stimulation of the human aromatase promoter by forskolin in bovine granulosa cells [76]. We investigated the activation of the two NRE elements in the rat corpus luteum throughout pregnancy using gel shift analysis and found that their activation is developmentally regulated and correlate with luteal aromatase expression (Fig. 23.5B).

GATA (TGATAA)—this element was described first in the human aromatase proximal promoter as part of a silencer element [95]. The GATA response element is conserved in several species, including the mouse, rat, and human [45]. In rat luteal and granulosa cells GATA is recognized by GATA-4 [45, 96]. GATA-4 binding is stimulated in vitro and in vivo by FSH [45, 96] and by PGE2 [97]. In addition, GATA-4 silencing blunts FSH-induced aromatase expression, suggesting that GATA-4 mediates at least in part the stimulatory effect of FSH in granulosa cells [96]. As shown in Fig. 23.5, the activation of this GATA binding site in the corpus luteum follows a pattern similar to that of aromatase expression, suggesting that GATA-4 participates in the regulation of aromatase in luteal cells.

AP-3 (TAACCACA)—we recently described the presence of an AP-3 binding site in the proximal promoter of the rat aromatase gene [77]. AP-3 binding sites are also present in the promoter of humans and mice and it seems to be necessary for full activation of the aromatase promoter [77]. In rats, this AP-3 binding site interacts with luteal nuclear extracts obtained from pregnant rats but not with ovarian extracts of immature rats or of immature rats treated with PMSG (Fig. 23.5B). In PMSG-treated immature rats, AP-3 binding activity increases after treatment with hCG, suggesting that AP-3 is active only in luteal cells [77]. In pregnant rats, luteal protein binding to AP-3 correlates with variations in aromatase expression (Fig. 23.5B). These results suggest that AP-3 activation may play a role in the regulation of luteal aromatase expression. The protein that recognizes AP-3 has a molecular weight of approximately 48 kDa [77]; however, its identity is not known yet.

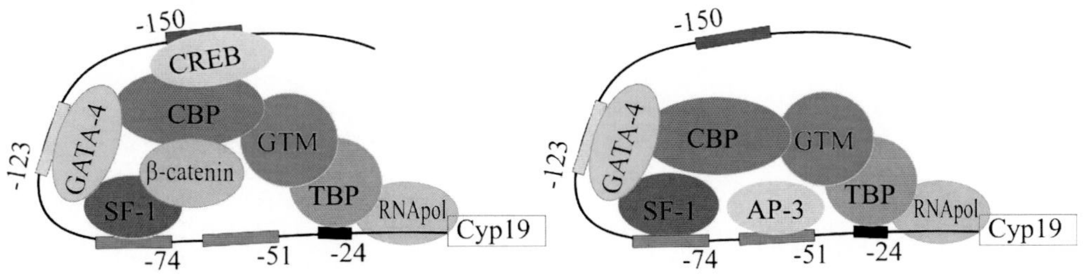

Fig. 23.6 Scheme depicting hypothetical multiunit complexes on the aromatase promoter in granulosa and luteal cells. The participation and binding of CREB, GATA-4, β-catenin, and AP-3 have been shown using in vitro approaches such as gel shift and gene reporter assays. The involvement of SF-1 has been inferred from knockout and overexpression experiments. Interactions between CREB, SF-1, and GATA-4 with CBP are deduced from experiments in cells other than granulosa or luteal cells. The spatial binding of transcription factors on the DNA does not necessarily represent an in vivo situation but highlights the complexity of this promoter and the differences between granulosa and luteal cells on the activation of the aromatase gene. GTM: general transcriptional machinery; TBP: TATA binding protein; RNApol: RNA polymerase

23.4 Summary

In the rat ovary, aromatase expression result from the activation of its proximal promoter. The cAMP/PKA/CREB pathway is considered to be the primary signaling cascade through which this promoter is regulated. In rat luteal cells, the proximal promoter is not controlled in a cAMP-dependent manner. Of the elements found in this region, GATA and NRE are used by both cell types. On the other hand, CLS is active only in granulosa cells, whereas the AP-3 binding site is activated exclusively in the luteal cells (Fig. 23.5B). It is clear then that there is a change in the composition of the transcription complex formed on the aromatase promoter during the transformation of granulosa cells into luteal cells (Fig. 23.6). CREB is the principal regulatory component in the regulation of aromatase gene, which is stimulated by FSH-activated PKA. FSH stimulation of aromatase expression is also mediated by activation of Akt, which seems to relieve aromatase from a repressive effect of FOXO1, although the mechanism by which FOXO1 blocks aromatase expression is not known. FSH also stimulates binding of GATA-4 and SF-1 to the aromatase promoter. The stimulatory effect of FSH on aromatase expression is potentiated by IGF-1 and androstenedione, probably by enhancing FSH intracellular signaling. IGF-I also seems to assist FSH in the repression of FOXO1. CREB interaction with cofactors such as the CREB binding protein (CBP) leads to the assembly of the general transcription machinery and the recruitment of the TATA binding protein to the aromatase promoter. GATA-4 and SF-1 participate in the activation of the promoter by binding to NRE and GATA and by interacting with cofactors such as CBP and β-catenin, respectively.

In luteal cells, CLS is inactive and it seems that CREB does not participate in the regulation of the aromatase gene. This is a striking feature of the regulation of aromatase in luteal cells, especially if we consider that cAMP is the major regulator of the proximal promoter across species. It remains to be determined which intracellular signaling pathway is involved in the upregulation of aromatase in luteal cells during the second half of pregnancy in rats. Since activation of the PRL receptor at this time stimulates luteal aromatase and because Akt can be activated by PRL [98], it is possible that Akt participate in the regulation of aromatase in luteal cells. Binding assays suggest that luteal aromatase activation is controlled by the GATA, NRE, and AP-3 response elements. Except for GATA, the proteins that recognize these elements in luteal cells are still under investigation. Since GATA is recognized by GATA-4 in luteal cells, and the activity of this transcription factor can be modulated by Akt [99], it is possible to postulate that the PRL/PI3/Akt/GATA-4 pathway is involved in the activation of the aromatase gene in the corpus luteum.

23.5 Glossary of Terms and Acronyms

Akt: oncogene of the retrovirus AKT8 with activity serine/threonine-specific protein kinase, also termed protein kinase-B (PKB)

AP-3: activating transcription factor-3 binding site

BMP: bone morphogenetic protein;

BRCA1: breast cancer-1 early onset tumor suppressor gene

C/EBPβ: CCAAT/enhancer-binding protein-β

cAMP: cyclic Adenosine monophosphate

CBP: CREB-binding protein

CLS: cAMP-responsive element-like sequence

CRE: cAMP response element

CREB: CRE binding protein

CREM: cAMP-responsive element modulator

Ctnnb1: β-catenin

EGF-1: epidermal growth factor-1

ER: estrogen receptor

ERKs: extracellular regulated kinases

FOXO1: forkhead box-O1

FSH: follicle stimulating hormone

GATA: a family of transcription factors that contain two zinc finger motif and binds to the DNA sequence (A/T)GATA(A/G)

GDF: growth differentiation factor

hCG: human chorionic gonadotropin

ICER: inducible cAMP early repressor

IGF-1: insulin growth factor-1

IGFBP-4: IGF-binding protein

JunB: protooncogene homolog B

LH-R: LH receptor

LRH-1: liver receptor homolog-1

MAPKs: mitogen-activated protein kinases

MEK: Mitogen-activated protein kinase kinase, also termed MAP2K

NRE: nuclear response element

PGE2: prostaglandin E2

PGF2α: prostaglandin F2α

PI3K: phosphatidylinositol-3 kinase

PKA: protein kinase-A

PKB: protein kinase-B, also termed Akt

PKC: protein kinase-C

PMSG: pregnant mare serum gonadotropin

PRL: prolactin

SF-1: steroidogenic factor-1

TBP: TATA binding protein

TNF-α: tumor necrosis factor-α

References

1. Simpson ER. Models of aromatase insufficiency. Semin Reprod Med 2004; 22:25–30.
2. Schomberg DW, Couse JF, Mukherjee A, et al. Targeted disruption of the estrogen receptor-alpha gene in female mice: characterization of ovarian responses and phenotype in the adult. Endocrinology 1999; 140:2733–44.
3. George FW, Wilson JD. Conversion of androgen to estrogen by the human fetal ovary. J Clin Endocrinol Metab 1978; 47:550–5.
4. George FW, Ojeda SR. Vasoactive intestinal peptide enhances aromatase activity in the neonatal rat ovary before development of primary follicles or responsiveness to follicle-stimulating hormone. Proc Natl Acad Sci U S A 1987; 84:5803–7.
5. Fisher CR, Graves KH, Parlow AF, et al. Characterization of mice deficient in aromatase (ArKO) because of targeted disruption of the cyp19 gene. Proc Natl Acad Sci U S A 1998; 95:6965–70.
6. Guigon CJ, Mazaud S, Forest MG, et al. Unaltered development of the initial follicular waves and normal pubertal onset in female rats after neonatal deletion of the follicular reserve. Endocrinology 2003; 144:3651–62.
7. Turner KJ, Macpherson S, Millar MR, et al. Development and validation of a new monoclonal antibody to mammalian aromatase. J Endocrinol 2002; 172:21–30.
8. Otsuka F, Yamamoto S, Erickson GF, et al. Bone morphogenetic protein-15 inhibits follicle-stimulating hormone (FSH) action by suppressing FSH receptor expression. J Biol Chem 2001; 276:11387–92.
9. Spicer LJ, Aad PY, Allen D, et al. Growth differentiation factor-9 has divergent effects on proliferation and steroidogenesis of bovine granulosa cells. J Endocrinol 2006; 189:329–39.
10. Vitt UA, Hayashi M, Klein C, et al. Growth differentiation factor-9 stimulates proliferation but suppresses the follicle-stimulating hormone-induced differentiation of cultured granulosa cells from small antral and preovulatory rat follicles. Biol Reprod 2000; 62:370–7.
11. Elvin JA, Yan C, Wang P, et al. Molecular characterization of the follicle defects in the growth differentiation factor 9-deficient ovary. Mol Endocrinol 1999; 13:1018–34.
12. Phillips KW, Goldsworthy SM, Bennett LM, et al. Brca1 is expressed independently of hormonal stimulation in the mouse ovary. Lab Invest 1997; 76:419–25.
13. Hu Y, Ghosh S, Amleh A, et al. Modulation of aromatase expression by BRCA1: a possible link to tissue-specific tumor suppression. Oncogene 2005; 24:8343–8.
14. Lu M, Chen D, Lin Z, et al. BRCA1 negatively regulates the cancer-associated aromatase promoters I.3 and II in breast adipose fibroblasts and malignant epithelial cells. J Clin Endocrinol Metab 2006; 91:4514–9.
15. Diaz FJ, Wigglesworth K, Eppig JJ. Oocytes determine cumulus cell lineage in mouse ovarian follicles. J Cell Sci 2007; 120: 1330–40.
16. Fitzpatrick SL, Richards JS. Regulation of cytochrome P450 aromatase messenger ribonucleic acid and activity by steroids and gonadotropins in rat granulosa cells. Endocrinology 1991; 129:1452–62.
17. Adashi EY, Hsueh AJ. Estrogens augment the stimulation of ovarian aromatase activity by follicle-stimulating hormone in cultured rat granulosa cells. J Biol Chem 1982; 257:6077–83.
18. Wang H, Eriksson H, Sahlin L. Estrogen receptors alpha and beta in the female reproductive tract of the rat during the estrous cycle. Biol Reprod 2000; 63:1331–40.
19. Couse JF, Yates MM, Deroo BJ, et al. Estrogen receptor-beta is critical to granulosa cell differentiation and the ovulatory response to gonadotropins. Endocrinology 2005; 146:3247–62.
20. Tetsuka M, Hillier SG. Differential regulation of aromatase and androgen receptor in granulosa cells. J Steroid Biochem Mol Biol 1997; 61:233–9.
21. Hillier SG, de Zwart FA. Androgen/antiandrogen modulation of cyclic AMP-induced steroidogenesis during granulosa cell differentiation in tissue culture. Mol Cell Endocrinol 1982; 28: 347–61.

22. El-Hefnawy T, Zeleznik AJ. Synergism between FSH and activin in the regulation of proliferating cell nuclear antigen (PCNA) and cyclin D2 expression in rat granulosa cells. Endocrinology 2001; 142:4357–62.
23. Tetsuka M, Hillier SG. Androgen receptor gene expression in rat granulosa cells: the role of follicle-stimulating hormone and steroid hormones. Endocrinology 1996; 137:4392–7.
24. Shiina H, Matsumoto T, Sato T, et al. Premature ovarian failure in androgen receptor-deficient mice. Proc Natl Acad Sci U S A 2006; 103:224–9.
25. Davoren JB, Hsueh JW, Li CH. Somatomedin C augments FSH-induced differentiation of cultured rat granulosa cells. Am J Physiol 1985; 249:E26–33.
26. Dorrington JH, Bendell JJ, Chuma A, et al. Actions of growth factors in the follicle. J Steroid Biochem 1987; 27:405–11.
27. Steinkampf MP, Mendelson CR, Simpson ER. Effects of epidermal growth factor and insulin-like growth factor I on the levels of mRNA encoding aromatase cytochrome P-450 of human ovarian granulosa cells. Mol Cell Endocrinol 1988; 59:93–9.
28. Zhou J, Kumar TR, Matzuk MM, et al. Insulin-like growth factor I regulates gonadotropin responsiveness in the murine ovary. Mol Endocrinol 1997; 11:1924–33.
29. Liu XJ, Malkowski M, Guo Y, et al. Development of specific antibodies to rat insulin-like growth factor-binding proteins (IGFBP-2 to -6): analysis of IGFBP production by rat granulosa cells. Endocrinology 1993; 132:1176–83.
30. Mason HD, Cwyfan-Hughes S, Holly JM, et al. Potent inhibition of human ovarian steroidogenesis by insulin-like growth factor binding protein-4 (IGFBP-4). J Clin Endocrinol Metab 1998; 83: 284–7.
31. Erickson GF, Garzo VG, Magoffin DA. Insulin-like growth factor-I regulates aromatase activity in human granulosa and granulosa luteal cells. J Clin Endocrinol Metab 1989; 69:716–24.
32. Hickey GJ, Krasnow JS, Beattie WG, et al. Aromatase cytochrome P450 in rat ovarian granulosa cells before and after luteinization: adenosine 3′,5′-monophosphate-dependent and independent regulation. Cloning and sequencing of rat aromatase cDNA and 5′ genomic DNA. Mol Endocrinol 1990; 4:3–12.
33. Fitzpatrick SL, Richards JS. Cis-acting elements of the rat aromatase promoter required for cyclic adenosine 3′,5′-monophosphate induction in ovarian granulosa cells and constitutive expression in R2C Leydig cells. Mol Endocrinol 1993; 7:341–54.
34. Fitzpatrick SL, Richards JS. Identification of a cyclic adenosine 3′,5′-monophosphate-response element in the rat aromatase promoter that is required for transcriptional activation in rat granulosa cells and R2C leydig cells. Mol Endocrinol 1994; 8:1309–19.
35. Hunzicker-Dunn M, Maizels ET. FSH signaling pathways in immature granulosa cells that regulate target gene expression: branching out from protein kinase A. Cell Signal 2006; 18:1351–9.
36. Zeleznik AJ, Saxena D, Little-Ihrig L. Protein kinase B is obligatory for follicle-stimulating hormone-induced granulosa cell differentiation. Endocrinology 2003; 144:3985–94.
37. Park Y, Maizels ET, Feiger ZJ, et al. Induction of cyclin D2 in rat granulosa cells requires FSH-dependent relief from FOXO1 repression coupled with positive signals from Smad. J Biol Chem 2005; 280:9135–48.
38. Richards JS, Sharma SC, Falender AE, et al. Expression of FKHR, FKHRL1, and AFX genes in the rodent ovary: evidence for regulation by IGF-I, estrogen, and the gonadotropins. Mol Endocrinol 2002; 16:580–99.
39. Shi F, LaPolt PS. Relationship between FoxO1 protein levels and follicular development, atresia, and luteinization in the rat ovary. J Endocrinol 2003; 179:195–203.
40. Kooijman R. Regulation of apoptosis by insulin-like growth factor (IGF)-I. Cytokine & Growth Factor Rev 2006; 17:305–3.
41. Cardona-Gomez GP, Mendez P, DonCarlos LL, et al. Interactions of estrogen and insulin-like growth factor-I in the brain: molecular mechanisms and functional implications. J Steroid Biochem Mol Biol 2002; 83:211–7.
42. Driggers PH, Segars JH. Estrogen action and cytoplasmic signaling pathways. Part II: the role of growth factors and phosphorylation in estrogen signaling. TEM 2002; 13:422–7.
43. McDonald CA, Millena AC, Reddy S, et al. Follicle-stimulating hormone-induced aromatase in immature rat Sertoli cells requires an active phosphatidylinositol 3-kinase pathway and is inhibited via the mitogen-activated protein kinase signaling pathway. Mol Endocrinol 2006; 20:608–18.
44. Gibori G, Khan I, Warshaw ML, et al. Placental-derived regulators and the complex control of luteal cell function. Recent Prog Horm Res 1988; 44:377–429.
45. Stocco C. In Vivo and in vitro inhibition of cyp19 gene expression by prostaglandin F2a in murine luteal cells: implication of GATA-4. Endocrinology 2004; 145:4957–66.
46. Donadeu FX, Ascoli M. The differential effects of the gonadotropin receptors on aromatase expression in primary cultures of immature rat granulosa cells are highly dependent on the density of receptors expressed and the activation of the inositol phosphate cascade. Endocrinology 2005; 146:3907–16.
47. Mukherjee A, Urban J, Sassone-Corsi P, et al. Gonadotropins regulate inducible cyclic adenosine 3′,5′-monophosphate early repressor in the rat ovary: implications for inhibin alpha subunit gene expression. Mol Endocrinol 1998; 12:785–800.
48. Fitzpatrick SL, Carlone DL, Robker RL, et al. Expression of aromatase in the ovary: down-regulation of mRNA by the ovulatory luteinizing hormone surge. Steroids 1997; 62:197–206.
49. Cameron MR, Foster JS, Bukovsky A, et al. Activation of mitogen-activated protein kinases by gonadotropins and cyclic adenosine 5′-monophosphates in porcine granulosa cells. Biol Reprod 1996; 55:111–9.
50. Salvador LM, Maizels E, Hales DB, et al. Acute signaling by the LH receptor is independent of protein kinase C activation. Endocrinology 2002; 143:2986–94.
51. Seger R, Hanoch T, Rosenberg R, et al. The ERK signaling cascade inhibits gonadotropin-stimulated steroidogenesis. J Biol Chem 2001; 276:13957–64.
52. Andric N, Ascoli M. A delayed gonadotropin-dependent and growth factor-mediated activation of the extracellular signal-regulated kinase 1/2 cascade negatively regulates aromatase expression in granulosa cells. Mol Endocrinol 2006; 20:3308–20.
53. Sirois J, Richards JS. Transcriptional regulation of the rat prostaglandin endoperoxide synthase 2 gene in granulosa cells. Evidence for the role of a cis-acting C/EBP beta promoter element. J Biol Chem 1993; 268:21931–8.
54. Sterneck E, Tessarollo L, Johnson PF. An essential role for C/EBPbeta in female reproduction. Genes Dev 1997; 11:2153–62.
55. Yang S, Fang Z, Suzuki T, et al. Regulation of aromatase P450 expression in endometriotic and endometrial stromal cells by CCAAT/enhancer binding proteins (C/EBPs): decreased C/EBPbeta in endometriosis is associated with over expression of aromatase. J Clin Endocrinol Metab 2002; 87:2336–45.
56. Burkart AD, Mukherjee A, Mayo KE. Mechanism of repression of the inhibin alpha-subunit gene by inducible 3′,5′-cyclic adenosine monophosphate early repressor. Mol Endocrinol 2006; 20:584–97.
57. Morales V, Gonzalez-Robayna I, Hernandez I, et al. The inducible isoform of CREM (inducible cAMP early repressor, ICER) is a repressor of CYP19 rat ovarian promoter. J Endocrinol 2003; 179:417–25.
58. Morales V, Gonzalez-Robayna I, Santana MP, et al. Tumor necrosis factor-alpha activates transcription of inducible repressor form of 3′,5′-cyclic adenosine 5′-monophosphate-responsive element binding modulator and represses P450 aromatase and

inhibin alpha-subunit expression in rat ovarian granulosa cells by a p44/42 mitogen-activated protein kinase-dependent mechanism. Endocrinology 2006; 147:5932–9.
59. Molina CA, Foulkes NS, Lalli E, et al. Inducibility and negative autoregulation of CREM: an alternative promoter directs the expression of ICER, an early response repressor. Cell 1993; 75:875–86.
60. Niehof M, Kubicka S, Zender L, et al. Autoregulation enables different pathways to control CCAAT/enhancer binding protein beta (C/EBP beta) transcription. J Mol Biol 2001; 309:855–68.
61. Gonzalez-Robayna IJ, Alliston TN, Buse P, et al. Functional and subcellular changes in the A-kinase-signaling pathway: relation to aromatase and Sgk expression during the transition of granulosa cells to luteal cells. Mol Endocrinol 1999; 13:1318–37.
62. Krasnow JS, Hickey GJ, Richards JS. Regulation of aromatase mRNA and estradiol biosynthesis in rat ovarian granulosa and luteal cells by prolactin. Mol Endocrinol 1990; 4:13–2.
63. Aten RF, Kolodecik TR, Behrman HR. A cell adhesion receptor antiserum abolishes, whereas laminin and fibronectin glycoprotein components of extracellular matrix promote, luteinization of cultured rat granulosa cells. Endocrinology 1995; 136:1753–8.
64. Stocco C, Telleria C, Gibori G. The molecular control of corpus luteum formation, function, and regression. Endocr Rev 2007; 28:117–49.
65. Morishige WK, Pepe GJ, Rothchild I. Serum luteinizing hormone, prolactin and progesterone levels during pregnancy in the rat. Endocrinology 1973; 92:1527–30.
66. Hickey GJ, Oonk RB, Hall PF, et al. Aromatase cytochrome P450 and cholesterol side-chain cleavage cytochrome P450 in corpora lutea of pregnant rats: diverse regulation by peptide and steroid hormones. Endocrinology 1989; 125:1673–82.
67. Doody KJ, Lorence MC, Mason JI, et al. Expression of messenger ribonucleic acid species encoding steroidogenic enzymes in human follicles and corpora lutea throughout the menstrual cycle. J Clin Endocrinol Metab 1990; 70:1041–5.
68. Voss AK, Fortune JE. Levels of messenger ribonucleic acid for cytochrome P450 17 alpha- hydroxylase and P450 aromatase in preovulatory bovine follicles decrease after the luteinizing hormone surge. Endocrinology 1993; 132:2239–45.
69. Vanselow J, Furbass R, Rehbock F, et al. Cattle and sheep use different promoters to direct the expression of the aromatase cytochrome P450 encoding gene, Cyp19, during pregnancy. Domest Anim Endocrinol 2004; 27:99–114.
70. Bulun SE, Takayama K, Suzuki T, et al. Organization of the human aromatase p450 (Cyp19) gene. Semin Reprod Med 2004; 22:5–9.
71. Sasano H, Harada N. Intratumoral aromatase in human breast, endometrial, and ovarian malignancies. Endocr Rev 1998; 19: 593–607.
72. Zhao Y, Agarwal VR, Mendelson CR, et al. Estrogen biosynthesis proximal to a breast tumor is stimulated by PGE2 via cyclic AMP, leading to activation of promoter II of the Cyp19 (aromatase) gene. Endocrinology 1996; 137:5739–42.
73. Lanzino M, Catalano S, Genissel C, et al. Aromatase messenger RNA is derived from the proximal promoter of the aromatase gene in Leydig, Sertoli, and germ cells of the rat testis. Biol Reprod 2001; 64:1439–43.
74. Kato J, Yamada-Mouri N, Hirata S. Structure of aromatase mRNA in the rat brain. J Steroid Biochem Mol Biol 1997; 61:381–5.
75. Hinshelwood MM, Smith ME, Murry BA, et al. A 278 bp region just upstream of the human CYP19 (aromatase) gene mediates ovary-specific expression in transgenic mice. Endocrinology 2000; 141:2050–3.
76. Hinshelwood MM, Repa JJ, ,a.Shelton JM, et al. Expression of LRH-1 and SF-1 in the mouse ovary: localization in different cell types correlates with differing function. Mol Cell Endocrinol 2003; 207:39–45.
77. Stocco C, Kwintkiewicz J, Cai Z. Identification of regulatory elements in the Cyp19 proximal promoter in rat luteal cells. J Mol Endocrinol 2007; 39:211–21.
78. Michael MD, Michael LF, Simpson ER. A CRE-like sequence that binds CREB and contributes to cAMP-dependent regulation of the proximal promoter of the human aromatase P450 (CYP19) gene. Mol Cell Endocrinol 1997; 134:147–56.
79. Somers JP, DeLoia JA, Zeleznik AJ. Adenovirus-directed expression of a nonphosphorylatable mutant of CREB (cAMP response element-binding protein) adversely affects the survival, but not the differentiation, of rat granulosa cells. Mol Endocrinol 1999; 13:1364–72.
80. Mukherjee A, Park-Sarge OK, Mayo KE. Gonadotropins induce rapid phosphorylation of the 3′,5′-cyclic adenosine monophosphate response element binding protein in ovarian granulosa cells. Endocrinology 1996; 137:3234–45.
81. Carlone DL, Richards JS. Functional interactions, phosphorylation, and levels of 3′,5′-cyclic adenosine monophosphate-regulatory element binding protein and steroidogenic factor-1 mediate hormone-regulated and constitutive expression of aromatase in gonadal cells. Mol Endocrinol 1997; 11:292–304.
82. Hinshelwood MM, Michael MD, Simpson ER. The 5′-flanking region of the ovarian promoter of the bovine CYP19 gene contains a deletion in a cyclic adenosine 3′,5′-monophosphate-like responsive sequence. Endocrinology 1997; 138:3704–10.
83. Silva JM, Price CA. Insulin and IGF-I are necessary for FSH-induced cytochrome P450 aromatase but not cytochrome P450 side-chain cleavage gene expression in oestrogenic bovine granulosa cells in vitro. J Endocrinol 2002; 174:499–507.
84. Fayard E, Auwerx J, Schoonjans K. LRH-1: an orphan nuclear receptor involved in development, metabolism and steroidogenesis. Trends cell Biol 2004; 14:250–60.
85. Lynch JP, Lala DS, Peluso JJ, et al. Steroidogenic factor 1, an orphan nuclear receptor, regulates the expression of the rat aromatase gene in gonadal tissues. Mol Endocrinol 1993; 7:776–86.
86. Falender AE, Lanz R, Malenfant D, et al. Differential expression of steroidogenic factor-1 and FTF/LRH-1 in the rodent ovary. Endocrinology 2003; 144:3598–610.
87. Hinshelwood MM, Shelton JM, Richardson JA, et al. Temporal and spatial expression of liver receptor homologue-1 (LRH-1) during embryogenesis suggests a potential role in gonadal development. Dev Dyn 2005; 234:159–68.
88. Pezzi V, Sirianni R, Chimento A, et al. Differential expression of Sf-1/Ad4bp and Lrh-1/Ftf in the rat testis: Lrh-1 as a potential regulator of testicular aromatase expression. Endocrinology 2004; 145:2186–96.
89. Clyne CD, Speed CJ, Zhou J, et al. Liver receptor homologue-1 (LRH-1) regulates expression of aromatase in preadipocytes. J Biol Chem 2002; 277:20591–7.
90. Saxena D, Safi R, Little-Ihrig L, et al. Liver receptor homolog-1 stimulates the progesterone biosynthetic pathway during follicle-stimulating hormone-induced granulosa cell differentiation. Endocrinology 2004; 145:3821–9.
91. Jeyasuria P, Ikeda Y, Jamin SP, et al. Cell-specific knockout of steroidogenic factor 1 reveals its essential roles in gonadal function. Mol Endocrinol 2004; 18:1610–9.
92. Britt KL, Drummond AE, Dyson M, et al. The ovarian phenotype of the aromatase knockout (ArKO) mouse. J Steroid Biochem Mol Biol 2001; 79:181–5.
93. Saxena D, Escamilla-Hernandez R, Little-Ihrig L, et al. Liver Receptor Homolog-1 (LRH-1) and Steroidogenic Factor-1 (SF-1) Have Similar Actions on Rat Granulosa Cell Steroidogenesis. Endocrinology 2007; 148:9.
94. Parakh TN, Hernandez JA, Grammer JC, et al. Follicle-stimulating hormone/cAMP regulation of aromatase gene expression requires beta-catenin. Proc Natl Acad Sci U S A 2006; 103:12435–40.

95. Jin T, Zhang X, Li H, Goss PE. Characterization of a novel silencer element in the human aromatase gene PII promoter. Breast Cancer Res Treat 2000; 62:151–9.
96. Kwintkiewicz J, Cai Z, Stocco C. Follicle-stimulating hormone-induced activation of Gata4 contributes in the up-regulation of Cyp19 expression in rat granulosa cells. Mol Endocrinol 2007; 21:933–47.
97. Cai Z, Kwintkiewicz J, Young ME, et al. Prostaglandin E2 increases cyp19 expression in rat granulosa cells: implication of GATA-4. Mol Cell Endocrinol 2007; 263:181–9.
98. Tessier C, Prigent-Tessier A, Ferguson-Gottschall S, et al. PRL antiapoptotic effect in the rat decidua involves the PI3K/protein kinase B-mediated inhibition of caspase-3 activity. Endocrinology 2001; 142:4086–94.
99. Naito AT, Tominaga A, Oyamada M, et al. Early stage-specific inhibitions of cardiomyocyte differentiation and expression of Csx/Nkx-2.5 and GATA-4 by phosphatidylinositol 3-kinase inhibitor LY294002. Exp cell Res 2003; 291:56–69.

Chapter 24

Epigenetic Mechanisms of Ovarian Gene Regulation

Holly A. LaVoie

24.1 Introduction

Epigenetics refers to alterations in gene function that are heritable between generations of cells or animals and that do not involve a change in the DNA sequence itself [1]. Such alterations occur at the level of chromatin and include post-translational modifications of histone tails and methylation of DNA. In a more narrow sense, epigenetics can be used to describe chromatin altering events within a single cell or groups of cells that leads to changes in gene function in response to hormones or growth factors. This chapter will briefly describe the most common epigenetic marks and review studies of chromatin alterations as they relate to gene function in cells of the developing ovarian follicle and corpus luteum.

24.2 Histone Modifications

Chromatin is comprised of DNA, RNA, and proteins [2]. The DNA is packaged into nucleosomes that allow a massive amount of DNA to be compacted into the small nucleus of a cell, but at the same time restrict transcriptional access. The nucleosome consists of 146 bp of DNA wrapped around a histone core formed by two molecules each of histones H2A, H2B, H3, and H4. A single histone H1 molecule is outside the core bound to linker DNA. The core histones have protruding charged amino-terminal tails that are subject to post-translational covalent modifications. These modifications include acetylation, methylation, phosphorylation, ubiquitination, and sumoylation [3]. The pattern of histone modifications makes up a histone code [4]. It is proposed that this code takes the form of binary switches and modification cassettes that are read by regulatory proteins [5]. Collectively some groups of modifications promote local transcription, whereas others inhibit it. Modifications that promote transcription recruit chromatin-remodeling ATPase complexes to open chromatin. These ATP-dependent enzyme complexes lead to nucleosome repositioning and increased access to DNA [6]. The following will highlight modifications of histones H3 and H4.

24.2.1 Acetylation and Deacetylation

Acetylation occurs on specific histone lysine residues and occurs primarily in the tail regions. The addition of an acetyl group to a lysine of the histone tail neutralizes a positive charge resulting in alterations to the chromatin fiber that increase accessibility of the transcriptional machinery to DNA, facilitating transcription [4, 7]. Histone acetyltransferases (HATs) catalyze the acetylation reaction. Although all the core histones can undergo this modification, histones H3 and H4 have been studied the most intensively. Figure 24.1 shows a summary of the reported histone H3 modifications including acetylation [5, 8, 9].

In the tail region of histone H3, acetylation occurs at lysines at amino acid positions 9, 14, 18, and 23 with other potential lysine acetylation sites possible [5, 9]. In the tail of histone H4 positions 5, 8, 12, and 16 can be acetylated and there are other possible lysine sites as well [5, 9]. Numerous HATs exist in mammals including the coactivators p300, CBP (CREB-binding protein), PCAF (p300/CBP-associated factor), TAF250, and GCN5 [10]. These acetyltransferases are recruited to specific gene promoters by transcription factors associated with the DNA [11].

Deacetylation of histone lysine residues is catalyzed by histone deacetylases (HDACs) and important for turning off genes and maintaining some genes in a repressed state [12]. Mammalian histone deacetylases fall into four classes (I–IV) based on their homology to yeast HDACs [12, 13]. Class I

H.A. LaVoie (✉)
Department of Cell Biology and Anatomy, University of South Carolina School of Medicine, Columbia, SC, USA
e-mail: HLAVOIE@gw.med.sc.edu

P.J. Chedrese (ed.), *Reproductive Endocrinology*,
DOI 10.1007/978-0-387-88186-7_24,

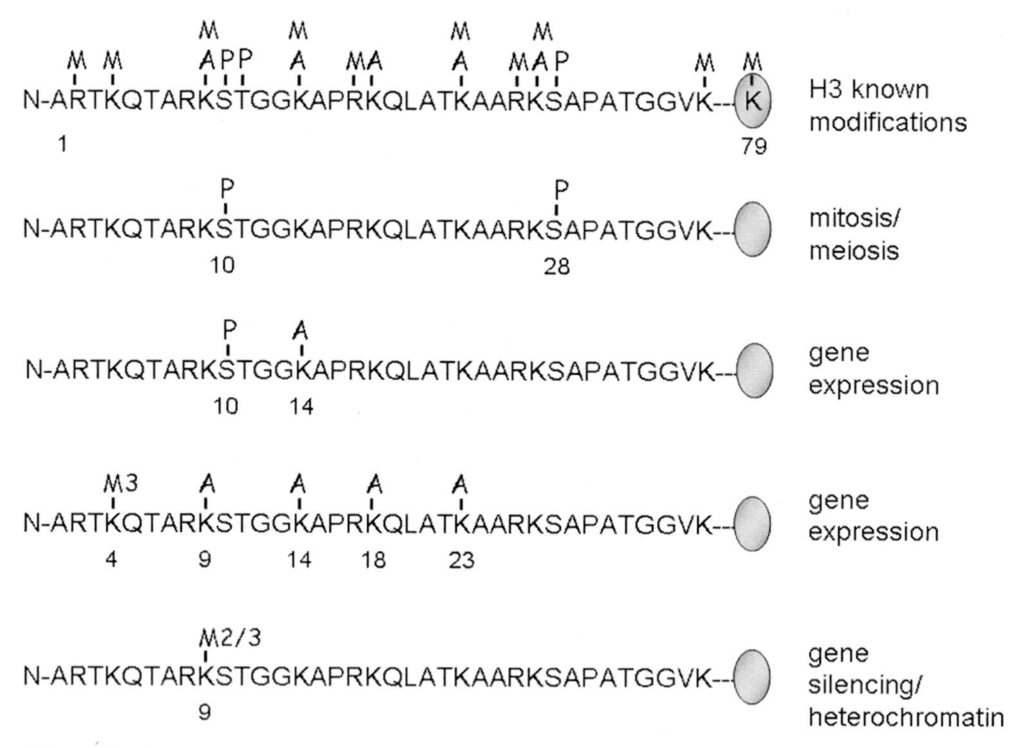

Fig. 24.1 Histone H3 modifications and the histone code. The *top row* summarizes the known modifications of histone H3. M, A, and P above the amino acid symbols represent methylation, acetylation, and phosphorylation, respectively. M2/3 represents di- or trimethylation. The *bottom four rows* show examples of H3 modifications and their corresponding functions

and II HDACs are sensitive to inhibition by trichostatin A (TSA). Class III HDACs require NAD^+ as a cofactor. Little is known about Class IV. Class I is comprised of HDAC1, 2, 3, and 8. Class II is comprised of HDAC4, 5, 6, 7, 9, and 10. Class III is comprised of sirtuin family members SIRT1-7, but only 1, 2, 3, and 5 have demonstrated significant deacetylase activity [14]. HDAC11 is the sole member of Class IV [15]. HDACs do not bind DNA but rather are recruited by specific transcription factors and frequently exist as part of multiprotein complexes associated with chromatin. For example, HDACs 1 and 2 are found in three main complexes known as the Sin3, NuRD, and CoREST complexes [16].

The abundance, activity, and availability of HATs and HDACs are regulated [17, 18]. The activities of both groups of enzymes are regulated by post-translational modifications such as phosphorylation and acetylation and by protein–protein interactions. Select Class II members can shuttle between the nucleus and cytoplasm altering their access to chromatin.

24.2.2 Methylation and Demethylation

Methylation of histones can stimulate or inhibit transcription depending on the particular residue modified and the placement of methyl groups [3, 19]. Within the histone tails, methylation occurs on unacetylated lysines and is carried out by a family of histone methyltransferases (HMTs). Most lysine HMTs possess a conserved SET domain and are specific in their substrate recognition [20]. Lysines can be mono-, di-, or trimethylated. In addition, histone arginines can be mono- or dimethylated and the enzymes responsible are called protein arginine methyltransferases (PRMTs). Lysines 4 (K4) and 9 (K9) of H3 and arginine 3 (R3) of H4 have been well studied. Dimethylation of K4 of H3 is associated with both active and inactive genes, whereas trimethylation of this residue is associated with active genes only [3]. Di- and tri-methylation of K9 of H3 is typically linked with gene silencing or repression through recruitment of heterochromatin 1 protein (HP1) and can lead to DNA methylation [7, 21]. Interestingly, data show multi-methylation of K4 of H3 can block the binding of the deacetylase complex NuRD and methylation of K9 of H3 and thus help maintain active genes [10]. Asymmetric dimethylation of R3 of H4 is permissive to transcription, whereas symmetric dimethylation is repressive [19]. In some cases, histone deacetylases have been found in complexes containing methyltransferases; such complexes are proposed to deacetylate a lysine residue then methylate it to silence the gene [3, 7].

Until recently histone methylation was believed to be irreversible or reversed at a very slow rate [22]. Histone demethylation is now believed to be a dynamic event. LSD1 was the first identified histone demethylase with specificity for mono- and dimethylated K4 of H3 [23]. The search for demethylases capable of recognizing and demethylating trimethylated lysines recently yielded characterization of the Jumonji domain family of histone demethylases [22, 24]. Jumonji family members JMJD2A-D can demethylate both di- and trimethylated K9 of H3. Jumonji members falling in the JARID1 subfamily can demethylate di- and trimethylated K4 of H3 [22]. With candidates for lysine demethylases in hand, the search for arginine demethylases continued. Experiments showed that arginine demethylation could be reversed by a deimination reaction through the intermediate citrulline [23]. Just recently, Jumonji family member JMJD6 was identified as the first histone arginine demethylase [25]. This is a rapidly evolving area and new advances will greatly further our understanding of the role of specific histone methylation events in transcription.

24.2.3 Phosphorylation

Just like methylation, phosphorylation of specific histone tail residues has been associated with different functions. Phosphorylation occurs on selected serines and threonines in the histone tails introducing a negative charge that can locally alter chromatin accessibility [26]. Phosphorylation of serines 10 and 28 and threonine 11 of histone H3 is associated with chromatin condensation during mitosis and meiosis [27]. In contrast, H3 serine 10 (S10) phosphorylation during interphase has been implicated in transcriptional activation [27]. The phosphorylation of histone H3 S10 is associated with the transcription of immediate early genes such as *FOS* in mitogen-activated cells and this mark can synergize with lysine acetylation [28]. Phosphorylated S10 of H3 acts as a preferential substrate for HATs p300 and PCAF and fosters K14 acetylation [3, 29]. Phosphorylation of H3 S10 has been shown to occur on H3 preacetylated at lysines 9 and 14 as well [27]. S10 phosphorylation can inhibit K9 methylation, which in turn would allow easier access for HATs [3]. Additional roles for phosphorylation are evident in other core histones [30]. Core histone variant H2A.X is phosphorylated at serine 139 during DNA repair processes. H2B undergoes phosphorylation at serine 14 during apoptosis.

H3 phosphorylation is regulated by specific kinases. During mitosis Aurora kinases phosphorylate S10 and S28 of histone H3 [31]. During mitosis in breast cancer cells, p21-activated kinase-1 (Pak1) phosphorylates H3 S10 [32]. Mitogen- and stress-activated protein kinases 1 and 2 (MSK1/2) are the major kinases mediating non-mitotic H3 S10 phosphorylation [31]. RSK2 (ribosomal S6 kinase 2) is also a possible H3 kinase, but studies with this kinase have not been conclusive [27, 31]. H3 S10 phosphorylation has also been shown to be protein kinase A (PKA)-dependent in ovarian granulosa cells, and there is in vitro evidence that the catalytic subunit of PKA can phosphorylate histone H3 directly [33, 34]. Dephosphorylation of histones occurs by general cellular phosphatases such as type 1 (PP1) and type 2A (PP2A) [27].

24.2.4 Ubiquitination, ADP-Ribosylation, and Sumoylation

Other less well characterized histone modifications include ubiquitination, ADP-ribosylation, and sumoylation. Ubiquitination of H3 is associated with nucleosome loosening in preparation for histone removal and H2A ubiquitination may be required for some H3 methylation events [21]. ADP-ribosylation is not well characterized but mono-ADP-ribosylation of histones increases with damage to DNA [35]. Sumoylation of histones in yeast is associated with transcriptional repression [36].

24.2.5 Bromo and Chromo Domain Proteins

Many chromatin modifying enzymes such as HATs, HMTs, and chromatin remodeling ATPases contain one of two distinct chromatin binding domains referred to as bromodomains and chromodomains [10]. These domains are important for protein recognition of histone modifications and thus act to interpret the histone code. The bromodomain can bind acetylated lysines in histone tails and may act to tether HATs to the chromatin to propagate local acetylation. Likewise, chromatin remodeling ATPases can be directed to specific acetylated lysines to form and stabilize protein complexes. Bromodomains are mostly associated with facilitating transcription. Chromodomains recognize methylated lysines within the histone tails. Chromodomain proteins can be negative or positive regulators of transcription depending on which specific methylated residue is recognized. The chromodomain of HP-1 binds to methylated K9 of H3 facilitating chromatin compaction.

24.3 Detection of Histone Modifications

24.3.1 Chromatin Immunoprecipitation Assay

The chromatin immunoprecipitation (ChIP) assay has facilitated the study of the protein–DNA interactions including the detection of modified histones associated with DNA.

By using antibodies that recognize modified histones (e.g., acetylated K9 and K14 H3, phosphorylated S10 H3) the relative measurement of modified histones in specific regions of the DNA in response to hormone treatments can be assessed. A more detailed protocol can be obtained from the reviews [37, 38]. For the ChIP assay, formaldehyde is used to crosslink proteins to DNA in cells. Cells are lysed and DNA fragments in the 200–1,000 bp range are generated by sonication or enzymatic digestion of DNA. Antibodies to modified histones, transcription factors, or other chromatin associated proteins are incubated with the sheared DNA and immunoprecipitated. Protein–DNA crosslinks are reversed and the DNA is purified and used for PCR with primers to specific regions of a gene (Fig. 24.2). These primers may include sequences within the gene itself or various locations in its flanking regions. Frequently, the 5′-flanking region regulatory region of a gene is amplified. Data are presented typically as autoradiograms of gels showing the amplified products from samples with and without treatment. The application of real-time PCR with ChIP allows for better quantification of amplified products. The controls for this assay should include performing the procedure with a non-specific antibody pool such as normal IgG of the same species. Other controls used for normalization are to amplify the region of interest from an aliquot of the sonicated DNA prior to immunoprecipitation (input DNA) to show the input DNA did not vary between treatments. However, since the ChIP assay is subject to numerous steps that could result in procedural losses of template, amplification of another unregulated DNA region from the same immunoprecipitated DNA is an appropriate control. The experimental product is then normalized for its input DNA product or to another unregulated DNA product amplified from the same ChIP reaction. An increase in the amount of amplified target DNA indicates that more of that protein (e.g., modified histone) was associated with that region of DNA.

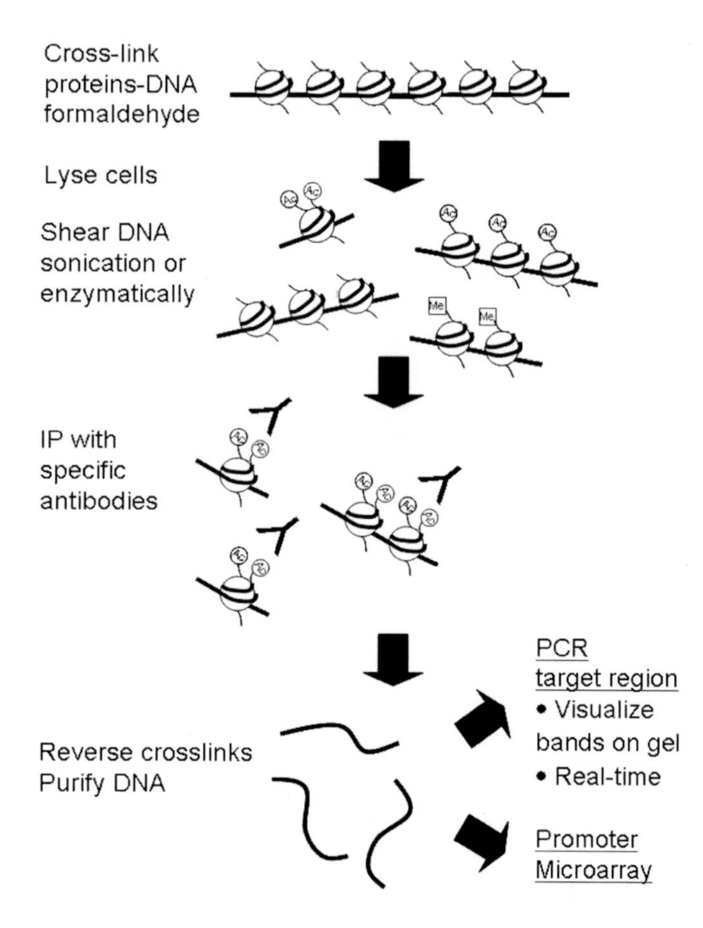

Fig. 24.2 Overview of chromatin immunoprecipitation (IP) assay used for detecting modified histones within DNA. Antibodies recognizing specifically modified histone molecules are used. DNA is quantified by PCR or large scale modifications can be determined using intergenic microarrays

The ChIP assay has been used primarily to evaluate specific small groups of genes. One application of this assay has been expanded to include the ChIP-chip (or ChIP on chip) assay that utilizes the immunoprecipitated DNA for microarray analysis [39]. The microarrays typically have intergenic DNA or open reading frames. The coupling of these two techniques allows one to evaluate protein–DNA interactions on a genomic scale. For example, the technique has been used to show methylation of H3 lysine 4 occurs in the coding regions of active genes of yeast [40].

24.3.2 DNA Methylation

A more stable epigenetic mark is DNA methylation. Methylation occurs mainly on cytosines of CpG dinucleotides (an adjacent C and G on the same strand). Clusters of CpG dinucleotides are called CpG islands and have been reported to be associated with up to 76% of mammalian gene promoters [41]. Active genes, such as tissue-specifically expressed and housekeeping genes, tend to have low methylation of their CpG islands. For example, the follicle-stimulating hormone (FSH) receptor gene promoter is hypomethylated in Sertoli cells where it is expressed compared to tissues that do not express the gene [42]. The pattern of DNA methylation is heritable between cells and generations. DNA methylation patterns in the embryo are erased and then reestablished as cells differentiate [43]. In primordial germ cells DNA methylation patterns are erased and reestablished in a sex-specific pattern during oogenesis and spermatogenesis [44]. This later point is particularly important for the imprinting. Imprinting refers to the sex-specific DNA methylation pattern that occurs in a relatively small number of genes. The DNA in or surrounding an imprinted gene will be methylated in either the maternally- or paternally-derived chromosome, limiting expression of the gene to one copy only [45]. Primarily, two families of DNA cytosine-5 methyltransferases DNMT1 and DNMT3 participate in DNA methylation in mammals [41]. Only recently have data supported a role for DNA methylation in differential activation of genes in adult tissues [43].

24.4 The Dynamic Ovarian Follicle and the Study of Histone Modifications

In the ovarian follicle, the precise temporal activation and inactivation of different gene pools is coordinated by hormones and growth factors and is required for successful ovulation and corpus luteum (CL) formation (for reviews see [46, 47]). Prior to the midcycle gonadotropin surge, theca cells of developing antral follicles possess luteinizing hormone (LH) receptors and granulosa cells express FSH receptors. LH stimulates theca cell androgen production. FSH stimulates granulosa cell proliferation and aromatase activity converting thecal androgens into estradiol. Under the influence of FSH and estradiol, mural granulosa cells of the periovulatory follicle acquire LH receptors allowing them to respond to LH and undergo terminal differentiation following the midcycle gonadotropin surge. The LH surge transiently decreases LH receptor and aromatase expression and activates genes required for increased progesterone production associated with luteinization including *STAR*, *CYP11A*, and *HSD3B*. The functional CL maintains the expression of these genes at high levels until physiological regression occurs.

Unique combinations of hormone-activated transcription factors control the temporal regulation of these numerous genes. Existing and regulated histone modifications will lead to recruitment of chromatin remodeling enzymes and the transcriptional machinery. The following section will summarize our current knowledge of histone modifications and chromatin remodeling of genes important to follicular development and CL function.

Early studies of ovarian histone modifications—in 1999, Demanno [48] first demonstrated that FSH and cyclic AMP but not epidermal growth factor (EGF) led to the phosphorylation of histone H1 and histone H3 in cultured rat granulosa cells. It was subsequently shown that FSH mediated phosphorylation of histone H3 S10 in a PKA-dependent manner [33]. Follicular S10 phosphorylation peaked during proestrus in rats, was shown to be also stimulated by estradiol, and was indirectly linked to mitosis [49]. FSH-induction of genes encoding serum-glucocorticoid kinase, inhibin alpha subunit, and c-Fos in granulosa cells was associated with increased H3 S10 phosphorylation and K14 acetylation within their promoters [33]. Under the same conditions the progesterone receptor gene promoter did not exhibit altered H3 acetylation or phosphorylation. Studies of histone modifications and their association with transcription of other genes have followed.

Steroidogenic acute regulatory protein (STAR) gene—STAR is responsible for the rate-limiting step in steroidogenesis, the transfer of cholesterol from the outer to the inner mitochondrial membrane [50], and is discussed more extensively elsewhere in this text. Expression of the *STAR* gene is tightly regulated in granulosa cells with the transcript being nearly undetectable in maturing granulosa cells of the follicle and then being robustly activated by the midcycle gonadotropin or LH surge [51]. Using hormonally primed monkeys, Christenson [52] examined histone modifications in granulosa cells before and after luteinization induced by hCG injection. Luteinized granulosa cells exhibited a 32- to 206-fold increase in acetylated H3 associated with the primate *STAR* promoter when compared with non-luteinized granulosa cells. High levels of acetylated H3 within the human *STAR* promoter were also present in human luteinized granulosa cells isolated from women undergoing assisted reproduction [52].

Further evaluation of the *STAR* gene in mice during the early post-hCG response (0–4 h) yielded a wealth of information about the temporal changes in transcription factor recruitment and histone modifications of the gene in granulosa cells [53]. Transcription factors GATA4, CEBPβ, SF-1, and cofactor CBP associated with the proximal *STAR* promoter after hCG stimulus and in temporal agreement with increased nascent *STAR* RNA. Methylation of H3 K9 within the proximal *STAR* promoter decreased immediately after treatment. H3 acetylation remained unaltered within the same *STAR* promoter region in the same time frame. In other words, a reduction in the silencing mark methylated H3 K9 resulted from hCG facilitating *STAR* transcription.

Another study focused on *STAR* gene repression in granulosa cells [54]. In cultured porcine granulosa cells, FSH can stimulate *STAR* transcription and EGF can suppress this FSH effect. Concordantly, FSH increased acetylation of histone H3 within the proximal promoter of the endogenous *STAR* gene and EGF also blocked this FSH effect. Under similar conditions, methylation of H3 K9 was unchanged. EGF may therefore block FSH recruitment of HATs or increase that of HDACs.

The above studies suggest that *STAR* expression in granulosa cells may be repressed at the level of chromatin either by silencing methylation or by hypoacetylation until a strong stimulus like the LH surge relieves repression. A study in Leydig cells showed HDAC1/2 and Sin3A interact with Sp3 to form a corepressor complex on a repressive region of the *STAR* promoter [55], raising the possibility that a similar mechanism may occur in undifferentiated granulosa cells. Another study showed nucleosome-remodeling ATPase SNF2L to be induced in granulosa cells by hCG and to be required for STAR gene expression [56]. More in vivo studies are needed to understand the complex regulation of the STAR gene that occurs following the LH surge.

LH receptor gene—the LH receptor is increased in granulosa cells during follicular development, downregulated by the LH surge, and expressed subsequently in some luteal cells [46]. Studies in non-ovarian cells have provided evidence that histone-associated enzymes can mediate repression of LH receptor transcription. The LH receptor gene is activated by transcription factors Sp1 and Sp3 [57]. Repression of the gene occurs when a complex containing HDAC1/HDAC2/Sin3A associates with Sp1 or Sp3 bound to its proximal DNA element [58]. HDAC inhibitor TSA increased H3 and H4 acetylation within the proximal LH receptor gene promoter and recruited RNA polymerase II, thereby promoting transcription. These studies suggest a mechanism by which the LH receptor gene might be kept silent at the early stages of granulosa cell development or be downregulated following the LH surge and serve as a starting point for studies in ovarian cells.

Relaxin gene—relaxin mRNA is abundantly expressed in the rat CL during days 15–20 of pregnancy. Its expression is regulated by rat placental lactogen (rPRL) and prolactin via STAT proteins [59, 60]. Relaxin expression is low before day 10 of pregnancy and at the end of pregnancy (days 22–23). Soloff [61] evaluated acetylated-H3, RNA polymerase II, and STAT protein association with the rat relaxin gene 5′-flanking region. H3 acetylation within the proximal promoter during high relaxin expression (day 15) was greater than during low relaxin expression (day 6). Higher levels of acetylated H3 were accompanied by an increase in RNA polymerase II binding to the proximal promoter and a increased binding of STAT5a approximately 4 kb upstream of the region of acetylation.

Low-density lipoprotein (LDL) receptor gene—the LDL receptor is responsible for the uptake of blood-borne LDL cholesterol and is expressed in theca, luteinized granulosa, and luteal cells [62, 63, 64]. In non-rodent species, LDL cholesterol is the primary source for steroidogenesis [65]. The LDL receptor gene promoter is stimulated by transcription factors Sp1 and SREBP, and antagonized by Kruppel-like factor 13 (KLF13) [66]. KLF-13 repression of the LDL receptor gene is antagonized by HDAC inhibitors, suggesting KLF13 recruits HDACs or antagonizes HAT activity at the promoter.

Niemann-Pick C1 gene—the Niemann-Pick C1 (*NPC1*) gene codes for a molecule mediating the intracellular transport of LDL-derived cholesterol and is abundant in theca, luteal, and luteinized granulosa cells [67, 68]. Cyclic AMP analogue increases *NPC1* mRNA expression in cultured luteinized porcine granulosa cells through a mechanism that likely involves CREB [68]. Cyclic AMP stimulated gene expression in granulosa cells and increased CBP, acetylated H3, and phosphorylated H3 affiliation with the *NPC1* promoter [69]. In addition, cholesterol-deprivation of granulosa cells in culture led to increased SREBP and acetylated histone H3 association with the sterol-responsive region of the *NPC-1* gene promoter [70].

Inhibin alpha subunit gene—ovarian inhibin suppresses pituitary FSH, can act as a tumor suppressor, and has paracrine functions in the ovary [71, 72]. As mentioned above H3 within the inhibin alpha subunit gene promoter is both phosphorylated and acetylated in granulosa cells in response to FSH [33]. In transfected HEK tsa201 cells, inhibin alpha subunit gene promoter constructs demonstrated PKA-dependent increases in H4 acetylation [73]. Within the rat granulosa GRMO2 cell line adenylate cyclase activator forskolin decreased SF-1 association with and increased the association of CREB and LRH-1, cofactors CBP and SRC-1, and acetylated histone H4 with the proximal inhibin alpha subunit gene promoter [74].

Thecal genes—the synthesis of androgens by theca cells depends on the coordinated activities of several genes including *STAR*, *CYP11A*, and *CYP17* among others [72]. The antiepileptic drug valproate, which has HDAC inhibitor properties, alters steroid hormone secretion from ovarian follicles and can create symptoms of polycystic ovary syndrome (PCOS) in women [75, 76]. Valproate increased total H3 and H4 acetylation in cultured human theca cells [77]. Valproate augmented cAMP-driven androstenedione and DHEA biosynthesis in theca cells from normal and PCOS women, as well as 17α-hydroxyprogesterone in normal theca cells. This increase in androgen production was accompanied by the ability of valproate to augment *CYP11A* and *CYP17* mRNA synthesis and gene promoter activity. A second HDAC inhibitor, butyric acid, was also able to enhance forskolin-stimulated DHEA production by theca cells. Comparison of gene profiles between valproate-treated theca cells and theca cells from PCOS patients showed similar gene profiles [78]. These studies indicate that drug-induced increases in histone acetylation affect a subset of genes involved in androgen biosynthesis (e.g., *CYP11A* and *CYP17*) and that acetylation of the regulatory regions of these genes may account for their activation under physiological conditions.

24.5 Summary

Gene repression or silencing in the ovary is critical for some genes such as the LH receptor and *STAR* at certain times of the cycle. As described above there are several mechanisms by which gene expression may be suppressed, including silencing methylation, HDAC recruitment, and HAT inhibition or dissociation. Methylation of genes silent during the follicular phase (e.g., *STAR* in granulosa cells) could be a general mechanism by which genes associated with terminal differentiation are suppressed until activated by the LH surge. The induction of chromatin remodeling ATPases such as SNF2L participates in post-LH surge gene induction as well. The study of epigenetic modifiers in ovarian somatic cells is still at its beginning stage and future studies are needed to understand the cycle-specific regulation of genes mediating ovarian cell proliferation and differentiation.

24.6 Glossary of Terms and Acronyms

Aurora kinases: a family of serine/threonine kinases important for cell division

CBP/p300: CREB-binding protein and related E1A binding protein p300

CEBPβ: CCAAT enhancer binding protein beta

ChIP: chromatin immunoprecipitation

CL: corpus luteum

CoREST: corepressor for RE1 silencing transcription factor (REST)

***CYP11A*:** gene coding for P450scc

***CYP17*:** gene coding for cytochrome P450 17α-hydroxylase/17,20-lyase

DNMT1 and DNMT3: DNA cytosine-5 methyltransferases 1 and 3

***FOS*:** gene coding for c-Fos

GATA: family of transcription factors that contain a two-zinc-finger motif and bind to the DNA sequence (A/T)GATA(A/G)

GCN5: an histone acetyltransferase

GRMO2: rat granulosa stable cell line

H2A.X: core histone 2A variant

HATs: histone acetyltransferases

HDAC1/2: histone deacetylases 1 and 2

HDACs: histone deacetylases

HMTs: histone methyltransferases

HP1: heterochromatin 1 protein

***HSD3B*:** gene(s) coding for 3β-HSD

JMJD2A-D: histone demethylases of the Jumonji family

JMJD6: histone demethylase of the Jumonji family

KLF13: Kruppel-like factor 13

LDL: low-density lipoprotein

LSD1: lysine specific demethylase 1

MSK1/2: mitogen- and stress-activated protein kinases 1 and 2

***NPC1*:** Niemann-Pick C1 gene

NuRD: nucleosome remodeling and deacetylase

Pak1: p21-activated kinase-1

PCAF: p300/CBP-associated factor

PCOS: polycystic ovary syndrome

PKA: protein kinase A

PP1: phosphatase type 1

PP2A: phosphatase type 2A

PRMTs: protein arginine methyltransferases

rPRL: rat placental lactogen

RSK2: ribosomal S6 kinase 2

SET domain: a domain found in some lysine methyltransferases, derived from *Drosophilia* proteins designated suppressor of variegation, enhancer of zeste and trithorax

SF-1: transcription factor steroidogenic factor 1, also termed NR5A1

Sin3: a SWI independent corepressor found in complexes with HDACs

Sin3A: Sin3 family member A

SNF2L: nucleosome-remodeling ATPase

Sp1 and Sp3: transcription factor specificity proteins 1 and 3

SRC-1: steroid receptor coactivator

***STAR*:** gene coding for steroidogenic acute regulatory (STAR) protein

STAT: signal transducers and activators of transcription

SWI: SWItch proteins, part of chromatin remodeling ATPase complexes

TAF250: an histone acetyltransferase

TSA: trichostatin A, an HDAC inhibitor

References

1. Jaenisch R, Bird A. Epigenetic regulation of gene expression: how the genome integrates intrinsic and environmental signals. Nat Genet 2003; 33:245–54.

2. Lodish H, Berk A, Zipursky LS, et al. Regulation of transcription initiation. Molecular Cell Biology. New York: W.H. Freeman, 2000.
3. Fischle W, Wang Y, Allis CD. Histone and chromatin cross-talk. Curr Opin Cell Biol 2003; 15:172–83.
4. Strahl BD, Allis CD. The language of covalent histone modifications. Nature 2000; 403:41–5.
5. Fischle W, Wang Y, Allis CD. Binary switches and modification cassettes in histone biology and beyond. Nature 2003; 425:475–9.
6. Eberharter A, Becker PB. ATP-dependent nucleosome remodelling: factors and functions. J Cell Sci 2004; 117:3707–11.
7. Eberharter A, Becker PB. Histone acetylation: a switch between repressive and permissive chromatin. Second in review series on chromatin dynamics. EMBO Rep 2002; 3:224–9.
8. Rice JC, Allis CD. Histone methylation versus histone acetylation: new insights into epigenetic regulation. Curr Opin Cell Biol 2001; 13:263–73.
9. Santos-Rosa H, Caldas C. Chromatin modifier enzymes, the histone code and cancer. Eur J Cancer 2005; 41:2381–402.
10. de lC, X, Lois S, Sanchez-Molina S, et al. Do protein motifs read the histone code? Bioessays 2005; 27:164–75.
11. Utley RT, Ikeda K, Grant PA, et al. Transcriptional activators direct histone acetyltransferase complexes to nucleosomes. Nature 1998; 394:498–502.
12. de Ruijter AJ, van Gennip AH, Caron HN, et al. Histone deacetylases (HDACs): characterization of the classical HDAC family. Biochem J 2003; 370:737–49.
13. Yang XJ, Seto E. The Rpd3/Hda1 family of lysine deacetylases: from bacteria and yeast to mice and men. Nat Rev Mol Cell Biol 2008; 9:206–18.
14. Haigis MC, Guarente LP. Mammalian sirtuins–emerging roles in physiology, aging, and calorie restriction. Genes Dev 2006; 20:2913–21.
15. Gao L, Cueto MA, Asselbergs F, et al. Cloning and functional characterization of HDAC11, a novel member of the human histone deacetylase family. J Biol Chem 2002; 277:25748–55.
16. Grozinger CM, Schreiber SL. Deacetylase enzymes: biological functions and the use of small-molecule inhibitors. Chem Biol 2002; 9:3–16.
17. Legube G, Trouche D. Regulating histone acetyltransferases and deacetylases. EMBO Rep 2003; 4:944–7.
18. Sengupta N, Seto E. Regulation of histone deacetylase activities. J Cell Biochem 2004; 93:57–67.
19. Wysocka J, Allis CD, Coonrod S. Histone arginine methylation and its dynamic regulation. Front Biosci 2006; 11:344–55.
20. Lee DY, Teyssier C, Strahl BD, et al. Role of protein methylation in regulation of transcription. Endocr Rev 2005; 26:147–70.
21. He H, Lehming N. Global effects of histone modifications. Brief Funct Genomic Proteomic 2003; 2:234–43.
22. Agger K, Christensen J, Cloos PA, et al. The emerging functions of histone demethylases. Curr Opin Genet Dev 2008; 22:115–40.
23. Bannister AJ, Kouzarides T. Reversing histone methylation. Nature 2005; 436:1103–6.
24. Takeuchi T, Watanabe Y, Takano-Shimizu T, et al. Roles of jumonji and jumonji family genes in chromatin regulation and development. Dev Dyn 2006; 235:2449–59.
25. Chang B, Chen Y, Zhao Y et al. JMJD6 is a histone arginine demethylase. Science 2007; 318:444–7.
26. Workman JL, Kingston RE. Alteration of nucleosome structure as a mechanism of transcriptional regulation. Annu Rev Biochem 1998; 67:545–79.
27. Nowak SJ, Corces VG. Phosphorylation of histone H3: a balancing act between chromosome condensation and transcriptional activation. Trends Genet 2004; 20:214–20.
28. Cheung P, Tanner KG, Cheung WL, et al. Synergistic coupling of histone H3 phosphorylation and acetylation in response to epidermal growth factor stimulation. Mol Cell 2000; 5:905–15.
29. Lo WS, Trievel RC, Rojas JR, et al. Phosphorylation of serine 10 in histone H3 is functionally linked in vitro and in vivo to Gcn5-mediated acetylation at lysine 14. Mol Cell 2000; 5: 917–26.
30. Wang GG, Allis CD, Chi P. Chromatin remodeling and cancer, Part I: Covalent histone modifications. Trends Mol Med 2007; 13: 363–72.
31. Clayton AL, Mahadevan LC. MAP kinase-mediated phosphoacetylation of histone H3 and inducible gene regulation. FEBS Lett 2003; 546:51–8.
32. Li F, Adam L, Vadlamudi RK, et al. p21-activated kinase 1 interacts with and phosphorylates histone H3 in breast cancer cells. EMBO Rep 2002; 3:767–73.
33. Salvador LM, Park Y, Cottom J, et al. Follicle-stimulating hormone stimulates protein kinase A-mediated histone H3 phosphorylation and acetylation leading to select gene activation in ovarian granulosa cells. J Biol Chem 2001; 276:40146–55.
34. Taylor SS. The in vitro phosphorylation of chromatin by the catalytic subunit of cAMP-dependent protein kinase. J Biol Chem 1982; 257:6056–63.
35. Hassa PO, Haenni SS, Elser M, et al. Nuclear ADP-ribosylation reactions in mammalian cells: where are we today and where are we going? Microbiol Mol Biol Rev 2006; 70:789–829.
36. Nathan D, Ingvarsdottir K, Sterner DE, et al. Histone sumoylation is a negative regulator in Saccharomyces cerevisiae and shows dynamic interplay with positive-acting histone modifications. Genes Dev 2006; 20:966–76.
37. Kuo MH, Allis CD. In vivo cross-linking and immunoprecipitation for studying dynamic Protein:DNA associations in a chromatin environment. Methods 1999; 19:425–33.
38. Das PM, Ramachandran K, vanWert J, et al. Chromatin immunoprecipitation assay. Biotechniques 2004; 37:961–9.
39. Hanlon SE, Lieb JD. Progress and challenges in profiling the dynamics of chromatin and transcription factor binding with DNA microarrays. Curr Opin Genet Dev 2004; 14:697–705.
40. Bernstein BE, Humphrey EL, Erlich RL, et al. Methylation of histone H3 Lys 4 in coding regions of active genes. Proc Natl Acad Sci U S A 2002; 99:8695–700.
41. Goll MG, Bestor TH. Eukaryotic cytosine methyltransferases. Annu Rev Biochem 2005; 74:481–514.
42. Griswold MD, Kim JS. Site-specific methylation of the promoter alters deoxyribonucleic acid-protein interactions and prevents follicle-stimulating hormone receptor gene transcription. Biol Reprod 2001; 64:602–10.
43. Nagase H, Ghosh S. Epigenetics: differential DNA methylation in mammalian somatic tissues. FEBS J 2008; 275:1617–23.
44. Schaefer CB, Ooi SK, Bestor TH, et al. Epigenetic decisions in mammalian germ cells. Science 2007; 316:398–9.
45. Reik W, Dean W, Walter J. Epigenetic reprogramming in mammalian development. Science 2001; 293:1089–93.
46. Richards JS. Hormonal control of gene expression in the ovary. Endocr Rev 1994; 15:725–51.
47. Richards JS. Perspective: the ovarian follicle–a perspective in 2001. Endocrinology 2001; 142:2184–93.
48. DeManno DA, Cottom JE, Kline MP, et al. Follicle-stimulating hormone promotes histone H3 phosphorylation on serine-10. Mol Endocrinol 1999; 13:91–105.
49. Ruiz-Cortes ZT, Kimmins S, Monaco L, et al. Estrogen mediates phosphorylation of histone H3 in ovarian follicle and mammary epithelial tumor cells via the mitotic kinase, Aurora B. Mol Endocrinol 2005; 19:2991–3000.
50. Clark BJ, Wells J, King SR, et al. The purification, cloning, and expression of a novel luteinizing hormone-induced mitochondrial protein in MA-10 mouse Leydig tumor cells. Characterization of the steroidogenic acute regulatory protein (StAR). J Biol Chem 1994; 269:28314–22.

51. Kiriakidou M, McAllister JM, Sugawara T, et al. Expression of steroidogenic acute regulatory protein (StAR) in the human ovary. J Clin Endocrinol Metab 1996; 81:4122–8.
52. Christenson LK, Stouffer RL, Strauss JF, III. Quantitative analysis of the hormone-induced hyperacetylation of histone H3 associated with the steroidogenic acute regulatory protein gene promoter. J Biol Chem 2001; 276:27392–9.
53. Hiroi H, Christenson LK, Chang L, et al. Temporal and spatial changes in transcription factor binding and histone modifications at the steroidogenic acute regulatory protein (StAR) locus associated with StAR transcription. Mol Endocrinol 2004; 18:791–806.
54. Rusovici R, Hui YY, LaVoie HA. Epidermal growth factor-mediated inhibition of follicle-stimulating hormone-stimulated StAR gene expression in porcine granulosa cells is associated with reduced histone H3 acetylation, Biol Reprod 2005; 72:862–71.
55. Clem BF, Clark BJ. Association of the mSin3A-histone deacetylase 1/2 corepressor complex with the mouse steroidogenic acute regulatory protein gene. Mol Endocrinol 2006; 20:100–13.
56. Lazzaro MA, Pepin D, Pescador N, et al. The imitation switch protein SNF2L regulates steroidogenic acute regulatory protein expression during terminal differentiation of ovarian granulosa cells. Mol Endocrinol 2006; 20:2406–17.
57. Zhang Y, Dufau ML. Dual mechanisms of regulation of transcription of luteinizing hormone receptor gene by nuclear orphan receptors and histone deacetylase complexes. J Steroid Biochem Mol Biol 2003; 85:401–14.
58. Zhang Y, Dufau ML. Silencing of transcription of the human luteinizing hormone receptor gene by histone deacetylase-mSin3A complex. J Biol Chem 2002; 277:33431–8.
59. Soloff MS, Shaw AR, Gentry LE, et al. Demonstration of relaxin precursors in pregnant rat ovaries with antisera against bacterially expressed rat prorelaxin. Endocrinology 1992; 130:1844–51.
60. Peters CA, Maizels ET, Robertson MC, et al. Induction of relaxin messenger RNA expression in response to prolactin receptor activation requires protein kinase C delta signaling. Mol Endocrinol 2000; 14:576–90.
61. Soloff MS, Gal S, Hoare S, et al. Cloning, characterization, and expression of the rat relaxin gene. Gene 2003; 323:149–55.
62. Garmey JC, Guthrie HD, Garrett WM, et al. Localization and expression of low-density lipoprotein receptor, steroidogenic acute regulatory protein, cytochrome P450 side-chain cleavage and P450 17-alpha-hydroxylase/C17-20 lyase in developing swine follicles: in situ molecular hybridization and immunocytochemical studies. Mol Cell Endocrinol 2000; 170:57–65.
63. Golos TG, August AM, Strauss JF, III. Expression of low density lipoprotein receptor in cultured human granulosa cells: regulation by human chorionic gonadotropin, cyclic AMP, and sterol. J Lipid Res 1986; 27:1089–96.
64. LaVoie HA, Benoit AM, Garmey JC, et al. Coordinate developmental expression of genes regulating sterol economy and cholesterol side-chain cleavage in the porcine ovary. Biol Reprod 1997; 57:402–7.
65. Grummer RR, Carroll DJ. A review of lipoprotein cholesterol metabolism: importance to ovarian function. J Anim Sci 1988; 66:3160–73.
66. Natesampillai S, Fernandez-Zapico ME, Urrutia R, et al. A novel functional interaction between the Sp1-like protein KLF13 and SREBP-Sp1 activation complex underlies regulation of low density lipoprotein receptor promoter function. J Biol Chem 2006; 281:3040–7.
67. Watari H, Blanchette-Mackie EJ, Dwyer NK, et al. NPC1-containing compartment of human granulosa-lutein cells: a role in the intracellular trafficking of cholesterol supporting steroidogenesis. Exp Cell Res 2000; 255:56–66.
68. Gevry N, Lacroix D, Song JH, et al. Porcine Niemann Pick-C1 protein is expressed in steroidogenic tissues and modulated by cAMP. Endocrinology 2002; 143:708–16.
69. Gevry NY, Lalli E, Sassone-Corsi P, et al. Regulation of niemann-pick c1 gene expression by the 3′5′-cyclic adenosine monophosphate pathway in steroidogenic cells. Mol Endocrinol 2003; 17:704–15.
70. Gevry N, Schoonjans K, Guay F, et al. Cholesterol supply and sterol regulatory element binding proteins modulate transcription of the Niemann-Pick C1 gene in steroidogenic tissues. J Lipid Res 2008; 49:1024–33
71. Bernard DJ, Chapman SC, Woodruff TK. Mechanisms of inhibin signal transduction. Recent Prog Horm Res 2001; 56:417–50.
72. Wood JR, Strauss JF, III. Multiple signal transduction pathways regulate ovarian steroidogenesis. Rev Endocr Metab Disord 2002; 3:33–46.
73. Ito M, Park Y, Weck J, et al. Synergistic activation of the inhibin alpha-promoter by steroidogenic factor-1 and cyclic adenosine 3′,5′-monophosphate. Mol Endocrinol 2000; 14:66–81.
74. Weck J, Mayo KE. Switching of NR5A proteins associated with the inhibin alpha-subunit gene promoter after activation of the gene in granulosa cells. Mol Endocrinol 2006; 20:1090–103.
75. Gregoraszczuk E, Wojtowicz AK, Tauboll E, et al. Valproate-induced alterations in testosterone, estradiol and progesterone secretion from porcine follicular cells isolated from small- and medium-sized ovarian follicles. Seizure 2000; 9:480–5.
76. Isojarvi JI, Laatikainen TJ, Pakarinen AJ, et al. Polycystic ovaries and hyperandrogenism in women taking valproate for epilepsy. N Engl J Med 1993; 329:1383–8.
77. Nelson-DeGrave VL, Wickenheisser JK, Cockrell JE, et al. Valproate potentiates androgen biosynthesis in human ovarian theca cells. Endocrinology 2004; 145:799–808.
78. Wood JR, Nelson-DeGrave VL, Jansen E, et al. Valproate-induced alterations in human theca cell gene expression: clues to the association between valproate use and metabolic side effects. Physiol Genomics 2005; 20:233–43.

Chapter 25

Ovarian Function and Failure: The Role of the Oocyte and Its Molecules

Loro L. Kujjo and Gloria I. Perez

25.1 Introduction

The ovaries are complex organs by virtue of their primary function and are also genetically unique, since they have a mixture of both somatic cells and germ cells. At birth, each ovarian germ cell or oocyte is enclosed by a specialized population of somatic (pregranulosa) cells to form the follicle, the most basic functional unit of the female gonads. Most follicles present in the ovaries of neonates exist in a state of growth arrest and are referred to as primordial follicles. Although the number of these follicles endowed in the ovaries at birth varies among species (from 2×10^4 to 4×10^4 in mice to 1×10^6–2×10^6 in humans), this stockpile of oocytes is non-renewable in all species and must provide for the entire reproductive needs of the female throughout adult life. By the age of 50 years, the ovaries in most women are exhausted and menopause ensues as a direct consequence of ovarian senescence. This heralds a major life change for women, highlighted by the abrupt loss of ovarian estrogens, which leads to increased risk for a number of debilitating health problems, including cardiovascular disease, osteoporosis, neurological dysfunction, and macular degeneration. In women undergoing adjuvant cancer therapies the stockpile of oocytes is subjected to accelerated depletion resulting in early ovarian failure, loss of reproductive potential, and premature onset of menopause-related health problems. Since oocytes are the prime players influencing ovarian function and reproductive lifespan, the following pages will focus on the major molecules and factors determining the survival and quality of female germ cells.

L.L. Kujjo (✉)
Department of Physiology, Biomedical and Physical Sciences, Michigan State University, East Lansing, MI, USA

25.2 Oocyte Quality and Its Role in Fertility

Women experience a progressive decline in fecundity as they pass through the reproductive years, with the maternal age at birth of the last child averaging 41–42 years in all populations studied [1, 2, 3]. Additionally, in 1% of women, ovarian dysgenesis occurs before the onset of puberty, or ovarian function ceases before 40 years; these conditions being collectively referred to as premature ovarian failure (POF) [4]. On average, though, in normal healthy females optimal fertility is maintained until 30 years of age and then decreases sharply, attributed mainly to a decline in the quality of oocytes [5, 6]. Therefore, the trend to delay childbearing world wide in the last 30 years has increased the risk of infertility [3].

The quality or developmental competence of the oocyte is acquired during folliculogenesis, as it grows, and during the period of its maturation. Throughout the process of oogenesis, oocytes accumulate gene transcripts and translation products that are later mobilized to direct the initial stages of embryogenesis [7]. These maternally supplied factors are responsible for orchestrating numerous developmental processes including crucial early embryonic events such as axis formation, cell fate determination, and embryonic genome activation; all the processes occurring entirely by post-transcriptional mechanisms [8]. Therefore, there is no doubt that oocyte quality affects early embryonic survival, since products of its gene expression control many processes central to the early development of the whole organism; examples include the establishment and maintenance of pregnancy, fetal development, and even predisposition to certain adult diseases.

Besides being modulated by intrinsic factors such as age and genetic defects, the quality of the oocyte is dependent also on extrinsic factors, such as stimulation agents, culture conditions, and nutrition. There is consensus on the point that proper oocyte handling in vitro, including the environment in the incubator, constitutes part of the pertinent determinants for successful fertilization and embryo development. These

P.J. Chedrese (ed.), *Reproductive Endocrinology*,
DOI 10.1007/978-0-387-88186-7_25, © Springer Science+Business Media, LLC 2009

elements have been thoroughly addressed in several recent publications [9, 10, 11, 12, 13]. However, the question of whether hormonal stimulation of the ovary imparts a negative effect on the quality of oocytes retrieved remains controversial and needs clarification.

Data from donor–recipient models in mice show that the uterine milieu is not the only tissue adversely affected by ovarian hyperstimulation [14]. Potentially detrimental effects of ovarian stimulation do also affect oocyte and embryo quality, manifesting in several forms such as reduced proportions of normal embryos [15, 16, 17, 18] and decreased implantation rates and postimplantation development [15]. Some other studies in murine have confirmed this negative effect, mainly showing that in vitro blastocysts of stimulated mice hatch later than oocytes of naturally cycling controls [19]. Additionally, chromosome analysis revealed an increased number of abnormalities in oocytes from hyperstimulated donors, compared with spontaneously ovulating donors [20].

Human studies in this area show less consistent results. For example, data from some studies suggest that controlled ovarian hyperstimulation leads to reduced implantation rates [21, 22]. However, this pathophysiology was possibly caused by diminished uterine receptivity rather than by inferior quality of the oocytes retrieved. Elevated E_2 levels post-hyperstimulation have been also suggested to exert a deleterious effect on the embryo and thought to explain the lower implantation rates [23]. Other studies, however, did not confirm the lower implantation rates in cases of hyperstimulation [24]. Even in OHSS (ovarian hyper-stimulation syndrome), no reduction in implantation rates were found, though lower fertilization rates were observed [25]. Moreover, it has also been reported that a high number of oocytes obtained after ovarian hyperstimulation for in vitro fertilization or intracytoplasmic sperm injection is not associated with decreased pregnancy outcome [26]. Therefore, the data do not support the concept that the quality of oocytes is impaired in women with a strong ovarian response.

One probable reason for the disparity between the rodent and human data is the lack of reliable biochemical and/or morphological markers that can be used to distinguish a good quality oocyte. Currently, assessment of oocyte quality relies on morphology, which by itself is a difficult task and often dependent on subjective judgments. Furthermore, studies do not usually agree on the final parameter for determining oocyte quality; hence, one queries: is it based on fertilization rates, or blastocyst development, or implantation rates? Several scientists believe that when the health of the uterus and the risk of pregnancy complications are not intervening factors to be considered, classic oocyte quality indicators like morphology, fertilization rate, embryo quality, and implantation rate should always be correlated with delivery of live and healthy offspring.

25.3 In Search of Other Markers of Oocyte Quality

Because of the difficulties in assessing oocyte morphology, the search for other alternative determinants of its quality has created a great deal of interest to identify specific factors. As such, some research efforts are being focused on cumulus cells (CC), which are considered as ideal surrogates for assessment of oocyte developmental potential [27]. While earlier studies speculated that a set of cumulus genes may determine oocyte maturation, fertilization potential, and embryo quality [27], recent gene expression data by Assou [28] have provided some molecular evidence for mediations by the CC in embryo development.

There are already data from sibling human oocytes suggesting that the quality of embryos improve when oocytes are allowed to interact with CC, hence indicative of an improvement of cytoplasmic maturation [29]. It appears that CC have a protective and beneficial effect on embryo development [30, 31]. However, there are also data suggesting just the opposite [32]. Whatever the case, under certain circumstances, there is no doubt that CC play a significant role in controlling the fate of the oocyte. For example, in this chapter, we discuss data showing that ceramide (see under 25.4) concentration in the CC correlates with low fertility and increased rates of oocyte apoptosis.

Efforts toward identifying a blueprint of oocyte quality from genes and proteins in CC should take into account two phenomena: first, the majority of oocytes are destined to die rather than survive; and second, that most probably, the oocyte is the generator of signals sent to the CC to program the latter on how to proceed. A reflection of this communication strategy can be seen in the increased wave of oocyte apoptosis that occurs in mice with advanced age; here the process appears to be directed by the oocyte itself [33]. The oocyte transmits the message to the executioners (the CC) to go ahead and kill. Thus, we speculate that the chances of spotting a bad oocyte based on cumulus cell gene expression are quite high (many bad oocytes in a pool); as such, CC genomic/proteomic analyses may provide a valuable means. However, the situation can turn more complicated when trying to do the opposite; in other words, finding a CC pattern(s) that could be used to distinguish a good oocyte. Our prediction is that in a group of oocytes, the chances for emergence of a good one are minimal; and therefore within a group, the genomic signature of the good oocyte can easily be masked, unless of course, the analysis is performed in CC from individual oocytes.

So, there are no doubts that genome wide analyses of the mouse oocyte transcriptome can accelerate our understanding of the molecular mechanisms underlining oocyte competence. However, it is worth emphasizing that functional genomics need to be complemented with proteomics when monitoring gene expression because proteins being the functional molecules in a cell reflect functional differences in gene transcription, translation, and post-translational modifications. This multifaceted approach is particularly important since transcription profiles and actual cellular protein content sometimes lack correlation [34, 35]. Besides, epigenetic influences do introduce considerable variation.

25.4 Aging: Mitochondria, DNA Check Point, and Lipids

There is general consensus that oocyte quality decreases with advanced age [1, 3, 6, 36]. The aging process by itself is a complex phenomenon, driven by as yet incompletely described genetic programs. While many factors are thought to modulate cellular and organismal life span, only a few are recognized as being prominently involved. Among these are mitochondrial respiration and genomic stability. As such, eukaryotic cells have developed specific mechanisms for monitoring and responding to environmental and genetic perturbations. The checkpoints that monitor mitochondrial integrity and the genomic DNA represent two of the most important surveillance mechanisms available to cells for maintaining homeostasis [37, 38]. In addition, several studies have also presented evidence for the involvement of the sphingolipid second messenger, ceramide [39], in cellular senescence and organismal aging in diverse species [40, 41, 42]. Consequently, the significance of mitochondria, genomic DNA checkpoint, and ceramide with respect to their relation to oocyte quality and aging will be discussed in the following paragraphs.

Mitochondria have been implicated in both general body aging [43, 44] and aging of female reproductive tissues [45, 46, 47]. As women age, the success rates of in vitro fertilization fall and evidence suggests that a concurrent increase in mitochondrial defects may be in part responsible for this decline in fertility [48]. A glimpse of this phenomenon is seen where increased maternal age has been associated with increased rate of mitochondrial mutations in the oocytes [49], decreased mitochondrial metabolic activity [50], and inefficient ATP production through mitochondrial respiration [51]. Furthermore, the level of mtDNA deletions is reportedly significantly higher in oocytes than in embryos, strongly suggesting that mtDNA integrity plays a role in determining the fertilizability of oocytes [52, 53]. These data are supported by studies in which the transfer of either pure mitochondria or mitochondria-enriched cytoplasts to mouse oocytes resulted in lower rates of apoptosis [54] and significant increases in ATP production [55]. In addition, transfer of donor ooplasm to recipient oocytes in ART clinics resulted in improved in vitro embryo development as evident by the decrease in their fragmentation rates [56]. These findings, along with others, have attracted increased interest in researching the role of mitochondria in infertility and embryo development.

In oocytes as well as in many other cell types, mitochondria are principal determinants of cell fate, not only for their well established control of energy production via oxidative phosphorylation but also by virtue of their direct role in the regulation of cell death [57, 58]. Compared to the other cell types though, germ cells are in a unique status in regards to their mitochondrial complement. For example, germ cell mitochondria contain denser matrixes and fewer cristae. The number of cristae in a mitochondrion has been directly associated with the levels of ATP production [46]. Hence, the low surface to volume ratio of cristae in germ cell mitochondria limits the ability to generate ATP, a fact documented in human oocytes found to contain relatively low levels of ATP [51]. Other mitochondrial ultrastructures such as reduced organelle size and less complex internal structure typify the oocyte mitochondria as morphologically primitive or immature compared with those of somatic cells [59]. In spite of this disparity, metabolic evidence suggests that the mitochondria of oocytes and early embryo are constitutively active and that maintenance of the low-level activity is necessary and sufficient for ongoing development. Additionally, mitochondria found in oocytes have a limited number of replication cycles prior to the blastocyst stage [60]. Inevitably, the developing embryo is entirely dependent on the population of mitochondria present at the time of ovulation.

In addition to the preceding facts, all mitochondria in a zygote are exclusively maternally derived through the ooplasm; the few mitochondria that enter the oocyte with the sperm at fertilization are reportedly degraded [61] or diluted out during sequential cleavage [62] and cannot be detected by the blastocyst stage. Therefore, given that mitochondria are (1) maternally inherited, (2) principal sites of oxidative damage, (3) required for cytoplasmic energy production in all embryonic cells, and (4) important regulators of preimplantation embryo development, it is feasible that any inherited or acquired mitochondrial dysfunction in the oocyte can be detrimental to postimplantation development.

To investigate the role of mitochondria in the developmental compromises observed with female aging, Thouas [63, 64, 65] developed a mouse model of mitochondrial dysfunction. They found that blastocyst development in vitro is relatively resistant to low degrees of mitochondrial injury induced by fluorophore photosensitization in oocytes from young mice. In contrast, aged oocytes were more developmentally sensitive to mitochondrial damage than pubertal oocytes. The authors concluded that this was possibly due to an age-related mitochondrial energy deficiency. As further evidence, Thouas identified delayed developmental effects directly related to mitochondrial deficiencies caused by advanced aging.

Since accumulation of mtDNA deletions in oocytes may contribute to mitochondrial dysfunction and impaired ATP production as well, reproductive age in women constitutes a significance factor in this context. In fact, it has been reported that loss of mitochondrial activity in oocytes obtained from older women may be associated with a lower rate of both pregnancy and embryonic development [51].

More recently, in unpublished studies of mouse and human oocytes, we found that there was a global decline in mitochondrial function in the aged oocytes. This decline was accompanied by changes in mitochondrial structure as analyzed by electron tomography. The aged oocytes were more highly susceptible to spontaneous fragmentation in vitro, had a lower ratio of highly to lowly polarized mitochondria, and exhibited a significant decrease in ROS and ATP content. These results indicate that alterations of mitochondrial function precede changes that result in oocyte demise and thus may play a role in determining the quality of the oocyte.

Mitochondrial function is sensitive to environmental stressors, including those induced by in vitro manipulations and conditions [66]. For example, a significant increase in the proportion of mtDNA deletions have been observed in stimulated oocytes and embryos from rhesus macaques, compared with levels of mtDNA deletions in immature, unstimulated oocytes [67]. Also, data we have recently obtained using human unfertilized oocytes support the tenet that the environment in which the oocytes mature affects their final quality. For instance, conditions such as polycystic ovary syndrome, pelvic inflammatory disease, and endometriosis, all of which impact ovarian function, resulted in a trend toward an aberrant increase in the ratio of mitochondrial membrane potential. In addition, advanced aging causes further depression of mitochondrial function and oocyte developmental competence [68, 69].

Oocyte mitochondria are therefore physiological regulators of early embryo development and potential sites of pathological insults that may perturb the oocyte and subsequent viability of the preimplantation embryo. Particular research focus should be directed at the specific biological effects of clinical and laboratory interventions such as ovarian stimulation and in vitro manipulations and culture; the emphasis should be on reducing the potential for mitochondrial insult.

It is therefore plausible to conclude that any type of low-level biochemical interference affecting mitochondrial function in the oocyte do translate into negative developmental outcomes, especially in conjunction with differentiation events. However, under certain circumstances, the mitochondrial phenotype can be rescued as we discovered to be the case in oocytes from FVB mice [70]. The oocytes from these mice exhibit a marked reduction in ROS content, indicative of diminished metabolic activity. Further to this, analysis of the oocytes mitochondrial ultrastructure revealed a myriad of abnormalities in mitochondrial membrane composition and cristae. The most striking of these were the onion-like whorling of the peripheral cristae and the frequent sites of outer mitochondrial membrane rupture. All of these alterations led to a marked elevation in apoptosis susceptibility in female germ cells. Despite all these, the FVB oocyte phenotype can be rescued by fertilization, probably a benefit of calcium waves triggered by the sperm [70].

The preceding data are indicative of tremendous plasticity in mitochondrial ultrastructure and function and might explain why the pathophysiology of the oocyte mitochondria has not yet been directly identified as a causal factor in female infertility. In the case of the FVB mice, the females are fertile despite having oocytes with multiple mitochondrial phenotypes. In humans, naturally occurring mtDNA mutations that result in an array of systemic and tissue-specific pathologies have been identified [71, 72] but none of those has been directly linked to infertility. Therefore, the generalization that mitochondrial dysfunction represents a potential epigenetic factor in the etiology of clinical subfertility in females has to be made with caution, in the absence of direct evidence from either animal or human oocytes. Nonetheless, all those studies reinforce the value of mitochondrial function as a potentially valuable indicator of oocyte developmental competence. The data also suggest that the quality of embryonic and fetal nutrition in utero may have downstream heritable influences on adult health [65], a point that still needs further confirmation.

Mitochondrial dysfunction may not only compromise developmental processes in the embryo but also trigger apoptosis [73, 74]. This dual role for mitochondria (to maintain life or to commit to cell death) may well represent a quality control system that will determine whether the early embryo proceeds further in development or is quickly eliminated. It is therefore not surprising that a very delicate balance in mitochondrial function must be maintained to have a high quality oocyte or embryo. Aberrant increases or decreases in the balance of mitochondrial polarization indicate a disparity in mitochondrial activation, in both cases seemingly leading to an increased incidence of oocyte fragmentation.

However, whether during early development mitochondria have a regulatory role that is distinct from their metabolic contribution has only recently been investigated relative to other functions, such as their ability to sequester and release calcium, modify proteins, or initiate apoptosis [75].

Due to the magnitude of cell death that normally occurs within the female gonad during both fetal development and post-natal life, the ovary has proven to be an excellent model for studying the role of cell death genes in a physiological setting of endocrine-regulated apoptosis. A relatively detailed blueprint of specific genes and pathways involved in signaling for, and executing, oocyte apoptosis following disruption of either mitochondrial or genomic integrity has been formulated over the past decade or so [70, 76, 77, 78, 79]. This topic has been recently reviewed [80, 81] and therefore we refer the reader to those publications. Nevertheless, we want to mention some more recent findings where we demonstrated that in the oocytes from FVB mice (mentioned earlier), mitochondria are directly responsible for the enhancement of cell death. In those studies [70], microinjection of FVB mitochondria into B6C3F1 oocytes, which have an inherently low basal rate of death, increased the incidence of apoptosis by nearly fivefold. We interpreted these results as indicating that FVB mitochondria are supplying pro-apoptotic factors to the B6C3F1 oocytes, most likely by virtue of their "leaky" outer membrane structure. Consistent with this, mitochondria collected from FVB oocytes released approximately 20% more cytochrome *c* in short-term incubations than did mitochondria of B6C3F1 oocytes. However, direct microinjection of cytochrome *c* into B6C3F1 oocytes did not, by itself, induce apoptosis, suggesting that the FVB mitochondria must be supplying additional pro-apoptotic factors. To this end, microinjection of recombinant Smac/DIABLO increased apoptosis in B6C3F1 oocytes, and co-injecting cytochrome *c* synergistically enhanced this effect.

Besides dysfunctions in metabolic activity, abnormal distribution pattern of mitochondria has been associated with reduced developmental competence in mice [82] and humans [50, 83]. In oocytes from aging B6CBAF1 mice, mitochondrial condensation of an atypical nature has been observed and interpreted as evidence of apoptosis [64]. Interestingly, the number of MII oocytes possessing the abnormal mitochondrial clusters increased threefold between 12 and 40 weeks of age. This mitochondrial abnormality was also pronounced in oocytes that had undergone prolonged in vitro aging [84]. More recently, we observed that uneven aggregation of mitochondria consistently occurs in mouse and bovine oocytes of poor quality (Perez, unpublished observations). Collectively, all those findings suggest that abnormal mitochondrial distribution in oocytes at MII stage is a cause of developmental retardation. A logical conclusion would have been to declare that the mitochondrial distribution in MII oocytes could be used as an important criterion for selection of good oocytes and prediction of embryonic developmental competence during incubation. Paradoxically though, we also found the same atypical aggregations of mitochondria in oocytes derived from FVB mice that do not comparatively exhibit fertility problems, yet their oocytes are prone to apoptosis. Therefore, we hypothesize that the aggregation of mitochondria is not a permanent genetic impression but possibly epigenetic in nature. Oocytes with this pattern can be rescued, for example, with spikes of Ca at the time of fertilization. Other mechanisms, still unknown, might be equally potent in reversing these abnormal mitochondrial patterns.

The effects of aging on physical and biochemical interactions between mitochondria and other important ooplasmic organelles merit further investigation. Unfortunately, assessment of mitochondrial membrane potential as well as localization of the organelles in a clinical setting is unfeasible, as it necessitates the developing embryo to take up potentially harmful dyes; some of those dyes have been shown to increase the rates of cell death in murine blastocysts [73]. Non-invasive measurements of nutrient consumption, and hence indirectly assessing mitochondrial function, as suggested by Gardner [85], and the development of other non-invasive assessments of mitochondrial function are likely to prove fruitful in selecting the most viable embryos for transfer in the IVF setting for both young and maternally aged women. Such a tool would be also useful for understanding the age-related alterations in mitochondrial gene expression and protein expression that ultimately regulate physiological processes.

So far we have focused our attention on mitochondria and aerobic respiration. However, the exploitation of other metabolic pathways by the oocyte to achieve developmental competence cannot be ignored. Basal glycolytic activity in mature oocytes is correlated with increased embryonic development [86]. However, relying for too long on glycolysis leads to overproduction of lactate, resulting in reduced embryo viability [87]. The contribution of the glycolytic pathway to oocyte quality has been recently reviewed by Krisher [88].

More recently, Nutt [89], using extracts from *Xenopus laevis* oocytes, showed that generation of NADPH through the pentose-phosphate-pathway is critical for oocyte survival and that the target of this regulation is caspase-2, previously shown to be required for oocyte death in mice. Pentose-phosphate-pathway-mediated inhibition of cell death (driven by glucose-6-phosphate) is effected via the inhibitory phosphorylation of caspase-2 by calcium/calmodulin-dependent protein kinase II (CaMKII). These data, which are fully consistent with the reported requirement for caspase-2 in mouse oocyte apoptosis, link the operation of a specific metabolic pathway to the direct CaMKII-mediated regulation of caspase-2, thereby providing insight into the control of germ-cell life and death.

Coupling metabolic state and cell survival may provide a mechanism for regulating cell numbers during aging at both the cellular and organismal level. Nutt [89] noted an age-related decrease in the activity of G6P dehydrogenase in murine oocytes, which could contribute to their decreased viability. Although it is attractive to speculate that nutrient stockpiles serve as a timer for oocyte survival, changes in pentose-phosphate-pathway flux might also result from age- or hormone-related alterations in enzyme activities. Nutt

[89] also suggested that their data might help to explain why oocytes are lost through apoptosis as females age, thereby providing potential therapeutic targets for the maintenance of oocyte viability and fertility. However, other mechanisms might also exist, since caspase-2 knockout females do not show prolonged ovarian function with advanced age [90].

In addition to the mitochondrial check-point, cells must also maintain DNA integrity to avoid cellular transformation or death. In this regard, eukaryotic cells have developed highly sophisticated responses to DNA double-strand breaks (DDSB), including repair pathways involving homologous recombination driven by the *Rad51* gene product [91, 92]. Formation and repair of DDSB by homologous recombination ensures correct pairing and subsequent segregation of homologous chromosomes during the first meiotic division. However, irreparable or excessive DDSB activate the DNA surveillance checkpoint and trigger apoptosis [37]. It is therefore not totally surprising that defects on this pathway result in infertility [70, 93, 94].

We recently demonstrated that in mice of the AKR/J background a defect in DNA repair in oocytes leads to high rates of spontaneous apoptosis and infertility [70]. This phenotype was obviously prominently due to deficiency of Rad51, since we were able to minimize it by microinjection of the protein; a marked reduction in apoptosis followed. As shown more recently by Kuznetsov [93] deficiency of RAD51C in mice results in early prophase I arrest in males, and sister chromatid separation at metaphase II in females, leading to infertility in both sexes. In our unpublished observations in mice, we have observed that fertility decline due to advanced age may be also partly attributable to a decreased ability of their oocytes to repair DNA damage.

Interestingly, the oocytes recovered from aged females had a tendency toward higher rates of aneuploidy [95]. Age-related decline in oocyte quality has been associated with aneuploidy linked to the age-associated increase of spontaneous abortions and numerical meiotic division errors [96, 97]. Whereas the primary cause of such anomalies and other factors determining oocyte quality remain to be elucidated, there is ample data indicating that the vast majority of human aneuploidies result from maternal MI errors [98, 99, 100]. It has been proposed that mono-oriented and misaligned univalents can bypass the spindle assembly checkpoint (SAC) in oocytes owing to the large volume of this cell, which makes it refractory to residual checkpoint signaling [101]. In contrast, recently data from Kouznetsova [102], using mouse oocytes, argue that chromosomes present as unpaired univalents, despite a conserved side-by-side kinetochore arrangement, succeed in establishing a bipolar orientation, enabling them to satisfy the requirements of the SAC which remains fully functional. Despite the controversy, and although many questions still remain unanswered, the contributions of the DNA checkpoint to oocyte quality and fertility are significant.

Sphingolipids are also implicated in both development and aging at the organismal [40] and the tissue level [103]. Currently many different types of lipids are known to exist in cells, and their functions are defined beyond being considered primarily as structural only: originally ascribed to only forming bilayers that compartmentalize cellular organelles and functions, tremendous appreciation is currently growing for some of these lipid classes and their critical roles as bioactive second messengers. For example, the main membrane lipids, phosphatidylcholine, phosphatidylethanolamine, phosphatidylserine, sphingomyelin, and cholesterol serve structural or storage functions, whereas lipids that are present in low quantities such as diacylglycerol, phosphatidic acid, ceramide, or phosphoinositides function as signaling molecules [104] primarily due to their ability to interact with specific protein targets, and thereby influence the activity/function of those targets.

Ceramide signaling is a complex network and has been implicated in various cellular processes such as necrosis [105], survival and proliferation [106], differentiation [107], and aging [108]. In the human ovary, an organ whose function declines precipitously with advancing age [109, 110], ceramide levels have been shown to increase during the years immediately preceding the menopause [111], when a woman's endowed pool of germ cells (oocytes) has been nearly exhausted [109]. It is interesting to note here that in aging mice the increase in ceramide is negatively correlated to a decline in fertility. Not surprisingly, we recently identified ceramide as one of the key molecules signaling the accelerated incidence of apoptosis in oocytes of aged female mice [33]. We uncovered a novel role for the intercellular trafficking of ceramide as a key step in the process. The oocytes from aged females are particularly sensitive to a cytosolic ceramide spike released by CC and translocated via gap junction-dependent communication. Considering that the rate of oocyte depletion from the human ovaries accelerates dramatically in the 10 or so years prior to menopause [109], it is logical to speculate that similar events may be at work in the ovaries of women as they age. But we are not aware of any studies analyzing lipid content of human oocytes and/or the changes in lipids according to maternal age.

In further studies we also discovered that the spatial location of ceramide pools (within the oocyte) is also a key determinant of whether the cell dies or survives [34]. In a recent review, van Blitterswijk [112] highlighted the physicochemical aspects of ceramide signaling, and they stressed the need to understand where in the cell and on which membrane-leaflet ceramide is located.

The de novo synthesis of ceramide commences in the endoplasmic reticulum (ER) and continues in the Golgi apparatus, and possibly other compartments such as the plasma membrane and mitochondria. Because of the various locations of ceramide within the cell, the significance of its intracellular transport is worth mentioning. Sphingolipid-binding proteins are now described as being responsible for moving specific sphingolipids from one membrane to another. So far, the ceramide transport protein (CERT), also known as the Good-pasture antigen binding protein (GPBP), is the only ceramide-specific transport protein known [113]. It is described as functioning in the non-vesicular transport of ceramide from the ER to the Golgi apparatus.

Hanada unraveled the biochemical aspects of CERT function in several elegant studies [113, 114]. However, the relevance of CERT in the context of development and physiological functions of the body remained unknown until just recently, when Rao [115] demonstrated that in Drosophila, CERT function is essential for normal oxidative stress response and lifespan.

The relevance of this model to this chapter is that the majority of the phenotypes observed in the mutant flies, lacking a functional CERT, have amazing similarities with the changes we and others have observed in the aged oocyte. For example, the mutant flies became susceptible to reactive oxygen species, they developed metabolic imbalance including decreased ATP levels; their ceramide levels were decreased by 70% and the flies died early in life. In aged oocytes we and others have seen that with advanced age the oocytes become more susceptible to ROS [64] and other insults [34]; they experience reduced ATP levels [51], and low levels of total ceramide [33] which precede an increased rate of apoptosis. Similarities between changes observed in the mutant flies lacking CERT and changes in the oocyte with age prompted us to hypothesize that most probably the aging oocyte will shift to low expression levels of CERT. This concept is currently under investigation in our laboratory.

Understanding the cellular and molecular mechanisms that maintain physiological levels of ceramide in oocytes in vivo has implications for the therapeutic management of infertility and also, perhaps, of the aging processes itself.

25.5 Summary

A mature healthy viable oocyte is the product of a complex and as yet poorly understood growth and differentiation processes. Future work will hopefully expose the molecular blueprint that will allow us to identify a fully competent oocyte. In the meantime, one can postulate that perhaps the best oocyte is not the one with the best mitochondria and the best quality DNA, but one that has the capacity to activate relevant metabolic pathways and/or DNA repair mechanisms whenever the need arises. The best oocyte is therefore the one that assures its survival against all odds, always determined to survive and protect the genetic information that it carries. Even so, they all succumb to the negative impacts of advanced maternal age. This remains a fundamental query in aging research.

25.6 Glossary of Terms and Acronyms

ART: assisted reproductive technology

ATP: adenosine 5'-triphosphate

CaMKII: calcium/calmodulin-dependent protein kinase II

CC: cumulus cells

CERT: ceramide transport protein

DDSB: DNA double strand breaks

E_2: estradiol

ER: endoplasmic reticulum

G6P: glucose-6-phosphate

GPBP: good-pasture antigen binding protein

IVF: in vitro fertilization

MI: metaphase I

MII: metaphase II

mtDNA: mitochondrial DNA

NADPH: nicotinamide adenine dinucleotide phosphate

OHSS: ovarian hyper-stimulation syndrome

POF: premature ovarian failure

Rad51: protein involved in repair of DNA.

ROS: reactive oxygen species

SAC: spindle assembly checkpoint

Smac/DIABLO: mitochondrial protein

References

1. Ruman J, Klein J, Sauer MV. Understanding the effects of age on female fertility. Minerva Ginecol 2003; 55:117–27.
2. Sauer MV. The impact of age on reproductive potential: lessons learned from oocyte donation. Maturitas 1998; 30:221–5.
3. Ottolenghi C, Uda M, Hamatani T, et al. Aging of oocyte, ovary, and human reproduction. Ann N Y Acad Sci 2004; 1034:117–31.
4. Lobo RA. Potential options for preservation of fertility in women. N Engl J Med 2005; 353:64–73.
5. Faddy MJ. Follicle dynamics during ovarian ageing. Mol Cell Endocrinol 2000; 163:43–8.
6. Gougeon A. The biological aspects of risks of infertility due to age: the female side. Rev Epidemiol Sante Publique 2005; 53:2S37–45.

7. Schultz GA, Heyner S. Gene expression in pre-implantation mammalian embryos. Mutat Res 1992; 296:17–31.
8. Latham KE, Schultz RM. Embryonic genome activation. Front Biosci 2001; 6:D748–59.
9. Balaban B, Urman B. Effect of oocyte morphology on embryo development and implantation. Reprod Biomed Online 2006; 12:608–15.
10. Choi WJ, Banerjee J, Falcone T, et al. Oxidative stress and tumor necrosis factor-alpha-induced alterations in metaphase II mouse oocyte spindle structure. Fertil Steril 2007; 81:1220–31.
11. Higdon HL 3rd, Blackhurst DW, Boone WR. Incubator management in an assisted reproductive technology laboratory. Fertil Steril 2007; 89:703–10.
12. Mikkelsen AL. Strategies in human in-vitro maturation and their clinical outcome. Reprod Biomed Online 2005; 10:593–9.
13. Yeung QS, Briton-Jones CM, Tjer GC, et al. The efficacy of test tube warming devices used during oocyte retrieval for IVF. J Assist Reprod Genet 2004; 21:355–60.
14. Fossum GT, Davidson A, Paulson RJ. Ovarian hyperstimulation inhibits embryo implantation in the mouse. J In Vitro Fert Embryo Transf 1989; 6:7–10.
15. Ertzeid G, Storeng R. The impact of ovarian stimulation on implantation and fetal development in mice. Hum Reprod 2001; 16:221–5.
16. Katz-Jaffe MG, Trounson AO, Cram DS. Chromosome 21 mosaic human preimplantation embryos predominantly arise from diploid conceptions. Fertil Steril 2005; 84:634–43.
17. Munne S, Magli C, Adler A, et al. Treatment-related chromosome abnormalities in human embryos. Hum Reprod 1997; 12:780–4.
18. Ziebe S, Lundin K, Janssens R, et al. Influence of ovarian stimulation with HP-hMG or recombinant FSH on embryo quality parameters in patients undergoing IVF. Hum Reprod 2007; 22:2404–13.
19. Van der Auwera I, D'Hooghe T. Superovulation of female mice delays embryonic and fetal development. Hum Reprod 2001; 16:1237–43.
20. Spielmann H, Vogel R. Genotoxic and embryotoxic effects of gonadotropin hyperstimulated ovulation on murine oocytes, preimplantation embryos and term fetuses. Ann Ist Super Sanita 1993; 29:35–9.
21. Check JH, Choe JK, Katsoff D, et al. Controlled ovarian hyperstimulation adversely affects implantation following in vitro fertilization-embryo transfer. J Assist Reprod Genet 1999; 16:416–20.
22. Simon C, Cano F, Valbuena D, et al. Clinical evidence for a detrimental effect on uterine receptivity of high serum oestradiol concentrations in high and normal responder patients. Hum Reprod 1995; 10:2432–7.
23. Valbuena D, Martin J, de Pablo JL, et al. Increasing levels of estradiol are deleterious to embryonic implantation because they directly affect the embryo. Fertil Steril 2001;76:962–8.
24. van Kooij RJ, Looman CW, Habbema JD, et al. Age-dependent decrease in embryo implantation rate after in vitro fertilization. Fertil Steril 1996; 66:769–75.
25. Aboulghar MA, Mansour RT, Serour GI, et al. Oocyte quality in patients with severe ovarian hyperstimulation syndrome. Fertil Steril 1997; 68:1017–21.
26. Kok JD, Looman CW, Weima SM, et al. A high number of oocytes obtained after ovarian hyperstimulation for in vitro fertilization or intracytoplasmic sperm injection is not associated with decreased pregnancy outcome. Fertil Steril 2006; 85: 918–24.
27. McKenzie LJ, Pangas SA, Carson SA, et al. Human cumulus granulosa cell gene expression: a predictor of fertilization and embryo selection in women undergoing IVF. Hum Reprod 2004; 19:2869–74.
28. Assou S, Anahory T, Pantesco V, et al. The human cumulus–oocyte complex gene-expression profile. Hum Reprod 2006; 21:1705–19.
29. Hassan HA. Cumulus cell contribution to cytoplasmic maturation and oocyte developmental competence in vitro. J Assist Reprod Genet 2001; 18:539–43.
30. Ebner T, Moser M, Sommergruber M, et al. Incomplete denudation of oocytes prior to ICSI enhances embryo quality and blastocyst development. Hum Reprod 2006; 21:2972–7.
31. Magier S, van der Ven HH, Diedrich K, et al. Significance of cumulus oophorus in in-vitro fertilization and oocyte viability and fertility. Hum Reprod 1990; 5:847–52.
32. Perez GI, Tilly JL. Cumulus cells are required for the increased apoptotic potential in oocytes of aged mice. Hum Reprod 1997; 12:2781–3.
33. Perez GI, Jurisicova A, Matikainen T, et al. A central role for ceramide in the age-related acceleration of apoptosis in the female germline. Faseb J 2005; 19:860–2.
34. Jurisicova A, Lee HJ, D'Estaing SG, et al. Molecular requirements for doxorubicin-mediated death in murine oocytes. Cell Death Differ 2006; 13:1466–74.
35. Lopez MF. Better approaches to finding the needle in a haystack: optimizing proteome analysis through automation. Electrophoresis 2000; 21:1082–93.
36. Tarlatzis BC, Zepiridis L. Perimenopausal conception. Ann N Y Acad Sci 2003; 997:93–104.
37. Li L, Zou L. Sensing, signaling, and responding to DNA damage: organization of the checkpoint pathways in mammalian cells. J Cell Biochem 2005; 94:298–306.
38. Singh KK. Mitochondria damage checkpoint, aging, and cancer. Ann N Y Acad Sci 2006; 1067:182–90.
39. Kolesnick RN, Kronke M. Regulation of ceramide production and apoptosis. Annu Rev Physiol 1998; 60:643–65.
40. Cutler RG, Mattson MP. Sphingomyelin and ceramide as regulators of development and lifespan. Mech Ageing Dev 2001; 122:895–908.
41. Obeid LM, Hannun YA. Ceramide, stress, and a "LAG" in aging. Sci Aging Knowledge Environ 2003; 39:PE27.
42. Cutler RG, Kelly J, Storie K, et al. Involvement of oxidative stress-induced abnormalities in ceramide and cholesterol metabolism in brain aging and Alzheimer's disease. Proc Natl Acad Sci U S A 2004; 101:2070–5.
43. Ames BN, Shigenaga MK, Hagen TM. Mitochondrial decay in aging. Biochim Biophys Acta 1995;1271:165–70.
44. Sastre J, Borras C, Garcia-Sala D, et al. Mitochondrial damage in aging and apoptosis. Ann N Y Acad Sci 2002; 959:448–51.
45. Janny L, Menezo YJ. Maternal age effect on early human embryonic development and blastocyst formation. Mol Reprod Dev 1996; 45:31–7.
46. Jansen RP, de Boer K. The bottleneck: mitochondrial imperatives in oogenesis and ovarian follicular fate. Mol Cell Endocrinol 1998; 145:81–8.
47. Kirkwood TB. Ovarian ageing and the general biology of senescence. Maturitas 1998; 30:105–11.
48. Keefe DL. Aging and infertility in women. Med Health R I 1997; 80:403–5.
49. Keefe DL, Niven-Fairchild T, Powell S, et al. Mitochondrial deoxyribonucleic acid deletions in oocytes and reproductive aging in women. Fertil Steril 1995; 64:577–83.
50. Wilding M, Dale B, Marino M, et al. Mitochondrial aggregation patterns and activity in human oocytes and preimplantation embryos. Hum Reprod 2001; 16:909–17.
51. Van Blerkom J, Davis PW, Lee J. ATP content of human oocytes and developmental potential and outcome after in-vitro fertilization and embryo transfer. Hum Reprod 1995; 10: 415–24.

52. Brenner CA, Wolny YM, Barritt JA, et al. Mitochondrial DNA deletion in human oocytes and embryos. Mol Hum Reprod 1998; 4:887–92.
53. Barritt JA, Brenner CA, Cohen J, et al. Mitochondrial DNA rearrangements in human oocytes and embryos. Mol Hum Reprod 1999; 5:927–33.
54. Perez GI, Trbovich AM, Gosden RG, et al. Mitochondria and the death of oocytes. Nature 2000; 403:500–1.
55. Van Blerkom J, Sinclair J, Davis P. Mitochondrial transfer between oocytes: potential applications of mitochondrial donation and the issue of heteroplasmy. Hum Reprod 1998; 13: 2857–68.
56. Cohen J, Scott R, Alikani M, et al. Ooplasmic transfer in mature human oocytes. Mol Hum Reprod 1998; 4:269–80.
57. Danial NN, Korsmeyer SJ. Cell death: critical control points. Cell 2004; 116:205–19.
58. Wang X. The expanding role of mitochondria in apoptosis. Genes Dev 2001; 15:2922–33.
59. Dumollard R, Duchen M, Carroll J. The role of mitochondrial function in the oocyte and embryo. Curr Top Dev Biol 2007; 77:21–49.
60. McConnell JM, Petrie L. Mitochondrial DNA turnover occurs during preimplantation development and can be modulated by environmental factors. Reprod Biomed Online 2004; 9:418–24.
61. Sutovsky P, Moreno RD, Ramalho-Santos J, et al. Ubiquitinated sperm mitochondria, selective proteolysis, and the regulation of mitochondrial inheritance in mammalian embryos. Biol Reprod 2000; 63:582–90.
62. Cummins JM. Fertilization and elimination of the paternal mitochondrial genome. Hum Reprod 2000; 15:92–101.
63. Thouas GA, Trounson AO, Wolvetang EJ, et al. Mitochondrial dysfunction in mouse oocytes results in preimplantation embryo arrest in vitro. Biol Reprod 2004; 71:1936–42.
64. Thouas GA, Trounson AO, Jones GM. Effect of female age on mouse oocyte developmental competence following mitochondrial injury. Biol Reprod 2005; 73:366–73.
65. Thouas GA, Trounson AO, Jones GM. Developmental effects of sublethal mitochondrial injury in mouse oocytes. Biol Reprod 2006; 74:969–77.
66. Manoli I, Alesci S, Blackman MR, et al. Mitochondria as key components of the stress response. Trends Endocrinol Metab 2007; 18:190–8.
67. Gibson TC, Kubisch HM, Brenner CA. Mitochondrial DNA deletions in rhesus macaque oocytes and embryos. Mol Hum Reprod 2005; 11:785–9.
68. Meng Q, Wong YT, Chen J, et al. Age-related changes in mitochondrial function and antioxidative enzyme activity in fischer 344 rats. Mech Ageing Dev 2007; 128:286–92.
69. Eichenlaub-Ritter U, Vogt E, Yin H, et al. Spindles, mitochondria and redox potential in ageing oocytes. Reprod Biomed Online 2004; 8:45–58.
70. Perez GI, Acton BM, Jurisicova A, et al. Genetic variance modifies apoptosis susceptibility in mature oocytes via alterations in DNA repair capacity and mitochondrial ultrastructure. Cell Death Differ 2007; 14:524–33.
71. Thorburn DR, Dahl HH. Mitochondrial disorders: genetics, counseling, prenatal diagnosis and reproductive options. Am J Med Genet 2001; 106:102–14.
72. Christodoulou J. Genetic defects causing mitochondrial respiratory chain disorders and disease. Hum Reprod 2000; 15:28–43.
73. Acton BM, Jurisicova A, Jurisica I, et al. Alterations in mitochondrial membrane potential during preimplantation stages of mouse and human embryo development. Mol Hum Reprod 2004; 10: 23–32.
74. Van Blerkom J, Cox H, Davis P. Regulatory roles for mitochondria in the peri-implantation mouse blastocyst: possible origins and developmental significance of differential DeltaPsim. Reproduction 2006; 131:961–76.
75. Van Blerkom J. Mitochondria in human oogenesis and preimplantation embryogenesis: engines of metabolism, ionic regulation and developmental competence. Reproduction 2004; 128: 269–80.
76. Takai Y, Matikainen T, Jurisicova A, et al. Caspase-12 compensates for lack of caspase-2 and caspase-3 in female germ cells. Apoptosis 2007; 12:791–800.
77. Takai Y, Canning J, Perez GI, et al. Bax, caspase-2, and caspase-3. are required for ovarian follicle loss caused by 4-vinylcyclohexene diepoxide exposure of female mice in vivo. Endocrinology 2003; 144:69–74.
78. Morita Y, Perez GI, Paris F, et al. Oocyte apoptosis is suppressed by disruption of the acid sphingomyelinase gene or by sphingosine-1-phosphate therapy. Nat Med 2000; 6:1109–14.
79. Hanoux V, Pairault C, Bakalska M, et al. Caspase-2 involvement during ionizing radiation-induced oocyte death in the mouse ovary. Cell Death Differ 2007; 14:671–81.
80. Kim MR, Tilly JL. Current concepts in Bcl-2. family member regulation of female germ cell development and survival. Biochim Biophys Acta 2004; 1644:205–10.
81. Tilly JL. Commuting the death sentence: how oocytes strive to survive. Nat Rev Mol Cell Biol 2001; 2:838–48.
82. Nagai S, Mabuchi T, Hirata S, et al. Correlation of abnormal mitochondrial distribution in mouse oocytes with reduced developmental competence. Tohoku J Exp Med 2006; 210: 137–44.
83. Chao HT, Lee SY, Lee HM, et al. Repeated ovarian stimulations induce oxidative damage and mitochondrial DNA mutations in mouse ovaries. Ann N Y Acad Sci 2005; 1042:148–56.
84. Tarin JJ, Perez-Albala S, Cano A. Cellular and morphological traits of oocytes retrieved from aging mice after exogenous ovarian stimulation. Biol Reprod 2001; 65:141–50.
85. Gardner DK, Lane M, Stevens J, et al. Noninvasive assessment of human embryo nutrient consumption as a measure of developmental potential. Fertil Steril 2001; 76:1175–80.
86. Swain JE, Bormann CL, Clark SG, et al. Use of energy substrates by various stage preimplantation pig embryos produced in vivo and in vitro. Reproduction 2002; 123:253–60.
87. Lane M, Gardner DK. Amino acids and vitamins prevent culture-induced metabolic perturbations and associated loss of viability of mouse blastocysts. Hum Reprod 1998; 13:991–7.
88. Krisher RL. The effect of oocyte quality on development. J Anim Sci 2004; 82:E14–23.
89. Nutt LK, Margolis SS, Jensen M, et al. Metabolic regulation of oocyte cell death through the CaMKII-mediated phosphorylation of caspase-2. Cell 2005; 123:89–103.
90. Bergeron L, Perez GI, Macdonald G, et al. Defects in regulation of apoptosis in caspase-2-deficient mice. Genes Dev 1998; 12:1304–14.
91. Karran P. DNA double strand break repair in mammalian cells. Curr Opin Genet Dev 2000; 10:144–50.
92. Thacker J. The RAD51 gene family, genetic instability and cancer. Cancer Lett 2005; 219:125–35.
93. Kuznetsov S, Pellegrini M, Shuda K, et al. RAD51C deficiency in mice results in early prophase I arrest in males and sister chromatid separation at metaphase II in females. J Cell Biol 2007; 17:581–92.
94. Bannister LA, Pezza RJ, Donaldson JR, et al. A dominant, recombination-defective allele of Dmc1 causing male-specific sterility. PLoS Biol 2007; 5:e105.
95. Perez GI, Jurisicova A, Wise L, et al. Absence of the proapoptotic Bax protein extends fertility and alleviates age-related health complications in female mice. Proc Natl Acad Sci U S A 2007; 104:5229–34.

96. Eichenlaub-Ritter U. Genetics of oocyte ageing. Maturitas 1998; 30:143–69.
97. Baird DT, Collins J, Egozcue J, et al. Fertility and ageing. Hum Reprod Update 2005; 11:261–76.
98. Thomas NS, Ennis S, Sharp AJ, et al. Maternal sex chromosome non-disjunction: evidence for X chromosome-specific risk factors. Hum Mol Genet 2001; 10:243–50.
99. Hassold T, Hunt P. To err (meiotically) is human: the genesis of human aneuploidy. Nat Rev Genet 2001; 2:280–91.
100. Hall H, Hunt P, Hassold T. Meiosis and sex chromosome aneuploidy: how meiotic errors cause aneuploidy; how aneuploidy causes meiotic errors. Curr Opin Genet Dev 2006;16: 323–9.
101. LeMaire-Adkins R, Radke K, Hunt PA. Lack of checkpoint control at the metaphase/anaphase transition: a mechanism of meiotic nondisjunction in mammalian females. J Cell Biol 1997; 139:1611–9.
102. Kouznetsova A, Lister L, Nordenskjold M, et al. Bi-orientation of achiasmatic chromosomes in meiosis I oocytes contributes to aneuploidy in mice. Nat Genet 2007; 39:966–8.
103. Lightle SA, Oakley JI, Nikolova-Karakashian MN. Activation of sphingolipid turnover and chronic generation of ceramide and sphingosine in liver during aging. Mech Ageing Dev 2000; 120:111–25.
104. Cho W, Stahelin RV. Membrane-protein interactions in cell signaling and membrane trafficking. Annu Rev Biophys Biomol Struct 2005; 34:119–51.
105. Hetz CA, Hunn M, Rojas P, et al. Caspase-dependent initiation of apoptosis and necrosis by the Fas receptor in lymphoid cells: onset of necrosis is associated with delayed ceramide increase. J Cell Sci 2002; 115:4671–83.
106. Adam D, Heinrich M, Kabelitz D, et al. Ceramide: does it matter for T cells? Trends Immunol 2002; 23:1–4.
107. Okazaki T, Bielawska A, Bell RM, et al. Role of ceramide as a lipid mediator of 1 alpha,25-dihydroxyvitamin D3-induced HL-60 cell differentiation. J Biol Chem 1990; 265:15823–31.
108. Venable ME, Webb-Froehlich LM, Sloan EF, et al. Shift in sphingolipid metabolism leads to an accumulation of ceramide in senescence. Mech Ageing Dev 2006; 127:473–80.
109. Richardson SJ, Senikas V, Nelson JF. Follicular depletion during the menopausal transition: evidence for accelerated loss and ultimate exhaustion. J Clin Endocrinol Metab 1987; 65:1231–7.
110. Faddy MJ, Gosden RG, Gougeon A, et al. Accelerated disappearance of ovarian follicles in mid-life: implications for forecasting menopause. Hum Reprod 1992; 7:1342–6.
111. Diatlovitskaia EV, Andreasian GO, Malykh Ia N. Human ovarian ceramides and gangliosides in aging. Biokhimiia 1995; 60: 1302–6.
112. van Blitterswijk WJ, van der Luit AH, Caan W, et al. Sphingolipids related to apoptosis from the point of view of membrane structure and topology. Biochem Soc Trans 2001; 29: 819–24.
113. Hanada K, Kumagai K, Tomishige N, et al. CERT and intracellular trafficking of ceramide. Biochim Biophys Acta 2007; 1771:644–53.
114. Kawano M, Kumagai K, Nishijima M, et al. Efficient trafficking of ceramide from the endoplasmic reticulum to the Golgi apparatus requires a VAMP-associated protein-interacting FFAT motif of CERT. J Biol Chem 2006; 281:30279–88.
115. Rao RP, Yuan C, Allegood JC, et al. Ceramide transfer protein function is essential for normal oxidative stress response and lifespan. Proc Natl Acad Sci U S A 2007; 104:11364–9.

Chapter 26

Molecular Control of Corpus Luteum Function

Carlos Stocco

26.1 Introduction

Follicles and corpora lutea are the major endocrine components of the ovary. These are ephemeral structures that develop and disintegrate periodically. Follicle development begins with the recruitment of several primordial follicles. In primates, only one of these follicles will develop all the way to the preovulatory state and ovulate, whereas the rest become atresic. After the release of the oocyte at ovulation, the selected follicle differentiates into a corpus luteum. Thus, the corpus luteum can be viewed as the last differentiation step of the selected follicle. After a short period, the corpus luteum regresses, allowing a new cycle to begin. The corpus luteum's most prominent product is progesterone, which is critical for the establishment and maintenance of pregnancy. Therefore, if implantation occurs, complex mechanisms are activated to maintain luteal function during pregnancy. In this chapter, one perspective of the formation, regulation, demise, and rescue of the corpus luteum will be presented.

26.2 Luteinization

26.2.1 *Intracellular and Genomic Effects of the LH surge*

Follicle-stimulating hormone (FSH) increases estradiol production by the selected follicle and also induces the expression of the luteinizing hormone (LH) receptor (LH-R), which reaches maximal levels of expression in the granulosa cells of preovulatory follicles [1]. During the middle of the menstrual cycle, the increasing levels of estradiol cause an increase in LH (and FSH) secretion, which is known as the LH surge. In the mature preovulatory follicle, the activation of LH-R by the preovulatory LH surge induces ovulation and initiates the process of differentiation of granulosa and theca cells into luteal cells through a process termed luteinization.

The LH-R signals via membrane-associated heterotrimeric GTP-binding proteins (G-proteins), which are composed of α-, β-, and γ-subunits. The specificity of G-protein coupling to a given receptor is typically defined in terms of the class of the Gα subunit (Gs, Gi/o, Gq/11, G12). Particularly, the LH-R is coupled to the stimulatory guanine nucleotide-binding protein Gαs (Fig. 26.1), although it has been shown that the LH-R may also couple with Gi, G13, and Gq/11 [2]. The LH-R is capable of generating two major signals: one activates the adenylyl cyclase (AC) to produce cAMP, and the other activates phospholipase-C (PLC), mobilizes phosphoinositides, and increases intracellular calcium (Ca^{2+}) [3]. Gαs is responsible for activating the AC that produces cAMP, which activates the cAMP-dependent protein kinase-A (PKA) [4]. Once PKA is activated, its moves to the nucleus where phosphorylates a number of transcription factors, including the cAMP regulatory element-binding protein (CREB) [5]. CREB is phosphorylated at serine 133 by PKA [5], which allows the recruitment of coactivators such as the CREB binding protein (CBP/p300) [6].

In granulosa and luteal cells, as in all cells, cAMP is rapidly degraded by phosphodiesterases. In granulosa cells, the phosphodiesterase, PDE4D, regulates cAMP levels [7]. PDE4D knockout mice exhibit reduced ovulation rates and litter size due to the altered responsiveness of the granulosa cells to the LH surge [8]. Moreover, many follicles fail to reach the preovulatory stage due to premature luteinization, suggesting that the lack of PDE4D primarily permits accumulation of cAMP such that follicles exhibit premature luteinization.

The molecular mechanism by which LH stimulates the PLC/phosphoinositides pathway has not been clearly determined. In rat granulosa cells, it has been demonstrated that only cells with a high density of receptors respond to LH

C. Stocco (✉)
Department of Obstetrics Gynecology and Reproductive Sciences, Yale University School of Medicine, New Haven, CT, USA
e-mail: carlos.stocco@yale.edu

P.J. Chedrese (ed.), *Reproductive Endocrinology*,
DOI 10.1007/978-0-387-88186-7_26, © Springer Science+Business Media, LLC 2009

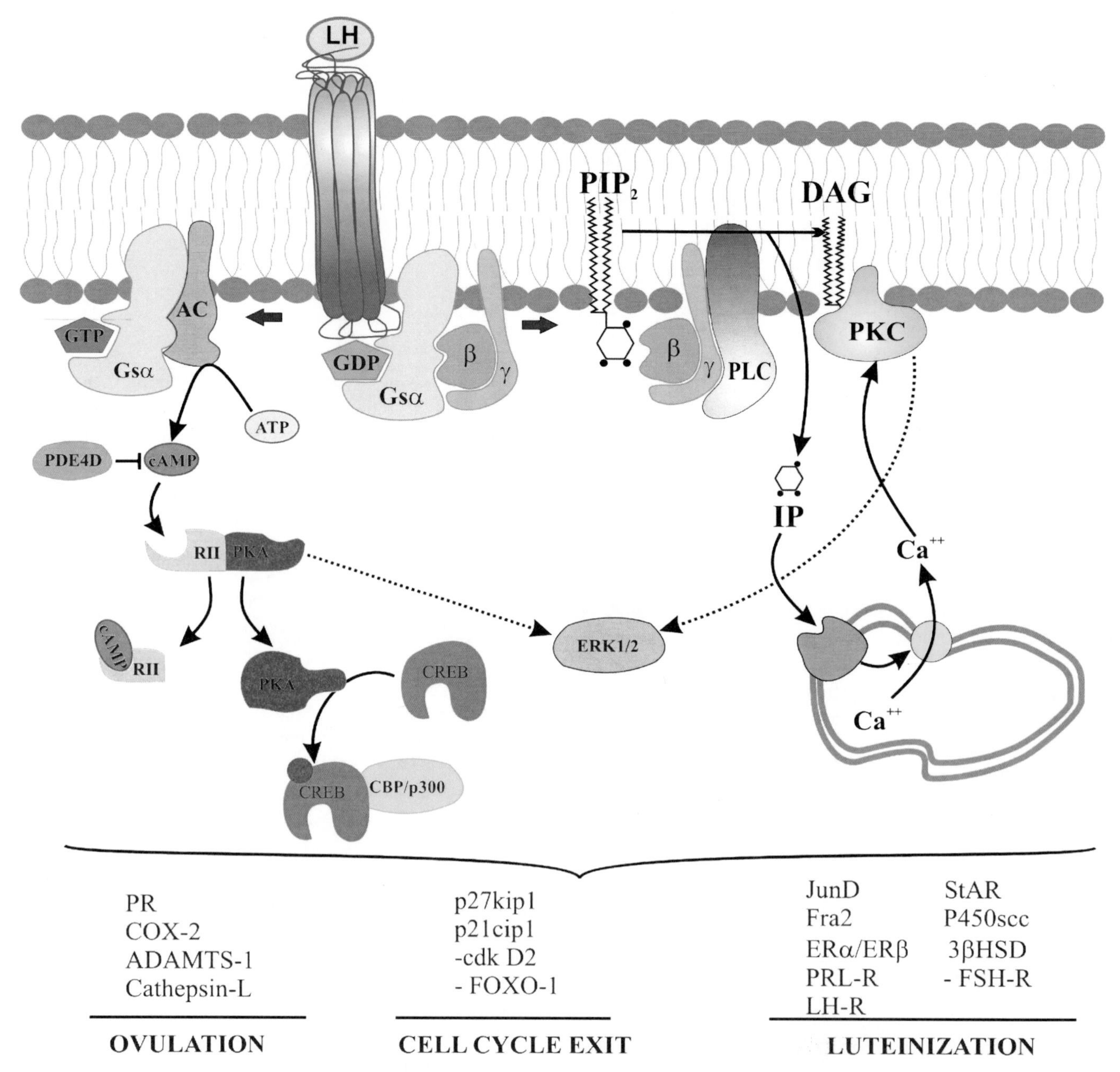

Fig. 26.1 Intracellular signaling and genomics effects of the LH surge leading to luteinization. The LH-R signals via Gαs, which activates the adenylyl cyclase (AC) to produce cAMP. cAMP activates cAMP-dependent protein kinase A (PKA). Once PKA is activated, its catalytic unit moves to the nucleus where it phosphorylates transcription factors such as cAMP regulatory element binding protein (CREB). The phorphorylation of CREB facilitates the recruitment of coactivators such as CBP or p300. Simultaneously, cAMP is rapidly degraded by the action of phosphodiesterases E4D (PDE4D). LH also stimulates the phospholipase C (PLC) pathway, probably mediated by the release of the βγ subunits after the activation of Gα. PLC-β cleaves the inositol 1,4,5-trisphosphate (InsP3) moiety from phosphatidylinositol 4, 5-bisphosphate (PIP2) with the retention of 1,2-diacylglycerol (DAG) in the membrane. InsP3 enters the cytoplasm and binds to receptors on the endoplasmic reticulum, resulting in the mobilization of intracellular calcium stores. DAG, in concert with calcium, activates PKC. In preovulatory granulosa cells, LH causes a delayed and sustained stimulation of the extracellular regulated kinase (ERK1/2 or MAPK) pathway in a PKA- and PKC-dependent manner. These signaling pathways will ultimately lead to the expression of several genes involved in ovulation, exit of the proliferative cell cycle, and luteinization

by activating the inositol phosphate cascade in addition to the cAMP [9]. This observation suggests that only granulosa cells from preovulatory follicles will respond to LH with an increase in phosphoinositides production. It has been proposed that the release of the G-protein β- and γ-subunits after the activation of Gα leads to the activation of PLC-β [10]. PLC-β cleaves the inositol 1,4,5-trisphosphate (IP3) moiety from phosphatidylinositol 4,5-bisphosphate (PIP2) with the retention of 1,2-diacylglycerol (DAG) in the membrane [2] (Fig. 26.1). IP3 enters the cytoplasm and binds to receptors on the endoplasmic reticulum, resulting in the mobilization of intracellular Ca^{2+} stores. DAG, in concert with Ca^{2+}, activates protein kinase-C (PKC) [11, 12, 13]. A role for PKC in mediating the effect of LH is supported by the fact that LH action can be mimicked by suboptimal doses of LH plus a PKC activator [14].

Luteinizing hormone also stimulates the extracellular-regulated kinase (ERK1/2 or MAPK) pathway in granulosa cells [15]. Remarkably, the effect of LH on ERK1/2 phosphorylation can be rapid and transient, or delayed and sustained. The early increase in ERK1/2 phosphorylation elicited by the activation of the LH receptor is PKA-dependent but does not depend on PKC activation [15]. This rapid and transient activation of the ERK1/2 occurs in immature granulosa cells, which express low levels of LH receptor. In preovulatory mural granulosa cells, which express high levels of LH-R, LH provokes a delayed and more sustained activation of ERK1/2 that is PKA- and PKC-dependent [16], which is an indirect mechanism because it is mediated by the release of epidermal growth-like factors (EGF) [16]. In fact, it is known that LH stimulates the expression of EGF in granulosa cells [17]. LH activates the transcription of the transcription factor Jun, Fos, and Fra [18].

Genes affected by the LH surge. LH-R activation in the preovulatory follicle results in the induction of a broad range of genes that are crucial for ovulation and differentiation of the granulosa cells [19]. Some of these genes are transiently induced, whereas others remain constitutively expressed in luteal cells. Between the transiently induced genes are progesterone receptor (PR) [20], cyclooxygenase-2 (COX-2) [21], CCAAT/Enhancer Binding Protein-β (C/EBP?) [22], and early growth response protein-1 (Egr-1) [23].

Mice deficient in PR, C/EBPβ, or COX-2 are infertile because preovulatory follicles fail to ovulate. COX-2 and PR knockout female mice do not ovulate even in response to exogenous hormones, but they do form corpora lutea containing trapped oocytes, suggesting that luteinization can occur in the absence of these molecules [24]. In contrast, C/EBPβ-null mice release fertilizable eggs in response to gonadotropin stimulation, yet luteinization does not take place [22]. These observations suggest a key role for C/EBPβ in corpus luteum formation. The genes targeted by C/EBPβ that are specifically involved in follicular luteinization are not yet known. Interestingly, once luteinization occurs, C/EBPβ is no longer expressed in the corpus luteum, suggesting that the role of this transcription factor is limited to the process of luteinization and not involved in the maintenance of luteal function.

Disruption of the Egr-1 gene also leads to infertility. In Egr-1 knockout animals infertility is due to diminished production of LH in the pituitary gland [25] and the lack of LH-R expression in the ovary [26], suggesting that Egr-1 controls not only the expression of LH but also the capacity of granulosa cells to respond to this hormone. Consequently, in the absence of Egr-1, luteinization of the granulosa cells does not occur even when animals are treated with gonadotropins [26].

26.2.2 Granulosa Cell Exit from the Proliferative Cycle

Follicle-stimulating hormone stimulates proliferation of granulosa cells, resulting in an enormous increase in follicle size. After the LH surge, granulosa cells stop dividing and exit the proliferative cycle. Luteal cells are found arrested predominantly at the G_0/G_1 phase of the cell cycle [27]. Cyclin-dependent kinases (Cdks) and several proteins that either stimulate or inhibit their activities regulate the G_1 phase of the cell cycle, governing the transition between proliferation and quiescence [28]. Progression through G_1 is controlled largely by Cdks 4/6 and 2 in association with cyclins D and E, respectively. The entry of cells into the S-phase involves the cooperation of Cdk 4/6 and D-type cyclins, D1, D2, and D3, with Cdk2/cyclin-E. These kinases phosphorylate the retinoblastoma protein (pRb), which in turn activates diverse genes required for S-phase entry and progression [29]. Accordingly, in mice and human luteal cells, only unphosphorylated pRb is found, and consistent with this, they express the Cdk inhibitor protein $p27^{Kip1}$ but not E2F-1, which is normally expressed in proliferating cells [30].

Cessation of cell proliferation during luteinization is associated with a progressive loss of positive cell cycle regulators, including cyclins and Cdk2, and with increased expression of the Cdk inhibitors $p21^{cip1}$ and $p27^{kip130}$. The LH surge silences the expression of cyclin-D2 and induces the expression of $p21^{cip1}$ very rapidly; however, its stimulatory effect on $p27^{kip1}$ is delayed by 12–24 hours [31]. Deletion of $p21^{cip1}$ causes no detectable defects in luteal cell function and fertility, whereas $p27^{kip1}$ knockout ovaries show a proliferation of luteal cells [32]. These observations suggest that $p27^{kip1}$ is a major limiting factor for the successful exit of luteal cells from the proliferative cycle during luteinization. However, luteal cells obtained from the $p27^{kip1}$ and $p21^{cip1}$-double knockout mice undergo more prolonged proliferation relative to those in $p27^{kip1}$-single knockout mice [33], suggesting that $p27^{kip1}$ and $p21^{cip1}$ synergistically cooperate in the exit of granulosa cells from the proliferative cycle.

Interestingly, p27-deficient granulosa cells complete the differentiation program and become luteal cells, and express luteal cells markers such as the cytochrome-P450 side chain cleavage (P450scc) and produce progesterone, but can do so without withdrawing from the cell cycle [32]. This ability of luteal cells to uncouple differentiation and growth arrest in the absence of p27 suggests that the exit from the proliferative cycle is not an obligatory step for the differentiation of granulosa into luteal cells.

26.2.3 Role of the Oocyte in Luteinization

Oocyte-derived regulatory molecules act within the follicle to inhibit premature luteinization and limit progesterone biosynthesis [34]. In vitro studies demonstrated that members of the transforming growth factor (TGF) superfamily, including TGF-β, activins, inhibins, bone morphogenetic proteins (BMPs), growth/differentiation factors (GDFs), and anti-Müllerian hormone [35], mediates the anti-luteinizing effect of the oocyte [36]. The TGFs signals through a serine–threonine kinase cascade that results in the cytoplasmic to nuclear translocation of intracellular effectors proteins termed "SMAD" [37]. In agreement with a role of the TGF/SMAD signaling pathway in preventing premature luteinization, in SMAD4 and activin-α conditional knockout mice granulosa cells luteinize prematurely [38]. The ovaries of these mutant animals contain a large number of corpora lutea and high levels of luteal markers and produce large amounts of progesterone [38]. This evidence suggests that SMAD4 and activin-α prevent premature luteinization of granulosa cells during follicle development.

26.2.4 Structural Changes

Luteal cell types—two types of cells can be found in the corpus luteum: large and small luteal cells [39, 40]. Noteworthy, although large and small luteal cells are the main functional components of the corpus luteum, they represent less than the 30% of the cells that form this gland. The rest is composed by pericytes, endothelial, and immune cells. In ruminants, humans, and rodents, small and large luteal cells differ in their basal rates of progesterone secretion; the large cells produce more progesterone than the small cells [39, 40]. In humans and domestic animals, LH increases secretion of progesterone from small luteal cells but not large cells [40]. In contrast, in rats, both cells respond to LH with an increase in progesterone secretion [39]. It is believed that the origin of the large luteal cells is the granulosa cells, whereas the theca cells differentiate into small luteal cells; however, although this hypothesis is well supported in domestic animals [40], this does not seem to be the case in rodents [39]. In most species, there is considerable mixing of large and small cell types during the reorganization of the follicle into the corpus luteum, leading to close contact between the two cell types. Primates are an exception with the two cell populations remaining relatively separate and therefore called granulosa-luteal cells and theca-luteal cells, respectively [41]. Thus, in primates, the two cell populations can be easily distinguished in tissue sections by their location. Furthermore, theca-luteal cells are the primary source of androgens [42], while granulosa-luteal cells are the site of estrogen synthesis [43], suggesting that the two-cell model of estrogen biosynthesis invoked to explain follicular estrogen production is preserved in the corpora lutea of primates.

Tissue remodeling and vascularization—along with the differentiation of granulosa and theca cells into large and small luteal cells, alterations at the tissue level are needed to form the vascular network of the corpus luteum. These alterations include degradation of the follicular basement membrane, the construction of the extracellular matrix (ECM), and the development of new capillaries. The formation of luteal microvasculature is essential for the transfer of the high amounts of progesterone produced by luteal cells to the systemic circulation. At the end of this process, each luteal cell is in direct contact with several capillaries, giving the corpus luteum one of the highest rates of blood flow in the organism [44].

Degradation of basement membrane—extensive angiogenesis is already present in the theca layer of preovulatory follicles as a result of the coordinated production of angiogenic factors by granulosa and theca cells. In follicles, however, blood vessels cannot cross the barrier represented by the basement membrane. After ovulation, the basement membrane between the granulosa and theca layers dissolves and thecal capillaries expand by sprouting into the avascular granulosa cell layer (Fig. 26.2).

Laminin, collagen type-IV, fibronectin, and proteoglycans are the major ECM components of the basement membrane of the follicle [45]. These components of the basement membrane are cleaved by matrix metalloproteinases (MMP) family of proteins, MMP-2, MMP-9, and MMP-19, which appear to be important during corpus luteum formation because they cleave collagen type-IV [46]. In the rat ovary, MMP-2 is present in granulosa, theca, and luteal cells, whereas MMP-9 localization is restricted to the plasma membrane of the luteal cells [47]. In mice, granulosa and theca cell expression of MMP-19 is increased five- to tenfold following activation of the LH-R [48].

Luteal extracellular matrix—the ECM consists of proteinaceous (collagen, fibronectin, and laminin) and non-proteinaceous (proteoglycans and glycoproteins) components that provide tissue-specific extracellular architecture to which granulosa and luteal cells attach. Laminin and collagen IV are scarce in the granulosa layer of follicles but become abundantly expressed in the corpora lutea of primates and rodents [49, 50]. In the rat corpus luteum, collagen type IV and laminin are detected in the granulosa layer approximately

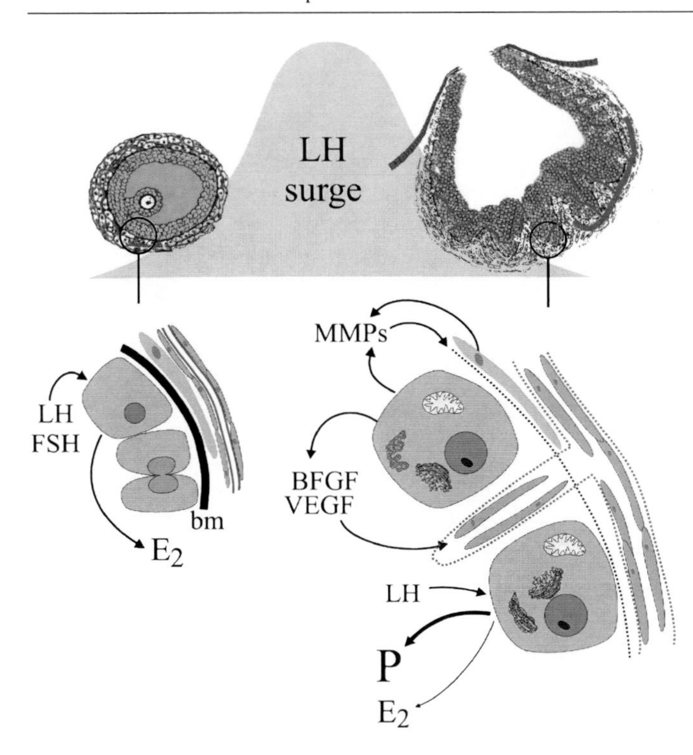

Fig. 26.2 Structural and vascular changes during luteinization. The LH surge leads to ovulation and the transformation of the ovulated follicles into the corpus luteum. In preovulatory follicles, two well-defined compartments are present: the theca layer, which is highly vascularized, and the granulosa layer, which is devoid of blood vessels. Granulosa cells proliferate under the action of FSH; however, as they differentiate into luteal cells, they exit the proliferative cell cycle. Along with the differentiation of granulosa and theca cells into luteal cells, degradation of the follicular basement membrane (bm), which allows the invasion and development of new capillaries, take place. These processes are orchestrated by luteinizing theca and granulosa cells through the production of angiogenic factors such as vascular endothelial growth factor (VEGF) and basic fibroblast growth factor (BFGF) as well as matrix metalloproteinases (MMPs), which degrade the basement membrane of the follicle. Follicles produce limited quantities of estrogens; in contrast, luteal cells produce large amounts of progesterone. In some species, luteal cells also produce estradiol

6 hours after ovulation [50, 51]. Cells interact with the ECM through cell surface receptors named integrins [52]. Integrin expression in luteal cells increases during luteinization due to the stimulatory action of LH [53, 54]. The interaction of integrins and components of ECM has an important role to play in modulating survival, growth, and steroidogenesis of luteal cells in several species. For instance, differentiation of granulosa cells into luteal cells in vitro is promoted by fibronectin or laminin, and blocked by an antibody against integrin [55]. Thus, the ECM acts not only as a "scaffold" but also modulates luteal function through the presence of cell surface receptors in the cells.

The ECM is constantly renewed, and equilibrium between synthesis and degradation of the ECM is critical. Luteal cells express not only MMP proteins but also tissue inhibitors of metalloproteinases (TIMP), which inhibit the activity of MMP. The ratio of active MMP:TIMP is important to maintain an ECM microenvironment favorable to the differentiation of follicle-derived cells into luteal cells. If the balance MMP:TIMP is altered to favor MMP activity, the normal development and function of the corpus luteum are impaired. This is, in fact, the phenotype observed in TIMP-1 knockout mice in which progesterone levels are significantly lower when compared to wild-type animals [56].

Development of blood vessels—the development of new blood vessels involves endothelial cell proliferation, expansion of the capillaries, and development of the capillary lumen (maturation). The dense capillary network formed during luteinization efficiently supplies nutrients, hormones, and lipoprotein-bound cholesterol to the luteal cells, and provides a mechanism for speedy and efficient output of progesterone from the corpus luteum. Soon after ovulation, pericytes derived from the theca compartment are the first vascular cells to invade the developing luteal parenchyma. In the rat corpus luteum, new capillaries can be found within 16 hours after ovulation [50, 51]. The molecular regulation of angiogenesis in the corpus luteum is complex, with a growing list of regulators that include vascular endothelial growth factor (VEGF), basic fibroblast growth factor (bFGF), endocrine gland-derived VEGF (EG-VEGF), and angiopoietins.

VEGF—VEGF stimulates differentiation, survival, migration, proliferation, tubulogenesis, and vascular permeability in endothelial cells. In primates and rodents, VEGF is expressed in the granulosa compartment only at the preovulatory stage; after ovulation, luteal cells continue expressing VEGF [57, 58]. In humans, VEGF expression peaks in the early luteal phase in association with the initial development of the capillary plexus [59]. Luteal VEGF is stimulated by LH, estradiol, and IGF-1 [60, 61, 62].

The importance of VEGF in the vascularization of the corpus luteum has been clearly demonstrated by using antibodies to VEGF [63], antibodies to VEGF receptors [64], or soluble truncated VEGF receptors to inhibit VEGF bioactivity [65]. Administration of these compounds to rodents before ovulation inhibits luteal angiogenesis and luteal development [64, 66]. Furthermore, when an anti-VEGF receptor antibody is injected in early-pregnant mice, blood vessels within the already formed corpus luteum regress as a consequence of the removal of endothelial cells that detach from the vascular basement membrane, suggesting that VEGF is not only required for luteinization but is also critical to maintain the functionality of luteal blood vessels during pregnancy [67].

EG-VEGF—endocrine gland-derived vascular endothelial growth factor, or prokineticin-1 (PK-1), is an angiogenic regulator expressed specifically in steroidogenic glands [68]. Similar to VEGF, EG-VEGF promotes proliferation, survival, and chemotaxis of endothelial cells; however, EG-VEGF actions are limited to endocrine tissues [69, 70]. Within the ovary, EG-VEGF is highly expressed in developing follicles and the corpus luteum [68]. Luteal expression of EG-VEGF is low in the early stage of luteal development but is upregulated during the mid-luteal to late luteal phase at a time when VEGF expression is considerably reduced [59]. The pattern of expression of VEGF and EG-VEGF supports the notion that VEGF activity is more important for the creation of the luteal capillary plexus, whereas both factors are needed to maintain the structure of the capillaries in the mid to late luteal phase.

bFGF—this was the first angiogenic factor identified in the ovary [71]. bFGF is produced by luteal steroidogenic and endothelial cells in rats [72] and humans [73]. bFGFs have been found to stimulate proliferation and motility of luteal endothelial cells [74], and treatment with antibodies to bFGF suppresses the endothelial cell proliferation activity of luteal extracts from cows, sheep, and pigs [75]; however, deletion of the bFGF gene in mice does not result in disruption or loss of fertility [76].

Angiopoietins—angiopoietins (Ang) are also critical for angiogenesis and blood vessel integrity. Ang-1 and Ang-2 bind to a common receptor named Tie-2; however, whereas Ang-1 stimulates sprouting and maturation of blood vessels, Ang-2 inhibits this effect, acting as a competitive inhibitor by binding Tie-2 without activating intracellular signaling pathways [77]. Tie-2 or Ang-1 knockout mice are embryonic lethal, with the most prominent defects involving the vasculature; however, Ang-2 knockout mice develop normally and are fertile [78], suggesting a more important role for Tie-2 and Ang-1 in angiogenesis. The Tie-2 receptor is localized exclusively in luteal endothelial cells and is highest during the early luteal phase and during luteal rescue [79]. Ang-1 is expressed uniformly in luteal and endothelial cells throughout the human corpus luteum. In contrast, Ang-2 is strongly expressed in a minority of individual luteal and endothelial cells found either singly or in clusters [79].

LH-R activation increases Ang-1 expression in human luteal cells [79] and in macaque granulosa cells [80]. Ang-1 and Ang-2 expression is low during the early to mid-luteal phase followed by a transient peak during corpus luteum regression during the late luteal phase [79, 80]. This is in contrast to the expression of VEGF, which peaks during the mid- to mid-late luteal phase before declining during the late luteal phase [81]. In rats, VEGF is abundant within the center of the developing corpus luteum, including regions where blood vessels have not yet developed [82]. In contrast, Ang-1 transcripts are associated with blood vessels and appear to follow or coincide with, rather than precede, vessel growth in the early rat corpus luteum. The pattern of Ang-2 expression, however, suggests that this factor plays an early role at the sites of vessel invasion. Initially, Ang-2 transcripts are clustered in close association with blood vessels in the theca interna of the late preovulatory follicle; after the LH surge, Ang-2 becomes abundant at the front of vessels invading the developing corpus luteum [82]. These expression patterns suggest that Ang-2 may collaborate with VEGF at the front of invading vascular sprouts by blocking the stabilizing or maturing function of Ang-1 and thus allowing vessels to remain in a plastic state where they may be more responsive to the sprouting signal provided by VEGF. After tube formation, Ang-1 recruits peri-endothelial support cells to promote vessel maturation and to maintain vessel integrity [82].

26.3 Luteal Steroidogenesis

Granulosa cells display limited steroidogenic activity, primarily converting androgens to estrogens. Luteal cells in contrast produce a large amount of progesterone. As all steroids, progesterone is synthesized from cholesterol. Therefore, regulation of the uptake and storage of cholesterol plays an integral part in luteal progesterone synthesis. Thus, during luteinization, luteal cells acquire the capacity to uptake and store cholesterol and express several proteins associated with the mobilization of intracellular cholesterol.

Cholesterol uptake—although there are three potential sources of cholesterol that could contribute to the pool needed for luteal steroidogenesis (de novo synthesis, hydrolysis of stored cholesterol esters, and exogenous lipoproteins; see the top left part of Fig. 26.3), it is well accepted that lipoproteins are the major source of cholesterol in luteal cells. Thus, luteal cholesterol is provided via either the endocytosis of cholesterol-rich low-density lipoproteins (LDL) through the LDL-receptor (LDL-R) pathway, or the selective uptake of cholesterol esters from high-density lipoproteins (HDL) through the scavenger receptor BI (SR-BI) pathway [83]. Accordingly, HDL and LDL stimulate progesterone production in human [84] and rat [85] luteal cells in culture.

Low-density lipoproteins are processed via the LDL-R, where the intact lipoprotein is internalized and then degraded in lysosomes. The cholesteryl esters delivered via this pathway are hydrolyzed by lysosomal acid lipase, and the released unesterified cholesterol is available for steroid synthesis or re-esterified into cholesteryl esters for storage in lipid droplets [86]. Unlike the LDL-R, in which the entire lipoprotein is internalized, HDL binds to SR-BI, and the core cholesterol ester is delivered to the plasma membrane without the concomitant uptake and degradation of the entire HDL particle, leaving the lipoprotein at the cell surface [87]. The mechanism by which cholesteryl esters are selectively transferred to cells by SR-BI is not fully understood, but it has been proposed that SR-BI forms a hydrophobic "channel" through which the cholesteryl esters in SR-BI-bound HDL move down a concentration gradient into the plasma membrane. The expression of SR-BI increases several folds during the development of the corpora lutea in vivo and

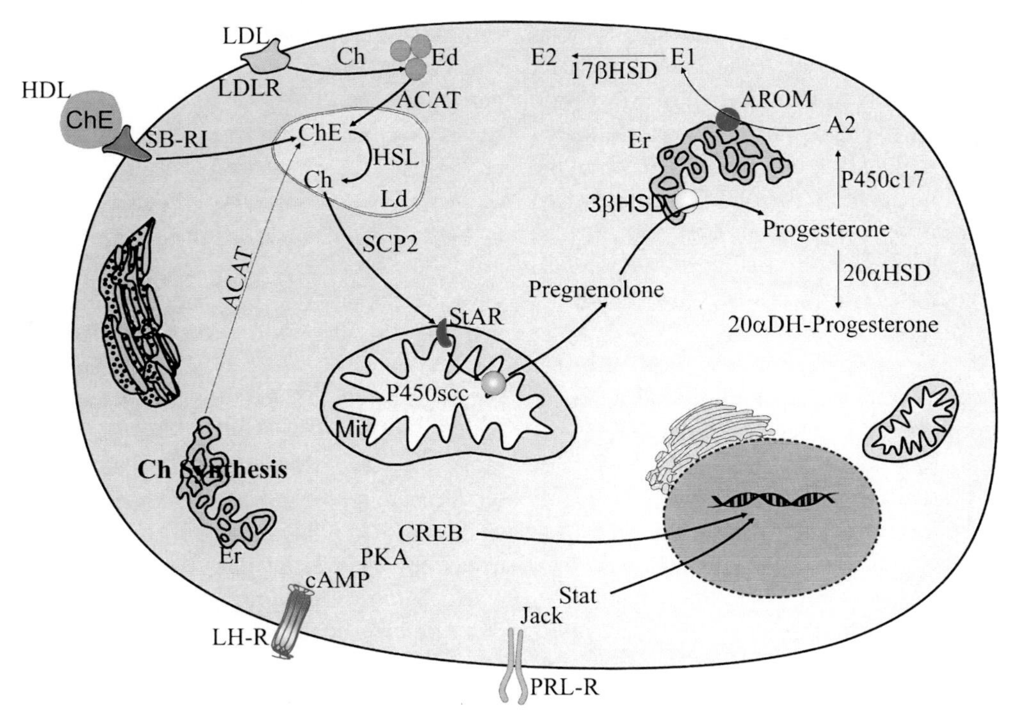

Fig. 26.3 Synthesis of progesterone, 20α-dehydro-progesterone, and estradiol in the corpus luteum. Cholesterol (Ch) needed for luteal steroidogenesis enters luteal cells through two routes: the LDLR and endosome (Ed) transfer pathway and the HDL/SR-B1/lipid droplet (Ld) transfer pathway. Lipid droplets are very extensive in luteal cells and comprise 95% cholesterol esters (ChE) due to high acyl-CoA:cholesterol acyltransferase (ACAT) activity. Ch can also be synthesized de novo in the endoplasmic reticulum (Er). Ch is transferred to the mitochondria (Mit.) by sterol carrier protein 2 (SCP2) after activation of hormone-sensitive lipase (HSL), which hydrolyzes ChE to free Ch. Steroidogenic acute activator protein (StAR) facilitates the movement of cholesterol from the outer mitochondrial membrane to the inner mitochondrial membrane. Once Ch reaches the inner mitochondrial membrane, it is converted to pregnenolone by the cytochrome P450 side-chain cleavage (P450scc). Conversion of pregnenolone into progesterone occurs in the Er catalyzed by the 3βhydroxysteroid dehydrogenase /$\Delta^5\Delta^4$ isomerase (3βHSD). In rodents, the level of progesterone secreted by the corpus luteum also depends on the expression of 20αHSD that catabolizes progesterone into the inactive progestin, 20α-dihydroprogesterone. In rodents and primates, the corpus luteum converts progestins into androstenedione (A2) by action of P45017α-hydroxylase/C17–20 lyase (P450c17). A2 is converted into estrone (E1) by the cytochrome P450 aromatase (P450arom). 17β-HSD-1/7 catalyzes the conversion of the weak estrogen E1 into estradiol (E2), which is a very active estrogen. LH and PRL stimulate progesterone synthesis at several points.

during luteinization of granulosa cells in vitro [88]. In rat luteal cells, SR-BI gene expression is stimulated by estradiol [89]. SR-BI-null mice are infertile although they synthesize normal amounts of progesterone during pseudopregnancy [90]. Abnormal oocyte maturation and embryonic development seem to be the cause of infertility in SR-BI-knockout mice. These mutant animals, however, have a reduced reserve of cholesterol in the corpus luteum, suggesting that normal ovarian lipid stores are not essential for the production of adequate amounts of progesterone. It is also possible that endogenous cholesterol synthesis or SR-BI-independent pathways of cholesterol uptake increase and compensate for the absence of SR-BI. In fact, luteal cells are able to produce cholesterol de novo from acetyl coenzyme A [91].

Cholesteryl esters derived by the HDL/SR-BI pathway must be hydrolyzed by ester hydrolases before the unesterified cholesterol can be transferred to mitochondria for steroid production. This step is catalyzed by hormone-sensitive lipase (HSL), which is a multifunctional enzyme that hydrolyzes triacylglycerol and cholesterol esters. HSL is expressed in the rat corpus luteum and is regulated at both the mRNA expression and mRNA translation levels by LH and PRL [92]. HSL knockout males are sterile because of oligospermia [93]; however, the effect of this mutation of ovarian function has not been examined. In contrast, the enzyme hepatic lipase (HL), which facilitates selective uptake and mobilization of cholesterol of HDL [94], is necessary for optimal progesterone production in luteal cells [95]. HL knockout mice become pregnant; however, these mutants have fewer corpora lutea, lower levels of progesterone, and smaller litters.

Intracellular cholesterol mobilization—due to its hydrophobic nature, cholesterol cannot freely diffuse in the cytosol and reach the mitochondria where the first enzymatic step in the synthesis of progesterone takes place. Several proteins have been shown to participate in the intracellular movement of cholesterol in steroidogenic cells. Among these proteins, sterol carrier protein-2 (SCP-2) is considered to play a major role in the cholesterol mobilization within the cytoplasm [96]. SCP-2 is increased in luteal cells by LH [97] and estradiol [98]. Moreover, SCP-2 expression during pregnancy closely parallels progesterone production by the corpus luteum [98]. Once cholesterol reaches the outer mitochondrial membrane, cholesterol is transported to the inner mitochondrial membrane through the aqueous intermembrane space by the action of at least two proteins, the steroidogenic acute regulatory protein (StAR) and the peripheral-type benzodiazepine receptor (PBR).

Steroidogenic acute regulatory protein facilitates the movement of cholesterol from the outer mitochondrial membrane to the inner mitochondrial membrane, but the mechanism of action of StAR is not fully understood, and at least four models of StAR's action have been proposed [99]. In follicles, StAR expression is restricted to theca cells; after ovulation, this protein becomes highly expressed in human [100] and rat [101] luteal cells. In these cells, StAR expression increases rapidly after LH administration [102]. In StAR-null mice, ovaries appear normal at birth; however, after puberty, ovaries exhibit impaired follicular maturation and contain abundant lipid deposits primarily in ovarian stromal cells [103]. These alterations lead to infertility and premature ovarian failure.

The PBR is a high-affinity cholesterol-binding protein found in between outer and inner mitochondrial membranes [104]. In this location, PBR could function as a pore, allowing for the translocation of cholesterol from the outer to the inner mitochondrial membrane [105]. PBR is expressed ubiquitously but is most abundant in steroidogenic cells, including luteal cells [106]; however, since PBR-knockout mice are embryonic lethal [107], the precise role of PBR in ovarian steroidogenesis remains unclear.

Progesterone biosynthesis—once cholesterol reaches the inner mitochondrial membrane, the transformation of cholesterol into steroid hormones begins. The synthesis of progesterone is the most simple of all the steroidogenic pathways, and only two enzymes are involved (Fig. 26.3). The first modification to the cholesterol molecule occurs in the inner mitochondrial membrane where the molecule is converted to pregnenolone by the cytochrome P450 side-chain cleavage (P450scc). Conversion of pregnenolone into progesterone occurs in the plasma reticulum and is catalyzed by the 3βhydroxysteroid dehydrogenase/$\Delta^5\Delta^4$ isomerase (3βHSD). The expression of both enzymes [108] and of the organelles that house them [109] increases during corpus luteum formation, allowing luteal cells to synthesize large amounts of progesterone. Both enzymes remain highly expressed in the corpus luteum throughout pregnancy [110].

In rodents, the level of progesterone secreted by the corpus luteum does not depend only on the amount of progesterone synthesized by the luteal cells but also on the expression of the enzyme 20αHSD that catabolizes progesterone into the inactive progestin, 20α-dihydroprogesterone (20α-DHP). Once 20αHSD becomes expressed in the corpus luteum, progesterone secretion drops, and 20α-DHP becomes the major steroid secreted by luteal cells [111]. Due to the detrimental effect of 20αHSD on luteal progesterone secretion, little or no 20αHSD activity is found in the rat corpus luteum throughout pregnancy; however, elevated activity is found just before parturition concomitant with the rapid decrease in the concentration of progesterone in serum [112]. In rodents, this decrease in serum progesterone is essential for parturition to occur. Accordingly, mice deficient in 20αHSD sustain high progesterone levels and display a delay of several days in parturition [113]. Whereas 20αHSD is one of the most important players in the regulation of luteal progesterone secretion in rodents, the role of this enzyme in the control of progesterone production and/or action in other species needs further investigation.

Biosynthesis of androgens and estradiol by the corpus luteum—in rodents as in primates, the corpus luteum produces significant quantities of androgens and estrogens. The conversion of progestins into androgens is mediated by the enzyme P45017α-hydroxylase/C17–20 lyase (P450c17 or CYP17). Although P450c17 is expressed in the corpora lutea of primates and rodents, the distribution of this enzyme differs between species; thus, while rat large and small luteal cells express P450c17 [39], in humans only theca-derived small luteal cells produce androgens [114]. Androgens are converted into estrogens by action of the cytochrome P450 aromatase, which is expressed in the corpora lutea of primates and rodents [114, 115].

26.4 Hormonal Regulation

One of the most important changes that take place during luteinization is the alteration in the cellular responsiveness to external signals that allows luteal cells to respond to a new set of hormones. In general, corpus luteum function is controlled by the interaction of tropic hormones secreted by the pituitary, the corpus luteum itself, the decidua, and the placenta; however, the kind of hormones secreted by these tissues and their combination differ between species.

26.4.1 Role of Luteinizing Hormone

In humans and non-human primates, LH is essential and sufficient for the stimulation and maintenance of luteal function [116]. In monkeys treated with a gonadotropin-releasing hormone (GnRH) antagonist (which blocks pituitary LH secretion), there is a rapid and sustained fall in serum progesterone production by the corpus luteum [117]. During the luteal phase, LH is secreted from the pituitary in a pulsatile manner due to the intermittent secretion of GnRH by the hypothalamus [118]. During the early luteal phase, pulses of low amplitude and high frequency have been observed, whereas by the mid-luteal phase, LH pulses are less frequent but of greater amplitude [118]. A progesterone peak occurs after each LH pulse during the mid-late luteal phase [119].

In primates, the steroidogenic actions of LH are directed primarily at the delivery of cholesterol to the mitochondrial cholesterol side-chain cleavage system by increasing the expression of StAR [120] and by stimulating the expression of SR-BI [121]. LH also stimulates the expression of 3βHSD, aromatase, estrogen receptor β [122], and genes involved in tissue remodeling [123] and in the maintenance of luteal vasculature [124].

In rodents, LH has been shown to increase SR-BI and HDL uptake [125] and to stimulate the synthesis of androgens [126]. Because LH is essential for follicular maturation and ovulation [127], knockout animals for LH or the LH receptor have not provided information regarding the role of LH in corpus luteum function.

Luteinizing hormone receptor—in humans, the LH-R is highly expressed by luteal cells [128]; however, the stimulus that maintains the expression of this receptor is not known. In contrast, it is clear that in rodents the upregulation of the LH receptor in the corpus luteum is due to prolactin (PRL) [129]. The preovulatory LH surge causes activation first and thereafter a transient desensitization of the LH receptor. A marked downregulation of LH-R protein and mRNA follows desensitization [130]. A model for such desensitization involving ADP-ribosylation factor 6 (ARF6) and arrestin 2 was recently proposed [131]. The downregulation of LH-R mRNA that occurs under these conditions is not due to decreased transcription of the LH-R gene but rather to increased degradation of LH-R mRNA [132]. Menon and colleagues have identified and purified a rat ovarian protein, designated LH-R binding protein (LRBP), that binds to a region of the open reading frame of the rat LH receptor mRNA and enhances its degradation [133]. The LH-R is also activated by human chorionic gonadotropin (hCG). In the event of fertilization, hCG secreted by the syncytiotrophoblast cells of the placenta stimulates the LH receptor in the corpus luteum, resulting in the production of progesterone essential for the maintenance of pregnancy (see below, Rescue of corpus luteum function during pregnancy).

26.4.2 Role of Prolactin (PRL)

In rodents, LH also enhances progesterone synthesis, but is not sufficient to sustain steroidogenesis by itself and depends on the previous exposure of the corpus luteum to PRL [134]. Rats, mice, hamsters, and other small rodents have a short, incomplete estrous cycle that lacks a spontaneous luteal phase. In the absence of mating, the corpora lutea of these animals are short-lived and non-functional. The activation of the luteal phase in these species depends on the neuroendocrine reflex provided by the male at the time of mating, which initiates biphasic surges of PRL secretion from the pituitary [135].

In the rat corpus luteum, PRL increases LH receptor expression [136] and also facilitates the luteotrophic effect of estradiol (see below) by stimulating estrogen receptor expression [137]. PRL not only enables luteal cells to respond to the luteotropic stimuli of LH and estradiol but also indirectly stimulates progesterone production by increasing the uptake of cholesterol. Thus, in the rat corpus luteum, PRL stimulates lipoprotein binding and enhances cholesterol uptake [85]. PRL also regulates the expression of P450scc and 3βHSD [138, 139] and enhances the expression of the two enzymes involved in estradiol biosynthesis: aromatase [115] and 17βHSD type-7 [140]. PRL also has a protective role in luteal cells via the stimulation of superoxide dismutase, an enzyme that scavenges free radicals [141], and the inhibition of annexin-5 expression, a protein associated with cell death [142]. However, one of the main luteotrophic actions of PRL is the inhibition of the 20αHSD enzyme [143]. In PRL and PRL-receptor (PRL-R) null mice, ovulation, fertilization, and corpus luteum formation occur normally; however, the corpus luteum fails to organize appropriately and degenerates rapidly [144]. One of the major defects found in the PRL-R null mice is the premature expression of luteal 20αHSD [144], leading to decreased secretion of progesterone and involution of the corpus luteum. As a consequence, PRL and PRL-R null mice are sterile because uterine decidualization and embryo implantation fail to take place.

In rodents, PRL is secreted from the pituitary only until mid-pregnancy [134]. After mid-pregnancy, the decidua and the placenta secrete PRL and placental lactogens (PLs) I and II [145]. Both PL-I and PL-II bind to the PRL receptor and sustain luteal function until parturition [145]. Noteworthy, human decidua cells also secrete PRL [146]; however, although human decidual PRL is crucial for implantation, it does not affect luteal function [147].

PRL receptors—in the rat, the PRL-R is expressed in two variant forms, long (PRL-RL) and short (PRL-RS), both of which increase during luteinization [148]. The PRL-R does not have tyrosine kinase activity, but it is associated with a protein kinase named Jak2, which is activated by phosphorylation upon PRL binding and phosphorylates the transcription factors Stat5a and Stat5b [149]. Stat5b is the predominant isoform in luteal cells, and its expression increases during luteinization [148]. Stat5b is essential for luteal function since Stat5b-deficient mice abort beyond day

7 of pregnancy, a phenomenon that can be partially prevented by treatment with progesterone [150]. Because abortion occurs after day 7 of pregnancy, it appears that the major role of Stat5b is not in luteinization but rather in corpus luteum maintenance. In contrast, Stat5a/b double mutant mice have normal developing follicles and ovulation, yet few corpora lutea are evident in their ovaries, indicating that both Stat5 molecules are necessary for the normal development and survival of the corpus luteum [151]. Stat5a/b deletion leads to extensive expression of the progesterone-metabolizing enzyme 20αHSD [151], which explains the low circulating levels of progesterone found in these animals. This deficiency is partially corrected in the double knockout for 20αHSD and Stat5b [113], supporting the concept that activation of Stat5b is important in suppressing 20αHSD gene expression.

26.4.3 Role of Estradiol

In rodents, estradiol is a potent luteotropic hormone that stimulates progesterone biosynthesis, vascularization, and hypertrophy of the corpus luteum [152]. In fact, it has been demonstrated that estradiol mediates LH actions on the rat corpus luteum and that estradiol can maintain the corpus luteum in the absence of LH [153, 154]. Thus, the main luteotropic factors in the rat are PRL and estradiol, and these two hormones synergize in the stimulation of progesterone secretion.

Estradiol induces a remarkable proliferation of vascular endothelial cells apparently due to the stimulation of VEGF and VEGF receptor expression [63]. Estradiol also stimulates overall protein synthesis [155] and enhances cholesterol supply in luteal cells by stimulating cholesterol synthesis [156], uptake of cholesterol from circulation [157], and intracellular cholesterol mobilization [98]. These effects of estradiol are due, at least in part, to transcriptional stimulation of the SR-BI gene [158]. In rabbits, estradiol is the main luteotrophic hormone, and has been linked with elevated production of StAR and with cholesterol accumulation [159].

In contrast, in primates, estradiol has been related to the process of involution of the corpus luteum. For instance, intraovarian administration of estradiol induces premature functional luteolysis in monkeys [160]. However, whether estradiol acts directly in luteal cells or indirectly by reducing pituitary LH release and therefore depriving the corpus luteum of adequate gonadotropin support is still unclear.

Estrogen receptor—estrogen receptors are expressed in two distinct forms: ERα and ERβ. ERβ mRNA and protein are present in the corpora lutea of monkeys and humans [122, 161]. In the monkey, ERβ expression is low during the first half of the luteal phase and increases at mid-late luteal phase at the onset of luteal regression [122]. Moreover, progesterone seems to inhibit ERβ expression in the primate corpus luteum [122]. This inhibitory effect of progesterone and the pattern of ERβ expression support a role for locally produced estrogens in functional luteolysis in primates.

The rodent corpus luteum expresses both estrogen receptors [162], although ERα is found at levels tenfold higher than ERβ [163]. In contrast, in granulosa cells ERβ is the predominant receptor [164]. Granulosa ERβ expression is downregulated by the LH surge and can be mimicked in vivo and in vitro by activation of the LH receptor [164]. Generation of mice lacking ERα or ERβ has demonstrated that while estradiol-17β (estradiol) is critical for corpus luteum formation and maintenance, estradiol can act through either receptor to carry out its actions. Although ERβ or ERα knockout mice present severe fertility problems, both show normal corpus luteum development [165]. In contrast, aromatase knockout and ERαβ double-knockout mice do not form corpora lutea [166]. Taken together, these findings indicate that estradiol is indeed needed for corpus luteum formation and maintenance in rodents and that estradiol likely acts non-selectively through ERα or ERβ.

26.4.4 Role of Progesterone

A number of findings suggest that progesterone can promote its own secretion in many species; however, the cellular mechanisms and specific actions of progesterone in the corpus luteum remain controversial. In humans, it has been suggested that progesterone may contribute to maintaining luteal function during early pregnancy [167]. Moreover, progesterone depletion causes premature functional and structural degeneration of the macaque corpus luteum [168]. In the rat, progesterone also has luteotrophic effects. For instance, progesterone seems to stimulate its own production in vivo [169] and in vitro [170]. Progesterone also appears to participate in the downregulation of 20αHSD [171] and to protect the rat corpus luteum from cell death [172].

Progesterone receptors (PR)—the corpus luteum of most species, including humans and ruminants, with the exception of rodents, expresses PR. PR can be expressed in two major isoforms: PR-A and PR-B. PR-A is a truncated protein lacking the first 164 amino acids from the N-terminal domain; otherwise, the two PRs have an identical amino acid sequence. PR expression is rapidly induced by LH in preovulatory follicles and in cultured granulosa cells of several species [173, 174]. In primates, both PR-A and PR-B remain expressed in the corpus luteum, where PR-B is the predominant isoform [175]. In contrast, in mice and rats, PR expression decreases rapidly after ovulation, and the receptor is not expressed in luteal cells throughout pregnancy [176].

In rodents, the induction of PR expression in granulosa cells is a central event in ovulation; however, the role of PR in luteinization is not clear. This is because PR-null mice treated with gonadotropin are able to form corpora lutea, which contain trapped oocytes [24]. Nevertheless, the absence of PR does not preclude progesterone action via other pathways. For instance, Cai [177] established the expression of membrane PR in the rat corpus luteum. The authors demonstrated that rat luteal membranes contain proteins that bind progesterone with high affinity and specificity and that these proteins localize exclusively on luteal microsomal fractions. However, the function and signaling of membrane PR in luteal cell formation and function remain unknown.

26.5 Luteolysis

Luteolysis is defined as the process whereby the corpus luteum ceases to function, i.e., to secrete progesterone. Luteolysis plays an important role in reproduction because progesterone suppresses FSH and LH secretion, thus preventing the development of new preovulatory follicles. Therefore, luteolysis terminates the female reproductive cycle by decreasing progesterone secretion and allowing the continuation of follicular development. In fact, luteolysis appears to have evolved as a mechanism to increase reproductive efficiency; thus, if the female does not conceive following ovulation, removal of the corpus luteum permits a new ovarian cycle to begin. In this way, after a relatively short interval of time, a new opportunity is provided for the female to conceive. In rodents, as in many other species in which the corpus luteum in the only source of progesterone during pregnancy, luteolysis plays an even more important role because circulating progesterone must fall to allow parturition to occur. During normal luteolysis, two closely related events occur. First, the capacity to synthesize and secrete progesterone is lost (functional regression) followed by the death of luteal cells and the loss of vascular integrity of the gland (structural regression).

26.5.1 Factors Involved in Functional Regression

Prostaglandin F2α—early studies demonstrated that hysterectomy in cyclic guinea pigs causes abnormal persistence of the corpus luteum. Similar effects were subsequently observed in sheep, cows, pigs, mares, hamsters, rabbits, and rats [178]. The factor that mediates the luteolytic effect of the uterus was identified as prostaglandin F2α (PGF2α) [179]. The role of PGF2α in the demise of corpus luteum function at the end of pregnancy in rodents was clearly demonstrated in mice lacking the PGF2α receptor. In these mutant animals, luteal progesterone production at the end of pregnancy is extended, and consequently, parturition does not occur [180]. A similar phenotype has been observed in mice defective in enzymes involved in prostaglandin biosynthesis such as cytosolic phospholipase A2 (cPLA2), which catalyzes the formation of arachidonic acid from membrane phospholipids, and COX-1, which allows the conversion of arachidonic acid to prostaglandins [181].

Although not discussed here, it is well established that neurohypophysial oxytocin stimulates the release of PGF2α from the endometrium in large domestic ruminants. PGF2α in turn increases ovarian oxytocin secretion, forming a feed-forward loop that leads to luteolysis and parturition [178]. Oxytocin is mainly associated with the initiation of parturition, especially by facilitating uterine contraction. Oxytocin, however, is also synthesized by the corpora lutea of several species, including humans, ruminants, and rodents, and may contribute to the progress of luteolysis by stimulating the production of PGF2α in the ovary [178].

In contrast to the clear role of the uterus in the initiation of luteolysis in ruminants and rodents, the uterus has no effect on the lifespan of the corpus luteum in primates. Luteal regression in primates seems to be due to a reduction in the responsiveness of the aging corpus luteum to LH [116, 182]. In fact, luteal lifespan in monkeys is prolonged beyond the expected time of spontaneous luteal regression by treatment with exponentially increasing doses of LH [182] but not by maintaining mid-cycle LH levels. A second hypothesis suggests that luteolysis is initiated by autocrine and/or paracrine mechanisms [178, 183]. In this case, factors synthesized within the ovary or the corpus luteum may initiate luteolysis near the end of the menstrual cycle. Estradiol, PGF2α, and oxytocin have been proposed to act as local luteolytic agents in the primate corpus luteum.

In monkeys, estradiol injected into the ovary containing the corpus luteum induces a premature decline in circulating progesterone [184]; however, no changes in progesterone levels occur if estradiol is injected into the contralateral ovary. Moreover, estrogen receptor-β expression increases toward the end of the luteal phase [122]. Similarly, PGF2α induces luteal regression when given as a single injection directly into the human corpus luteum but not when injected into the adjacent stroma [185]. Moreover, in monkeys and humans, the secretion of PGF2α increases toward the end of the menstrual cycle only in the ovary bearing the corpus luteum [186]. The PGF2α receptor is expressed in primate luteal cells [187], and PGF2α inhibits the stimulatory effect of LH on progesterone secretion in the corpora lutea of mid- and late luteal phases but not in earlier stages [188]. The stage-dependent resistance to PGF2α-induced luteolysis is common among farm animals, but the mechanism responsible for this is not clear [189]. Oxytocin injected into the human corpus luteum causes a fall in serum progesterone and increases PGF2α production [190]. Moreover, the oxytocin-induced decrease in serum progesterone can be prevented with an inhibitor of prostaglandin biosynthesis [190]. Similar to PGF2α, the effect of oxytocin on the corpus luteum depends on the age of the gland [191]. Human luteal cells produce oxytocin [192] and possess oxytocin receptors [192]. These findings suggest that ovarian and/or neural oxytocin signals may induce local production of PGF2α, which explains why luteolysis can proceed in the absence of the uterus in primates.

Prostaglandin F2α signaling—in luteal cells, PGF2α signals through a Gq coupled receptor that has been detected in the corpora lutea of multiple mammalian species [193]. Activation of this receptor causes PLC-mediated generation of inositol triphosphate (InsP3) and diacylglycerol (DAG) followed by an increase in intracellular calcium and PKC activity [194]. In addition to PLC, there is evidence that in luteal cells PGF2α activates the phospholipase D pathway, producing phosphatidic acid in addition to DAG [195] and the MAPK signaling cascade [196, 197].

Effect of PGF2α on luteal steroidogenesis—in rodents, one of the main actions of PGF2α is to stimulate the inactivation of progesterone. This effect is mediated by the induction of the 20αHSD enzyme [198]. Consequently, at the end of pregnancy, the rat corpus luteum principally secretes 20αDH-progesterone instead of progesterone. PGF2α administration to pregnant rats induces a rapid increase in the expression of the 20α-HSD gene [199]. Moreover, the massive expression of the 20αHSD gene at the end of pregnancy does not take place in mice lacking the PGF2α receptor [199]. The stimulatory effect of PGF2α on the expression of the 20α-HSD genes is mediated by the activation of the transcription factor JunD, which in turn induces the expression of Nur77 [196].

Prostaglandin F2α also reduces cholesterol transport in the rat ovary by decreasing SCP-2 [200] and StAR expression [201]. Inhibition of StAR by PGF2α has been associated with increased expression of DAX-1 [202], c-Fos [203], and ying yang-1 [204] transcription factors. Administration of PGF2α on day 19 of pregnancy also reduces 3βHSD activity [170]. However, this inhibitory effect of PGF2α does not appear to be at the level of gene expression [205]. The physiological significance of the downregulation of protein involved in cholesterol transport and processing is not clear since the drop in progesterone secretion at the end of pregnancy in rodents is not due to the decrease in progesterone biosynthesis but rather to the catabolism of progesterone.

Anti-luteotrophic effect of PGF2α—in rodents, PGF2α prevents both LH and PRL stimulation of progesterone biosynthesis. The anti-LH action of PGF2α involves two interrelated actions, the blockage of LH-induced cAMP accumulation and the inhibition of the response of the luteal cells to cAMP [206]. PGF2α also inhibits the expression of the PRL receptors and curtails PRL signaling through the Jak/Stat pathway [207, 208]. Thus, PGF2α stimulates the expression of suppressors of cytokine signaling-3 (SOCS-3) that inhibits the Jak/Stat signaling pathway and prevents PRL-induced Stat5 activation [208]. PGF2α also curtails progesterone synthesis by suppressing estradiol production by inhibiting luteal aromatase expression [209]. Accordingly, the normal decrease in luteal aromatase expression at the end of pregnancy does not occur in PGF2α receptor-knockout mice [209].

PGF2α and Endothelin-1—endothelins (ET-1, ET-2, and ET-3) constitute a family of peptides derived from distinct genes produced mainly by endothelial cells. Particularly, ET-1 is secreted by endothelial cells and acts locally in a paracrine/autocrine manner. The corpus luteum is a highly vascular tissue and 50% of the cells in the mature gland are endothelial cells; therefore, it is an important site of ET-1 synthesis. Luteal ET-1 levels are hormonally regulated throughout the reproductive cycle with a marked increase toward the initiation of luteolysis [210]. Moreover, PGF2α increases endothelin-1 (ET-1) production from endothelial cells in the corpus luteum [210]. Elevation of ET-1 during luteolysis participates in terminating progesterone production by the corpus luteum by inhibiting basal and LH-stimulated progesterone biosynthesis [211].

Role of reactive oxygen species—reactive oxygen species (ROS) are the partially-reduced products of oxygen produced in all aerobic cells. The principal ROS are the superoxide anion (O_2^-), hydrogen peroxide (H_2O_2), hypochlorous acid (HClO), the hydroxyl radical (OH), lipid hydroperoxides, singlet oxygen (O_2^1), and the peroxynitrite radical (NOO^-). These compounds can be beneficial to organisms as cell regulators, but in high doses they become cytotoxic, often leading to cell death. There are several mechanisms of protection against ROS. These mechanisms include specific degradative enzymes such as catalase and superoxide dismutase (SOD) that eliminate H_2O_2 and O_2^-, respectively, and antioxidant vitamins such as vitamins C and E. In particular, luteal cells contain conspicuously high levels of lutein, a lipid-soluble catotenoid similar to vitamin E.

Prostglandin F2α-induced and naturally occurring functional luteal regression is associated with accumulation of ROS and/or with a decrease in protective enzymes and antioxidant vitamins [212, 213]. PGF2α treatment rapidly depletes luteal antioxidants such as vitamin C [214], increases superoxide radicals [215, 216], and downregulates genes involved in the elimination of free radicals such as catalase and SOD [212]. Although PGF2α has been shown to stimulate production of ROS by rat luteal cells in vitro [217] and to cause an increase in hydrogen peroxide in the regressing rat corpus luteum in vivo [218], immune cells seem to be the main source of ROS during luteolysis. For instance, neutrophil activation blocks luteal function in a catalase-sensitive manner [219]. Non-steroidogenic cells, most probably leukocytes, produce ROS in response to PGF2α via a PKC-activated pathway [215]. Moreover, activated neutrophils have been shown to inhibit gonadotropin action and progesterone synthesis by producing ROS [219]. On the other hand, progesterone inhibits O_2^- production by mononuclear phagocytes in pseudopregnant rats [220]. During regression, phagocytes are activated to produce O_2^- in the corpus luteum, but the mechanism by which this occurs is unclear. Low concentrations of progesterone or PGF2α could be some of the factors that contribute to the activation of phagocytes [220].

In the corpus luteum, generation of ROS causes the invasion of leukocytes [216] and the stimulation of ROS scavengers, such as superoxide dismutase and catalase, prevents the invasion of leukocytes and the decrease in serum progesterone induced by PGF2α [216]. The finding that pretreatment with either superoxide dismutase or catalase, which are large molecules that do not easily penetrate cell membranes, prevents luteal regression triggered by PGF2α suggests that neutrophils are the major source of ROS during functional luteal regression [216].

26.5.2 Structural Regression

The structural regression of the corpus luteum is grossly characterized by a decrease in the size and weight of the gland, which eventually becomes a scar within the ovarian stroma known as corpus albicans. The corpus albicans is eventually reabsorbed and replaced by ovarian stroma. The major event that causes the structural regression of the corpus luteum is death by apoptosis of luteal and vascular cells.

There are two major apoptotic signaling pathways: the death-receptor-mediated, or extrinsic, pathway and the mitochondrial, or intrinsic, pathway [221]. In the extrinsic pathway, the signal is provided by the interaction between a ligand and "death receptors," including Fas and tumor necrosis factor receptor (TNFR-3, TNFR-4, and TNF-5). The ligand for Fas, FasL, initiates the activation of initiator caspases (e.g., caspase-8). The intrinsic apoptotic signaling cascade, on the other hand, is activated by stress stimuli that change the mitochondrial permeability, leading to the release of cytochrome c, which binds to Apaf-1 (apoptotic protease-activating factor) and procaspase-9 to form a complex named the apoptosome leading to caspase-9 activation. Both pathways lead to activation of the final effectors of the system or executioner caspases such as caspase-3, caspase-6, and caspase-7. These caspases cleave a variety of intracellular polypeptides, including major structural elements of the cytoplasm such as actin, components of the DNA repair machinery such as poly (ADP-ribose) polymerase (PARP), a number of protein kinases, and the inhibitor of caspase-activated DNase (ICAD) among others [222]. Apoptosis is tightly regulated by members of the Bcl-2 family of proteins, which can either block (e.g., Bcl-2, Bcl-xL, Mcl-1) or promote (e.g., Bad, Bax, Bak, Bid, Puma) cell death mainly by regulating the permeability of the mitochondria [223].

Caspase-3 has been localized in the corpora lutea of rats [224] and mice [225]. The importance of caspase-3 as a mediator of apoptosis in luteal regression has been demonstrated by studies performed with caspase-3 null mice. The corpora lutea of these animals show attenuated rates of apoptosis and delay in the process of involution [225]; yet the corpora lutea of caspase-3 null mice finally involute, indicating that caspase-3 is not the sole factor leading to cell death in this gland.

In rats, PGF2α-induced luteal apoptosis is associated with the upregulation of caspase-9, caspase-8, and caspase-3 activities as well as an increase in the expression of Bax and FasL [226]. In this system, a caspase-8 inhibitor, but not an inhibitor of caspase-9, completely prevents the activation of caspase-3 induced by PGF2α. This observation suggests a greater importance of the extrinsic pathway in mediating PGF2α-induced apoptosis of rat luteal cells.

Role of immune cells—immune cells are normal constituents of the ovary and serve essential roles in the regulation of luteal function [227]. The number of macrophages and neutrophils present in the theca increases five- to eight-fold during ovulation [228, 229] and migrate into the forming corpora lutea [230]. Immune cells also increases during luteolysis in a variety of species, including rodents [231] and humans [229], suggesting a fundamental role, or their cytokine products, in the luteolytic process. PGF2α induces leukocyte accumulation in the rat corpus luteum [216]. Moreover, lymphocytes are immunohistochemically detected only in discrete regions of the corpus luteum where apoptosis occurs [232]. Macrophage accumulation in the regressing corpus luteum is believed to be in response to chemotactic factors such as monocyte chemoattractant protein-1 (MCP-1). MCP-1 has been identified in luteal cells [233] and its expression increases in the rat corpus luteum close to luteolysis [234]. Macrophages participate in luteal regression by phagocytosis of apoptotic luteal cells and cell remnants [235] and by secreting oxygen radicals [236].

Role of cytokines—macrophages and lymphocytes also promote luteolysis through the secretion of cytokines such as tumor necrosis factor-α (TNF)-α, interferon γ (IFN-γ), and Fas.

Tumor necrosis factor-α is a pleiotropic cytokine known to initiate apoptosis. In human and rodent corpora lutea, TNF-α can be immunologically localized in macrophages and luteal cells, and its expression increases during luteolysis [237]. TNF-α treatment increases DNA fragmentation in luteal cells [237] and in luteal vascular endothelial cells [238]. Moreover, TNF-α inhibits gonadotropin-supported progesterone accumulation [239] and reduces StAR expression [240] in rat luteal cells.

Interferon γ in combination with TNF-α reduces progesterone production in mouse luteal cells [241]. IFN-γ can be detected in luteal tissue collected around natural luteolysis and after induced luteolysis [242]. Since IFN-γ upregulates TNF-α receptors in a variety of cell types [243], luteal cells may become more sensitive to the action of TNF-α in the presence of IFN-γ. Moreover, TNF-α and IFN-γ stimulate PGF2α synthesis [244], suggesting that once progesterone production decreases, which leads to the invasion of immune cells, intraluteal positive feedback is initiated, leading to an increase in the production of luteal PGF2α.

Fas ligand (Fas L) is a member of the tumor necrosis factor superfamily that primarily engages its membrane receptor (Fas) to induce apoptosis [245]. Fas and FasL have been described in the rat corpus luteum during luteal regression [232, 246]. Binding of FasL to Fas initiates the activation of initiator caspases (e.g., caspase-8). In mice, administration of either a Fas-activating antibody or PGF2α induces activation of luteal caspase-8 and caspase-3 together with DNA fragmentation. The PGF2α receptor appears not to be directly coupled to caspase-8. Therefore, PGF2α may initiate luteal apoptosis, at least in part, by increasing the bioactivity or bioavailability of FasL (Fig. 26.4). Indeed, FasL has been shown to be highly expressed in the regressing corpora lutea of rats [246], probably due to the recruitment of immune cells [247]. Progesterone can suppress Fas expression in a large variety of cells, including rodent luteal cells. Moreover, Takahashi and collaborators [172] have proposed that progesterone may suppress luteal cell apoptosis by inhibiting the expression of Fas.

26.6 Rescue of the Corpus Luteum

In non-pregnant primates, the functional lifespan of the corpus luteum is about 2 weeks, which corresponds to the luteal phase of the cycle. However, if conception and implantation take place, luteolysis must be prevented to provide adequate levels of progesterone for maintenance of the pregnancy. In humans, progesterone from the corpus luteum is required only during early pregnancy until placental production of progesterone begins, which occurs between the fifth and seventh weeks of gestation. The extension of the lifespan of the corpus luteum during pregnancy in primates is caused by the implanting embryo [248]. At implantation, the syncytiotrophoblast of the placenta secretes chorionic gonadotrophin

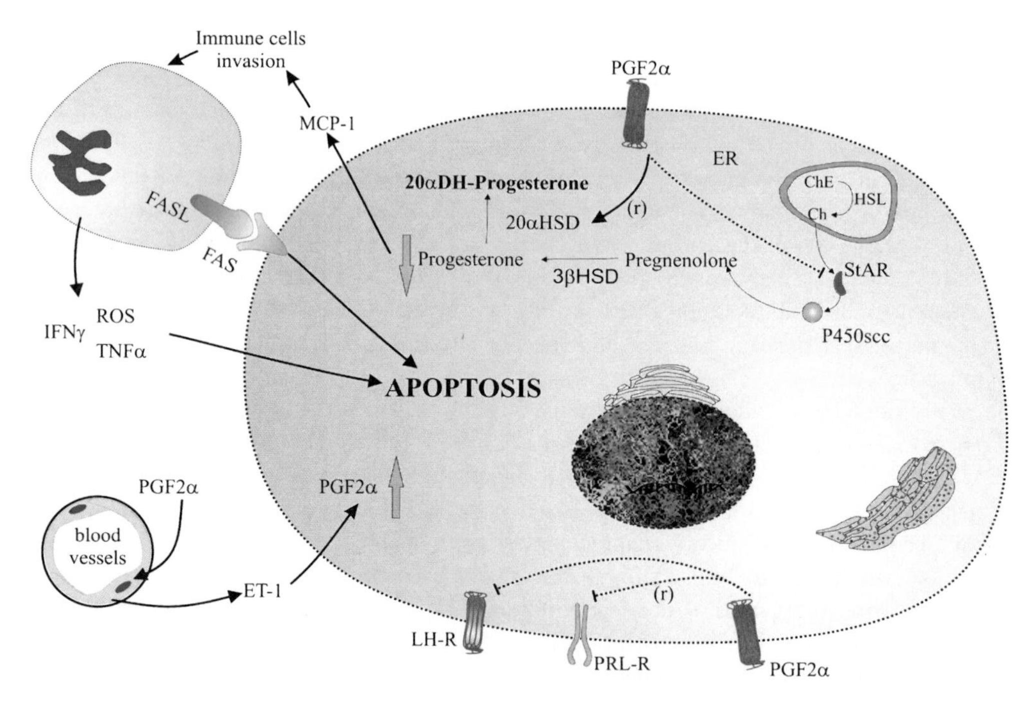

Fig. 26.4 An integrative model for luteal regression. In this model, PGF2α stimulates 20αHSD, decreases StAR expression, and blocks PRL and LH luteotrophic actions, leading to a decline in progesterone production. This allows an increase in Fas expression and invasion of immune cells that were under the repressive control of progesterone. The immune cell invasion results in an increase in cytokines such as FasL, TNF-α, and IFN-γ and superoxide radical production. The action of the cytokines leads to the initiation of apoptosis. Apoptotic luteal cells are then cleared by immune cells. In ruminants, PGF2α effects are also mediated by stimulation of ET-1 in endothelial cells. ET-1, TNF-α, and IFN-γ stimulate luteal PGF2α production, further stimulating the progress of luteolysis. The sign (r) denotes effects mainly observed in rodents

(CG), which binds to the same receptors as LH, thus extending luteotrophic signals beyond the normal menstrual cycle. In monkeys, administration of human CG mimics the patterns of circulating levels of progesterone found in early pregnancy [182]. In rats and mice, if mating does not occur, the ovulated follicle reorganizes into the corpus luteum of the cycle, which is a small, minimally developed structure that secretes very little progesterone and begins to regress after only 3 days. In these species, the development of the corpus luteum into a structure that can secrete sufficient progesterone to allow decidualization and implantation requires mating. The physical stimulus of mating induces secretion of PRL from the pituitary, and this hormone stimulates development of the corpus luteum of pregnancy [249]. Alternatively, in the event of a sterile mating or artificial vaginal stimulation, an artificial state known as pseudopregnancy is instated. In pseudopregnant animals, the corpus luteum regresses after 10–12 days, a period 3–4 times longer than the length of the normal cycle.

26.7 Glossary of Terms and Acronyms

17βHSD: 17β-hydroxysteroid dehydrogenase

20αDHP: 20α-dyhydroprogesterone

20αHSD: 20α-hydroxysteroid dehydrogenase

3βHSD: 3β-hydroxysteroid dehydrogenase

AC: adenylyl cyclase

Ang: angiopoietins;

Apaf-1: apoptotic protease-activating factor

AR: androgen receptor

ARF6: ADP ribosylation factor 6

bFGF: basic fibroblast growth factor

C/EBP: CATT/enhancer binding protein

CAM-1: intracellular adhesion molecule-1

cAMP: cyclic AMP

CBP: CREB-binding protein

Cdk: cyclin dependent kinase

CG: human chorionic gonadotropin

CL: corpus luteum;

COX-2: cyclooxygenase-2

CRE: cAMP response element

CREB: CRE binding protein

DAG: diacylglycerol

ECM: extracellular matrix

EG-VEGF: endocrine gland-derived VEGF

ER: estrogen receptor

Erg-1: early growth response factor-1

ERK: extracellular regulated kinase

FasL: fas ligand

FSH: follicle stimulating hormone

FSH-R: FSH receptor

GTP: guanosine-5′-triphosphate

HDL: high density lipoproteins

HSL: hormone-sensitive lipase

IFN-γ: interferon γ

IGF-1: insulin growth factor-1

InsP3: inositol triphosphate

Jak: janus kinase

Ld: lipid droplet

LDL: low density lipoproteins

LH-R: LH receptor

LIP: liver-enriched

LRBP: LH-R binding protein inhibitory protein

MAPK: mitogen activated protein kinase

MCP-1: monocyte chemoattractant protein-1

MMPs: matrix metalloproteinases

NO: nitric oxide

P450scc: cytochrome P450 cholesterol side-chain cleavage

PARP: poly ADP-ribose polymerase

PBR: peripheral-type benzodiazepine receptor

PDE: phosphodiesterase

PGF2α: prostaglandin F2α

PI3K: phosphatidylinositol-3-kinase

PKA: protein kinase A

PKC: protein kinase C

PLC: phospholipase C

PLs: placental lactogens

PR: progesterone receptor

pRb: retinoblastoma

PRL: prolactin; LH, luteinizing hormone

RAR: retinoid acid receptor

ROS: reactive oxygen species

SCP-2: sterol carrier protein-2

SF-1: steroidogenic factor-1

SOCS-3: suppressor of cytokine signaling-3

StAR: steroidogenic acute regulatory protein

Stat: signal transduction and activation of transcription

TGF: transforming growth factor

TIMPS: tissue inhibitors of metalloproteinases

TNF-α: tumor necrosis factor-α

VEGF: vascular endothelial growth factor

α2 M: α2-macroglobulin

References

1. Mukherjee A, Urban J, Sassone-Corsi P, et al. Gonadotropins regulate inducible cyclic adenosine 3′,5′-monophosphate early repressor in the rat ovary: implications for inhibin α-subunit gene expression. Mol Endocrinol 1998; 12:785–800.
2. Ascoli M, Fanelli F, Segaloff DL. The lutropin/choriogonadotropin receptor, a 2002 perspective. Endocr Rev 2002; 23:141–74.
3. Davis JS, West LA, Weakland LL, et al. Human chorionic gonadotropin activates the inositol 1,4,5-trisphosphate- Ca2+ intracellular signalling system in bovine luteal cells. FEBS Lett 1986; 208:287–91.
4. Richards JS. New signaling pathways for hormones and cyclic adenosine 3′,5′-monophosphate action in endocrine cells. Mol Endocrinol 2001; 15:209–18.
5. Mukherjee A, Park-Sarge OK, Mayo KE. Gonadotropins induce rapid phosphorylation of the 3′,5′-cyclic adenosine monophosphate response element binding protein in ovarian granulosa cells. Endocrinology 1996; 137:3234–45.
6. Arias J, Alberts AS, Brindle P, et al. Activation of cAMP and mitogen responsive genes relies on a common nuclear factor. Nature 1994; 370:226–9.
7. Tsafriri A, Chun SY, Zhang R, et al. Oocyte maturation involves compartmentalization and opposing changes of cAMP levels in follicular somatic and germ cells: studies using selective phosphodiesterase inhibitors. Dev Biol 1996; 178:393–402.
8. Park JY, Richard F, Chun SY, et al. Phosphodiesterase regulation is critical for the differentiation and pattern of gene expression in granulosa cells of the ovarian follicle. Mol Endocrinol 2003; 17:1117–30.
9. Donadeu FX, Ascoli M. The differential effects of the gonadotropin receptors on aromatase expression in primary cultures of immature rat granulosa cells are highly dependent on the density of receptors expressed and the activation of the inositol phosphate cascade. Endocrinology 2005; 146:3907–16.
10. Kosugi S, Mori T, Shenker A. An anionic residue at position 564 is important for maintaining the inactive conformation of the human lutropin/choriogonadotropin receptor. Mol Pharmacol 1998; 53:894–901.
11. Flores JA, Aguirre C, Sharma OP, et al. Luteinizing hormone (LH) stimulates both intracellular calcium ion ([Ca2+]i) mobilization and transmembrane cation influx in single ovarian (granulosa) cells: recruitment as a cellular mechanism of LH-[Ca2+]i dose response. Endocrinology 1998; 139:3606–12.
12. Gudermann T, Birnbaumer M, Birnbaumer L. Evidence for dual coupling of the murine luteinizing hormone receptor to adenylyl cyclase and phosphoinositide breakdown and Ca2+ mobilization. Studies with the cloned murine luteinizing hormone receptor expressed in L cells. J Biol Chem 1992; 267:4479–88.
13. Lee PS, Buchan AM, Hsueh AJ, et al. Intracellular calcium mobilization in response to the activation of human wild-type and chimeric gonadotropin receptors. Endocrinology 2002; 143:1732–40.
14. Morris JK, Richards JS. Luteinizing hormone induces prostaglandin endoperoxide synthase-2 and luteinization in vitro by A-kinase and C-kinase pathways. Endocrinology 1995; 136:1549–58.
15. Salvador LM, Maizels E, Hales DB, et al. Acute signaling by the LH receptor is independent of protein kinase C activation. Endocrinology 2002; 143:2986–94.
16. Andric N, Ascoli M. A delayed gonadotropin-dependent and growth factor-mediated activation of the extracellular signal-regulated kinase 1/2 cascade negatively regulates aromatase expression in granulosa cells. Mol Endocrinol 2006; 20:3308–20.
17. Park JY, Su YQ, Ariga M, et al. EGF-like growth factors as mediators of LH action in the ovulatory follicle. Science 2004; 303:682–4.
18. Sharma SC, Richards JS. Regulation of AP1 (Jun/Fos) factor expression and activation in ovarian granulosa cells. Relation of JunD and Fra2 to terminal differentiation. J Biol Chem 2000; 275:33718–28.
19. Espey LL, Richards JS. Temporal and spatial patterns of ovarian gene transcription following an ovulatory dose of gonadotropin in the rat. Biol Reprod 2002; 67:1662–70.
20. Natraj U, Richards JS. Hormonal regulation, localization, and functional activity of the progesterone receptor in granulosa cells of rat preovulatory follicles. Endocrinology 1993; 133:761–9.
21. Lim H, Paria BC, Das SK, et al. Multiple female reproductive failures in cyclooxygenase 2-deficient mice. Cell 1997; 91: 197–208.
22. Sterneck E, Tessarollo L, Johnson PF. An essential role for C/EBPβ in female reproduction. Genes Dev 1997; 11: 2153–62.
23. Espey LL, Ujioka T, Russell DL, et al. Induction of early growth response protein-1 gene expression in the rat ovary in response to an ovulatory dose of human chorionic gonadotropin. Endocrinology 2000; 141:2385–91.
24. Lydon JP, DeMayo FJ, Conneely OM, et al. Reproductive phenotpes of the progesterone receptor null mutant mouse. J Steroid Biochem Mol Biol 1996; 56:67–77.
25. Lee SL, Sadovsky Y, Swirnoff AH, et al. Luteinizing hormone deficiency and female infertility in mice lacking the transcription factor NGFI-A (Egr-1). Science 1996; 273:1219–21.
26. Topilko P, Schneider-Maunoury S, Levi G, et al. Multiple pituitary and ovarian defects in Krox-24 (NGFI-A, Egr-1)-targeted mice. Mol Endocrinol 1998; 12:107–22.
27. Green C, Chatterjee R, McGarrigle HH, et al. p107 is active in the nucleolus in non-dividing human granulosa lutein cells. J Mol Endocrinol 2000; 25:275–86.
28. Morgan DO. Principles of CDK regulation. Nature 1995; 374:131–4.
29. Stevaux O, Dyson NJ. A revised picture of the E2F transcriptional network and RB function. Curr Opin Cell Biol 2002; 14:684–91.

30. Hampl A, Pachernik J, Dvorak P. Levels and interactions of p27, cyclin D3, and CDK4 during the formation and maintenance of the corpus luteum in mice. Biol Reprod 2000; 62: 1393–401.
31. Richards JS, Russell DL, Robker RL, et al. Molecular mechanisms of ovulation and luteinization. Mol Cell Endocrinol 1998; 145:47–54.
32. Tong W, Kiyokawa H, Soos TJ, et al. The absence of p27Kip1, an inhibitor of G1 cyclin-dependent kinases, uncouples differentiation and growth arrest during the granulosa-luteal transition. Cell Growth Differ 1998; 9:787–94.
33. Jirawatnotai S, Moons DS, Stocco CO, et al. The cyclin-dependent kinase inhibitors p27Kip1 and p21Cip1 cooperate to restrict proliferative life span in differentiating ovarian cells. J Biol Chem 2003; 278:17021–7.
34. el-Fouly MA, Cook B, Nekola M, et al. Role of the ovum in follicular luteinization. Endocrinology 1970; 87:286–93.
35. Miyazawa K, Shinozaki M, Hara T, et al. Two major Smad pathways in TGF-β superfamily signalling. Genes Cells 2002; 7: 1191–204.
36. Otsuka F, Yamamoto S, Erickson GF, et al. Bone morphogenetic protein-15 inhibits follicle-stimulating hormone (FSH) action by suppressing FSH receptor expression. J Biol Chem 2001; 276:11387–92.
37. Pangas SA, Matzuk MM. Genetic models for transforming growth factor β superfamily signaling in ovarian follicle development. Mol Cell Endocrinol 2004; 225:83–91.
38. Pangas SA, Li X, Robertson EJ, et al. Premature luteinization and cumulus cell defects in ovarian-specific Smad4 knockout mice. Mol Endocrinol 2006; 20:1406–22.
39. Nelson SE, McLean MP, Jayatilak PG, et al. Isolation, characterization, and culture of cell subpopulations forming the pregnant rat corpus luteum. Endocrinology 1992; 130:954–66.
40. Niswender GD, Juengel JL, Silva PJ, et al. Mechanisms controlling the function and life span of the corpus luteum. Physiol Rev 2000; 80:1–29.
41. Devoto L, Kohen P, Vega M, et al. Control of human luteal steroidogenesis. Mol Cell Endocrinol 2002; 186:137–41.
42. Sanders SL, Stouffer RL, Brannian JD. Androgen production by monkey luteal cell subpopulations at different stages of the menstrual cycle. J Clin Endocrinol Metab 1996; 81:591–6.
43. Ohara A, Mori T, Taii S, et al. Functional differentiation in steroidogenesis of two types of luteal cells isolated from mature human corpora lutea of menstrual cycle. J Clin Endocrinol Metab 1987; 65:1192–200.
44. Fraser HM, Dickson SE, Lunn SF, et al. Suppression of luteal angiogenesis in the primate after neutralization of vascular endothelial growth factor. Endocrinology 2000; 141: 995–1000.
45. Berkholtz CB, Lai BE, Woodruff TK, et al. Distribution of extracellular matrix proteins type I collagen, type IV collagen, fibronectin, and laminin in mouse folliculogenesis. Histochem Cell Biol 2006; 126:583–92.
46. Luck MR, Zhao Y. Identification and measurement of collagen in the bovine corpus luteum and its relationship with ascorbic acid and tissue development. J Reprod Fertil 1993; 99:647–52.
47. Bagavandoss P. Differential distribution of gelatinases and tissue inhibitor of metalloproteinase-1 in the rat ovary. J Endocrinol 1998; 158:221–8.
48. Hagglund AC, Ny A, Leonardsson G, Ny T. Regulation and localization of matrix metalloproteinases and tissue inhibitors of metalloproteinases in the mouse ovary during gonadotropin-induced ovulation. Endocrinology 1999; 140:4351–8.
49. Irving-Rodgers HF, Friden BE, Morris SE, et al. Extracellular matrix of the human cyclic corpus luteum. Mol Hum Reprod 2006; 12:525–34.
50. Matsushima T, Fukuda Y, Tsukada K, et al. The extracellular matrices and vascularization of the developing corpus luteum in rats. J Submicrosc Cytol Pathol 1996; 28:441–55.
51. Tsukada K, Matsushima T, Yamanaka N. Neovascularization of the corpus luteum of rats during the estrus cycle. Pathol Int 1996; 46:408–16.
52. Irving-Rodgers HF, Rodgers RJ. Extracellular matrix in ovarian follicular development and disease. Cell Tissue Res 2005; 322:89–98.
53. Honda T, Fujiwara H, Yamada S, et al. Integrin α5 is expressed on human luteinizing granulosa cells during corpus luteum formation, and its expression is enhanced by human chorionic gonadotrophin in vitro. Mol Hum Reprod 1997; 3: 979–84.
54. Rolaki A, Coukos G, Loutradis D, et al. Luteogenic hormones act through a vascular endothelial growth factor-dependent mechanism to up-regulate α5-β1 and α-β3 integrins, promoting the migration and survival of human luteinized granulosa cells. Am J Pathol 2007; 170:1561–72.
55. Aten RF, Kolodecik TR, Behrman HR. A cell adhesion receptor antiserum abolishes, whereas laminin and fibronectin glycoprotein components of extracellular matrix promote, luteinization of cultured rat granulosa cells. Endocrinology 1995; 136:1753–8.
56. Nothnick WB. Tissue inhibitor of metalloproteinase-1 (TIMP-1) deficient mice display reduced serum progesterone levels during corpus luteum development. Endocrinology 2003; 144:5–8.
57. Phillips HS, Hains J, Leung DW, et al. Vascular endothelial growth factor is expressed in rat corpus luteum. Endocrinology 1990; 127:965–7.
58. Kamat BR, Brown LF, Manseau EJ, et al. Expression of vascular permeability factor/vascular endothelial growth factor by human granulosa and theca lutein cells. Role in corpus luteum development. Am J Pathol 1995; 146:157–65.
59. Fraser HM, Bell J, Wilson H, et al. Localization and quantification of cyclic changes in the expression of endocrine gland vascular endothelial growth factor in the human corpus luteum. J Clin Endocrinol Metab 2005; 90:427–34.
60. Schams D, Kosmann M, Berisha B, et al. Stimulatory and synergistic effects of luteinising hormone and insulin like growth factor 1 on the secretion of vascular endothelial growth factor and progesterone of cultured bovine granulosa cells. Exp Clin Endocrinol Diabetes 2001; 109:155–62.
61. Martinez-Chequer JC, Stouffer RL, Hazzard TM, et al. Insulin-like growth factors-1 and -2, but not hypoxia, synergize with gonadotropin hormone to promote vascular endothelial growth factor-A secretion by monkey granulosa cells from preovulatory follicles. Biol Reprod 2003; 68:1112–8.
62. Sugino N, Kashida S, Takiguchi S, et al. Expression of vascular endothelial growth factor (VEGF) receptors in rat corpus luteum: regulation by oestradiol during mid-pregnancy. Reproduction 2001; 122:875–81.
63. Kashida S, Sugino N, Takiguchi S, et al. Regulation and role of vascular endothelial growth factor in the corpus luteum during mid-pregnancy in rats. Biol Reprod 2001; 64:317–23.
64. Zimmermann RC, Hartman T, Bohlen P, et al. Preovulatory treatment of mice with anti-VEGF receptor 2 antibody inhibits angiogenesis in corpora lutea. Microvasc Res 2001; 62:15–25.
65. Hazzard TM, Rohan RM, Molskness TA, et al. Injection of antiangiogenic agents into the macaque preovulatory follicle: disruption of corpus luteum development and function. Endocrine 2002; 17:199–206.
66. Ferrara N, Chen H, Davis-Smyth T, et al. Vascular endothelial growth factor is essential for corpus luteum angiogenesis. Nat Med 1998; 4:336–40.
67. Pauli SA, Tang H, Wang J, et al. The vascular endothelial growth factor (VEGF)/VEGF receptor 2 pathway is critical for blood

vessel survival in corpora lutea of pregnancy in the rodent. Endocrinology 2005; 146:1301–11.
68. LeCouter J, Kowalski J, Foster J, et al. Identification of an angiogenic mitogen selective for endocrine gland endothelium. Nature 2001; 412:877–84.
69. Ferrara N, LeCouter J, Lin R, et al. EG-VEGF and Bv8: a novel family of tissue-restricted angiogenic factors. Biochim Biophys Acta 2004; 1654:69–78.
70. LeCouter J, Lin R, Ferrara N. Endocrine gland-derived VEGF and the emerging hypothesis of organ-specific regulation of angiogenesis. Nat Med 2002; 8:913–7.
71. Gospodarowicz D, Cheng J, Lui GM, et al. Corpus luteum angiogenic factor is related to fibroblast growth factor. Endocrinology 1985; 117:2383–91.
72. Guthridge M, Bertolini J, Cowling J, et al. Localization of bFGF mRNA in cyclic rat ovary, diethylstilbesterol primed rat ovary, and cultured rat granulosa cells. Growth Factors 1992; 7:15–25.
73. Yamamoto S, Konishi I, Nanbu K, et al. Immunohistochemical localization of basic fibroblast growth factor (bFGF) during folliculogenesis in the human ovary. Gynecol Endocrinol 1997; 11:223–30.
74. Gospodarowicz D, Massoglia S, Cheng J, et al. Effect of fibroblast growth factor and lipoproteins on the proliferation of endothelial cells derived from bovine adrenal cortex, brain cortex, and corpus luteum capillaries. J Cell Physiol 1986; 127:121–36.
75. Reynolds LP, Redmer DA. Growth and development of the corpus luteum. J Reprod Fertil Suppl 1999; 54:181–91.
76. Ortega S, Ittmann M, Tsang SH, et al. Neuronal defects and delayed wound healing in mice lacking fibroblast growth factor 2. Proc Natl Acad Sci U S A 1998; 95:5672–7.
77. Fiedler U, Augustin HG. Angiopoietins: a link between angiogenesis and inflammation. Trends Immunol 2006; 27:552–8.
78. Dumont DJ, Gradwohl G, Fong GH, et al. Dominant-negative and targeted null mutations in the endothelial receptor tyrosine kinase, tek, reveal a critical role in vasculogenesis of the embryo. Genes Dev 1994; 8:1897–909.
79. Wulff C, Wilson H, Largue P, et al. Angiogenesis in the human corpus luteum: localization and changes in angiopoietins, tie-2, and vascular endothelial growth factor messenger ribonucleic acid. J Clin Endocrinol Metab 2000; 85:4302–9.
80. Hazzard TM, Molskness TA, Chaffin CL, et al. Vascular endothelial growth factor (VEGF) and angiopoietin regulation by gonadotrophin and steroids in macaque granulosa cells during the peri-ovulatory interval. Mol Hum Reprod 1999; 5:1115–21.
81. Hazzard TM, Christenson LK, Stouffer RL. Changes in expression of vascular endothelial growth factor and angiopoietin-1 and -2 in the macaque corpus luteum during the menstrual cycle. Mol Hum Reprod 2000; 6:993–8.
82. Maisonpierre PC, Suri C, Jones PF, et al. Angiopoietin-2, a natural antagonist for Tie2 that disrupts in vivo angiogenesis. Science 1997; 277:55–60.
83. Azhar S, Reaven E. Scavenger receptor class BI and selective cholesteryl ester uptake: partners in the regulation of steroidogenesis. Mol Cell Endocrinol 2002; 195:1–26.
84. Ragoobir J, Abayasekara DR, Bruckdorfer KR, et al. Stimulation of progesterone production in human granulosa-lutein cells by lipoproteins: evidence for cholesterol-independent actions of high-density lipoproteins. J Endocrinol 2002; 173:103–11.
85. Menon M, Peegel H, Menon KM. Lipoprotein augmentation of human chorionic gonadotropin and prolactin stimulated progesterone synthesis by rat luteal cells. J Steroid Biochem 1985; 22:79–84.
86. Kraemer FB. Adrenal cholesterol utilization. Mol Cell Endocrinol 2007; 265–266:42–5.
87. Connelly MA, Williams DL. SR-BI and HDL cholesteryl ester metabolism. Endocr Res 2004; 30:697–703.
88. Miranda-Jimenez L, Murphy BD. Lipoprotein receptor expression during luteinization of the ovarian follicle. Am J Physiol 2007; 293:E1053–61.
89. Landschulz KT, Pathak RK, Rigotti A, et al. Regulation of scavenger receptor, class B, type I, a high density lipoprotein receptor, in liver and steroidogenic tissues of the rat. J Clin Invest 1996; 98:984–95.
90. Trigatti B, Rayburn H, Vinals M, et al. Influence of the high density lipoprotein receptor SR-BI on reproductive and cardiovascular pathophysiology. Proc Natl Acad Sci U S A 1999; 96: 9322–7.
91. Azhar S, Khan I, Puryear T, et al. Luteal cell 3-hydroxy-3-methylglutaryl coenzyme-A reductase activity and cholesterol metabolism throughout pregnancy in the rat. Endocrinology 1988; 123:1495–503.
92. Aten RF, Kolodecik TR, Macdonald GJ, et al. Modulation of cholesteryl ester hydrolase messenger ribonucleic acid levels, protein levels, and activity in the rat corpus luteum. Biol Reprod 1995; 53:1110–7.
93. Osuga J, Ishibashi S, Oka T, et al. Targeted disruption of hormone-sensitive lipase results in male sterility and adipocyte hypertrophy, but not in obesity. Proc Natl Acad Sci U S A 2000; 97:787–92.
94. Connelly PW, Hegele RA. Hepatic lipase deficiency. Crit Rev Clin Lab Sci 1998; 35:547–72.
95. Wade RL, Van Andel RA, Rice SG, et al. Hepatic lipase deficiency attenuates mouse ovarian progesterone production leading to decreased ovulation and reduced litter size. Biol Reprod 2002; 66:1076–82.
96. Seedorf U, Ellinghaus P, Roch Nofer J. Sterol carrier protein-2. Biochim Biophys Acta 2000; 1486:45–54.
97. Rennert H, Amsterdam A, Billheimer JT, et al. Regulated expression of sterol carrier protein 2 in the ovary: a key role for cyclic AMP. Biochemistry 1991; 30:11280–5.
98. McLean MP, Puryear TK, Khan I, et al. Estradiol regulation of sterol carrier protein-2 independent of cytochrome P450 side-chain cleavage expression in the rat corpus luteum. Endocrinology 1989; 125:1337–44.
99. Miller WL. Mechanism of StAR's regulation of mitochondrial cholesterol import. Mol Cell Endocrinol 2007; 265:46–50.
100. Kiriakidou M, McAllister JM, Sugawara T, et al. Expression of steroidogenic acute regulatory protein (StAR) in the human ovary. J Clin Endocrinol Metab 1996; 81:4122–8.
101. Mizutani T, Sonoda Y, Minegishi T, et al. Molecular cloning, characterization and cellular distribution of rat steroidogenic acute regulatory protein (StAR) in the ovary. Life Sci 1997; 61:1497–506.
102. Sandhoff TW, McLean MP. Hormonal regulation of steroidogenic acute regulatory (StAR) protein messenger ribonucleic acid expression in the rat ovary. Endocrine 1996; 4:259.
103. Hasegawa T, Zhao L, Caron KM, et al. Developmental roles of the steroidogenic acute regulatory protein (StAR) as revealed by StAR knockout mice. Mol Endocrinol 2000; 14:1462–71.
104. Li H, Yao Z, Degenhardt B, et al. Cholesterol binding at the cholesterol recognition/ interaction amino acid consensus (CRAC) of the peripheral-type benzodiazepine receptor and inhibition of steroidogenesis by an HIV TAT-CRAC peptide. Proc Natl Acad Sci U S A 2001; 98:1267–72.
105. Culty M, Li H, Boujrad N, et al. In vitro studies on the role of the peripheral-type benzodiazepine receptor in steroidogenesis. J Steroid Biochem Mol Biol 1999; 69:123–30.
106. Sridaran R, Philip GH, Li H, et al. GnRH agonist treatment decreases progesterone synthesis, luteal peripheral benzodiazepine receptor mRNA, ligand binding and steroidogenic acute regulatory protein expression during pregnancy. J Mol Endocrinol 1999; 22:45–54.

107. Lacapere JJ, Papadopoulos V. Peripheral-type benzodiazepine receptor: structure and function of a cholesterol-binding protein in steroid and bile acid biosynthesis. Steroids 2003; 68:569–85.
108. Oonk RB, Krasnow JS, Beattie WG, Richards JS. Cyclic AMP-dependent and -independent regulation of cholesterol side chain cleavage cytochrome P-450 (P-450scc) in rat ovarian granulosa cells and corpora lutea. cDNA and deduced amino acid sequence of rat P-450scc. J Biol Chem 1989; 264:21934–42.
109. Enders AC. Cytology of the corpus luteum. Biol Reprod 1973; 8:158–82.
110. Goldring NB, Durica JM, Lifka J, et al. Cholesterol side-chain cleavage P450 messenger ribonucleic acid: evidence for hormonal regulation in rat ovarian follicles and constitutive expression in corpora lutea. Endocrinology 1987; 120:1942–50.
111. Wiest WG, Kidwell WR, Balogh K, Jr. Progesterone catabolism in the rat ovary: a regulatory mechanism for progestational potency during pregnancy. Endocrinology 1968; 82:844–59.
112. Mao J, Duan WR, Albarracin CT, et al. Isolation and characterization of a rat luteal cDNA encoding 20 α-hydroxysteroid dehydrogenase. Biochem Biophys Res Commun 1994; 201:1289–95.
113. Piekorz RP, Gingras S, Hoffmeyer A, et al. Regulation of progesterone levels during pregnancy and parturition by signal transducer and activator of transcription 5 and 20α-hydroxysteroid dehydrogenase. Mol Endocrinol 2005; 19:431–40.
114. Sanders SL, Stouffer RL. Localization of steroidogenic enzymes in macaque luteal tissue during the menstrual cycle and simulated early pregnancy: immunohistochemical evidence supporting the two-cell model for estrogen production in the primate corpus luteum. Biol Reprod 1997; 56:1077–87.
115. Hickey GJ, Oonk RB, Hall PF, et al. Aromatase cytochrome P450 and cholesterol side-chain cleavage cytochrome P450 in corpora lutea of pregnant rats: diverse regulation by peptide and steroid hormones. Endocrinology 1989; 125:1673–82.
116. Stouffer RL. Progesterone as a mediator of gonadotrophin action in the corpus luteum: beyond steroidogenesis. Hum Reprod Update 2003; 9:99–117.
117. Hutchison JS, Zeleznik AJ. The rhesus monkey corpus luteum is dependent on pituitary gonadotropin secretion throughout the luteal phase of the menstrual cycle. Endocrinology 1984; 115:1780–6.
118. Ellinwood WE, Norman RL, Spies HG. Changing frequency of pulsatile luteinizing hormone and progesterone secretion during the luteal phase of the menstrual cycle of rhesus monkeys. Biol Reprod 1984; 31:714–22.
119. Filicori M, Butler JP, Crowley WF, Jr. Neuroendocrine regulation of the corpus luteum in the human. Evidence for pulsatile progesterone secretion. J Clin Invest 1984; 73:1638–47.
120. Devoto L, Vega M, Kohen P, et al. Molecular regulation of progesterone secretion by the human corpus luteum throughout the menstrual cycle. J Reprod Immunol 2002; 55:11–20.
121. Xu J, Stouffer RL, Searles RP, et al. Discovery of LH-regulated genes in the primate corpus luteum. Mol Hum Reprod 2005; 11:151–9.
122. Duffy DM, Chaffin CL, Stouffer RL. Expression of estrogen receptor α and β in the rhesus monkey corpus luteum during the menstrual cycle: regulation by luteinizing hormone and progesterone. Endocrinology 2000; 141:1711–7.
123. Young KA, Stouffer RL. Gonadotropin and steroid regulation of matrix metalloproteinases and their endogenous tissue inhibitors in the developed corpus luteum of the rhesus monkey during the menstrual cycle. Biol Reprod 2004; 70:244–52.
124. Ravindranath N, Little-Ihrig L, Phillips HS, et al. Vascular endothelial growth factor messenger ribonucleic acid expression in the primate ovary. Endocrinology 1992; 131:254–60.
125. Chen Z, Menon KM. Expression of high density lipoprotein-binding protein messenger ribonucleic acid in the rat ovary and its regulation by gonadotropin. Endocrinology 1994; 134: 2360–6.
126. Risk M, Shehu A, Mao J, et al. Cloning and characterization of a 5′ regulatory region of the prolactin receptor-associated protein/17β-hydroxysteroid dehydrogenase 7 gene. Endocrinology 2005; 146:2807–16.
127. Lei ZM, Mishra S, Zou W, et al. Targeted disruption of luteinizing hormone/human chorionic gonadotropin receptor gene. Mol Endocrinol 2001; 15:184–200.
128. Duncan WC, McNeilly AS, Fraser HM, et al. Luteinizing hormone receptor in the human corpus luteum: lack of down- regulation during maternal recognition of pregnancy. Hum Reprod 1996; 11:2291–7.
129. Segaloff DL, Wang HY, Richards JS. Hormonal regulation of luteinizing hormone/chorionic gonadotropin receptor mRNA in rat ovarian cells during follicular development and luteinization. Mol Endocrinol 1990; 4:1856–65.
130. Peegel H, Randolph J, Jr., Midgley AR, et al. In situ hybridization of luteinizing hormone/human chorionic gonadotropin receptor messenger ribonucleic acid during hormone-induced down-regulation and the subsequent recovery in rat corpus luteum. Endocrinology 1994; 135:1044–51.
131. Hunzicker-Dunn M, Gurevich VV, Casanova JE, et al. ARF6: a newly appreciated player in G protein-coupled receptor desensitization. FEBS Lett 2002; 52:3–8.
132. Lu DL, Peegel H, Mosier SM, et al. Loss of lutropin/human choriogonadotropin receptor messenger ribonucleic acid during ligand-induced down-regulation occurs post transcriptionally. Endocrinology 1993; 132:235–40.
133. Nair AK, Menon KM. Isolation and characterization of a novel trans-factor for luteinizing hormone receptor mRNA from ovary. J Biol Chem 2004; 279:14937–44.
134. Risk M, Gibori G. Mechanisms of luteal cell regulation by prolactin. In: Horseman ND, editor. Prolactin. New York: Kluwer Academic, 2001:265–95.
135. Smith MS. Role of prolactin in regulating gonadotropin secretion and gonad function in female rats. Fed Proc 1980; 39:2571–6.
136. Richards JS, Williams JJ. Luteal cell receptor content for prolactin (PRL) and luteinizing hormone (LH): regulation by LH and PRL. Endocrinology 1976; 99:1571–81.
137. Frasor J, Barkai U, Zhong L, et al. PRL-induced ERα gene expression is mediated by Janus kinase 2 (Jak2) while signal transducer and activator of transcription 5b (Stat5b) phosphorylation involves Jak2 and a second tyrosine kinase. Mol Endocrinol 2001; 15:1941–52.
138. Martel C, Gagne D, Couet J, et al. Rapid modulation of ovarian 3β-hydroxysteroid dehydrogenase/$\Delta 5 \Delta 4$ isomerase gene expression by prolactin and human chorionic gonadotropin in the hypophysectomized rat. Mol Cell Endocrinol 1994; 99:63–71.
139. Jones PB, Valk CA, Hsueh AJ. Regulation of progestin biosynthetic enzymes in cultured rat granulosa cells: effects of prolactin, β2-adrenergic agonist, human chorionic gonadotropin and gonadotropin releasing hormone. Biol Reprod 1983; 29:572–85.
140. Duan WR, Parmer TG, Albarracin CT, et al. PRAP, a prolactin receptor associated protein: its gene expression and regulation in the corpus luteum. Endocrinology 1997; 138:3216–21.
141. Stocco C, Callegari E, Gibori G. Opposite effect of prolactin and prostaglandin F_{2a} on the expression of luteal genes as revealed by rat cDNA expression array. Endocrinology 2001; 142:4158–61.
142. Kawaminami M, Shibata Y, Yaji A, et al. Prolactin inhibits annexin 5 expression and apoptosis in the corpus luteum of pseudopregnant rats: involvement of local gonadotropin-releasing hormone. Endocrinology 2003; 144:3625–31.
143. Albarracin CT, Parmer TG, Duan WR, et al. Identification of a major prolactin-regulated protein as 20 α-hydroxysteroid dehydrogenase: coordinate regulation of its activity, protein content,

and messenger ribonucleic acid expression. Endocrinology 1994; 134:2453–60.
144. Grosdemouge I, Bachelot A, Lucas A, et al. Effects of deletion of the prolactin receptor on ovarian gene expression. Reprod Biol Endocrinol 2003; 1:12.
145. Linzer DI, Fisher SJ. The placenta and the prolactin family of hormones: regulation of the physiology of pregnancy. Mol Endocrinol 1999; 13:837–40.
146. Golander A, Hurley T, Barrett J, et al. Prolactin synthesis by human chorion-decidual tissue: a possible source of prolactin in the amniotic fluid. Science 1978; 202:311–3.
147. Jabbour HN, Critchley HO. Potential roles of decidual prolactin in early pregnancy. Reproduction 2001; 121:197–205.
148. Russell DL, Richards JS. Differentiation-dependent prolactin responsiveness and stat (signal transducers and activators of transcription) signaling in rat ovarian cells. Mol Endocrinol 1999; 13:2049–64.
149. Levy DE, Darnell JE, Jr. Stats: transcriptional control and biological impact. Nat Rev Mol Cell Biol 2002; 3:651–62.
150. Udy GB, Towers RP, Snell RG, et al. Requirement of STAT5b for sexual dimorphism of body growth rates and liver gene expression. Proc Natl Acad Sci U S A 1997; 94:7239–44.
151. Teglund S, McKay C, Schuetz E, et al. Stat5a and Stat5b proteins have essential and nonessential, or redundant, roles in cytokine responses. Cell 1998; 93:841–50.
152. Gibori G, Khan I, Warshaw ML, et al. Placental-derived regulators and the complex control of luteal cell function. Recent Prog Horm Res 1988; 44:377–429.
153. Gibori G, Keyes PL, Richards JS. A role for intraluteal estrogen in the mediation of luteinizing hormone action on the rat corpus luteum during pregnancy. Endocrinology 1978; 103:162–9.
154. Gibori G, Rodway R, Rothchild I. The luteotrophic effect of estrogen in the rat: prevention by estradiol of the luteolytic effect of an antiserum to luteinizing hormone in the pregnant rat. Endocrinology 1977; 101:1683–9.
155. McLean MP, Khan I, Puryear TK, et al. Induction and repression of specific estradiol sensitive proteins in the rat corpus luteum. Chin J Physiol 1990; 33:353–66.
156. Puryear TK, McLean MP, Khan I, et al. Mechanism for control of hydroxymethylglutaryl-coenzyme A reductase and cytochrome P-450 side chain cleavage message and enzyme in the corpus luteum. Endocrinology 1990; 126:2910–8.
157. Gibori G, Chen YD, Khan I, et al. Regulation of luteal cell lipoprotein receptors, sterol contents, and steroidogenesis by estradiol in the pregnant rat. Endocrinology 1984; 114: 609–17.
158. Lopez D, Sanchez MD, Shea-Eaton W, et al. Estrogen activates the high-density lipoprotein receptor gene via binding to estrogen response elements and interaction with sterol regulatory element binding protein-1A. Endocrinology 2002; 143:2155–68.
159. Cok SJ, Hay RV, Holt JA. Estrogen-mediated mitochondrial cholesterol transport and metabolism to pregnenolone in the rabbit luteinized ovary. Biol Reprod 1997; 57:360–6.
160. Butler WR, Hotchkiss J, Knobil E. Functional luteolysis in the rhesus monkey: ovarian estrogen and progesterone during the luteal phase of the menstrual cycle. Endocrinology 1975; 96:1509–12.
161. Misao R, Nakanishi Y, Sun WS, et al. Expression of oestrogen receptor α and β mRNA in corpus luteum of human subjects. Mol Hum Reprod 1999; 5:17–21.
162. Telleria CM, Zhong L, Deb S, et al. Differential expression of the estrogen receptors α and β in the rat corpus luteum of pregnancy: regulation by prolactin and placental lactogens. Endocrinology 1998; 139:2432–42.
163. Frasor J, Park K, Byers M, et al. Differential roles for signal transducers and activators of transcription 5a and 5b in PRL stimulation of ERα and ERβ transcription. Mol Endocrinol 2001; 15:2172–81.
164. Fitzpatrick SL, Funkhouser JM, Sindoni DM, et al. Expression of estrogen receptor-β protein in rodent ovary. Endocrinology 1999; 140:2581–91.
165. Krege JH, Hodgin JB, Couse JF, et al. Generation and reproductive phenotypes of mice lacking estrogen receptor β. Proc Natl Acad Sci U S A 1998; 95:15677–82.
166. Couse JF, Hewitt SC, Bunch DO, et al. Postnatal sex reversal of the ovaries in mice lacking estrogen receptors α and β. Science 1999; 286:2328–31.
167. Ottander U, Hosokawa K, Liu K, et al. A putative stimulatory role of progesterone acting via progesterone receptors in the steroidogenic cells of the human corpus luteum. Biol Reprod 2000; 62:655–63.
168. Duffy DM, Hess DL, Stouffer RL. Acute administration of a 3 β-hydroxysteroid dehydrogenase inhibitor to rhesus monkeys at the midluteal phase of the menstrual cycle: evidence for possible autocrine regulation of the primate corpus luteum by progesterone. J Clin Endocrinol Metab 1994; 79:1587–94.
169. Telleria CM, Deis RP. Effect of RU486 on ovarian progesterone production at pro-oestrus and during pregnancy: a possible dual regulation of the biosynthesis of progesterone. J Reprod Fertil 1994; 102:379–84.
170. Telleria CM, Stocco CO, Stati AO, et al. Progesterone receptor is not required for progesterone action in the rat corpus luteum of pregnancy. Steroids 1999; 64:760–6.
171. Sugino N, Telleria CM, Gibori G. Progesterone inhibits 20α-hydroxysteroid dehydrogenase expression in the rat corpus luteum through the glucocorticoid receptor. Endocrinology 1997; 138:4497–500.
172. Kuranaga E, Kanuka H, Hirabayashi K, Suzuki M, Nishihara M, Takahashi M. Progesterone is a cell death suppressor that downregulates Fas expression in rat corpus luteum. FEBS Lett 2000; 466:279–82.
173. Hild-Petito S, Stouffer RL, Brenner RM. Immunocytochemical localization of estradiol and progesterone receptors in the monkey ovary throughout the menstrual cycle. Endocrinology 1988; 123:2896–905.
174. Park OK, Mayo KE. Transient expression of progesterone receptor messenger RNA in ovarian granulosa cells after the preovulatory luteinizing hormone surge. Mol Endocrinol 1991; 5:967–78.
175. Duffy DM, Wells TR, Haluska GJ, et al. The ratio of progesterone receptor isoforms changes in the monkey corpus luteum during the luteal phase of the menstrual cycle. Biol Reprod 1997; 57:693–9.
176. Park-Sarge OK, Parmer TG, Gu Y, et al. Does the rat corpus luteum express the progesterone receptor gene? Endocrinology 1995; 136:1537–43.
177. Cai Z, Stocco C. Expression and regulation of progestin membrane receptors in the rat corpus luteum. Endocrinology 2005; 146:5522–32.
178. McCracken JA, Custer EE, Lamsa JC. Luteolysis: a neuroendocrine-mediated event. Physiol Rev 1999; 79:263–323.
179. Strauss JF 3rd, Sokoloski J, Caploe P, et al. On the role of prostaglandins in parturition in the rat. Endocrinology 1975; 96:1040–3.
180. Sugimoto Y, Yamasaki A, Segi E, et al. Failure of parturition in mice lacking the prostaglandin F receptor. Science 1997; 277:681–3.
181. Bonventre JV, Huang Z, Taheri MR, et al. Reduced fertility and postischaemic brain injury in mice deficient in cytosolic phospholipase A2. Nature 1997; 390:622–5.
182. Zeleznik AJ. In vivo responses of the primate corpus luteum to luteinizing hormone and chorionic gonadotropin. Proc Natl Acad Sci U S A 1998; 95:11002–7.

183. Maas S, Jarry H, Teichmann A. Paracrine actions of oxytocin, prostaglandin F2α, and estradiol within the human corpus luteum. J Clin Endocrinol Metab 1992; 74:306–12.
184. Karsch FJ, Sutton GP. An intra-ovarian site for the luteolytic action of estrogen in the rhesus monkey. Endocrinology 1976; 98:553–61.
185. Bennegard B, Hahlin M, Wennberg E, et al. Local luteolytic effect of prostaglandin F2α in the human corpus luteum. Fertil Steril 1991; 56:1070–6.
186. Auletta FJ, Kamps DL, Wesley M, et al. Luteolysis in the rhesus monkey: ovarian venous estrogen, progesterone, and prostaglandin F2α-metabolite. Prostaglandins 1984; 27:299–310.
187. Ristimaki A, Jaatinen R, Ritvos O. Regulation of prostaglandin F2α receptor expression in cultured human granulosa-luteal cells. Endocrinology 1997; 138:191–5.
188. Patwardhan VV, Lanthier A. Effect of prostaglandin F2α on the hCG-stimulated progesterone production by human corpora lutea. Prostaglandins 1984; 27:465–73.
189. Berisha B, Schams D. Ovarian function in ruminants. Domes Anim Endocrinol 2005; 29:305–17.
190. Bennegard-Eden B, Hahlin M, Kindahl H. Interaction between oxytocin and prostaglandin F2α in human corpus luteum? Hum Reprod 1995; 10:2320–4.
191. Khan-Dawood FS, Huang JC, Dawood MY. Baboon corpus luteum oxytocin: an intragonadal peptide modulator of luteal function. Am J Obstet Gynecol 1988; 158:882–91.
192. Khan-Dawood FS, Dawood MY. Human ovaries contain immunoreactive oxytocin. J Clin Endocrinol Metab 1983; 57:1129–32.
193. Anderson LE, Wu YL, Tsai SJ, et al. Prostaglandin F2α receptor in the corpus luteum: recent information on the gene, messenger ribonucleic acid, and protein. Biol Reprod 2001; 64: 1041–7.
194. Narumiya S, Sugimoto Y, Ushikubi F. Prostanoid receptors: structures, properties, and functions. Physiol Rev 1999; 79: 1193–226.
195. Yamamoto H, Endo T, Kiya T, et al. Activation of phospholipase D by prostaglandin F2α in rat luteal cells and effects of inhibitors of arachidonic acid metabolism. Prostaglandins 1995; 50: 201–11.
196. Stocco C, Lau LF, Gibori G. A calcium/calmodulin-dependent activation of ERK1/2 mediates JunD phosphorylation and induction of *nur77* and *20a-hsd* genes by prostaglandin F_{2a} in ovarian cells. J Biol Chem 2002; 277:3293–302.
197. Chen D, Fong HW, Davis JS. Induction of c-fos and c-jun messenger ribonucleic acid expression by prostaglandin F2α is mediated by a protein kinase C-dependent extracellular signal-regulated kinase mitogen-activated protein kinase pathway in bovine luteal cells. Endocrinology 2001; 142:887–95.
198. Strauss JF 3rd, Stambaugh RL. Induction of 20α-hydroxysteroid dehydrogenase in rat corpora lutea of pregnancy by prostaglandin F2α. Prostaglandins 1974; 5:73–85.
199. Stocco CO, Zhong L, Sugimoto Y, et al. Prostaglandin F_{2a} induced expression of 20a-hydroxysteroid dehydrogenase involves the transcription factor NUR77. J Biol Chem 2000; 275:37202–11.
200. Colles SM, Woodford JK, Moncecchi D, et al. Cholesterol interaction with recombinant human sterol carrier protein-2. Lipids 1995; 30:795–803.
201. Sandhoff TW, McLean MP. Prostaglandin F2a reduces steroidogenic acute regulatory (StAR) protein messenger ribonucleic acid expression in the rat ovary. Endocrine 1996; 5:183–90.
202. Sandhoff TW, McLean MP. Repression of the rat steroidogenic acute regulatory (StAR) protein gene by PGF2α is modulated by the negative transcription factor DAX- 1. Endocrine 1999; 10: 83–91.
203. Shea-Eaton W, Sandhoff TW, Lopez D, et al. Transcriptional repression of the rat steroidogenic acute regulatory (StAR) protein gene by the AP-1 family member c-Fos. Mol Cell Endocrinol 2002; 188:161–70.
204. Shea-Eaton WK, Trinidad MJ, Lopez D, et al. Sterol regulatory element binding protein-1a regulation of the steroidogenic acute regulatory protein gene. Endocrinology 2001; 142: 1525–33.
205. Stocco CO, Chedrese J, Deis RP. Luteal expression of cytochrome P450 side-chain cleavage, steroidogenic acute regulatory protein, 3b-hydroxysteroid dehydrogenase, and 20a-hydroxysteroid dehydrogenase genes in late pregnant rats: effect of luteinizing hormone and RU486. Biol Reprod 2001; 65:1114–9.
206. Behrman HR, Grinwich DL, Hichens M. Studies on the mechanism of PGF2α and gonadotropin interactions on LH receptor function in corpora lutea during luteolysis. Adv Prostaglandin Thromboxane Res 1976; 2:655–66.
207. Stocco C, Djiane J, Gibori G. Prostaglandin (PG) F2a and prolactin signaling: PGF2a-mediated inhibition of prolactin receptor expression in the Corpus luteum. Endocrinology 2003; 144:3301–5.
208. Curlewis JD, Tam SP, Lau P, et al. A prostaglandin (PG) F2a analog induces suppressors of cytokine signaling-3 expression in the corpus luteum of the pregnant rat: a potential new mechanism in luteolysis. Endocrinology 2002; 143:3984–93.
209. Stocco C. In vivo and in vitro inhibition of cyp19 gene expression by prostaglandin F_{2a} in murine luteal cells: implication of GATA-4. Endocrinology 2004; 145:4957–66.
210. Girsh E, Wang W, Mamluk R, et al. Regulation of endothelin-1 expression in the bovine corpus luteum: elevation by prostaglandin F2α. Endocrinology 1996; 137:5191–6.
211. Hinckley ST, Milvae RA. Endothelin-1 mediates prostaglandin F2α-induced luteal regression in the ewe. Biol Reprod 2001; 64:1619–23.
212. Foyouzi N, Cai Z, Sugimoto Y, et al. Changes in the expression of steroidogenic and antioxidant genes in the mouse corpus luteum during luteolysis. Biol Reprod 2005; 72:1134–41.
213. Behrman HR, Kodaman PH, Preston SL, et al. Oxidative stress and the ovary. J Soc Gynecol Investig 2001; 8:S40–2.
214. Musicki B, Kodaman PH, Aten RF, et al. Endocrine regulation of ascorbic acid transport and secretion in luteal cells. Biol Reprod 1996; 54:399–406.
215. Aten RF, Kolodecik TR, Rossi MJ, et al. Prostaglandin F2α treatment in vivo, but not in vitro, stimulates protein kinase C-activated superoxide production by nonsteroidogenic cells of the rat corpus luteum. Biol Reprod 1998; 59:1069–76.
216. Minegishi K, Tanaka M, Nishimura O, et al. Reactive oxygen species mediate leukocyte-endothelium interactions in prostaglandin F2α-induced luteolysis in rats. Am J Physiol 2002; 283:E1308–15.
217. Tanaka M, Miyazaki T, Tanigaki S, et al. Participation of reactive oxygen species in PGF2α-induced apoptosis in rat luteal cells. J Reprod Fertil 2000; 120:239–45.
218. Riley JC, Behrman HR. In vivo generation of hydrogen peroxide in the rat corpus luteum during luteolysis. Endocrinology 1991; 128:1749–53.
219. Pepperell JR, Wolcott K, Behrman HR. Effects of neutrophils in rat luteal cells. Endocrinology 1992; 130:1001–8.
220. Sugino N, Shimamura K, Tamura H, et al. Progesterone inhibits superoxide radical production by mononuclear phagocytes in pseudopregnant rats. Endocrinology 1996; 137:749–54.
221. Gupta S. Molecular signaling in death receptor and mitochondrial pathways of apoptosis (Review). Int J Oncol 2003; 22:15–20.
222. Earnshaw WC, Martins LM, Kaufmann SH. Mammalian caspases: structure, activation, substrates, and functions during apoptosis. Annu Rev Biochem 1999; 68:383–424.

223. Borner C. The Bcl-2 protein family: sensors and checkpoints for life-or-death decisions. Mol Immunol 2003; 39:615–47.
224. Slot KA, Voorendt M, de Boer-Brouwer M, et al. Estrous cycle dependent changes in expression and distribution of Fas, Fas ligand, Bcl-2, Bax, and pro- and active caspase-3 in the rat ovary. J Endocrinol 2006; 188:179–92.
225. Carambula SF, Matikainen T, Lynch MP, et al. Caspase-3 is a pivotal mediator of apoptosis during regression of the ovarian corpus luteum. Endocrinology 2002; 143:1495–501.
226. Yadav VK, Lakshmi G, Medhamurthy R. Prostaglandin F2α-mediated activation of apoptotic signaling cascades in the corpus luteum during apoptosis: involvement of caspase-activated DNase. J Biol Chem 2005; 280:10357–67.
227. Pate JL, Landis Keyes P. Immune cells in the corpus luteum: friends or foes? Reproduction 2001; 122:665–76.
228. Adashi EY. The potential relevance of cytokines to ovarian physiology: the emerging role of resident ovarian cells of the white blood cell series. Endocr Rev 1990; 11:454–64.
229. Brannstrom M, Pascoe V, Norman RJ, et al. Localization of leukocyte subsets in the follicle wall and in the corpus luteum throughout the human menstrual cycle. Fertil Steril 1994; 61:488–95.
230. Hehnke KE, Christenson LK, Ford SP, et al. Macrophage infiltration into the porcine corpus luteum during prostaglandin F2α-induced luteolysis. Biol Reprod 1994; 50:10–5.
231. Brannstrom M, Giesecke L, Moore IC, et al. Leukocyte subpopulations in the rat corpus luteum during pregnancy and pseudopregnancy. Biol Reprod 1994; 50:1161–7.
232. Kuranaga E, Kanuka H, Bannai M, et al. Fas/Fas ligand system in prolactin-induced apoptosis in rat corpus luteum: possible role of luteal immune cells. Biochem Biophys Res Commun 1999; 260:167–73.
233. Hosang K, Knoke I, Klaudiny J, et al. Porcine luteal cells express monocyte chemoattractant protein-1 (MCP- 1): analysis by polymerase chain reaction and cDNA cloning. Biochemical and biophysical research communications 1994; 199:962–8.
234. Townson DH, Warren JS, Flory CM, et al. Expression of monocyte chemoattractant protein-1 in the corpus luteum of the rat. Biol Reprod 1996; 54:513–20.
235. Paavola LG. The corpus luteum of the guinea pig. IV. Fine structure of macrophages during pregnancy and postpartum luteolysis, and the phagocytosis of luteal cells. Am J Anat 1979; 154: 337–64.
236. Bagavandoss P, Wiggins RC, Kunkel SL, et al. Tumor necrosis factor production and accumulation of inflammatory cells in the corpus luteum of pseudopregnancy and pregnancy in rabbits. Biol Reprod 1990; 42:367–76.
237. Abdo M, Hisheh S, Dharmarajan A. Role of tumor necrosis factor-α and the modulating effect of the caspases in rat corpus luteum apoptosis. Biol Reprod 2003; 68:1241–8.
238. Pru JK, Lynch MP, Davis JS, et al. Signaling mechanisms in tumor necrosis factor α-induced death of microvascular endothelial cells of the corpus luteum. Reprod Biol Endocrinol 2003; 1:17.
239. Adashi EY, Resnick CE, Packman JN, et al. Cytokine-mediated regulation of ovarian function: tumor necrosis factor α inhibits gonadotropin-supported progesterone accumulation by differentiating and luteinized murine granulosa cells. Am J Obstet Gynecol 1990; 162:889–96.
240. Chen YJ, Feng Q, Liu YX. Expression of the steroidogenic acute regulatory protein and luteinizing hormone receptor and their regulation by tumor necrosis factor α in rat corpora lutea. Biol Reprod 1999; 60:419–27.
241. Jo T, Tomiyama T, Ohashi K, et al. Apoptosis of cultured mouse luteal cells induced by tumor necrosis factor-α and interferon-γ. Anat Rec 1995; 241:70–6.
242. Penny LA, Armstrong D, Bramley TA, et al. Immune cells and cytokine production in the bovine corpus luteum throughout the oestrous cycle and after induced luteolysis. J Reprod Fertil 1999; 115:87–96.
243. Bebo BF, Jr., Linthicum DS. Expression of mRNA for 55-kDa and 75-kDa tumor necrosis factor (TNF) receptors in mouse cerebrovascular endothelium: effects of interleukin-1 β, interferon-γ and TNF-α on cultured cells. J Neuroimmunol 1995; 62:161–7.
244. Kurusu S, Sakaguchi S, Kawaminami M. Regulation of luteal prostaglandin F2α production and its relevance to cell death: an in vitro study using rat dispersed luteal cells. Prostaglandins Other Lipid Mediat 2007; 83:250–6.
245. Nagata S, Golstein P. The Fas death factor. Science 1995; 267:1449–56.
246. Roughton SA, Lareu RR, Bittles AH, et al. Fas and Fas ligand messenger ribonucleic acid and protein expression in the rat corpus luteum during apoptosis-mediated luteolysis. Biol Reprod 1999; 60:797–804.
247. Kuranaga E, Kanuka H, Furuhata Y, et al. Requirement of the Fas ligand-expressing luteal immune cells for regression of corpus luteum. FEBS Lett 2000; 472:137–42.
248. Goodman AL, Hodgen GD. Corpus luteum–conceptus–follicle relationships during the fertile cycle in rhesus monkeys: pregnancy maintenance despite early luteal removal. J Clin Endocrinol Metab 1979; 49:469–71.
249. Freeman ME, Neill JD. The pattern of prolactin secretion during pseudopregnancy in the rat: a daily nocturnal surge. Endocrinology 1972; 90:1292–4.

Chapter 27

The Molecular Landscape of Spermatogonial Stem Cell Renewal, Meiotic Sex Chromosome Inactivation, and Spermatic Head Shaping

Laura L. Tres and Abraham L. Kierszenbaum

27.1 Introduction

Spermatogenesis is a hormonally regulated process involving three sequential events: (1) the mitotic amplification of the spermatogonial cell progeny, (2) the completion of meiosis by the spermatocyte progeny, and (3) spermiogenesis, the gradual morphogenesis of the spermatid progeny. Mitosis, meiosis and spermiogenesis coexist in the seminiferous epithelium in association with a post-mitotic stable population of somatic Sertoli cells. Cell components of each spermatogonial, spermatocyte, and spermatid cell progeny remain connected by intercellular cytoplasmic bridges. Intercellular bridges are disrupted upon completion of spermiogenesis leading to the release in the seminiferous tubular lumen of single mature spermatids transported to the epididymal duct for acquisition of fertilizing activity. Several key cell cycle regulators have been shown to operate during the mitotic amplification of the spermatogonial progeny. During meiotic prophase, autosomal bivalents are engaged in prominent ribosomal RNA and non-ribosomal RNA transcriptional activity, in contrast with the transcriptional silencing of the condensed XY chromosomes. An autosomal bivalent is a synapsed (conjoined) chromosomal pair, excluding the sex chromosomes X and Y, observed during meiotic prophase I. Each member of a chromosomal bivalent (autosomes and X-Y) consists of two sister chromatids that will disjoin (separate) upon completion of meiosis to produce a haploid genome (spermatid). During spermiogenesis, gradual genetic inactivation of the spermatid genome correlates with spermatid head shaping. The acrosome-acroplaxome-manchette complex is emerging as a significant player in spermatid head shaping as well as in the assembly of the sperm head–tail coupling apparatus and the development of the outer dense fiber-axoneme-containing sperm tail. The acroplaxome is a cytoskeletal plate bordered by a desmosome-like marginal ring fastening the descending recess of the acrosomal sac to the nuclear envelope of the spermatid. The manchette is a transient microtubular-containing structure developed beneath the acroplaxome and encircling the elongating spermatid nucleus. This chapter is restricted to recent developments in the bioregulation of the spermatogonial stem cell progeny, the process of transcriptional inactivation of the XY bivalent, and the steps leading to spermatid head shaping. These are three relevant aspects that, when disrupted, can lead to male infertility.

L.L. Tres (✉)
Department of Cell Biology and Anatomy, The Sophie Davis School of Biomedical Education, The City University of New York Medical School, New York, NY, USA
e-mail: tres@med.cuny.edu

27.2 General Organization and Dynamics of the Seminiferous Epithelium

The seminiferous epithelium consists of several cellular layers representing the coexistence of overlapping spermatogenic cell progenies. Each spermatogenic cell progeny starts periodically from a spermatogonial reserve stem cell. A given cross-section of the mammalian seminiferous epithelium depicts a combination of cell types, called a *cellular association*, representing the coexistence of cellular members corresponding to preceding and subsequent spermatogenic cell progenies. Each cellular association corresponds to a *stage* of the cyclic process of spermatogenesis. A *cycle* is defined by the time it takes for a sequence of cellular associations (or *stages of the cycle*) to change at a particular point of the seminiferous tubule. The number of cellular associations is constant for any given species. In man, one cycle lasts 16 days and consists of six cellular associations. It takes four cycles (64 days) to develop testicular sperm from the starting spermatogonial stem cell. The alignment of successive cellular associations along the length of a seminiferous tubule of rodents is known as *spermatogenic wave* [1].

The topography of the seminiferous epithelium and the space allocation to the developing spermatogenic cells are

P.J. Chedrese (ed.), *Reproductive Endocrinology*,
DOI 10.1007/978-0-387-88186-7_27, © Springer Science+Business Media, LLC 2009

controlled by Sertoli cells. Developing spermatogenic cells occupy niches and crypts in Sertoli cells. Like a mechanical and functional bridge, a Sertoli cell spans from the seminiferous tubular wall to the lumen of the seminiferous tubule. This bridge-like arrangement is further refined by basal tight or occluding junctions linking adjacent Sertoli cells. Consequently, the seminiferous epithelium becomes divided into a *basal compartment*—below the intersertolian tight junctions—and an *adluminal compartment*—above the tight junctions. The spermatogonial cell progeny occupies niches in the basal compartment while members of the spermatocyte progeny are housed in the adluminal compartment, more specifically in *niches* within Sertoli cells. Maturing spermatids occupy *crypts* at the luminal surfaces of Sertoli cells.

27.3 Extrinsic and Intrinsic Bioregulation of the Spermatogonial Cell Progeny

Spermatogonial stem cells are always present in the seminiferous epithelium and undergo a self-renewal process throughout the lifespan of the animal. One of the cells derived from the self-renewal cycle enters the spermatogenic pathway, while the other remains quiescent as an undifferentiated reserve cell. The molecular regulation of the spermatogonial self-renewal cycle and the commitment of the derived cell products to either spermatogenesis or a temporary quiescent state are essential for the steady state output of sperm upon completion of each spermatogenic cycle. Extrinsic and intrinsic aspects of the molecular regulation of the spermatogonial stem cell progeny have attracted considerable interest in recent years.

Members of the conjoined spermatogonial cell progeny are directly in contact with Sertoli cells and the wall of the seminiferous tubule. Members of the progeny are readily accessible to regulatory and signaling molecules derived from Sertoli cells or transported across the basal lamina. Under the influence of follicle stimulating hormone (FSH), Sertoli cells secrete glial cell line-derived neurotrophic factor (GDNF), a small protein shown to promote the survival and differentiation of enteric neurons and prevention of apoptosis of motor neurons induced by axotomy. GDNF is a member of the transforming growth factor-β superfamily. GDNF is also involved in spermatogonial stem cell self-renewal [2]. A lack of GDNF expression decreases the self-renewal of spermatogonial stem cells, whereas overexpression of GDNF in transgenic mice stimulates the proliferation of undifferentiated (reserve) spermatogonia. How does GDNF exert its stimulatory growth function? GDNF appears to exert its effect through the RET/GDNF family receptor α1 (GFRα1) [3]. RET (rearranged during transfection) is a protooncogene tyrosine kinase receptor whose ligands are members of the GDNF family, including GDNF, neurturin, artemin, and persephin. In human embryos, RET is expressed in a cranial population of neural crest cells and in the developing nervous and urogenital systems. When Tyr-1062 in the intracellular region of RET is phosphorylated, several adaptor and effector proteins bind to the phosphotyrosine residue and activate several intracellular anti-apoptotic signaling pathways. RET Y1062F homozygous mice, in which Tyr-1062 was replaced by phenylalanine, showed atrophic testes determined by a decrease in the RET-expressing spermatogonial cell progeny [3]. These observations suggest that RET signaling mediated by phospho-Tyr-1062 is essential for self-renewal and differentiation of the spermatogonial stem cell progeny. How adaptor and effector proteins trigger a signaling cascade leading to the maintenance of spermatogonial cell progeny needs to be determined.

Glial cell line-derived neurotrophic factor effects on the spermatogonial cell progeny are mediated by an extrinsic mechanism initiated in Sertoli cells. An intrinsic mechanism can also determine which spermatogonial cell resulting from the self-renewal cycle may enter spermatogenesis while the other remains as a reserve cell. An intrinsic mechanism involves the transcriptional repressor Plzf encoded by the *Zfp145* gene. Plzf (for promyelocytic leukemia zinc-finger) belongs to the POK (POZ [Poxviruses and Zinc-finger] and Krüppel) family of transcriptional repressors [4]. The N-terminal POZ domain, common to several zinc finger-containing transcription factors, facilitates protein–protein interactions during hematopoiesis, neurogenesis, adipogenesis, osteoclastogenesis, and muscle differentiation. Plzf participates in chromatin remodeling by recruiting DNA histone acetylases and nuclear co-repressors and exerting growth-suppressive activities associated with the buildup of cells in the G0/G1 phase of the cell cycle.

Plzf expression is first detected in primordial germinal cells of the fetal testis and reaches a maximum around 1 week postnatal when prespermatogonia (also called gonocytes) exit their cell cycle arrest to initiate a mitotic proliferative cycle of the spermatogonial cell progeny [reviewed in 5]. Reserve spermatogonia can be distinguished from mitotically-dividing spermatogonia using cyclin D1, a marker of mitotically active spermatogonia, and cyclin D2, a marker of undifferentiated spermatogonia progressing into the mitotic amplification cycle to generate type A spermatogonia [6]. Plzf is detected in quiescent non-differentiating prespermatogonial stem cells not expressing cyclin D1 and cyclin D2. A genetic model system that validates the role of Plzf in the spermatogonial stem cell progeny is the Plzf-null/luxoid mutant mice characterized by a defect in spermatogonial self-renewal [7, 8]. Inactivation of Plzf expression results in an initial proliferation peak followed by

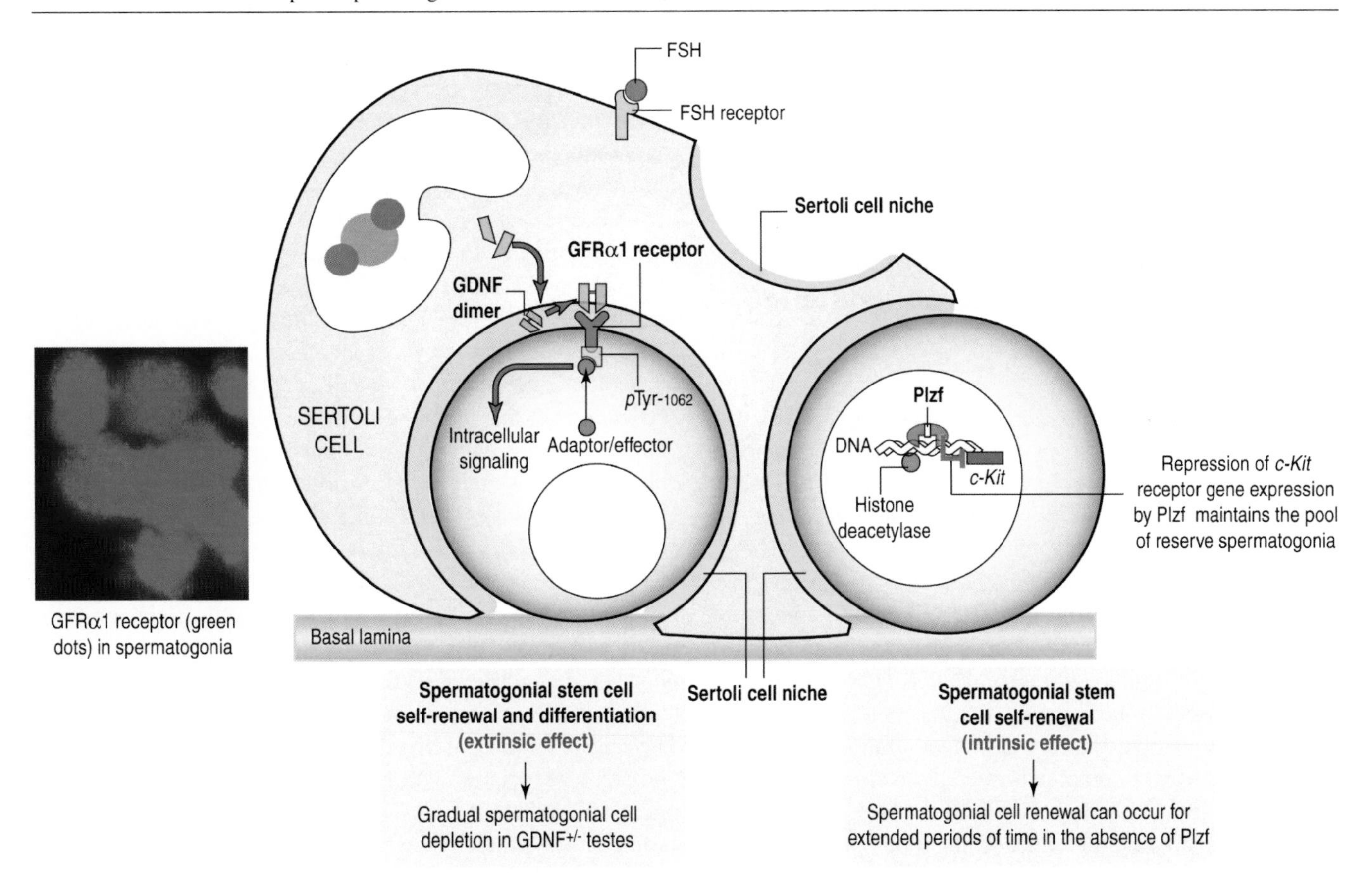

Fig. 27.1 Extrinsic effect of GDNF and intrinsic effect of Plzf on spermatogonial stem cell self-renewal and differentiation. Sertoli cells lodges members of the spermatogonial cell progeny in niches adjacent to the basal lamina of the seminiferous tubule. Under the influence of FSH, GDNF is produced and released by Sertoli cells. GDNF binds to the GRFα1 receptor on spermatogonial cell surfaces (shown by indirect immunofluorescence in the inserted panel) and induces the phosphorylation of Tyr-1062 present in the intracellular domain of the receptor. Phosphorylated Tyr-1062 enables binding of adapor/effector proteins to GRFα1, a mechanism that triggers an intracellular signaling cascade leading to spermatogonial stem cell self-renewal and eventual mitotic amplification. In contrast, an intrinsic effect mediated by the transcription factor Plzf enables epigenetic remodeling changes in spermatogonial cell chromatin mediated by histone deacetylase and other proteins to regulate self-renewal. Plzf also represses the gene encoding c-Kit receptor to prevent the differentiation of spermatogonial stem cells and maintain the pool of reserve spermatogonial stem cells. Depletion of the spermatogonial stem cell progeny occurs after an extended period of time in the absence of GDNF and Plzf. These observations suggest that additional intrinsic and/or extrinsic mechanisms may operate during a limited interval to maintain the spermatogonial stem cell progeny when GDNF and Plzf are defective

a decrease in the spermatogonial mitotic cell amplification. A progressive loss of spermatogonia with age, starting about 2 weeks postnatally, correlates with an increase in apoptosis leading to a gradual disruption of spermatogenesis with a consequent decrease in sperm production. It appears that the function of Plzf as a cell cycle regulator is to maintain the population of undifferentiated spermatogonia, a key step for feeding constantly mitotic spermatogonial cell progenies into the meiotic and spermiogenic steps of spermatogenesis. It remains to be determined whether Plzf can operate either directly or indirectly through the recruitment of developmentally regulated genes. Within this context, it has been reported that Plzf represses the expression of the *c-kit* receptor gene, thus preventing its ligand stem cell factor from stimulating the differentiation of spermatogonial stem cells. By repressing *c-kit* gene expression, Plzf maintains the pool of reserve spermatogonial stem cells [9]. Figure 27.1 provides a summary of the mechanism of action of GDNF and Plzf in the regulation of the spermatogonial cell progeny.

Available evidence indicates that mutations in genes encoding Plzf and GDNF lead to a defect in mammalian spermatogonial cell self-renewal. An intriguing observation is that a loss of Plzf depletes the spermatogonial cell progeny within a 2-month period. This time span suggests that spermatogonial cell self-renewal can proceed for a limited time in the absence of Plzf. GDNF is secreted by Sertoli cells under the influence of FSH. $GDNF^{-/-}$ exhibits embryonic lethality and most of the available data derive from testes of the heterozygous $GDNF^{+/-}$ mouse mutant because the homozygous $GDNF^{-/-}$ is lethal. The Sertoli cell, regarded as a niche cell, can support and regulate spermatogonial stem cell self-renewal. The mechanical and regulatory role of Sertoli cells of spermatogenesis is not surprising. In fact, spermatogenesis in vitro requires a Sertoli cell–spermatogenic cell partnership for spermatogenic cells to develop [10, 11].

27.4 Molecular Aspects Chromosome Transcription and Inactivation During Meiotic Prophase

The mitotic amplification cycle by spermatogonia is followed by the commitment of conjoined type B spermatogonia to meiosis. One of the primary objectives of meiosis is to generate haploid spermatids after two consecutive meiotic cell divisions. Homologous chromosomes orient and align with each other during meiotic prophase I to form bivalents and establish synapsis mediated by a protein-containing synaptonemal complex and undergo genetic recombination. Numerous DNA double-strand breaks (DSB) indicate recombination between homologous chromosomes. The repair of meiotic DSB involves homologous recombination proteins, including the Mre11 complex formed by the proteins Mre11, Rad50, and Nbs1. Molecular details of the Mre11 complex in several organisms and the effect of mutations of the complex have been recently reviewed [12]. The molecular components of the synaptonemal complex have been described [13].

Two significant molecular aspects of meiotic prophase I not seen during mitotic prophase are the nucleolar organization by autosomal bivalents and meiotic sex chromosome inactivation (MSCI). Meiotic autosomal nucleolar organization (MANO) begins at leptotene, when meiotic chromosomes, each formed by conjoined sister chromatids, start to become visible. MANO is completed during pachytene, when transcriptionally active chromatin loops of the synapsed autosomal bivalents, each stabilized by the synaptonemal complex, are also engaged in non-ribosomal RNA synthesis [reviewed in 14]. Contrasting with the substantial transcriptional activity of the uncondensed autosomal bivalents is the transcriptional inactivation of the condensing XY bivalent (MSCI).

The X and Y chromosomes are transcriptionally active during spermatogonial mitotic amplification. Upon entering meiotic prophase I, X and Y chromosomes are still transcriptionally active at zygotene. Autosomal bivalents are transcriptionally active as synapsis is progressing during zygotene and remain transcriptionally active during the early part of pachytene. During the zygotene-early pachytene transition, the synapsis between autosomal bivalents is gradually completed and nucleolar and non-nucleolar RNA transcription are at a maximum. In contrast, the partially paired XY bivalent, formed by chromosomes of different length, becomes transcriptionally silent, condenses, and is visualized as a peripheral nuclear mass, called the XY body, a state that persists throughout the rest of pachytene. X and Y chromosomes synapse extensively through a synaptonemal complex along their homologous pairing segments during late zygotene-early pachytene. The pairing segments have the characteristics of the autosomal-type synaptonemal complex consisting of two axial elements with a central element in a zipper-like fashion. In the unpaired region, axial core segments of the X and Y chromosomes, an equivalent to the lateral element of the synaptonemal complex, diverge from each other and remain unsynapsed [14]. Chromatin loops associate with the components of the synaptonemal complex and axial cores. Most genetic inactivation of the X and Y chromosomes is maintained during spermiogenesis [15].

The molecular events leading to MSCI of the XY bivalent have been described. Briefly, a key element in MSCI is the histone H2a variant H2AX. H2AX is abundant in testis, is prevalent in the nucleosome in meiotic cells, plays a role in the DNA-damage response following DSB of DNA, is phosphorylated at serine-139 to form γH2AX, and recruits proteins of the DNA-repair pathway to the sites of DNA breaks. The relevance of H2AX is demonstrated by a complete meiotic arrest associated with MSCI failure in H2AX-null male mice [16]. Whether H2AX phosphorylation at serine-139 is essential for MSCI needs to be determined. During male meiotic prophase, H2AX phosphorylation is detected during leptotene, when DNA DSB take place [17] and when X and Y chromosome synapse during the late zygotene–early pachytene transition [16], marking the onset of MSCI.

H2AX phosphorylation is determined by the DNA-repair protein ATR (ataxia telangectasia and Rad3 related), a member of the PI3-kinase-like family. ATR co-localizes with phosphorylated H2AX on the XY bivalent when MSCI starts and remains there until H2AX dephosphorylates during the diplotene-metaphase I transition [17]. ATR is targeted to the XY bivalent by the tumor suppressor protein BRCA1 (breast cancer 1).

BRCA1 and ATR associate with the unsynapsed axial cores of the XY bivalent before MSCI starts. ATR relocates from the axial cores to the chromatin loops concurrent with the visualization of phosphorylated H2AX along the loops. BRCA1 and ATR are not seen along the paired region of the XY bivalent, thus indicating that BRCA1 and ATR associate with the unsynapsed axial cores of the XY pair. The association of BRCA1, ATR, and phosphorylated H2AX with unsynapsed axial cores is not restricted to the XY bivalent; it is also observed in the unsynapsed regions of meiotic chromosome 16 resulting from a reciprocal translocation to the sex chromosomes [18]. Similar to H2AX-null mice, MSCI is defective in *Brca1* mutant mice. In the *Brca1* mutant mouse, H2AX phosphorylation is visualized throughout the nucleus of primary spermatocytes but not on the XY bivalent. The aberrant distribution of phosphorylated H2AX

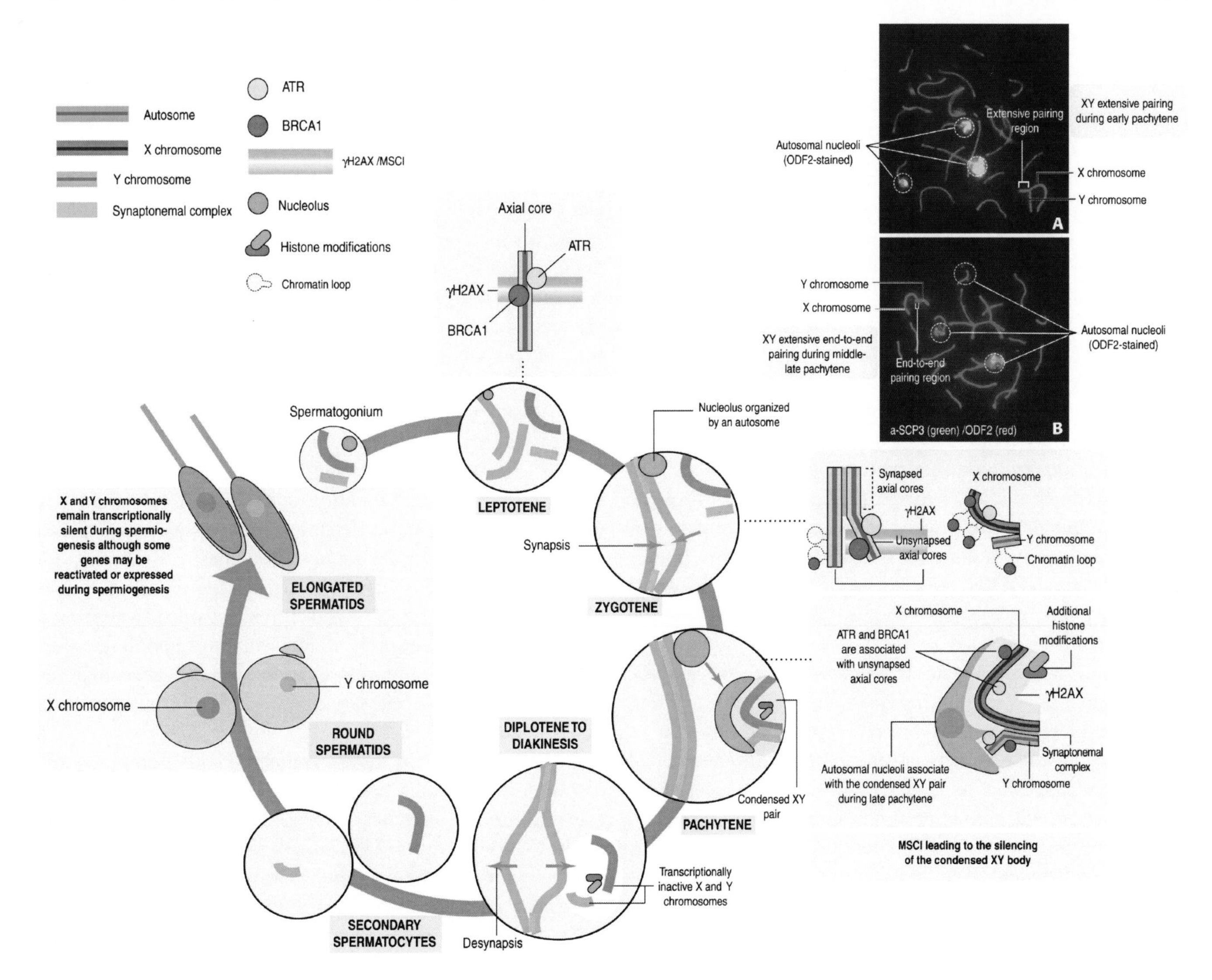

Fig. 27.2 Autosomal and XY chromosomal transcriptional activities during spermatogenesis and distribution of ATR, BRCA1, and γH2AX sites in meiotic chromosomes. Autosomes and X and Y chromosomes are engaged in ribosomal and non-ribosomal transcriptional activity during the mitotic amplification of the spermatogonial stem cell progeny. Non-ribosomal transcription by autosomes and X and Y chromosomes persists during leptotene and early zygotene. Nucleolar function is a relevant aspect of meiotic autosomal nucleolar organizing activity (MANO). X and Y chromosomes are not involved in nucleolar organization. ATR and BRCA1 are associated with the unsynapsed segments of the axial cores of autosomes and X and Y chromosomes, concurrent with chromatin displaying phosphorylated H2AX. H2AX phosphorylation occurs in response to meiotic DNA double strand breaks. X and Y chromosomes synapse extensively during late zygotene-early pachytene through a synaptonemal complex. ATR and BRCA1 are present in the unsynapsed axial cores of the XY bivalent. The chromosomal axial cores will become the lateral elements of the synaptonemal complex when synapsis occurs. H2AX phosphorylation, facilitated by ATR transported by BRCA1 to the XY chromatin, determines MSCI and transcription inactivation of the XY condensed pair adjacent to the intact nuclear envelope. As MSCI occurs, nucleolar masses detach from autosomal bivalents involved in MANO, migrate, and associate with the condensed XY pair. Simultaneous histone modifications maintain the MSCI condition during diplotene-diakinesis of meiotic prophase I when ATR, BRCA1, and γH2AX are no longer visible. Genetic inactivation and condensation of X and Y chromosomes (postmeiotic sex chromosome repression) persists during spermiogenesis, although some genetic activity has been detected. In contrast, autosomes remain transcriptionally active in round spermatids until somatic histones are replaced by transient proteins and later by the more permanent arginine- and lysine-rich protamines during spermatid elongation. Inserts **A** and **B** show the lateral elements stained with anti-SCP3 (synaptonemal complex 3 protein; *green*) of the synapsed chromosomes and associated nucleolar masses to autosomal bivalents (stained with anti-ODF2; *red*). Note in **A** the extensive pairing of the XY bivalent during early pachytene and the end-to-end pairing of the XY bivalent during middle-late pachytene in **B**

has been interpreted as consequence of a disrupted distribution of ATR, an event dependent on the presence of BRCA1.

A general conclusion is that unsynapsed autosomal and XY chromosome regions are silenced during meiotic prophase and that the BRCA1-ATR-phosphorylated H2AX complex [19], and presumably other chromosomal proteins, contribute to genetic silencing. An unanswered question is whether the unsynapsed segments of autosomal bivalents during zygotene, known to be transcriptionally active, operate independently of the ATR-BRCA1-phosphorylated H2AX silencing mechanism or are controlled by an undetermined mechanism that prevents silencing of specific chromosomal segments.

Although the BRCA1-ATR protein complex accounts for the phosphorylation of H2AX, a determinant of MSCI, other post-translationally modified proteins are associated with the chromatin of the XY bivalent [reviewed in 20]. Further studies should determine whether the emerging catalog of post-translationally modified histones (for example, ubiquitylated histone H2A, H2AFY, and other modified histones) are directly involved in determining and/or maintaining the MSCI condition.

A largely neglected issue is the significance of MANO, a process that starts during the leptotene–zygotene transition and is completed during pachytene. A puzzling observation is the detachment of nucleolar masses from the autosomal bivalent putative sites and their migration and segregation in association with the transcriptionally silent XY bivalent undergoing MSCI. Whether the relocation of nucleolar masses to the condensed and transcriptionally silent XY bivalent has a role in MSCI needs to be determined. A recent study has demonstrated the presence of the protein ODF2 in the nucleus of primary spermatocytes and in the autosomal nucleoli in particular [21]. ODF2 is a major component of the outer dense fibers of the sperm tail, the acroplaxome (see below), and the centrosome of somatic and spermatogenic cells. This observation is relevant to the finding that mice expressing a truncated form of ODF2 exibit preimplantation embryonic lethality [22]. Whether defective ODF2 expression in the embryo affects nucleolar function needs to be determined. Figure 27.2 provides a summary of the relevant aspect of MSCI and MANO during male meiotic prophase (mouse).

27.5 Molecular Aspects of Spermatid Head Shaping

Normal spermatid head shaping is regarded as essential for male fertility. Recent studies have called attention to the role of the acrosome-acroplaxome-manchette complex in sperm head shaping [reviewed in 23, 24]. The acrosome assembles from proacrosomal vesicles derived from the Golgi and is essential for fertilization. The acroplaxome is a cytoskeletal plate outlined by a desmosome-like marginal ring anchoring the developing acrosome to the spermatid nuclear envelope [25]. The manchette is a transient microtubule-containing structure assembled caudally to the acroplaxome and encircling the elongating spermatid nucleus. One of the functions of the manchette is the intramanchette transport of cargos to the developing spermatid tail and an involvement in nucleo-cytoplasmic exchange during the condensation and silencing of the spermatid nucleus [26]. The acrosome and acroplaxome persist as sperm head components; the manchette instead disappears upon completion of spermatid head elongation. X and Y chromosomes appear to exist in a transcriptionally repressed state during spermiogenesis, a continuation of the MSCI state achieved during meiotic prophase. However, very few X- and Y-linked genes are expressed during spermiogenesis. Examples are the Y-encoded genes *Ssty1* and *Ssty2* (spermiogenesis specific transcript on the Y 1 and 2, respectively), whose lack is associated with sperm head abnormalities and male infertility [27]. The precise nature of sperm head abnormalities in mouse with reduced *Ssty* gene expression has not been determined.

A large number of mouse mutants lacking an acrosome display round-headed spermatids and sperm and are infertile. Acrosome biogenesis starts early in spermiogenesis when Golgi-derived proacrosomal vesicles tether along the acroplaxome attached to the spermatid nuclear envelope. Proacrosomal vesicles fuse with each other to form the acrosome sac. The final size and shape of the acroplaxome involves a steady-state vesicular shuttling from the Golgi to the acrosome and vesicular retrival by the acroplaxome. A lack of acrosome development in Hrb-deficient mice [28] and GOPC-deficient mice [29] correlates with the development of round-headed sperm and infertility. Hrb protein (also called Rab or hRip) binds to the cytosolic surface of proacrosomal transporting vesicles. A lack of Hrb prevents vesicles from fusing and forming the acrosome. GOPC is associated with the trans-Golgi region in round spermatids and a lack of GOPC prevents vesicle transport from the Golgi to the acrosome.

The transport mechanism of Golgi-derived proacrosomal vesicles to the acroplaxome involves motor proteins using F-actin and/or microtubule tracks. Myosin-Va utilizes F-actin and kinesin uses microtubule tracks. The presence of myosin-Va-decorated proacrosomal vesicles in the Golgi apparatus and in vesicles aligned along the acroplaxome in Hrb mutant and wild-type spermatids suggest a molecular motor-mediated vesicle transport. An F-actin-based myosin Va motor-driven mechanism may coexist with a microtubule-based kinesin motor, KIFC1, which was recently reported to be involved in acrosome biogenesis and vesicle transport [30]. Members of the membrane-bound Rab GTPase protein family, in particular Rab27a and Rab27b, are known to facilitate the interaction of myosin-Va with the vesicles [reviewed 31]. Essentially, the transport of proacrosome vesicles during acrosome biogenesis may share intracellular traffic features of melanosomes in mouse melanocytes [reviewed in 32], consisting in shifting cargos from microtubule to F-actin tracks and vice versa. It is likely that the existence of F-actin and microtubule tracks and the corresponding motors represent a mechanism necessary for the appropriate and timely delivery of Golgi-derived proacrosomal vesicles to the acroplaxome and to the head–tail coupling apparatus and the developing tail through intramanchette transport. F-actin coexists with microtubules in the manchette [33]. The capture of proacrosomal vesicles by myosin-Va involves an adaptor protein associated with the proacrosomal vesicular membrane [22]. Further studies should determine how microtubule-base molecular motors capture proacrosomal vesicles and transport them to the acroplaxome site and along microtubules of the manchette.

Taken together, these findings suggest that the acroplaxome plays a significant role in acrosome biogenesis and in determining the assembly site of the manchette. Data from a number of mutants show that abnormal spermatid head shape is associated with alteration in the position and shape of the acrosome-acroplaxome-manchette complex. A relevant aspect is the potential role that the acrosome-acroplaxome-manchette complex plays in spermatid head shaping. The acroplaxome consists of a malleable F-actin/keratin 5-containing cytoskeletal plate and a desmosome-like marginal ring encircling the spermatid

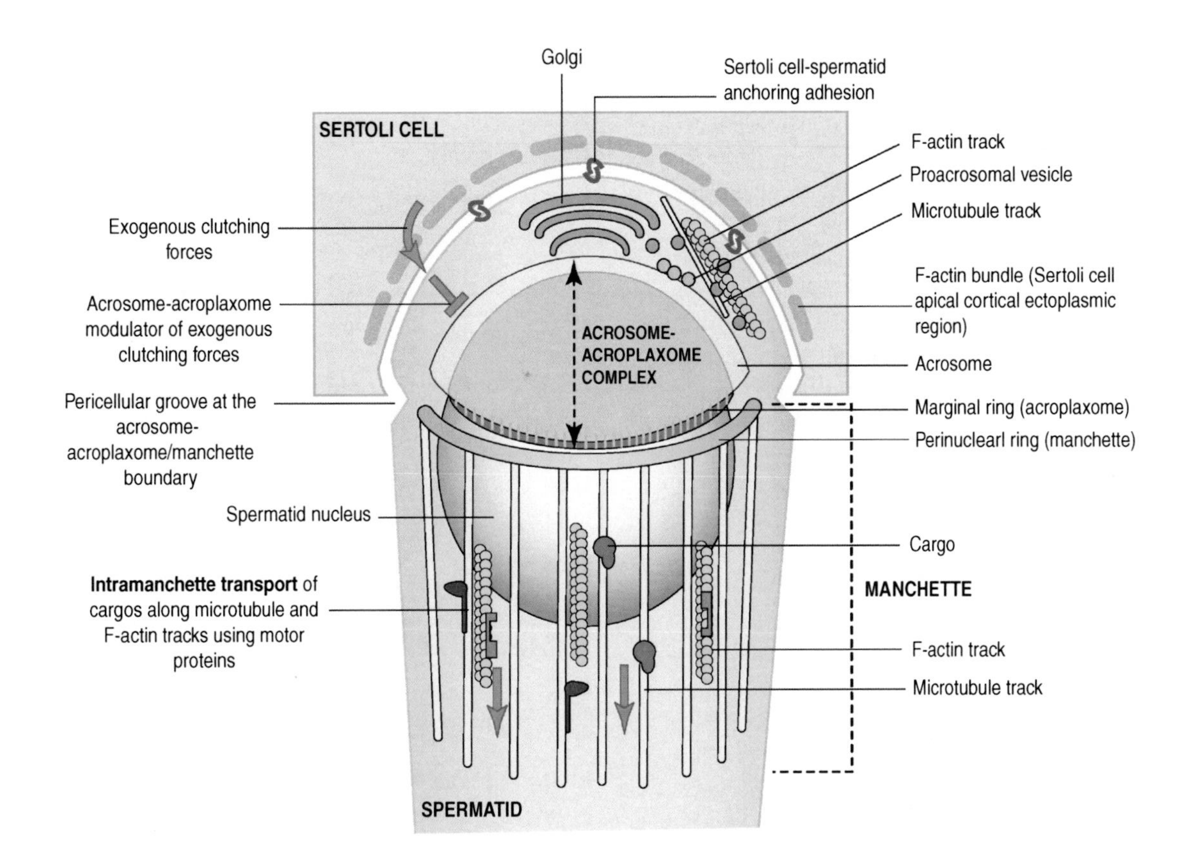

Fig. 27.3 The acrosome-acroplaxome-manchette complex during spermatid head shaping. The elongating spermatid head is partially surrounded by Sertoli cell F-actin hoops encircling the upper region of the cell. Exogenous clutching forces applied by the F-actin hoops are modulated inside the spermatid by the acrosome-acroplaxome complex. The Sertoli cell-spermatid relationship is stabilized by cell adhesion molecules. Golgi-derived proacrosomal vesicles are transported by motor proteins along F-actin and microtubule tracks to the acroplaxome where vesicular fusion determines the formation of the acrosome sac. The acroplaxome, an F-actin/keratin 5-containing cytoskeletal plate, links the developing acrosome to the spermatid nuclear envelope. An acroplaxome desmosome-like marginal ring links the caudal recess of the acrosome to the nuclear envelope and its subjacent nuclear lamina. As the acrosome-acroplaxome complex advances in its development, the perinuclear ring of the manchette develops adjacent to the marginal ring of the acroplaxome. A peri-cellular groove marks the acroplaxome/manchette ring boundary. Microtubules and associated F-actin enable the intramanchette transport of cargos to the developing head–tail coupling apparatus and developing spermatid tail (not shown). A gradual reduction in the diameter of the marginal ring of the acroplaxome and the subjacent perinuclear ring of the manchette steers the apical-to-caudal elongation of the spermatid head

nucleus. The acroplaxome plate and marginal ring anchor the developing acrosome to the spermatid nucleus. During spermatid elongation, the two overlapping rings (the marginal rings of the acroplaxome and perinuclear ring of the manchette) reduce their diameter to fit, in a sleeve-like fashion, the decreasing diameter of the elongating spermatid nucleus. The acroplaxome may provide a planar scaffold to modulate exogenous constriction forces generated by Sertoli cell F-actin hoops during spermatid head elongation. The dynamics of the F-actin cytoskeleton, one of the components of the acroplaxome and Sertoli cell hoops, can be regulated by tyrosine kinases, which target cortactin, an F-actin-associated protein. Tyrosine phosphorylation of cortactin correlates with a reduction in the crosslinking properties of F-actin, a mechanism that can modify the F-actin scaffold in response to exogenous clutching forces exerted by Sertoli cell F-actin hoops. Phosphorylated cortactin and the tyrosine kinase Fer are present in the acroplaxome, thus supporting a mechanism of F-actin remodeling during spermatid head shaping [24]. Keratin 5, an additional component of the acroplaxome, may also undergo dynamic reorganization during spermatid head elongation by an undetermined mechanism. Further studies are required to elucidate additional details of the composition of the acroplaxome and its marginal ring as well as the coexistence of additional structural and functional molecules involved in endowing the acroplaxome with malleable properties required for spermatid head shaping. Figure 27.3 provides a summary of the topography of the developing spermatid head and possible function of the acrosome-acroplaxome-manchette complex during spermatid head shaping.

27.6 Concluding Remarks

Ongoing research has provided insights on major aspects of spermatogenesis, including the dynamics of the spermatogonial stem cell renewal cycle, the selective transcriptional activities of autosomal and XY bivalents highlighted by the MSCI process, and the participation of the acrosome-acroplaxome-manchette complex in spermatid head shaping. These three aspects of spermatogenesis have far reaching consequences on understanding causes of male infertility that may occur at any of the three steps of spermatogenesis: spermatogonial mitotic amplification, meiosis, and spermiogenesis. There are several aspects that still remain unclear. If a mechanism exists to dictate the entry of a stem cell into a spermatogenic cycle while a daughter stem cell remains as reserve, what mechanism regulates the consecutive rounds of mitotic amplification of the spermatogonial progeny before an eventual entry to meiotic prophase I? If H2AX phosphorylation is triggered by the BRCA1-ATR complex associated with the unsynapsed chromosomal segments to silence selective chromosomal regions, what type of genes are present in these segments and what role do they play in meiosis? How do selective genes manage to escape the condition of MSCI during spermiogenesis and ensure normal sperm development? If spermatid head remodeling depends on the pliability of the acroplaxome, how complex is the composition of the acroplaxome plate and its marginal ring and what dynamic mechanisms participate in the remodeling of the acroplaxome during spermatid head shaping? These are some of the questions left to be address in our quest of a better understanding of the major causes of male infertility.

27.7 Glossary of Terms and Acronyms

ATR: DNA repair protein, member of the PI3-kinase-like family

***Brca1*:** breast cancer 1 gene

***c-kit*:** cellular homolog of the feline sarcoma viral oncogene v-kit

cyclin D1 and D2: cell cycle-regulatory genes

DSB: double-strand breaks

FSH: follicle stimulating hormone

GDNF: glial cell line-derived neurotrophic factor

GOPC: Golgi-associated PDZ and coiled-coil motif containing

H2AFY: H2A histone family, member Y (also known as histone macroH2A1)

H2AX: variant of the histone H2a

Hrb: Asn-Pro-Phe (NPF) motif-containing protein (also called Rab or hRip)

hRip: human immunodeficiency virus Rev-interacting protein.

KIFC: kinesin family member C

MANO: meiotic autosomal nucleolar organization

MCSI: meiotic sex chromosome inactivation

Mre11: meiotic recombination 11 protein

Nbs1: Nijmegen breakage syndrome 1

ODF2: Outer dense fiber 2

Plzf: promyelocytic leukemia zinc-finger, a transcriptional repressor encoded by the *Zfp145* gene

POK: Poxviruses and zinc-finger (POZ) and Krüppel family of transcription repressors

POZ: Poxviruses and zinc-finger

Rab: member of the Ras superfamily of monomeric G proteins

Rad3: a DNA helicase repair protein

Rad50: Mre11-interacting protein with binding affinity to double stranded DNA.

RET: protooncogene tyrosine kinase receptor that binds members of the GDNF family

Ssty 1 and Ssty 2: Y-linked spermiogenesis specific transcript

Zfp145: zinc finger protein 145 gene

References

1. Kierszenbaum AL. Spermatogenesis. In: Kierszenbaum AL. Histology and Cell Biology: An Introduction to Pathology, Second edition. Philadelphia:Mosby, 2007:569–96.
2. Meng X, Lindahl M, Hyvönen ME, et al. Regulation of cell fate decision of undifferentiated spermatogonia by GDNF. Science2000; 287:1489–93.
3. Jijiwa M, Kawai K, Fukihara J, et al. GDNF-mediated signaling via RET tyrosine 1062 is essential for maintenance of spermatogonial stem cells. Genes Cells 2008; 13:365–74.
4. Costoya JA. Functional analysis of the role of POK transcriptional repressor. Brief Funct Genomic Proteomic 2007; 6:8–18.
5. Tres LL, Kierszenbaum AL. The ADAM-integrin-tetraspanin complex in fetal and postnatal testicular cords. Birth Defects Res C Embryo Today 2005; 75:130–41.
6. Beumer TL, Roepers-Gajadien HL, Gademan IS, et al. Involvement of the D-type cyclins in germ cell proliferation and differentiation in the mouse. Biol Reprod 2000; 63:1893–98.
7. Costoya JA, Hobbs RM, Barna M, et al. Essential role of Plzf in maintenance of spermatogonial stem cells. Nat Genet 2004; 36:653–9.
8. Buaas FW, Kirsh AL, Sharma M, et al. Plzf is required in adult male germ cells for stem cell self-renewal. Nat Genet 2004; 36:647–52.
9. Filipponi D, Hobbs RM, Ottolenghi S, et al. Repression of kit expression by Plzf in germ cells. Mol Cell Biol 2007; 27: 6770–81.
10. Kierszenbaum AL. Mammalian spermatogenesis in vivo and in vitro: a partnership of spermatogenic and somatic cell lineages. Endocr Rev 1994; 15:116–34.
11. Marh J, Tres LL, Yamazaki Y, et al. Mouse round spermatids developed in vitro from preexisting spermatocytes can produce normal offspring by nuclear injection into in vivo-developed mature oocytes. Biol Reprod 2003; 69:169–76.
12. Borde V. The multiple roles of the Mre11 complex for meiotic recombination. Chromosome Res 2007; 15:551–63.
13. Page SL, Hawley RS. The genetics and molecular biology of the synaptonemal complex. Annu Rev Cell Dev Biol 2004; 20:525–58.
14. Tres LL. XY chromosomal bivalent: nucleolar attraction. Mol Reprod Dev 2005; 72:1–6.
15. Namekawa SH, Park PJ, Zhang LF, et al. Postmeiotic sex chromatin in the male germline of mice. Curr Biol 2006; 16:660–7.
16. Fernandez-Capetillo O, Mahadevaiah SK, Celeste A, et al. H2AX is required for chromatin remodeling and inactivation of sex chromosomes in male mouse meiosis. Dev Cell 2003; 4: 497–508.
17. Mahadevaiah SK, Turner JM, Baudat F, et al. Recombinational DNA double-strand breaks in mice precede synapsis. Nat Genet 2001; 27:271–6.
18. Turner JM, Mahadevaiah SK, Fernandez-Capetillo O, et al. Silencing of unsynapsed meiotic chromosomes in the mouse. Nat Genet 2005; 37:41–7.
19. Turner JM, Aprelikova O, Xu X, et al. BRCA1, histone H2AX phosphorylation, and male meiotic sex chromosome inactivation. Curr Biol 2004; 14:2135–42.
20. Turner JMA. Meiotic sex chromosome inactivation. Development 2007; 134:1823–31.
21. Rivkin E, Tres LL, Kierszenbaum AL. Genomic origin, processing and developmental expression of testicular outer dense fiber 2 (ODF2) transcripts and a novel nucleolar localization of ODF2 protein. Mol Reprod Dev 2008; 75:1591–602.
22. Salmon NA, Reijo Pera RA, Xu EY. A gene trap knockout of the abundant sperm tail protein, outer dense fiber 2, results in preimplantation lethality. Genesis 2006; 44:515–22.
23. Kierszenbaum AL, Tres LL. The acrosome-acroplaxome-manchette complex and the shaping of the spermatid head. Arch Histol Cytol 2004; 67:271–84.
24. Kierszenbaum AL, Rivkin E, Tres LL. Molecular biology of sperm head shaping. Soc Reprod Fertil Suppl 2007; 65:33–43.
25. Kierszenbaum AL, Rivkin E, Tres LL. Acroplaxome, an F-actin-keratin-containing plate, anchors the acrosome to the nucleus during shaping of the spermatid head. Mol Biol Cell 2003; 14: 4628–40.
26. Kierszenbaum AL. Intramanchette transport (IMT): managing the making of the spermatid head, centrosome, and tail. Mol Reprod Dev 2002; 63:1–4.
27. Touré A, Szot M, Mahadevaiah SK, et al. A new deletion of the mouse Y chromosome long arm associated with the loss of Ssty expression, abnormal sperm development and sterility. Genetics 2004; 166:901–12.
28. Kang-Decker N, Mantchev GT, Juneja SC, et al. Lack of acrosome formation in Hrb-deficient mice. Science 2001; 294:1531–3.
29. Yao R, Ito C, Natsume Y, et al. Lack of acrosome formation in mice lacking a Golgi protein, GOPC. Proc Natl Acad Sci U S A 2002; 99:11211–6.
30. Yang WX, Sperry AO. C-terminal kinesin motor KIFC1 participates in acrosome biogenesis and vesicle transport. Biol Reprod 2003; 69:1719–29.
31. Langford GM. Myosin-V, a versatile motor for short-range vesicle transport. Traffic 2002; 3:859–65.
32. Seabra MC, Mules EH, Hume AN. Rab GTPases, intracellular traffic and disease. Trends Mol Med 2002; 8:23–30.
33. Kierszenbaum AL, Rivkin E, Tres LL. The actin-based motor myosin Va is a component of the acroplaxome, an acrosome-nuclear envelope junctional plate, and of manchette-associated vesicles. Cytogenet Genome Res 2003; 103:337–44.

Appendix A

Glossary of Molecular Term

Adaptor protein: a small protein with few binding modules that bridges two other proteins.

Adaptors: short double-stranded synthetic oligonucleotides that can be ligated on to the ends of double-stranded DNA fragments to generate known restriction enzyme sites at each end.

Adenine: a purine base of DNA and RNA.

Allele or allelomorph: alternative forms of a gene occupying the same locus on homologous chromosomes. It is one of a series of alternative forms of a given gene, differing in DNA sequence, and affecting the functioning of a single product (RNA and/or protein). If more than two alleles have been identified in a population, the locus is said to show *multiple allelism*.

Allosteric proteins: are enzymatic proteins with at least two binding sites, one for the substrate and one or more for regulatory ligands. Allosteric proteins are involved in feedback regulation, in which one binding site can affect another by changing the protein conformation.

Amber mutation: a mutant nucleotide sequence of AUG that causes premature termination of protein synthesis.

Aminoacyl-tRNA: a tRNA molecule carrying an amino acid, which is to be used to extend a growing polypeptide chain during protein synthesis.

Amphipathic: a molecule containing both hydrophobic and hydrophilic properties.

Amplification refractory mutation system (ARMS): a PCR system, which, due to the primer used, will allow amplification of a single specific allele.

Anaphase: a stage of mitosis during which sister chromatids separate to the spindle poles.

Aneuploidy: a change in the number of chromosomes.

Anti-codon: a sequence of three nucleotides in transfer RNA that is complementary to a codon on mRNA.

Antisense: strand of DNA complementary to the sense strand.

Apoptosis: also known as programmed cell death is an energy requiring process that leads to cell shrinkage and fragmentation and condensation of chromatin. Unlike necrosis, apoptosis does not elicit an inflammatory response. Apoptosis is an important feature of the predictable life cycle of many cells.

Arrestins: family of proteins that uncouple receptors from the G-proteins and serve as adaptors to couple the receptor to clathrin-coated pits, inducing endocytosis.

ATPase: a class of enzymes that catalyzes the decomposition of adenosine triphosphate (ATP) into adenosine diphosphate (ADP) and a free phosphate ion. This dephosphorylation reaction releases energy, which, in most cases, the enzyme harnesses to drive other chemical reactions that would not otherwise occur.

Auto-antibodies: antibodies produced by an individual against antigens present in that individual.

Auto-antigen: antigens that elicit auto-antibodies.

Auto-immune disease: disease resulting from the generation of auto-antibodies.

Autonomously replicating sequence (ARS): a DNA sequence that allows a plasmid DNA molecule to replicate in yeast.

Autoradiography: detection of radioactively labeled molecules on X-ray film.

Autosome: any chromosome that is not a sex chromosome.

Bacteriophage (phage): a virus that infects bacteria.

Basepair (bp): a pair of complementary nucleotide bases in a duplex DNA or RNA molecule.

Blunt ends: ends of a duplex DNA molecule that have no overhangs due to restriction with an enzyme that generates blunt ends or due to removal of the overhangs by single-strand specific nucleases.

Ca^{2+}-CaM: complexes of Ca^{2+} bound to calmodulin.

Calcineurin: a phosphatase with broad specificity for phosphoproteins, including tyrosyl phosphoproteins. Calcineurin is involved in the signal transduction pathways that regulate interleukin production in T cells and is an intracellular target for the immunosuppressive effects of cyclosporine.

Calmodulin (CaM): is a small acidic protein that contains multiple copies of a helix-loop-helix motif that binds Ca^{2+} with high affinity. CaM serves as an intracellular Ca^{2+} receptor forming the complex Ca^{2+}-CaM.

Calpain: an acronym for calcium-activated, papain-like protease.

Cap: structure at the 5′ end of mRNAs containing a methylated guanine residue.

Capsid: the outer protein coat of a virus.

Cell line: a population of cells grown in culture. Cell lines are usually clonal and have the capacity for indefinite passage in culture.

Chimera: recombinant DNA construction that combines elements from distinct genes to express a protein that is a mixture of these elements. Often used in receptor field to study distinct domains involved in signal transduction.

Chromatid: one-half of a replicated chromosome.

Chromatin: protein-DNA complexes that make up chromosomes.

Chromosome walking: sequencing cloned DNA fragments directionally along a chromosome.

Cis-acting element: specific DNA sequences within a gene that binds transcription factors and confer a defined regulatory response. They act "in cis" because regions are on the same DNA molecule as the gene they regulate.

Clone: a population of cells or DNA molecules that came from a single progenitor.

Codon: the DNA or corresponding RNA sequence of three basepairs that codes for a particular amino acid or termination signal.

Complementary DNA (cDNA): a DNA copy of a messenger RNA generated by reverse transcriptase.

Concatemer: multiple copies of a DNA sequence covalently joined together in tandem arrays.

Concordance: members of a twin pair exhibiting the same trait.

Confluent cells: is a term used to define when all cells in culture are in contact with other cells and no space in the dish is left uncovered.

Consensus sequence: a DNA or amino acid sequence that specifies the most commonly found DNA base or amino acid at each position in a sequence of similar DNA or amino acid sequences.

Cooperativity: property of certain receptors in which ligand binding to a subset of receptors alters affinity for ligand binding to remaining receptors. May be positive or negative denoting increase or decrease, respectively, in subsequent affinity.

Cos site: cohesive ends of a phage genome.

Cosmid: a plasmid DNA containing Cos sites to enable it to be packaged into phage particles.

Cotig: blocks of DNA that are shown to be immediately adjacent because they contain common sequences in their overlapping region. A term used in physical, as opposed to genetic, mapping.

Cross-talk: interaction between two or more distinct intracellular signaling pathways that affect the same response and interact to produce and additive or synergistic response, or that activation of one pathway may inhibit the response stimulated by the other.

Cytoplasm: contents of a cell outside the nucleus.

Cytosine: a pyrimidine base of RNA and DNA.

Cytoskeleton: network of microtubutes that support the cellular structure and are used for directed transport of many cell constituents.

Dedifferentiation: irreversible loss of specialized functions that a cell expresses in vivo.

Degeneration of the genetic code: it refers to the fact that more than one codon can specify a particular amino acid.

Denaturing gradient gel electrophoresis (DGGE): a method of screening for mutations in a DNA sequence that depends on the idea that mismatched DNA falls apart (denatures) more easily than perfectly matched double-stranded DNA.

Desensitization: diminished responsiveness of a receptor pathway. This can be homologous, when is caused by a ligand for the receptor itself, or heterologous when is caused by activation of some other pathway.

Diploid: refers to a cell or organism containing two complete sets of homologous chromosomes.

Discontinuous replication: refers to synthesis of DNA in short fragments (Okazaki fragments) which are eventually joined together to form a complete DNA strand.

DNA polymerase: an enzyme that adds nucleotides to a growing chain of DNA in a 5′ to 3′ direction during DNA replication using a DNA strand as a template.

DNase footprinting: a technique for the detection of a region of DNA that is complexed with proteins. This technique is based in the principle that proteins protect that region of DNA from DNase digestion.

Docking protein: a larger and more versatile version of the adaptor protein that contains many binding modules and acts as a working platform for many intracellular signaling transducers.

Domain: refers to a discrete region of the protein that has a specific function associated with it.

Dominant allele: the allele observed in the phenotype of a heterozygote.

Dominant negative effects: effects caused by the expression of a mutant form of a gene that obliterates the normal function of the normal form of the gene, even though the normal gene is expressed.

Dominant negative mutation: a mutation generating a mutant gene product whose characteristics are dominant over the wild-type product.

Down-regulation: actual loss of receptors that may follow prolonged exposure to agonists. Down-regulation often occurs through receptor internalization from the cell surface to the cell interior.

Duplex: a complex of two complementary strands of nucleic acids.

Effector: enzyme or ion channel, such as adenylyl cyclase, that generates the second messenger in a receptor-activated signal transduction pathway.

Endoplasmic reticulum: continuous membrane network within the cytoplasm of cells.

Enhancer: DNA domain in the regulatory region of a gene that increases transcription independently of its orientation or site in the promoter.

Epigenetics: is the study of heritable changes in gene expression or function that occur without changes in the DNA sequence itself. Epigenetic changes are formally defined as an alteration of phenotype without change in the genotype, and are having an increasingly important role in development and in the etiology of a variety of disease.

Episome: genetic information expressed outside of the genome

Epitope: the part of an antigen that binds to the antigen binding region of an antibody.

Euchromatin: chromatin that is poorly condensed so allowing transcriptional activity.

Eukaryote: an organism comprising cells that have a true nucleus.

Excision repair: a system that allows the removal of a damaged piece of single-stranded DNA from a DNA duplex and replacement with a repaired DNA sequence copied from the undamaged DNA strand.

Exon: a portion of a gene that is transcribed into sequences found in the mature cytoplasmic RNA and that is flanked by introns. Introns, in contrast, are spliced out from the initial transcript.

Expressed sequence tags: short cDNA sequences that, when mapped to specific chromosomal regions, serve as useful anchors in the physical mapping of unknown genes.

Expression cloning: a method for cloning receptors that involves use of recombinant DNA techniques to express all (or a portion) of receptor protein. Requires assay for expressed protein such as binding of antibody (if available), binding of ligand, or agonist-stimulated signal transduction (e.g. change in ion current in *Xenopus* oocyte in which mRNA has been injected).

Farwestern blot: a technique involving transfer of protein molecules, after size fractionation on gel electrophoresis, to filter papers to analyse their interaction with other protein probes that are not antibodies.

Fluoresce *in situ* hybridization (FISH): a technique used to visualize locations on chromosomes, which hybridize to specific nucleotide probes. It is based on the use of large, fluorescent DNA probes to bind to specific sites on metaphase or interphase chromosomes and thereby physically locate the specific DNA on the chromosome.

Fluorescence-activated cell sorting (FACS): a technique to separate cells from a population on the basis of their expression of specific antigens.

***fos*:** oncogene of the FBJ murine osteogenic sarcoma virus. Its cellular counterpart, the proto-oncogene c-*fos*, belong to the family of early response genes activated in the conversion of quiescent serum-starved 3T3 fibroblast to growing cells. c-*fos* dimerize with c-*jun* (c-*fos*/c-*jun* heterodimer) to form the trancription factor AP-1 that bind to the AP-1-binding site (TGACTCA) present in the regulatory region of several PKC-stimulated genes.

Frame shift: refers to a mutation that shifts the reading frame of triplet codons in a gene during translation of mRNA.

G-banding: technique to visualize the band patterns of chromosome on staining with Giemsa.

Gene knock-out: ablation of a portion of a gene accomplished by introducing mutant genes into cells and screening/selecting for replacement of the normal sequence by the mutant sequence (by homologous recombination). When performed using murine embryonic stem cells, these cells can be used to generate gene "knock out" mice.

Gene mapping: a map of different genetic loci, constructed based on their physical position with respect to each other.

Gene: a unit of heredity that specifies an RNA or mRNA. A gene also contain intronic regions, which are not translated and regions that control transcription.

General transcription factors (GTFS): proteins that are required to allow RNA polymerase to transcribe a gene at the basal level of transcription. This level of gene expression is then further activated by tissue-specific transcription factors.

Genotype: the actual genetic make up of a cell or organism.

Genotype: the genetic constitution of an individual.

G-proteins: proteins that bind GTP with high affinity and specificity. The superfamily includes heterotrimeric G proteins involved in signal transduction by membrane receptors and small G proteins homologous to *ras.*

Growth curve: a plot of cell number against time in a proliferating cell culture. *Lag phase:* before growth is initiated; *log phase:* the period of exponential growth; *plateau:* stable cell count achieved when cells stop growing, usually after achieving high density.

Guanine: a purine base of RNA or DNA.

Haploid: refers to a cell or organism containing only one copy member of a homologous chromosome pair.

Hemizygote: refers to a diploid cell or organism that contains only one allele of a gene due to loss of one chromosome of a homologous chromosome pair.

Heterochromatin: chromatin regions that are condensed and so generally transcriptionally inactive.

Heterodimer: a complex of two non-identical moieties, e.g. proteins.

Heterogenous nuclear RNA (hnRNA): the primary transcript generated from transcription of a gene prior to splicing etc. associated with maturation.

Heterozygote: refers to a diploid cell or organism that contains different alleles of a gene at one locus on homologous chromosomes.

Heterozygous: an individual that carries a pair of non-identical allele.

Histones: a group of conserved basic proteins that structure DNA in the eukaryotic cell into basic chromatin structure.

Homodimer: a complex of two identical moieties, e.g. proteins.

Homologous recombination: genetic recombination between nucleic acid sequences that have extensive regions of identical sequence.

Homology cloning: a method for cloning receptors (and other entities) that relies on conserved regions shared by members of a gene family. Particularly powerful when PCR is used as initial step.

Homozygous: an individual carries a pair of identical alleles.

Hybridization: refers to the ability of complementary single-stranded DNA or RNA molecules to form a duplex. Screening technique that exploits the complementarity of nucleic acid sequences and the hydrogen bonds formed between complementary G-C and A-T pairs. A labeled probe, either synthetic single-stranded DNA, denatured double-stranded DNA, or cRNA – can bind to its complementary single-stranded DNA or RNA sequence. The specificity of the combination is controlled by the stringency of the hybridization. High stringency = high temperature, low salt, polar solvent. Disruption of complementary nucleic acid strands is called denaturation.

Hydrophilic: effectively interacts with water.

Hydrophobic: ineffectively interacts with water.

Immunogen: a compound able to elicit an antibody response.

Immunoglobulin: a family of proteins that function as antibodies.

Imprinting: a change in gene function that occurs early in development, in the egg or sperm such that maternal and paternal copies may differ.

Interphase: period of the cell cycle that span between one mitosis and the next.

Intron: the segments within the coding region of a gene that are not present in the fully mature RNA. Introns are removed after transcription by splicing during the formation of mature cytoplasmic RNA.

Jak: Janus kinase (after the two-faced Roman god). Some authors refer it as "just another kinase." Jak are cytoplasmic tyrosine kinases that activate latent gene regulatory proteins called STATs.

***jun*:** oncogene of the avian sarcoma virus-17.

Karyotype: the number and characteristics of the full chromosome set of an organism.

Kinase: a phosphorylation enzyme. Kinases transfer the ATP γ-phosphate group to the amino acids tyrosine, serine or threonine, and to the 5′ end of nucleotide chains.

Klenow: the largest fragment of the *Escherichia coli* DNA polymerase complex. The Klenow fragment is devoid of the exonucleolitic activity of the DNA polymerase, but it retains its polymerase activity.

Knockout: the ability to remove a specific gene in a cell or organism by molecular techniques.

Lagging strand: refers to the strand of newly synthesized DNA that is made up from joining of Okazaki fragments.

Leading strand: refers to the strand of newly synthesized DNA that is continuously synthesized during DNA replication.

Leucine zipper: a leucine-rich domain of a protein that allows protein-protein interaction.

Ligand: a molecule that binds to another macromolecule such as a receptor.

Ligase: an enzyme that joins the ends of two duplexes of DNA.

Linkage: the tendency for two genes in close proximity on a chromosome to be inherited together.

Liposome: small vesicle made up of a lipid bilayer. Liposomes are used as carriers to introduce foreign DNA into cell.

Locus: the position of a gene on a chromosome.

Lod score: odds that two markers are truly linked/odds that two markers are not linked. The higher the lod score, the more likely that the two markers (one usually the presence of disease) are linked.

Logarithm of the odds score (LOD): a value for the likelihood of two loci being within a measurable distance from each other.

Long terminal repeat (LTR): the direct repeats at the ends of the proviral DNA.

Lytic cycle: the events associated with virus infection during a productive infection.

Major histocompatibility complex (MHC): a family of genes that are involved in mediating T-cell immune responses.

Maternal inheritance: preferential carriage of a gene by the maternal parent.

Meiosis: eukaryotic cell division during which two sequential divisions generate cells containing a haploid complement of chromosomes.

Messenger RNA (mRNA): the mature transcript from a gene transcribed by RNA polymerase that specifies the order of amino acids during mRNA translation to protein.

Metaphase: a stage in mitosis when the parental and newly synthesized chromosomes are maximally condensed but prior to their segregation to opposite spindle poles.

Methylation of DNA: an epigenetic modification of the DNA molecule consisting in the transfer of methyl groups ($-CH_3$) to a cytosine residue, which is converted into 5-methylcytosine, by a DNA methylase. Methylation of DNA directly switches off gene expression.

Microsatellites: simple sequence repeats.

Microtubules: filamentous protein structures that make up the cytoskeleton of the cell.

Minisatellites: variable number of tandem repeats.

Mitosis: the mechanism by which a cell undergoes nuclear division to generate two identical daughter cells with equal complements of chromosomes.

Mobility shift assay: technique to analyze the interaction between known DNA sequences and protein. It is based on the slow electophoretic migration rate of the DNA-protein complex compared with DNA alone.

Monosomy: a condition in which one member of a chromosome pair is missing.

Monosomy: the presence of only one chromosome from a pair in a cell's nucleous. In partial monosomy only a portion of the chromosome has one copy, while the rest has two copies.

Monozygotic: genetically identical due to originating from the same fertilized cell.

Mosaicism: the presence of two populations of cells with different genotypes in an individual who has developed from a single fertilized egg. Mosaicism may result from a mutation during development, which is propagated to only a subset of the adult cells.

Mutation: a transmissible change in nucleotide sequence considered as a genetic mistake. Mutations lead to a change or loss of normal function encoded by that nucleotide sequence.

Neutral substitution: a substitution in the DNA sequence that results in a change in the amino acid without change in the function of the protein.

Nondisjunction: failure of chromosomes to properly segregate during meiosis or during the mitotic anaphase. Nondisjunction generates cells with abnormal number of chromosomes that can result in genetic diseases or abnormalities.

Non-disjunction: the process by which duplicate chromosomes fail to separate during cell division resulting in one daughter cell containing both duplicate chromosomes.

Nonsense mutation: a mutation resulting in the premature termination during protein synthesis.

Northern blot analysis: a technique for detection of RNA, which is separated by size by electrophoresis and transferred to a solid nitrocellulose or nylon support. Specific RNA sequences are detected by hybridization of the support with defined labeled probes. Although Northern blot is a qualitative technique, by densitometric analysis of the autoradiograms it is possible to quantitate steady-state levels of mRNA.

Nucleosome: a subunit of chromatin comprising a core of histone proteins with approximately 146 basepairs of DNA wrapped around.

Okazaki fragment: short segments of DNA synthesized during lagging strand DNA synthesis.

Oncogene: a mutated gene which is normally involved in the correct control of cell division such that disruption of the normal gene function leads to cell immortalization and transformation.

Open reading frame (ORF): a series of triplet codons in the coding region of a gene that lie between the signals to start and stop translation.

Organelle: a membrane-bound compart-ment of a eukaryotic cell.

Origin of replication: a specific site on DNA at which DNA replication starts.

Orthologous: this term is used to describe two similar genes in two different species that originated from a common ancestor. Alternatively, orthologous is used to describe any two genes in two different species with very similar functions.

PDGF: platelet-derived growth factor.

Phage: virus consisting of a nucleic acid core surrounded by a protein coat. The common lambda (k) phage acts by infecting bacteria with its DNA, which is replicated as part of the bacterial genome. After replication, the phage lyses the bacteria cell to form a plaque or clear area on a bacterial lawn.

Phenotype: the observable characteristics of a cell or organism resulting from the expression of the cell's genotype.

Phenotype: the observable expression of a genotype as a morphological, biochemical or molecular trait. A phenotype can be either normal or abnormal in a given individual. Abnormal phenotypes are caused by genetic disorders.

Phosphatase, protein tyrosine: an enzyme that cleaves phosphate bound covalently to tyrosine residues of proteins.

Plaque: the clear area on a lawn of bacteria or cells due to virus infection.

Plasmid: a circular DNA molecule capable of self-replication in a cell. Closed circular piece of DNA, which usually carries a resistance factor and can multiply independently in bacteria. Used for cloning cDNAs and for some genes, although limited somewhat by the size of the DNA insert.

Platelet-derived growth factor (PDGF): a dimeric peptide (~30Kd) stored in blood platelets and released during the clotting reaction.

Polar molecule: molecule soluble in water.

Polyadenylation: addition of tracts of polyadenylic acid to the ends of transcribed RNA molecules.

Polymerase chain reaction (PCR): a technique to amplify a target DNA sequence by multiple rounds of DNA synthesis. PCR is a powerful method to amplify DNA or RNA sequences for easier detection or analysis, it requires known DNA sequences from which short oligonucleotide primers (one for each strand) can be synthesized. Using a special *Taq* DNA polymerase that retains activity at high temperatures, the PCR technique consists of repetitive rounds of primer hybridization, DNA chain elongation with Taq polymerase, and DNA denaturation.

Polymorphism: mean different form. In genetics, polymorphis refers to two or more phenotypes products of allelic genes.

Position effect: refers to the differences in levels of expression of a gene due to its position in the chromatin.

Positional cloning: the identification of a gene largely by finding its location by genetic and physical mapping strategies.

Primer: a short nucleotide sequence that provides the starting point for polymerases to copy a nucleotide sequence and make a double strand.

Prokaryotes: cells that lack nuclei and membrane-limited organelles.

Promoter: a DNA sequence that targets RNA polymerase to a gene for transcription.

Protein kinase: enzyme catalyzing the addition of phosphate to a nucleic acid or aprotein, usually involving ATP as the phosphate donor. In proteins phosphorylation may occur on serine, threonine, or tyrosine, the latter catalyzed by protein tyrosine kinases.

Proteomics: the study of the protein complement of the genome, also called proteome. The terms proteomics and proteome, mirror the terms "genomics" and "genome," which describe the entire collection of genes in an organism. Although the definition of proteomics varies depending on the authors, it is agreed that can be defined as:

Protooncogene: a normal gene that, when mutated, contributes to transformation of a normal cell into a cancer cell.

Provirus: the double-stranded DNA copy of a retrovirus.

Pseudogene: a duplicated nonfunctional gene.

Pulsed field gel electrophoresis (PFGE): electrophoretic technique used to separate very large molecules of DNA by periodically altering the direction of the electric field through which the samples are migrating.

Purine: an organic base containing two heterocyclic rings that occurs in nucleic acids.

Pyrimidine: an organic base containing one heterocyclic ring that occurs in nucleic acids.

Q-banding: a technique to visualize the band patterns of chromosome on staining with quinacrine.

Quiescent: refers to a cell that has exited the cell cycle and is resting.

***ras*:** oncogene of the Harvey (*ras*H) and Kristen (*ras*K) rat sarcoma viruses. These genes, which are frequently activated in human tumors, encode a 21 kD G-protein.

Reading frame: any one of three ways that a specific nucleotide sequence can be read in triplets.

Receptor mediated endocytosis: a process in which ligand binding to receptor promotes receptor clustering and internalization. Used by LDL and transferrin receptors to convey ligand to cell interior.

Recessive allele: the allele masked by the dominant allele in a heterozygote due to the absence or comparative inactivity of the product of the recessive allele.

Replication forks: region of DNA resulting from DNA replication in which the parental DNA strands are displaced and DNA polymerase copies the parental template.

Reporter gene: a gene encoding a product that can be easily measured when introduced into cell by transfection, eg. luciferase, cathecol methil transferase.

Residue: term used to define the unit of a polymer, such as a monosaccharide, an amino acid, or a nucleotide.

Restriction enzyme: a bacterial endonuclease that cuts DNA at specific sequences. Each enzyme cuts at its own unique sequence or site, usually a 4- or 6-base pair (bp) motif.

Restriction fragment length polymorphism (RFLP): refers to heritable differences in the length of DNA fragments from a specific region of DNA generated by restriction enzymes due to DNA sequence differences.

Restriction map: Diagrammatic representation of a DNA molecule indicating the sites of cleavage by various restriction endonucleases.

Retrovirus: an RNA virus that replicates by first converting its RNA genome to a double-stranded DNA copy using reverse transcriptase.

Reverse transcriptase: viral enzyme with the unique property of synthesizing a cDNA molecule from an RNA template.

Reverse transcription: enzymatic reaction carried by the reverse transcriptase.

Reverse transcription-polymerase chain reaction (RT-PCR): amplification of RNA by PCR after copying of the RNA to cDNA by reverse transcription.

Ribosomal RNA (rRNA): structural molecules of RNA present in the ribosome.

Ribosome: organelle that serves as a biochemical machine to translate mRNA into protein. Ribosomes are constituted by a complex of different proteins associated with structural RNA molecules called ribosomal RNA (rRNA).

RNA polymerase: an enzyme that makes a RNA copy from a DNA template.

RNA splicing: removal of introns from transcribed RNA to generate a mature mRNA.

S1-nuclease: an enzyme that specifically degrades single-stranded DNA molecules.

Scatchard plot: a method for plotting equilibrium binding data that allows calculation of receptor number and affinity.

Second messenger: small intracytoplasmic molecule released in response to an extracellular signal that convey the signaling information to the interior of the cells, eg. cAMP, IP_3, and Ca^{2+}.

Secondary structure: regular folding of polymers. In proteins, secondary structures, such as α helices and β sheets are formed by hydrogen bonds between side groups of polar amino acids.

Semi-conservative replication: the generation of daughter duplexes of DNA, which contain one parental and one newly synthesized strand of DNA.

SH2 (SRC Homology 2) domains: peptide sequences similar to a motif found in the SRC protein kinase. SH2 domains can bind to certain phosphorylated tyrosine residues.

SH3 (SRC Homology 3) domains: peptide sequences similar to another motif found in the src protein kinase. SH3 domains can bind certain proline-rich motifs found in proteins.

Signal sequence: a short amino acid sequence that targets a protein to a specific cellular localization.

Silencer element: DNA domain in the regulatory region of a gene that acts to decrease transcription independently of its orientation or site in the promoter.

Silent mutations: changes in the DNA molecule without apparent effects. Since the same amino acid can be encoded by a different DNA sequence, a silent mutation may render a change that do not cause any change in the amino acid sequence. A silent mutation can also be produced by an alteration in the DNA that cause the change of an amino acid that does not change the function of the protein, which are called *neutral substitutions.*

Simple sequence repeats (SSRS): also called microsatellites. Tandemly repeated DNA sequences with 2–8 base pairs in the repeat unit. Very frequent variation in the number of repeat units allows corresponding chromosomes to be distinguished from each other.

Single-stranded conformational polymorphism (SSCP): a method of screening for mutations in a DNA sequence that depends on the altered mobility of single-stranded DNA. SSCP can detect a single base alteration in a DNA sequence.

Somatic cell hybrid: a fusion between two different cell types.

Somatic cell: a cell other than a haploid sex cell.

Southern blot: a technique in which DNA is separated by size through electrophoresis, transferred to a solid nitrocellulose support, and specific sequences detected by hybridization with specific probes.

Southwestern blot: a technique of transfer protein molecules after size fractionation on gels to filter papers to analyze their interaction with DNA probes.

S-phase: stage of the eukaryotic cell cycle at which DNA synthesis occurs.

Spindle: a microtubule structure of the nucleus that is involved in organizing replicated chromosomes during cell division.

***src (pronounced "sark")*:** name of the first described retroviral oncogene (*v-src*), from the chicken Rous sarcoma retrovirus and its precursor (*c-src*), which encode a membrane-associated protein kinase.

STAT: signal transducers and activators of transcription. STAT are inactive proteins located in the cytoplasm. When the cytokines receptors are activated, they recruit and activate STATs, which then migrate to the nucleus and activate gene transcription.

Stop codons: DNA triplet codon that terminates translation of an mRNA.

Synergism or potentiation: two or more hormones are said to act synergistically when the response to their simultaneous administration is greater than the sum of the responses to each when given alone. For example, both growth hormone and cortisol modestly increase lipolysis in adipocytes. When given simultaneously, however, glycerol production is nearly twice as great as the sum of the effects of each.

TATA box DNA sequence found in many eukaryotic promoters that binds the TATA binding protein in order to recruit RNA polymerase for transcription.

TATA-associated factors: (TAFS): proteins that bind to TBP tightly and, together with TBP, make up the transcription factor IID.

TATA-binding protein (TBP): protein that binds to the TATA sequence in the promoter region of genes transcribed by RNA polymerase II.

T-cell receptor: membrane protein complexes that are expressed on T-lymphocytes and recognize specific antigens when associated with MHC molecules.

Telophase: final stage of cell division when the nuclear membrane re-forms around replicated chromosomes.

- The identification of all the proteins made in a given cell, tissue or organisms;

- The outline of the precise three-dimensional structure of the proteins.
- The protein complement encoded by a genome;
- The study of how proteins join forces to form networks similar to electrical circuits; and

Tolerance: reduced ability to mount an immune response to specific antigens.

***Torpedo californica*:** a specie of electric ray native to the eastern Pacific ocean that was used as a model to isolate and characterize the nicotinic cholinergic ligand-gated channel receptor.

Transcription factor (or trans-acting factors): proteins that bind directly to DNA and influence the transcription rate of specific genes. These proteins include the steroid and thyroid hormone receptors and many other nuclear proteins. These factors act "in trans" because they are encoded on genes separate from the ones they regulate.

Transcription: process by which a DNA template is copied to RNA by RNA-polymerase. RNA-polymerase binds to DNA at the promoter region and transcribes RNA complementary to the DNA template. Both coding regions (exons) and noncoding regions (introns) are transcribed. The transcriptional process is regulated by a number of proteins, called transcription factors, which bind to the DNA of the gene to be transcribed.

Transfection: the introduction of foreign DNA into cells in culture. Mammalian cells can take up DNA if they are treated in culture with chemical or physical procedures that disrupt the cell membrane, such as calcium phosphate, DEAE-Dextran or electroporation. Although only a minute fraction of the DNA is incorporated into chromosomes, it is usually sufficient to test its presence in a transient transfection assay, which is conducted within a few days time. Using transient transfection assays it is possible to assess functionality of the introduced gene, such as responsiveness to regulatory molecules. Stable transfections require the integration of the introduced DNA into the genome and depend on a selectable marker to detect and amplify a rare event, which is used to generate stable cell lines, transgenic animals or for potential gene therapy.

Transfer RNA (tRNA): small RNAs that function to transport specific amino acids to a growing polypeptide chain on the ribosome during translation.

Transformation: permanent alteration of the cell phenotype occurs after an irreversible genetic change. Transformed cell lines have increased growth rate, infinite life span, high plating efficiency, and often exhibit tumorigenicity.

Translation: the mechanism by which mRNA is used as a template to synthesize protein on the ribosome.

Transmembrane (membrane-spanning) domain: domain of about 20 contiguous and relatively hydrophobic amino acids predicted to span a membrane bilayer. Receptors coupled to G proteins have seven putative membrane-spanning domains; many other receptors appear to have only one (e.g. EGF receptor).

Transposon: mobile genetic element, which can insert at random into plasmids or the bacterial chromosome independently of the host cell recombination system.

Trk: tropomyosin-related kinase receptor.

Tumour suppressor gene: a gene that negatively regulates cell division such that mutation in these genes results in uncontrolled cell division and turnout progression.

UPD: Uniparental disomy.

Uracil: a pyrimidine base that replaces the DNA base thymine in RNA molecules.

Variable number tandem repeats (VNTR): tandemly repeated DNA sequences with 10–100 base pairs in the repeat unit. Frequent variation in the number of repeat units allows corresponding chromosomes to be distinguished from each other. VNTR are also called minisatellites.

Vector: a DNA molecule in which DNA sequences can be cloned.

Western blot: transfer of protein molecules, after size fractionation by electrophoresis, on gels to filter papers for analysis with antibodies.

Wobble hypothesis: refers to the ability of a tRNA molecule to recognize more than one codon by relatively free pairing between the third base of the codon and first base of the anti-codon.

X-inactivation: refers to the inactivation of one of the X chromosomes in female somatic cells.

Yeast artificial chromosomes (YACS): plasmid DNA which contains DNA sequences that allow plasmid maintenance in yeast cells and allow cloning of very large regions of DNA. YACS replicate as chromosomes in yeast because they contain sequences that define the ends of the chromosome (telomeres) and sequences that allow the chromosome to separate appropriately at mitosis (centromeres).

Appendix B

Glossary of Terms and Acronyms

%GC: percentage of guanine and cytosine

ΨU: pseudouridilic acid

[Ca2+]i: intracellular calcium concentration

17βHSD: 17β-hydroxysteroid dehydrogenase

20αDHP: 20α-dyhydroprogesterone

20αHSD: 20α-hydroxysteroid dehydrogenase

2-deoxy-D-ribose: deoxyribose

3'-UT: 3' untranslated region

^{32}P: radioactive isotope

3βHSD: 3β-hydroxysteroid dehydrogenase

3β-HSD: 3β-hydroxysteroid dehydrogenase/Δ4-Δ5 isomerase

3αHP: 3α-hydroxy-4-pregnen-20-one

3β-HSD: 3β-hydroxysteroid dehydrogenase/Δ4-Δ5 isomerase

5'-UT: 5' untranslated region

5′AMP: adenosine monophosphate

5-HT$_3$ receptor: 5-hydroxytryptamine type 3 (serotonin) receptor is a serotonin receptor subclass that contains selective channels for sodium, potassium and calcium

5-HT$_3$: 5-hydroxytriptamine

5α-DHT: 5α-dihydrotestosterone

5α-DHT: 5α-dihydrotestosterone

7-TMS: seven transmembrane segment

9 K-PGR: 9-keto-PGE2-reductase

A: adenine or adenosine

AA: arachidonic acid

AAV: adeno-associated viruses

ABP: androgen-binding protein, a glycosylated dimeric protein secreted by the Sertoli cells homologous to steroid hormone-binding globulin

AC: Adenylate cyclase

AC: adenylyl cyclase

ACh: Acetylcholine

ACTH: adrenocorticotropic hormone

Ad5: type of adenovirus

Ad-ER-DN: adenoviral vector that expresses a dominant negative ER mutant

ADH: antidiuretic hormone

Ad-LacZ: adenovirus expressing the marker gene β-lactamase

ADX: adrenalectomized

AF-1, -2 and -5: activation function-1,-2 and -5.

aFGF: acidic fibroblast growth factor

AGRP: agouti-related protein

AII: angiotensin-II

AIS: androgen insensitivity syndrome

AKAP121: A-kinase anchor protein

Akt: oncogene of the retrovirus AKT8 with activity serine/threonine-specific protein kinase, also termed protein kinase-B (PKB)

Akt: protein kinase-B, also termed PKB

ALDH-1: aldehyde dehydrogenase class 1 gene

ALK4: activin receptor-like kinase 4

Allopregnanolone: 3α,5α-tetrahydroprogesterone, also called THP

ALS: amyotrophic lateral sclerosis

Amenorrhea: absence of menstruation

AMH: anti-Müllerian hormone or Müllerian-inhibiting substance (MIS)

AMP: adenosine 5-monophosphate

AMPA: α-Amino-3-hydroxy-5-methyl-4-isoxazole-propionic acid; agonist for the AMPA subtype of glutamate receptors

amp^r: ampicillin resistance gene

Ang: angiopoietins;

ANGPT1: angiopoietin 1

ANGPT2: angiopoietin 2

ANGPT3: angiopoietin 3

ANGPT4: angiopoietin 4

ANP: atrial natriuretic peptides

AP-1: activating protein-1, Jun-Fos heterodimer

AP-1: activator protein-1

AP-2: activating protein-2

AP-3: activating transcription factor-3 binding site

Apaf-1: apoptotic protease-activating factor

AR: androgen receptor

AR-A and AR-B: androgen receptor isoforms

ARA: AR associated proteins

ARE: androgen response element

ARF6: ADP ribosylation factor 6

ARKO: androgen receptor knockout mouse

ART: assisted reproductive technology

ASI: androgen sensitivity index

Asn: asparagine

ATP: adenosine 5'-triphosphate

ATP: adenosine triphosphate

ATR: DNA repair protein, member of the PI3-kinase-like family

Aurora kinases: a family of serine/threonine kinases important for cell division

AVPV: anteroventral periventricular

α2 M: α2-macroglobulin

αT3-1: ovarian cell line

BAC: bacterial artificial chromosome

BAD: Bcl-2-associated death promoter

Bcl-2: proto-oncogene

BDNF: brain-derived neurotrophic factor

bFGF: basic fibroblast growth factor

Bfl-1/A1: Bcl-2 family member

bHLH: basic helix-loop-helix

BLAST: basic local alignment search tools

BMI: body mass index

BMP: bone morphogenetic protein

BNP: B-type natriuretic peptide, also known as brain natriuretic peptide or GC-B

BOX2: amino acid sequence [V/L]E[V/L]L present in the single chain cytokine receptors required in some cases for full activation of Jak2

bp: base pairs

***Brca1*:** breast cancer 1 gene

BRCA1: breast cancer-1 early onset tumor suppressor gene

BRE: TFIIB recognition element

BSP: bisulfite sequencing PCR

Bulbar: describe any bulb-shaped organ of the body

bZip: DNA binding domain of the leucine zippers family of heterodimeric proteins

βARK: β-adrenergic receptor kinase

C/EBP: CAAT box/enhancer binding protein

C/EBP: CAAT enhancing binding protein

C/EBP: CATT/enhancer binding protein

C/EBP: CCAAT/enhancer binding protein

C/EBP: transcription factor CCAAT/Enhancer Binding Protein

C/EBPβ: CCAAT/enhancer-binding protein-β

C: cytosine or cytidine

Ca^{2+}: calcium ions

Ca^{2+}-CaM: Ca^{2+} bound to CaM

$CaCl_2$: calcium chloride

CAIS: complete androgen insensitivity syndrome

CaM: calmodulin

CAM-1: intracellular adhesion molecule-1

CaMK: Ca^{2+}-mediated kinase

CaMKII: calcium/calmodulin-dependent protein kinase II

cAMP: cyclic adenosine monophosphate

cAMP: cyclic AMP

Caov-3: ovarian cell line

CAP: capped mRNA

CAP: catabolic activator protein

CAR: coxsackie-adenovirus receptor

CART: cocaine- and amphetamine-regulated transcript

CBF: CAAT box-binding factor

CBP/p300: CREB-binding protein and related E1A binding protein p300

CBP: CREB binding protein

CC: cumulus cells

CD/5-FCyt: cytosine deaminase plus 5-fluorocytosine

CDD: conserved domain database

Cdk: cyclin dependent kinase

cDNA: complementary deoxyribonucleic acid

CEBPβ: CCAAT enhancer binding protein beta

CERT: ceramide transport protein

***c-fos*:** cellular protooncogene of the transforming gene of the FBJ and FBR osteosarcome viruses

***c-fos*:** cellular protooncogene of the transforming gene of the Finkel-Biskis-Jinkins murine osteosarcoma viruses

c-Fos: product of the proto-oncogene c-*fos* that dimerize with c-Jun (c-Fos/c-Jun heterodimer) to form the trancription factor AP-1

c-Fos: transcription factor member of the AP-1 family of transcription factors. cFos is product of the cellular protooncogene c-*fos*, homologue of v-FOS, identified as the transforming gene of the Finkel-Biskis-Jinkins murine osteosarcoma viruses

c-Fos: transcription factor product of the cellular protooncogene homologue of v-FOS identified as the transforming gene of the Finkel-Biskis-Jinkins murine osteosarcoma virus; member of the AP-1 family

CG: human chorionic gonadotropin

CGA: common glycoprotein alpha

CGH: comparative genomic hybridization

cGMP: cyclic guanosine monophosphate

cGMP-PK: cGMP-dependent protein kinase

ChIP: chromatin immunoprecipitation

CHO cells: Chinese hamster ovary cells

c-IAP1: caspase inhibitor

c-IAP2: caspase inhibitor

***c-jun*:** cellular potooncogene of the transforming gene of avian sarcoma virus

c-Jun: product of the proto-oncogene c-*jun* that dimerize with c-Fos to form the transcription factor AP-1.

c-Jun: transcription factor product of the cellular protooncogene homologue of v-Jun identified as the transforming gene of avian sarcoma virus 17; member of the AP-1 family

***c-kit*:** cellular homolog of the feline sarcoma viral oncogene v-kit

Cl^-: chlorine ion

CL: corpus luteum

Clozapine: an atypical antipsychotic agent principally described as a D_2 receptor antagonist, but also serves as a serotonin antagonist and can affect other receptors types as well

CLS: cAMP-responsive element-like sequence

CMC: chemical mismatch cleavage

CMP: cytidine 5'-monophosphate

CMV: cytomegalovirus

CNP: C-type natriuretic peptide

CNS: central nervous system; consists of the spinal cord, cranial nerves and the brain

CNTF: ciliary neurotrophic factor

CO: carbon monoxide

CO_2: carbon dioxide

CoREST: corepressor for RE1 silencing transcription factor (REST)

COUP-TF: chicken ovoalbumin upstream promoter transcription factor

COX: cyclooxygenases

COX-2: cyclooxygenase-2

CRAds: conditionally replicative adenoviruses

CRE: cAMP response element

CRE: cAMP response element

CREB: cAMP response element binding protein

CREB: transcription factor cyclic AMP response element binding protein

CREM: cAMP-responsive element modulator

CREM: cAMP-responsive gene modulator

CREM: CRE modulator

CRF: corticotropin releasing factor

CRH: corticotropin-releasing hormone

cRNA: complementary RNA

***Cro* repressor:** dimeric protein composed of identical subunits

CRS: cAMP response sequences

CsCL: cesium chloride

CSF: colony-stimulating factor

CT: computed topography

Ct: threshold cycle

CTCF: CTC-binding factor

CTF: CAAT binding transcription factor

Ctnnb1: β-catenin

cyclin D1 and D2: cell cycle-regulatory genes

***CYP11A*:** gene coding for P450scc

Cyp11AS: aldosterone synthase, also known as Cyp11B2

***CYP17*:** gene coding for cytochrome P450 17α-hydroxylase/17,20-lyase

Cα, Cβ, C: isotypes of the PKA C-subunit

D: dihydrouridine

D_1 receptor: dopamine type 1 receptor

DAG: 1,2-diacylglycerol

dAMP: deoxyadenosine monophosphate

dATP: deoxyadenosine triphosphate

DAX-1: dosage-sensitive sex reversal, adrenal hypoplasia critical region, on chromosome X, gene 1

dbB: diagonal band of Broca

DBD: DNA-binding domain

dCMP: deoxycytidine monophosphate

dCMP: deoxyribose cytidine 5'-monophosphate

dCTP: deoxycytidine triphosphate

ddATP: dideoxyadenosine triphosphate

DDBJ: DNA database of japan

ddCTP: dideoxycytidine triphosphate

ddGTP: dideoxyguanosine triphosphate

ddNTP: dideoxynucleoside triphosphate

DDSB: DNA double strand breaks

ddTTP: dideoxythymidine triphosphate

DGGE: denaturing gradient gel electrophoresis

DGLA: dihomo-gamma-linoleic acid

dGMP: deoxyguanosine monophosphate

dGTP: deoxyguanosine triphosphate

DHEA: dehydroepiandrosterone

DHEA-S: DHEA-sulfate

DNA: deoxyribonucleic acid

DNMT: DNA methyltransferases

DNMT1 and DNMT3: DNA cytosine-5 methyltransferases 1 and 3

dNTP: deoxynucleotide triphosphate

Dopamine: catecholamine neurotransmitter important in the regulation of movement

DP: PGD receptor

DPE: downstream promoter element

D-ribose: ribose

DSB: double-strand breaks

dsDNA: double stranded DNA

dsRNA: double-stranded RNA

dTMP: deoxythymidine monophosphate

dTTP: deoxythymidine triphosphate

dUTP: deoxyuridine triphosphate

E_2: estradiol

E_2B: estradiol benzoate

EBI: European Bioinformatics Institute

E-box: sequence CACGTG that binds members of the basic helix-loop-helix

eCG: equine chorionic gonadotropin

ECM: extracellular matrix

EDN1: endothelin-1

EDN2: endothelin-2

EDN3: endothelin-3

EDNRA: type A endothelin receptors

EDNRB: type B endothelin receptors

EEG: electroencephalography

EFA: essential fatty acids

EFO-21: ovarian cancer cell line

EFO-27: ovarian cancer cell line

EGF: epidermal growth factor

EGF: epidermal growth factor also known as urogastrone

EGF-1: epidermal growth factor-1

EGF-R: EGF receptor

EGFR: epidermal growth factor receptor

EG-VEGF: endocrine gland-derived VEGF

EL: extracellular loop

ELISA: enzyme-linked immunoabsorbent assays

ELT3: cell line derived from Eker rat leiomyoma

EMBL: European Molecular Biology Laboratory

EMBOSS: The European Molecular Biology Open Software Suite

EMC: enzyme mismatch cleavage

EMG: electromyography

EP1-4: PGE receptors with four subtypes

Eph: ephrins

ER: endoplasmic reticulum

ER: estrogen receptor

ERE: estrogen receptor element

ERE: estrogens response element

Erg-1: early growth response factor-1

ERI-536: estrogen receptor mutant

ERK: extracellular signal-regulated kinase

ERK: extracellular-signal-regulated kinase (ERK) pathways.

ES: embryonic stem cells

ESTs: expressed sequence tag

ExPASy: expert protein analysis system

FAD: flavin adenine dinucleotide

$FADH_2$: reduced FAD

FasL: fas ligand

FGF: fibroblast growth factor or heparin-binding growth factors

Finasteride: 5α-reductase inhibitor marketed under the brand name Proscar

FISH: fluorescence in-situ hybridization

Fluoxetine: antidepressant SSRI marketed under the brand name Prozac

Fluxomics: identification of the dynamic changes of molecules within a cell over time.

Fos: transcription factor expressed by the c-*fos* gene

FOXL2: transcription factor Forkhead box protein L2

FOXO1: forkhead box-O1

FP: PGF receptor

Fra: transcription factor Fos-related antigen

FRET: fluorescence resonance energy transfer

FRP: FSH releasing proteins

FSH: follicle stimulating hormone

FSHR: follicle stimulating hormone receptor

G: guanine or guanosine

G6P: glucose-6-phosphate

GABA: γ-aminobutyric acid type A; amino acid that is an inhibitory neurotransmitter

$GABA_A$ receptor: γ-aminobutyric acid type A receptor; multisubunit GABA receptor subtype that forms a chloride channel activated by GABA. It is the major inhibitory receptor in the brain and is modulated by the binding of agents such as neurosteroids

GalNAc: N-acetylgalactosamine

GAP: GnRH-associated peptide region

GAP: GTPase-activating protein

GAPDH: glyceral-dehyde-3-phosphate-dehydrogenase

GAS: interferon-gamma activated sequence promoter

GATA: a family of transcription factors that contain two zinc finger motif and binds to the DNA sequence (A/T)GATA(A/G)

GATA4 and GATA6: members of a family of transcription factors that contain a two-zinc-finger motif and bind to the DNA sequence (A/T)GATA(A/G)

GC: guanylate cyclase

GCN5: an histone acetyltransferase

GCV: ganciclovir

GDF: growth differentiation factor

GDF-9: growth differentiation factor-9

GDNF: glial cell line-derived neurotrophic factor

GDP: guanosine diphosphate

GDX: gonadectomized

GEF: Ras guanine nucleotide exchange factor

Genomics: the analysis of gene expression and regulation in cell, tissue or organs under given conditions ([2,47], in Chapter 10) including how the genes interact with each other and with the environment. It has the potential to revolutionize the practice of medicine.

GGH3: somatolactotroph cells

GH: growth hormone

GHIH: GH inhibitory hormone

GHRH: growth hormone releasing hormone

GlcNAc: N-acetylglucosamine

GLCs: granulosa luteal cells

Glutamate receptor: Family of receptors that include metabotropic glutamate receptors and ionotropic receptors defined by binding of kainate, AMPA or NMDA

Glycomics: identification of all carbohydrates in a cell or tissues.

Glycoproteomics: a branch of proteomics that identifies, catalogs, and characterizes glycoproteins.

GM-CSF: granulocyte-macrophage colony-stimulating factor

GMP: guanosine 5-monophosphate

GMP: guanosine monophosphate

GnRH: gonadotropin-releasing hormone

GnRH: gonadotropin-releasing hormone; secreted by hypothalamic neurons, that stimulates the release of LH and FSH from pituitary gonadotrophes

GnRH-a: gonadotroping releasing hormone agonist

GnRHR: gonadotropin-releasing hormone receptor

GnSE: GnRHR-specific enhancer

GOPC: Golgi-associated PDZ and coiled-coil motif containing

GOS: global ocean sampling expedition

GPBP: good-pasture antigen binding protein

GPCR: G-protein-coupled receptor

GPR54: G-protein-coupled receptor-54

GR: glucocorticoid receptor

GRAS: GnRHR activating sequence

Grb2: growth factor receptor-bound protein-2

GRE/PRE: glucocorticoid responsive element/ progesterone responsive element

GRE: glucocorticoids response element

GRK: G-protein-linked receptor kinase

GRKs: G protein-coupled receptor kinases

GRMO2: rat granulosa stable cell line

GTC: guanidinium isothiocyanate

GTFs: general transcription factors

GTP: guanosine triphosphate

GTP: guanosine-5′-triphosphate

Gynecomastia: abnormal overdevelopment of the male breasts

$G_{\alpha i}$: G-protein inhibitory α-subunit

$G_{\alpha s}$: G-protein stimulatory α-subunit

H2A.X: core histone 2A variant

H2AFY: H2A histone family, member Y (also known as histone macroH2A1)

H2AX: variant of the histone H2a

HAC: human artificial chromosome

HATs: histone acetyltransferases

HB-EGF: heparin-binding EGF

hCG: chorionic gonadotropin hormone

hCG: human chorionic gonadotropin

hCG: human Chorionic Gonadotropin

hCG: human chorionic gonadotropin;

HDAC1/2: histone deacetylases 1 and 2

HDACs: histone deacetylases

HDL: high density lipoprotein

HEK-293 cells: human embryonic kidney cells

HGF: hepatocyte growth factor

hGL: human granulosa-luteal cells

HMG: high mobility group of proteins

hMG: human menopausal gonadotropin

HMG-box: homologous DNA binding domain of the HMG proteins

HMG-CoA reductase: 3-hydroxy-3-methyl-glutaryl-CoA reductase

HMTs: histone methyltransferases

HNF: hepatocyte nuclear factor

hnRNA: heterogeneous nuclear RNA

HP1: heterochromatin 1 protein

HPL: human placental lactogen

HPLC: high-performance liquid chromatography

HPO_4^{2-}: monohydrogen phosphate ion

Hrb: Asn-Pro-Phe (NPF) motif-containing protein (also called Rab or hRip)

HRE: hormone response element

hRip: human immunodeficiency virus Rev-interacting protein.

HSD3B: gene(s) coding for 3β-HSD

HSL: hormone-sensitive lipase

HSP: heat shock protein

HSP70: heat shock protein-70

HSV: herpes simplex virus

HSV-tk: herpes simplex virus-thymidine kinase

HTH: helix-turn-helix

Hypospadia: failure of the distal urethra to develop normally, resulting in a ventral urinary meatus

i.c.v.: intracerebroventricular

I: inosine

ICER: inducible cAMP early repressor

ICM: inner cell mass

IEVT: immortalized extravillous trophoblast

IFN-γ: interferon γ

IGF: insuline-like growth factor

IGF-1: insulin like growth factor-1

IGFBP-4: IGF-binding protein

IGFBPs: insulin like growth factor binding protein

IGF-I: insulin like growth factor-I

IGF-I: insulin-like growth factor 1

IGF-I: insulin-like growth factor-I

IGF-I: insulin-like growth factor-I or somatomedin-C

IGF-II: insulin-like growth factor-II

IkB: inhibitor of kappa B

IL: interleukin

IL: intracellular loop

Inr: Initiator sequence

InsP3: inositol triphosphate

Interactomics: identification of protein-protein interactions but also include interactions between all molecules within a cell.

IP: PGI receptor

IP3: inositol 1,4,5-triphosphate

IP_3: inositol triphosphate

IPF: insulin promoter factor

IPTG: isoprppyl-β-D thiogalactopyranoside

IRE: interferon response element

IRS-1: insulin receptor substrate-1

IU: international units

IUGR: intrauterine growth restriction

IVF: in vitro fertilization

JAK/STAT: Janus tyrosine kinases/signal transduction and activators of transcription

Jak: Janus kinase

JEG-3: placental cell line

JMJD2A-D: histone demethylases of the Jumonji family

JMJD6: histone demethylase of the Jumonji family

JNK: Jun N-terminal kinase

Jun: transcription factor expressed by the c-*jun* gene

JunB: protooncogene homolog B

JunB: transcription factor related to c-Jun

JunD: transcription factor related to c-Jun

K^+: potassium ion

KAF: keratinocyte autocrine factor or amphiregulin

KCL: potassium chloride

KGF: keratocyte growth factor

KID: kinase inducible domain

KIFC: kinesin family member C

KiSS-1: cognate ligands of the GPR54

KLF13: Kruppel-like factor 13

LacZ′: β-galactosidase gene of *E.coli*

LBD: ligand-binding domain

LbT2: gonadotroph cell line

Ld: lipid droplet

LDL: low-density lipoprotein

LEP: human/primate leptin

Lep: rodent leptin

***Lep^{ob}/Lep^{ob}*:** obese, genetically leptin deficient, mouse

$LEPR_L$: long isoform, primate leptin receptor

$LEPR_S$: short isoform, primate leptin receptor

LGIC: ligand-gated ion channel

LGIC: ligand-gated ion channels

LH: luteinizing hormone

LH-R: LH receptor

LHR: luteinizing hormone receptor

LIF: leukemia inhibitory factor

LIP: liver-enriched

Lipoid CAH: congenital Lipoid Adrenal Hyperplasia

LLC: large luteal steroidogenic cells

LM-15: human leiomyoma cell line

LR: leptin receptor

LRa: a short LR isoform

LRb: long LR isoform

LRBP: LH-R binding protein inhibitory protein

LRH-1: liver receptor homolog-1

LRH-1: transcription factor Liver Receptor Homologue-1

LRR: leucine-rich repeat

LSD1: lysine specific demethylase 1

LT: leukotrienes

L-type: long lasting VSCC

MA-10: stable cell line originated from a Leydig cell tumor

mAch: muscarinic acetylcholine receptor

MAD/MAX: group of proteins of the bHLH family that can form heterodimers with myc and regulate transcription

MAIS: mild AIS

MANO: meiotic autosomal nucleolar organization

MAP: mitogen activated protein

MAPK: mitogen activated protein kinase

MAPK: mitogen-activated protein kinase

MAPKs: mitogen-activated protein kinases

MARKS: myristolated alanine-rich C-kinase substrate

MAS: marker-assisted selection

MC1: melanocortin receptor

MCH: melanin-concentrating-hormone

MCP-1: monocyte chemoatractant protein-1

MCP-1: monocyte chemoattractant protein-1

MCS: multiple cloning sites also called polylinkers

M-CSF: macrophage colony-stimulating factor

MCSI: meiotic sex chromosome inactivation

MDR1: multidrug resistance gene

MEK 1: threonine and tyrosine recognition kinase

MEK: MAP-kinase-kinase or MAPKK

MEK: Mitogen-activated protein kinase kinase, also termed MAP2K

Metabolomics: identification and measurement of all small metabolites in a cell or tissue.

MG: methylguanosine

Mg^{2+}: magnesium ion

mGC: membrane-associated guanylate cyclase

MHC: major histocompatibility complex

MI: metaphase I

MI: methylinosine

microRNA: micro RNA

MII: metaphase II

miRNA: micro RNAs

MIS: Müllerian-inhibiting substance

MLN64: StAR homologue

mLTC-1: murine Leydig tumor cells

MMPs: matrix metalloproteinases

MOI: multiplicity of infection

Mosaicism: the presence of two populations of cells with different genotypes in one individual originated from a single fertilized egg

MR: mineralocorticoid receptor

MR: mineralocorticoids

MRE: mineralocorticoids response element

Mre11: meiotic recombination 11 protein

MRI: magnetic resonance imaging

mRNA: messenger RNA

MSH: melanocyte-stimulating hormone

MSK1/2: mitogen- and stress-activated protein kinases 1 and 2

MSP: methylation specific PCR

MT1-R: melatonin receptor subtype

MT2-R: melatonin receptor subtype

mtDNA: mitochondrial DNA

MTF-1: metal-responsive transcription factor-1

Myc: transcription factor expressed by a gene originally described in the avian MC29 myelocytomatosis virus (v-*myc*). A homologous gene (c-*myc*) is located in the long arm of the human chromosome 8.

Na^+: sodium ion

nACh-R: nicotinic acetylcholine receptor

NADP: nicotinamide adenine dinucleotide phosphate

NaOH: sodium hydroxide

Nbs1: Nijmegen breakage syndrome 1

NCBI: National Center for Biotechnology Information

N-COR: nuclear receptor co-repressor

NF kB: or E-box nuclear factor kappa B

NF-1: nuclear factor-1.

NF-kB: nuclear factor-kappa B

NF-Y: nuclear factor-Y

NGF: nerve growth factor

NH^{4+}: ammonium

NIH : National Institutes of Health

NMDA receptor: excititatory glutamate receptor subtype

NMDA: *N*-methyl-D-aspartic acid

nnAchR: neuronal nicotinic acetylcholine receptor

NO: nitric oxide

***NPC1*:** Niemann-Pick C1 gene

NPY: neuropeptide Y

NR4A: transcription factor nuclear receptor 4A subgroup

NR5A1: transcription factor SF-1

NR5A2: SF-1 family member LRH-1

NRE: negative regulatory element

NRE: nuclear response element

NSAID: non-steroidal anti-inflammatory drugs

NT-3: neurotropin-3

NT-4/5: neurotrophin 4/5

NT-6: neurotropin-6

N-type: neither long-lasting nor transient VSCC

Nur77, Nurr1 and Nor1: family members of NR4A

NuRD: nucleosome remodeling and deacetylase

***ob*:** original name for *Lep* gene (*obese*)

ODF2: Outer dense fiber 2

OE: ovarian epithelial cells

OH: hydroxil group

OHSS: ovarian hyper-stimulation syndrome

OKdb: Ovarian Kaleidoscope database

Olanzapine: similar drug to clozapine (above)

oligo-dT: oligonucleotide chain of deoxythimidines

ORX: orexins

OSE: ovarian surface epithelial cells

OT: oxytocin

OVCAR-3: ovarian cell line

OVX: ovariectomized

p300: E1A-binding protein p300, also termed EP300

P4: progesterone

P450arom: cytochrome P450 aromatase

P450c17: cytochrome P450 17α-hydroxylase/17,20-lyase

P450scc: cytochrome P450 cholesterol side chain cleavage

PA: eicosapentenoic acid

PACAP: pituitary adenylate cyclase-activating polypeptide

PAGE: polyacrilamide gel electrophoresis

PAI-I: plasminogen activator inhibitor

PAIS: partial AIS

Pak1: p21-activated kinase-1

PARP: poly ADP-ribose polymerase

PAX-8: transcription factor expressed by a member of the paired box (PAX) family of genes that encode proteins that contain a paired box domain, an octapeptide, and a paired-type homeodomain. It is expressed in the thyroid and the kidney and binds thyroglobulin and thyroid peroxidase genes promoters

PBR: peripheral-type benzodiazepine receptor

PCAF: p300/CBP-associated factor

PCOS: polycystic ovary syndrome

PCR: polymerase chain reaction

PDE: nucleotide phosphodiesterase

PDE: phosphodiesterase

PDGFs: platelet-derived growth factors

PDGFs: platelet-derived growth factors

PDKI: phosphatidylinositol-dependent protein kinase

PG: prostaglandins

PGDH: 15 - hydroxyprostaglandin dehydrogenase

PGE2: prostaglandin E2

PGF2α: prostaglandin F2α

PGFS: prostaglandin F synthase

PGG2: hydroperoxy endoperoxide prostaglandin G_2

PGH2: prostaglandin H_2

PGI: prostacyclins

PGT: prostaglandin transporter

PH: Pleckstrin homology

Pharmacogenomics: the combination of pharmacology and genomics that deals with analysis of the genome and its products (RNA and proteins) related to drug responses.

Phosphoproteomics: a branch of proteomics that identifies, catalogs, and characterizes phosphorylated proteins.

$PI(3,4)P_2$: phosphatidylinositol 3,4-biphosphate

$PI(3,4,5)P_3$: phosphatidylinositol 3,4,5-triphosphate

$PI(4,5)P_2$: phosphatidylinositol 4,5-biphosphate

PI/PKC: phosphatidylinositol/protein kinase-C

P_i: inorganic phosphate

PI: phosphatidylinositol

PI3K: phosphatidylinositol-3 kinase

PIAS: protein inhibitors of activated STATs

Piezoelectricity: the ability of some materials mainly crystals and certain ceramics to generate an electric charge in response to a mechanical stress. If the material is not shortcircuited, the applied charge induces a voltage across the material.

PI-PLC: phosphatidilinositol-specific phospholipase-C

PIR: protein information resources

Pit-1: a transcription factor expressed specifically in the pituitary gland; member of the POU-homeodomain family

Pitx-1: pan pituitary homeobox transcription factor

PKA: cAMP-dependent protein kinase-A

PKA: protein kinase A

PKA: protein kinase-A

PKB: protein kinase-B, also termed Akt

PKC: protein kinase C

PKC: protein kinase-C

PKG: cGMP-dependent protein kinase

PLA2: phospholipase-A2

PLC: phospholipase-C

PLs: placental lactogens

Plzf: promyelocytic leukemia zinc-finger, a transcriptional repressor encoded by the *Zfp145* gene

PMSG: pregnant mare serum gonadotropin

PNS: peripheral nervous system; consists of nerves outside the CNS that connect it with tissues, muscles and organs in the body

POA: preoptic area; structure localized by the optic chiasm and the anterior commissure. The medial POA contains the sexually dimorphic nucleus, which is generally larger in males. The medial POA is particularly essential for male sexual behavior and function, but is also involved with estrus cycle regulation, lordosis and maternal behavior in the female

POA-AH: preoptic area of anterior hypothalamus

POF: premature ovarian failure

POK: Poxviruses and zinc-finger (POZ) and Krüppel family of transcription repressors

poly-A: poly-adenylated

Poly-A: poly-adenylated tail at the 3' end of the mRNA

POMC: proopiomelanocortin

POU: acronym derived from the homeodomain proteins Pit-1, Oct and Unc-86

POZ: Poxviruses and zinc-finger

PP1: phosphatase type 1

PP2A: phosphatase type 2A

PR: progesterone

PR: progesterone receptor

pRb: retinoblastoma

PRE: progesterone response element

Pregnanolone: 3α,5β-tetrahydroprogesterone

pre-miRNA: precursor of miRNA

pri-miRNA: primary transcript of an miRNA gene

PRL: prolactin

PRMTs: protein arginine methyltransferases

Protein microarray: a substrate of glass or silicon on which different molecules of protein have been affixed at separate locations in an ordered manner thus forming a microscopic array. These are used to identify protein-protein interactions, substrates of protein kinases, or the targets of biologically active small molecules. The main use of the protein microarrays, also termed protein chip, is to determine the presence and/or the amount of proteins in biological samples, e.g. blood. One common protein microarray is the antibody microarray, where antibodies (most frequently monoclonal) are spotted onto the protein chip and are used as capture-molecules to detect proteins from cell lysates. There are several types of protein chips; however the most common are glass slide chips and nano-well arrays.

Proteomics: complete identification of proteins and protein expression patterns of a cell or tissue through two-dimensional gel electrophoresis or other multi-dimensional protein separation techniques and mass spectrometry. It is a large-scale study of proteins, particularly of their structure and function under given conditions ([2], in Chapter 10). This term was created to make an analogy with genomics ([47], in Chapter 10). Proteomics is much more complicated than genomics: while the genome is a rather constant entity, the proteome differs from cell to cell and is constantly changing through its biochemical interactions with the genome and the environment. One organism has radically different protein expression in different parts of its body, during different stages of its life cycle and under different environmental conditions ([47], in Chapter 10).

PTH: parathyroid-stimulating hormone

PTP: phosphotyrosine phosphatase-like receptors

PTT: protein truncation test

qPCR: quantitative real time polymerase chain reaction

R2A: PKA regulatory subunit

Rab: member of the Ras superfamily of monomeric G proteins

RACK: receptor for activated C-kinase

Rad3: a DNA helicase repair protein

Rad50: Mre11-interacting protein with binding affinity to double stranded DNA.

Rad51: protein involved in repair of DNA.

RAF: MAP-kinase-kinase-kinase

RAR: retinoic acid and 9-*cis*-retinoic acid

RAR: retinoic acid receptor

***ras*:** oncogene of the Harvey (*ras*H) and Kristen (*ras*K) rat sarcoma viruses

Ras: oncogene of the Harvey (*ras*H) and Kristen (*ras*K) rat sarcoma viruses. These genes, which are frequently activated in human tumors, encode a 21 kD G-protein.

Ras: small G-protein, first identified as product of the *ras* oncogene

RBM-8A: RNA-binding motif protein-8A

RER: rough endoplasmic reticulum

RET: protooncogene tyrosine kinase receptor that binds members of the GDNF family

RFLP: restriction fragment length polymorphism

rFSH: recombinant follicle stimulating hormone

RH: Ras homology domain

RIA: radioimmunoassay

RID: nuclear receptor interaction domain

RISC: RNA-induced silencing complex

RNA: ribonucleic acid

RNAi: RNA interference

RNAses: enzymes that specifically degrade RNA

ROC: receptor operated Ca^{2+} channel

ROS: reactive oxygen species

rPRL: rat placental lactogen

RRE: retinoic acid response element

rRNA: ribosomal RNA

RSK2: ribosomal S6 kinase 2

RT: ribothymidine

RT-PCR: reverse transcriptase polymerase chain reaction

RU486: synthetic antiprogestin (also known as mifepristone) that is classically regarded as an inhibitor of the progesterone receptor (though it has anti-glucocorticoid activity as well), and used as an emergency contraceptive

s.c.: subcutaneous

S6: ribosomal S6 kinase

SAC: spindle assembly checkpoint

SAGE: Serial Analysis of Gene Expression.

SBMA: X-linked spinal and bulbar muscular atrophy or Kennedy Disease

SBSS: selective brain steroidogenic stimulant; recent term to describe drugs that elicit neurosteroidogenesis in the brain, like fluoxetine ([101], in Chapter 20)

SCC1: SF-1 binding element

SCC2: SF-1 binding element

SCF/KL: stem cell factor/Kit ligand

SCP-2: sterol carrier protein-2

SDS-PAGE: sodium dodecyl sulfate polyacrilamide gel electrophoresis

SET domain: a domain found in some lysine methyltransferases, derived from *Drosophilia* proteins designated suppressor of variegation, enhancer of zeste and trithorax

SF-1: transcription factor steroidogenic factor 1, also termed NR5A1

SFE: SMAD-binding element

SFK: Src-family kinases

sGC: cytoplasmatic soluble guanylate cyclase

SH2 and SH3: src homology region-2 and -3 respectively

SH2: Src Homology (SH) region 2, a phosphotyrosine-binding domain originally described in proteins of the Rous sarcoma virus (*src*) oncogene family of tyrosine kinases

SHBG: steroid hormone-binding globulin, a glycosylated dimeric protein homologous to the androgen-binding protein secreted by the Sertoli cells of the testes

SIB: Swiss Institute of Bioinformatics

Sin3: a negative regulator of transcription in yeast also known as SDII

Sin3: a SWI independent corepressor found in complexes with HDACs

Sin3A: Sin3 family member A

siRNA: short/small interfering RNA

siRNA: silencing RNA

SK-OV-3: ovarian cell line

SLC: small luteal steroidogenic cells

Smac/DIABLO: mitochondrial protein

SMAD: Small Mothers Against Decapentaplegic

SMRT: silencing mediator of retinoid and thyroid receptors

SNF2L: nucleosome-remodeling ATPase

snoRNA: small nucleolar RNA

SNP: single nucleotide polymorphism

snRNA: small nuclear RNA

SO_4^{2-}: sulfate ion

SOCS: suppressors of cytokine signaling

SOCS-3: suppressor of cytokine signaling-3

solLR: soluble leptin receptor

SOX: SRY box

Sp1 and Sp3: transcription factor specificity proteins 1 and 3

Sp1: selective promoter-1

Sp1: transcription factor stimulatory protein 1

sPLA$_2$: secreted phopholipase-A$_2$

SR-BI: scavenger type receptor, class B, type 1

***src (pronounced "sark")*:** retroviral oncogene (*v-src*) from the chicken Rous sarcoma retrovirus and its precursor (*c-src*)

Src: name of the first described retroviral oncogene (*v-src*), from the chicken Rous sarcoma retrovirus and its precursor (*c-src*), which encode a membrane-associated protein kinase.

SRC: steroid receptor coactivator

SRD5A1: 5α-reductase-1 gene

SRD5A2: 5α-reductase-2 gene

SRE: putative sterol response element or serum responsive element

SREBP: transcription factor sterol regulatory element binding protein

SRF: serum response factor

SRIF: somatotropin release-inhibiting factor or somatostatins also called GHIH

SRY: sex determining region of the Y chromosome

SSCP: single-stranded conformational polymorphism

SSRI: selective serotonin-reuptake inhibitor; class of drugs used to treat depression and obsessive-compulsive disorder that extend the action of serotonin by inhibiting reabsorption into neurons

ssRNA: single-stranded RNA

***Ssty 1 and Ssty 2*:** Y-linked spermiogenesis specific transcript

***STAR*:** gene coding for steroidogenic acute regulatory (STAR) protein

StAR: steroidogenic acute regulatory protein, also known as StarD1; mediates the rate-limiting delivery of cholesterol to P450scc to initiate steroid synthesis

StarD1: START domain-containing protein 1(StAR)

START: StAR-related lipid transfer domain

STATS: signal transducers and activators of transcription

STK: serine-threonine kinase

SULT: sulfotransferase

SURG-1: sequence underlying responsiveness to GnRH element

SVOG-4m: human ovarian granulosa-luteal cell line TLCs: theca luteal cells

SVOG-4o: human ovarian granulosa-luteal cell line

SWI: SWItch proteins, part of chromatin remodeling ATPase complexes

System biology: the strategy of pursuing integration of complex data about biological interactions from diverse experimental sources using interdisciplinary tools and high-throughput experiments and bioinformatics.

σ_1: metabotropic sigma type 1 glycine receptor

T: thymine or thymidine

T$_3$: triiodothyronine

T$_4$: thyroxine

TAF250: an histone acetyltransferase

TBP: TATA-binding protein

TCF/LEF: T-cell factor/lymphoid enhancer factor

TE-671: neuronal cell line

TF: transcription factor

TFIID: transcription factor IID

TGF-α: transforming growth factor alpha

TGF: transforming growth factor

TGF-β: transforming growth factor-β

TGGE: temperature gradient gel electrophoresis

THDOC: 3α,5α-tetrahydrodeoxycorticosterone

THR: thyroid hormone receptor

Tie1: angiopoietin receptor type 1

Tie2: angiopoietin receptor type 2

TIMPS: tissue inhibitors of metalloproteinases

TK: tyrosine kinase

TK-GCV: thymidine kinase – ganciclovir

Tm: melting temperature

TM: trans-membrane

TNF: tumor necrosis factor

TNF-α: tumor necrosis factor-α

TPA: 12-*O*-tetradecanoyl-13-acetate

TR: thyroid hormone

TR: thyroid hormone receptor

TRAIL: TNF apoptosis-inducing ligand

Transcriptomic: measurement of gene expression in whole cells or tissue by DNA microarray or SAGE.

Trap: VEGF antagonist

TRE: TPA response element

TRH: thyrotropin-releasing hormone

Trk: tropomyosin-related kinase receptor

tRNA: tranfer RNA

TRPV1: transient receptor potential vanilloid of subtype-1, also called vanilloid type 1 receptor

TSA: trichostatin A, an HDAC inhibitor

TSE: tissue specific element

TSH: thyrotropin, or thyroid-stimulating hormone

TSH-R: TSH receptor

TTF-1: thyroid transcription factor

T-type: transient type VSCC

TX: thromboxanes

Tx: transcription start site

TxA: thromboxane receptor

U: uracil or uridine

UMP: uridine monophosphate

UniMES: UniProt metagenomic and environmental sequence database

UniParc: UniProt archive

UniProt: universal protein resource

UniProtKB: UniProt knowledgebase

UniRef: UniProt reference clusters

USF: Upstream stimulatory factor

USF-1 and USF-2: upstream stimulating factors 1 and 2, respectively

USF-1/2: upstream regulatory factor-1/2

UTP: uridine triphosphate

UTR: untranslated region

UV: ultra-violet

VDAC: voltage-dependent anion channel

VD-R: vitamin D receptor

VDR: vitamin-D

VEGFA or VEGF: vascular endothelial growth factor

VEGFR1 or Flt-1: vascular endothelial growth factor receptor type 1

VEGFR2 or Flk-1/KDR: vascular endothelial growth factor receptor type 2

VGCR: voltage-gated channel receptors

VIP: vasoactive intestinal peptide

VIP: vasointestinal peptide

VMN: ventromedial nucleus; structure in the middle hypothalamus essential for female sexual behavior

VNTR: variable number tandem repeats

VSCC: voltage-sensitive Ca^{2+} channel

VTA: ventral tegmental area; region in the mesencephalon from which originates mesocortical and mesolimbic dopaminergic neurons involved in the dopamine mesolimbic reward pathway

wHTH: winged HTH

YAC: yeast artificial chromosome

YY1: transcription factor Yin Yang 1

ψ: pseudouridine

Zfp145: zinc finger protein 145 gene

Index

D

P

T

Printed by Books on Demand, Germany